# A SIGNAL THEORETIC INTRODUCTION TO RANDOM PROCESSES

# A SIGNAL THEORETIC INTRODUCTION TO RANDOM PROCESSES

**ROY M. HOWARD**
Department of Electrical Engineering & Computing
Curtin University of Technology
Perth, Australia

WILEY

Published by John Wiley & Sons, Inc., Hoboken, New Jersey
Published simultaneously in Canada



***Library of Congress Cataloging-in-Publication Data:***

Howard, Roy M.
A signal theoretic introduction to random processes / Roy M. Howard.
pages cm
Includes bibliographical references and index.
ISBN 978-1-119-04677-6 (cloth)
1. Signal processing. 2. Signal theory (Telecommunication) 3. Stochastic processes.
4. Random noise theory. I. Title.
TK5102.9.H695 2015
003′.54–dc23

2014049440

Set in 10/12pt Times by SPi Global, Pondicherry, India

Printed in the United States of America

10 9 8 7 6 5 4 3 2 1

1 2016

# ABOUT THE AUTHOR

Clarity follows from rigour
All research is preliminary

Roy M. Howard has been awarded a BE and a Ph.D in Electrical Engineering, as well as a BA (Mathematics and Philosophy), by the University of Western Australia, Perth, Australia. To date, his academic career has been at the University of Western Australia and then in the School of Electrical and Computer Engineering at Curtin University, Perth, Australia. Over recent years, he has been a regular visitor to the Technical University of Darmstadt, Darmstadt, Germany. He is the author of *Principles of Random Signal Analysis and Low Noise Design: The Power Spectral Density and its Applications*, Wiley, 2002. He has expertise in low-noise amplifier design, signal theory, and modeling and characterization of random phenomena.

# CONTENTS

# PREFACE

This book has arisen from research, over a 30 year period, related to modeling and characterizing random phenomena. From this research I have formed the view, which underpins the book, that an introduction to random processes should be grounded in signal theory as well as probability theory; results for random processes should be established on the finite interval; and mathematical rigor is of fundamental importance.

Writing of material for the book commenced around 2005 with an initial goal of extending my book *Principles of Random Signal Analysis and Low Noise Design: The Power Spectral Density and its Applications*, Wiley, 2002. Around 2006 it became clear that a new book was required and, when time permitted, research and writing for this book was undertaken. Significant impetus for the book was given with a Sabbatical taken between August 2011 and January 2012, which was spent at The Technical University of Darmstadt, Darmstadt, Germany.

The book assumes a prior introduction to probability theory and random variable theory and is suited to final year Electrical, Electronic, and Communications Engineering students, postgraduate students, and researchers with an interest in characterizing randomness.

The following people have contributed to the book in various ways. First, Prof. A. Zoubir, Head of the Signal Processing Group, The Technical University of Darmstadt, Darmstadt, Germany, has generously supported my career and visits to The Technical University of Darmstadt where a significant amount of writing and research for the book was undertaken. Second, Prof. K. Fynn, Head of School, Electrical Engineering and Computing, Curtin University, Perth, Australia, for support for my research and travel that has facilitated the writing of the book. Third, as Head of the Electrical and Computer Engineering Department, Curtin University,

Prof. S. Nordholm has provided good collegial support for the book. Fourth, the staff at John Wiley (Susanne Steitz-Filler, Senior Editor; Sari Friedman, Senior Editorial Assistant; F. Pascal Raj, Assistant Account Manager) have supported and managed the publication of the book. I also wish to note the contribution of the staff at a countable number of cafes (countable, but large) for their indirect support whilst a significant level of writing and editing was in progress.

Finally, one can only lament the era of unparalleled wealth and knowledge where important scholarly endeavors increasingly are not supported and where many academic values, which underpin important aspects of a society that one can value, have been seriously eroded. For those who have supported, in this context, my endeavors I am thankful.

Roy M. Howard
August 2014

# 1

# A SIGNAL THEORETIC INTRODUCTION TO RANDOM PROCESSES

## 1.1 INTRODUCTION

Consistent with the nature of the physical universe, the technical world, and the social world and at the usual levels of observation, randomness is ubiquitous. Consistent with this, randomness has received significant attention over human history and prehistory. The appearance of gods, and associated offerings to such deities, in early civilizations was, in part, an attempt to understand the randomness inherent in nature and to have control over this variability. As our understanding of the nature of the physical universe has expanded, interest in random phenomena, and its characterization, has increased significantly. Today, the accumulated knowledge on characterizing randomness is vast. Cohen (2005) provides a good overview of part of the more recent scientific endeavor to understand and characterize random phenomena. Meyer (2009) provides a historical perspective on the mathematical development of stochastic process theory from the 1940s.

Random phenomena occur widely in the physical, social, and technical worlds, and well-known examples include the variation, with time, in measures of the weather (wind velocity, temperature, humidity, etc.), economic activity (inflation rate, stock indexes, currency exchange rates, etc.), an individual's physical state (heart rate, blood pressure, feeling of well-being, etc.), and technical entities (the information flow to a mobile communication device, the fuel economy of a car, etc.). Other technical examples include the lifetime of a set product manufactured by a manufacturer, variations in the queue length of data packets waiting to be forwarded at a given

*A Signal Theoretic Introduction to Random Processes*, First Edition. Roy M. Howard.

network node, the effective information rate on a given communication channel, the voltage variation between the terminals of a conductive medium due to the random movement of electrons, unwanted signal variations at the receiver of a communication system or at the output of a sensor, etc. Random phenomena generally inhibit the performance of a system and, for example, limit the transmission distance for electronic/photonic communication systems, the sensitivity of sensors for monitoring phenomenon in the natural world, the timing accuracy of all timing reference sources, etc. Randomness is not always detrimental to system operation, and the introduction of a random signal component to a system can, in certain circumstances, enhance system performance, and stochastic resonance is a common term associated with such an outcome.

## 1.2 MOTIVATION

The transformation of the world to the information age has been underpinned by advances in electrical, electronic, and photonic technology. Such technology is based on controlling the random movement of electrons and photons. As electrons have a mass of $9.11 \times 10^{-31}$ kg, they, potentially, can move with very high velocities (of the order of $10^5$ m/s), between collisions, when subject to an electric field and/or due to ambient thermal energy. (Such velocities should not be confused with the average drift velocity of electrons in a conducting media—of the order of $10^{-4}\text{m/s}$ in a copper conductor.) One extraordinary achievement of the modern era has been the development of sophisticated devices and systems with robust well-defined performance based on controlling electrons. Figure 1.1 provides a perspective.

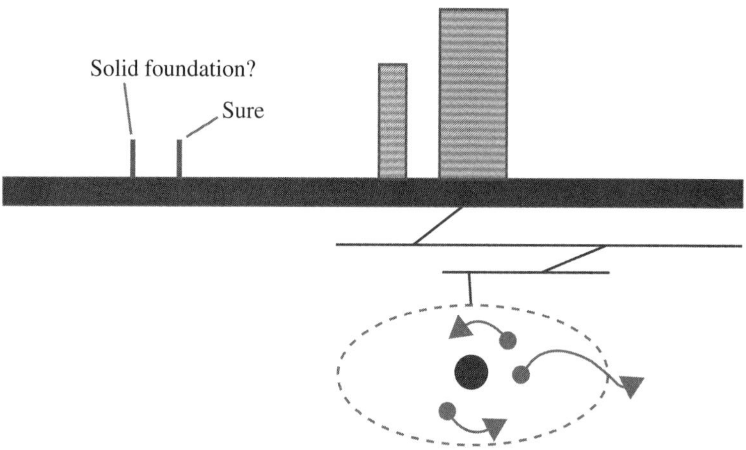


**FIGURE 1.1** On the foundation of modern electrical and electronic technology.

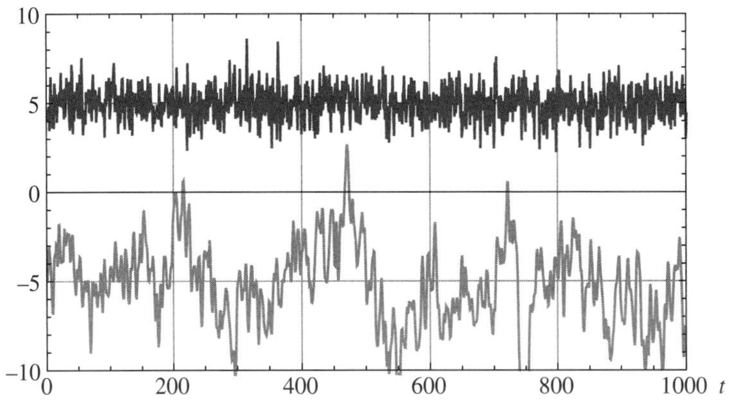


**FIGURE 1.2** Upper: white Gaussian noise signal—offset by 5. Lower: $1/f$ noise signal—offset by −5.

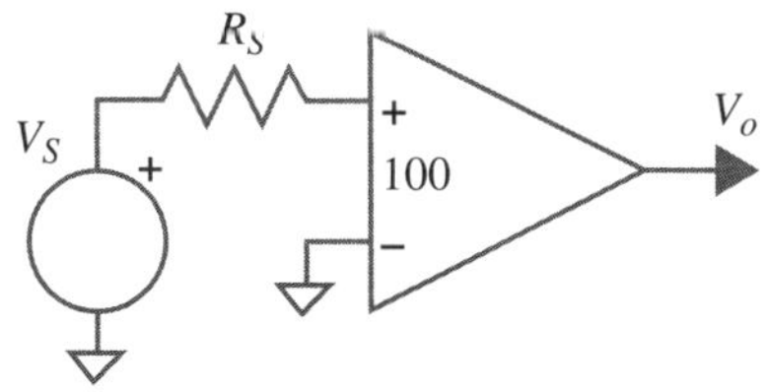


**FIGURE 1.3** Schematic diagram of a signal source connected to an amplifier.

The following examples illustrate, in small part, the reality of engineering in the context of the nature of our physical world and the diverse nature of the manifestation of randomness.

First, the voltage at a node in an electrical/electronic system will exhibit randomness consistent, for example, with white Gaussian noise or $1/f$ noise. The nature of such randomness is illustrated in Figure 1.2. Note the *regular* nature of a white Gaussian noise signal in comparison to a $1/f$ noise signal. The irregularity of $1/f$ noise is consistent with the nature of randomness in complex systems with multiple sources of noise, and such noise has been found in signals from a diverse range of systems including graphene, electronic devices, the human heart and brain, the human response to stimuli, phenomena in the natural world, and economic activity. The spectrum of many forms of music is consistent with a $1/f$ noise spectrum, and $1/f$ noise is often a signature of complexity. The introduction, and cited references, in Grigolini et al. (2009) provides a good overview of the diverse sources of such noise.

Second, as an example of the importance of randomness in electrical engineering, consider the random movement of electrons that leads to a signal at the output of an electronic amplifier, illustrated in Figure 1.3, having the form illustrated in Figure 1.4. In a communication context, the noise floor (the output noise signal in the absence of an input signal) of an amplifier limits the distance information can be transmitted and

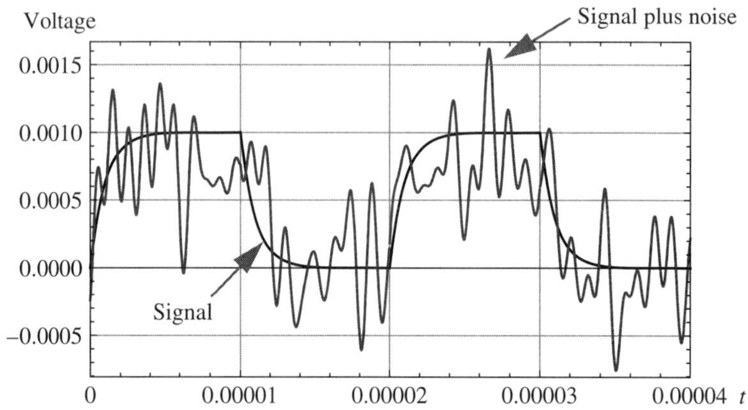


**FIGURE 1.4** Example of a signal corrupted by the noise inherent in an amplifier, at the amplifier output, for the case where the maximum signal level is $\sqrt{10}$ times the rms noise level.

recovered. In a sensor context, the amplifier noise limits the sensitivity of the sensor. The effect of the noise on a signal depends on the nature of the noise, the bandwidth of the noise, the noise level, the signal form, and the information required from the signal. Modelling and characterization of noise is fundamental to designing amplifiers to minimize the noise generated and for subsequent processing that, potentially, can limit the effect of the introduced noise.

Third, the randomness of the movement of electrons, and the randomness inherent in the natural environment (e.g., temperature variations), results in the drift of all electronic-based timing references. This necessitates, for example, circuitry, or algorithms, to maintain synchronization and, hence, reliable communication, between the nodes in a communication network. The first-order model of an electronic oscillator is

$$x(t) = A_o \sin(2\pi f_o t + \phi(t)) \tag{1.1}$$

where $f_o$ is the oscillator frequency and $\phi(t)$ is the phase variation due to the influence of random electrical and thermal variations. As a first-order model, the phase $\phi(t)$ variation will exhibit a random walk behavior as illustrated in Figure 1.5.

Noise in a system often leads to the delay or advance of the timing of a signal when compared with the zero noise case. This leads to jitter of the time a signal crosses a set threshold as illustrated in Figure 1.6. When the jitter is associated with a clock signal, which serves as a timing reference for a system, constraints on the switching rate, for reliable system operation, result. In a communication context, jitter in the timing reference at a receiver leads to an increase in the probability of error for the received information.

Fourth, the time a signal first reaches a set threshold level is called the first passage time, and the first passage time is of interest in many areas including the spread of diseases (when does a measure of the disease reach a set level for the first time),

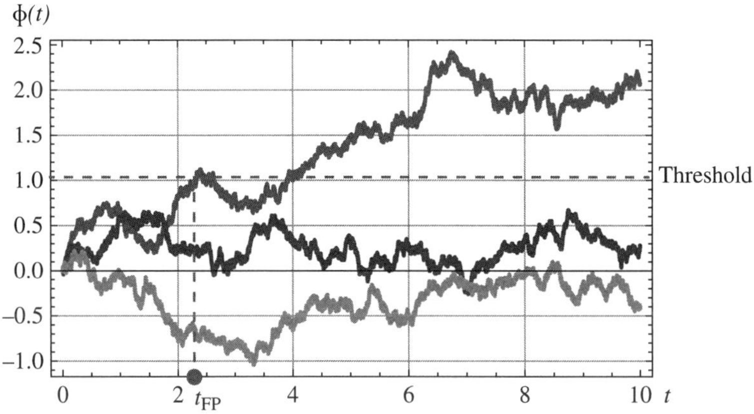


**FIGURE 1.5** Three trajectories consistent with the random-walk behavior of the phase of an oscillator (arbitrary units). The time $t_{FP}$ is the time the upper random walk first reaches the threshold level of unity.

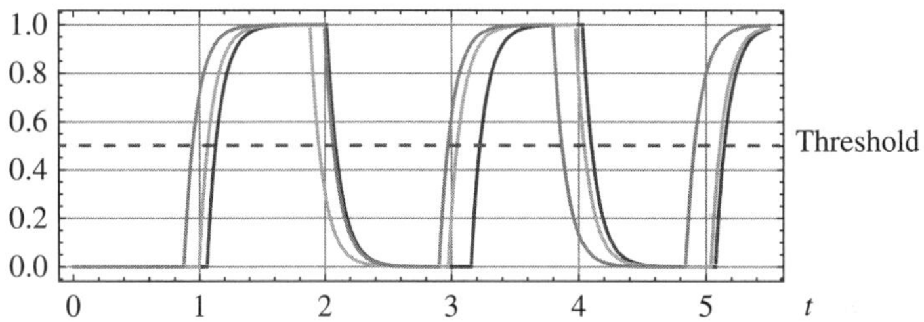


**FIGURE 1.6** Three jittered binary signals.

finance (when does a set measure of economic activity reach a set level for the first time), neuronal firing (a neuron fires when the collective inputs reach a set level), etc. The first passage time for a random walk is illustrated in Figure 1.5.

Fifth, in many contexts, our intuition and experience are consistent with events occurring at random and at a regular rate. Examples include the arrival of people at a supermarket queue, the timing of incoming phone calls/emails during a set period of the day, etc. In a technical context, examples include the arrival of photons on a photodetector, the crossing times of electrons in a PN junction, the arrival of data packets at a network node, etc. The simplest model for such timing is the Poisson point process (random timing of points but at a set rate), and examples of times defined by such a process are shown in Figure 1.7. Such point processes underpin, for example, the examples noted above.

Sixth, the efficient and inexpensive conveyance of information underpins, in significant measure, the latest transformation of our world and is based on agreed protocols with respect to encoding of information in signals. Efficient communication requires the use of signals that are spectrally efficient with a unit of information taking

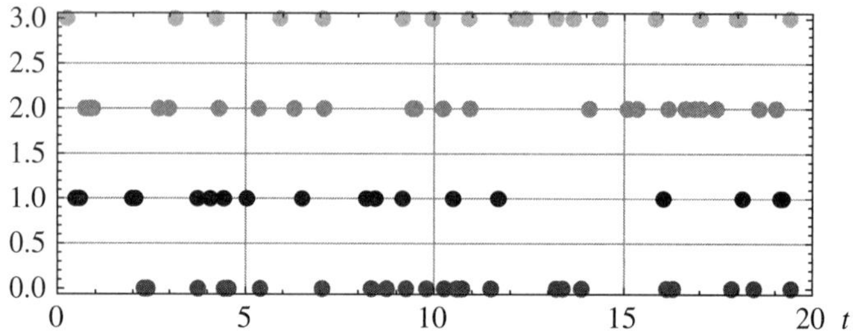


**FIGURE 1.7** Four examples of the times defined by a Poisson point random process with a rate of one point per unit time.

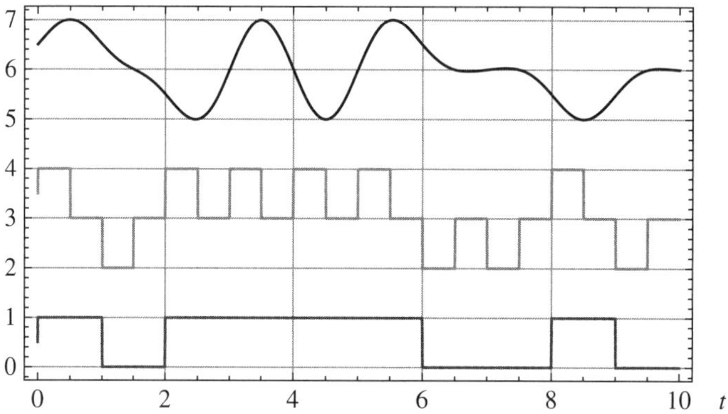


**FIGURE 1.8** Three signalling waveforms for encoding the data 1, 0, 1, 1, 1, 1, 0, 0, 1, 0, at a rate of 1 bit per second. Bottom: signalling with rectangular pulses—reference level of zero. Middle: signalling with return to zero pulses and polar coding—reference level of three. Top: signalling with raised cosine pulses using bipolar coding—reference level of six.

the smallest possible amount of the available capacity of a communication channel. Three different signals, each with different spectral efficiency but encoding the same information, are shown in Figure 1.8. Further, the nature of the signals used has implications for the ability of the receiver to recover the sent information in the context of noise and interference. Signal theory, random process theory, and methods of characterizing the spectral content of communication signals, underpin modern communication.

### 1.2.1 Usefulness of Randomness

Randomness is not always detrimental to the outcomes of a system. One example is illustrated in Figure 1.9 where the irregularity of the main traffic flow facilitates the movement of traffic from the right.

Another application where noise is useful is illustrated in Figures 1.10 and 1.11 and arises when an input level is to be estimated from samples provided by a device, for

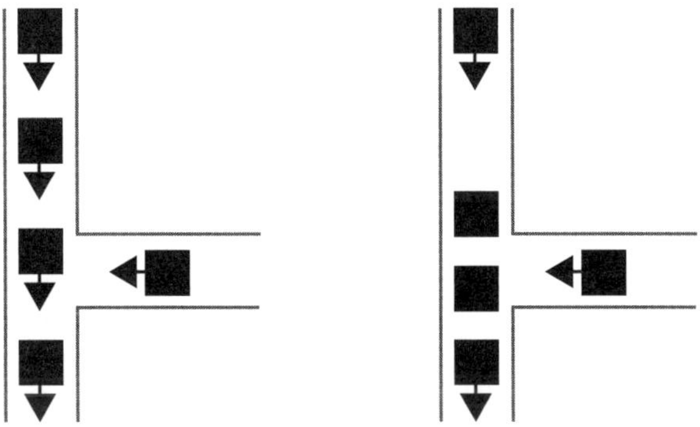

**FIGURE 1.9** Regular and irregular traffic flow.

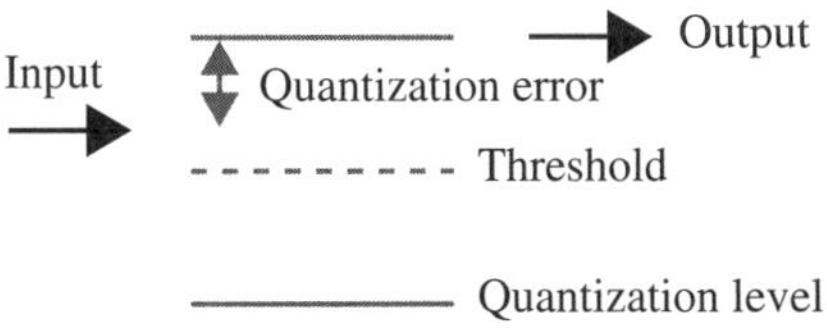


**FIGURE 1.10** Illustration of the quantization error arising from sampling a signal with finite amplitude resolution.

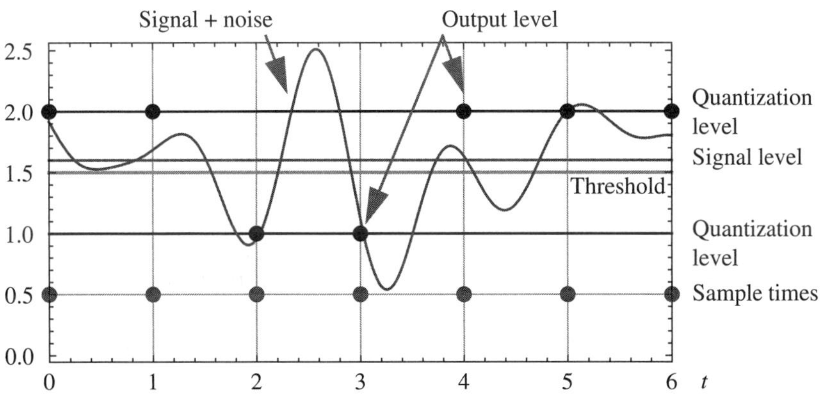


**FIGURE 1.11** Illustration of the output values arising from sampling a constant signal level plus additive noise. The sampling rate is unity.

example, an analogue to digital converter, that produces quantized levels. In the absence of noise, the output level will differ from the input level by the quantization error. When noise, with a variation greater than the quantization resolution, is present, it is possible to recover the input level accurately by averaging output values taken at times greater than the correlation time of the noise. The resolution can be improved according to $1/\sqrt{N}$ where $N$ is the number of data values averaged.

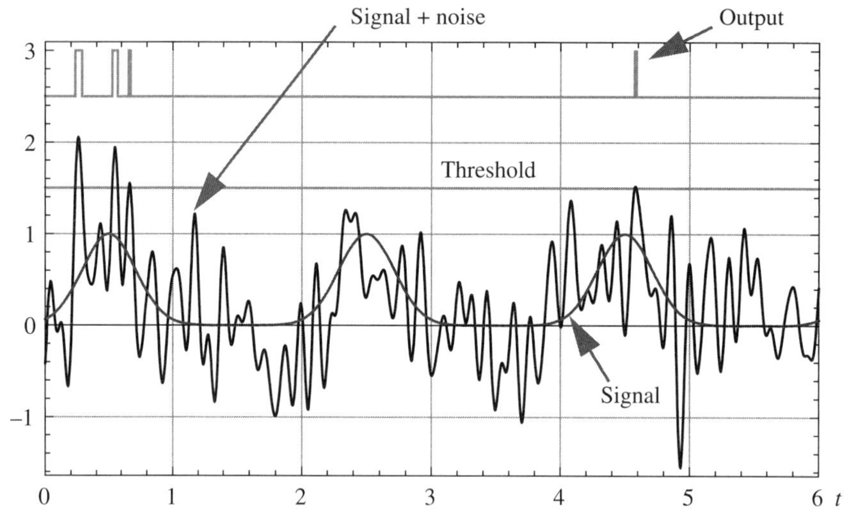


**FIGURE 1.12** Illustration of the beneficial effect of additive noise on the detection of a pulse train signal that is below a threshold level. The output level (offset) is a pulse when the signal plus noise is above the threshold.

Additive noise, or the noise inherent in a system, can aid subthreshold detection of a periodic signal as illustrated in Figure 1.12. Such detection occurs, for example, in neuronal networks. The phrase *stochastic resonance* is widely used when noise aids the detection of a signal.

### 1.2.2 Engineering

The first requirement for engineering, in the context of random phenomena, is the modelling of the random phenomena. The second requirement is the characterization of the random phenomena such that measures of system performance can be ascertained and the effects of the randomness minimized or utilized as appropriate. The goal of this book is to provide a theoretical basis that leads to the facilitation of both the modelling and the characterization of the prototypical random phenomena encountered in electrical engineering.

## 1.3 BOOK OVERVIEW

The theory for modelling and characterizing random phenomena is vast, and a single book can, at best, provide a modest introduction in a specific area. This book, and the approach taken for introducing and characterizing random phenomena, has arisen out of long-term research in the electronics and communications field. The theory and examples included are consistent with the broad electrical/electronic/communications engineering discipline and include the prototypical random phenomena of these disciplines: the random walk, Brownian motion, the random telegraph signal, the Poisson point process, the Poisson counting process, shot noise, white noise, $1/f$ noise, signalling random processes (which underpin most forms of communication), jitter, random clustering, and birth–death random processes.

The rationale for the book is threefold: first, random process theory should be grounded in signal theory as well as probability theory. Second, results for random processes should be established on the finite interval and results for the infinite interval obtained by taking an appropriate limit. Third, attention to mathematical rigor provides clarity and facilitates understanding, and such rigor is well suited to the modelling and characterization of random phenomena. Detailed proofs of results have been included to provide a comprehensive treatment of material at a moderate mathematical level. The book assumes a prior introduction to probability theory and random variable theory.

The approach taken, with a strong mathematical and signal theory basis, provides a foundation for random process theory that facilitates the continued development of random process theory in the context of many unsolved problems and the increasing importance of modelling and characterizing random phenomena. The modelling of random processes on the finite interval allows transient as well as steady-state results to be obtained. Importantly, it allows the use of finite-dimensional functions for characterizing random phenomena, and this facilitates the development of random process theory. The use of a signal theory basis provides a general framework for defining functions used for characterizing random phenomena including the autocorrelation function and the power spectral density. This has pedagogical value as the definitions have a signal basis rather than a random process basis and is less problematic for students. The use of a signal basis set approach for defining the power spectral density, which is the most widely used measure for characterizing random phenomena, provides a simple and natural interpretation of this function for the general case and for the usual case where a sinusoidal basis set is assumed.

Chapters 2–5 provide the necessary mathematical theory, background signal theory, random variable theory, and random process theory for subsequent discussion of random processes. Chapter 6 details the prototypical random processes that are fundamental to electrical, electronic, and communication engineering. Chapters 7–10 provide a basis for the characterization of random phenomena: Chapter 7 provides a general overview, Chapter 8 details probability mass function/probability density function evolution, Chapter 9 details the autocorrelation function, and Chapter 10 details the power spectral density. Chapter 11 provides an introduction to order statistics, and this provides the background for a discussion of the Poisson point random process in Chapter 12. Chapter 13 provides an introduction to birth–death random processes, while Chapter 14 provides an introduction to first passage time theory.

# 2

# BACKGROUND: MATHEMATICS

## 2.1 INTRODUCTION

This chapter details relevant mathematical theory, and mathematical results, to provide the basis for analysis in subsequent chapters and to ensure a coherent and systematic development of material. A summary of mathematical notation is detailed in the notation table in the reference section at the end of the book.

A basic introduction to set theory, and function theory, is provided, and this is followed by an introduction to measure theory. Two types of measure are considered. The first, Lebesgue measure, underpins Lebesgue integration, and this is fundamental to signal theory, which is discussed in Chapter 3. The second, Lebesgue–Stieltjes measure, provides the basis for Lebesgue–Stieltjes integration. Lebesgue–Stieltjes integration underpins Cramer's representation of a signal, which is discussed in Chapter 3. The discussion, in general, complements that of Howard (2002, ch. 2).

## 2.2 SET THEORY

### 2.2.1 Basic Definitions

**Definition: Set** A set is a collection of distinct entities.
The notation $\{\alpha_1, \alpha_2, \ldots, \alpha_N\}$ is used for the set of distinct entities $\alpha_1, \alpha_2, \ldots, \alpha_N$. The notation $\{x : f(x)\}$ is used for the set of elements $x$ for which the property $f(x)$ is true. The notation $x \in S$ means that the entity, denoted $x$, is an element of the set $S$.

*A Signal Theoretic Introduction to Random Processes*, First Edition. Roy M. Howard.

The empty set {} is denoted by Ø. The complement of a set $S$, denoted $S^C$, is defined as $S^C = \{x : x \notin S\}$ where $S$ is usually a subset of a larger set—often the *universal set*. The union and intersection of two sets are defined as follows:

$$\begin{aligned} A \cup B &= \{x : x \in A \text{ or } x \in B\} \\ A \cap B &= \{x : x \in A \text{ and } x \in B\} \end{aligned} \tag{2.1}$$

***2.2.1.1 Numbering and Notation*** Number theory (e.g., Feferman, 1964; Kranz, 2005, ch. 2) provides one of the foundations of modern analysis, and the following notation is widely used for the set of natural numbers, integer numbers, and rational numbers:

$$\begin{aligned} \boldsymbol{N} &= \{1,2,3,\ldots\} \\ \boldsymbol{Z} &= \{\ldots,-3,-2,-1,0,1,2,3,\ldots\} \\ \boldsymbol{Q} &= \{p/q : \ p,q \in \boldsymbol{Z}, q \neq 0, \gcd(p,q) = 1\} \end{aligned} \tag{2.2}$$

where gcd is the greatest common divisor function.

The set of positive integers, $\boldsymbol{Z}^+$, is defined as being equal to $\boldsymbol{N}$ and, depending on the context, may include the number 0.

The set of rational numbers, however, is not *complete* in the sense that it does not include useful numbers such as the length of the hypotenuse of a right triangle whose sides have unity length or the area of a circle of unit radius. By augmenting the set of rational numbers with irrational numbers, the set of real numbers, denoted $\boldsymbol{R}$, can be defined.

The set of complex numbers, denoted $\boldsymbol{C}$, is the set of possible ordered pairs that can be generated from real numbers, that is,

$$\boldsymbol{C} = \{(\alpha,\beta) : \ \alpha,\beta \in \boldsymbol{R}\} \tag{2.3}$$

When representing a complex number in the plane, the notation $(x,y) = x + jy$ is used where $j = (0,1)$. The algebra of complex numbers is governed by the rules of vector addition and scalar multiplication, that is,

$$\begin{aligned} &(x_1,y_1) + (x_2,y_2) = (x_1 + x_2, y_1 + y_2) \\ &a(x_1,y_1) = (ax_1, ay_1) \quad a \in \boldsymbol{R} \\ &(x_1,y_1)(x_2,y_2) = (x_1x_2 - y_1y_2, x_1y_2 + y_1x_2) \end{aligned} \tag{2.4}$$

From these definitions, the familiar result of $j^2 = -1$, or $j = \sqrt{-1}$, follows. The conjugate of a complex number $(x, y)$, by definition, is $(x, -y)$.

***2.2.1.2 Countability***

**Definition: Countable and Uncountable Sets** A set is a countable set if each element of the set can be associated, uniquely, with a natural number, that is, an element of $\boldsymbol{N}$ (Sprecher, 1970, p. 29). If such an association is not possible, then the set is an uncountable set.

The sets $\boldsymbol{N}$, $\boldsymbol{Z}$, and $\boldsymbol{Q}$ are countable sets. The sets $\boldsymbol{R}$ and $\boldsymbol{C}$ are uncountable sets.

### 2.2.2 Infinity

The term *infinity*, written $\infty$, does not denote a number but is notation for that which is *arbitrarily large*. Consider a sequence of numbers $x_1, x_2, \ldots$ where $x_i > x_{i-1}$ and which has no upper bound. The notation $\lim_{i \to \infty} x_i = \infty$ has the meaning

$$\forall I_o \in \mathbf{Z}^+, \exists x_i \in S, \text{ such that } x_i > I_o \quad S = \{x_1, x_2, \ldots\} \tag{2.5}$$

***2.2.2.1 Infinite Set Operations*** The following notation is widely used for the union and intersection of an infinite number of sets: $\bigcup_{i=1}^{\infty} A_i$ and $\bigcap_{i=1}^{\infty} A_i$. The following definitions apply:

$$\bigcup_{i=1}^{\infty} A_i = \lim_{N \to \infty} \bigcup_{i=1}^{N} A_i \qquad \bigcap_{i=1}^{\infty} A_i = \lim_{N \to \infty} \bigcap_{i=1}^{N} A_i \tag{2.6}$$

### 2.2.3 Supremum and Infimum

**Definition: Supremum and Infimum** The supremum of a set $A$ of real numbers, denoted sup($A$), is the least upper bound of that set and is such that (Marsden and Hoffman, 1993, p. 45)

$$\begin{aligned} &\sup(A) \geq x \quad \forall x \in A \\ &\forall \varepsilon > 0, \exists x \in A \text{ such that } \sup(A) - x < \varepsilon \end{aligned} \tag{2.7}$$

The infimum of a set $A$ of real numbers, denoted inf($A$), is the greatest lower bound of that set and is such that

$$\begin{aligned} &\inf(A) \leq x \quad \forall x \in A \\ &\forall \varepsilon > 0, \exists x \in A \text{ such that } x - \inf(A) < \varepsilon \end{aligned} \tag{2.8}$$

## 2.3 FUNCTION THEORY

Random phenomena define signals, and accordingly, signal and function theory are of fundamental importance. The following provides a definition for a function, defines common functions, and defines some useful properties of functions.

### 2.3.1 Function Definition

**Definition: Function** A function $f$ is a mapping, as illustrated in Figure 2.1, from a set $S_D$, called the domain, to a set $S_R$, called the range, such that only one element in the range is associated with each element in the domain. Such a function is written $f : S_D \to S_R$. If $x \in S_D$ and $y \in S_R$, with $x$ mapping to $y$ under $f$, then the notation $y = f(x)$ is used.

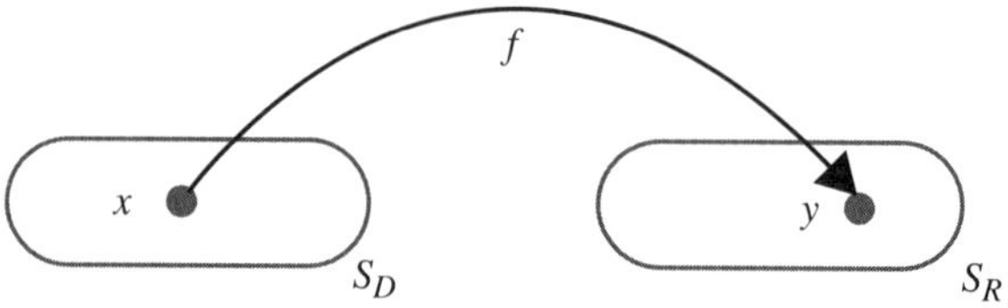


**FIGURE 2.1** Illustration of the mapping of a function.

### 2.3.2 Common Functions

The following are widely used functions:

**Definition: Kronecker Delta** The Kronecker delta function, $\delta_K$, is defined according to

$$\delta_K : \boldsymbol{R} \to \{0,1\} \quad \delta_K(x) = \begin{cases} 0 & x \neq 0 \\ 1 & x = 0 \end{cases} \tag{2.9}$$

**Definition: Sign or Signum Function** The sign, or signum, function, sgn, is defined according to

$$\text{sgn} : \boldsymbol{R} \to \{-1, 0, 1\} \quad \text{sgn}(x) = \begin{cases} -1 & x < 0 \\ 0 & x = 0 \\ 1 & x > 0 \end{cases} \tag{2.10}$$

**Definition: Unit Step Function** The unit step function, $u$, is defined according to

$$u : \boldsymbol{R} \to \{0,1\} \quad u(x) = \begin{cases} 0 & x < 0 \\ 1 & x \geq 0 \end{cases} \tag{2.11}$$

**Definition: Indicator Function** The indicator function, $I_A$, for a set $A \subset \boldsymbol{R}$ is defined according to

$$I_A : \boldsymbol{R} \to \{0,1\} \quad I_A(x) = \begin{cases} 1 & x \in A \\ 0 & x \notin A \end{cases} \tag{2.12}$$

One example of an indicator function is illustrated in Figure 2.2. Note that indicator functions are measurable functions (to be defined later).

**Definition: Sinc Function** The sinc function is defined according to

$$\text{sinc} : \boldsymbol{R} \to \boldsymbol{R} \quad \text{sinc}(x) = \frac{\sin(\pi x)}{\pi x} \tag{2.13}$$

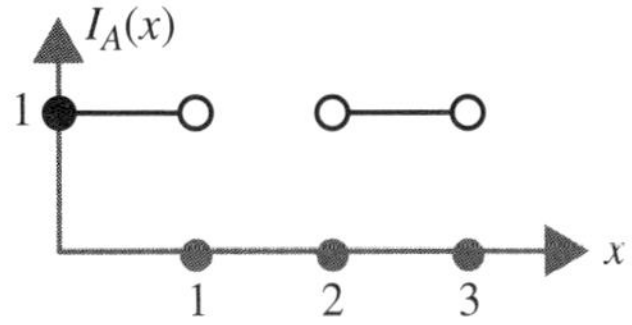


**FIGURE 2.2** An indicator function for the set $A = [0, 1) \cup (2, 3)$.

**Definition: Gaussian Pulse with Defined FWHM** A Gaussian pulse, $g : \boldsymbol{R} \to \boldsymbol{R}$, with unit area and a full width half maximum (FWHM) of $x_{\text{fwhm}}$, by definition, is

$$\begin{aligned} g(x) &= \frac{1}{\sqrt{2\pi}\sigma} \cdot \exp\left(-x^2/2\sigma^2\right) \quad \sigma^2 = x_{\text{fwhm}}^2/8\ln(2) \\ &= \frac{2\sqrt{\ln(2)}}{\sqrt{\pi}x_{\text{fwhm}}} \cdot \exp\left[-4\ln(2)x^2/x_{\text{fwhm}}^2\right] \end{aligned} \tag{2.14}$$

**Definition: Error Function** There are two alternative definitions for the error function:

$$\text{erf}_1(x) = \frac{1}{\sqrt{2\pi}} \int_0^x e^{-\lambda^2/2} d\lambda \quad \text{erf}_2(x) = \frac{2}{\sqrt{\pi}} \int_0^x e^{-\lambda^2} d\lambda \tag{2.15}$$

The following properties hold: $\text{erf}_1(\infty) = 0.5$, $\text{erf}_2(\infty) = 1$, $\text{erf}_1(x) = -\text{erf}_1(-x)$, and

$$\text{erf}_1(x) = 0.5\,\text{erf}_2\left[\frac{x}{\sqrt{2}}\right] \quad \text{erf}_2(x) = 2\,\text{erf}_1\left[\sqrt{2}x\right] \tag{2.16}$$

$$\int_{-\infty}^{x} \frac{e^{-(\lambda-\mu)^2/2\sigma^2}}{\sqrt{2\pi}\sigma} d\lambda = 0.5 + \text{erf}_1\left[\frac{x-\mu}{\sigma}\right] = 0.5 + 0.5\,\text{erf}_2\left[\frac{x-\mu}{\sqrt{2}\sigma}\right] \tag{2.17}$$

**Definition: Sine Integral Function** The sine integral function is defined according to

$$\text{Si} : \boldsymbol{R} \to \boldsymbol{R} \quad \text{Si}(x) = \int_0^x \frac{\sin(\lambda)}{\lambda} d\lambda \quad \begin{aligned} &\text{Si}(-x) = -\text{Si}(x) \\ &\text{Si}(\infty) = \pi/2 \\ &\text{Si}(-\infty) = -\pi/2 \end{aligned} \tag{2.18}$$

**Definition: Gamma Function** The gamma function $\Gamma$ is widely used in the field of probability and statistics and has the definition

$$\Gamma(n) = \int_0^\infty t^{n-1} e^{-t} dt \quad n \geq 1 \tag{2.19}$$

For the positive integer case,

$$\Gamma(n) = (n-1)! \quad n \geq 1, n \in \mathbf{Z}^+ \tag{2.20}$$

### 2.3.3 Function Properties

**Definition: Bounded Variation** A function $f : \boldsymbol{R} \to \boldsymbol{R}$ has bounded variation if, for all intervals $[\alpha, \beta]$, and for all finite sets of ordered points $\{\alpha_1, \ldots, \alpha_N\}$ from such intervals, it is the case that

$$\sum_{i=1}^{N-1} \left| f(\alpha_{i+1}) - f(\alpha_i) \right| \tag{2.21}$$

is finite.

**Definition: Right Continuous** A function $f : \boldsymbol{R} \to \boldsymbol{R}$ is right continuous at a point $x_o$ if for all $\varepsilon > 0$, there exists a number $\delta_o > 0$ such that

$$|f(x_o + \delta) - f(x_o)| < \varepsilon \quad 0 < \delta < \delta_o \tag{2.22}$$

**Definition: Piecewise Continuity at a Point—Left and Right Function Limits** A function $f : \boldsymbol{R} \to \boldsymbol{R}$ is piecewise continuous at a point $x_o$ if the left and right limits, $f(x_o^-)$ and $f(x_o^+)$, exist, that is,

$$\forall \varepsilon > 0 \ \exists \delta_o > 0 \ \text{ such that } \ 0 < \delta < \delta_o \ \Rightarrow \ \left| f(x_o + \delta) - f(x_o^+) \right| < \varepsilon \tag{2.23}$$

$$\forall \varepsilon > 0 \ \exists \delta_o > 0 \ \text{ such that } \ 0 < \delta < \delta_o \ \Rightarrow \ \left| f(x_o^-) - f(x_o - \delta) \right| < \varepsilon \tag{2.24}$$

and $f(x_o) \in \{f(x_o^-), f(x_o^+)\}$.

**Definition: Continuity at a Point** A function $f : \boldsymbol{R} \to \boldsymbol{R}$ is continuous at a point $x_o$ if $f(x_o^-) = f(x_o^+) = f(x_o)$.

**Definition: Piecewise Differentiability at a Point** A function $f$ is piecewise differentiable at $x_o$ if the left and right hand derivatives defined according to

$$\begin{aligned} f'(x_o^+) &= \lim_{\delta \to 0} \left[ \frac{f(x_o + \delta) - f(x_o^+)}{\delta} \right] \quad \delta > 0 \\ f'(x_o^-) &= \lim_{\delta \to 0} \left[ \frac{f(x_o^-) - f(x_o - \delta)}{\delta} \right] \quad \delta > 0 \end{aligned} \tag{2.25}$$

exist. The assumption in these definitions is that left and right hand limits $f\left(x_o^+\right)$ and $f\left(x_o^-\right)$ also exist. As for the case of piecewise continuity, the additional constraint $f(x_o) \in \left\{f\left(x_o^-\right), f\left(x_o^+\right)\right\}$ is included in the definition.

Piecewise differentiability at a point $x_o$ constrains a function, for the case where $f'\left(x_o^-\right)$ and $f'\left(x_o^+\right)$ are nonzero, such that it can be approximated by a first-order Taylor series expansions on either side of the point, that is,

$$f(x) \approx f\left(x_o^+\right) + (x - x_o)f'\left(x_o^+\right) \qquad x_o < x < x_o + \delta, \delta > 0 \tag{2.26}$$

$$f(x) \approx f\left(x_o^-\right) + (x - x_o)f'\left(x_o^-\right) \qquad x_o - \delta < x < x_o, \delta > 0 \tag{2.27}$$

**Definition: Differentiability at a Point** A function $f$ is differentiable at a point $x_o$ if $f\left(x_o^+\right) = f\left(x_o^-\right)$ and $f'\left(x_o^+\right) = f'\left(x_o^-\right)$.

**Definition: Monotonically Increasing Function** A function $f : \boldsymbol{R} \to \boldsymbol{R}$ is monotonically increasing on an interval $I$ if

$$f(x_2) \geq f(x_1) \qquad \forall x_1, x_2 \in I, x_2 > x_1 \tag{2.28}$$

A monotonically increasing function is not necessarily continuous at a point. However, the following can be proved.

**Theorem 2.1 Convergence Property: Monotonically Increasing Function**
If a function $f : \boldsymbol{R} \to \boldsymbol{R}$ is monotonically increasing on an open interval $I$, then, for all points $\alpha \in I$, it is the case that $\lim_{i \to \infty} f(\alpha - \delta/i)$ and $\lim_{i \to \infty} f(\alpha + \delta/i)$ exist for $\delta > 0$ and with $\alpha \pm \delta \in I$. The following notation is used:

$$\lim_{i \to \infty} f\left(\alpha - \frac{\delta}{i}\right) = f(\alpha^-) \qquad \lim_{i \to \infty} f\left(\alpha + \frac{\delta}{i}\right) = f(\alpha^+) \tag{2.29}$$

**Proof**
The proof is detailed in Appendix 2.A.

***2.3.3.1 Interchange of Limit and Function*** Continuity allows the interchange of the limit and function operations.

**Theorem 2.2 Interchanging the Order of the Limit and Function Operators**
If $f : \boldsymbol{R} \to \boldsymbol{R}$ and $g : \boldsymbol{R} \to \boldsymbol{R}$ are continuous, respectively, in neighborhoods of $g(x_o)$ and $x_o$, then, as illustrated in Figure 2.3,

$$\lim_{x \to x_o} f(g(x)) = f\left[\lim_{x \to x_o} g(x)\right] = f(g(x_o)) \tag{2.30}$$

Hence, if $f : \boldsymbol{R} \to \boldsymbol{R}$ is continuous in a neighborhood of $x_o$, then

$$\lim_{x \to x_o} f(x) = f\left[\lim_{x \to x_o} x\right] = f(x_o) \tag{2.31}$$

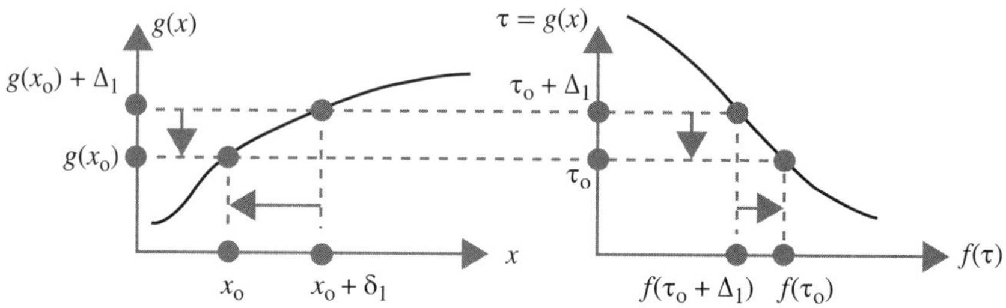


**FIGURE 2.3** Illustration of the nature of the convergence of $g(x)$ and $f(g(x))$ as $x \to x_o$.

**Proof**

The proof is detailed in Appendix 2.B.

*2.3.3.2 Little Oh Function* The little Oh function is widely used and is used in the discussion of the Poisson random process.

**Definition: Little Oh Function** A function $o: \mathbf{R} \to \mathbf{R}$, called *little oh*, has the characteristic

$$\lim_{x \to 0} \frac{o(x)}{x} = 0 \tag{2.32}$$

which is consistent with

$$\forall \varepsilon > 0, \exists \delta > 0 \text{ such that } 0 < x < \delta \;\Rightarrow\; \left| \frac{o(x)}{x} \right| < \varepsilon \tag{2.33}$$

A *little oh* function has the property that it decreases to zero at a rate faster than the function $f(x) = x$. For example, $o(x) = x^{1+\alpha}$, $\alpha > 0$, is a *little oh* function as

$$\lim_{x \to 0} \frac{o(x)}{x} = \lim_{x \to 0} x^{\alpha} = 0 \tag{2.34}$$

## 2.4 MEASURE THEORY

Underpinning Lebesgue integration and the Lebesgue–Stieltjes integral is the concept of measure. The measure operator is a mapping based on a sigma algebra. Cohn (1980), for example, provides a useful introduction to measure theory.

### 2.4.1 Sigma Algebra

**Definition: Sigma Algebra** A set $\Omega$ is a sigma algebra on a set $X$ if it is a set of the possible subsets of $X$ and if it has the following properties:

- $\Omega$ is nonempty and contains $\emptyset$ and $X$.
- $\Omega$ is closed under complementation, that is, if $A \in \Omega$, then $A^C \in \Omega$.

- $\Omega$ is closed under a countable union of sets, that is, if $A_i \in \Omega, i \in \boldsymbol{Z}^+$, then $\bigcup_{i=1}^{\infty} A_i \in \Omega$.

As an example, if $X = \{a,b,c\}$, then

$$\Omega = \{\emptyset,\{a\},\{b\},\{c\},\{a,b\},\{a,c\},\{b,c\},\{a,b,c\}\} \tag{2.35}$$

**Definition: Borel Sigma Algebra on the Real Line** A Borel sigma algebra, denoted $\Omega_B$, on $\boldsymbol{R}$ is a sigma algebra defined by the open sets of $\boldsymbol{R}$.

As a point $x$ of the real line can be defined as

$$x = \bigcap_{i=1}^{\infty} \left( x - \frac{1}{i}, x + \frac{1}{i} \right) \tag{2.36}$$

it follows that all points of $\boldsymbol{R}$ are in a Borel sigma algebra of $\boldsymbol{R}$.

### 2.4.2 Measure

The concept of measure underpins important areas of analysis. Different measures can be defined, and two of importance to signal theory are the Lebesgue measure and the Lebesgue–Stieltjes measure.

**Definition: Measure** A measure, $\mu$, is a mapping from a sigma algebra to the real line, that is, $\mu : \Omega \to \boldsymbol{R}$, which has the following properties:

$$\mu(\emptyset) = 0 \tag{2.37}$$

$$\text{If } A_1, A_2 \in \Omega \text{ and } A_1 \subseteq A_2 \text{ then } \mu(A_1) \le \mu(A_2) \tag{2.38}$$

$$\text{Countable sub-additivity: if } A_1, \ldots, A_i, \ldots \in \Omega \text{ then } \mu\left[\bigcup_{i=1}^{\infty} A_i\right] \le \sum_{i=1}^{\infty} \mu(A_i) \tag{2.39}$$

These properties are illustrated in Figure 2.4 for a measure operator that is defined by the area of a set.

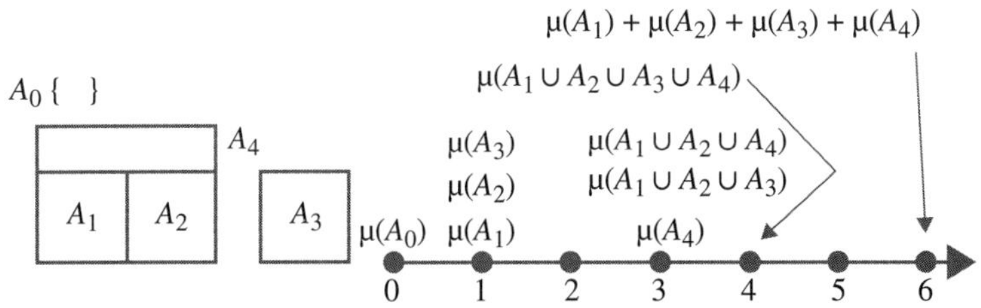


**FIGURE 2.4** Illustration of the properties that a measure must satisfy. The measure used is defined by the set area.

**Definition: Outer Measure** An outer measure $\mu * : \Omega \to \boldsymbol{R}$ of a set $A$ is the greatest lower bound of the measures of the sets that cover $A$, that is,

$$\mu*(A) = \inf\left\{ \sum_{j=1}^{\infty} \mu\left[B_{ij}\right] :\ A \subseteq \bigcup_{j=1}^{\infty} B_{ij}, B_{ij} \subseteq \Omega, i \in \boldsymbol{Z}^{+} \right\} \tag{2.40}$$

If the sets that cover $A$, denoted $A_1, A_2, \ldots$, can be ordered such that $A_{i+1} \subseteq A_i$, then

$$\mu*(A) = \lim_{i \to \infty} \mu[A_i] \tag{2.41}$$

This result also holds for a suitable subcollection, $A_1, A_2, \ldots, A_{i+1} \subseteq A_i$, of the sets that cover $A$.

*2.4.2.1 Notes* In general, subadditivity does not imply additivity, that is, for disjoint sets $A_1, \ldots, A_i \ldots \in \Omega$, it does not follow, in general, that $\mu\left[\bigcup_{i=1}^{\infty} A_i\right] = \sum_{i=1}^{\infty} \mu(A_i)$.

The measure operation is an operation to establish an indication—that is, a measure—of the *length* of a set. The direct analogy of length follows for Lebesgue measure of an interval. Consistent with the concept of *length*, the measure operation on any set yields a number that is greater than, or equal to, zero.

### 2.4.3 Lebesgue Measure

The Lebesgue measure of an interval of the real line is consistent with our understanding of the length of such an interval. Formally:

**Definition: Lebesgue Measure for an Open Interval** The Lebesgue measure, denoted $m$, is a mapping from the Borel sigma algebra of $\boldsymbol{R}$ to the positive real line and is defined according to

$$m : \Omega_{\mathrm{B}} \to \boldsymbol{R}^{+} \qquad m[(\alpha, \beta)] = \beta - \alpha \qquad m[\{\}] = 0 \tag{2.42}$$

**Definition: Lebesgue Measure for an Arbitrary Set** For an arbitrary set $E \in \Omega_{\mathrm{B}}$, the Lebesgue measure $m : \Omega_{\mathrm{B}} \to \boldsymbol{R}^{+}$ is the outer measure:

$$m[E] = \inf\left\{ \sum_{j=1}^{\infty} m\left[I_{ij}\right] : E \subset \bigcup_{j=1}^{\infty} I_{ij}, I_{ij} = \left(\alpha_{ij}, \beta_{ij}\right), i \in \boldsymbol{Z}^{+} \right\} \tag{2.43}$$

The Lebesgue measure of an arbitrary set is the minimum measure of all possible sets, based on open intervals, which cover $E$. If a suitable subcollection of sets,

$A_1, A_2, \ldots$, based on open intervals and which cover $E$, can be ordered such that $A_{i+1} \subseteq A_i$, then

$$m(E) = \lim_{i \to \infty} m(A_i) \tag{2.44}$$

***2.4.3.1 Lebesgue Measurable Sets*** A problem arises with the outer Lebesgue measure definition as given by Equation 2.43 in that it is possible to construct two disjoint sets whose outer Lebesgue measures are not additive (Jain and Gupta, 1986, pp. 83f ). To avoid this situation, the Lebesgue measure is restricted to Lebesgue measurable sets or to simplify notation, measurable sets.

**Definition: Lebesgue Measurable Set** A set $E$ of real numbers is a Lebesgue measurable set, or simply a measurable set, if it can be approximated arbitrarily closely by an open set and a closed set, that is, if $\forall \varepsilon > 0$, there exist an open set $O$ and a closed set $C$ such that

$$E \subseteq O \qquad C \subseteq E \tag{2.45}$$

and

$$m(O \cap E^C) < \varepsilon \qquad m(E \cap C^C) < \varepsilon \tag{2.46}$$

These relationships imply

$$m(O) - m(E) < \varepsilon \qquad m(E) - m(C) < \varepsilon \tag{2.47}$$

and are illustrated in Figure 2.5.

The following is a more formal definition of a Lebesgue measurable set:

**Definition: Formal Definition—The Caratheodory Approach** A set $E$ is Lebesgue measurable if for every set $A \in \Omega_B$, it is the case that

$$m(A) = m(A \cap E) + m(A \cap E^C) \tag{2.48}$$

This definition states that a set $E$ is measurable if it, and its complement $E^C$, partitions all Borel sets into disjoint sets whose measures are additive. This partitioning is illustrated in Figure 2.6.

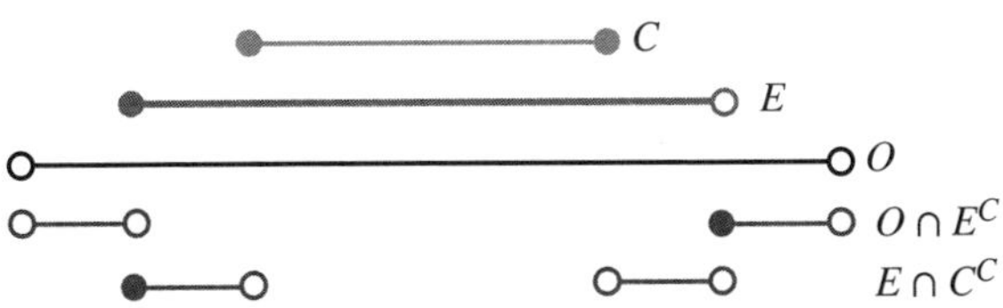


**FIGURE 2.5** Illustration of the approximation of an arbitrary set by an open and closed set.

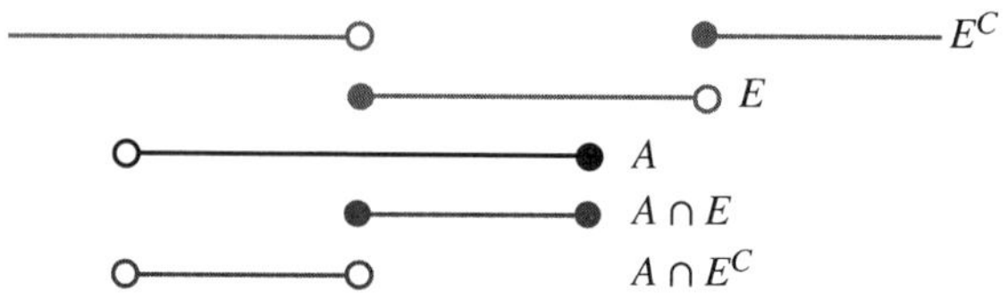


**FIGURE 2.6** Illustration of partitioning of an arbitrary set $A$ by a set $E$ and its complement.

A rationale for the definition may be found by considering two disjoint sets $A_1$ and $A_2$ whose outer measures are not additive. Then with $A = A_1 \cup A_2$ and $E = A_1$, the definition, as given by Equation 2.48, implies it should be the case that

$$m(A_1 \cup A_2) = m(A_1) + m(A_2) \tag{2.49}$$

As this does not hold, by assumption, it is the case that $A_1$ is not Lebesgue measurable. Repeating the argument with $E = A_2$ implies that $A_2$ is not Lebesgue measurable.

***2.4.3.2 Validity of Lebesgue Measure*** To be a valid measure, the properties stated in Equations 2.37–2.39 need to be verified. The following two theorems provide relevant details.

**Theorem 2.3 Measurable Sets Imply Additivity for Lebesgue Measure**
If $E_1, \ldots, E_i, \ldots$ are disjoint measurable sets, then

$$m\left(\bigcup_{i=1}^{\infty} E_i\right) = \sum_{i=1}^{\infty} m(E_i) \tag{2.50}$$

**Proof**
Consider two disjoint measurable sets $E_1$ and $E_2$. Measurability, consistent with Equation 2.48 and with $A = E_1 \cup E_2$, $E = E_1$, implies

$$m(E_1 \cup E_2) = m((E_1 \cup E_2) \cap E_1) + m\left((E_1 \cup E_2) \cap E_1^C\right) \tag{2.51}$$

Disjointness implies

$$m(E_1 \cup E_2) = m(E_1) + m(E_2) \tag{2.52}$$

The general case can then be proved by induction.

**Theorem 2.4 Properties of the Lebesgue Measure**
If $A_1, A_2$ are Lebesgue measurable sets and $A_1 \subseteq A_2$, then $m(A_1) \leq m(A_2)$.

**Proof**

First, it is the case that $A_2 = A_2 \cap \left[A_1 \cup A_1^C\right]$, and if $A_1 \subseteq A_2$, it then follows that

$$\begin{aligned} A_2 = A_2 \cap \left[A_1 \cup A_1^C\right] &\Rightarrow A_2 = \left[A_2 \cap A_1\right] \cup \left[A_2 \cap A_1^C\right] \\ &\Rightarrow A_2 = A_1 \cup \left[A_2 \cap A_1^C\right] \end{aligned} \tag{2.53}$$

As $A_1$ and $A_2 \cap A_1^C$ are disjoint sets, it follows from Theorem 2.3 that

$$\begin{aligned} m(A_2) &= m\left(A_1 \cup \left[A_2 \cap A_1^C\right]\right) \\ &= m(A_1) + m\left(A_2 \cap A_1^C\right) \geq m(A_1) \end{aligned} \tag{2.54}$$

### *2.4.3.3 Lebesgue Measure for Basic Sets*

**Theorem 2.5 Lebesgue Measure of a Point**

The Lebesgue measure of a point, as defined by the set $\{\alpha\}$, is zero, that is,

$$m[\{\alpha\}] = 0 \tag{2.55}$$

**Proof**

Consider the result $\{\alpha\} = \bigcap_{i=1}^{\infty} \left(\alpha - \frac{1}{i}, \alpha + \frac{1}{i}\right)$. It then follows that

$$\begin{aligned} m[\{\alpha\}] &= \inf\left\{ m(I_i):\ I_i = \left(\alpha - \frac{1}{i}, \alpha + \frac{1}{i}\right), i \in \mathbf{Z}^+ \right\} = \inf\left\{ \frac{1}{2i}:\ i \in \mathbf{Z}^+ \right\} \\ &= 0 \end{aligned} \tag{2.56}$$

**Theorem 2.6 Lebesgue Measure of a Closed and Half-Open Set**

The Lebesgue measures of the sets $(\alpha, \beta)$, $[\alpha, \beta)$, $(\alpha, \beta]$, and $[\alpha, \beta]$ are identical and equal to $\beta - \alpha$, that is,

$$m[(\alpha, \beta)] = m[[\alpha, \beta)] = m[(\alpha, \beta]] = m[[\alpha, \beta]] = \beta - \alpha \tag{2.57}$$

**Proof**

Consider the set $[\alpha, \beta)$, which is the union of the two disjoint sets $\{\alpha\}$ and $(\alpha, \beta)$. Consistent with Theorems 2.3 and 2.5, the Lebesgue measure of $[\alpha, \beta\}$ equals the Lebesgue measure of the open interval $(\alpha, \beta)$ which, by definition, is $\beta - \alpha$.

**Definition: Zero Measure** A set $E$ is said to have zero measure if $m(E) = 0$. The measure of a countable set of points has zero measure. For example, $m(\boldsymbol{Q}) = 0$.

**Definition: Almost Everywhere (a.e.)** A property is said to hold *almost everywhere* if it holds everywhere except on a set of points that have zero measure.

## 2.5 MEASURABLE FUNCTIONS

The Lebesgue integration of a function presupposes that the function is measurable.

**Definition: Measurable Function** A function $f:\boldsymbol{R}\rightarrow\boldsymbol{C}$ is a Lebesgue measurable function, or simply a measurable function, if for any open set, $O$, of $\boldsymbol{C}$, the inverse image defined by $f^{-1}(O)=\{x:\ f(x)\in O\}$ is a Lebesgue measurable set. Figure 2.7 illustrates this definition for the real function case.

Jain and Gupta (1986, ch. 4) and Burk (1998, ch. 4) are useful references for more detailed information on measurable functions.

### 2.5.1 Simple or Elementary Functions

The use of simple functions facilitates the definition of the Lebesgue integral.

**Definition: Simple or Elementary Function** A function $s:\boldsymbol{R}\rightarrow\boldsymbol{R}$ is a simple, or elementary, function if it takes only a finite number of values $s_1, \ldots, s_N$ and is a measurable function. Accordingly, the function is a mapping $s:\boldsymbol{R}\rightarrow\{s_1,\ldots,s_N\}$.

An example of a simple function is illustrated in Figure 2.8. A simple function can be written as a summation of indicator functions according to

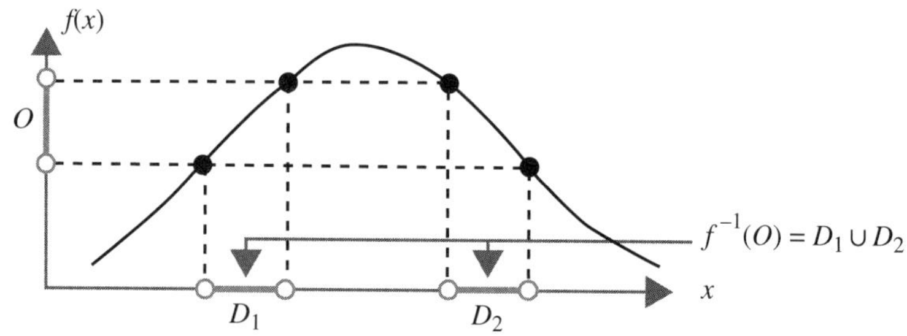


**FIGURE 2.7** Illustration for the definition of a measurable function assuming a real function $f:\boldsymbol{R}\rightarrow\boldsymbol{R}$.

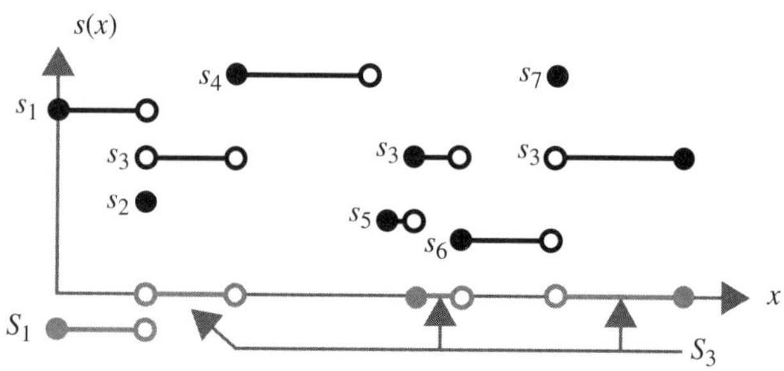


**FIGURE 2.8** An example of a simple function.

$$s(x) = \sum_{i=1}^{N} s_i I_{S_i}(x) \qquad I_{S_i}(x) = \begin{cases} 1 & x \in S_i \\ 0 & x \notin S_i \end{cases} \tag{2.58}$$

where $S_i = \{x:\ s(x) = s_i\}$. For a simple function $s : \boldsymbol{R} \to \{s_1, \ldots, s_N\}$, the sets $S_i = \{x:\ s(x) = s_i\},\ i \in \{1, \ldots, N\}$ form a partition of $\boldsymbol{R}$.

**Theorem 2.7 From Simple Function to a Step Function**
If a simple function results in a partition of the real line into a countable number of intervals, then it is possible to write the simple function as a summation of a countable number of step functions, that is, it is possible to find pairs of numbers $(a_1, x_1), \ldots, (a_i, x_i), \ldots$ and write

$$s(x) = \sum_{i=1}^{\infty} a_i u(x - x_i) \tag{2.59}$$

**Proof**
The proof is detailed in Appendix 2.C.

***2.5.1.1 Approximation of a Measurable Function by Simple Functions*** For Lebesgue integration theory, for example, it is of interest if a measurable function can be approximated by a simple function. The following theorems specify that this is the case.

**Theorem 2.8 Measurable Function as the Limit of Simple Functions**
If $f : \boldsymbol{R} \to \boldsymbol{R}$ is a measurable function and $f(x) \ge 0, x \in \boldsymbol{R}$, then there exists a sequence of monotonically increasing simple functions $s_i : \boldsymbol{R} \to \boldsymbol{R},\ i \in \boldsymbol{Z}^+$ such that

$$\begin{aligned} &s_{i+1}(x) \ge s_i(x), \quad s_i(x) \le f(x) \quad x \in \boldsymbol{R} \\ &\lim_{i \to \infty} s_i(x) = f(x) \quad x \in \boldsymbol{R}, \text{a.e.} \end{aligned} \tag{2.60}$$

**Proof**
The proof is detailed in Appendix 2.D.

This theorem states that a positive measurable function can be approximated, arbitrarily closely, by a simple function at all points of the real line except, possibly, on a set of zero measure. The nature of the simple function approximations is illustrated in Figure 2.9 and is based on partitioning the range of the function *f* with an increasingly fine resolution.

As an example, consider the function defined on the interval [0,1] according to

$$f(x) = \begin{cases} 0 & x \in \boldsymbol{Q} \\ x & x \notin \boldsymbol{Q} \end{cases} \tag{2.61}$$

For a simple function approximation, with an error less than $\varepsilon$, define the levels, and corresponding sets, according to

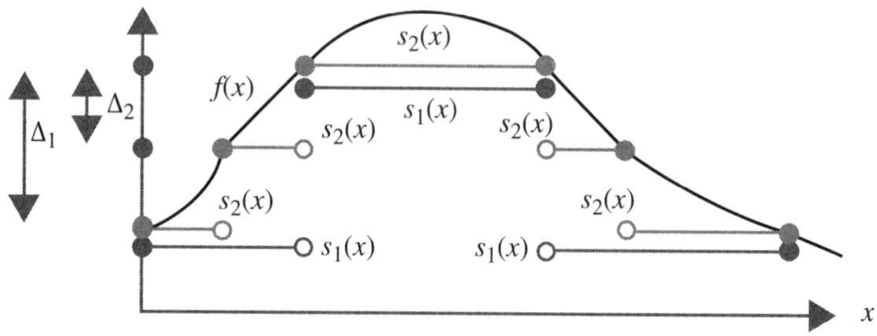


**FIGURE 2.9** Illustration of the partitioning of the range of $f$, with increasingly fine resolution, to create a sequence of simple function approximations to $f$. For clarity, the simple function $s_1$, with resolution $\Delta_1$, is shown slightly below the simple function $s_2$ with resolution $\Delta_2$.

$$\begin{aligned}
&s_0 = 0 \quad S_0 = \boldsymbol{Q} \cap [0,1] \\
&s_1 = 0 \quad S_1 = \boldsymbol{Q}^C \cap [0,\varepsilon) \\
&s_2 = \varepsilon \quad S_2 = \boldsymbol{Q}^C \cap [\varepsilon, 2\varepsilon) \quad \cdots \\
&s_{\lfloor 1/\varepsilon \rfloor + 1} = \lfloor 1/\varepsilon \rfloor \varepsilon \quad S_{\lfloor 1/\varepsilon \rfloor + 1} = \boldsymbol{Q}^C \cap [\lfloor 1/\varepsilon \rfloor \varepsilon, \lfloor 1/\varepsilon \rfloor \varepsilon + \varepsilon) \cap [0,1]
\end{aligned} \tag{2.62}$$

etc. The simple function approximation for $x \in [0,1]$ then is

$$s(x) = s_0 I_{S_0}(x) + \sum_{i=1}^{\lfloor 1/\varepsilon \rfloor + 1} (i-1)\varepsilon I_{S_i}(x) \qquad I_{S_i}(x) = \begin{cases} 0 & x \notin S_i \\ 1 & x \in S_i \end{cases} \tag{2.63}$$

**2.5.1.2 *Extension to Nonnegative Functions*** Theorem 2.8 states that a positive function can be approximated, arbitrarily closely, by a simple function. It is important to extend this result to the general case where a function takes on both positive and negative values. The following definition facilitates this:

**Definition: Positive and Negative Components of a Function** The positive component, $f^+ : \boldsymbol{R} \to \boldsymbol{R}$, and negative component, $f^- : \boldsymbol{R} \to \boldsymbol{R}$, of a function $f : \boldsymbol{R} \to \boldsymbol{R}$ are defined according to

$$\begin{aligned}
f^+(x) &= \max\{f(x), 0\} \quad x \in \boldsymbol{R} \\
f^-(x) &= f^+(x) - f(x) \quad x \in \boldsymbol{R}
\end{aligned} \tag{2.64}$$

Clearly,

$$f(x) = f^+(x) - f^-(x) \tag{2.65}$$

Further, both functions are positive, that is, $f^+(x) \geq 0,\ x \in \boldsymbol{R}$ and $f^-(x) \geq 0,\ x \in \boldsymbol{R}$. The functions $f^+$ and $f^-$ are illustrated in Figure 2.10.

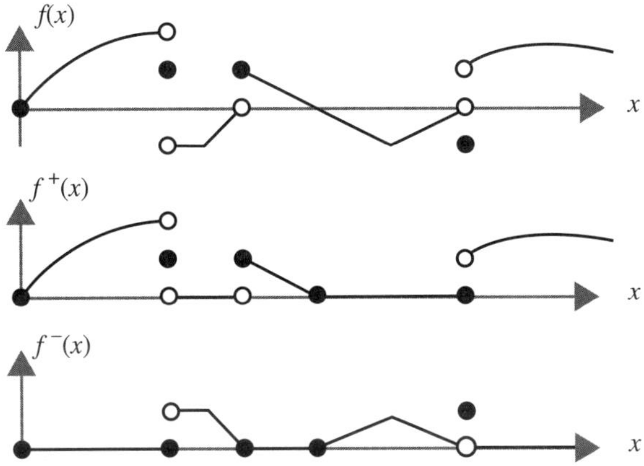


**FIGURE 2.10** Illustration of the positive and negative components of a function.

**Theorem 2.9 Approximation of an Arbitrary Measurable Function**
If $f : \boldsymbol{R} \to \boldsymbol{R}$ is a measurable function, then there exists a sequence of simple functions $s_i : \boldsymbol{R} \to \boldsymbol{R},\ \ i \in \boldsymbol{Z}^+$ such that

$$\begin{aligned} &|s_{i+1}(x)| \geq |s_i(x)|, |s_i(x)| \leq |f(x)| \quad x \in \boldsymbol{R} \\ &\lim_{i \to \infty} s_i(x) = f(x) \quad x \in \boldsymbol{R}, \text{a.e.} \end{aligned} \tag{2.66}$$

Thus, an arbitrary measurable function can be approximated, arbitrarily closely, by a simple function except, possibly, on a set of zero measure.

**Proof**
As both $f^+$ and $f^-$ are positive functions, it follows that $f^+$ and $f^-$ can be approximated, arbitrarily closely, by a sequence of simple functions in a manner consistent with Theorem 2.8. The required result follows, consistent with Equation 2.65, as $f(x) = f^+(x) - f^-(x)$.

***2.5.1.3 Approximating a Right Continuous Function*** It is of interest if a simple function approximation to a function can be replaced by a step function approximation with a finite number of terms. In general, this cannot be done as the function, illustrated in Figure 2.11 and defined according to

$$f(x) = \begin{cases} (-1)^i & x \in \left[1 + \dfrac{2^i - 1}{2^i}, 1 + \dfrac{2^{i+1} - 1}{2^{i+1}}\right) \quad i \in \{0,1,\ldots\} \\ 0 & \text{elsewhere} \end{cases} \tag{2.67}$$

demonstrates. This function is a simple function and oscillates between the levels $-1$ and 1 on the interval $[1,2)$ with increasing frequency as the point 2 is approached.

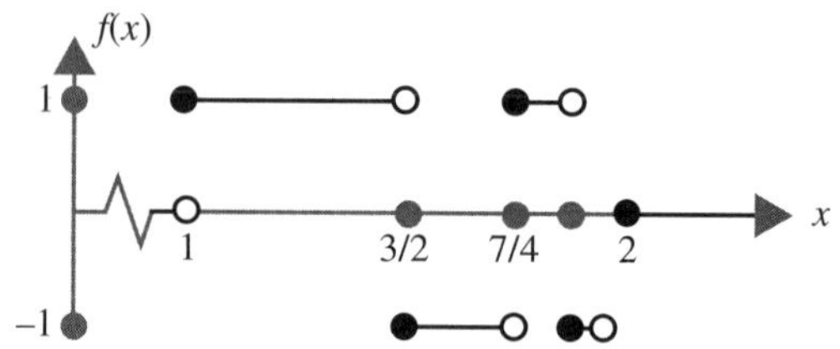


**FIGURE 2.11** Graph of a function that cannot be approximated by a finite number of step functions.

Consistent with Theorem 2.7, it can be approximated by a countable number of step functions. It cannot, however, be approximated by a finite number of step functions as, for example, on the interval $[2-\delta_1, 2+\delta_2)$ and for all $\delta_1, \delta_2 > 0$, the function is such that

$$\max\{f(x): \ x \in [2-\delta_1, 2+\delta_2)\} - \min\{f(x): \ x \in [2-\delta_1, 2+\delta_2)\} = 2 \tag{2.68}$$

However, the following can be proved.

**Theorem 2.10 Approximating a Right Cont. Function by Step Functions**
On a finite interval $[\alpha, \beta]$, a right continuous function $f: \boldsymbol{R} \to \boldsymbol{R}$, with a finite number of discontinuities on any given finite interval, can be approximated, arbitrarily closely, by a finite number of step functions, that is, $\forall \varepsilon > 0$, there exist a positive integer $N$ and pairs of numbers $\{(a_0, x_0), \ldots, (a_N, x_N)\}$, such that

$$\left| f(x) - \sum_{i=0}^{N} a_i u(x - x_i) \right| < \varepsilon \quad x \in [\alpha, \beta] \tag{2.69}$$

**Proof**
The proof is detailed in Appendix 2.E.

Right continuity, and a finite number of discontinuities on a finite interval, is consistent with a finite path length.

## 2.6 LEBESGUE INTEGRATION

Lebesgue integration underpins a significant level of signal theory and its applications. A detailed discussion of Lebesgue integration can be found in the books of Burk (1998, ch. 5), Jain and Gupta (1986, ch. 5), Titchmarsh (1939, pp. 332f), and Debnath and Mikusinski (1999, ch. 2).

### 2.6.1 The Lebesgue Integral

The following two equivalent approaches for defining a Lebesgue integral provide an introduction to Lebesgue integration.

**2.6.1.1 *The Lebesgue Integral: Definition via Partition of Range*** Consider a positive bounded measurable function $f: \boldsymbol{R} \to \boldsymbol{R}$ on an interval $(\alpha, \beta)$ where the function is bounded according to

$$f_L \le f(x) \le f_U \quad x \in (\alpha, \beta) \tag{2.70}$$

The range of $f$ can be partitioned by the $N+1$ numbers $f_0, f_1, \ldots, f_N$ such that

$$f_L = f_0 < f_1 < \ldots < f_{N-1} < f_N = f_U \tag{2.71}$$

and the sets $E_0, E_1, \ldots, E_N$ can then be defined according to

$$\begin{aligned} E_i &= \{x: \; f_i \le f(x) < f_{i+1}\} \quad i \in \{0, \ldots, N-1\} \\ E_N &= \{x: \; f(x) = f_N\} \end{aligned} \tag{2.72}$$

The measurability of $f$ guarantees the existence of the sets $E_0, \ldots, E_N$. As illustrated in Figure 2.12, the area under the function $f$ over the interval $(\alpha, \beta)$ can be approximated by the lower and upper sums defined according to

$$S_L = \sum_{i=0}^{N-1} f_i m(E_i) \qquad S_U = \sum_{i=0}^{N-1} f_{i+1} m(E_i) \tag{2.73}$$

Clearly, $S_L < S_U$. As the number of levels, $N+1$, demarcating the range of $f$ increases, $f_{i+1} - f_i$ tends toward zero for $i \in \{0, \ldots, N-1\}$ and $S_L$ and $S_U$ converge to the same number. This number is defined as the Lebesgue integral of the function $f$ over the interval $(\alpha, \beta)$.

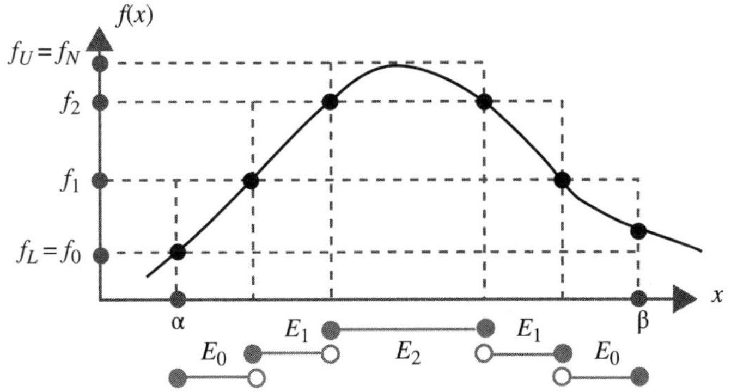


**FIGURE 2.12** Illustration of the partition of the range of $f$, and the sets partitioning the domain of $f$, for the case of $N = 3$.

**Definition: Lebesgue Integral of a Bounded Measurable Function** The Lebesgue integral of a positive bounded measurable function $f$ over a set $E$ is written as $\int_E f$ and is defined as

$$\int_E f = \lim_{N\to\infty} \sum_{i=0}^{N-1} f_i m(E_i) = \lim_{N\to\infty} \sum_{i=0}^{N-1} f_{i+1} m(E_i) \tag{2.74}$$

*2.6.1.2 The Lebesgue Integral: Definition via Simple Functions* The following is an equivalent approach to defining the Lebesgue integral and is based on the following definition.

**Definition: Lebesgue Integral of a Simple Function** The Lebesgue integral of a simple function $s:\boldsymbol{R}\to\{s_1,\ldots,s_N\}$ defined according to

$$s(x) = \sum_{i=1}^{N} s_i I_{S_i}(x) \quad S_i = \{x:\ s(x) = s_i\}, I_{S_i}(x) = \begin{cases} 0 & x\notin S_i \\ 1 & x\in S_i \end{cases} \tag{2.75}$$

is

$$\int_{\boldsymbol{R}} s = \sum_{i=1}^{N} s_i m(S_i) \tag{2.76}$$

As a positive measurable function, consistent with Theorem 2.8, can be approximated arbitrarily closely by a simple function, the following definition can be made.

**Definition: Lebesgue Integral for a Positive Function** The Lebesgue integral of a positive measurable function $f$ over a set $E$ is

$$\int_E f(x)dx = \sup\left\{ \int_E s(x)dx,\ \ s(x) \le f(x), x\in R,\ \ s \ \text{simple} \right\} \tag{2.77}$$

Alternatively, consider an ordered sequence $\{s_i\}_{i=1}^{\infty}$ of simple functions that approximate a measurable function $f$ over a set $E$ according to

$$s_{i+1}(x) \ge s_i(x) \quad s_i(x) \le f(x) \quad \lim_{i\to\infty} s_i(x) = f(x) \tag{2.78}$$

The Lebesgue integral of $f$ over $E$ then is

$$\int_E f(x)dx = \lim_{i\to\infty} \int_E s_i(x)dx = \lim_{i\to\infty} \sum_{j=1}^{N_i} s_{ij} m(S_{ij}) \tag{2.79}$$

where

$$s_i(x) = \sum_{j=1}^{N_i} s_{ij} I_{S_{ij}}(x) \qquad S_{ij} = \{x: \ s_i(x) = s_{ij}\}, I_{S_{ij}}(x) = \begin{cases} 0 & x \notin S_{ij} \\ 1 & x \in S_{ij} \end{cases} \tag{2.80}$$

***2.6.1.3 Lebesgue Integral of an Arbitrary Measurable Function*** The Lebesgue integral can readily be extended to an arbitrary measurable function through use of the positive and negative components of a function as defined by Equation 2.64 and according to $f(x) = f^+(x) - f^-(x)$. As both components are positive, the following definition is valid.

**Definition: General Case—Lebesgue Integral of a Measurable Function** For an arbitrary measurable function $f : \boldsymbol{R} \to \boldsymbol{R}$, the Lebesgue integral of $f$ over a set $E$ is

$$\int_E f(x)dx = \int_E f^+(x)dx - \int_E f^-(x)dx \tag{2.81}$$

The first integral yields a positive contribution, and the second a negative contribution, and overall, the result is consistent with the interpretation of an integral as an area under a function.

***2.6.1.4 Notes*** First, a sufficient condition for the existence of a Riemann integral on a finite interval is boundedness and continuity almost everywhere. If a function is bounded and Riemann integrable over [α, β], then it is also Lebesgue Integrable and the two integrals are the same (Burk, 1998, pp. 181–182; Jain and Gupta, 1986, p. 136). The usual notation for this equality is

$$\int_\alpha^\beta f(x)dx = \int_{[\alpha,\beta]} f \tag{2.82}$$

It is useful to use both notations for a Lebesgue integral as given in this equation, and both forms are used in subsequent analysis.

Second, if $m(E) = 0$, then $\int_E f = 0$ and this result is independent of $f$. As an example, the Lebesgue integral of the function illustrated in Figure 2.13 is zero.

Third, for any measurable function $f$, it follows that

$$\int_{[\alpha,\beta]} f = \int_{[\alpha,\beta]\cap \boldsymbol{Q}} f + \int_{[\alpha,\beta]\cap \boldsymbol{Q}^C} f = \int_{[\alpha,\beta]\cap \boldsymbol{Q}^C} f \tag{2.83}$$

as the measure of $[\alpha,\beta] \cap \boldsymbol{Q}$ is zero.

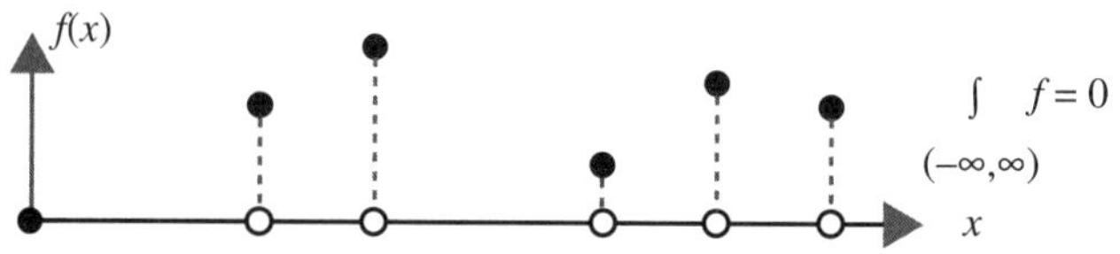


**FIGURE 2.13** Illustration of a function that takes on non-zero values only on a set with zero measure.

***2.6.1.5 Example*** Consider the functions

$$f(x)=\begin{cases} f_1 & x\in \boldsymbol{Q} \\ f_2 & x\in \boldsymbol{Q}^C \end{cases} \qquad s(x)=\begin{cases} f_1 & x\in S_1 \\ f_2 & x\in S_2 \end{cases} \tag{2.84}$$

where $S_1=[\alpha,\beta]\cap \boldsymbol{Q}$ and $S_2=[\alpha,\beta]\cap \boldsymbol{Q}^C$. The simple function $s$ equals $f$ over the interval $[\alpha, \beta]$, and the Lebesgue integral of $f$ over the interval $[\alpha, \beta]$ is

$$\int_{[\alpha,\beta]} f = f_1\cdot m(S_1)+f_2\cdot m(S_2)=f_2(\beta-\alpha) \tag{2.85}$$

### 2.6.2 Demarcation of Signal Space

The definition of the Lebesgue integral leads to the following definitions.

**Definition: Locally Integrable Functions** A function $f$ is locally integrable if the Lebesgue integral of its magnitude, over all finite intervals, is finite, that is, if

$$\int_{[\alpha,\beta],\,\alpha,\,\beta\in \boldsymbol{R}} |f| < \infty \tag{2.86}$$

**Definition: Set of Lebesgue Integrable Functions** The set of Lebesgue integrable functions, denoted $\boldsymbol{L}$, is defined according to

$$\boldsymbol{L}=\left\{ f:\boldsymbol{R}\to \boldsymbol{C},\ \int_{\boldsymbol{R}} |f| < \infty \right\} \tag{2.87}$$

**Definition: Set of Squared Lebesgue Integrable Functions** The set of squared Lebesgue integrable functions, denoted $\boldsymbol{L}^2$, is defined according to

$$\boldsymbol{L}^2=\left\{ f:\boldsymbol{R}\to \boldsymbol{C},\ \int_{\boldsymbol{R}} |f|^2 < \infty \right\} \tag{2.88}$$

Clearly, $\boldsymbol{L}^2$ is the set of functions with finite energy on the infinite interval.

***2.6.2.1 Notation*** For integration over the interval $[\alpha, \beta]$, the following sets can be defined:

$$L^p[\alpha,\beta] = \left\{ f : [\alpha,\beta] \to \boldsymbol{C}, \quad \int_{[\alpha,\beta]} |f|^p < \infty \right\} \qquad p \in \boldsymbol{R}^+ \tag{2.89}$$

When $p = 1$, the superscript is omitted.

***2.6.2.2 Demarcation of Signal Space*** The earlier definitions, and the distinction between measurable and nonmeasurable signals, lead to the demarcation of signal space as illustrated in Figure 2.14.

### 2.6.3 Miscellaneous Results

The following results are useful for subsequent analysis.

**Theorem 2.11 Area Associated with the Tail and a Neighborhood**
If $f \in \boldsymbol{L}$, then the area under the tail of $f$, the area associated with the neighborhood of any point, and the area under $f$ in the neighborhood of a point where $f$ is unbounded, can be made arbitrarily small, that is, $\forall \varepsilon > 0$, there exist $T_o > 0$, $\delta > 0$, and $f_o > 0$ such that

$$\int_{T_o}^{\infty} |f(x)|dx < \varepsilon \quad \int_{-\infty}^{-T_o} |f(x)|dx < \varepsilon \quad \int_{x_o-\delta}^{x_o+\delta} |f(x)|dx < \varepsilon \quad x_o \in \boldsymbol{R}$$
$$\int_E |f(x)|dx < \varepsilon \quad E = \{x : |f(x)| > f_o\} \tag{2.90}$$

**Proof**
See Howard (2002, ch. 2).

**Theorem 2.12 Approximation by a Simple Function**
If $f \in \boldsymbol{L}$, then for all $\varepsilon > 0$, there exists a simple function $s : \boldsymbol{R} \to \boldsymbol{C}$ such that

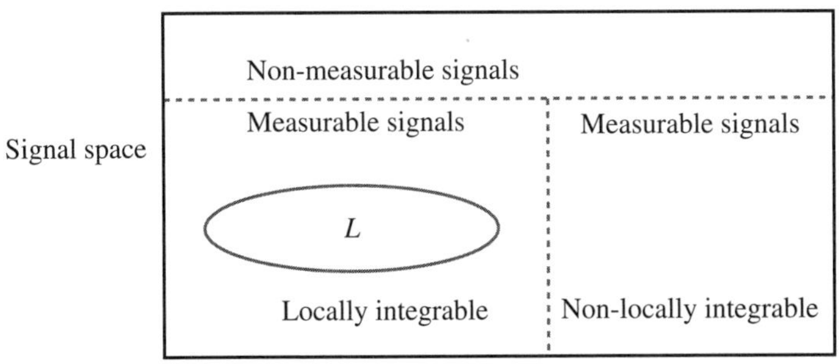


**FIGURE 2.14** Demarcation of signal space.

$$\left|\int_R [f-s]\right| < \varepsilon \quad \int_R |f-s| < \varepsilon \tag{2.91}$$

where $s(x) = \sum_{i=1}^{N} s_i I_{S_i}(x)$ and $S_i = \{x: \ s(x) = s_i\}$.

**Proof**

The first result is implicit in the definition of the Lebesgue integral, and the result that the area under the tail of a function $f \in \boldsymbol{L}$ can be made arbitrarily small as stated in Theorem 2.11. See, for example, Jain and Gupta (1986, pp. 130f) and Titchmarsh (1939, pp. 332f). The second result follows from Theorems 2.9 and 2.11.

**Theorem 2.13 Relationships between Signals in $L$ and $L^2$**

If $f \in \boldsymbol{L}^2[\alpha, \beta]$, then $f \in \boldsymbol{L}[\alpha, \beta]$. If $f \in \boldsymbol{L}^2$, it is not necessarily the case that $f \in \boldsymbol{L}$. If $f \in \boldsymbol{L}$, it is not necessarily the case that $f \in \boldsymbol{L}^2$.

**Proof**

Assume $f \in \boldsymbol{L}^2[\alpha, \beta]$. With $f(x) = u(x) + jv(x)$, it follows that

$$\int_{\alpha}^{\beta} \left[u^2(x) + v^2(x)\right] dx < \infty \tag{2.92}$$

Clearly, $u, v \in \boldsymbol{L}^2[\alpha, \beta]$. Define the sets $E_1$ and $E_2$ according to

$$\begin{aligned} E_1 &= \{x: |u(x)| \le 1\} \quad m(E_1) \le \beta - \alpha \\ E_2 &= \{x: |u(x)| > 1\} \quad m(E_2) \le \beta - \alpha \end{aligned} \tag{2.93}$$

It then follows that

$$\begin{aligned} u^2(x) \le |u(x)| \le 1 \quad x \in E_1 \\ 1 < |u(x)| < u^2(x) \quad x \in E_2 \end{aligned} \tag{2.94}$$

and, hence,

$$\int_{E_2} |u(x)| dx < \int_{E_2} u^2(x) dx < \infty \qquad \int_{E_1} |u(x)| dx \le \int_{E_1} dx = m(E_1) \le \beta - \alpha \tag{2.95}$$

which proves that $u \in \boldsymbol{L}[\alpha, \beta]$. A similar argument can be used to show $v \in \boldsymbol{L}[\alpha, \beta]$ and, hence, $f \in L[\alpha, \beta]$.

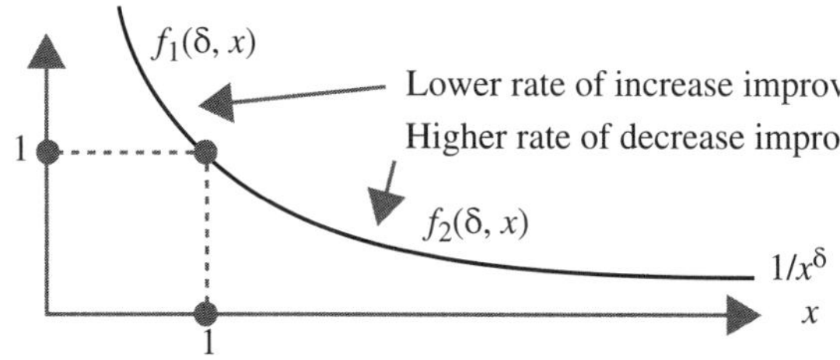


**FIGURE 2.15** Illustration of how the rate of increase, or the rate of decrease, of a function affects the integrability of a function.

Figure 2.15 illustrates two functions, $f_1$ and $f_2$, defined according to

$$f_1(\delta,x)=\begin{cases}0 & x\le 0, x>1\\ \dfrac{1}{x^\delta} & 0<x\le 1\end{cases}\qquad f_2(\delta,x)=\begin{cases}0 & x\le 1\\ \dfrac{1}{x^\delta} & x>1\end{cases}\tag{2.96}$$

which provide proof of the second and third stated results. Using standard integration results, it follows, for $0.5\le\delta<1$, that $f_1(\delta,x)\in\boldsymbol{L}$ but $f_1(\delta,x)\notin\boldsymbol{L}^2$. Also, for $0.5<\delta\le 1$, it follows that $f_2(\delta,x)\in\boldsymbol{L}^2$ but $f_2(\delta,x)\notin\boldsymbol{L}$.

**Theorem 2.14 Boundedness and Integrability**
If $f$ is bounded and $f\in\boldsymbol{L}$, then $f\in\boldsymbol{L}^2$. If $f$ is bounded on $[\alpha,\beta]$ and $f\in\boldsymbol{L}[\alpha,\beta]$, then $f\in\boldsymbol{L}^2[\alpha,\beta]$.

**Proof**
Assume $|f(x)|<f_{\max}$ for $\forall x\in\boldsymbol{R}$. It then follows that $|f(x)|/f_{\max}<1$ for $\forall x\in\boldsymbol{R}$, which implies $0<|f|^2/f_{\max}^2<|f|/f_{\max}$. From this result, it follows that $\int_{\boldsymbol{R}}|f|^2<f_{\max}\int_{\boldsymbol{R}}|f|<\infty$ and, hence, $f\in\boldsymbol{L}^2$. The result for the finite interval case follows in an analogous manner.

***2.6.3.1 Schwarz's Inequality*** Schwarz's inequality is a general relationship that applies to any two elements of an inner product space (Debnath and Mikusinski, 1999, p. 90; Kreyszig, 1978, p. 137). The following specific forms are useful for subsequent material.

**Theorem 2.15 Schwarz's Inequality**
If $f,g\in\boldsymbol{L}^2[\alpha,\beta]$, then

$$\left|\int_\alpha^\beta f(x)g^*(x)dx\right|\le\sqrt{\int_\alpha^\beta|f(x)|^2dx}\sqrt{\int_\alpha^\beta|g(x)|^2dx}\tag{2.97}$$

If $f, g \in \boldsymbol{L}^2$, then

$$\left|\int_{-\infty}^{\infty} f(x)g^*(x)dx\right| \le \sqrt{\int_{-\infty}^{\infty} |f(x)|^2 dx}\sqrt{\int_{-\infty}^{\infty} |g(x)|^2 dx} \tag{2.98}$$

If $\sum_{i=1}^{\infty} |f_i|^2$ and $\sum_{i=1}^{\infty} |g_i|^2$ are finite, then

$$\left|\sum_{i=1}^{\infty} f_i g_i^*\right| \le \sqrt{\sum_{i=1}^{\infty} |f_i|^2}\sqrt{\sum_{i=1}^{\infty} |g_i|^2} \tag{2.99}$$

**Proof**

Consider the first result and assume $f, g \in \boldsymbol{L}^2[\alpha, \beta]$. Equality holds if either, or both, of $f$, $g$ is zero almost everywhere. Assume $g$ is non-zero on a set of non-zero measure, and consider the inequality

$$\begin{aligned} 0 &\le \int_{\alpha}^{\beta} |f(x) - kg(x)|^2 dx \\ &= \int_{\alpha}^{\beta} |f(x)|^2 dx + |k|^2 \int_{\alpha}^{\beta} |g(x)|^2 dx - k^* \int_{\alpha}^{\beta} f(x)g^*(x)dx - k\int_{\alpha}^{\beta} f^*(x)g(x)dx \end{aligned} \tag{2.100}$$

which is valid for any $k \in \boldsymbol{C}$. For the case where $k = \int_{\alpha}^{\beta} f(x)g^*(x)dx \Big/ \int_{\alpha}^{\beta} |g(x)|^2 dx$, it follows that

$$0 \le \int_{\alpha}^{\beta} |f(x)|^2 dx - \left|\int_{\alpha}^{\beta} f(x)g^*(x)dx\right|^2 \Big/ \int_{\alpha}^{\beta} |g(x)|^2 dx \tag{2.101}$$

and the required result directly follows. The proofs for the other two stated results follow in an analogous manner.

***2.6.3.2 Interchanging Integration Order and Summation Order*** It is often the case that the order of integration in multiple integrals needs to be interchanged. The Fubini–Tonelli theorems, summarized below, specify when this interchange is valid (Champeney, 1987, p. 18).

**Theorem 2.16 Fubini–Tonelli Theorem**
If $f : \boldsymbol{R}^2 \to \boldsymbol{C}$ is measurable and defined almost everywhere, and one of the integrals

$$\int\int |f(x,y)|dxdy \qquad \int\int |f(x,y)|dydx \tag{2.102}$$

can be shown to be finite, then $f$ is integrable over the $x$, $y$ plane and

$$\int f(x,y)d(x,y) = \int\left[\int f(x,y)dxdy = \int\left[\int f(x,y)dydx \tag{2.103}$$

This result can be generalized to higher dimensions in a straightforward manner.

An analogous result follows for interchanging the order of summation in double summations.

**Theorem 2.17 Absolute Convergence Implies Summation Interchange**
If $\sum_{i=1}^{\infty}\sum_{j=1}^{\infty}|x_{ij}|$ is finite, then the following summations are equal:

$$\sum_{i=1}^{\infty}\left[\sum_{j=1}^{\infty}x_{ij}\right] = \sum_{j=1}^{\infty}\left[\sum_{i=1}^{\infty}x_{ij}\right] \tag{2.104}$$

**Proof**
See, for example, Hirschman (1962, pp. 119–121).

## 2.7 CONVERGENCE

The standard forms of convergence for a sequence of functions include pointwise convergence, uniform convergence, and convergence in the mean. A useful reference is Champeney (1987, ch. 4). For notational convenience, a sequence of functions $\{f_n : \boldsymbol{R} \to \boldsymbol{C}\}_{n=1}^{\infty}$, or $\{f_n : \boldsymbol{R} \to \boldsymbol{R}\}_{n=1}^{\infty}$, is written as $\{f_n\}$.

**Definition: Pointwise and Uniform Convergence** A sequence of functions $\{f_n\}$ converges pointwise to a function $f : \boldsymbol{R} \to \boldsymbol{C}$ on a set $E$, that is, $\lim_{n\to\infty} f_n(x) = f(x)$, if for all $x \in E$, it is the case that

$$\forall \varepsilon > 0 \ \ \exists N_o > 0 \ \text{ such that } \ \forall n > N_o \ \ |f_n(x) - f(x)| < \varepsilon \tag{2.105}$$

where $N_o$, in general, depends on $x$ and $\varepsilon$. When $N_o$ is independent of the value of $x \in E$, the convergence is said to be uniform over $E$.

**Definition: Convergence in the Mean** A sequence of functions $\{f_n\}$ converges to a function $f$ *in the mean* on a set $E$ if

$$\lim_{n\to\infty}\int_E |f_n(x)-f(x)|dx=0 \tag{2.106}$$

that is,

$$\forall\varepsilon>0\ \ \exists N_o>0 \quad \text{such that} \quad \forall n>N_o \quad \int_E |f_n(x)-f(x)|dx<\varepsilon \tag{2.107}$$

### 2.7.1 Dominated and Monotone Convergence

In analysis, it is often useful to be able to interchange the order of the limit and integral operations. The two important results, which give sufficient conditions for such an interchange, are the monotone and dominated convergence theorems (Champeney, 1987, p. 26).

**Theorem 2.18 Monotone Convergence Theorem**
If $\{f_n\}$ is a sequence of nonnegative functions such that $f_n\in\boldsymbol{L}, f_n(x)\le f_{n+1}(x)$ for all $n$ and all $x$, except on a set of zero measure, and $\int f_n<A_o$, $A_o\in\boldsymbol{R}^+$ for all $n$, then there exists a function $f\in\boldsymbol{L}$ such that

$$\lim_{n\to\infty} f_n(x)=f(x) \qquad \text{pointwise a.e.} \tag{2.108}$$

$$\lim_{n\to\infty}\int f_n(x)dx=\int \lim_{n\to\infty} f_n(x)dx=\int f(x)dx \tag{2.109}$$

**Theorem 2.19 Dominated Convergence Theorem**
If $\{f_n\}$ is a sequence of Lebesgue integrable functions, that is, $f_n\in\boldsymbol{L}$ for all $n$, $\lim_{n\to\infty} f_n(x)=f(x)$ pointwise almost everywhere, and there exists a function $g\in\boldsymbol{L}$ with the property $|f_n|<|g|$ almost everywhere, and for all $n$, then

$$\lim_{n\to\infty}\int f_n(x)dx=\int \lim_{n\to\infty} f_n(x)dx=\int f(x)dx \tag{2.110}$$

that is, $\forall\varepsilon>0$, there exists a constant $N_o$ such that for all $n>N_o$, it is the case that

$$\left|\int [f_n(x)-f(x)]dx\right|<\varepsilon \tag{2.111}$$

The functions associated with dominated convergence are illustrated in Figure 2.16.
Dominated convergence implies convergence in the mean.

**Theorem 2.20 Dominated Convergence Implies Convergence in the Mean**
Consider a sequence of functions $\{f_n\}$, where $f_n\in\boldsymbol{L}$ and $\lim_{n\to\infty} f_n(x)=f(x)$ pointwise almost everywhere on a set $E$. Dominated convergence of $\{f_n\}$ implies convergence in the mean, that is,

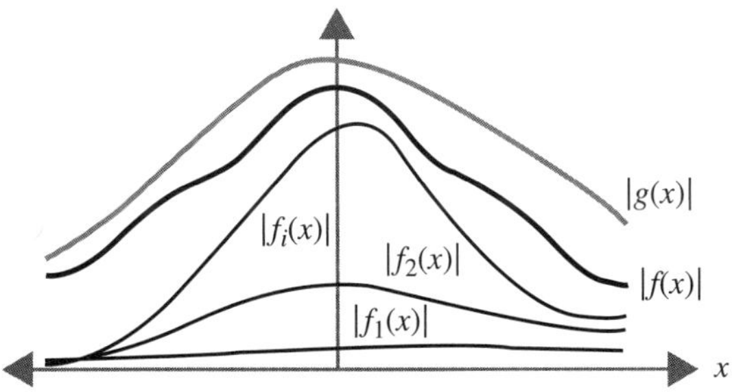


**FIGURE 2.16** Illustration of functions that are consistent with dominated convergence.

$$\lim_{n\to\infty}\int_E |f_n(x)-f(x)dx| = 0 \tag{2.112}$$

**Proof**

Dominated convergence implies that for all $\varepsilon > 0$, there exists a constant $N_o$ and for all $n > N_o$ and for all subsets $E_i$ of $E$, it is the case that

$$\left|\int_{E_i} [f_n(x)-f(x)]dx\right| < \varepsilon/2 \tag{2.113}$$

For all $n > N_o$, the set $E$ can be partitioned according to

$$\begin{aligned} E_1 &= \{x \in E:\ f_n(x)-f(x) \geq 0\} \\ E_2 &= \{x \in E:\ f_n(x)-f(x) < 0\} \end{aligned} \tag{2.114}$$

whereupon it follows that

$$\int_{E_1} [f_n(x)-f(x)]dx < \varepsilon/2 \qquad \int_{E_2} -[f_n(x)-f(x)]dx < \varepsilon/2 \tag{2.115}$$

The required result then follows, namely,

$$\int_E |f_n(x)-f(x)|dx < \varepsilon \tag{2.116}$$

## 2.8 LEBESGUE–STIELTJES MEASURE

The Lebesgue–Stieltjes measure is a generalization of the Lebesgue measure. This measure underpins Lebesgue–Stieltjes integration, and this finds application in probability and random process theory. References for Lebesgue–Stieltjes measure and

Lebesgue–Stieltjes integration include Carter and van Brunet (2000) and Haaser and Sullivan (1991).

### 2.8.1 Lebesgue–Stieltjes Measure: Monotonic Function Case

Lebesgue–Stieltjes measure is first defined assuming a monotonic function.

***2.8.1.1 Lebesgue–Stieltjes Measure: Open Interval*** The basis for the Lebesgue–Stieltjes measure is a measure defined over an open interval.

**Definition: Lebesgue–Stieltjes Measure over an Open Interval** The Lebesgue–Stieltjes measure over an open interval (α, β), as determined by a monotonically increasing function $g:\boldsymbol{R}\rightarrow\boldsymbol{R}$, is denoted $m_g$ and is defined according to

$$m_g[(\alpha,\beta)] = g(\beta^-) - g(\alpha^+) \tag{2.117}$$

The Lebesgue–Stieltjes measure is illustrated in Figure 2.17 and equals the change in the function $g$ over the open interval (α, β).

***2.8.1.2 Notes*** For the case of a function $g(x) = x$, the Lebesgue–Stieltjes measure defaults to the Lebesgue measure as

$$m_g[(\alpha,\beta)] = \beta - \alpha = m[(\alpha,\beta)] \tag{2.118}$$

***2.8.1.3 Lebesgue–Stieltjes Measure over an Arbitrary Set*** The aforementioned definition is the basis for the general definition of the Lebesgue–Stieltjes measure.

**Definition: Lebesgue–Stieltjes Measure over an Arbitrary Set** The Lebesgue–Stieltjes measure defined by a monotonically increasing function $g:\boldsymbol{R}\rightarrow\boldsymbol{R}$, and denoted $m_g$, is a mapping from the Borel sigma algebra of $\boldsymbol{R}$ to the positive real line, that is, $m_g:\Omega_B\rightarrow\boldsymbol{R}$. For an arbitrary set $E\subseteq\Omega_B$, the definition is

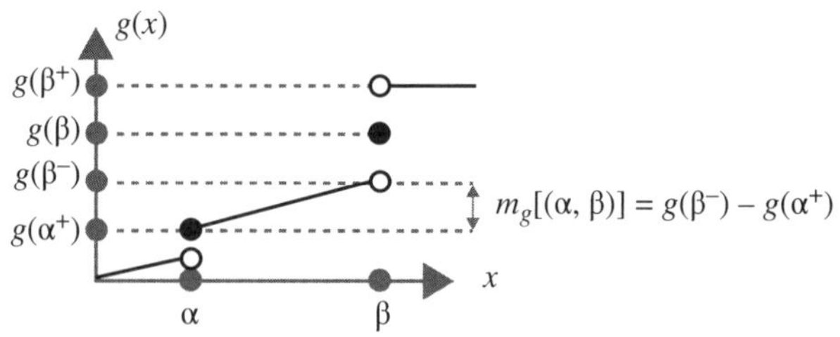


**FIGURE 2.17** Illustration of the Lebesgue–Stieltjes measure, over the open interval (α, β), for a monotonically increasing function $g$.

$$\begin{aligned} m_g[E] &= \inf\left\{ \sum_{i=1}^{\infty} m_g(I_i) : E \subseteq \bigcup_{i=1}^{\infty} I_i, I_i = (\alpha_i, \beta_i), \alpha_i, \beta_i \in \boldsymbol{R} \right\} \\ m_g[\{\}] &= 0 \end{aligned} \quad (2.119)$$

The Lebesgue–Stieltjes measure is an outer measure and, for a set $E$, is the minimum of the sum of Lebesgue–Stieltjes measures defined by open intervals that cover the set $E$. If a suitable subcollection of sets, $A_1, A_2, \ldots$, based on open intervals and which cover $E$, can be ordered such that $A_{i+1} \subseteq A_i$, then

$$m_g[E] = \lim_{i\to\infty} m_g[A_i] \quad (2.120)$$

It is easy to prove that the Lebesgue–Stieltjes measure satisfies the three properties, as required and given by Equations 2.37–2.39.

The following important property holds for the Lebesgue–Stieltjes measure.

**Theorem 2.21 Measure of Sum Equals Sum of Measures**
If a monotonically increasing function $g$ can be decomposed into a finite summation of monotonically increasing functions $g_1, \ldots, g_N$, then the Lebesgue–Stieltjes measure defined by $g$ over a set $E$ equals the sum of the Lebesgue–Stieltjes measures over the set $E$ defined by the functions $g_1, \ldots, g_N$, that is,

$$m_g[E] = \sum_{k=1}^{N} m_{g_k}[E] \qquad g(x) = \sum_{k=1}^{N} g_k(x) \quad x \in E \quad (2.121)$$

**Proof**
Consider the definition given by Equation 2.120. The ability to interchange the order of the limit and summation operations for a finite sum yields the required result, that is,

$$\begin{aligned} m_g[E] &= \lim_{i\to\infty} m_g[A_i] \qquad\qquad A_i = (\alpha_i, \beta_i) \\ &= \lim_{i\to\infty} \left[g(\beta_i^-) - g(\alpha_i^+)\right] = \lim_{i\to\infty}\left[\sum_{k=1}^{N} g_k(\beta_i^-) - g_k(\alpha_i^+)\right] \\ &= \sum_{k=1}^{N} \lim_{i\to\infty} \left[g_k(\beta_i^-) - g_k(\alpha_i^+)\right] = \sum_{k=1}^{N} m_{g_k}[E] \end{aligned} \quad (2.122)$$

In general, $A_i = (\alpha_{i1}, \beta_{i1}) \cup (\alpha_{i2}, \beta_{i2}) \cup \cdots$, and for this case, the proof follows in an analogous manner assuming disjoint open intervals and the additivity of the Lebesgue–Stieltjes measure over disjoint sets.

***2.8.1.4 Lebesgue–Stieltjes Measurable Set*** As the Lebesgue–Stieltjes measure is a generalization of the Lebesgue measure, it is possible to construct two disjoint sets whose Lebesgue–Stieltjes measures are not additive. In a similar manner to that used for the Lebesgue measure, the following definition is used to restrict the class of sets to those with the property of additive measure over disjoint sets.

**Definition: Lebesgue–Stieltjes Measurable Set** A set $E \subseteq \boldsymbol{R}$ is Lebesgue–Stieltjes measurable with respect to a monotonically increasing function $g : \boldsymbol{R} \to \boldsymbol{R}$ if, for all sets $A \subseteq \Omega_B$, it is the case that

$$m_g[A] = m_g[A \cap E] + m_g\left[A \cap E^C\right] \tag{2.123}$$

Thus, a set $E$ is Lebesgue–Stieltjes measurable if it, and its complement $E^C$, partitions all Borel sets into disjoint sets whose Lebesgue–Stieltjes measures are additive.

**Theorem 2.22 Additive Property of Lebesgue–Stieltjes Measure**
If $E_1, \ldots, E_N, \ldots$ are disjoint Lebesgue–Stieltjes measurable sets with respect to $g : \boldsymbol{R} \to \boldsymbol{R}$, then

$$m_g\left[\bigcup_{i=1}^{\infty} E_i\right] = \sum_{i=1}^{\infty} m_g[E_i] \tag{2.124}$$

**Proof**
Consider a set $A = E_1 \cup E_2$, where $E_1$ and $E_2$ are disjoint, and a set $E = E_1$. From the definition of a Lebesgue–Stieltjes measurable set, it follows that

$$\begin{aligned} m_g[E_1 \cup E_2] &= m_g[(E_1 \cup E_2) \cap E_1] + m_g\left[(E_1 \cup E_2) \cap E_1^C\right] \\ &= m_g[E_1] + m_g[E_2] \end{aligned} \tag{2.125}$$

as required. The result for the general case follows by induction.

***2.8.1.5 Lebesgue–Stieltjes Measure of Elementary Sets*** The following two theorems state fundamental results for the Lebesgue–Stieltjes measure of elementary sets comprising of a point and open and closed intervals.

**Theorem 2.23 Lebesgue–Stieltjes Measure at a Point**
If $g : \boldsymbol{R} \to \boldsymbol{R}$ is monotonically increasing, then the Lebesgue–Stieltjes measure defined by $g$ at a point $\alpha$ is

$$m_g[\{\alpha\}] = g(\alpha^+) - g(\alpha^-) \tag{2.126}$$

This result is illustrated in Figure 2.18.

**Proof**
Consider the result $\{\alpha\} = \bigcap_{i=1}^{\infty} \left(\alpha - \frac{1}{i}, \alpha + \frac{1}{i}\right)$. It then follows that

$$\begin{aligned} m_g[\{\alpha\}] &= \inf\left\{m_g[I_i] : \ I_i = (\alpha - 1/i, \alpha + 1/i), i \in \boldsymbol{Z}^+\right\} \\ &= \inf\left\{g[(\alpha + 1/i)^-] - g[(\alpha - 1/i)^+] : \ i \in \boldsymbol{Z}^+\right\} \\ &= g(\alpha^+) - g(\alpha^-) \end{aligned} \tag{2.127}$$

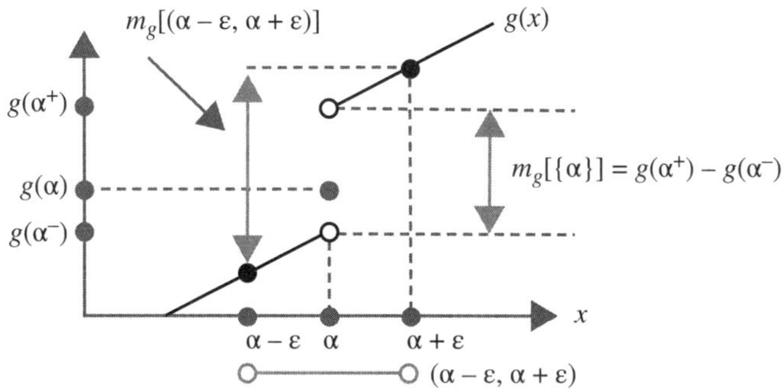


**FIGURE 2.18** Illustration of the Lebesgue–Stieltjes measure at a point.

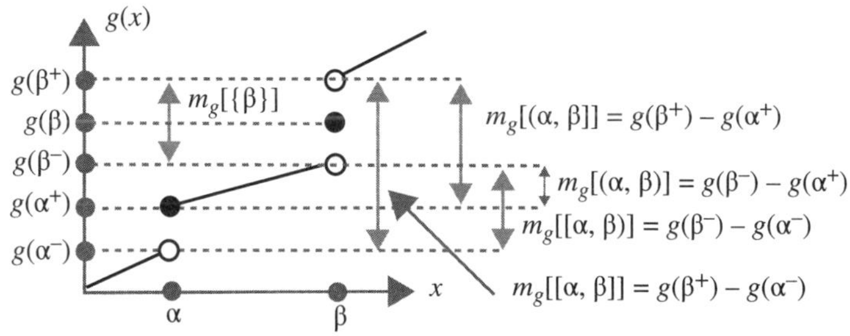


**FIGURE 2.19** Illustration of the Lebesgue–Stieltjes measure for a monotonically increasing function $g$ over the intervals $(\alpha, \beta)$, $(\alpha, \beta]$, $[\alpha, \beta)$, and $[\alpha, \beta]$.

**Theorem 2.24 Lebesgue–Stieltjes Measure over Open and Closed Intervals**
If $g: \boldsymbol{R} \to \boldsymbol{R}$ is monotonically increasing, then the Lebesgue–Stieltjes measures defined by $g$, over the intervals $(\alpha, \beta)$, $(\alpha, \beta]$, $[\alpha, \beta)$, and $[\alpha, \beta]$, are

$$\begin{aligned} m_g[(\alpha,\beta)] &= g(\beta^-) - g(\alpha^+) \quad & m_g[[\alpha,\beta]] &= g(\beta^+) - g(\alpha^-) \\ m_g[(\alpha,\beta]] &= g(\beta^+) - g(\alpha^+) \quad & m_g[[\alpha,\beta)] &= g(\beta^-) - g(\alpha^-) \end{aligned} \tag{2.128}$$

These measures are illustrated in Figure 2.19.

**Proof**
The result for the interval $(\alpha, \beta)$ follows by definition.

For the interval $[\alpha, \beta]$, the additivity property of the Lebesgue–Stieltjes measure for disjoint sets can be used as follows:

$$\begin{aligned} m[[\alpha,\beta]] &= m_g[\{\alpha\}] + m_g[(\alpha,\beta)] + m_g[\{\beta\}] \\ &= [g(\alpha^+) - g(\alpha^-)] + [g(\beta^-) - g(\alpha^+)] + [g(\beta^+) - g(\beta^-)] \\ &= g(\beta^+) - g(\alpha^-) \end{aligned} \tag{2.129}$$

A similar approach can be used for the intervals $[\alpha,\beta)$ and $(\alpha,\beta]$. An alternative approach for the interval $[\alpha,\beta)$ is to consider the equivalence

$$[\alpha,\beta) = \bigcap_{i=1}^{\infty}\left(\alpha-\frac{1}{i},\beta\right) \tag{2.130}$$

It then follows that

$$\begin{aligned}\inf\left\{m_g[I_i]:\ I_i=\left(\alpha-\frac{1}{i},\beta\right), i\in \boldsymbol{Z}^+\right\} &= \inf\left\{g(\beta^-)-g\left(\left(\alpha-\frac{1}{i}\right)^+\right), i\in \boldsymbol{Z}^+\right\}\\ &= g(\beta^-)-g(\alpha^-)\end{aligned} \tag{2.131}$$

**2.8.1.6 *Example*** Consider $g(x)=\begin{cases}1 & x>x_o\\ 0 & x<x_o\end{cases}$. It is the case that $m_g[\{x_o\}]=1$, $m_g[(-\infty,x_o)]=m_g[(x_o,\infty)]=0$, $m_g[(-\infty,x_o]]=1$, and $m_g[[x_o,\infty)]=1$.

**2.8.1.7 *Alternative Definition for Lebesgue–Stieltjes Measure*** It is common to define the Lebesgue–Stieltjes measure, for a monotonically increasing function $g:\boldsymbol{R}\to\boldsymbol{R}$, on the half-open interval $(\alpha,\beta]$ rather than the open interval. For this case, the definition is

$$m_g[(\alpha,\beta]] = g(\beta^+)-g(\alpha^+) \tag{2.132}$$

The two definitions are equivalent as

$$\begin{aligned}m_g[(\alpha,\beta]] &= m_g[(\alpha,\beta)]+m_g[\{\beta\}] = [g(\beta^-)-g(\alpha^+)]+[g(\beta^+)-g(\beta^-)]\\ &= g(\beta^+)-g(\alpha^+)\end{aligned} \tag{2.133}$$

### 2.8.2 Lebesgue–Stieltjes Measure: Decreasing Function

A measure, by definition, is a mapping from sets to the positive real line, and the Lebesgue–Stieltjes measure establishes the positive change of an increasing function over a set. The issue is how to define a Lebesgue–Stieltjes measure for the case of a nonincreasing function where the change is zero or negative. The approach is to define the Lebesgue–Stieltjes measure as the magnitude of the change in the function over the specified interval as illustrated in Figure 2.20 for the case of a monotonically decreasing function. The definitions are as follows.

**Definition: Lebesgue–Stieltjes Measure—Decreasing Function, Open Interval**
The Lebesgue–Stieltjes measure over an open interval $(\alpha,\beta)$, and for a monotonically decreasing function $g:\boldsymbol{R}\to\boldsymbol{R}$, is defined according to

$$m_g[(\alpha,\beta)] = g(\alpha^+)-g(\beta^-) \tag{2.134}$$

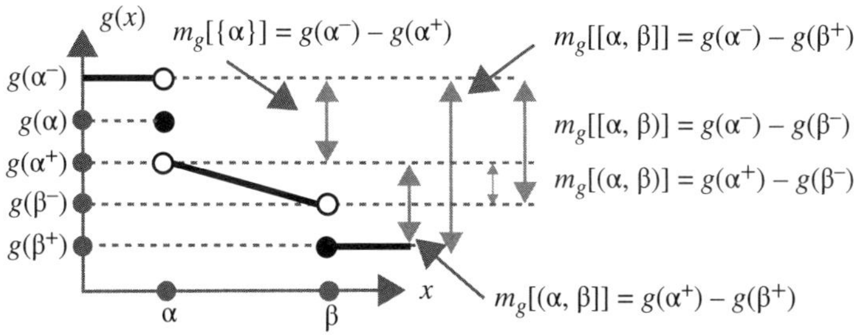


**FIGURE 2.20** Illustration of the measure defined by a monotonically decreasing function $g$ at a point and over the intervals $(\alpha, \beta)$, $(\alpha, \beta]$, $[\alpha, \beta)$, and $[\alpha, \beta]$.

The definition is such as to define the magnitude of the change of the function over the open interval.

**Definition: Lebesgue–Stieltjes Measure—Decreasing Function, Arbitrary Set**
The Lebesgue–Stieltjes measure, denoted $m_g$, defined by a monotonically decreasing function $g:\boldsymbol{R}\to\boldsymbol{R}$ over a Lebesgue–Stieltjes measurable set $E$, is

$$m_g[E] = \inf\left\{\sum_{i=1}^{\infty} m_g[I_i]:\ E\subseteq\bigcup_{i=1}^{\infty} I_i,\ \ I_i=(\alpha_i,\beta_i),\alpha_i,\beta_i\in\boldsymbol{R}\right\} \tag{2.135}$$

where $m_g$ is defined by Equation 2.134.

Using these definitions, the Lebesgue–Stieltjes measure for closed and half-open intervals can be specified.

**Theorem 2.25 Lebesgue–Stieltjes Measure: Point and Intervals**
If $g:\boldsymbol{R}\to\boldsymbol{R}$ is monotonically decreasing, then the Lebesgue–Stieltjes measures defined by $g$ at a point $\alpha$, and over the intervals $(\alpha, \beta)$, $(\alpha,\beta]$, $[\alpha,\beta)$, and $[\alpha, \beta]$, respectively, are

$$m_g[\{\alpha\}] = g(\alpha^-) - g(\alpha^+) \tag{2.136}$$

$$\begin{aligned} m_g[(\alpha,\beta)] &= g(\alpha^+) - g(\beta^-) \quad & m_g[[\alpha,\beta]] &= g(\alpha^-) - g(\beta^+) \\ m_g[(\alpha,\beta]] &= g(\alpha^+) - g(\beta^+) \quad & m_g[[\alpha,\beta)] &= g(\alpha^-) - g(\beta^-) \end{aligned} \tag{2.137}$$

These results are illustrated in Figure 2.20.

**Proof**
The proof follows in an analogous manner to that of Theorems 2.23 and 2.24.

### 2.8.3 Lebesgue–Stieltjes Measure: General Case

Consider the general case, as illustrated in Figure 2.21, of a function $g:\boldsymbol{R}\to\boldsymbol{R}$ that both increases and decreases. To define the Lebesgue–Stieltjes measure for such a function, the approach is to decompose the function into its increasing, $g_I:\boldsymbol{R}\to\boldsymbol{R}$,

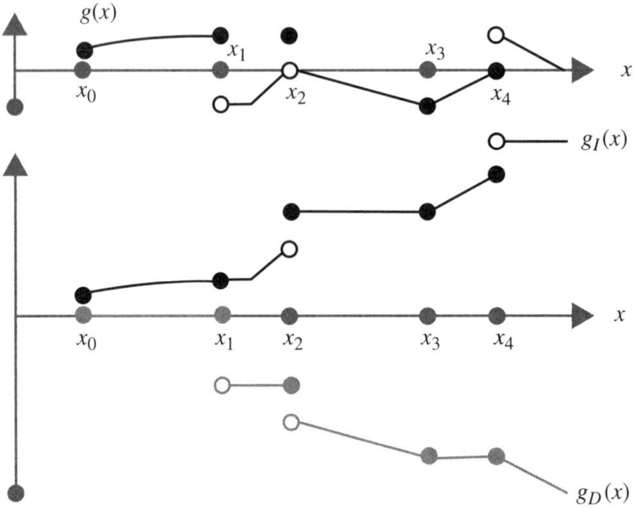


**FIGURE 2.21** Illustration of the increasing and decreasing components of a function.

and decreasing, $g_D:\boldsymbol{R}\rightarrow\boldsymbol{R}$, components and to then use the defined measures for these individual components.

The decomposition of a function into its increasing and decreasing function components is illustrated in Figure 2.21. Such a decomposition can readily be achieved by writing the function in a manner as specified in the following theorem:

**Theorem 2.26 Function Form to Facilitate Decomposition**
Assume a bounded function $g:\boldsymbol{R}\rightarrow\boldsymbol{R}$ has, at most, a finite number of changes, as defined by a change from being an increasing to a decreasing function, on all finite intervals. These changes are assumed to be determined by discontinuities in the function, or a change in the sign of its derivative, or points where $\text{sgn}[g(x+\Delta)-g(x)]\neq\text{sgn}[g(x)-g(x-\Delta)]$ as $\Delta\rightarrow 0$. The points of change are denoted $x_1, x_2, \ldots, x_N$. For the interval $[x_0, x_{N+1}]$, and with $x_1, x_2, \ldots, x_N \in (x_0, x_{N+1})$, such a function can be written as

$$
\begin{aligned}
&g(x)=g(x_0)+\left[g(x)-g\left(x_0^+\right)\right] \quad x_0\leq x<x_1\\
&g(x)=g(x_0)+\left[g\left(x_1^-\right)-g\left(x_0^+\right)\right]+\left[g(x_1)-\left[g\left(x_1^-\right)\right] \quad x=x_1\\
&g(x)=g(x_0)+\sum_{k=1}^{n(x)}\left[g\left(x_k^-\right)-g\left(x_{k-1}^+\right)\right]+\sum_{k=1}^{n(x)}\left[g(x_k)-g\left(x_k^-\right)\right]+\\
&\qquad\sum_{k=1}^{n(x)}\left[g\left(x_k^+\right)-g(x_k)\right]+g(x)-g\left(x_{n(x)}^+\right) \quad \begin{cases}x\in(x_i,x_{i+1})\\ i\in\{1,2,\ldots,N\}\end{cases} \qquad (2.138)\\
&g(x)=g(x_0)+\sum_{k=1}^{n(x)}\left[g\left(x_k^-\right)-g\left(x_{k-1}^+\right)\right]+\sum_{k=1}^{n(x)-1}\left[g(x_k)-g\left(x_k^-\right)\right]+\\
&\qquad\sum_{k=1}^{n(x)-1}\left[g\left(x_k^+\right)-g(x_k)\right]+g\left(x_{n(x)}\right)-g\left(x_{n(x)}^-\right) \quad \begin{cases}x=x_{n(x)}\\ n(x)\in\{2,\ldots,N\}\end{cases}
\end{aligned}
$$

where $g(x_0^+) = g(x_0)$ and $n(x)$ is the number of points of change in the interval $(x_0, x]$ and this is defined by

$$n(x) = \sum_{i=1}^{N} u[x - x_i] \tag{2.139}$$

The first summation in Equation 2.138 defines the changes in the intervals $(x_{k-1}, x_k), k \in \{1, \ldots, n(x)\}$, while the second and third summations define the change at the point of discontinuities $x_1, \ldots, x_{n(x)}$.

**Proof**

The proof follows from the construction of the component functions that specify the change in the function over intervals as determined by the points of change and the change in the function at the points of change.

The stated restrictions avoid the problems associated with a function such as $x\cos(1/x)$ which changes from increasing to decreasing an infinite number of times on any interval that includes the origin.

**Theorem 2.27 Decomposition into Increasing and Decreasing Components**

The increasing $g_I : \boldsymbol{R} \to \boldsymbol{R}$ and decreasing $g_D : \boldsymbol{R} \to \boldsymbol{R}$ components of a function, which satisfy the conditions stated in Theorem 2.27, are defined, using a generic symbol $G$, according to

$$G(x) = H_0[g(x_0)] + H\left[g(x) - g(x_0^+)\right] \qquad x_0 \le x < x_1$$

$$G(x) = H_0[g(x_0)] + H\left[g(x_1^-) - g(x_0^+)\right] + H\left[g(x_1) - g(x_1^-)\right] \quad x = x_1$$

$$\begin{aligned} G(x) = H_0[g(x_0)] &+ \sum_{k=1}^{n(x)} H\left[g(x_k^-) - g(x_{k-1}^+)\right] + \sum_{k=1}^{n(x)} H\left[g(x_k) - g(x_k^-)\right] \\ &+ \sum_{k=1}^{n(x)} H\left[g(x_k^+) - g(x_k)\right] + H\left[g(x) - g\left(x_{n(x)}^+\right)\right] \quad \begin{cases} x \in (x_i, x_{i+1}) \\ i \in \{1, 2, \ldots, N\} \end{cases} \end{aligned} \tag{2.140}$$

$$\begin{aligned} G(x) = H_0[g(x_0)] &+ \sum_{k=1}^{n(x)} H\left[g(x_k^-) - g(x_{k-1}^+)\right] + \sum_{k=1}^{n(x)-1} H\left[g(x_k) - g(x_k^-)\right] \\ &+ \sum_{k=1}^{n(x)-1} H\left[g(x_k^+) - g(x_k)\right] + H\left[g\left(x_{n(x)}\right) - g\left(x_{n(x)}^-\right)\right] \begin{cases} x = x_{n(x)} \\ n(x) \in \{2, \ldots, N\} \end{cases} \end{aligned}$$

where $H_0[g(x_0)] = g(x_0)u\left[g(x_1^-) - g(x_0)\right]$, $H(q) = qu(q)$ for the case of the increasing function $g_I$ and $H_0[g(x_0)] = g(x_0)\left(1 - u\left[g(x_1^-) - g(x_0)\right]\right)$, $H[q] = q(1 - u(q))$ for the decreasing function $g_D$.

By construction

$$g(x) = g_I(x) + g_D(x) \tag{2.141}$$

and for all sets of points $(\alpha_1, \alpha_2, \ldots, \alpha_N)$, such that $\alpha_1 \le \alpha_2 \le \cdots \le \alpha_N$, it is the case that

$$g_I(\alpha_1) \le \cdots \le g_I(\alpha_N) \qquad g_D(\alpha_1) \ge \cdots \ge g_D(\alpha_N) \tag{2.142}$$

**Proof**

The proof follows from Theorem 2.26 and by noting that the function $H$ determines whether the component of the function it is acting on is increasing or decreasing.

***2.8.3.1 Lebesgue–Stieltjes Measure: General Case*** The Lebesgue–Stieltjes measure for the general case is defined consistent with the net change in the function as determined by the change due to the increasing component of a function minus the change due to the decreasing component.

For a function with both increasing and decreasing components, there are two approaches for defining the Lebesgue–Stieltjes measure over an arbitrary set $E$. First,

$$m_g[E] = m_{g_I}[E] - m_{g_D}[E] \tag{2.143}$$

where the appropriate Lebesgue–Stieltjes measure definition is used for the increasing, $g_I$, and decreasing, $g_D$, component functions. Note that the increasing and decreasing component functions are changing on disjoint sets. If one component is changing on a set $E_1$, then the other has zero change over $E_1$.

The alternative approach is to write the function in terms of the difference of two increasing functions according to

$$g(x) = g_I(x) - [-g_D(x)] \tag{2.144}$$

which leads to

$$m_g[E] = m_{g_I}[E] - m_{-g_D}[E] \tag{2.145}$$

where the Lebesgue–Stieltjes measure definition used is that for an increasing function.

Naturally, the two approaches lead to the same result: the Lebesgue–Stieltjes measure defined by a function $g$ over a set $E$ is the difference between the Lebesgue–Stieltjes measures of $g_I$ and $-g_D$ and gives the magnitude of the change of the function $g$ as a set is traversed after taking into account the increasing and decreasing changes. Formally:

**Definition: Lebesgue–Stieltjes Measure for a Set—General Case** Consider a function $g : \boldsymbol{R} \to \boldsymbol{R}$ that satisfies the conditions such that its increasing and decreasing components are well defined. The Lebesgue–Stieltjes measure of $g$, denoted $m_g$, over a Lebesgue–Stieltjes measurable set $E$, is the change in the increasing component of the function over the set $E$ minus the change in the decreasing component over the set $E$:

$$m_g[E] = \inf\left\{\sum_{i=1}^{\infty} m_{g_I}[I_i]:\ E \subset \bigcup_{i=1}^{\infty} I_i,\ \ I_i = (\alpha_i, \beta_i), \alpha_i, \beta_i \in \boldsymbol{R}\right\}$$
$$- \inf\left\{\sum_{i=1}^{\infty} m_{-g_D}[I_i]:\ E \subset \bigcup_{i=1}^{\infty} I_i,\ \ I_i = (\alpha_i, \beta_i), \alpha_i, \beta_i \in \boldsymbol{R}\right\} \quad (2.146)$$

An alternative, but equivalent, definition is:

**Definition: Lebesgue–Stieltjes Measure for a Set—General Case** The Lebesgue–Stieltjes measure of $g: \boldsymbol{R} \to \boldsymbol{R}$, over a Lebesgue–Stieltjes measurable set $E$, is the sum of the changes in the function when the function is increasing minus the sum of the changes when the function is decreasing:

$$m_g[E] = m_g[E_I] - m_{-g}[E_D]$$
$$= \inf\left\{\sum_{i=1}^{\infty} m_g[I_i]:\ E_I \subset \bigcup_{i=1}^{\infty} I_i,\ \ I_i = (\alpha_i, \beta_i), \alpha_i, \beta_i \in \boldsymbol{R}\right\}$$
$$- \inf\left\{\sum_{i=1}^{\infty} m_{-g}[I_i]:\ E_D \subset \bigcup_{i=1}^{\infty} I_i,\ \ I_i = (\alpha_i, \beta_i), \alpha_i, \beta_i \in \boldsymbol{R}\right\} \quad (2.147)$$

where

$$E = E_I \cup E_D \qquad E_I \cap E_D = \emptyset \quad (2.148)$$

and $g$ is constant or increasing on $E_I$ and $g$ is constant or decreasing on $E_D$.

An example of the Lebesgue–Stieltjes measure for a function, with both increasing and decreasing components, is illustrated in Figure 2.22.

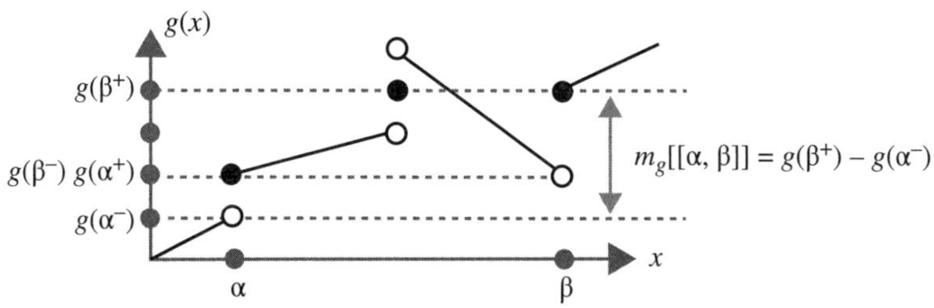


**FIGURE 2.22** Illustration of the Lebesgue–Stieltjes measure for a function $g$, with both increasing and decreasing components, over the interval $[\alpha, \beta]$.

## 2.9 LEBESGUE–STIELTJES INTEGRATION

The Lebesgue–Stieltjes integral is useful when a summation of changes, or a weighted summation of changes, in a function is of interest. One application of the integral is in probability theory:

### 2.9.1 Motivation

For a random variable $X$, with outcomes that take on specific discrete values with nonzero probability and also outcomes that form an uncountable set, the cumulative distribution function

$$F_X(x) = P[X \leq x] \tag{2.149}$$

is well defined. Here, $P$ is the probability operator. Consider the case, for example, where the cumulative distribution function has the form illustrated in Figure 2.23.

The possibility of either a discrete outcome or an outcome from an uncountable set results in the probability of an outcome not being completely specified by a probability mass function or a probability density function. As the expectation of a random variable is important, and this is readily defined for the case of a discrete random variable or a continuous random variable, the question arises as how to define this operator for the mixed case with both discrete and continuous outcomes. The expectation of a random variable $X$, denoted $\boldsymbol{E}[X]$, is the summation, over all values taken by the random variable, of the value taken times the probability of occurrence of the value. For the mixed case being considered, the expectation is

$$\boldsymbol{E}[X] = \lim_{\Delta \to 0} \sum_{i=-\infty}^{\infty} i\Delta \cdot [F_X((i+1)\Delta) - F_X(i\Delta)] \tag{2.150}$$

The limit of this summation is the Lebesgue–Stieltjes integral

$$\int_{\boldsymbol{R}} x dF(x) \tag{2.151}$$

where the notation $dF(x)$ is to be defined.

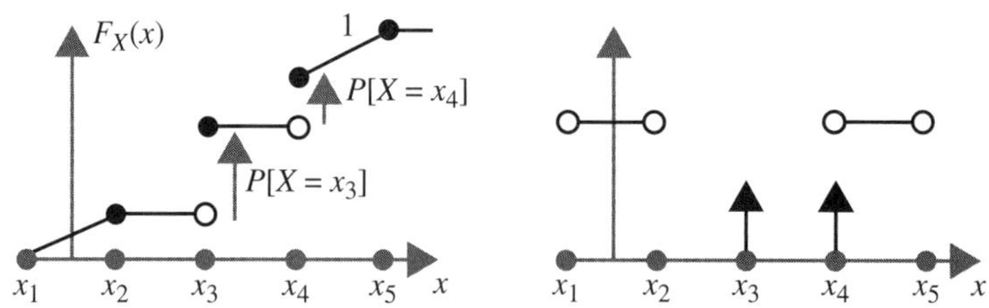


**FIGURE 2.23** Left: illustration of the cumulative distribution function of a random variable that takes on specific discrete values with nonzero probability and also has outcomes that form an uncountable set. Right: probability mass function for the discrete outcomes and probability density function for the outcomes defined on $(x_1, x_2)$ and $(x_4, x_5)$.

### 2.9.2 Lebesgue–Stieltjes Integral

The Lebesgue–Stieltjes integral is best introduced by considering a simple function.

#### *2.9.2.1 Step Function Case*

**Definition: Lebesgue–Stieltjes Integral of a Simple Function** The Lebesgue–Stieltjes integral of a simple function $s:\boldsymbol{R}\to\boldsymbol{R}$, with respect to the Lebesgue–Stieltjes measure defined by an increasing function $g:\boldsymbol{R}\to\boldsymbol{R}$, is

$$\int_{\boldsymbol{R}} s(x)dg(x)=\sum_{i=1}^{N} s_i m_g[S_i] \quad (2.152)$$

where

$$s(x)=\sum_{i=1}^{N} s_i I_{S_i}(x) \quad I_{S_i}(x)=\begin{cases}1 & x\in S_i \\ 0 & x\notin S_i\end{cases} \quad S_i=\{x:\ s(x)=s_i\} \quad (2.153)$$

This result holds irrespective of whether the step levels are positive or negative and equals the sum of the changes of $g$ over the sets defined by the simple function and weighted by the levels of the simple function.

As an example, consider the increasing function

$$g(x)=\begin{cases}0 & x<0 \\ x & 0\le x<\beta \\ \beta & x\ge\beta\end{cases} \quad (2.154)$$

and the step function $s$ illustrated in Figure 2.24. The step function partitions the real line into the intervals

$$S_0=(-\infty,0), S_1=[0,\alpha), S_2=[\alpha,\beta), S_3=\{\beta\}, S_4=(\beta,\infty) \quad (2.155)$$

and the Lebesgue–Stieltjes integral of $s$ with respect to $g$ is

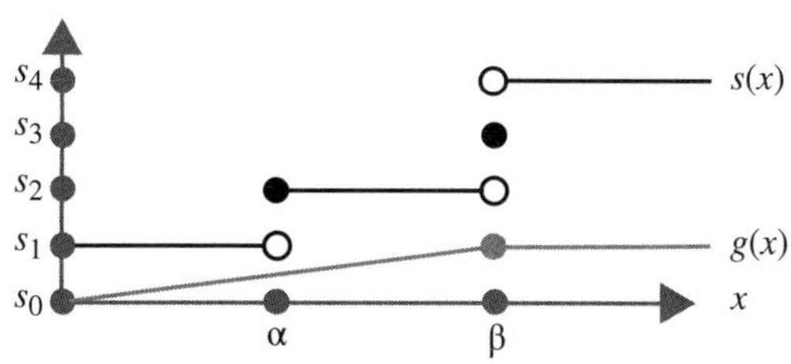


**FIGURE 2.24** Graph of the function $g$ and the step function $s$.

$$\int_{(-\infty,\infty)} s(x)dg(x) = s_0 m_g[S_0] + s_1 m_g[S_1] + s_2 m_g[S_2] + s_3 m_g[S_3] + s_4 m_g[S_4]$$
$$= s_0 \cdot 0 + s_1 \cdot \alpha + s_2 \cdot (\beta - \alpha) + s_3 \cdot 0 + s_4 \cdot 0 = \alpha(s_1 - s_2) + \beta s_2 \quad (2.156)$$

***2.9.2.2 Positive Function Case*** Consistent with Theorem 2.9, there exists a sequence of simple functions that approximate a measurable function *f*. This is the basis for the definition of the Lebesgue–Stieltjes integral of a function.

**Definition: Lebesgue–Stieltjes Integral—Positive Measurable Function Case** The Lebesgue–Stieltjes integral of a measurable function $f: \boldsymbol{R} \to \boldsymbol{R}$, which is positive, that is, $f(x) \geq 0, \ x \in \boldsymbol{R}$, with respect to the Lebesgue–Stieltjes measure defined by an increasing function $g: \boldsymbol{R} \to \boldsymbol{R}$, is

$$\int_{\boldsymbol{R}} f(x)dg(x) = \sup\left\{\int_{\boldsymbol{R}} s(x)dg(x): \quad s(x) \leq f(x), x \in \boldsymbol{R}, \quad s \text{ simple}\right\} \quad (2.157)$$

Alternatively,

$$\int_{\boldsymbol{R}} f(x)dg(x) = \lim_{i\to\infty} \int_{\boldsymbol{R}} s_i(x)dg(x) = \lim_{i\to\infty} \sum_{j=1}^{N_i} s_{ij} m_g[S_{ij}] \quad (2.158)$$

where $s_i: \boldsymbol{R} \to \{s_{i1}, \ldots, s_{iN_i}\}, \ i \in \boldsymbol{Z}^+$ is a sequence of simple functions, consistent with Equation 2.60, which converge to *f*, and where the *i*th simple function has the form

$$s_i(x) = \sum_{j=1}^{N_i} s_{ij} I_{S_{ij}}(x) \qquad S_{ij} = \{x: \quad s_i(x) = s_{ij}\} \quad (2.159)$$

The Lebesgue–Stieltjes integral can readily be extended to an arbitrary measurable function through use of the positive and negative components of a function consistent with $f(x) = f^+(x) - f^-(x)$. As both components are positive, the following definition is valid.

**Definition: Lebesgue–Stieltjes Integral of a Measurable Function** The Lebesgue–Stieltjes integral of a measurable function $f: \boldsymbol{R} \to \boldsymbol{R}$, with respect to the Lebesgue–Stieltjes measure defined by an increasing function $g: \boldsymbol{R} \to \boldsymbol{R}$, is

$$\int_{\boldsymbol{R}} f(x)dg(x) = \int_{\boldsymbol{R}} f^+(x)dg(x) - \int_{\boldsymbol{R}} f^-(x)dg(x) \quad (2.160)$$

The first integral yields a positive component; the second a negative component.

***2.9.2.3 Lebesgue–Stieltjes Integral: General Case*** The Lebesgue–Stieltjes integral can be extended to an arbitrary function $g$ by decomposing this function into its increasing and decreasing components denoted, consistent with the previous notation, as $g_I$ and $g_D$. As the Lebesgue–Stieltjes integral establishes the net change in a function $g$ on an interval, and weighted by the function $f$ on that interval, the following definition holds.

**Definition: Lebesgue–Stieltjes Integral—General Case** The Lebesgue–Stieltjes integral of a measurable function $f:\boldsymbol{R}\to\boldsymbol{R}$, with respect to the Lebesgue–Stieltjes measure defined by a function $g:\boldsymbol{R}\to\boldsymbol{R}$, is

$$\begin{aligned}\int_{\boldsymbol{R}} f(x)dg(x) &= \int_{\boldsymbol{R}} f^{+}(x)dg_I(x)-\int_{\boldsymbol{R}} f^{-}(x)dg_I(x)\\ &\quad -\left[\int_{\boldsymbol{R}} f^{+}(x)dg_D(x)-\int_{\boldsymbol{R}} f^{-}(x)dg_D(x)\right]=I_1+I_2+I_3+I_4\end{aligned} \tag{2.161}$$

The contributions are as follows: the integral $I_1$ is for the positive component of $f$ and the Lebesgue–Stieltjes measure defined by the increasing component of $g$; the integral is positive. The integral $I_2$ is for the negative component of $f$ and the Lebesgue–Stieltjes measure defined by the increasing component of $g$; the integral is negative. The integral $I_3$ is for the positive component of $f$ and the Lebesgue–Stieltjes measure defined by the decreasing component of $g$; the integral is negative. The integral $I_4$ is for the negative component of $f$ and the Lebesgue–Stieltjes measure defined by the decreasing component of $g$; the integral is positive.

***2.9.2.4 Modified Definition:*** The following modified definition is based on a modified interpretation of the Lebesgue–Stieltjes measure and the goal of a Lebesgue–Stieltjes integral to establish the net change in a function $g$ on an interval when weighted by a function $f$.

**Definition: Modified Lebesgue–Stieltjes Integral Definition:** The Lebesgue–Stieltjes integral of a measurable function $f:\boldsymbol{R}\to\boldsymbol{R}$, with respect to the Lebesgue–Stieltjes measure defined by a measurable function $g:\boldsymbol{R}\to\boldsymbol{R}$, is

$$\int_{\boldsymbol{R}} f(x)dg(x)=\sup\left\{\int_{\boldsymbol{R}} s(x)dg(x):\quad |s(x)|\le|f(x)|, x\in\boldsymbol{R},\ s\text{ simple}\right\} \tag{2.162}$$

where the measures $dg(x)$ and $m_g[S_{ij}]$ are defined as the net change in $g$ over the appropriate set consistent with $m_g[(\alpha,\beta)]=g(\beta^{-})-g(\alpha^{+})$. Alternatively,

$$\int_E f(x)dg(x)=\lim_{i\to\infty}\int_E s_i(x)dg(x)=\lim_{i\to\infty}\sum_{j=1}^{N_i} s_{ij}m_g\left[S_{ij}\right] \tag{2.163}$$

where $s_i : \boldsymbol{R} \to \boldsymbol{R},\ \ i \in \boldsymbol{Z}^+$ is a sequence of simple functions

$$s_i(x) = \sum_{j=1}^{N_i} s_{ij} I_{S_{ij}}(x) \qquad S_{ij}(x) = \{x:\ \ s_i(x) = s_{ij}\} \tag{2.164}$$

such that when $f(x) \geq 0$,

$$s_{i+1}(x) \geq s_i(x),\ s_i(x) \leq f(x) \qquad \lim_{i \to \infty} s_i(x) = f(x) \tag{2.165}$$

and when $f(x) < 0$,

$$s_{i+1}(x) \leq s_i(x),\ s_i(x) \geq f(x) \qquad \lim_{i \to \infty} s_i(x) = f(x) \tag{2.166}$$

***2.9.2.5 Notation*** The Lebesgue–Stieltjes integral of a measurable function $f : \boldsymbol{R} \to \boldsymbol{R}$, with respect to the Lebesgue–Stieltjes measure defined by a function $g : \boldsymbol{R} \to \boldsymbol{R}$, is written, as has already been noted, as

$$\int_{\boldsymbol{R}} f(x) dg(x) \ \text{ or } \ \int_{\boldsymbol{R}} f dg \tag{2.167}$$

The integral is called a Lebesgue–Stieltjes integral of $f$ with respect to $g$. The function $f$ is called the integrand and the function $g$ is called the integrator.

***2.9.2.6 Example 1*** Consider a right continuous function $g(x) = g_1, x < \alpha$, $g(x) = g_2, x \geq \alpha$, as illustrated in Figure 2.25. Consider the Lebesgue–Stieltjes integral of a constant function $f(x) = f_1$ over the intervals $[0,1], [0,\alpha), [0,\alpha], [\alpha,1], (\alpha,1]$ assuming $0 < \alpha < 1$:

$$\begin{aligned} \int_{[0,1]} f(x) dg(x) &= s_1 m_g[S_1] \ \text{ where } \ s_1 = f_1, S_1 = [0,1] \\ &= f_1 \cdot (g(1^+) - g(0^-)) = f_1 \cdot (g_2 - g_1) \end{aligned} \tag{2.168}$$

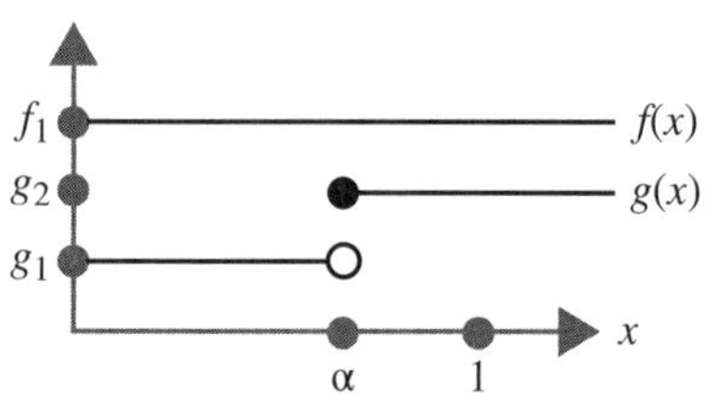


**FIGURE 2.25** Graph of the function $g$ and the function $f$.

$$\int_{[0,\alpha)} f(x)dg(x) = s_1 m_g[S_1] \quad \text{where} \quad s_1 = f_1, S_1 = [0,\alpha)$$
$$= f_1 \cdot [g(\alpha^-) - g(0^-)] = f_1 \cdot 0 = 0 \tag{2.169}$$

$$\begin{aligned}\int_{[0,\alpha]} f(x)dg(x) &= s_1 m_g[S_1] \quad \text{where} \quad s_1 = f_1, S_1 = [0,\alpha] = [0,\alpha) \cup \{\alpha\} \\ &= f_1 \left(m_g[[0,\alpha)] + m_g[\{\alpha\}]\right) = f_1[(g(\alpha^-) - g(0^-)) + g(\alpha^+) - g(\alpha^-)] \\ &= f_1(g_2 - g_1)\end{aligned} \tag{2.170}$$

$$\int_{[\alpha,1]} f(x)dg(x) = s_1 m_g[S_1] \quad \text{where} \quad s_1 = f_1, S_1 = [\alpha,1]$$
$$= f_1 \cdot m_g[[\alpha,1]] = f_1 \cdot (g(1^+) - g(\alpha^-)) = f_1 \cdot (g_2 - g_1) \tag{2.171}$$

$$\int_{(\alpha,1]} f(x)dg(x) = s_1 m_g[S_1] \quad \text{where} \quad s_1 = f_1, S_1 = (\alpha, 1]$$
$$= f_1 \cdot m_g[(\alpha, 1]] = f_1 \cdot (g(1^+) - g(\alpha^+)) = 0 \tag{2.172}$$

***2.9.2.7 Example 2*** Consider a right continuous function $g(x) = g_1, x < \alpha, g(x) = g_2$, $x \geq \alpha$, as illustrated in Figure 2.26, where $0 < \alpha < 1$, and a simple function defined according to $f(x) = f_1, x < x_o, f(x) = f_2, x \geq x_o$, also illustrated in Figure 2.26. Consider, first, the case where $x_o \neq \alpha$. By definition of a Lebesgue–Stieltjes integral,

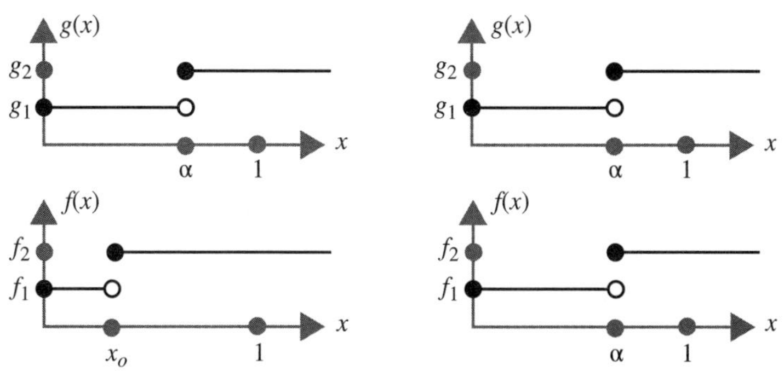


**FIGURE 2.26** Graphs of the function $g$ and $f$. Left: the case where $x_o < \alpha$. Right: the case where $x_o = \alpha$.

$$\int_{[0,1]} f(x)dg(x) = s_1 m_g[S_1] + s_2 m_g[S_2] \quad \text{where} \begin{cases} s_1 = f_1, S_1 = [0, x_o) \\ s_2 = f_2, S_2 = [x_o, 1] \end{cases}$$
$$= \begin{cases} f_1 \cdot 0 + f_2 \cdot (g_2 - g_1) = f_2 \cdot (g_2 - g_1) & 0 < x_o < \alpha < 1 \\ f_1 \cdot (g_2 - g_1) + f_2 \cdot 0 = f_1 \cdot (g_2 - g_1) & 0 < \alpha < x_o < 1 \end{cases} \tag{2.173}$$

For the case where, $x_o = \alpha$ as illustrated in the right part of Figure 2.26, it follows that

$$\int_{[0,1]} f(x)dg(x) = s_1 m_g[S_1] + s_2 m_g[S_2] \quad \text{where} \begin{cases} s_1 = f_1, S_1 = [0, \alpha) \\ s_2 = f_2, S_2 = [\alpha, 1] \end{cases}$$
$$= f_1 \cdot m_g[[0, \alpha)] + f_2 \cdot m_g[[\alpha, 1]]$$
$$= f_1 \cdot (g(\alpha^-) - g(0^-)) + f_2 \cdot m_g[g(1^+) - g(\alpha^-)] = f_2 \cdot (g_2 - g_1) \tag{2.174}$$

***2.9.2.8 Example 3*** Consider the case of

$$f(x) = \begin{cases} f_1 & x \in \boldsymbol{Q} \\ x & x \notin \boldsymbol{Q} \end{cases} \tag{2.175}$$

whereupon

$$\int_{[0,1]} f(x)dg(x) = \int_{\boldsymbol{Q} \cap [0,1]} f_1 dg(x) + \int_{[0,1] \cap \boldsymbol{Q}^C} x dg(x)$$
$$= f_1 \cdot m_g[\boldsymbol{Q} \cap [0,1]] + \lim_{i \to \infty} \sum_{k=1}^{i} \frac{k-1}{i} \cdot m_g\left[\boldsymbol{Q}^C \cap \left[\frac{k-1}{i}, \frac{k}{i}\right]\right] \tag{2.176}$$

For the case where $g$ is a step function as illustrated in Figure 2.26, $g(x) = g_1, x < \alpha$, $g(x) = g_2, x \geq \alpha$, and $\alpha \in (0,1), \alpha \in \boldsymbol{Q}$, the integral equals $f_1(g_2 - g_1)$. For the case where $\alpha \in (0,1), \alpha \in [0,1] \cap \boldsymbol{Q}^C$, the integral equals $\alpha(g_2 - g_1)$.

### 2.9.3 Lebesgue–Stieltjes Integrals: Specific Cases

#### *2.9.3.1 Lebesgue–Stieltjes Integral over a Point*

**Theorem 2.28 Lebesgue–Stieltjes Integral over a Point**
The Lebesgue–Stieltjes integral of a measurable function $f: \boldsymbol{R} \to \boldsymbol{R}$, with respect to the Lebesgue–Stieltjes measure defined by an increasing function $g: \boldsymbol{R} \to \boldsymbol{R}$, over a point $\alpha$, is

$$\int_{\{\alpha\}} f(x)dg(x) = [g(\alpha^+) - g(\alpha^-)]f(\alpha) \tag{2.177}$$

and equals the change of measure of the function $g$ at the point, as given by $g(\alpha^+)-g(\alpha^-)$, weighted by the value of the function $f$ at the point.

If $g:\boldsymbol{R}\to\boldsymbol{R}$ is continuous at a point $\alpha$, then $\int_{\{\alpha\}} f(x)dg(x)=0$.

**Proof**

The stated result follows from the definition of a Lebesgue–Stieltjes integral, according to Equation 2.158:

$$\begin{aligned}\int_{\{\alpha\}} f(x)dg(x) &= \lim_{i\to\infty}\int_{\{\alpha\}} s_i(x)dg(x) = \lim_{i\to\infty} s_i(\alpha)m_g[\{\alpha\}] \\ &= \lim_{i\to\infty} s_i(\alpha)[g(\alpha^+)-g(\alpha^-)] = f(\alpha)[g(\alpha^+)-g(\alpha^-)]\end{aligned} \tag{2.178}$$

where $s_i:\boldsymbol{R}\to\{s_{i1},\ldots,s_{iN_i}\}$, $i\in\boldsymbol{Z}^+$ is a sequence of simple functions such that $s_{i+1}(x)\ge s_i(x), s_i(x)\le f(x), \lim_{i\to\infty} s_i(x)=f(x)$ for $x\in\boldsymbol{R}$ assuming the positive function case.

### *2.9.3.2 Lebesgue–Stieltjes Integral: Single-Step Integrator*

**Theorem 2.29 Lebesgue–Stieltjes Integral: Single-Step Integrator**

If $g(x)=g_1, x<\alpha$, $g(x)=g_2, x\ge\alpha$, as illustrated in Figure 2.27, then, for a measurable function $f:\boldsymbol{R}\to\boldsymbol{R}$, it is the case that

$$\int f(x)dg(x)=f(\alpha)[g_2-g_1]=f(\alpha)[g(\alpha)-g(\alpha^-)] \tag{2.179}$$

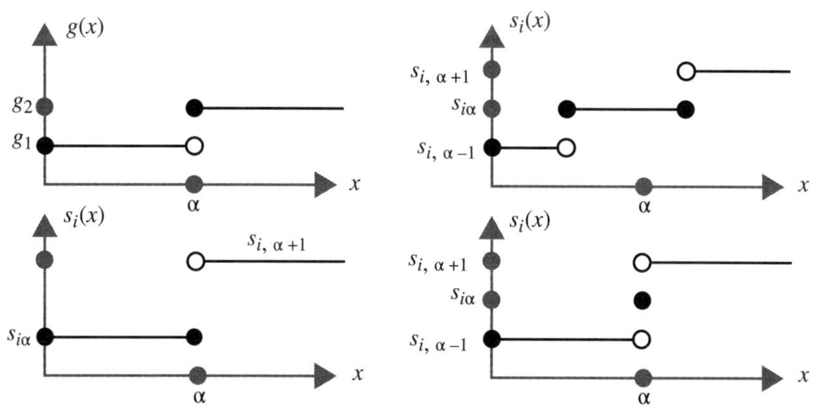


**FIGURE 2.27** Graph of the function $g$ and examples of simple functions.

**Proof**
Without loss of generality, assume $f(x) \geq 0, \ x \in \boldsymbol{R}$. Since $f: \boldsymbol{R} \to \boldsymbol{R}$ is measurable, according to Theorem 2.8, there exists a sequence of simple functions $s_i: \boldsymbol{R} \to \{s_{1i}, \ldots, s_{1N_i}\}, \ i \in \boldsymbol{Z}^+$ such that $s_{i+1}(x) \geq s_i(x), \ \lim_{i \to \infty} s_i(x) = f(x)$ for $x \in \boldsymbol{R}$. Consider the $i$th simple function $s_i$ which can be written as

$$s_i(x) = \sum_{j=1}^{N_i} s_{ij} I_{S_{ij}}(x) \qquad S_{ij} = \{x : s_i(x) = s_{ij}\} \tag{2.180}$$

Assume, consistent with the illustrations in Figure 2.27, that the set $S_{i\alpha}$, with the level $s_{i\alpha}$, contains the point $\alpha$. The Lebesgue–Stieltjes integral of $s_i$ with respect to the Lebesgue–Stieltjes measure defined by $g: \boldsymbol{R} \to \boldsymbol{R}$, according to Equation 2.152, then is

$$\int_{\boldsymbol{R}} s_i(x) dg(x) = \sum_{j=1}^{N_i} s_{ij} m_g [S_{ij}] = s_{i\alpha}(g_2 - g_1) = s_{i\alpha}[g(\alpha) - g(\alpha^-)] \tag{2.181}$$

Consistent with Theorem 2.8, $\lim_{i \to \infty} s_{i\alpha} = f(\alpha)$ and the required result follows.

### 2.9.3.3 *Lebesgue–Stieltjes Integral: Right Increasing Step Function*

**Theorem 2.30 Lebesgue–Stieltjes Integral: Right Increasing Step Function**
Consider a right increasing step function $g: \boldsymbol{R} \to \boldsymbol{R}$, as illustrated in Figure 2.28 and defined according to

$$g(x) = \sum_{i=1}^{N} g_i u(x - x_i) \tag{2.182}$$

The Lebesgue–Stieltjes integral of a measurable function $f: \boldsymbol{R} \to \boldsymbol{R}$, with respect to the Lebesgue–Stieltjes measure defined by $g$, is

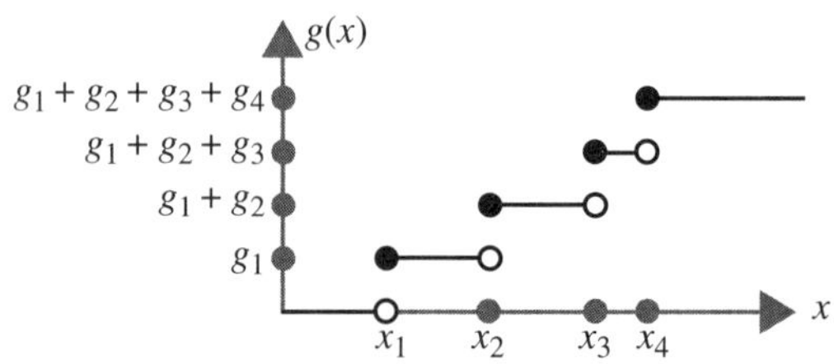


**FIGURE 2.28** Graph of the right increasing step function $g$.

$$\int_{R} f(x)dg(x) = \sum_{i=1}^{N} g_i f(x_i) \tag{2.183}$$

**Proof**

The result follows by writing the integral as

$$\begin{aligned}\int_{R} f(x)dg(x) = & \int_{(-\infty,x_1)} f(x)dg(x) + \int_{\{x_1\}} f(x)dg(x) + \int_{(x_1,x_2)} f(x)dg(x) + \cdots + \\ & \int_{\{x_N\}} f(x)dg(x) + \int_{(x_N,\infty)} f(x)dg(x)\end{aligned} \tag{2.184}$$

and using Theorem 2.28.

### *2.9.3.4 Integral: Continuous Function and Differentiable Measure Case*

**Theorem 2.31 Integral: Continuous Function and Differentiable Integrator**

The Lebesgue–Stieltjes integral of a measurable function $f: \boldsymbol{R} \to \boldsymbol{R}$, with respect to the Lebesgue–Stieltjes measure defined by a differentiable function $g: \boldsymbol{R} \to \boldsymbol{R}$, is

$$\int_{R} f(x)dg(x) = \int_{R} f(x)\frac{d}{dx}g(x)dx \tag{2.185}$$

**Proof**

Without loss of generality, assume $g$ is an increasing differentiable function and consider a sequence of simple functions $\{s_i\}$ that converge to $f$ consistent with Theorem 2.8, that is, $\lim_{i\to\infty} s_i(x) = f(x)$. Consider the following:

$$\begin{aligned}\int_{R} f(x)dg(x) &= \lim_{i\to\infty} \sum_{j=1}^{N_i} s_{ij} m_g\left[S_{ij}\right] \approx \lim_{i\to\infty} \sum_{j=1}^{N_i} s_{ij}\left[g\left(x_j + \frac{\Delta_j}{2}\right) - g\left(x_j - \frac{\Delta_j}{2}\right)\right] \\ &\approx \lim_{i\to\infty} \sum_{j=1}^{N_i} f(x_j)\left[\frac{g(x_j + \Delta_j/2) - g(x_j - \Delta_j/2)}{\Delta_j}\right]\Delta_j = \int_{R} f(x)\frac{d}{dx}g(x)dx\end{aligned} \tag{2.186}$$

Here, the $i$th simple function takes on the value $s_{ij}$ over the set $S_{ij}$, $x_j \in S_{ij}$, and the measure of $S_{ij}$ is $\Delta_j$. The last result assumes $\Delta_j \to 0$ as $i \to \infty$. For the case, for example, where $f$ is constant over a finite interval, the set $S_{ij}$ can be partitioned into smaller and smaller intervals to justify the result.

As an example, consider the Lebesgue–Stieltjes integral over the interval $[\alpha, \beta]$ for the case of $f(x) = 2x$ and $g(x) = x^2$:

$$\int_{[\alpha,\beta]} f(x)dg(x) = \int_{\alpha}^{\beta} f(x)\frac{d}{dx}g(x)dx = \int_{\alpha}^{\beta}(2x)(2x)dx = \left.\frac{4x^3}{3}\right|_{\alpha}^{\beta} = \frac{4\beta^3}{3} - \frac{4\alpha^3}{3} \qquad (2.187)$$

***2.9.3.5 Lebesgue–Stieltjes Integral of a Right Continuous Step Function*** The following theorem states a useful Lebesgue–Stieltjes integral result.

**Theorem 2.32 Lebesgue–Stieltjes Integral: Right Continuous Step Function** Consider the right continuous step function $s:[x_0,x_N) \to \boldsymbol{R}$, illustrated in Figure 2.29 and defined according to

$$s(x) = s_i \quad x \in S_i = [x_i, x_{i+1}) \qquad (2.188)$$

The Lebesgue–Stieltjes integral of this function, with respect to the Lebesgue–Stieltjes measure defined by an increasing function $g:\boldsymbol{R} \to \boldsymbol{R}$, is

$$\begin{aligned} \int_{[x_0,x_N)} s(x)dg(x) &= \sum_{i=0}^{N-1} s_i m_g[S_i] \\ &= s_0\left[g\left(x_1^-\right)-g\left(x_0^-\right)\right] + s_1\left[g\left(x_2^-\right)-g\left(x_1^-\right)\right] + \cdots + s_{N-1}\left[g\left(x_N^-\right)-g\left(x_{N-1}^-\right)\right] \end{aligned} \qquad (2.189)$$

**Proof**

The proof follows from the definition of the Lebesgue–Stieltjes integral of a step function as specified by Equation 2.152.

***2.9.3.6 Financial Interpretation of the Lebesgue–Stieltjes Integral*** Consider a signal arising from a Brownian motion random process, $B(t)$, as illustrated in Figure 2.30. This signal could represent, for example, the price fluctuations of a specific investment an investor has made. Assume the investment value is monitored at the times $t_0, t_1, \ldots, t_N$ where $t_0 = 0, t_N = T$, and $t_0 \le t_1 \le t_2 \le \cdots \le t_N$. Next, consider the case where the investor transacts shares in the investment at the times $\{t_0, t_1, \ldots, t_N\}$ and holds $f(t_i)$ shares at the time $t_i$, when the price is $B(t_i)$, as illustrated in Figure 2.30.

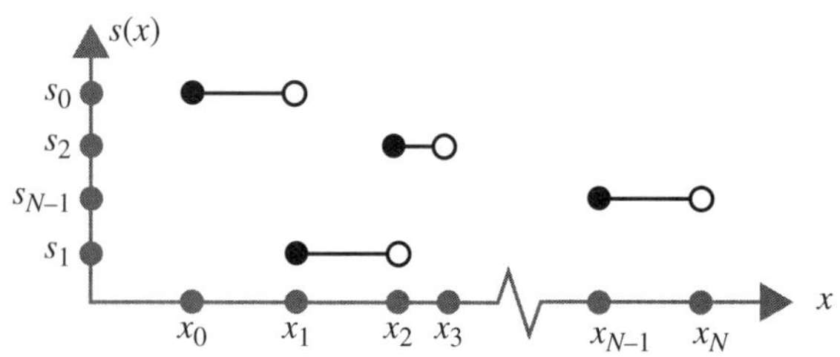


**FIGURE 2.29** Illustration of a simple function defined in Equation 2.188.

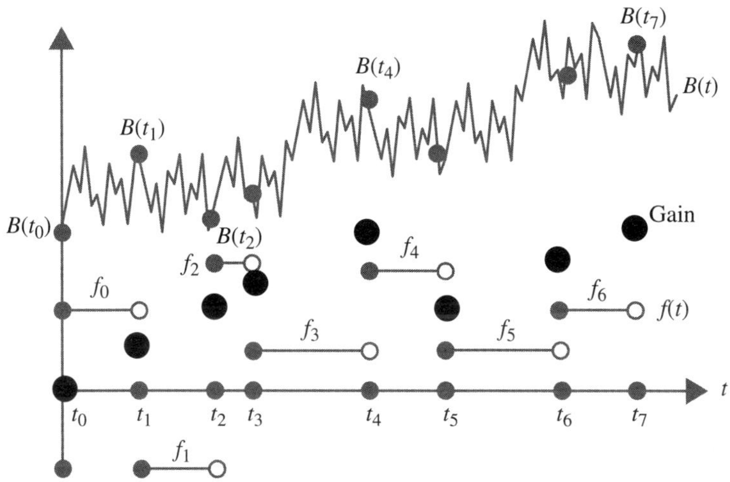


**FIGURE 2.30** Illustration of trading times, number of shares held, and financial gain, for an investment whose share price fluctuates in a manner as specified by $B(t)$.

The investor is, naturally, interested in his/her gain/loss on a time interval $[0, T]$, and this is given by the Lebesgue–Stieltjes integral

$$\begin{aligned}\text{gain}(T) &= \int_{[0,T]} f(t)dB(t) \\ &= f(t_0)\ [B(t_1)-B(t_0)] + f(t_1)\ [B(t_2)-B(t_1)] + \cdots + f(t_{N-1})\ [B(t_N)-B(t_{N-1})]\end{aligned}$$

## 2.10 MISCELLANEOUS RESULTS

**Theorem 2.33 Miscellaneous Results**
The following miscellaneous results are used in subsequent chapters.

$$\sum_{i=1}^{N} e^{-j2\pi i f/f_o} = e^{-j2\pi f/f_o} \sum_{i=0}^{N-1} e^{-j2\pi i f/f_o} = \begin{cases} e^{-j2\pi f/f_o}\left[\dfrac{1-e^{-j2\pi N f/f_o}}{1-e^{-j2\pi f/f_o}}\right] \\ N \qquad f = kf_o, k \in \mathbf{Z} \end{cases} \tag{2.190}$$

$$\begin{aligned}\left|\sum_{i=1}^{N} e^{-j2\pi i f/f_o}\right|^2 = \left|\sum_{i=0}^{N-1} e^{-j2\pi i f/f_o}\right|^2 &= \begin{cases} \dfrac{\sin(\pi N f/f_o)^2}{\sin(\pi f/f_o)^2} & -\infty < f < \infty \\ N^2 & f = kf_o, k \in \mathbf{Z} \end{cases} \\ &= N\left[1 + 2\sum_{i=1}^{N-1}\left(1-\frac{i}{N}\right)\cos(2\pi i f/f_o)\right]\end{aligned} \tag{2.191}$$

$$\lim_{i\to\infty}\frac{1}{N}\cdot\frac{\sin(\pi Nf/f_o)^2}{\sin(\pi f/f_o)^2}=\sum_{i=-\infty}^{\infty}f_o\delta(f-if_o) \quad (2.192)$$

$$\lim_{i\to\infty}T\,\text{sinc}^2(fT)=\delta(f) \quad (2.193)$$

Here, δ is the Dirac delta.

**Proof**
The proof of these results is detailed in Chapter 2 of Howard (2002).

## 2.11 PROBLEMS

**2.1** Find inf(*A*) and sup (*A*) for the set $A=\{1+\tanh(x):x\geq 1\}$.

**2.2** The first instance a function reaches a specified level (the first passage time) is often of interest. Consider the case of a linear signal plus a noise signal, which is assumed to be sinusoidal, and the first passage time to the level 1 defined according to

$$x_{FP}=\inf\{x:x+\varepsilon_o\sin(2\pi f_o x)=1, x>0\} \quad (2.194)$$

a. Determine $x_{FP}$ for the case of $\varepsilon_o=0.1$ and $f_o=10$. To do this, first, graph $x+\varepsilon_o\sin(2\pi f_o x)$ and use a numerical root solving package to find approximations to the solutions of $x+\varepsilon_o\sin(2\pi f_o x)=1$.

b. Consider the sequence of the first passage times defined by

$$x_{FP}(i)=\inf\{x:x+\varepsilon_o\sin(2\pi i f_o x)=1, x>0\}, i\in\{1,2,\ldots\} \quad (2.195)$$

What is $\lim_{i\to\infty}x_{FP}(i)$ for the case of $\varepsilon_o=0.1$ and $f_o=10$?

**2.3** A set *X* comprises of the sets

$$\{x:0\leq x<1\},\{x:1\leq x<2\},\{x:2\leq x<3\} \quad (2.196)$$

Specify the sigma algebra of *X*.

**2.4** Consider a set *X* comprising of a countable number of distinct entities. A measure μ, the counting measure, is defined on the sigma algebra of *X*, denoted $\Omega_X$, according to

$$\begin{aligned}\mu(A)&=\text{cardinality of }A\\&=\text{number of elements in }A\end{aligned} \quad (2.197)$$

for all $A\in\Omega_X$. Show that μ is a valid measure.

**2.5** Consider the sets

$$E_1 = [0,1), E_2 = [1,2), E_3 = (2,3), E_4 = (2,3] \tag{2.198}$$

Determine $m(E_1) + m(E_2) + m(E_3) + m(E_4)$ and $m(E_1 \cup E_2 \cup E_3 \cup E_4)$ where $m$ is the Lebesgue measure. Comment on the countable subadditivity property of the Lebesgue measure for this case.

**2.6** Determine the Lebesgue measure of a set $E$ defined according to

$$E = \bigcap_{i=1}^{\infty} \left[ a + \tanh(i), a + \frac{\pi}{1+i} \right) \tag{2.199}$$

**2.7** Consider the function $f(x) = \sin(2\pi f_o x)$ defined on the interval $[0, 2/f_o]$. Determine the inverse image $f^{-1}(O)$ for the case of $O = [0, 1/2)$. Is the function measurable?

**2.8** Define a sequence of simple function approximations, $s_1(x), s_2(x), \ldots$ to $f(x) = \sin(2\pi x)$ on the interval [0, 1/2] by partitioning the range of [0, 1] with a resolution $1/i$ for $i \in 1, 2, \ldots$. The simple functions are to be such that $s_i(x) \le s_{i+1}(x) \le f(x)$. For $i = 1$, use the levels 0, 1. For $i = 2$, use the levels 0, 1/2, 1, etc. Graph the simple function approximation for $i = 6$.

**2.9** Use the sequence of simple function approximations defined in the previous problem to establish a sequence of approximations to the integral

$$\int_0^{1/2} \sin(2\pi x)\, dx \tag{2.200}$$

Evaluate the relative error in the approximations for $i = 1, 2, 4, 8, 16, 100$.

**2.10** It is the case that

$$\int_0^{\infty} \operatorname{sinc}(kx)\, dx = \frac{\operatorname{sgn}(k)}{2k} \tag{2.201}$$

Consider the function

$$f(x) = \operatorname{sinc}(kx) \tag{2.202}$$

for a fixed value of $k$. Is $f \in \boldsymbol{L}$? Is $f \in \boldsymbol{L}^2$?

**2.11** The following sequence of functions

$$\{\delta_i(x) = i \operatorname{sinc}^2(ix)\}_{i=1}^{\infty} \tag{2.203}$$

is a suitable sequence for the Dirac delta. Define $\delta(x) = \lim_{i\to\infty} \delta_i(x)$. Show that

$$\lim_{i\to\infty} \int_{-\infty}^{\infty} \delta_i(x)dx \neq \int_{-\infty}^{\infty} \lim_{i\to\infty} \delta_i(x)dx = \int_{-\infty}^{\infty} \delta(x)dx \tag{2.204}$$

that is, show that the dominated convergence theorem does not hold for this sequence of functions. This result generally holds for such impulsive type of sequences.

**2.12** Determine the increasing and decreasing components of a function defined in Figure 2.31.

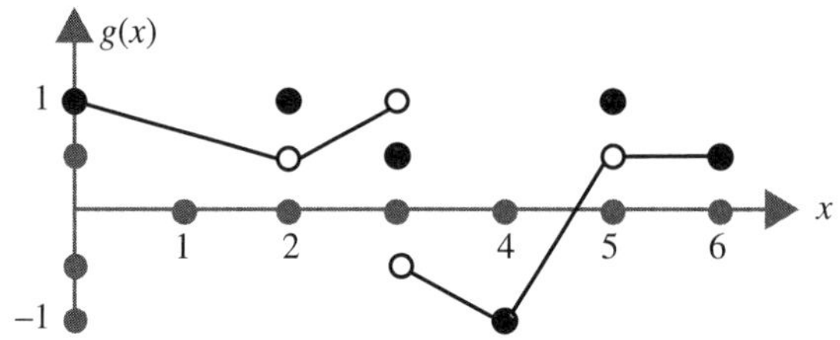


**FIGURE 2.31**

**2.13** Determine the Lebesgue–Stieltjes measure of the increasing and decreasing component functions of the function defined in Figure 2.31 over the sets $(0,2), \{2\}, (0,2], (2,3), \{3\}, (3,4), \{4\}, [3,4], (4,5), \{5\}, [4,5], (5,6)$, and $(0, 6)$. Hence, determine the Lebesgue–Stieltjes measure of the function $g$ over these sets.

**2.14** A random variable takes on values consistent with the distribution function $F$ defined in Figure 2.32. The expected value of this random variable is specified by the Lebesgue–Stieltjes integral:

$$\text{expected value} = \int_{(-\infty,\infty)} x dF(x) \tag{2.205}$$

Evaluate this expected value. For the specific case of $\alpha = -\beta$, what is the expected value?

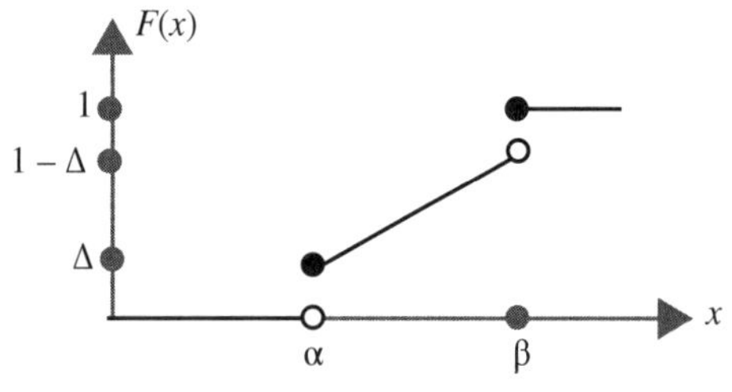


**FIGURE 2.32**

## APPENDIX 2.A PROOF OF THEOREM 2.1

Consider $\alpha$, $\delta$ fixed with $\alpha, \alpha-\delta \in I$, $j>i$, and $f(\alpha-\delta/j)-f(\alpha-\delta/i) \geq 0$ where the inequality is guaranteed by the monotonic nature of $f$. If there is equality for all $j,i>I_o$, $I_o \in \mathbf{Z}^+$, then convergence is guaranteed. For the case of $f(\alpha-\delta/j)-f(\alpha-\delta/i)>0$, consider the sequences of numbers illustrated in Figure 2.33 and, for each set value of $k \in \mathbf{Z}^+$, the sequence of numbers

$$\varepsilon_i(k) = f\left(\alpha - \frac{\delta}{(i+1)^k}\right) - f\left(\alpha - \frac{\delta}{i^k}\right) \quad i \in \mathbf{Z}^+ \tag{2.206}$$

The monotonicity of $f$ implies that $\varepsilon_i(k) \geq 0$, $i,k \in \mathbf{Z}^+$. For each value of $k \in \mathbf{Z}^+$, the summation $\sum_{i=1}^{\infty} \varepsilon_i(k)$ equals the change in the function as the argument changes from $\alpha-\delta$ to arbitrarily close to $\alpha$. As monotonicity requires that $f(\alpha)$ and $f(\alpha-\delta)$ are finite, it follows that each of the summations $\sum_{i=1}^{\infty} \varepsilon_i(1)$, $\sum_{i=1}^{\infty} \varepsilon_i(2)$, … converges, that is, for $k$ fixed, $\forall \varepsilon>0$, there exists an $I_o>0$ such that for $\forall i>I_o$, it is the case that $\varepsilon_i(k)<\varepsilon$. It then follows that $\forall \varepsilon>0$, there exists an $I_1>0$ such, for $\forall i,k>I_1$, that

$$f\left(\alpha - \frac{\delta}{(i+1)^k}\right) - f\left(\alpha - \frac{\delta}{i^k}\right) < \varepsilon \tag{2.207}$$

By appropriate selection of $i$ and $k$, it follows, for arbitrary $u, v \in \mathbf{Z}^+$ with $v>u>I_1$, that $f\left(\alpha - \frac{\delta}{v}\right) - f\left(\alpha - \frac{\delta}{u}\right) < \varepsilon$, which is sufficient for the existence of $\lim_{i\to\infty} f\left(\alpha - \frac{\delta}{i}\right)$. A similar argument justifies the existence of $\lim_{i\to\infty} f\left(\alpha + \frac{\delta}{i}\right)$.

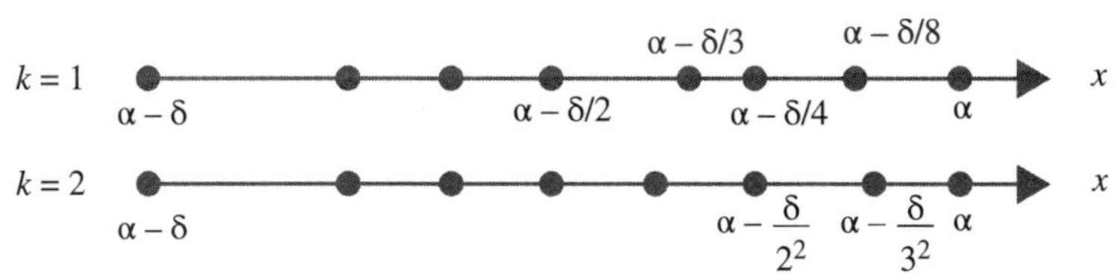


**FIGURE 2.33** Illustration of the sequences of numbers.

## APPENDIX 2.B PROOF OF THEOREM 2.2

First, continuity implies, for all $\varepsilon>0$, there exists a $\delta_o>0$ such that

$$|g(x_o+\delta)-g(x_o)|<\varepsilon \tag{2.208}$$

for all $|\delta| < \delta_o$. Thus, there is a sequence of numbers $\delta_1, \delta_2, \ldots, \delta_i, \ldots$, with $\lim_{i\to\infty} \delta_i = 0$, and a corresponding sequence $g(x_o+\delta_1),\ldots,g(x_o+\delta_i),\ldots$ that converges to $g(x_o)$. Thus, $f\left[\lim_{x\to x_o} g(x)\right] = f(g(x_o))$.

Second, corresponding to the sequence $g(x_o+\delta_1),\ldots,g(x_o+\delta_i),\ldots$ is a sequence $g(x_o)+\Delta_1,\ldots,g(x_o)+\Delta_i,\ldots$, where $g(x_o)+\Delta_i = g(x_o+\delta_i)$, which also converges to $g(x_o)$ and with $\lim_{i\to\infty} \Delta_i = 0$. The sequences $f[g(x_o+\delta_1)],\ldots,f[g(x_o+\delta_i)],\ldots$ and $f[g(x_o)+\Delta_1],\ldots,f[g(x_o)+\Delta_i],\ldots$ are equal and, from the continuity of $f$, have identical limits, that is, $\lim_{i\to\infty} f[g(x_o)+\Delta_i] = f(g(x_o))$. Thus, $\lim_{i\to\infty} f[g(x_o+\delta_i)] = f\left[\lim_{x\to x_o} g(x)\right]$ as required.

## APPENDIX 2.C PROOF OF THEOREM 2.7

First, consider the case of a simple function $s$ which takes the value of $s_1$ at a point $x_o$ and is zero everywhere else. Then consider the unit step functions, illustrated in Figure 2.34, which are the component functions of the function $f_{x_o}$ defined according to

$$\begin{aligned} f_{x_o}(x) = [u(x-x_o) - u(x-x_o-1)] + \left[u(x-x_o-1) - u\left(x-x_o-\frac{1}{2}\right)\right] + \cdots \\ + \left[u\left(x-x_o-\frac{1}{i}\right) - u\left(x-x_o-\frac{1}{i+1}\right)\right] + \cdots \end{aligned} \quad (2.209)$$

The function $f_{x_o}$ converges according to

$$f_{x_o}(x) = \begin{cases} 0 & x \neq x_o \\ 1 & x = x_o \end{cases} \quad (2.210)$$

and hence, the simple function $s$ is equal to a countable summation of unit step functions according to $s(x) = s_1 f_{x_o}(x)$.

Second, for the case where the level $s_i$ in the simple function defines the set $S_i$, which is a countable union of intervals, the results

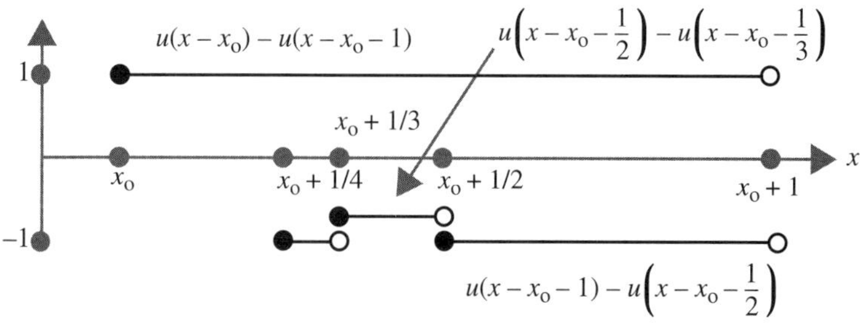


**FIGURE 2.34** Component functions of the function defined in Equation 2.209.

$$I_{[x_o,x_o+\alpha)}(x) = u[x-x_o] - u[x-(x_o+\alpha)] \tag{2.211}$$

$$I_{[x_o,x_o+\alpha]}(x) = u[x-x_o] - u[x-(x_o+\alpha)] + f_{x_o+\alpha}(x) \tag{2.212}$$

$$I_{(x_o,x_o+\alpha]}(x) = -f_{x_o}(x) + u[x-x_o] - u[x-(x_o+\alpha)] + f_{x_o+\alpha}(x) \tag{2.213}$$

$$I_{(x_o,x_o+\alpha)}(x) = -f_{x_o}(x) + u[x-x_o] - u[x-(x_o+\alpha)] \tag{2.214}$$

can be used to write the simple function as a summation of step functions. As the number of levels in a simple function is finite, a countable number of step functions can be used in any step function equivalence of a simple function that partitions the real line into a countable number of intervals.

## APPENDIX 2.D PROOF OF THEOREM 2.8

Assume $f: \boldsymbol{R} \to \boldsymbol{R}$ is a measurable function and $f(x) \geq 0, x \in \boldsymbol{R}$. Consider a fixed value $N \in \boldsymbol{Z}^+$ and the sets, as illustrated in Figure 2.35, defined according to

$$S_{iN} = \begin{cases} \left\{x : \dfrac{i}{2^N} \leq f(x) < \dfrac{i+1}{2^N}\right\} & i \in \{0, 1, \ldots, N2^N - 1\} \\ \{x : f(x) \geq N\} & i = N2^N \end{cases} \tag{2.215}$$

As $f$ is a measurable function, it follows, for any fixed value of $N$, that the sets $S_{iN}, i \in \{0, 1, \ldots, N2^N - 1\}$ are measurable sets. By construction, these sets are disjoint. Consider the simple function $s_N: \boldsymbol{R} \to \boldsymbol{R}$ defined according to

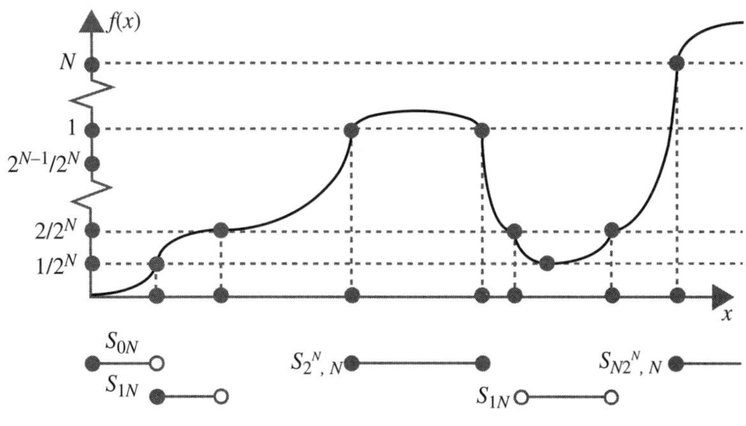


**FIGURE 2.35** Illustration of the sets defined by a function $f$.

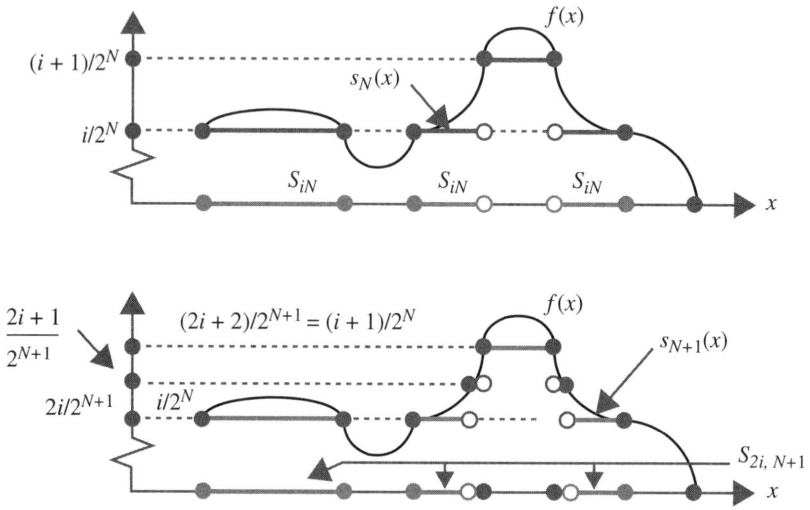


**FIGURE 2.36** Upper graph: illustration of *N*th simple function defined by *f*. Lower graph: illustration of (*N* + 1)th simple function defined by *f*.

$$s_N(x) = \sum_{i=0}^{N2^N} \frac{i}{2^N} I_{S_{iN}}(x) \qquad I_{S_{iN}}(x) = \begin{cases} 1 & x \in S_{iN} \\ 0 & x \notin S_{iN} \end{cases} \tag{2.216}$$

By construction, $s_N(x) \le f(x), x \in \boldsymbol{R}$. Consider the sequence of simple functions defined as $N$ takes on the values of $1, 2, \ldots,$ that is, consider the sequence $\{s_j(x)\}_{j=1}^{\infty}$. To show that $s_{j+1}(x) \ge s_j(x), x \in \boldsymbol{R}$, consider $j = N$ and the set $S_{iN}$ upon which $s_N(x) = i/2^N$ as illustrated in Figure 2.36. The same level, for the $(N+1)$th simple function, is when $s_{N+1}(x) = 2i/2^{N+1}$. Consistent with the illustration in Figure 2.36, it is the case that $s_{N+1}(x) \in \{i/2^N, (i+0.5)/2^N\}$ on the set $S_{iN}$ while $s_N(x) = i/2^N$. Thus, by construction, $s_{N+1}(x) \ge s_N(x), x \in \boldsymbol{R}$.

Finally, to prove that $\lim_{N\to\infty} s_N(x) = f(x), x \in \boldsymbol{R}$, assume an upper bound for $f$ of $N$. Then, by construction, it follows that

$$0 \le f(x) - s_N(x) < 1/2^N \tag{2.217}$$

Thus, for an arbitrary upper bound for $f$ of $N_B$, and for all $\varepsilon > 0$, there exists a value $N_o$ such that $\forall N > N_o$ it is the case that

$$f(x) - s_N(x) < \varepsilon \qquad x \in \boldsymbol{R},\ x : f(x) < N_B. \tag{2.218}$$

As the function $f$ may be unbounded at a countable number of points, the convergence holds almost everywhere and the theorem is proved.

## APPENDIX 2.E PROOF OF THEOREM 2.10

First, the necessity of the restriction of a finite number of discontinuities on a finite interval has been demonstrated by the function illustrated in Figure 2.11.

Assuming, at most, a finite number of discontinuities on any given finite interval, it follows, from the definition of right continuity, that $\forall\varepsilon > 0$, there exists a number $\delta_o > 0$ such that for all $0 < \delta < \delta_o$, it is the case that

$$|f(x_o+\delta)-f(x_o)| < \varepsilon \quad x_o \in [\alpha,\beta], 0<\delta<\delta_o \tag{2.219}$$

Hence, for a fixed value of $\varepsilon$, there exists a finite sequence of numbers $\delta_1, \ldots, \delta_N$ which partitions the interval $[\alpha, \beta]$ according to

$$[\alpha,\alpha+\delta_1), [\alpha+\delta_1,\alpha+\delta_1+\delta_2), \ldots, \left[\alpha+\sum_{i=1}^{N-1}\delta_i, \alpha+\sum_{i=1}^{N}\delta_i\right), \{\beta\} \tag{2.220}$$

where $\beta = \sum_{i=1}^{N}\delta_i$ and on the $i$th interval

$$\left| f\left(\alpha+\sum_{k=0}^{i-1}\delta_k+\delta\right) - f\left(\alpha+\sum_{k=0}^{i-1}\delta_k\right)\right| < \varepsilon \quad \begin{matrix} 0<\delta<\delta_i, \delta_0=0 \\ i\in\{1,\ldots,N\}\end{matrix} \tag{2.221}$$

This partition and the step approximation

$$\begin{aligned} f_A(x) = {} & f(\alpha)u(x-\alpha) + [f(\alpha+\delta_1)-f(\alpha)]u(x-\alpha-\delta_1) + \\ & [f(\alpha+\delta_1+\delta_2)-f(\alpha+\delta_1)]u(x-\alpha-\delta_1-\delta_2) + \cdots + \\ & \left[f\left(\alpha+\sum_{k=0}^{N-1}\delta_k\right) - f\left(\alpha+\sum_{k=0}^{N-2}\delta_k\right)\right] u\left[x-\alpha-\sum_{k=0}^{N-1}\delta_k\right] + \\ & \left[f(\beta)-f\left(\alpha+\sum_{k=0}^{N-1}\delta_k\right)\right]u(x-\beta) \end{aligned} \tag{2.222}$$

are illustrated in Figure 2.37. By construction, $|f(x)-f_A(x)| < \varepsilon$ for $x \in [\alpha, \beta]$ as required.

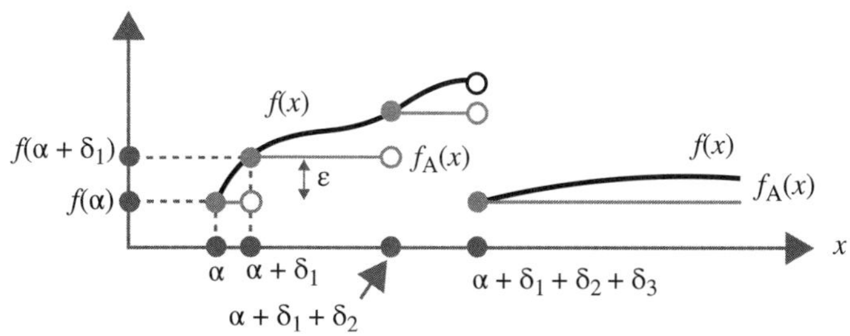


**FIGURE 2.37** Illustration of the partition of the interval [α, β], as determined by a right continuous function $f$, and the associated step approximation $f_A$ to $f$.

# 3

# BACKGROUND: SIGNAL THEORY

## 3.1 INTRODUCTION

This chapter details relevant signal theory and associated results that are important for subsequent chapters. Material on orthogonality, Dirichlet theory, and the Dirac Delta provide the basis for a discussion of Fourier theory and signal power. This is followed by a discussion of the power spectral density, and autocorrelation, functions. Finally, an introduction to the Cramer representation of a signal, and the Cramer transform, is given.

## 3.2 SIGNAL ORTHOGONALITY

Orthogonality and signal decomposition underpin a significant level of modern analysis, particularly, in the communications field. The following definitions are fundamental:

**Definition: Orthogonality** Two Lebesgue integrable signals $f_1 : \boldsymbol{R} \to \boldsymbol{C}$ and $f_2 : \boldsymbol{R} \to \boldsymbol{C}$ are orthogonal on an interval $[\alpha, \beta]$ if

$$\int_{\alpha}^{\beta} f_1(\lambda) f_2^*(\lambda)\, d\lambda = 0 \tag{3.1}$$

**Definition: Orthogonal Set** A set of functions $\{b_i : \boldsymbol{R} \rightarrow \boldsymbol{C},\ \ i \in \boldsymbol{Z}^+\}$ is an orthogonal set for $[\alpha, \beta]$ if

$$\int_{\alpha}^{\beta} b_i(\lambda) b_j^*(\lambda) d\lambda = \begin{cases} k_i & i = j, k_i > 0 \\ 0 & i \neq j \end{cases} \tag{3.2}$$

**Definition: Orthonormal Set** A set of functions $\{b_i : \boldsymbol{R} \rightarrow \boldsymbol{C},\ \ i \in \boldsymbol{Z}^+\}$ is an orthonormal set for $[\alpha, \beta]$ if

$$\int_{\alpha}^{\beta} b_i(\lambda) b_j^*(\lambda) d\lambda = \begin{cases} 1 & i = j \\ 0 & i \neq j \end{cases} \tag{3.3}$$

**Definition: Orthogonal/Orthonormal Basis Set** A set of functions $\{b_i : \boldsymbol{R} \rightarrow \boldsymbol{C},\ \ i \in \boldsymbol{Z}^+\}$, which satisfy the conditions specified in Equations 3.2/3.3, is an orthogonal/orthonormal basis set for $[\alpha, \beta]$ if any integrable function $f : \boldsymbol{R} \rightarrow \boldsymbol{C}$ defined on $[\alpha, \beta]$ can be written as

$$f(t) = \sum_{i=0}^{\infty} c_i b_i(t) \qquad t \in [\alpha, \beta], \text{a.e.} \tag{3.4}$$

for a set of real or complex coefficients $c_0, c_1, \ldots$.

### 3.2.1 Signal Decomposition

A basis set allows signal decomposition according to the following theorem:

**Theorem 3.1 Signal Decomposition: Orthogonal/Orthonormal Basis Set**
If $\{b_i :\ \ i \in \boldsymbol{Z}^+\}$ is an orthogonal/orthonormal basis set for $[\alpha, \beta]$, then an integrable signal $f : \boldsymbol{R} \rightarrow \boldsymbol{C}$ can be written as

$$f(t) = \sum_{i=0}^{\infty} c_i b_i(t) \tag{3.5}$$

where

$$c_i = \frac{\int_{\alpha}^{\beta} f(\lambda) b_i^*(\lambda) d\lambda}{\int_{\alpha}^{\beta} |b_i(\lambda)|^2 d\lambda} \qquad c_i = \int_{\alpha}^{\beta} f(\lambda) b_i^*(\lambda) d\lambda \tag{3.6}$$

The definition on the left is for the orthogonal case; the definition on the right is for the orthonormal case.

**Proof**

This important result is a special case of the more general result detailed in the following theorem:

**Theorem 3.2 Generalized Signal Decomposition**

If $\{b_i : \boldsymbol{R} \to \boldsymbol{C},\ i \in \boldsymbol{Z}^+\}$ is an orthogonal set for $[\alpha, \beta]$, with respect to a signal $g : \boldsymbol{R} \to \boldsymbol{C}$, that is,

$$\int_{\alpha}^{\beta} g(\lambda) b_i(\lambda) b_j^*(\lambda) d\lambda = 0 \qquad i \neq j \tag{3.7}$$

then an integrable signal $f : \boldsymbol{R} \to \boldsymbol{C}$ defined on $[\alpha, \beta]$ can be written as

$$f(t) = \sum_{i=0}^{\infty} c_i b_i(t) \qquad \text{where} \qquad c_i = \frac{\int_{\alpha}^{\beta} g(\lambda) f(\lambda) b_i^*(\lambda) d\lambda}{\int_{\alpha}^{\beta} g(\lambda) |b_i(\lambda)|^2 d\lambda} \tag{3.8}$$

**Proof**

With $f(t) = \sum_{i=0}^{\infty} c_i b_i(t)$ it follows that $f(t) g(t) b_i^*(t) = g(t) b_i^*(t) \sum_{k=0}^{\infty} c_k b_k(t)$ and

$$\begin{aligned} \int_{\alpha}^{\beta} g(\lambda) f(\lambda) b_i^*(\lambda) d\lambda &= c_i \int_{\alpha}^{\beta} g(\lambda) |b_i(\lambda)|^2 d\lambda + \sum_{\substack{k=0 \\ k \neq i}}^{\infty} c_k \int_{\alpha}^{\beta} g(\lambda) b_k(\lambda) b_i^*(\lambda) d\lambda \\ &= c_i \int_{\alpha}^{\beta} g(\lambda) |b_i(\lambda)|^2 d\lambda \end{aligned} \tag{3.9}$$

The required result then follows.

### 3.2.2 Generalization

The above definition of orthogonality arises from the more general framework of functional analysis and in particular Hilbert Space theory where an inner product is defined on a vector space. Useful references include Kreyszig (1978, ch. 3) and Debnath and Mikusinski (1999, ch. 3).

### 3.2.3 Example: Hermite Basis Set

For the infinite interval, a useful basis set is the orthonormal Hermite basis set defined according to

$$\left\{ b_i^N(t) = \frac{H_i(t)}{2^{i/2}\sqrt{i!}\pi^{1/4}} \cdot e^{-t^2/2} : i \in \{0,1,\ldots\} \right\} \tag{3.10}$$

where the Hermite polynomials $H_0, H_1, \ldots$ are defined according to

$$\begin{aligned} &H_0(t)=1 \quad H_1(t)=2t \quad H_2(t)=4t^2-2 \\ &H_3(t)=8t^3-12t \quad H_4(t)=16t^4-48t^2+12 \quad \cdots \end{aligned} \tag{3.11}$$

and by the recurrence relationship

$$H_i(t) = 2tH_{i-1}(t) - 2(i-1)H_{i-2}(t) \tag{3.12}$$

The first seven basis functions are graphed in Figure 3.1.

A basis decomposition for

$$f(t) = \frac{\sin(\pi t)}{\pi t} \tag{3.13}$$

and based on the 0th to 10th terms, is shown in Figure 3.2.

**3.2.3.1** ***Gaussian-Based Basis Set*** The following Gaussian-based basis set finds application, for example, in estimating the probability density function of random variables:

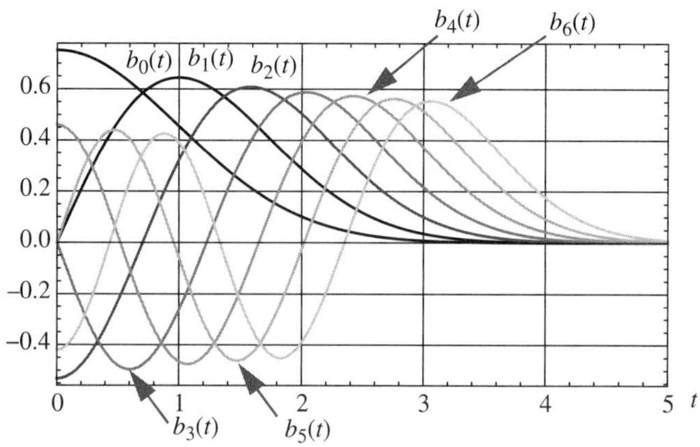


**FIGURE 3.1** Zeroth- to sixth-order Hermite basis functions.

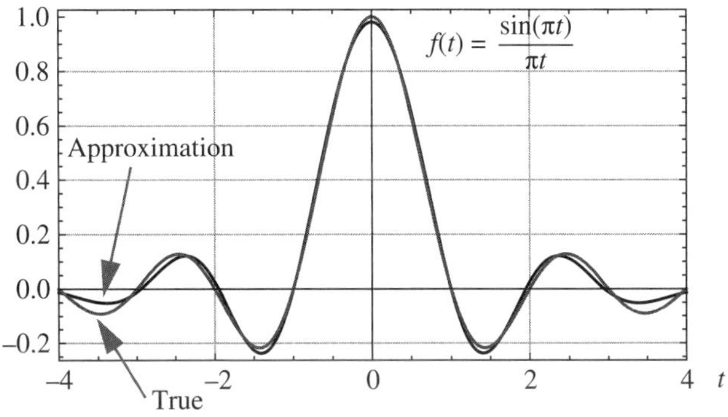


**FIGURE 3.2** Approximation to the sinc function by the first 11 terms of a decomposition based on a Hermite basis set. The coefficients are $c_0 = 0.750$, $c_1 = c_3 = c_5 = c_7 = c_9 = 0$, $c_2 = -0.511$, $c_4 = 0.350$, $c_6 = -0.110$, $c_8 = -0.096$, and $c_{10} = 0.067$.

**Theorem 3.3 Gaussian-Based Basis Set**
The set of functions, based on Gaussian functions with a mean $\mu$ and a variance $\sigma^2$,

$$\left\{\begin{aligned} b_0(t) &= \sqrt{2}\pi^{1/4}\sqrt{\sigma}H_0\left[\frac{t-\mu}{\sigma}\right]\cdot\frac{e^{-(t-\mu)^2/2\sigma^2}}{\sqrt{2\pi}\sigma},\ldots, \\ b_i(t) &= \frac{\pi^{1/4}\sqrt{\sigma}H_i[(t-\mu)/\sigma]}{2^{(i-1)/2}\sqrt{i!}}\cdot\frac{e^{-(t-\mu)^2/2\sigma^2}}{\sqrt{2\pi}\sigma},\ldots \end{aligned}\right\} \tag{3.14}$$

is an orthogonal basis set for $(-\infty, \infty)$.

**Proof**
The proof follows from the set of orthonormal signals defined by Equation 3.10.

## 3.3 THEORY FOR DIRICHLET POINTS

Consider the sequence of functions $\sin(2\pi\lambda u)/\pi u$, $\lambda \in \{1,2,3,\ldots\}$, as illustrated in Figure 3.3. Each function in the sequence has unit area. The following definitions of a Dirichlet point and Dirichlet value are based on this sequence of functions and are important in Fourier theory.

**Definition: Dirichlet Point** A point $t_o$ is a Dirichlet point of a locally integrable function, $f: \boldsymbol{R} \to \boldsymbol{C}$, if $\exists\delta_o > 0$ such that for $0 < \delta < \delta_o$ the following limit is finite:

$$\lim_{\lambda\to\infty}\int_{-\delta}^{\delta} f(t_o + u)\cdot\frac{\sin(2\pi\lambda u)}{\pi u}\cdot du \tag{3.15}$$

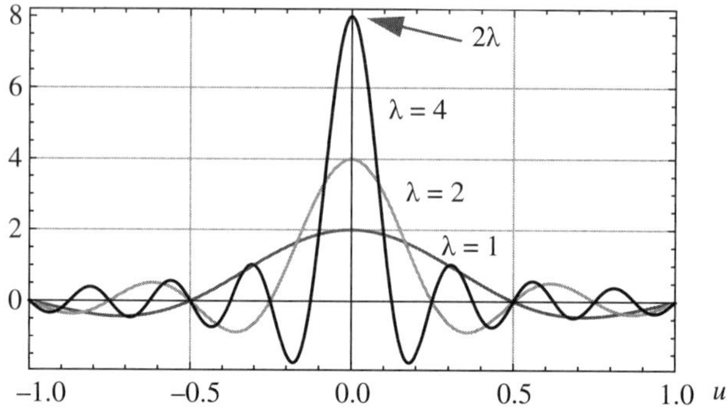


**FIGURE 3.3** Graph of $\sin(2\pi\lambda u)/\pi u$ for $\lambda = 1, 2, 4$. The height at $u = 0$ is $2\lambda$. The spacing between zero crossings is $1/2\lambda$ for $u > 0$. Each function has unity area.

**Definition: Dirichlet Value** If $t_o$ is a Dirichlet point of a locally integrable function, $f: \boldsymbol{R} \to \boldsymbol{C}$, then the value defined by the limit in Equation 3.15 is called the Dirichlet value of $f$ at $t_o$.

### 3.3.1 Existence of Dirichlet Points

Given the impulsive nature of the sequence of functions $\sin(2\pi\lambda u)/\pi u$, $\lambda \in \{1,2,3,\ldots\}$, it would be reasonable to conclude that

$$\lim_{\lambda\to\infty} \int_{-\delta}^{\delta} f(t_o + u) \cdot \frac{\sin(2\pi\lambda u)}{\pi u} \cdot du = f(t_o) \tag{3.16}$$

that is, the Dirichlet value of $f$ at $t_o$ exists and is equal to $f(t_o)$. However, there are no known necessary and sufficient conditions for the existence of a Dirichlet point (Champeney, 1987, p. 43). In fact, on any finite interval $(\alpha, \beta)$, there exist functions that are locally integrable but that have no Dirichlet points on that interval (Champeney, 1987, p. 38). As an illustration of the nonexistence of a Dirichlet point, consider the function $f(t) = \begin{cases} 0 & t = 0 \\ 1/\sqrt{|t|} & t \neq 0 \end{cases}$, which is continuous at all points except the origin and is integrable on all finite intervals. The limit associated with the potential Dirichlet value at $t = 0$ is infinite for all values of $\delta > 0$. To see this note, for a set $\delta_o > 0$ and for all $\lambda$ consistent with $\delta_o > 1/2\lambda$, that

$$\begin{aligned} \int_{-\delta_o}^{\delta_o} \frac{1}{\sqrt{|u|}} \cdot \frac{\sin(2\pi\lambda u)}{\pi u} du &> 2\sqrt{2\lambda} \int_0^{1/2\lambda} \frac{\sin(2\pi\lambda u)}{\pi u} du + 2 \int_{1/2\lambda}^{\delta_o} \frac{1}{\sqrt{u}} \cdot \frac{\sin(2\pi\lambda u)}{\pi u} du \\ &= \frac{2\sqrt{2\lambda}}{\pi} \cdot \mathrm{Si}(\pi) + 2 \int_{1/2\lambda}^{\delta_o} \frac{1}{\sqrt{u}} \cdot \frac{\sin(2\pi\lambda u)}{\pi u} du \end{aligned} \tag{3.17}$$

where $\mathrm{Si}(x) = \int_0^x \frac{\sin(x)}{x} dx$ is the sine integral. Hence, as $\lambda \to \infty$, the integral on the left-hand side becomes unbounded.

Further, when a Dirichlet point exists, its value may not coincide with the value of the function at the point. Consider the function $f(t) = \begin{cases} f_1 & t \in \boldsymbol{Q} \\ f_2 & t \in \boldsymbol{Q}^C \end{cases}$ and a point $t_o \in \boldsymbol{Q}$. It is the case, for all $\delta > 0$, that

$$\lim_{\lambda \to \infty} \int_{-\delta}^{\delta} f(t_o + u) \cdot \frac{\sin(2\pi\lambda u)}{\pi u} \cdot du = f_2 \tag{3.18}$$

as the integral is determined by the integrand values on $\boldsymbol{Q}^C$.

The following three theorems give sufficient conditions for the existence of a Dirichlet point (Champeney, 1987, pp. 37–42):

**Theorem 3.4 Square Integrability Implies Dirichlet Almost Everywhere**
If $f \in \boldsymbol{L}^2$, then all points, except, possibly, on a set of zero measure, are Dirichlet points and the Dirichlet value is equal to the value of $f$ at the point. If $t_o$ is a Dirichlet point, then $\exists \delta_o > 0$ such that

$$f(t_o) = \lim_{\lambda \to \infty} \int_{-\delta_o}^{\delta_o} f(t_o + u) \cdot \frac{\sin(2\pi\lambda u)}{\pi u} \cdot du \tag{3.19}$$

**Proof**
The proof has been detailed in Carleson (1966).

**Theorem 3.5 Differentiability Implies Dirichlet Point**
If $f$ is differentiable at a point, then that point is a Dirichlet point of $f$. Further, the Dirichlet value of $f$ at $t_o$ is $f(t_o)$.

**Proof**
The proof is detailed in Appendix 3.A.

**Theorem 3.6 Piecewise Smoothness Implies Dirichlet Point**
If $f$ is piecewise smooth (piecewise differentiable) at a point, then that point is a Dirichlet point of $f$. Further, the Dirichlet value of $f$ at $t_o$ is $\left[f\left(t_o^+\right) + f\left(t_o^-\right)\right]/2$, that is, $\exists \delta_o > 0$ such that

$$\frac{f\left(t_o^+\right) + f\left(t_o^-\right)}{2} = \lim_{\lambda \to \infty} \int_{t_o - \delta_o}^{t_o + \delta_o} f(u) \cdot \frac{\sin[2\pi\lambda(u - t_o)]}{\pi(u - t_o)} \cdot du \tag{3.20}$$

**Proof**

The proof follows in an analogous manner to the proof of Theorem 3.5.

## 3.4 DIRAC DELTA

The Dirac delta, denoted $\delta$, is widely used in analysis. It is often treated as a function but is a generalized function (e.g., Champeney, 1987, ch. 12) and its definition arises from the limit of a sequence of functions.

Consider a sequence of functions $\{\delta_i : \boldsymbol{R} \to \boldsymbol{R}\}_{i=1}^{\infty}$ that have the properties

$$\lim_{i\to\infty} \delta_i(t) = \begin{cases} \infty & t=0 \\ 0 & t \neq 0 \end{cases} \qquad \lim_{i\to\infty} \int_{-\infty}^{\infty} \delta_i(t)dt = 1 \tag{3.21}$$

If the following definition was made

$$\delta(t) = \lim_{i\to\infty} \delta_i(t) \tag{3.22}$$

then the following two results would hold: First, as $\delta$ is nonzero only on a set of zero measure, it follows that

$$\int_{-\infty}^{\infty} \delta(t)dt = 0 \tag{3.23}$$

Second, if $t_o$ is a Dirichlet point of $f \in \boldsymbol{L}$, then

$$\lim_{i\to\infty} \int_{-\infty}^{\infty} \delta_i(t-t_o)f(t)dt = \frac{f(t_o^+) + f(t_o^-)}{2} \tag{3.24}$$

but

$$\int_{-\infty}^{\infty} \delta(t-t_o)f(t)dt = 0 \tag{3.25}$$

The definition of the Dirac delta function is associated with the limit of the sequence of functions $\{\delta_i\}_{i=1}^{\infty}$ and not with the function definition $\delta(t) = \lim_{i\to\infty} \delta_i(t)$. For notational convenience $\lim_{i\to\infty} \int_{-\infty}^{\infty} \delta_i(t-t_o)f(t)dt$ is written as $\int_{-\infty}^{\infty} \delta(t-t_o)f(t)dt$ and $\lim_{i\to\infty} \int_{-\infty}^{\infty} \delta_i(t)dt = 1$ is written as $\int_{-\infty}^{\infty} \delta(t)dt = 1$. It is important to stress the distinction between the notation and the limit of the sequence of operations that the notation represents. A more detailed account of defining the Dirac delta, and associated distributions, in a manner consistent with the approach taken in this book, can be found in Howard (2012a).

## 3.5 FOURIER THEORY

Fourier theory is fundamental to many facets of science and engineering. The following sections state the definition of a Fourier series, the Fourier transform, and the inverse Fourier transform and detail relevant results including Parseval's theorem.

### 3.5.1 Fourier Series

The widely used sinusoidal basis set for the interval $[\alpha, \beta]$ is

$$\{1, \sin(2\pi f_o t)\}, \cos(2\pi f_o t), \sin(4\pi f_o t), \cos(4\pi f_o t), \ldots\} \tag{3.26}$$

where $f_o = 1/(\beta - \alpha)$ is the fundamental frequency. This basis set leads to a Fourier series decomposition as follows:

**Definition: Fourier Series** A Fourier series, $x_F$, on the interval $[\alpha, \beta]$ for a signal $x \in \boldsymbol{L}[\alpha, \beta]$ is defined according to

$$\begin{aligned} x_F(t) &= a_o + \sum_{i=1}^{\infty} a_i \cos(2\pi i f_o t) + b_i \sin(2\pi i f_o t) \\ &= \sum_{i=-\infty}^{\infty} c_i e^{j2\pi i f_o t} \quad t \in [\alpha, \beta] \end{aligned} \tag{3.27}$$

where

$$a_o = \frac{1}{\beta - \alpha} \int_\alpha^\beta x(t)dt \quad a_i = \frac{2}{\beta - \alpha} \int_\alpha^\beta x(t)\cos(2\pi i f_o t)dt \tag{3.28}$$

$$b_i = \frac{2}{\beta - \alpha} \int_\alpha^\beta x(t)\sin(2\pi i f_o t)dt \quad c_i = \frac{1}{\beta - \alpha} \int_\alpha^\beta x(t)^{-j2\pi i f_o t} dt \tag{3.29}$$

***3.5.1.1 Notes*** In general, all the coefficients are complex and from their definitions it follows that $a_{-i} = a_i$ and $b_{-i} = -b_i$ and for the case of real signals $c_{-i} = c_i^*$. Further,

$$c_i = \begin{cases} a_0 & i = 0 \\ 0.5(a_i - jb_i) & i \neq 0 \end{cases} \tag{3.30}$$

and it then follows that

$$|c_i|^2 = \begin{cases} |a_0|^2 & i=0 \\ 0.25\left(|a_i|^2 + |b_i|^2 - 2\mathrm{Im}\left[a_i b_i^*\right]\right) & i \neq 0 \end{cases} \tag{3.31}$$

The power in the sinusoidal components of the signal $x_F$ with a frequency $if_o$, $i \neq 0$, is given by $\left(|a_i|^2 + |b_i|^2\right)/2$. From Equation 3.31, and the results $a_{-i} = a_i$ and $b_{-i} = -b_i$, the important result

$$|c_{-i}|^2 + |c_i|^2 = \left(|a_i|^2 + |b_i|^2\right)/2 \tag{3.32}$$

follows. For the real signal case, $|c_{-i}|^2 = |c_i|^2$ but, in general, $|c_{-i}|^2 \neq |c_i|^2$.

#### 3.5.1.2 Convergence of a Fourier Series

Of importance is whether the Fourier series of a signal converges to the signal. For conditions that generally hold, this is the case:

**Theorem 3.7 Convergence of a Fourier Series**
If $x \in \boldsymbol{L}[\alpha, \beta]$ and is piecewise differentiable, or has bounded variation, on $[\alpha, \beta]$, then

$$x_F(t) = \begin{cases} x(t) & x \text{ continuous at } t, \quad t \in (\alpha, \beta) \\ 0.5[x(t^+) + x(t^-)] & x \text{ discontinuous at } t, \quad t \in (\alpha, \beta) \\ 0.5[x(\alpha^+) + x(\beta^-)] & t = \alpha, \ t = \beta \end{cases} \tag{3.33}$$

**Proof**
See Champeney (1987, pp. 156–157) and Tolstov (1962, p. 75f ).

### 3.5.2 Fourier Transform

The Fourier transform is widely used in characterizing signals and, hence, is important for random process theory.

**Definition: Fourier Transform** The Fourier transform of a signal $x$, denoted $X$ and evaluated over the intervals $[\alpha, \beta]$ and $(-\infty, \infty)$, is, respectively, defined according to

$$X([\alpha,\beta], f) = \int_{\alpha}^{\beta} x(t) e^{-j2\pi ft} dt \qquad X(f) = \int_{-\infty}^{\infty} x(t) e^{-j2\pi ft} dt \tag{3.34}$$

As the interval $[0, T]$ is primarily considered, the Fourier transform of a signal $x$, over this interval, is denoted $X(T, f)$.

Some useful Fourier transform pairs are detailed in the tables at the end of the book.

***3.5.2.1 Properties*** For the interval [0, $T$], the following properties hold for the Fourier transform. First, if $x \in \boldsymbol{L}[0,T]$, then $X$ exists. Second, for real signals,

$$X(T,-f)=X^*(T,f) \tag{3.35}$$

that is, the Fourier transform exhibits Hermitian symmetry. Third, the $i$th Fourier coefficient $c_i$ in the exponential Fourier series of a signal, as given by Equation 3.29, is related to the Fourier transform of the same signal according to

$$c_i=\frac{X(T,if_o)}{T} \qquad f_o=\frac{1}{T} \tag{3.36}$$

***3.5.2.2 Interpretation of Fourier Transform*** Consider the Fourier transform of a signal $x$ on the interval [0, $T$] and defined according to

$$X(T,f)=\int_0^T x(t)e^{-j2\pi ft}dt \tag{3.37}$$

At a set frequency, $f=f_1$, the Fourier transform gives a measure of the orthogonality of the signal with respect to the exponential signal $e^{j2\pi f_1 t}$.

***3.5.2.3 Example*** The Fourier transform of a sinusoid $x(t)=A\cos(2\pi f_C t+\phi)$ on the interval [0, $T$] is

$$\begin{aligned} X(T,f)=&\frac{ATe^{j\phi}e^{-j\pi(f-f_C)T}}{2}\cdot\operatorname{sinc}[(f-f_C)T] \\ &+\frac{ATe^{-j\phi}e^{-j\pi(f+f_C)T}}{2}\cdot\operatorname{sinc}[(f+f_C)T] \end{aligned} \tag{3.38}$$

The proof of this result is detailed in Appendix 3.C.

### 3.5.3 Inverse Fourier Transform

The definition of a Dirichlet point facilitates the statement of conditions related to the existence of the inverse Fourier transform as detailed in the following theorem:

**Theorem 3.8 Existence of Inverse Fourier Transform**
If $x \in \boldsymbol{L}[0,T]$, and if $x$ is piecewise differentiable at a point $t_o$, then the inverse Fourier transform at $t_o$ exists and is given by

$$\int_{-\infty}^{\infty} X(T,f)e^{j2\pi ft_o}df=\begin{cases} 0 & t_o \notin [0,T] \\ x(0^+)/2 & t_o=0 \\ \dfrac{x(t_o^+)+x(t_o^-)}{2} & t_o \in (0,T) \\ x(T^-)/2 & t_o=T \end{cases} \tag{3.39}$$

**Proof**
The proof is detailed in Appendix 3.B.

### 3.5.4 Parseval's Theorem

Parseval's theorem is fundamental to defining the power spectral density for the usual case where a sinusoidal basis set is assumed. A formal statement of this theorem is as follows (Champeney, 1987, p. 72; Titchmarsh, 1948, pp. 50, 69–76):

**Theorem 3.9 Parseval's Theorem**
With $X$ denoting the Fourier transform of $x$, the following results hold: First, if $x \in \boldsymbol{L}$ and $x \in \boldsymbol{L}^2$, then $X \in \boldsymbol{L}^2$ and

$$\int_{-\infty}^{\infty} |x(t)|^2 dt = \int_{-\infty}^{\infty} |X(f)|^2 df \tag{3.40}$$

Second, if $x \in \boldsymbol{L}^2[\alpha, \beta]$, then $x \in \boldsymbol{L}[\alpha, \beta]$ and

$$\int_{\alpha}^{\beta} |x(t)|^2 dt = \int_{-\infty}^{\infty} |X([\alpha,\beta],f)|^2 df \qquad \int_{0}^{T} |x(t)|^2 dt = \int_{-\infty}^{\infty} |X(T,f)|^2 df \tag{3.41}$$

**Proof**
A proof of this relationship, in a manner consistent with the theoretical framework being used, is detailed in Howard (2002, ch. 2).

## 3.6 SIGNAL POWER

In this section signal energy and signal power are defined and a fundamental result for the signal power, in terms of the power in its constituent signal components, is stated and proved. The use of the term power, while widely used, is misleading as the definition leads, in general, to a relative power measure. For example, in many communication and electronic applications, the signal being considered is a current or a voltage and the power definition leads to a result in amps squared or volts squared. The use of relative power measures, in general, is unproblematic. For example, when the signal to noise ratio (signal power divided by noise power) is to be established, provided both the signal and noise have the same units, for example, Watts or Volts squared, it does not matter whether relative or absolute power measures are used.

**Definition: Signal Energy** By definition, the energy in a signal $x \in \boldsymbol{L}^2[\alpha, \beta]$ over the interval $[\alpha, \beta]$ is

$$E = \int_{\alpha}^{\beta} |x(t)|^2 dt \tag{3.42}$$

**Definition: Average Signal Power** By definition, the average power in a signal $x \in \boldsymbol{L}^2[\alpha, \beta]$ over the interval [α, β] is

$$\bar{P} = \frac{1}{\beta - \alpha} \int_{\alpha}^{\beta} |x(t)|^2 dt \tag{3.43}$$

### 3.6.1 Sinusoidal Basis Set

The most widely used decomposition of a signal is through the use of a sinusoidal basis set, and this leads to a Fourier series. The following theorem details the important result that relates the average power of a signal to the power of its constituent sinusoidal components:

**Theorem 3.10 Power in a Signal: Sinusoidal Basis Set**
The average power in a bounded and piecewise differentiable signal $x \in \boldsymbol{L}[0, T]$, which has a Fourier series

$$x_F(t) = \sum_{i=-\infty}^{\infty} c_i e^{j2\pi i f_o t} \quad \begin{cases} t \in (0, T) \\ f_o = 1/T \end{cases} \quad c_i = \frac{1}{T} \int_0^T x(t) e^{-j2\pi i f_o t} dt \tag{3.44}$$

is

$$\bar{P} = \sum_{i=-\infty}^{\infty} |c_i|^2 \tag{3.45}$$

Thus, the average signal power equals the sum of the average power in the constituent exponential sinusoidal components.

**Proof**
First, boundedness and integrability of $x$ on $[0, T]$ implies $x \in \boldsymbol{L}^2[0, T]$ (Theorem 2.14) and the power definition $\bar{P} = \frac{1}{T} \int_0^T |x(t)|^2 dt$ is valid. Then

$$\bar{P} = \frac{1}{T} \int_0^T |x(t)|^2 dt = \frac{1}{T} \int_0^T \lim_{\lambda_o \to \infty} \left| \sum_{i=-\lambda_o}^{\lambda_o} c_i e^{j2\pi i f_o t} \right|^2 dt \tag{3.46}$$

Interchanging the order of the limit and integration operations is valid, consistent with the dominated convergence theorem (Theorem 2.19), as the summation converges to $|x(t)|^2$, which is Lebesgue integrable on $[0, T]$. As the integral of a finite sum equals the finite sum of the integrals, the following steps yield the required result:

$$\begin{aligned}\bar{P} &= \lim_{\lambda_o\to\infty}\frac{1}{T}\int_0^T \sum_{i=-\lambda_o}^{\lambda_o}\sum_{k=-\lambda_o}^{\lambda_o} c_i e^{j2\pi i f_o t} c_k^* \, e^{-j2\pi k f_o t}dt \\ &= \frac{1}{T}\sum_{i=-\infty}^{\infty}\sum_{k=-\infty}^{\infty} c_i c_k^* \int_0^T e^{j2\pi i f_o t} e^{-j2\pi k f_o t}dt = \frac{1}{T}\sum_{i=-\infty}^{\infty}|c_i|^2\int_0^T dt = \sum_{i=-\infty}^{\infty}|c_i|^2\end{aligned} \tag{3.47}$$

### 3.6.2 Arbitrary Basis Set

Consider an arbitrary interval $[\alpha, \beta]$ and an orthonormal basis set $\{b_0, b_1, \ldots\}$ for this interval. The following theorem states that the average power in a signal is equal to the sum of the power in the individual components as defined by the basis set:

**Theorem 3.11 Average Power in a Signal: General Case**
The average power in a signal $x \in \boldsymbol{L}^2[\alpha, \beta]$ on the interval $[\alpha, \beta]$, which has a decomposition

$$x(t) = \sum_{i=0}^{\infty} c_i b_i(t) \quad t \in [\alpha, \beta], c_i = \int_{\alpha}^{\beta} x(t) b_i^*(t)dt \tag{3.48}$$

is

$$\bar{P} = \frac{1}{\beta-\alpha}\sum_{i=0}^{\infty}|c_i|^2 \tag{3.49}$$

**Proof**
The proof follows in an analogous manner to the proof of Theorem 3.10. In brief,

$$\begin{aligned}\bar{P} &= \frac{1}{\beta-\alpha}\int_{\alpha}^{\beta}|x(t)|^2 dt = \frac{1}{\beta-\alpha}\sum_{i=0}^{\infty}\sum_{k=0}^{\infty} c_i c_k^* \int_{\alpha}^{\beta} b_i(t) b_k^*(t)dt \\ &= \frac{1}{\beta-\alpha}\sum_{i=0}^{\infty}|c_i|^2 \int_{\alpha}^{\beta} b_i(t) b_i^*(t) = \frac{1}{\beta-\alpha}\sum_{i=0}^{\infty}|c_i|^2\end{aligned} \tag{3.50}$$

***3.6.2.1 Note*** It is the orthogonality, or equivalently orthonormality, that guarantees the average power of a signal on an interval is the summation of the powers in the individual signal components.

## 3.7 THE POWER SPECTRAL DENSITY

The most widely used way to characterize a random process in engineering and the physical sciences is through the use of a power spectral density function. The definition for the power spectral density of a random process has its basis in the power spectral density of an individual signal, and this is discussed below in a manner consistent with Howard (2002, ch. 3).

The usual approach for defining the power spectral density is via an autocorrelation function. An equivalent, and more direct, approach is to define it based on a basis set decomposition. With this approach, a power spectral density should establish the powers in the individual signal components of a signal, assumed to be orthonormal or orthogonal, in a manner such that the sum of the powers of the signal components equals the total signal power. Consider the interval $[\alpha, \beta]$ and the basis set decomposition

$$x(t) = \sum_{i=0}^{\infty} c_i b_i(t) \quad t \in [\alpha, \beta],\ c_i = \int_{\alpha}^{\beta} x(t) b_i^*(t)\,dt \tag{3.51}$$

where the average signal power, consistent with Theorem 3.11, is

$$\bar{P} = \frac{1}{\beta - \alpha} \sum_{i=0}^{\infty} |c_i|^2 \tag{3.52}$$

Such a decomposition leads to the following general definition for a power spectral density function:

**Definition: Power Spectral Density** A power spectral density function $G: \boldsymbol{Z}^+ \to \boldsymbol{R}$ is defined, consistent with the illustration in Figure 3.4, according to

$$G(i) = \frac{1}{\beta - \alpha} \sum_{k=0}^{\infty} |c_k|^2 \delta_K(i-k) \quad \text{or} \quad G(i) = \frac{1}{\beta - \alpha} \sum_{k=0}^{\infty} |c_k|^2 \delta(i-k) \tag{3.53}$$

assuming $\delta$ is notation for the limit of an appropriate Dirac sequence of functions $\{\delta_r\}_{r=1}^{\infty}$ and $G(i)$ is notation for $\left\{ G_r(i) = \frac{1}{\beta - \alpha} \sum_{k=0}^{\infty} |c_k|^2 \delta_r(i-k) \right\}_{r=1}^{\infty}$.

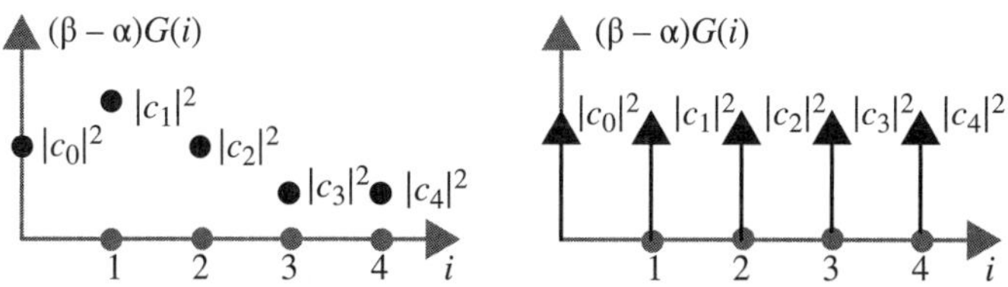


**FIGURE 3.4** Graphs of two possible forms for the power spectral density. Left: power spectral density based on the Kronecker delta. Right: power spectral density based on the Dirac delta.

With these definitions, the average power of the signal is given by the relationships

$$\bar{P} = \sum_{i=0}^{\infty} G(i) \quad \text{or} \quad \bar{P} = \int_{0}^{\infty} G(i)di \tag{3.54}$$

where $\int_{0}^{\infty} G(i)di$ is notation for the limit: $\lim_{r\to\infty} \int_{0}^{\infty} G_r(i)di$.

### 3.7.1 Energy Spectral Density

For signals with finite energy, an energy spectral density may be appropriate. This applies, in particular, to signals defined on the semi-infinite interval $[0, \infty]$.

Consider a signal with finite energy on an interval $[\alpha, \beta]$ and with the basis set decomposition

$$x(t) = \sum_{i=0}^{\infty} c_i b_i(t) \qquad t \in [\alpha, \beta], c_i = \int_{\alpha}^{\beta} x(t) b_i^*(t) dt \tag{3.55}$$

where the signal energy, consistent with Theorem 3.11, is

$$E_x = \sum_{i=0}^{\infty} |c_i|^2 \tag{3.56}$$

**Definition: Energy Spectral Density** The energy spectral density is defined according to

$$E(i) = \sum_{k=0}^{\infty} |c_k|^2 \delta_K(i-k) \quad \text{or} \quad E(i) = \sum_{k=0}^{\infty} |c_k|^2 \delta(i-k) \tag{3.57}$$

The energy of the signal is given by

$$E_x = \sum_{i=0}^{\infty} E(i) \quad \text{or} \quad E_x = \int_{0}^{\infty} E(i)di \tag{3.58}$$

### 3.7.2 Power Spectral Density: Sinusoidal Basis Set

The most widely used signal decomposition is based on a sinusoidal basis set and, for this case, it is usual to represent the power in the individual signal components according to their frequency and not their coefficient number. The notation $G(T, f), f \in \boldsymbol{R}$, is used for the power spectral density on the interval $[0, T]$.

Consider a signal defined on the interval [0, *T*], where the fundamental frequency is $f_o = 1/T$, and the signal decomposition, consistent with Equation 3.44, is

$$x(t) = \sum_{i=-\infty}^{\infty} c_i e^{j2\pi i f_o t} \qquad \begin{cases} t \in (0,T) \\ f_o = 1/T \end{cases} \qquad c_i = \frac{1}{T}\int_0^T x(t)e^{-j2\pi i f_o t}\,dt \tag{3.59}$$

Three definitions for the power spectral density can be made and these are discussed, in turn, below:

**Definition: Power Spectral Density: First Definition** Consistent with the definitions given for the general case in Equation 3.53, and as illustrated in Figure 3.5, the following power spectral definitions can be made:

$$G(T,f) = \sum_{k=-\infty}^{\infty} |c_k|^2 \delta_K(f - kf_o) \quad \text{or} \quad G(T,f) = \sum_{k=-\infty}^{\infty} |c_k|^2 \delta(f - kf_o) \tag{3.60}$$

The total signal power then is

$$\bar{P} = \sum_{i=-\infty}^{\infty} G(T, if_o) \quad \text{or} \quad \bar{P} = \int_{-\infty}^{\infty} G(T,f)\,df \tag{3.61}$$

Assuming a suitable sequence of functions, $\{\delta_r\}_{r=1}^{\infty}$, for the Dirac delta, in these relationships,

$$G(T,f) \text{ is notation for } \left\{ G_r(T,f) = \frac{1}{\beta - \alpha} \sum_{k=-\infty}^{\infty} |c_k|^2 \delta_r(f - kf_o) \right\}_{r=1}^{\infty}$$

$$\int_{-\infty}^{\infty} G(T,f)\,df \text{ is notation for the limit: } \lim_{r\to\infty} \int_{-\infty}^{\infty} G_r(T,f)\,df.$$

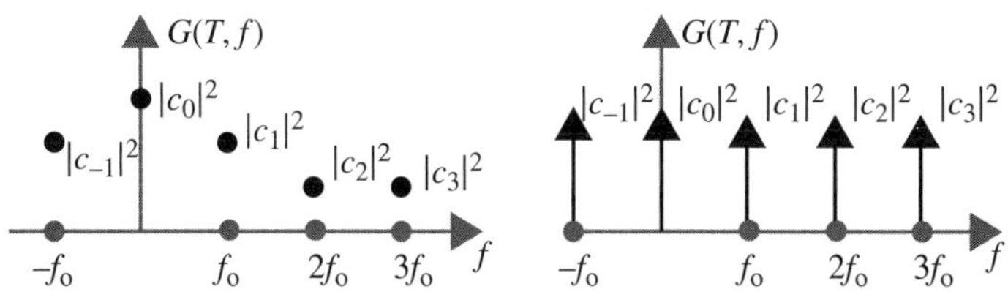


**FIGURE 3.5** Graphs of two possible forms for the power spectral density and for the sinusoidal case. Left: power spectral density based on the Kronecker delta. Right: power spectral density based on the Dirac delta.

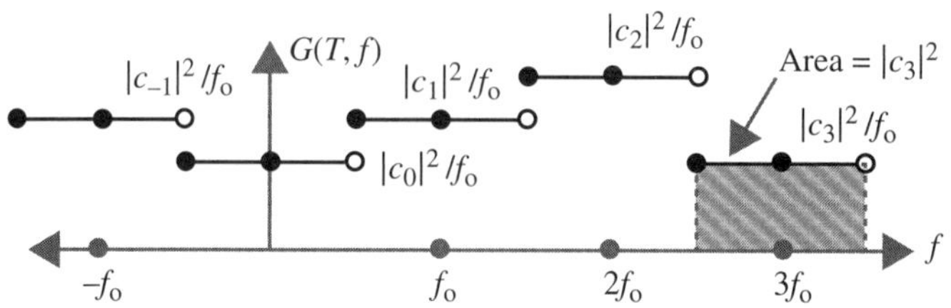


**FIGURE 3.6** Graph of a piecewise continuous power spectral density function for the interval [0,*T*].

#### 3.7.2.1 *Second Definition: Step Power Spectral Density*

A preferable definition for a power spectral density is one that leads to a continuous function and whose integral equals the total power. Toward this end, the following definition can be made:

**Definition: Power Spectral Density: Second Definition** A piecewise step definition for the power spectral density, as illustrated in Figure 3.6, can be made according to

$$\begin{aligned} G(T,f) &= \frac{\left|c_{\lfloor f/f_o+1/2\rfloor}\right|^2}{f_o} \quad f\in \mathbf{R} \\ &= \frac{|c_i|^2}{f_o} \quad f\in\left[if_o-\frac{f_o}{2}, if_o+\frac{f_o}{2}\right), i\in \mathbf{Z} \end{aligned} \tag{3.62}$$

Clearly, by construction, and as required,

$$\bar{P}=\int_{-\infty}^{\infty} G(T,f)df=\sum_{i=-\infty}^{\infty}|c_i|^2 \tag{3.63}$$

#### 3.7.2.2 *Third Definition: Continuous Power Spectral Density*

The basis for obtaining a continuous signal for the power spectral density is Parseval's relationship:

$$\int_0^T |x(t)|^2dt=\int_{-\infty}^{\infty}|X(T,f)|^2df \tag{3.64}$$

where $X(T,f)$ is the Fourier transform of the signal $x$ on the interval $[0, T]$. Scaling both integrals by $T$ yields

$$\bar{P}=\frac{1}{T}\int_0^T|x(t)|^2dt=\int_{-\infty}^{\infty}\frac{|X(T,f)|^2}{T}df=\int_{-\infty}^{\infty}G(T,f)df \tag{3.65}$$

and a power spectral density function, $G: \mathbf{R} \to \mathbf{R}$, can be defined as follows:

**Definition: Power Spectral Density: Third Definition** The power spectral density of a signal $x \in \mathbf{L}^2[0,T]$, evaluated on the interval [0, $T$], is defined according to

$$G(T,f) = \frac{|X(T,f)|^2}{T} \tag{3.66}$$

***3.7.2.3 Notes*** First, all the power spectral density definitions contain the same information and, thus, the definitions can be used interchangeably. Second, the power spectral density definition that should be used depends on the form that best facilitates analysis. Third, the power spectral density definition, as given by Equation 3.66, is the one that is commonly used. Finally, the third power spectral density definition results, as illustrated in Figure 3.7, in a continuous function passing through the midpoints of the step levels defined by the second power spectral density definition. To show this, consider the relationship (Eq. 3.36)

$$c_i = X(T, if_o)/T \quad f_o = 1/T \tag{3.67}$$

whereupon the third power spectral density function implies

$$G(T, if_o) = \frac{|X(T, if_o)|^2}{T} = \frac{|c_i|^2}{f_o} \tag{3.68}$$

***3.7.2.4 Interpretation of Continuous Power Spectral Density Function*** As a Fourier series decomposition of a signal on an interval [0, $T$] yields sinusoidal components with frequencies $f_o$, $2f_o$, …, it is reasonable to conclude that $G(T,f_x)$ should only be interpreted for $f_x = if_o, i \in \mathbf{Z}^+$. This is the case, and the interpretation of $G(T, if_o)$ for some integer value of $i$ is specified in the following theorem:

**Theorem 3.12 Interpretation of Power Spectral Density**
If $x \in \mathbf{L}^2[0,T]$ and is piecewise differentiable on [0, $T$], and the power spectral density of $x$ is defined according to

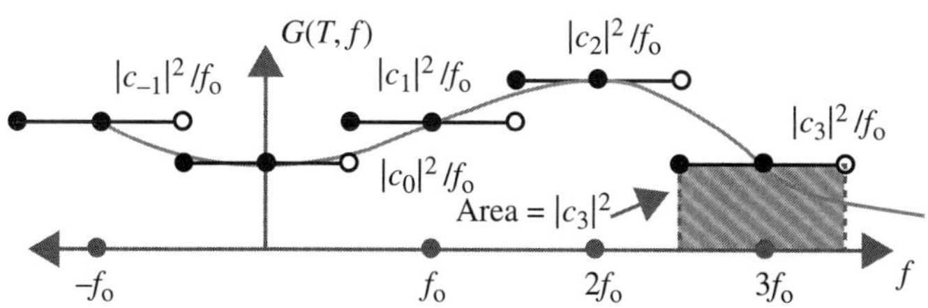


**FIGURE 3.7** The graph of the second and third power spectral density functions for the interval [0,$T$].

$$G(T,f) = \frac{|X(T,f)|^2}{T} \qquad f_o = \frac{1}{T} \tag{3.69}$$

then the average power in the sinusoidal components with a frequency $if_o$, and on the interval $[0, T]$, is given by

$$\begin{aligned} |c_0|^2 &= f_o G(T,0) \\ |c_{-i}|^2 + |c_i|^2 &= f_o[G(T,-if_o) + G(T,if_o)] \quad i \in \mathbf{Z}^+ \end{aligned} \tag{3.70}$$

**Proof**

From Equation 3.32 the power in the sinusoidal components with a frequency of $if_o$ Hz is $|c_{-i}|^2 + |c_i|^2$. The required result then follows from Equation 3.68.

***3.7.2.5 Example: Power Spectral Density of a Sinusoid*** Consider the case of the interval $[0, T]$, $f_o = 1/T$, and the signals

$$x_1(t) = A\cos(2\pi i f_o t) \qquad x_2(t) = A\sin(2\pi i f_o t) \qquad i \in \mathbf{Z}^+ \tag{3.71}$$

Using the results detailed in Appendix 3.C, the power spectral density functions, using the three definitions, respectively, are

$$G_1(T,f) = G_2(T,f) = \begin{cases} 0 & f \notin \{-if_o, if_o\} \\ \dfrac{A^2}{4} & f = -if_o, if_o \end{cases} \tag{3.72}$$

$$G_1(T,f) = G_2(T,f) = \begin{cases} 0 & \text{elsewhere} \\ \dfrac{A^2}{4f_o} & \begin{cases} (-i-0.5)f_o \le f < (-i+0.5)f_o \\ (i-0.5)f_o \le f < (i+0.5)f_o \end{cases} \end{cases} \tag{3.73}$$

$$\begin{aligned} G_{1,2}(T,f) = \frac{A^2T}{4}\Bigg[ &\text{sinc}^2\left[\left(\frac{f}{if_o}-1\right)i\right] + \text{sinc}^2\left[\left(\frac{f}{if_o}+1\right)i\right] \\ &\pm 2\,\text{sinc}\left[\left(\frac{f}{if_o}-1\right)i\right]\text{sinc}\left[\left(\frac{f}{if_o}+1\right)i\right]\Bigg] \end{aligned} \tag{3.74}$$

where, in Equation 3.74, the plus is for $G_1$ and the minus is for $G_2$. These respective power spectral densities are shown in Figures 3.8 and 3.9 for the case of the interval [0, 1] and 4 Hz sinusoids with unity amplitude. Note, for the third form, that the power spectral densities are zero at $f = 0, \pm 1, \pm 2, \pm 3, \pm 5, \ldots$ as required.

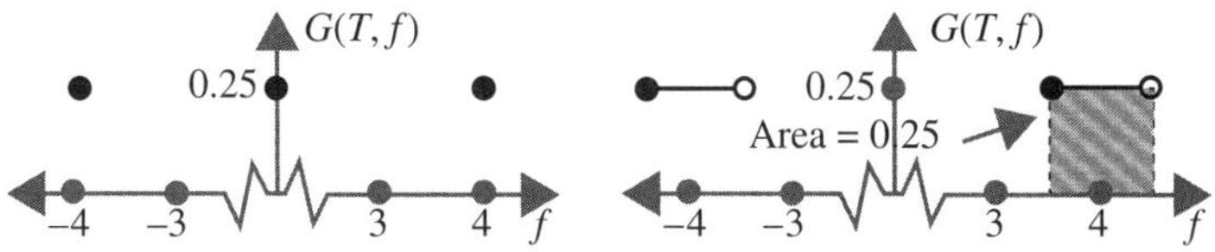


**FIGURE 3.8** First (left) and second (right) forms for the power spectral density for the signals $A\sin(2\pi i f_o t)$ and $A\cos(2\pi i f_o t)$. The signal parameters are $A = 1$, $T = 1$, $f_o = 1/T$, and $i = 4$.

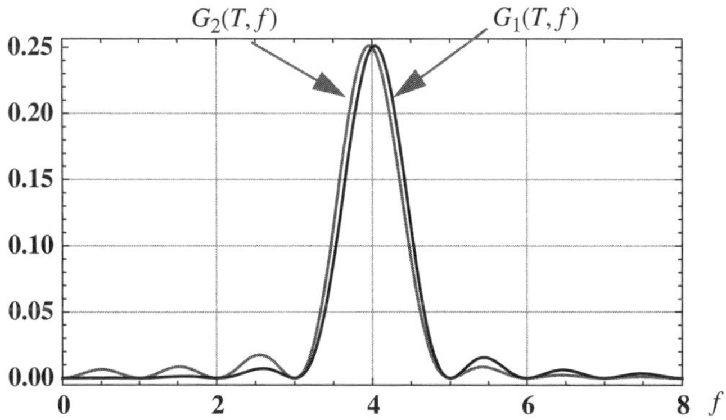


**FIGURE 3.9** Third forms for the power spectral density for the signals $A\sin(2\pi i f_o t)$ and $A\cos(2\pi i f_o t)$. The signal parameters are $A = 1$, $T = 1$, $f_o = 1/T$, and $i = 4$.

## 3.8 THE AUTOCORRELATION FUNCTION

The autocorrelation function is also widely used in characterizing random phenomena as it provides, in general, useful information about the correlatedness of a random process at two distinct times.

### 3.8.1 Definition of the Autocorrelation Function

**Definition: Definition of Autocorrelation Function** The autocorrelation function of a bounded function $x: \boldsymbol{R} \to \boldsymbol{C}$ on the interval $[0, T]$ is defined according to

$$R(T,t_1,t_2) = \begin{cases} x(t_1)x^*(t_2) & t_1,t_2 \in [0,T] \\ 0 & \text{elsewhere} \end{cases} \tag{3.75}$$

With the definition of a delay variable $\tau$ defined as $\tau = t_1 - t_2$ or $\tau = t_2 - t_1$, the following alternative autocorrelation function definitions can be made:

$$\begin{aligned} R_1(T,t,\tau) &= x(t)x^*(t-\tau) & t \in [0,T], \quad t-\tau \in [0,T] \\ R_2(T,t,\tau) &= x(t)x^*(t+\tau) & t \in [0,T], \quad t+\tau \in [0,T] \\ R_3(T,t,\tau) &= x(t-\tau)x^*(t) & t \in [0,T], \quad t-\tau \in [0,T] \\ R_4(T,t,\tau) &= x(t+\tau)x^*(t) & t \in [0,T], \quad t+\tau \in [0,T] \end{aligned} \tag{3.76}$$

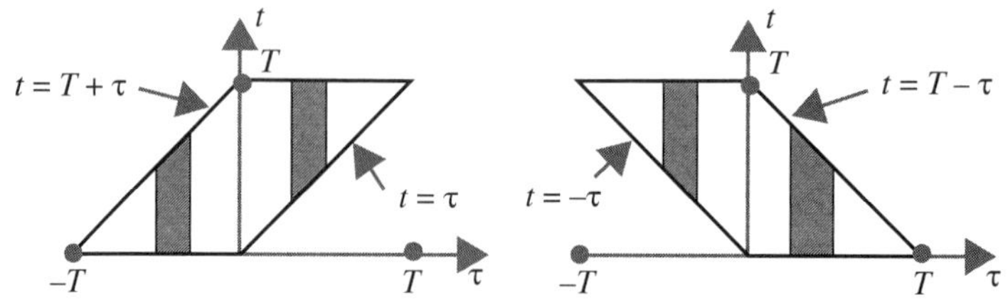


**FIGURE 3.10** Region where autocorrelation function is nonzero. Left: region for definition of $R_1$ (T, t, τ) or $R_3$ (T, t, τ). Right: region for definition of $R_2$ (T, t, τ) or $R_4$ (T, t, τ).

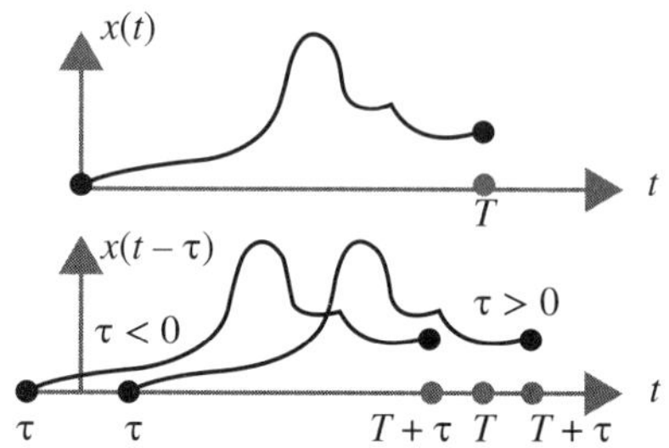


**FIGURE 3.11** A real signal and its shifted counterpart.

These autocorrelation functions are nonzero on the region of the $t$, $\tau$ plane as defined in Figure 3.10.

From the definitions, it follows that $R_1(T,t,\tau)=R_2(T,t,-\tau)$, $R_3(T,t,\tau)=R_4(T,t,-\tau)$, $R_1(T,t+\tau,\tau)=R_4(T,t,\tau)$, and $R_2(T,t-\tau,\tau)=R_3(T,t,\tau)$. As will be shown later, the definitions $R_1(T, t, \tau)$ and $R_4(T, t, \tau)$ lead to results that are consistent with the power spectral density definition stated in Equation 3.66 and, accordingly, either of these definitions is suitable for use. The following definition is appropriate:

**Definition: Definition of Autocorrelation Function** The autocorrelation function of a bounded function $x:\boldsymbol{R}\rightarrow\boldsymbol{C}$ on the interval $[0, T]$ is defined according to either

$$R(T,t,\tau)=\begin{cases} x(t)x^*(t-\tau) & t,t-\tau\in[0,T] \\ 0 & \text{elsewhere} \end{cases} \tag{3.77}$$

or

$$R(T,t,\tau)=\begin{cases} x(t+\tau)x^*(t) & t,t+\tau\in[0,T] \\ 0 & \text{elsewhere} \end{cases} \tag{3.78}$$

To provide an interpretation of the autocorrelation function, consider a signal $x$, and its shifted counterpart, as shown in Figure 3.11. For a fixed time $t$, the shift $\tau$ required before there is a significant change between $x(t)$ and $x^*(t-\tau)$, and hence

$x(t)x^*(t-\tau)$, is an indication of the interval over which the signal is correlated at the time $t$.

***3.8.1.1 Time Averaged Autocorrelation Function*** An average measure of the correlation time of a signal can be obtained through the time averaged autocorrelation function:

**Definition: Time Averaged Autocorrelation Function** For bounded and integrable signals on $[0, T]$, the time averaged autocorrelation function is defined according to

$$\bar{R}(T,\tau)=\frac{1}{T}\int_{t,t\pm\tau\in[0,T]} R(T,t,\tau)dt \tag{3.79}$$

For $R_1(T,t,\tau)=x(t)x^*(t-\tau)$ and $R_4(T,t,\tau)=x(t+\tau)x^*(t)$, the time averaged autocorrelation function is

$$\bar{R}(T,\tau)=\begin{cases}\dfrac{1}{T}\displaystyle\int_0^{T+\tau} x(t)x^*(t-\tau)dt & \tau<0\\[2ex] \dfrac{1}{T}\displaystyle\int_\tau^{T} x(t)x^*(t-\tau)dt & \tau>0\end{cases} \tag{3.80}$$

The integration limits arise from the illustration in Figure 3.10 for the case of $R_1(T, t, \tau)$ and, for the case of $R_4(T, t, \tau)$, after a change of variable.

**Definition: Alternative Time Averaged Autocorrelation Function** For the definitions $R_2(T,t,\tau)=x(t)x^*(t+\tau)$ and $R_3(T,t,\tau)=x(t-\tau)x^*(t)$, an alternative time averaged autocorrelation function can be defined:

$$\bar{R}_A(T,\tau)=\begin{cases}\dfrac{1}{T}\displaystyle\int_{-\tau}^{T} x(t)x^*(t+\tau)dt & \tau<0\\[2ex] \dfrac{1}{T}\displaystyle\int_0^{T-\tau} x(t)x^*(t+\tau)dt & \tau>0\end{cases} \tag{3.81}$$

Note that $\bar{R}_A(T,\tau)=\bar{R}(T,-\tau)$.

**Theorem 3.13 Boundedness of Averaged Autocorrelation Function**
For $T$ fixed, the following holds: First, if $x\in \boldsymbol{L}^2[0,T]$, then $\bar{R}(T,\tau)\in \boldsymbol{L}[-T,T]$. Second, if $x\in \boldsymbol{L}[0,T]$, and is bounded on $[0, T]$, then $\bar{R}(T,\tau)$ is bounded and, hence, $\bar{R}(T,\tau)\in \boldsymbol{L}[-T,T]$.

**Proof**

To prove the second result, boundedness and integrability ensures that

$$\bar{R}(T,\tau) \le \frac{1}{T}\int_0^{T+\tau} |x_{\max}|^2 dt < |x_{\max}|^2\left(1-\frac{|\tau|}{T}\right) \qquad -T<\tau<0 \tag{3.82}$$

where $x_{\max}$ is the maximum level of the magnitude of the signal. The same bound applies for $0 \le \tau < T$. Boundedness of $\bar{R}(T,\tau)$ for $\tau \in [-T,T]$ implies $\bar{R}(T,\tau) \in \boldsymbol{L}[-T,T]$ as required.

To prove the first result, note that the assumption of $x \in \boldsymbol{L}^2[0,T]$ implies, for $\tau$ fixed, that $x(t)x^*(t-\tau)$ is integrable over, for example, $[0,T+\tau]$ when $-T<\tau<0$. This follows from Schwarz's inequality (Theorem 2.15) according to

$$\int_0^{T+\tau} x(t)x^*(t-\tau)dt \le \sqrt{\int_0^{T+\tau} |x(t)|^2 dt}\sqrt{\int_0^{T+\tau} |x(t-\tau)|^2 dt} \qquad -T<\tau<0 \tag{3.83}$$

It then follows that $\bar{R}(T,\tau)$ is finite for $\tau \in [-T,T]$. Again, boundedness of $\bar{R}(T,\tau)$ for $\tau \in [-T,T]$ implies $\bar{R}(T,\tau) \in \boldsymbol{L}[-T,T]$ as required.

***3.8.1.2 Example: Autocorrelation Function of a Sinusoid*** Consider the case of the interval [0, $T$], $f_o = 1/T$, and the signals

$$x_1(t) = A\cos(2\pi i f_o t) \qquad x_2(t) = A\sin(2\pi i f_o t) \qquad i \in \boldsymbol{Z}^+ \tag{3.84}$$

Using the definition for the autocorrelation function, Equation 3.77, the respective autocorrelation functions for these signals are

$$R_1(T,t,\tau) = A^2\cos(2\pi i f_o t)\cos(2\pi i f_o(t-\tau)) \tag{3.85}$$

$$R_2(T,t,\tau) = A^2\sin(2\pi i f_o t)\sin(2\pi i f_o(t-\tau)) \tag{3.86}$$

The time averaged autocorrelation functions as given by Equation 3.80 then can be determined to be

$$\bar{R}_1(T,\tau), \bar{R}_2(T,\tau) = A^2\left[\left[1-\frac{|\tau|}{T}\right]\cdot\frac{\cos(2\pi i f_o \tau)}{2} \mp \frac{|\tau|}{2T}\cdot \text{sinc}(2 i f_o \tau)\right]$$
$$-T<\tau<T \tag{3.87}$$

where the negative sign is for $\bar{R}_1(T,\tau)$ and the positive sign for $\bar{R}_2(T,\tau)$. These time averaged autocorrelation functions are illustrated in Figure 3.12.

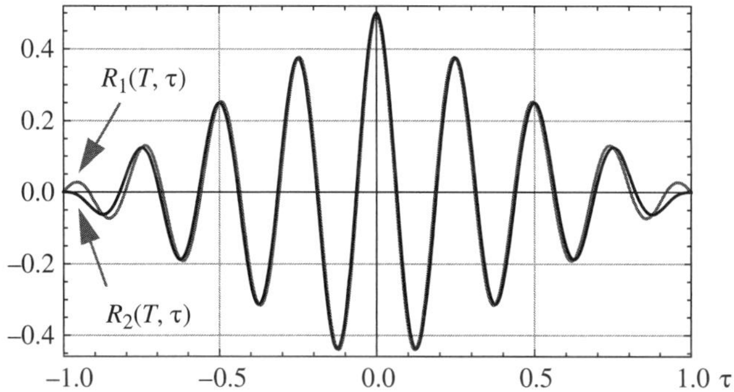


**FIGURE 3.12** Time averaged autocorrelation functions of $A\sin(2\pi i f_o t)$ and $A\cos(2\pi i f_o t)$ for the case of $A = 1$, $T = 1$, $f_o = 1/T$, and $i = 4$.

## 3.9 POWER SPECTRAL DENSITY–AUTOCORRELATION FUNCTION

The relationship between the power spectral density function, as given by $G(T,f) = |X(T,f)|^2/T$, and the time averaged autocorrelation function is via the Fourier transform according to the following theorem:

**Theorem 3.14 Power Spectral Density–Autocorrelation Relationship**
If $x \in \boldsymbol{L}^2[0,T]$, then the power spectral density–time averaged autocorrelation function relationships for $x$ are

$$G(T,f) = \int_{-T}^{T} \bar{R}(T,\tau)e^{-j2\pi f\tau}d\tau \qquad \bar{R}(T,\tau) = \int_{-\infty}^{\infty} G(T,f)e^{j2\pi f\tau}df \tag{3.88}$$

The additional requirement for the second integral is that $x$ be piecewise differentiable on $[0, T]$. This condition is a sufficient condition and the result holds almost everywhere.

**Proof**
The proof is detailed in Appendix 3.D.

### 3.9.1 Relationships for Alternative Autocorrelation Function

Denote the Fourier transform of the alternative time averaged autocorrelation function $\bar{R}_A(T,\tau)$ as $G_A(T,f)$, that is,

$$G_A(T,f) = \int_{-T}^{T} \bar{R}_A(T,\tau)e^{-j2\pi f\tau}d\tau \qquad \bar{R}_A(T,\tau) = \int_{-\infty}^{\infty} G_A(T,f)e^{j2\pi f\tau}df \tag{3.89}$$

The relationship $\bar{R}_A(T,\tau) = \bar{R}(T,-\tau)$ implies

$$G_A(T,f) = \int_{-T}^{T} \bar{R}(T,-\tau)e^{-j2\pi f\tau}d\tau = \int_{-T}^{T} \bar{R}(T,\zeta)e^{j2\pi f\zeta}d\zeta = G(T-f) \tag{3.90}$$

where a change of variable $\zeta = -\tau$ has been used. In summary:

**Theorem 3.15 Alternative Autocorrelation–PSD Relationship**
If $x \in \boldsymbol{L}^2[0,T]$, and is piecewise differentiable on [0, *T*], then the power spectral density–modified time averaged autocorrelation function relationships for this signal are

$$G(T,f) = \int_{-T}^{T} \bar{R}_A(T,\tau)e^{j2\pi f\tau}d\tau \qquad \bar{R}_A(T,\tau) = \int_{-\infty}^{\infty} G(T,f)e^{-j2\pi f\tau}df \tag{3.91}$$

For the real signal case where $X(T,f) = X^*(T,-f)$ (Hermitian symmetry), it follows that $G(T,f) = G(T,-f)$ and $G_A(T,f) = G(T,f)$.

**Proof**
The proof follows as detailed above.

***3.9.1.1 Note*** For the case of real signals, both time averaged autocorrelation functions lead, after a Fourier transform, to the signal power spectral density. This is not the case for complex signals.

## 3.10 RESULTS FOR THE INFINITE INTERVAL

For signals that are locally square integrable, that is, $x \in \boldsymbol{L}^2[0,T]$, the definitions for the time averaged autocorrelation function, and the power spectral density, are unproblematic for the finite interval [0, *T*]. However, when the definitions are extended to the infinite interval, additional constraints are required.

### 3.10.1 Average Power

Taking the limit as *T* tends toward infinity of the average power on the interval [0, *T*] yields a definition for the average signal power on the interval $(0,\infty)$, denoted $\bar{P}_\infty$:

**Definition: Power on Infinite Interval** By definition, the average power on the infinite interval $(0,\infty)$ is

$$\bar{P}_\infty = \lim_{T\to\infty} \frac{1}{T}\int_0^T |x(t)|^2 dt \tag{3.92}$$

### 3.10.2 The Power Spectral Density

A power spectral density definition for the infinite interval can be established by using Parseval's relationship on the definition for average power:

$$\bar{P}_{\infty} = \lim_{T\to\infty} \frac{1}{T}\int_{0}^{T} |x(t)|^2 dt = \lim_{T\to\infty} \int_{-\infty}^{\infty} \frac{|X(T,f)|^2}{T} df = \lim_{T\to\infty} \int_{-\infty}^{\infty} G(T,f) df \tag{3.93}$$

If it is possible to interchange the order of integration and limit operations in the last equation (e.g., consistent with the dominated or monotone convergence theorems), then $\bar{P}_{\infty}$ can be rewritten as

$$\bar{P}_{\infty} = \lim_{T\to\infty} \int_{-\infty}^{\infty} G(T,f) df = \int_{-\infty}^{\infty} \lim_{T\to\infty} G(T,f) df \tag{3.94}$$

and a power spectral density function, $G_{\infty}$, for the interval $[0, \infty]$ can be defined according to:

**Definition: Power Spectral Density on Infinite Interval** The power spectral density for the infinite interval is defined according to

$$G_{\infty}(f) = \lim_{T\to\infty} G(T,f) = \lim_{T\to\infty} \frac{|X(T,f)|^2}{T} \tag{3.95}$$

***3.10.2.1 Components with an Impulsive Power Spectral Density*** Many signals encountered in the physical sciences have periodic components, and, consistent with the sinusoidal case, as detailed in Equation 3.74, the power spectral densities of such signals become impulsive on a set of zero measure as $T$ approaches infinity. In general, impulsive signal components in the power spectral density result in the conditions specified in the dominated or monotone convergence theorems not being satisfied, and, in consequence, the interchange of limit and integral relationships in Equation 3.94 cannot be justified. Accordingly, it is not possible to define a power spectral density function for the infinite interval consistent with the definition specified in Equation 3.95 and with the requirement that the integral of the power spectral density equal the average signal power. However, the following definition is pragmatic:

**Definition: Power Spectral Density: Impulsive and Infinite Interval Case** For the case where a signal has a countable number of components that lead to impulsive components in the power spectral density on the infinite interval, the power spectral density is defined according to

$$G_{\infty}(f) = \lim_{T\to\infty} G(T,f) = \lim_{T\to\infty} \frac{|X(T,f)|^2}{T} \tag{3.96}$$

and the average signal power is

$$\bar{P}_{\infty} = \lim_{T \to \infty} \int_{-\infty}^{\infty} G(T,f)df \tag{3.97}$$

*3.10.2.2 Existence of Power Spectral Density: Infinite Interval* The following theorem states when the power spectral density, as defined by Equation 3.96, exists:

**Theorem 3.16 Existence of Power Spectral Density: Infinite Interval**
The power spectral density $G_{\infty}(f)$ exists if $\sup\{G(T,f):\ T \in \boldsymbol{R}^{+}\}$ is finite. A sufficient condition for this is for $x \in \boldsymbol{L}[0,T]$, for all $T \in \boldsymbol{R}^{+}$, and

$$\sup\left\{\frac{1}{T}\left[\int_{0}^{T}|x(t)|dt\right]^{2}\ :\ T \in \boldsymbol{R}^{+}\right\} < \infty \tag{3.98}$$

**Proof**
Consider

$$G_{\infty}(f) = \lim_{T \to \infty} G(T,f) = \lim_{T \to \infty} \frac{|X(T,f)|^{2}}{T} = \lim_{T \to \infty} \frac{1}{T}\left|\int_{0}^{T} x(t)e^{-j2\pi ft}dt\right|^{2} \tag{3.99}$$

which implies the bound:

$$G_{\infty}(f) \le \lim_{T \to \infty} \frac{1}{T}\left|\int_{0}^{T}|x(t)|dt\right|^{2} \tag{3.100}$$

provided $x \in \boldsymbol{L}[0,T]$ for all $T \in \boldsymbol{R}^{+}$.

*3.10.2.3 Notes* A sinusoidal signal with a frequency of $f_o$, for example, does not satisfy the conditions specified in Equation 3.98. When $f = \pm f_o$ its power spectral density, consistent with Equation 3.74, is infinite.

### 3.10.3 Integrated Spectrum

To overcome the inconsistencies inherent in the definition of the power spectral density function, $G(T,f)$, when it becomes impulsive as $T \to \infty$, a spectral distribution function $F$ is defined (Champeney, 1987, ch. 11; Parzen, 1962a, p. 110):

**Definition: Spectral Distribution Function** The spectral distribution function is defined as

$$F(f) = \lim_{T\to\infty} \int_{-\infty}^{f} G(T,\lambda)d\lambda \quad F(-\infty)=0 \tag{3.101}$$

From the definition, the average power over the interval $(0,\infty)$ is

$$\bar{P}_{\infty} = F(\infty) \tag{3.102}$$

and the average power due to exponential sinusoidal signal components with frequencies on the interval $(f_1, f_2]$ is given by

$$F(f_2) - F(f_1) \tag{3.103}$$

The integrated spectrum of a power spectral density function $G(T,f)$, which contains an *impulsive* component, is illustrated in Figure 3.13. With the spectral distribution function, the following power spectral density function definition can be made for the infinite interval:

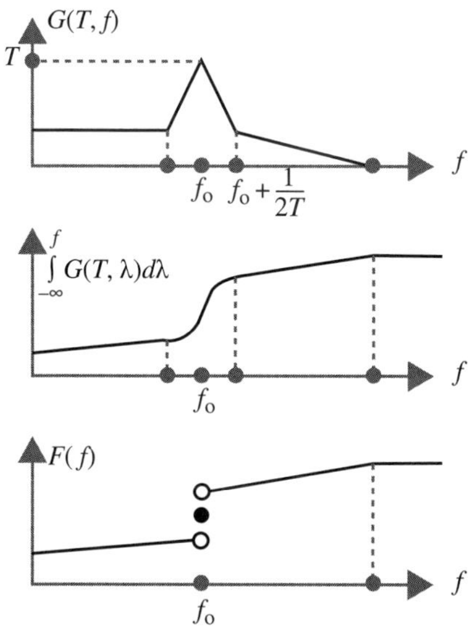


**FIGURE 3.13** Illustration of the spectral distribution function for the case where $G(T,f)$ has an *impulsive* component at $f=f_o$.

**Definition: Power Spectral Density: Infinite Interval** The power spectral density, $G_\infty$, for the infinite interval is defined according to

$$G_\infty(f) = \begin{cases} \dfrac{d}{df}F(f) & F \text{ differentiable at } f \\ \left[F(f_o^+) - F(f_o^-)\right]\delta(f - f_o) & F \text{ not differentiable at } f_o \end{cases} \tag{3.104}$$

When there are impulsive components in the power spectral density, the average power on the interval $[0, \infty)$ is not equal to the integral of $G_\infty(f)$ but is equal to $F(\infty)$.

For the case where there are no impulses in $G(T,f)$ as $T \to \infty$, and $F$ is differentiable, such that

$$F(f) = \int_{-\infty}^{f} \lim_{T\to\infty} G(T,\lambda)d\lambda = \int_{-\infty}^{f} G_\infty(\lambda)d\lambda \tag{3.105}$$

it follows that

$$G_\infty(f) = \frac{d}{df}F(f) \quad f \in \boldsymbol{R} \tag{3.106}$$

and the integral of the $G_\infty$ equals the average signal power.

### 3.10.4 Time Averaged Autocorrelation Function

For the infinite interval, the time averaged autocorrelation function is defined as a limit in the usual manner:

**Definition: Autocorrelation Function on Infinite Interval** The time averaged autocorrelation function for the infinite interval is defined according to

$$R_\infty(\tau) = \lim_{T\to\infty} \bar{R}(T,\tau) \tag{3.107}$$

#### *3.10.4.1 Existence of Autocorrelation Function*

**Theorem 3.17 Existence of Autocorrelation Function on Infinite Interval** If $x$ is locally integrable and is bounded, that is, $\sup\{|x(t)|: \ t \in [0, \infty)\} < \infty$, then $R_\infty(\tau)$ is finite for all values of $\tau$.

Alternatively, if $x \in \boldsymbol{L}^2[0,T]$ for all $T \in \boldsymbol{R}^+$ and $x$ has finite average power on the interval $[0, \infty)$, that is, if $\sup\left\{\dfrac{1}{T}\displaystyle\int_0^T |x(t)|^2 dt: \ T \in \boldsymbol{R}^+\right\} < \infty$, then $R_\infty(\tau)$ is finite for all values of $\tau$.

**Proof**
Consider the case of $\tau > 0$ and

$$\lim_{T\to\infty} \bar{R}(T,\tau) = \lim_{T\to\infty} \frac{1}{T}\int_{\tau}^{T} x(t)x^*(t-\tau)dt \tag{3.108}$$

Boundedness of $x$ on the infinite interval, consistent with $x_{max} = \sup\{|x(t)|:\ t\in[0,\infty)\}$ being finite, and local integrability, imply that the integral is finite for all $T$. Then

$$\lim_{T\to\infty} \bar{R}(T,\tau) \le \lim_{T\to\infty} \frac{1}{T}\int_{\tau}^{T} |x_{max}|^2 dt = \lim_{T\to\infty} |x_{max}|^2\left(1-\frac{\tau}{T}\right) = |x_{max}|^2 \tag{3.109}$$

Alternatively, assume $x \in \boldsymbol{L}^2[0,T]$, for all $T \in \boldsymbol{R}^+$, and

$$\sup\left\{\frac{1}{T}\int_0^T |x(t)|^2 dt:\ \ T\in \boldsymbol{R}^+\right\} < \infty \tag{3.110}$$

For the case of $\tau > 0$, Schwarz's inequality yields

$$\lim_{T\to\infty} \bar{R}(T,\tau) = \lim_{T\to\infty} \frac{1}{T}\int_{\tau}^{T} x(t)x^*(t-\tau)dt \le \lim_{T\to\infty} \frac{1}{T}\sqrt{\int_0^T |x(t)|^2 dt}\sqrt{\int_{\tau}^{T} |x(t-\tau)|^2 dt}$$

$$\le \lim_{T\to\infty} \frac{1}{T}\sqrt{\int_0^T |x(t)|^2 dt}\sqrt{\int_0^T |x(t)|^2 dt} \le \sup\left\{\frac{1}{T}\int_0^T |x(t)|^2 dt:\ \ T\in \boldsymbol{R}^+\right\} \tag{3.111}$$

The proof for the case of $\tau < 0$ follows in an analogous manner.

### 3.10.5 Power Spectral Density–Autocorrelation Relationship

The following theorem extends the result stated in Theorem 3.14, for the relationship between the power spectral density and the time averaged autocorrelation function, to the infinite interval case:

**Theorem 3.18 Power Spectral Density–Autocorrelation: Infinite Interval**
Assume $x \in \boldsymbol{L}^2[0,T]$ for all $T \in \boldsymbol{R}^+$ and the second relationship in Theorem 3.14 is valid for all $T$ almost everywhere. A sufficient condition for this is for $x$ to be piecewise differentiable on the interval $[0, T]$. Assume $G(T,f)$ is nonimpulsive and there

exists a function $B_G \in \boldsymbol{L}$ such that $G(T,f) \le B_G(f)$ for all $T$. Further, assume there exists a function $B_R \in \boldsymbol{L}$ such that $\bar{R}(T,\tau) \le B_R(\tau)$ for all $T$. With these assumptions it follows that the Fourier transform relationships

$$R_\infty(\tau) = \int_{-\infty}^{\infty} G_\infty(f)e^{j2\pi f\tau}df \quad G_\infty(f) = \int_{-\infty}^{\infty} R_\infty(\tau)e^{-j2\pi f\tau}d\tau \tag{3.112}$$

are valid.

These relationships are commonly called the Wiener–Khintchine relations (Champeney, 1987, p. 104; Parzen, 1962a, p. 110; Peebles, 2001, p. 228).

**Proof**

If $x \in \boldsymbol{L}^2[0,T]$ for all $T \in \boldsymbol{R}^+$, then Theorem 3.14 yields

$$G_\infty(f) = \lim_{T\to\infty} G(T,f) = \lim_{T\to\infty} \int_{-T}^{T} \bar{R}(T,\tau)e^{-j2\pi f\tau}d\tau \tag{3.113}$$

To interchange the limit and integration operations in this equation, first, note, consistent with Theorem 3.13, that $\bar{R} \in \boldsymbol{L}[-T,T]$. Second, assume there exists a function $B_R \in \boldsymbol{L}$ such that $\bar{R}(T,\tau) \le B_R(\tau)$ for all $T$. Then, from the dominated convergence theorem (Theorem 2.19), it follows that the order of the limit and integral operations can be interchanged to yield

$$G_\infty(f) = \int_{-\infty}^{\infty} \lim_{T\to\infty} \bar{R}(T,\tau)e^{-j2\pi f\tau}d\tau = \int_{-\infty}^{\infty} R_\infty(\tau)e^{-j2\pi f\tau}d\tau \tag{3.114}$$

To relate $R_\infty$ to $G_\infty$, consider

$$R_\infty(\tau) = \lim_{T\to\infty} \bar{R}(T,\tau) = \lim_{T\to\infty} \int_{-\infty}^{\infty} G(T,f)e^{j2\pi f\tau}df \tag{3.115}$$

Again it is necessary to interchange the limit and integration operations. However, and in general, the power spectral density function $G(T,f)$ will have impulsive components, and it is not possible to justify the interchange of the limit and integral operations. With the assumption of the existence of a function $B_G \in \boldsymbol{L}$ such that $G(T,f) \le B_G(f)$ for all $T$, it then follows, from the dominated convergence theorem (Theorem 2.19), that the order of the limit and integral operations can be interchanged to yield

$$R_\infty(\tau) = \int_{-\infty}^{\infty} \lim_{T\to\infty} G(T,f)e^{j2\pi f\tau}df = \int_{-\infty}^{\infty} G_\infty(f)e^{j2\pi f\tau}df \tag{3.116}$$

***3.10.5.1 Impulsive Case: Formal Definition of Wiener–Khintchine*** In general, the power spectral density on the infinite interval may contain impulsive components such that it is not possible to interchange the order of the limit and integral operations in the equation:

$$R_{\infty}(\tau) = \lim_{T\to\infty} \bar{R}(T,\tau) = \lim_{T\to\infty} \int_{-\infty}^{\infty} G(T,f)e^{j2\pi f\tau}df \tag{3.117}$$

For this case, a spectral distribution function, defined in Equation 3.101, is used:

$$F(f) = \lim_{T\to\infty} \int_{-\infty}^{f} G(T,\lambda)d\lambda \tag{3.118}$$

where $F(-\infty)=0$, $\bar{P}_{\infty}=F(\infty)$, and $F$ is a monotonically increasing function.

This function can be used as a basis for defining the autocorrelation function for the infinite interval, and this leads to a formal definition for the Wiener–Khintchine theorem (Champeney, 1987, p. 108; Grimmett and Stirzaker, 1992, p. 353; Parzen, 1962a, p. 110; Priestley, 1981, p. 218):

**Theorem 3.19 Wiener–Khintchine Theorem**
For the case where the signal has finite average power on the infinite interval, the autocorrelation function on the infinite interval is defined according to

$$R_{\infty}(\tau) = \int_{-\infty}^{\infty} e^{j2\pi f\tau}dF(f) \tag{3.119}$$

**Proof**
The proof is detailed in Appendix 3.E.

## 3.11 CONVERGENCE OF FOURIER COEFFICIENTS

The nature of the power spectral density of a signal on an interval $[0, T]$, consistent with Figure 3.7, is related to the nature of the Fourier series coefficients in a Fourier series decomposition of the signal on the interval. It is of interest to determine the nature of the change in the Fourier coefficients as the interval becomes increasingly large as this determines the nature of the integrability, for example, of the power spectral density.

### 3.11.1 Periodic Signal Case

Consider a bounded, locally integrable, and periodic signal $x:\boldsymbol{R}\to\boldsymbol{C}$, with period $T_p=1/f_p$, which can be written, using an exponential Fourier series, as

$$x(t) = \sum_{p=-\infty}^{\infty} d_p e^{j2\pi p f_p t} \qquad d_p = \frac{1}{T_p} \int_0^{T_p} x(t) e^{-j2\pi p f_p t} dt \tag{3.120}$$

Consider a Fourier series decomposition on the interval $[0, nT_p]$, $n \in \boldsymbol{Z}^+$, with a fundamental frequency $f_o = 1/nT_p = f_p/n$:

$$x(t) = \sum_{k=-\infty}^{\infty} c_k e^{j2\pi k f_o t} \qquad c_k = \frac{1}{nT_p} \int_0^{nT_p} x(t) e^{-j2\pi k f_o t} dt \tag{3.121}$$

The $k$th coefficient can be written as

$$c_k = \frac{1}{nT_p} \cdot \sum_{p=-\infty}^{\infty} d_p \int_0^{nT_p} \exp[j2\pi(np-k)f_o t] dt = \begin{cases} 0 & k \notin \{0, \pm n, \pm 2n, \pm 3n, \ldots\} \\ d_{k/n} & k \in \{0, \pm n, \pm 2n, \pm 3n, \ldots\} \end{cases} \tag{3.122}$$

after noting that boundedness of the signal guarantees the interchange of the summation and the integral.

The nature of the Fourier coefficient is illustrated in Figure 3.14. The zero coefficients arise as only sinusoids with frequencies that are an integer multiple of $f_p$ exist in the signal. Clearly, the nonzero coefficients do not decrease in value as the interval length increases.

### 3.11.2 Convergence of Fourier Coefficients to Zero

Consider an integrable signal $x: \boldsymbol{R} \to \boldsymbol{C}$, which is not necessarily periodic and which can be written as a Fourier series on an arbitrary interval $[0, T_o]$ according to

$$x(t) = \sum_{p=-\infty}^{\infty} d_p e^{j2\pi p f_o t} \quad t \in [0, T_o], f_o = \frac{1}{T_o}, d_p = \frac{1}{T_o} \int_0^{T_o} x(t) e^{-j2\pi p f_o t} dt \tag{3.123}$$

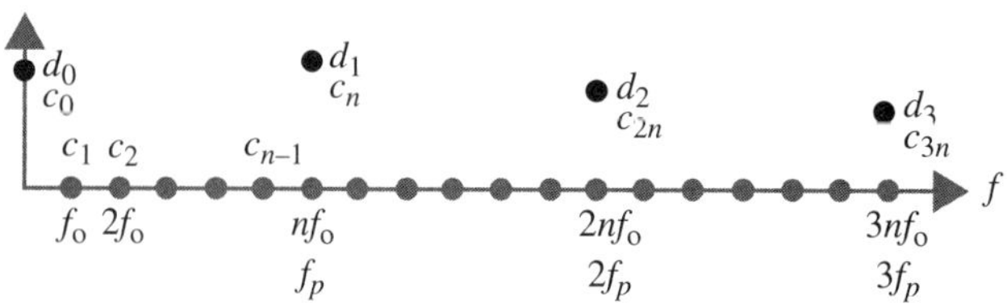


**FIGURE 3.14** Nature of the Fourier coefficients for the two cases of $[0, T_p]$ and $[0, nT_p]$.

On an interval $[0, nT_o]$, $n \in \mathbf{Z}^+$, the two coefficients $c_{nu}$, $c_{nv}$, corresponding to the set frequencies, respectively, of $uf_o$ and $vf_o$, in the decomposition

$$x(t) = \sum_{k=-\infty}^{\infty} c_k e^{j2\pi k(f_o/n)t} \qquad c_k = \frac{1}{nT_o}\int_0^{nT_o} x(t)e^{-j2\pi k(f_o/n)t}dt \qquad t \in [0, nT_o] \tag{3.124}$$

are

$$c_{nu} = \frac{1}{nT_o}\int_0^{nT_o} x(t)e^{-j2\pi u f_o t}dt \qquad c_{nv} = \frac{1}{nT_o}\int_0^{nT_o} x(t)e^{-j2\pi v f_o t}dt \tag{3.125}$$

It is of interest to know whether the coefficients, and hence the power in the associated set frequency components, decrease in magnitude as the interval length increases. The following theorem states conditions for $\lim_{n\to\infty}\left[c_{nu}c_{nv}^*\right] = 0$:

**Theorem 3.20 Convergence of Fourier Coefficients to Zero**

If a signal is locally integrable, and $\lim_{T\to\infty}\frac{1}{T}\int_0^T |x(t)|dt = 0$, then $\lim_{n\to\infty}\left|c_{nu}c_{nv}^*\right| = 0$. For the case of $u = v$, $\lim_{n\to\infty}|c_{nu}|^2 = 0$.

**Proof**
Consider

$$\begin{aligned} c_{nu}c_{nv}^* &= \left[\frac{1}{nT_o}\int_0^{nT_o} x(t)e^{-j2\pi u f_o t}dt\right]\left[\frac{1}{nT_o}\int_0^{nT_o} x^*(t)e^{j2\pi v f_o t}dt\right] \\ \Rightarrow \left|c_{nu}c_{nv}^*\right| &\leq \left[\frac{1}{nT_o}\int_0^{nT_o}|x(t)|dt\right]^2 \end{aligned} \tag{3.126}$$

If the sequence of numbers defined by $\frac{1}{nT_o}\int_0^{nT_o}|x(t)|dt$, as $n$ takes on the values of 1, 2, …, converges to zero, then the corresponding sequence defined by $\left[\frac{1}{nT_o}\int_0^{nT_o}|x(t)|dt\right]^2$ also converges to zero and the required result follows.

***3.11.2.1 Nature of Coefficients: Set Power in an Interval*** A common case with nonperiodic signals is for the signal power in a set frequency interval to converge to a set value as the length of the interval being considered increases. Consider a signal with the Fourier series representations, as given by Equations 3.123 and 3.124 for the respective intervals of $[0, T_o]$ and $[0, nT_o]$. The power in the sinusoidal components of the signal with frequencies in the range $pf_o$ to $[p+(q-1)]f_o$, $p,q \in \mathbf{Z}^+$ for these two intervals, respectively, are

$$P_1 = \sum_{k=p}^{p+q-1} |d_k|^2 \quad P_n = \sum_{k=np}^{(p+q)n-1} |c_k|^2 \tag{3.127}$$

For the case of $pf_o$ and $p+(q-1)]f_o$ being relatively close such that the power in each of the sinusoidal component signals in the interval $[pf_o, (p+(q-1))f_o]$ are approximately constant at the level of the component with a frequency $pf_o$, it follows that

$$P_1 \approx q|d_p|^2 \quad P_n \approx nq|c_{np}|^2 \tag{3.128}$$

and equal powers is consistent with $|c_{np}|^2 = |d_p|^2/n$. It then follows that constant power is a set frequency interval, as the interval duration is increased, is consistent with the magnitude of the Fourier coefficients decaying according to $1/\sqrt{n}$.

## 3.12 CRAMER'S REPRESENTATION AND TRANSFORM

For many signals of interest, for example, power signals and signals arising from stationary random processes where $x \notin \boldsymbol{L}$, the corresponding Fourier transform

$$X(f) = \int_{\boldsymbol{R}} x(t) e^{-j2\pi ft} dt \tag{3.129}$$

does not exist for one or more values of the argument $f$ as the integral is unbounded. The Cramer transform of a signal is valid for such signals on the infinite interval and represents a generalized harmonic characterization of a signal. A Cramer transform of a signal can distinguish between any two sinusoidal components of a signal and can establish the exact frequency of a sinusoidal component. Useful references include Priestley (1981), Yaglom (1962), and Wiener (1930).

### 3.12.1 Miscellaneous Mathematical Results

The following result is fundamental to the development of the theory related to the Cramer transform of a signal:

$$\frac{1}{\pi}\int_0^{\infty} \frac{\sin(p\lambda)}{\lambda} d\lambda = 0.5\,\mathrm{sgn}(p) = \begin{cases} 0.5 & p>0 \\ 0 & p=0 \\ -0.5 & p<0 \end{cases} \tag{3.130}$$

The following result from Chapter 2 (Theorem 2.30) is also important:

**Theorem 3.21 Lebesgue–Stieltjes Integral: Right Increasing Step Function** Consider an increasing step function $g:\boldsymbol{R}\to\boldsymbol{R}$, as illustrated in Figure 3.15 and defined according to

$$g(x)=\sum_{i=1}^{N}g_iu(x-x_i) \tag{3.131}$$

The Lebesgue–Stieltjes integral of a measurable function $f:\boldsymbol{R}\to\boldsymbol{R}$, with respect to the Lebesgue–Stieltjes measure defined by $g$, is

$$\int_{\boldsymbol{R}} f(x)dg(x)=\sum_{i=1}^{N}g_if(x_i) \tag{3.132}$$

### 3.12.2 Cramer Representation and Transform

**Definition: Cramer Representation and Transform** The Cramer representation of a signal $x:\boldsymbol{R}\to\boldsymbol{R}$ is defined as

$$x(t)=\int_{\boldsymbol{R}} e^{j2\pi ft}dX_C(f) \tag{3.133}$$

where the integrator function $X_C:\boldsymbol{R}\to\boldsymbol{C}$ is the Cramer transform:

$$X_C(f)=\frac{j}{2\pi}\int_{-\infty}^{\infty}\frac{e^{-j2\pi f\lambda}-1}{\lambda}\cdot x(\lambda)d\lambda \tag{3.134}$$

Alternative forms for the Cramer transform are

$$\begin{aligned} X_C(f)&=\frac{1}{\pi}\int_{-\infty}^{\infty}e^{-j\pi f\lambda}\cdot\frac{\sin(\pi f\lambda)}{\lambda}\cdot x(\lambda)d\lambda \\ X_C(f)&=f\int_{-\infty}^{\infty}e^{-j\pi f\lambda}\cdot \text{sinc}(f\lambda)\cdot x(\lambda)d\lambda \end{aligned} \tag{3.135}$$

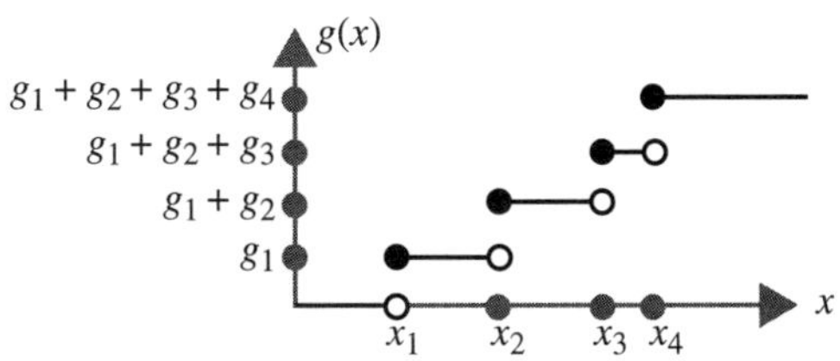


**FIGURE 3.15** Graph of the right increasing step function $g$.

The nature of the Cramer representation is defined by the transform $X_C$. The Cramer representation is also called the *spectral representation.*

An alternative but equivalent definition and transform is

$$y(t) = \int_R e^{j\omega t} dY_C(\omega) \quad Y_C(\omega) = \frac{j}{2\pi} \int_{-\infty}^{\infty} \frac{e^{-j\omega\lambda} - 1}{\lambda} \cdot y(\lambda) d\lambda \tag{3.136}$$

where $Y_C(\omega) = X_C(\omega/2\pi)$. Equivalence of the two definitions can be shown by considering $y(t) = x(t)$, which leads to $Y_C(\omega) = X_C(\omega/2\pi)$, and a change in variable, $\omega = 2\pi f$, in $y(t) = \int_R e^{j\omega t} dX_C(\omega/2\pi)$ leads to $y(t) = \int_R e^{j2\pi ft} dX_C(f) = x(t)$.

### 3.12.3 Initial Approach to the Cramer Transform

A useful starting point, with the goal of understanding the Cramer transform of a signal, is to consider the Fourier series decomposition of a signal $x$ on the interval $[0, T]$:

$$x(t) = \sum_{i=-\infty}^{\infty} c_i e^{j2\pi i f_o t} \quad f_o = \frac{1}{T}, \left| \sum_{i=-\infty}^{\infty} c_i \right| < \infty \tag{3.137}$$

and the Cramer representation of such a signal as specified in the following theorem:

**Theorem 3.22 Integrator Function for a Fourier Series Decomposition**
If a signal $x$ has a Fourier series decomposition given by Equation 3.137, then, with the definition of the integrator function according to

$$X_C(f) = \sum_{i=-\infty}^{\infty} c_i u(f - i f_o) \tag{3.138}$$

it is the case that

$$x(t) = \int_{-\infty}^{\infty} e^{j2\pi ft} dX_C(f) \tag{3.139}$$

The integrator function is illustrated in Figure 3.16.

**Proof**
This result follows from Theorem 3.21.

*3.12.3.1 Notes* The integrator function specified in Equation 3.138 is not unique. Any function that equals the specified function, apart from an added constant, is a valid integrator function.

The integrator function represents a decomposition of the signal in a manner consistent with a Fourier series. This provides useful intuition for the decomposition

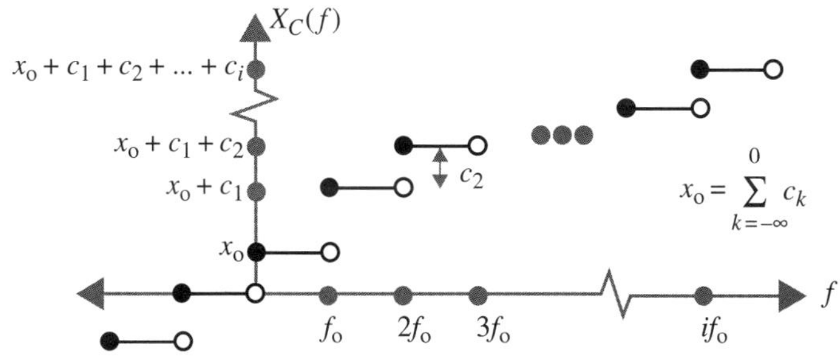


**FIGURE 3.16** Integrator function for a Fourier series representation of a signal. In general, the coefficients are complex.

associated with the Cramer representation. However, the precise decomposition associated with a Fourier series varies from this as detailed below.

### 3.12.4 The Cramer Transform

With the Cramer representation, as given by Equation 3.133, the question arises as to what is the correct form for the integrator function, or the Cramer transform, $X_C(f)$, such that

$$x(t) = \int_R e^{j2\pi ft} dX_C(f) = \lim_{\substack{f_o \to \infty \\ \Delta f \to 0}} \sum_{i=-\lfloor f_o/\Delta f \rfloor}^{\lfloor f_o/\Delta f \rfloor} e^{j2\pi i \Delta ft}[X_C((i+1)\Delta f) - X_C(\Delta f)] \qquad (3.140)$$

The summation form arises from the definition of the Lebesgue–Stieltjes integral and by using a step approximation to $e^{j2\pi ft}$, with respect to $f$, based on a resolution of $\Delta f$. The hypothesis is

$$X_C(f) = \frac{j}{2\pi} \int_R \frac{e^{-j2\pi f\lambda} - 1}{\lambda} \cdot x(\lambda) d\lambda \qquad (3.141)$$

First, the conditions for the existence of $X_C(f)$ are established. Second, the validity of the inversion formula, as given by Equation 3.140, is proved.

**Theorem 3.23 Existence of Cramer Transform**
The Cramer transform $X_C(f)$, as given by Equation 3.134, exists if

$$\lim_{T \to \infty} f \int_{-T}^{T} e^{-j\pi f\lambda} \cdot \text{sinc}(f\lambda) \cdot x(\lambda) d\lambda \qquad (3.142)$$

is finite. A sufficient condition for this is for $\frac{x(t)}{1+|t|} \in \boldsymbol{L}$.

**Proof**
The proof is detailed in Appendix 3.F.

**Theorem 3.24 Validity of Inversion of the Cramer Transform**
Provided $x$ is piecewise differentiable at $t$, the inverse of the Cramer transform $X_C(f)$, as given by the Cramer representation, is valid and

$$\int_R e^{j2\pi ft} dX_C(f) = \frac{x(t^+) + x(t^-)}{2} \tag{3.143}$$

The additional assumption here, consistent with Theorem 3.23, is that $\frac{x(t)}{1+|t|} \in \boldsymbol{L}$.

**Proof**
The proof is detailed in Appendix 3.G.

### *3.12.4.1 Alternative Form for $X_C(f)$*

**Theorem 3.25 Alternative Form $X_C(f)$**
Consider a Fourier series decomposition of a signal $x \in \boldsymbol{L}$ on $[-T, T]$:

$$x(t) = \sum_{i=-\infty}^{\infty} c_i(T) e^{j2\pi i f_o t} \quad f_o = \frac{1}{2T} \tag{3.144}$$

where equivalence almost everywhere is assumed. An alternative form for the Cramer transform, $X_C(f)$, is

$$X_C(f) = \lim_{T\to\infty} \sum_{i=-\infty}^{\infty} \frac{c_i(T)}{\pi} [\mathrm{Si}(i\pi) + \mathrm{Si}(2\pi f T - i\pi)] \tag{3.145}$$

where Si is the sine integral function.

**Proof**
The proof is detailed in Appendix 3.H.

**Theorem 3.26 Existence of Alternative Cramer Transform**
Assume $x \in \boldsymbol{L}[-T, T]$ for all $T$ and $x$ is piecewise differentiable such that the Fourier series representation, as given by Equation 3.144, is valid almost everywhere on $[-T, T]$. Sufficient conditions for the existence of $X_C(f)$, as given by Equation 3.145, over the interval $[-T, T]$ and the infinite interval, respectively, are

$$\sum_{i=1}^{\infty} \frac{|c_i(T)|}{i} < \infty \qquad \lim_{T\to\infty} \sum_{i=1}^{\infty} \frac{|c_i(T)|}{i} < \infty, \quad \lim_{T\to\infty} \sum_{i=1}^{\infty} c_i(T) < \infty \tag{3.146}$$

**Proof**
The proof is detailed in Appendix 3.I.

***3.12.4.2 Example*** Consider a square wave, as illustrated in Figure 3.17, whose Fourier series on the interval $[-kT_o, kT_o]$, $k \in \boldsymbol{Z}^+$, $f_o = 1/2kT_o$, is

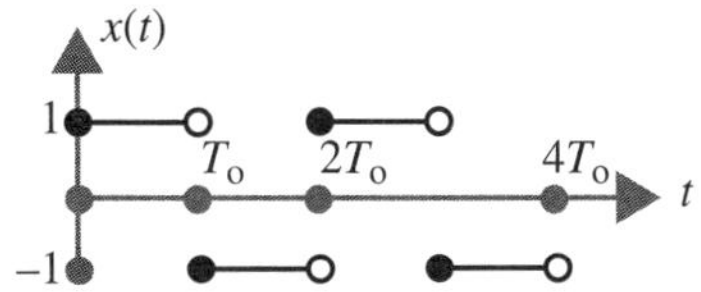


**FIGURE 3.17** Graph of square wave.

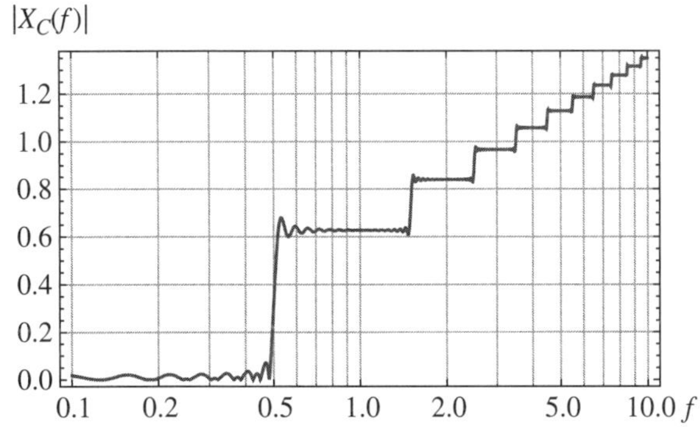


**FIGURE 3.18** The magnitude of the Cramer transform of a square wave for the case of $T_o = 1$ and $k = 16$.

$$x_F(t) = \frac{4}{\pi}\left[\sin(2\pi(kf_o)t) + \frac{\sin(2\pi 3(kf_o)t)}{3} + \frac{\sin(2\pi 5(kf_o)t)}{5} + \cdots\right] \quad (3.147)$$

The series $1 + \frac{1}{3} + \frac{1}{5} + \frac{1}{7} + \frac{1}{9} + \cdots$ does not converge but the series specified by Equation 3.146, that is, $1 + \frac{1}{3\cdot 3} + \frac{1}{5\cdot 5} + \frac{1}{7\cdot 7} + \cdots$, does for all values of $k$. The implication, consistent with Equation 3.146, is that the Cramer transform for a square wave, as given by Equation 3.145, exists on the infinite interval. A square wave does not satisfy the condition specified in Theorem 3.23 of $\frac{x(t)}{1+|t|} \in \boldsymbol{L}$. The magnitude of the Cramer transform of a square wave is shown in Figure 3.18 for the case of $k = 16$ and $T_o = 1$. Being an odd function the Cramer transform is imaginary.

### 3.12.5 Miscellaneous Results

***3.12.5.1 Real and Imaginary Parts of Transform*** Consider the transform in the form

$$X_C(f) = \frac{1}{\pi}\int_{-\infty}^{\infty} e^{-j\pi f\lambda}\cdot\frac{\sin(\pi f\lambda)}{\lambda}\cdot x(\lambda)d\lambda \quad (3.148)$$

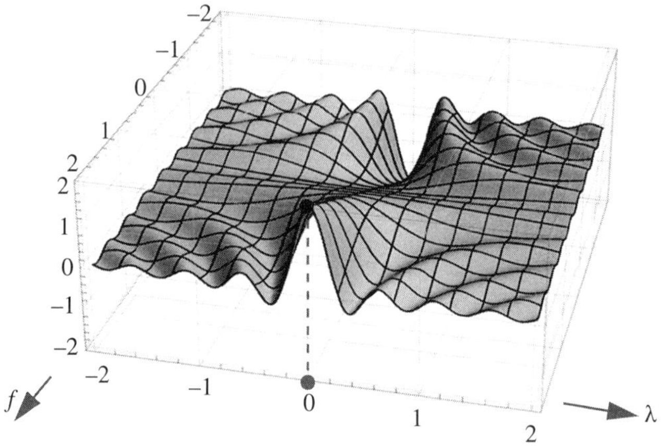


FIGURE 3.19 Graph of the real part of the kernel $K(f, \lambda)$.

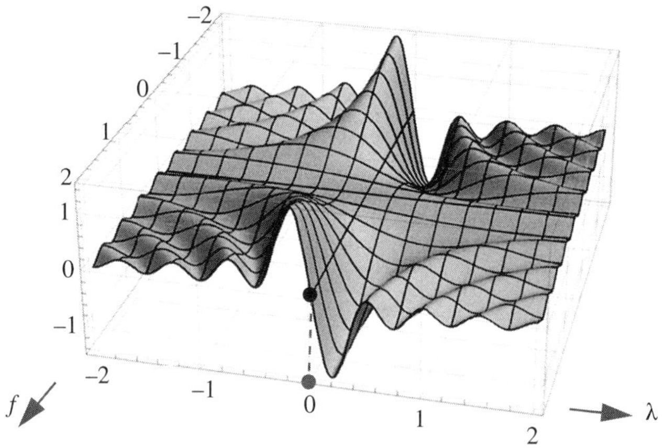


FIGURE 3.20 Graph of the imaginary part of the kernel $K(f, \lambda)$.

The real and imaginary parts of the kernel of the transform

$$K(f,\lambda)=\frac{e^{-j\pi f\lambda}}{\pi}\cdot\frac{\sin(\pi f\lambda)}{\lambda} \tag{3.149}$$

are shown in Figures 3.19 and 3.20.

***3.12.5.2 Basic Properties of Transform*** Basic properties of the Cramer transform include, first, $X_C(0)=0$. Second, the transform is linear. Third, for real signals,

$$X_C(-f)=-X_C^*(f) \tag{3.150}$$

*3.12.5.3 Cramer Transform: Generalized Harmonic Decomposition* The kernel $\frac{e^{-j\pi f\lambda}\cdot\sin(\pi f\lambda)}{\lambda}$ is such that it discriminates with respect to sinusoids, and this discrimination is independent of whether the sinusoids form part of a basis set or not. This is why the Cramer representation is a generalization of a Fourier series decomposition.

Specifically, $X_C(f)$ changes level as $f$ changes from below to above the frequency of a sinusoid. It is this change in $X_C(f)$ that allows the sinusoidal components of a signal to be isolated. Further, it is this change that leads to the interpretation of $X_C(f)$ as an *integrated spectrum*, and from this, a power spectral density can be defined. The details are provided in later subsections.

### 3.12.6 Transform of Common Signals

The following subsections detail the Cramer transform of common signals:

*3.12.6.1 Constant Signal* Consider a constant signal $x(t)=1$ for $t\in(-\infty,\infty)$ whereupon

$$X_C(f)=\frac{1}{\pi}\int_{-\infty}^{\infty}e^{-j\pi f\lambda}\cdot\frac{\sin(\pi f\lambda)}{\lambda}d\lambda=\frac{2}{\pi}\int_{0}^{\infty}\cos(\pi f\lambda)\cdot\frac{\sin(\pi f\lambda)}{\lambda}d\lambda$$

$$=\frac{1}{\pi}\int_{0}^{\infty}\frac{\sin(2\pi f\lambda)}{\lambda}d\lambda=\begin{Bmatrix}0.5 & f>0\\ 0 & f=0\\ -0.5 & f<0\end{Bmatrix}=0.5\,\text{sgn}(f) \tag{3.151}$$

The last result follows from Equation 3.130. The graph of this Cramer transform is shown in Figure 3.21.

*3.12.6.2 Unit Step Function* As detailed in Appendix 3.J, the Cramer transform of $u(t-t_o)$ on the interval $[-T,T]$, assuming $|t_o|<T$, $t_o\neq 0$, is

$$\begin{aligned}X_C(f)=&\frac{\text{sgn}(f)}{2\pi}[\text{Si}(2\pi|f|T)+[2u(-t_o)-1]\text{Si}(2\pi|f||t_o|)]\\&+\frac{j}{2\pi}\left[\text{Ci}(2\pi|f|T)-\text{Ci}(2\pi|f||t_o|)+\ln\left(\frac{|t_o|}{T}\right)\right]\end{aligned} \tag{3.152}$$

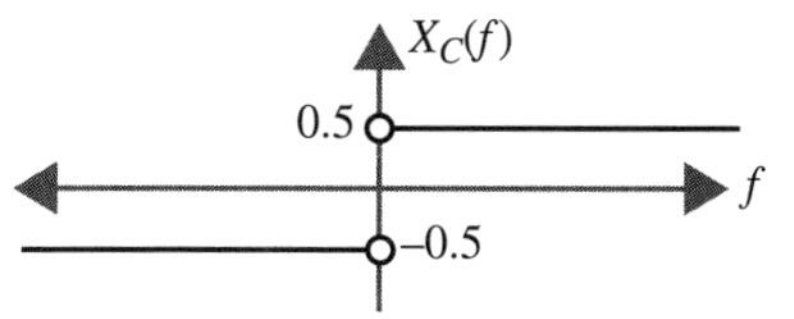


**FIGURE 3.21** Graph of the Cramer transform of a constant signal of unit amplitude.

where $\text{Ci}(x) = \gamma + \ln(x) + \int_0^x \frac{\cos(\lambda) - 1}{\lambda} d\lambda$ is the cosine integral function and $\gamma = 0.57721\ldots$ is Euler's constant.

For the case of $t_o = 0$, consistent with $u(t)$, the Cramer transform is

$$X_C(f) = \frac{1}{2\pi}[\text{sgn}(f)\text{Si}(2\pi|f|T) + j[\text{Ci}(2\pi|f|T) - \gamma - \ln(2\pi|f|T)]] \tag{3.153}$$

For the case of $t_o = -T$, that is, a constant signal on $[-T, T]$, it follows that

$$X_C(f) = \frac{\text{sgn}(f)}{\pi} \cdot \text{Si}(2\pi|f|T) = \frac{1}{\pi} \cdot \text{Si}(2\pi f T) \tag{3.154}$$

As $\text{Si}(\pm\infty) = \pm\pi/2$ and $\text{Si}(0) = 0$, this result is consistent with Equation 3.151.

The real and imaginary parts of the Cramer transform of $u(t - t_o)$, for the case of $t_o = 1$ and $T = 100$, are shown in Figure 3.22.

Note that the Cramer transform of $u(t - t_o)$ is not defined in the limit as $T \to \infty$. This arises as the imaginary part of the transform, as given by

$$\text{Im}[X_C(f)] = \frac{-1}{\pi} \int_{t_o}^{T} \frac{\sin(\pi f \lambda)^2}{\lambda} d\lambda \tag{3.155}$$

does not converge as $T \to \infty$.

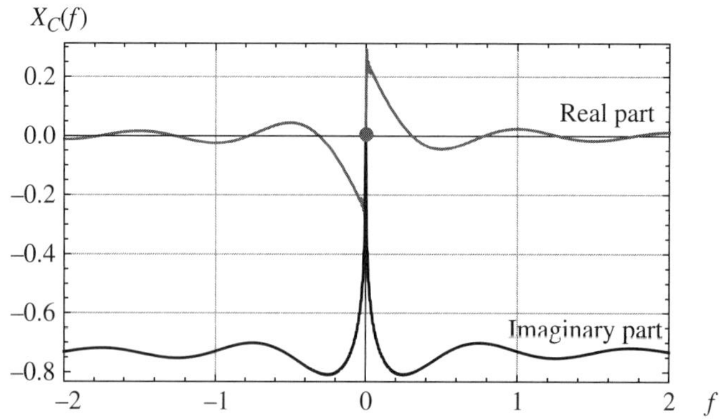


**FIGURE 3.22** Real and imaginary parts of the Cramer transform of $u(t - t_o)$ for the case of $t_o = 1$ and $T = 100$.

### *3.12.6.3 Complex Exponential and Sinusoidal Signals on Infinite Interval*

**Theorem 3.27 Sinusoids on Infinite Interval**
The Cramer transforms for the sinusoidal signals $x(t) = \sin(2\pi f_o t)$, $x(t) = \cos(2\pi f_o t)$, and $x(t) = \exp(j2\pi f_o t)$, on the infinite interval $(-\infty, \infty)$, are

$$X_C(f) = \begin{cases} j \cdot \dfrac{\operatorname{sgn}(f+f_o) - \operatorname{sgn}(f-f_o) - 2\operatorname{sgn}(f_o)}{4} & \text{sine} \\ \dfrac{\operatorname{sgn}(f+f_o) + \operatorname{sgn}(f-f_o)}{4} & \text{cosine} \\ \dfrac{\operatorname{sgn}(f-f_o) + \operatorname{sgn}(f_o)}{2} & \text{complex exponential} \end{cases} \tag{3.156}$$

The graph of these Cramer transforms is shown, respectively, in Figures 3.23, 3.24, and 3.25.

**Proof**
The proofs for these results are detailed in Appendix 3.K.

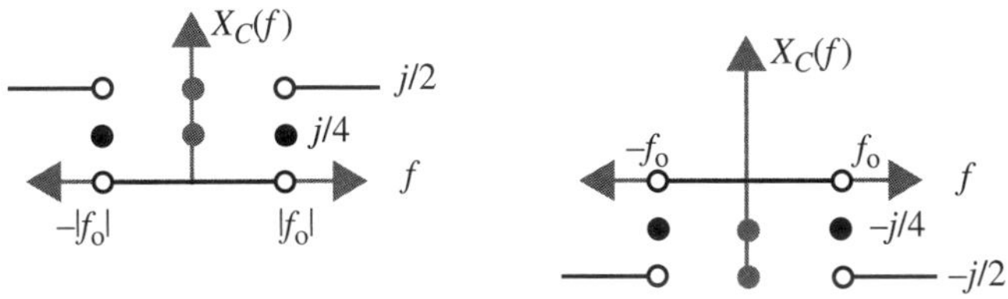


**FIGURE 3.23** Graph of the Cramer transform of a sine signal of unit amplitude and frequency $f_o$. Left: $f_o < 0$. Right: $f_o > 0$.

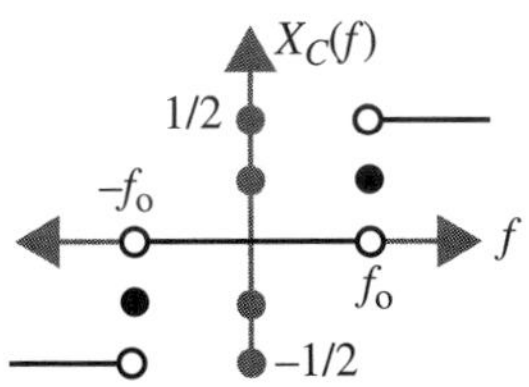


**FIGURE 3.24** Graph of the Cramer transform of a cosine signal of unit amplitude and frequency $f_o$. The graph is valid for $f_o$ both positive and negative.

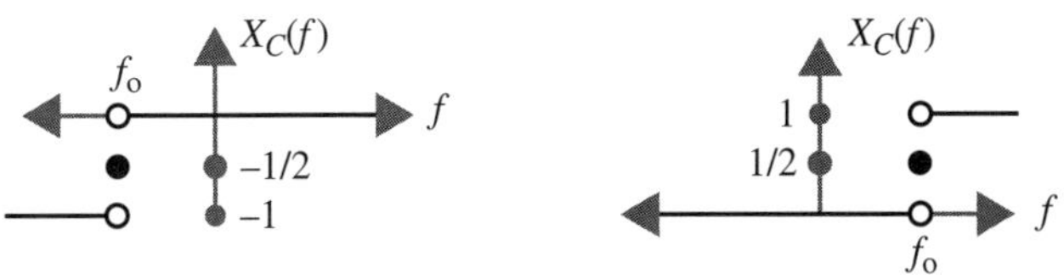


**FIGURE 3.25** Graph of the Cramer transform of a complex exponential signal of unit amplitude. Left: $f_o < 0$. Right: $f_o > 0$.

***3.12.6.4 Notes*** As $\exp(j2\pi f_o t) = \cos(2\pi f_o t) + j\sin(2\pi f_o t)$, it follows, from the linearity of the Cramer transform, that the Cramer transform of $\exp(j2\pi f_o t)$ is

$$X_C(f) = \frac{\mathrm{sgn}(f+f_o) + \mathrm{sgn}(f-f_o)}{4} + j\left[j \cdot \frac{\mathrm{sgn}(f+f_o) - \mathrm{sgn}(f-f_o) - 2\,\mathrm{sgn}(f_o)}{4}\right]$$
$$= \frac{\mathrm{sgn}(f-f_o) + \mathrm{sgn}(f_o)}{2} \qquad (3.157)$$

as required.

For the case of $f_o = 0$, consistent with a constant level signal, the Cramer transform is $X_C(f) = \mathrm{sgn}(f)/2$, which is consistent with the previously proved result stated in Equation 3.151.

### *3.12.6.5 Sinusoids on Finite Interval*

**Theorem 3.28 Sinusoids on Finite Interval**
For the interval $[-T, T]$, the Cramer transform of the signals $x(t) = \sin(2\pi f_o t)$, $x(t) = \cos(2\pi f_o t)$, and $x(t) = \exp(j2\pi f_o t)$, respectively, are

$$X_C(f) = \begin{cases} \frac{j}{2\pi}[\mathrm{Si}(2\pi(f+f_o)T) - \mathrm{Si}(2\pi(f-f_o)T) - 2\mathrm{Si}(2\pi f_o T)] & \text{sine} \\ \frac{1}{2\pi}[\mathrm{Si}(2\pi(f+f_o)T) + \mathrm{Si}(2\pi(f-f_o)T)] & \text{cosine} \\ \frac{1}{\pi}[\mathrm{Si}(2\pi f_o T) + \mathrm{Si}(2\pi(f-f_o)T)] & \text{exponential} \end{cases} \qquad (3.158)$$

**Proof**
The proofs for these results are detailed in Appendix 3.K.

### *3.12.6.6 Cramer Transform of a Fourier Series*

**Theorem 3.29 Cramer Transform of a Fourier Series Decomposition**
The Cramer transform of a signal on the interval $[-T, T]$, and assuming a Fourier series decomposition according to

$$x(t) = \sum_{i=-\infty}^{\infty} c_i e^{j2\pi i f_o t} \qquad c_i = \frac{1}{2T}\int_{-T}^{T} x(t)e^{-j2\pi i f_o t}\,dt, f_o = \frac{1}{2T} \qquad (3.159)$$

is

$$X_C(f) = \sum_{i=-\infty}^{\infty} \frac{c_i}{\pi} \cdot [\mathrm{Si}(i\pi) + \mathrm{Si}(2\pi f T - i\pi)] \qquad (3.160)$$

For $T$ large the Cramer transform can be approximated, as illustrated in Figure 3.26, according to

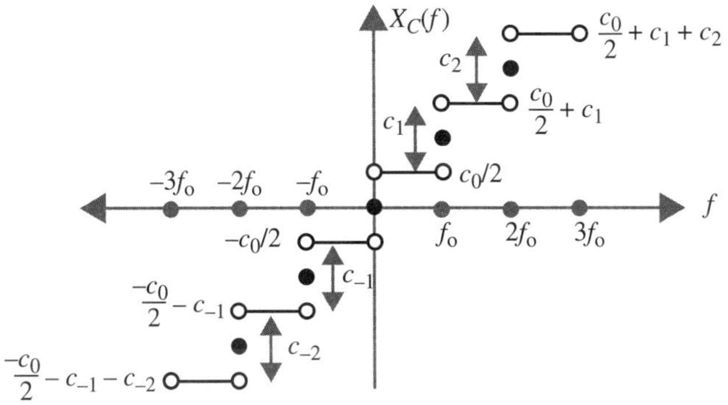


**FIGURE 3.26** Illustration of the approximation to the Cramer transform of a Fourier series representation of a signal. In general, the coefficients are complex.

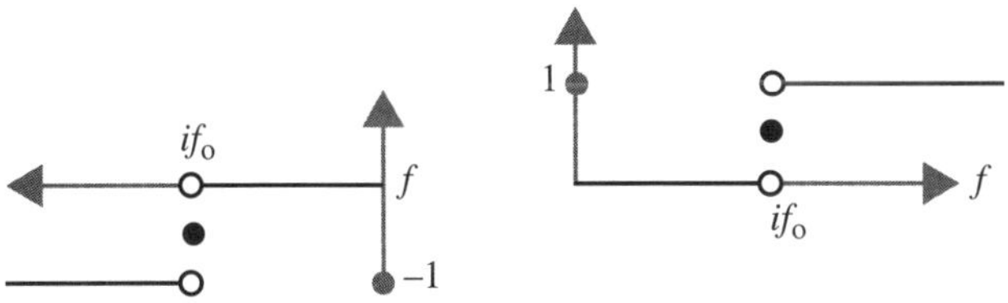


**FIGURE 3.27** Graph of $[\mathrm{sgn}(i) + \mathrm{sgn}(f - if_o)]/2$. Left: $i < 0$. Right: $i > 0$.

$$X_C(f) \approx \sum_{i=-\infty}^{\infty} \frac{c_i}{2} \cdot [\mathrm{sgn}(i) + \mathrm{sgn}(f - if_o)] \tag{3.161}$$

which can be written in the form

$$X_C(f) \approx \begin{cases} \sum_{i=0}^{\lfloor f/f_o \rfloor} \frac{c_i}{2} \cdot [\mathrm{sgn}(i) + \mathrm{sgn}(f - if_o)] & f > 0 \\ 0 & f = 0 \\ \sum_{i=\lfloor f/f_o \rfloor}^{0} \frac{c_i}{2} \cdot [\mathrm{sgn}(i) + \mathrm{sgn}(f - if_o)] & f < 0 \end{cases} \tag{3.162}$$

One component of the approximate transform, as defined by Equation 3.161, is shown in Figure 3.27.

**Proof**

From Theorem 3.28 and with $f_o = 1/2T$, the Cramer transform is

$$X_C(f) = \frac{c_0}{\pi} \cdot \mathrm{Si}(2\pi f T) + \sum_{i=-\infty, i \neq 0}^{\infty} \frac{c_i}{\pi} \cdot [\mathrm{Si}(i\pi) + \mathrm{Si}(2\pi f T - i\pi)] \tag{3.163}$$

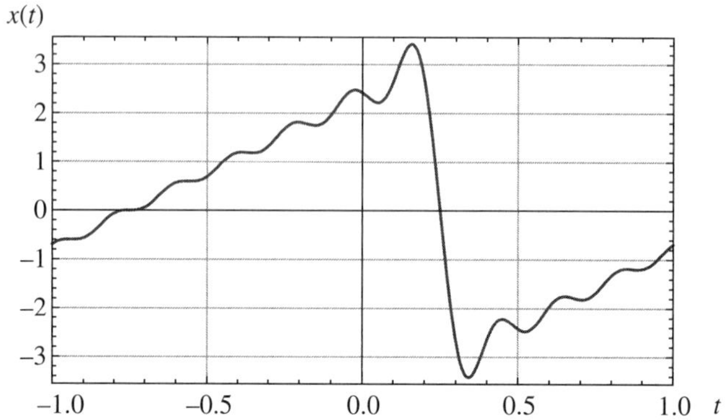


**FIGURE 3.28** Time domain signal as specified by Equations 3.167 and 3.168.

Consider the case of fixed $f = kf_o$, $k \in \mathbf{Z}$, and the approximation

$$\mathrm{Si}(i\pi) \approx \begin{cases} \pi/2 & i > 1 \\ -\pi/2 & i < -1 \end{cases} \tag{3.164}$$

This approximation is relatively poor for $i = \pm 1$ but improves as $|i|$ increases. It then follows that

$$\frac{c_i}{\pi} \cdot [\mathrm{Si}(i\pi) + \mathrm{Si}(2\pi f T - i\pi)] = \frac{c_i}{\pi} \cdot [\mathrm{Si}(i\pi) + \mathrm{Si}((k-i)\pi)]$$

$$\approx \begin{cases} -c_i & i < -1, k-i < -1 \quad f < (i-1)f_o \\ -c_i/2 & i < -1, k-i = 0 \quad f = if_o \\ 0 & i < -1, k-i > 1 \quad f > (i+1)f_o \\ 0 & i > 1, k-i < -1 \quad f < (i-1)f_o \\ c_i/2 & \mathrm{i} > 1, k-i = 0 \quad f = if_o \\ c_i & \mathrm{i} > 1, k-i > 1 \quad f > (i+1)f_o \end{cases} \tag{3.165}$$

The approximation has the form shown in Figure 3.27 and approaches this form as $T \to \infty, f_o \to 0$. Thus, as $T \to \infty, f_o \to 0$, it follows that

$$\frac{c_i}{\pi} \cdot [\mathrm{Si}(i\pi) + \mathrm{Si}(2\pi f T - i\pi)] \to \frac{c_i}{2} \cdot [\mathrm{sgn}(i) + \mathrm{sgn}(f - if_o)] \tag{3.166}$$

and the approximation specified in Equation 3.161 follows.

***3.12.6.7 Example*** Consider a signal shown in Figure 3.28 and defined according to

$$x(t) = \sum_{i=-10}^{10} c_i e^{j2\pi i f_o t} \quad f_o = \frac{1}{2T}, T = 1 \tag{3.167}$$

$$\begin{aligned} c_i &= \frac{\sin(\pi i/4)}{i} + \frac{j\cos(\pi i/4)}{i} \quad i > 0 \\ c_0 &= 0, c_{-i} = c_i^* \end{aligned} \tag{3.168}$$

The real and imaginary parts of the Cramer transform of this signal, as defined by Equation 3.160 and along with the approximation specified by Equation 3.161, are shown in Figures 3.29 and 3.30.

The discrepancy between the approximation and the true transform—most evident in the imaginary part—is due to the fact that the approximation

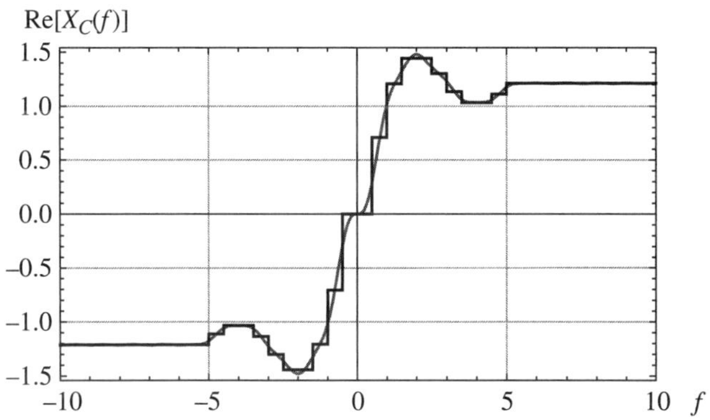


**FIGURE 3.29** Real part of Cramer transform of signal as given by Equation 3.160 and its approximation as given by Equation 3.161.

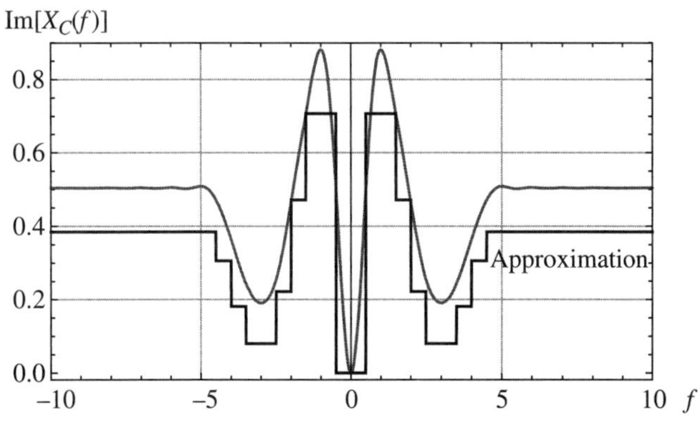


**FIGURE 3.30** Imaginary part of Cramer transform of signal as given by Equation 3.160 and its approximation as given by Equation 3.161.

$$\frac{\mathrm{Si}(2\pi fT - i\pi)}{\pi} \approx \frac{\mathrm{sgn}(f - if_o)}{2} \tag{3.169}$$

is relatively poor when $|f - if_o|T$ is not large.

### 3.12.7 Change in Transform

The following theorem provides the basis for interpreting the Cramer transform as an integrated spectrum:

**Theorem 3.30 Change in Cramer Transform**
The change in the Cramer transform between $f - \Delta f/2$ and $f + \Delta f/2$ has the following alternative forms:

$$\begin{aligned} X_C\left(f + \frac{\Delta f}{2}\right) - X_C\left(f - \frac{\Delta f}{2}\right) &= \frac{1}{\pi}\int_{-\infty}^{\infty} e^{-j2\pi f\lambda} \cdot \frac{\sin(\pi\Delta f\lambda)}{\lambda} \cdot x(\lambda)d\lambda \\ &= \Delta f \int_{-\infty}^{\infty} e^{-j2\pi f\lambda}\mathrm{sinc}(\Delta f\lambda)x(\lambda)d\lambda \end{aligned} \tag{3.170}$$

**Proof**
The proof is detailed in Appendix 3.L.

An interpretation of this result can be found by considering the response of a band-pass filter (Papoulis, 1965, p. 469):

#### *3.12.7.1 Band-Pass Filter Interpretation*

Consider an ideal one-sided band-pass filter with a transfer function as shown in Figure 3.31. For such a filter, and as detailed in Appendix 3.M, the impulse response is

$$h(t) = \Delta f e^{j2\pi f_o t}\mathrm{sinc}(\Delta f t) \tag{3.171}$$

For such a filter, the output signal $y$, in response to an input signal $x$, is

$$y(t) = \int_{-\infty}^{\infty} x(\lambda)h(t-\lambda)d\lambda = \Delta f \int_{-\infty}^{\infty} e^{j2\pi f_o(t-\lambda)}\mathrm{sinc}[\Delta f(t-\lambda)]x(\lambda)d\lambda \tag{3.172}$$

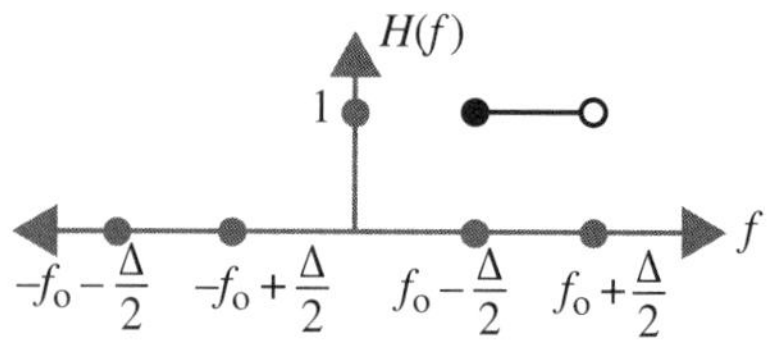


**FIGURE 3.31** Transfer function of an ideal band-pass filter with center frequency of $f_o$ and a bandwidth of $\Delta$.

The result stated in the following theorem then holds:

**Theorem 3.31 Band-Pass Filter Interpretation**
The change in the integrated spectrum between $f_o - \frac{\Delta f}{2}$ and $f_o + \frac{\Delta f}{2}$, as given by $X_C\left(f_o + \frac{\Delta f}{2}\right) - X_C\left(f_o - \frac{\Delta f}{2}\right)$, equals the time domain output, at $t = 0$, of an ideal one-sided band-pass filter with a transfer function as shown in Figure 3.31, that is,

$$X_C\left(f_o + \frac{\Delta f}{2}\right) - X_C\left(f_o - \frac{\Delta f}{2}\right) = y(0) = \Delta f \int_{-\infty}^{\infty} e^{-j2\pi f_o \lambda} \operatorname{sinc}(\Delta f \lambda) x(\lambda) d\lambda \qquad (3.173)$$

**Proof**
The proof is detailed in Appendix 3.M.

***3.12.7.2 Change in Integrated Spectrum*** That the Cramer transform $X_C$ is an *integrated spectrum* is evident from the following theorem:

**Theorem 3.32 Cramer Transform as an Integrated Spectrum**
Consider a signal $x$ that is integrable on the interval $[0, T]$ and whose Fourier transform on this interval is $X(T,f)$. The change in the Cramer transform between $f_1$ and $f_2$ equals the integral of the Fourier transform of the signal on the interval $[f_1, f_2]$, that is,

$$X_C(f_2) - X_C(f_1) = \int_{f_1}^{f_2} X(T, f) df \qquad (3.174)$$

This result also holds for the interval $[-T, T]$ assuming the Fourier transform is evaluated on that interval.

For the interval $[-T, T]$, and for the case of a Fourier series decomposition

$$x(t) = \sum_{i=-\infty}^{\infty} c_i e^{j2\pi i f_o t} \qquad c_i = \frac{1}{2T} \int_{-T}^{T} x(t) e^{-j2\pi i f_o t} dt, f_o = \frac{1}{2T} \qquad (3.175)$$

it is the case, for $0 < f_1 < f_2$, and assuming $T$ is sufficiently large such that Equation 3.161 is valid, that

$$X_C(f_2) - X_C(f_1) \approx \sum_{i=\lfloor f_1/f_o \rfloor}^{\lfloor f_2/f_o \rfloor} \frac{c_i}{2} [\operatorname{sgn}(f_2 - i f_o) - \operatorname{sgn}(f_1 - i f_o)] \qquad (3.176)$$

**Proof**
The proof of Equation 3.174 is detailed in Appendix 3.N. The result stated in Equation 3.176 follows from Equation 3.161.

### 3.12.8 Linear Filtering

The Cramer representation allows a simple relationship to be defined between the input and output signals of a linear system for a class of signals greater than that given by Fourier analysis. Consider linear filtering as illustrated in Figure 3.32 where the input and output relationships are given by the convolution integral

$$y(t) = \int_{-\infty}^{t} x(\lambda)h(t-\lambda)d\lambda \tag{3.177}$$

for the usual case of a causal impulse response.

**Theorem 3.33 Input–Output Relationship**
When the input and output signals associated with linear filtering, consistent with Figure 3.32, have the Cramer representation

$$x(t) = \int_{R} e^{j2\pi ft} dX_C(f) \qquad y(t) = \int_{R} e^{j2\pi ft} dY_C(f) \tag{3.178}$$

then it is the case that

$$dY_C(f) = H(f)dX_C(f) \tag{3.179}$$

The assumptions here are that $x$ is locally integrable, $h \in \mathbf{L}$, $x(t-\lambda)h(\lambda) \in \mathbf{L}[0,\infty]$ for all $t \in [0,\infty]$, and both $x$ and $h$ have bounded variation on all finite intervals such that the convolution integral is valid.

**Proof**
The proof is detailed in Appendix 3.O.

***3.12.8.1 Notes*** Intuitively, if the input to a linear system consists of a summation of sinusoids, then the output consists of a summation of sinusoids where their weighting is modified by the transfer function of the system. Equation 3.179 reflects this.

The result $dY_C(f) = H(f)dX_C(f)$ only holds for the infinite interval; it does not hold, for example, for the finite interval [0, $T$]. Useful references include Priestley (1981, p. 263f) and Yaglom (1962, p. 53).

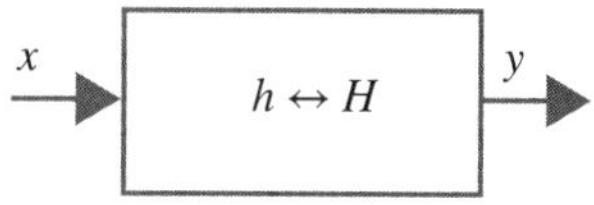


**FIGURE 3.32** Schematic diagram of a linear system with an impulse response function $h \in \mathbf{L}$ whose Fourier transform is denoted $H$.

### 3.12.9 Integrated Spectrum, Spectrum, and Power Spectrum

Consistent with Theorem 3.32, the Cramer transform $X_C(f)$ is an integrated spectrum, and from this function, the spectrum and power spectrum of a signal can be defined. Formally:

**Definition: Integrated Spectrum** The Cramer transform $X_C$, as specified by Equation 3.134, is the integrated spectrum of the signal $x$, that is, the integrated spectrum of a signal $x$ is

$$X_C(f) = \frac{j}{2\pi} \int_{-\infty}^{\infty} \frac{e^{-j2\pi f\lambda} - 1}{\lambda} \cdot x(\lambda) d\lambda \tag{3.180}$$

**Definition: Spectrum and Power Spectrum** The spectrum and power spectrum of a signal $x$, respectively, are defined according to

$$dX_C(f) = \lim_{df \to 0} [X_C(f + df/2) - X_C(f - df/2)] \tag{3.181}$$

$$|dX_C(f)|^2 = \left| \lim_{df \to 0} [X_C(f + df/2) - X_C(f - df/2)] \right|^2 \tag{3.182}$$

***3.12.9.1 Example*** To illustrate the definitions, and the interpretation implied by the names associated with the definitions, consider the case of a summation of sinusoids defined on the infinite interval $(-\infty, \infty)$:

$$x(t) = \sum_{i=1}^{\infty} A_i \sin(2\pi f_i t) \quad f_i > 0 \tag{3.183}$$

First, from Equation 3.156, the integrated spectrum $X_C$ is

$$X_C(f) = \frac{j}{4} \sum_{i=1}^{N} A_i [\text{sgn}(f + f_i) - \text{sgn}(f - f_i) - 2\,\text{sgn}(f_i)] \tag{3.184}$$

The graph of the integrated spectrum is illustrated in Figure 3.33. Second, the spectrum and power spectrum are given by

$$dX_C(f) = \sum_{i=1}^{N} \frac{jA_i}{2} \delta_K(f + f_i) + \frac{-jA_i}{2} \delta_K(f - f_i) \tag{3.185}$$

$$|dX_C(f)|^2 = \sum_{i=1}^{N} \frac{A_i^2}{4} \delta_K(f + f_i) + \frac{A_i^2}{4} \delta_K(f - f_i) \tag{3.186}$$

and are illustrated in Figure 3.34.

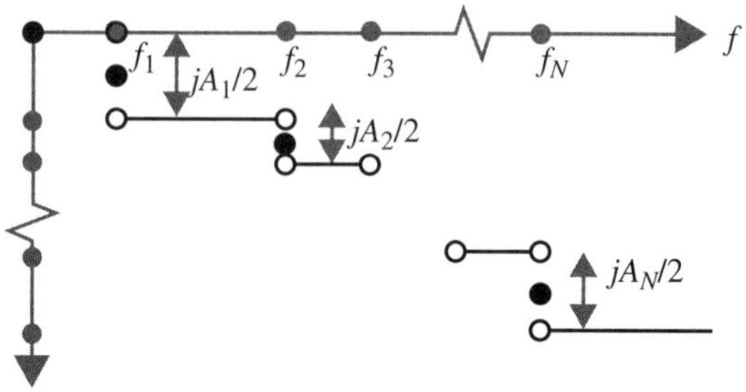


**FIGURE 3.33** Graph of integrated spectrum for $f > 0$ for the sum of sinusoids as defined by Equation 3.183.

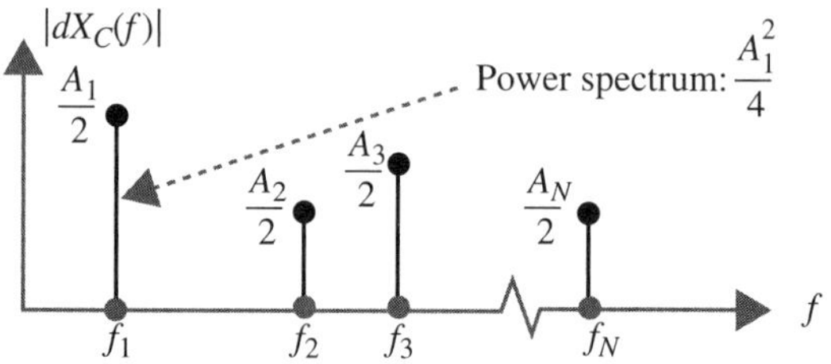


**FIGURE 3.34** Graph of the magnitude of the spectrum, and illustration of power spectrum, for $f > 0$ and for the case of a sum of sinusoids as defined by Equation 3.183.

**TABLE 3.1 Cramer transform of standard signals defined on $(-\infty, \infty)$**

| Signal: $x(t)$ | Cramer transform: $X_C(f)$ |
|---|---|
| 1 | $0.5\,\text{sgn}(f)$ |
| $\sin(2\pi f_o t)$ | $j\cdot\frac{\text{sgn}(f+f_o)-\text{sgn}(f-f_o)-2\text{sgn}(f_o)}{4}$ |
| $\cos(2\pi f_o t)$ | $\frac{\text{sgn}(f+f_o)+\text{sgn}(f-f_o)}{4}$ |
| $\exp(j2\pi f_o t)$ | $\frac{\text{sgn}(f_o)+\text{sgn}(f-f_o)}{2}$ |
| $\sum_{i=-\infty}^{\infty} c_i e^{j2\pi i f_o t}$ | $\sum_{i=-\infty}^{\infty} \frac{c_i}{2}\cdot[\text{sgn}(i)+\text{sgn}(f-if_o)]$ |

***3.12.9.2 A Note on the Power Spectrum*** Consider the power spectrum as defined by Equation 3.182. As the above example illustrates, this power spectrum identifies the power in the individual sinusoidal components of a signal. However, because such signal components, in general, are not orthogonal on a finite interval, for example, $[0, T]$, the sum of the individual signal powers, in general, will not equal the average signal power.

### 3.12.10 Cramer Transform of Standard Signals

The Cramer transform of selected signals is detailed in Table 3.1.

## 3.13 PROBLEMS

**3.1** Determine the intervals over which the set of functions

$$\{1, \cos(2\pi i f_o t), \sin(2\pi i f_o t) : \ i \in \{1,2,\ldots\}\} \tag{3.187}$$

is an orthogonal set. Assume $f_o > 0$.

**3.2** Consider an orthonormal basis set $\{b_0^N, b_1^N, \ldots\}$ defined on an interval $[a, b]$. Show that the set of functions $\{b_0, b_1, \ldots\}$

$$\begin{aligned} b_i(t) &= \frac{1}{\sqrt{k_1}} \cdot b_i^N\left(\frac{t-k_0}{k_1}\right) \qquad i \in \{0,1,\ldots\} \\ k_0 &= \alpha - \frac{a(\beta-\alpha)}{b-a}, k_1 = \frac{\beta-\alpha}{b-a} \end{aligned} \tag{3.188}$$

is an orthonormal set for the interval $[\alpha, \beta]$.

**3.3** An orthonormal basis set for the interval $[0, \infty]$, based on Laguerre polynomials, is defined according to

$$\left\{ b_i(t) = \frac{L_i(t)}{i!} e^{-t/2} : \ i \in \{0, 1,, \ldots\} \right\} \tag{3.189}$$

where $L_i$ is the $i$th Laguerre function:

$$\begin{aligned} &L_0(t) = 1 \quad L_1(t) = 1 - t \quad L_2(t) = 2 - 4t + t^2 \\ &L_n(t) = [2n - 1 - t]L_{n-1}(t) - (n-1)^2 L_{n-2}(t) \end{aligned} \tag{3.190}$$

a. Specify the zeroth- to fifth-order basis functions.
b. Graph the zeroth- to fifth-order basis functions.
c. Prove the individual orthonormality of $b_0(t)$ and $b_1(t)$.
d. Prove the orthogonality of $b_0(t)$ and $b_1(t)$ as well as $b_1(t)$ and $b_2(t)$.

**3.4** Consider the signal

$$f(t) = t^3 e^{-t} u(t) \tag{3.191}$$

and a scaled version of the Laguerre basis set defined in the previous problem according to

$$B_i(t) = \sqrt{2} b_i(2t) = \frac{\sqrt{2} L_i(2t)}{i!} e^{-t} \qquad i \in \{0,1,\ldots\} \tag{3.192}$$

a. Find the coefficients $c_0, c_1, \ldots$ in the decomposition

$$f(t) = \sum_{i=0}^{\infty} c_i B_i(t) \quad t > 0 \quad c_i = \int_0^{\infty} f(t) B_i(t) dt \tag{3.193}$$

b. Graph the decomposition to check the validity of the coefficients. Graph the constituent component signals.
c. Graph the energy spectral density of the signal with respect to the Laguerre basis set.

**3.5** Consider the set of Gaussian probability density functions defined according to

$$\left\{ \delta_i(t) = \frac{1}{\sqrt{2\pi}(\sigma/i)} \exp\left[\frac{-t^2}{2\sigma^2/i^2}\right] \right\}_{i=1}^{\infty} \quad \sigma > 0 \tag{3.194}$$

a. Show that this set of functions is a suitable sequence for the Dirac delta with

$$\delta(t) = \lim_{i \to \infty} \delta_i(t) = \begin{cases} 0 & t \neq 0 \\ \infty & t = 0 \end{cases} \tag{3.195}$$

$$\lim_{i \to \infty} \int_{-\infty}^{\infty} \delta_i(t) dt = 1 \tag{3.196}$$

b. Graph the first five functions in the sequence for the case of $\sigma = 1$.

**3.6** A first-order model for a digital signal is illustrated in Figure 3.35.

a. Specify an explicit analytical expression for this signal for $0 \le t < T$.
b. Consider the normalized case of $T = 1$, $W + \tau = 1/2$, and $A = 1$. Graph the signal for the three cases:

$$\begin{array}{ll} W = 9/20 & \tau = 1/20 \\ W = 3/8 & \tau = 1/8 \\ W = \tau = 1/4 & \end{array} \tag{3.197}$$

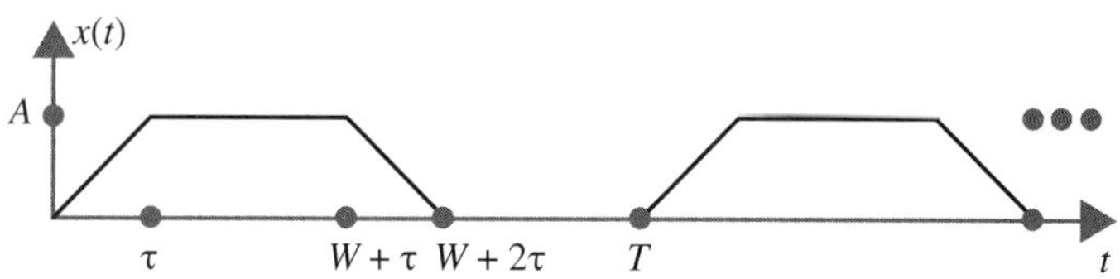


**FIGURE 3.35** Periodic digital pulse train.

c. Determine an analytical expression for the exponential Fourier series $x_F$ for the digital signal. Manipulate the expression for the $i$th coefficient so it is in the form

$$c_i = k_i e^{-j2\pi i f_o (W/2+\tau)} \cdot \text{sinc}(i f_o \tau) \cdot \text{sinc}[i f_o (W+\tau)] \tag{3.198}$$

where $k_i$ is a constant and $f_o = 1/T$.

d. For the three cases specified above, graph the Fourier series approximation to the digital signal based on the coefficients $c_{-16}$ to $c_{16}$.

e. For the three cases specified above, graph the magnitude of the coefficients in the Fourier series on a linear–log graph. Note the relationship between the pulse rise time and the number of significant terms in the Fourier series approximation to the digital signal.

f. Determine an analytical expression for the average power in the digital signal. Evaluate this power for the case of $T=1$, $W+\tau=1/2$, and $A=1$. Confirm that the average power is closely approximated by the sum of the magnitude of the Fourier coefficients squared by using $c_{-16}$ to $c_{16}$ for each of the three cases.

**3.7** Consider the digital pulse signal defined in Figure 3.36.

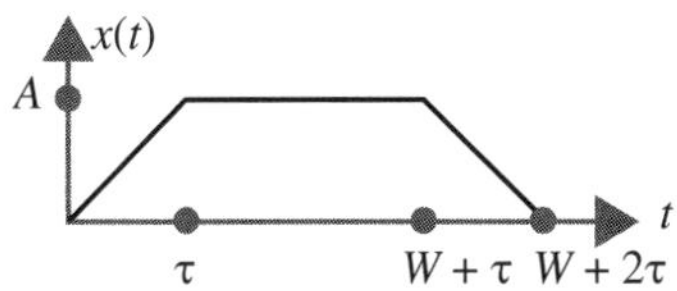


**FIGURE 3.36** Digital pulse signal.

a. Determine the Fourier transform of the signal over the interval $[0, \infty]$.

Write the Fourier transform in the form

$$X(f) = ke^{-j2\pi(W/2+\tau)f} \cdot \text{sinc}(\tau f) \cdot \text{sinc}[(W+\tau)f] \tag{3.199}$$

for an appropriately defined constant $k$.

b. Consider the normalized case of $T=1$, $W+\tau=1/2$, and $A=1$. Graph the signal and the magnitude of the Fourier transform (using a linear–log plot) for the three cases:

$$
\begin{aligned}
&W = 9/20 \quad \tau = 1/20 \\
&W = 3/8 \quad \tau = 1/8 \\
&W = \tau = 1/4
\end{aligned}
\tag{3.200}
$$

**3.8** Determine the Fourier transform of the signals:

$$
\begin{aligned}
&x_0(t) = \frac{e^{-t/\tau}}{\tau} \cdot u(t) \quad x_1(t) = \frac{te^{-t/\tau}}{\tau^2} \cdot u(t) \\
&x_2(t) = \frac{t^2}{2\tau^3} \cdot e^{-t/\tau} u(t)
\end{aligned}
\tag{3.201}
$$

**3.9** Consider the digital pulse signal in Figure 3.36. Determine the power spectral density over the interval [0, $T$], $T > W + 2\tau$, using a sinusoidal basis set. Specify the power spectral density using the three equivalent forms. Assume the normalized case of $T = 1$, $W + \tau = 1/2$, and $A = 1$. Assume $W = 9/20, \tau = 1/20$ and use linear–log graphs.

**3.10** Consider a signal defined by a sinusoidal pulse

$$
x(t) = \begin{cases} A\sin(2\pi f_o t) & 0 \le t < T/2 \\ 0 & T/2 \le t < T \end{cases}
\tag{3.202}
$$

on the interval [0, $T$], where $f_o = 1/T$.

a. Determine the autocorrelation function $R(T,t,\tau) = x(t+\tau)x(t)$ and the time averaged autocorrelation function $\bar{R}(T,\tau)$. Graph these functions for the normalized case of $A = 1$ and $T = 1$.
b. Determine the power spectral density of this signal on the interval [0, $T$]. Graph the power spectral density for the case of $A = 1$ and $T = 1$.
c. Show that the Fourier transform of the time averaged autocorrelation function equals the power spectral density function. Note the technical difficulty of the approach of determining the power spectral density via the time averaged autocorrelation function.

**3.11** Consider the sinusoid $x(t) = A\cos(2\pi f_o t)$ on the interval $[-T, T]$.

a. Graph the Cramer transform of this signal for the case of $A = 1$, $T = 10$, and $f_o = 1$.
b. Graph the spectrum and power spectrum of this signal as defined for the Cramer transform using resolutions of $df = 1/2T$, $df = 1/T$, and $df = 2/T$. Comment on the results.

## APPENDIX 3.A PROOF OF THEOREM 3.5

The point $t_o$ is a Dirichlet point of a differentiable function $f$, and the Dirichlet value of $f$ at $t_o$ is $f(t_o)$, if $\exists\delta_o > 0$ and

$$\forall\varepsilon > 0\ \exists\lambda_o > 0 \text{ such that } \forall\lambda > \lambda_o$$

$$\left| \int_{-\delta_o}^{\delta_o} f(t_o + u) \cdot \frac{\sin(2\pi\lambda u)}{\pi u} du - f(t_o) \right| < \varepsilon \tag{3.203}$$

First, as $f$ is differentiable at $t_o$, it follows, $\forall\varepsilon_1 > 0$, $\exists u_o$ such that for $-u_o < u < u_o$ it is the case that

$$\left| \frac{f(t_o + u) - f(t_o)}{u} - f'(t_o) \right| < \varepsilon_1 \tag{3.204}$$

which implies

$$f(t_o) + uf'(t_o) - |u|\varepsilon_1 < f(t_o + u) < f(t_o) + uf'(t_o) + |u|\varepsilon_1 \tag{3.205}$$

Hence, for $-u_o < u < u_o$, there exists a function $\varepsilon_1(u)$ such that

$$f(t_o + u) = f(t_o) + uf'(t_o) + u\varepsilon_1(u) \tag{3.206}$$

where $|\varepsilon_1(u)| < \varepsilon_1$.

Second, $\sin(2\pi\lambda u)/\pi u$ is an *impulsive* function with unit area as $\lambda \to \infty$. This is simply proved: Define $I(\lambda)$ as

$$I(\lambda) = \int_0^{\delta} \frac{\sin(2\pi\lambda u)}{\pi u} du \Rightarrow I(\lambda) = \int_0^{\lambda\delta} \frac{\sin(2\pi\zeta)}{\pi\zeta} d\zeta \tag{3.207}$$

where a change of variable $\zeta = \lambda u$ has been used. The definite integral (Spiegel, 1968, p. 96)

$$\int_0^{\infty} \frac{\sin(p\zeta)}{\zeta} d\zeta = \frac{\pi}{2} \quad p > 0 \tag{3.208}$$

yields the required result. It then follows, $\forall\varepsilon_2 > 0$, $\forall\delta_o > 0$, that there exists a constant $\lambda_o > 0$ such that for all $\lambda > \lambda_o$:

$$\left| \int_{-\delta_o}^{\delta_o} \frac{\sin(2\pi\lambda u)}{\pi u} du - 1 \right| < \varepsilon_2 \tag{3.209}$$

Third, choose $\delta_o = u_o$ and consider the analysis:

$$\begin{aligned} I &= \int_{-\delta_o}^{\delta_o} f(t_o+u)\cdot\frac{\sin(2\pi\lambda u)}{\pi u}du - f(t_o) \\ &= \int_{-\delta_o}^{\delta_o} f(t_o+u)\cdot\frac{\sin(2\pi\lambda u)}{\pi u}du - f(t_o)\left[\int_{-\delta_o}^{\delta_o}\frac{\sin(2\pi\lambda u)}{\pi u}du+\varepsilon_{22}\right] \\ &= \int_{-\delta_o}^{\delta_o}[f(t_o+u)-f(t_o)]\cdot\frac{\sin(2\pi\lambda u)}{\pi u}du-\varepsilon_{22}f(t_o) \end{aligned} \tag{3.210}$$

where $-\varepsilon_2 < \varepsilon_{22} < \varepsilon_2$. Substitution of Equation 3.206 into Equation 3.210 yields

$$\begin{aligned} I &= \int_{-\delta_o}^{\delta_o}[uf'(t_o)+u\varepsilon_1(u)]\cdot\frac{\sin(2\pi\lambda u)}{\pi u}du-\varepsilon_{22}f(t_o) \\ &= f'(t_o)\cdot\frac{\cos(2\pi\lambda u)}{2\pi^2\lambda}\Bigg|_{-\delta_o}^{\delta_o}+\int_{-\delta_o}^{\delta_o}\varepsilon_1(u)\cdot\frac{\sin(2\pi\lambda u)}{\pi}du-\varepsilon_{22}f(t_o) \end{aligned} \tag{3.211}$$

As $|\sin(2\pi\lambda u)|\le 1$ and $|\varepsilon_1(u)|<\varepsilon_1$ on $[-\delta_o,\delta_o]$, and $|\varepsilon_{22}|<\varepsilon_2$, it follows that

$$|I| \le 0+\frac{2\varepsilon_1\delta_o}{\pi}+\varepsilon_2|f(t_o)| \tag{3.212}$$

As $\varepsilon_1$, $\varepsilon_2$ can be made arbitrarily small, and as $\delta_o$ is fixed, the required result follows.

## APPENDIX 3.B PROOF OF THEOREM 3.8

Assume $x \in \boldsymbol{L}[0,T]$ which implies $X(T,f)$ is finite for all values of $f$. It then follows that the following integral is finite for all values of $t$ and $\lambda$:

$$I(\lambda,t) = \int_{-\lambda}^{\lambda} X(T,f)e^{j2\pi ft}df \tag{3.213}$$

With $x \in \boldsymbol{L}[0,T]$, it follows, from the definition of the Fourier transform, that

$$I(\lambda,t) = \int_{-\lambda}^{\lambda}\left[\int_0^T x(\tau)e^{-j2\pi f\tau}d\tau\right]e^{j2\pi ft}df \tag{3.214}$$

Since $x(\tau)e^{-j2\pi f\tau}e^{j2\pi ft}$ is absolutely integrable for $\tau\in[0,T]$ and $f\in[-\lambda,\lambda]$, it follows, from the Fubini–Tonelli theorem (Theorem 2.16), that the order of integration can be interchanged to yield

$$I(\lambda,t)=\int_0^T x(\tau)\left[\int_{-\lambda}^{\lambda} e^{-j2\pi f(\tau-t)}df\right]d\tau \tag{3.215}$$

Since sine is an odd function, it follows that

$$\int_{-\lambda}^{\lambda} e^{-j2\pi f(\tau-t)}df=\int_{-\lambda}^{\lambda}\cos[2\pi f(\tau-t)]df=\frac{\sin[2\pi\lambda(\tau-t)]}{\pi(\tau-t)} \tag{3.216}$$

which is valid for all $t,\tau$, including the case where $t\to\tau$. For this case, $\sin[2\pi\lambda(t-\tau)]/\pi(t-\tau)|_{t\to\tau}=2\lambda$. Thus,

$$y(t)=\lim_{\lambda\to\infty}I(\lambda,t)=\lim_{\lambda\to\infty}\int_0^T x(\tau)\frac{\sin[2\pi\lambda(\tau-t)]}{\pi(\tau-t)}d\tau \tag{3.217}$$

As $x$ has been assumed to be piecewise differentiable, it then follows, from Theorem 3.6, that

$$y(t)=\begin{cases}0 & t\notin[0,T]\\ x(0^+)/2 & t=0\\ \dfrac{x(t^+)+x(t^-)}{2} & t\in(0,T)\\ x(T^-)/2 & t=T\end{cases} \tag{3.218}$$

which is the required result.

## APPENDIX 3.C FOURIER TRANSFORM AND PSD OF A SINUSOID

Consider the signal $x(t)=A\cos(2\pi if_o t+\phi)$, where $\phi=0$ for the cosine case and $\phi=-\pi/2$ for the sine case. First, the $k$th coefficient in an exponential Fourier series decomposition for $x$ is

$$c_k=\frac{1}{T}\int_0^T A\cos(2\pi if_o t+\phi)e^{-j2\pi kf_o t}dt=\frac{A}{T}\int_0^T\left[\frac{e^{j(2\pi if_o t+\phi)}+e^{-j(2\pi if_o t+\phi)}}{2}\right]e^{-j2\pi kf_o t}dt$$

$$= \frac{A}{2T}\int_0^T \left[e^{j\phi}e^{j2\pi(i-k)f_o t} + e^{-j\phi}e^{-j2\pi(i+k)f_o t}\right]dt = \begin{cases} 0 & i \neq k, i \neq -k \\ \frac{Ae^{-j\phi}}{2} & i = -k \\ \frac{Ae^{j\phi}}{2} & i = k \end{cases} \tag{3.219}$$

Second, the Fourier transform for $x$ is

$$\begin{aligned} X(T,f) &= A\int_0^T \left[\frac{e^{j(2\pi i f_o t+\phi)} + e^{-j(2\pi i f_o t+\phi)}}{2}\right]e^{-j2\pi f t}dt \\ &= \frac{A}{2}\int_0^T \left[e^{j\phi}e^{-j2\pi(f-if_o)t} + e^{-j\phi}e^{-j2\pi(f+if_o)t}\right]dt \\ &= \begin{cases} \frac{ATe^{-j\phi}}{2}, \frac{ATe^{j\phi}}{2} & f = -if_o, if_o \\ \frac{A}{2}\left[e^{j\phi}\cdot\frac{e^{-j2\pi(f-if_o)t}}{-j2\pi(f-if_o)} + e^{-j\phi}\cdot\frac{e^{-j2\pi(f+if_o)t}}{-j2\pi(f+if_o)}\right]_0^T & f \neq \pm if_o \end{cases} \end{aligned} \tag{3.220}$$

Consider the term for the case where $f \neq \pm if_o$:

$$X(T,f) = \frac{-A}{2}\left[e^{j\phi}\cdot\frac{e^{-j2\pi(f-if_o)T}}{j2\pi(f-if_o)} + e^{-j\phi}\cdot\frac{e^{-j2\pi(f+if_o)T}}{j2\pi(f+if_o)} - \left[\frac{e^{j\phi}}{j2\pi(f-if_o)} + \frac{e^{-j\phi}}{j2\pi(f+if_o)}\right]\right] \tag{3.221}$$

Factoring out terms yields

$$\begin{aligned} X(T,f) = &\frac{-Ae^{j\phi}e^{-j\pi(f-if_o)T}}{2}\cdot\left[\frac{e^{-j\pi(f-if_o)T} - e^{j\pi(f-if_o)T}}{j2\pi(f-if_o)}\right] \\ &+ \frac{-Ae^{-j\phi}e^{-j\pi(f+if_o)T}}{2}\cdot\left[\frac{e^{-j\pi(f+if_o)T} - e^{j\pi(f+if_o)T}}{j2\pi(f+if_o)}\right] \end{aligned} \tag{3.222}$$

and, thus,

$$\begin{aligned} X(T,f) = &\frac{ATe^{j\phi}e^{-j\pi(f-if_o)T}}{2}\cdot\text{sinc}[(f-if_o)T] \\ &+ \frac{ATe^{-j\phi}e^{-j\pi(f+if_o)T}}{2}\cdot\text{sinc}[(f+if_o)T] \end{aligned} \tag{3.223}$$

For the case of $f_oT=1$ and $e^{\pm j\pi i f_o T}=(-1)^i$, it follows that

$$X(T,f)=\frac{(-1)^i ATe^{-j\pi fT}}{2}\left[e^{j\phi}\text{sinc}[(f-if_o)T]+e^{-j\phi}\text{sinc}[(f+if_o)T]\right] \tag{3.224}$$

As $T=1/f_o$, the final form follows:

$$X(T,f)=\frac{(-1)^i ATe^{-j\pi fT}}{2}\cdot\left[e^{j\phi}\text{sinc}\left[\left(\frac{f}{if_o}-1\right)i\right]+e^{-j\phi}\text{sinc}\left[\left(\frac{f}{if_o}+1\right)i\right]\right] \tag{3.225}$$

This result is also valid for the case where $f=-if_o, if_o$ as $\text{sinc}(0)=1$. Hence, for the cosine case where $\phi=0$,

$$X_1(T,f)=\frac{(-1)^i ATe^{-j\pi fT}}{2}\left[\text{sinc}\left[\left(\frac{f}{if_o}-1\right)i\right]+\text{sinc}\left[\left(\frac{f}{if_o}+1\right)i\right]\right] \tag{3.226}$$

and for the sine case where $\phi=-\pi/2$,

$$X_2(T,f)=\frac{(-1)^{i+1} jATe^{-j\pi fT}}{2}\left[\text{sinc}\left[\left(\frac{f}{if_o}-1\right)i\right]-\text{sinc}\left[\left(\frac{f}{if_o}+1\right)i\right]\right] \tag{3.227}$$

Hence, for the cosine and sine cases,

$$\begin{aligned}|X(T,f)|^2&=\frac{A^2T^2}{4}\left|\text{sinc}\left[\left(\frac{f}{if_o}-1\right)i\right]\pm\text{sinc}\left[\left(\frac{f}{if_o}+1\right)i\right]\right|^2\\&=\frac{A^2T^2}{4}\left[\text{sinc}^2\left[\left(\frac{f}{if_o}-1\right)i\right]+\text{sinc}^2\left[\left(\frac{f}{if_o}+1\right)i\right]\right.\\&\left.\quad\pm 2\text{sinc}\left[\left(\frac{f}{if_o}-1\right)i\right]\text{sinc}\left[\left(\frac{f}{if_o}+1\right)i\right]\right]\end{aligned} \tag{3.228}$$

and, thus,

$$\begin{aligned}G(T,f)=\frac{A^2T}{4}&\left[\text{sinc}^2\left[\left(\frac{f}{if_o}-1\right)i\right]+\text{sinc}^2\left[\left(\frac{f}{if_o}+1\right)i\right]\right.\\&\left.\pm 2\text{sinc}\left[\left(\frac{f}{if_o}-1\right)i\right]\text{sinc}\left[\left(\frac{f}{if_o}+1\right)i\right]\right]\end{aligned} \tag{3.229}$$

## APPENDIX 3.D PROOF OF THEOREM 3.14

Consider the integral $I$ defined as the Fourier transform of the time averaged autocorrelation function:

$$I = \int_{-T}^{T} \bar{R}(T,\tau)e^{-j2\pi f\tau}d\tau = \begin{cases} \int_{-T}^{0}\left[\frac{1}{T}\int_{0}^{T+\tau} x(t)x^*(t-\tau)dt\right]e^{-j2\pi f\tau}d\tau & \tau<0 \\ \int_{0}^{T}\left[\frac{1}{T}\int_{\tau}^{T} x(t)x^*(t-\tau)dt\right]e^{-j2\pi f\tau}d\tau & \tau>0 \end{cases} \tag{3.230}$$

By assumption, $x \in \boldsymbol{L}^2[0,T]$, which implies (Theorem 3.13) that the averaged autocorrelation function is finite and, thus, the inner integrals are finite. It then follows that the order of integration can be interchanged to yield (see Fig. 3.10 for the region of integration)

$$I = \frac{1}{T}\int_{0}^{T}\int_{t-T}^{t} x(t)x^*(t-\tau)e^{-j2\pi f\tau}d\tau dt \tag{3.231}$$

A change of variable $\xi = t - \tau$ for $\tau$ results in

$$\begin{aligned} I &= \frac{1}{T}\int_{0}^{T}\int_{0}^{T} x(t)x^*(\xi)e^{-j2\pi f(t-\xi)}d\xi dt \\ &= \frac{1}{T}\left[\int_{0}^{T} x(t)e^{-j2\pi ft}dt\right]\left[\int_{0}^{T} x^*(\xi)e^{j2\pi f\xi}d\xi\right] = \frac{|X(T,f)|^2}{T} = G(T,f) \end{aligned} \tag{3.232}$$

which is the first required result.

To prove the second result, assume $x$ is bounded and $x \in \boldsymbol{L}[0,T]$, which implies $x \in \boldsymbol{L}^2[0,T]$. Define the integral $I$ as

$$\begin{aligned} I &= \lim_{\lambda\to\infty}\int_{-\lambda}^{\lambda} G(T,f)e^{j2\pi f\tau}df \\ &= \lim_{\lambda\to\infty}\int_{-\lambda}^{\lambda}\left[\frac{1}{T}\left[\int_{0}^{T} x(t)e^{-j2\pi ft}dt\right]\left[\int_{0}^{T} x^*(\xi)e^{j2\pi f\xi}d\xi\right]\right]e^{j2\pi f\tau}df \end{aligned} \tag{3.233}$$

Interchanging the order of integration, which is valid as the integrands are finite and the regions of integration are finite, yields

$$I = \lim_{\lambda\to\infty} \frac{1}{T}\int_0^T\int_0^T x(t)x^*(\xi)\left[\int_{-\lambda}^{\lambda} e^{j2\pi f(\xi+\tau-t)}df\right]d\xi dt \tag{3.234}$$

The inner integral equals

$$\int_{-\lambda}^{\lambda} e^{j2\pi f(\xi+\tau-t)}df = 2\int_0^{\lambda} \cos[2\pi f(\xi+\tau-t)]df = \frac{\sin[2\pi\lambda(\xi+\tau-t)]}{\pi(\xi+\tau-t)} \tag{3.235}$$

This result is valid also for the case of $t-\xi\to\tau$ and for this case $\left.\frac{\sin[2\pi\lambda(\xi+\tau-t)]}{\pi(\xi+\tau-t)}\right|_{t-\xi\to\tau} = 2\lambda$.

Hence,

$$I = \lim_{\lambda\to\infty} \frac{1}{T}\int_0^T\int_0^T x(t)x^*(\xi)\cdot\frac{\sin[2\pi\lambda(\xi+\tau-t)]}{\pi(\xi+\tau-t)}d\xi dt = \lim_{\lambda\to\infty}\frac{1}{T}\int_0^T F(t,\tau)dt \tag{3.236}$$

where

$$F(t,\tau) = x(t)\int_0^T x^*(\xi)\cdot\frac{\sin[2\pi\lambda(\xi+\tau-t)]}{\pi(\xi+\tau-t)}d\xi \tag{3.237}$$

To determine whether the interchange of the limit and integral operations is valid in Equation 3.236, consider the case of $x(t)=1/t^{1/3}$, $t>0$. Clearly, $x\in \boldsymbol{L}[0,T]$. For the case of $\tau-t=0$, the integral in Equation 3.237 becomes infinite as $\lambda$ increases as the following shows:

$$\begin{aligned}\int_0^T x^*(\xi)\left[\frac{\sin(2\pi\lambda\xi)}{\pi\xi}\right]d\xi &> (2\lambda)^{1/3}\int_0^{1/2\lambda}\frac{\sin(2\pi\lambda\xi)}{\pi\xi}d\xi\\ &+\int_{1/2\lambda}^{T}\frac{1}{\xi^{1/3}}\left[\frac{\sin(2\pi\lambda\xi)}{\pi\xi}\right]d\xi\end{aligned} \tag{3.238}$$

The unboundedness follows as the first integral on the right-hand side of the equation is independent of $\lambda$, which can be seen by using a change of variable $p=\lambda\xi$. Accordingly, the dominated convergence theorem cannot be used to justify the interchange of the limit and integral operations in Equation 3.236.

However, for the case of piecewise differentiability, the integral is of a Dirichlet type and, from Theorem 3.6, is finite and bounded with a bound that is independent of $\lambda$. For this case, the dominated convergence theorem (Theorem 2.19) then justifies the interchange of the limit and integral operations, and it then follows, for all points almost everywhere and as required, that

$$I = \frac{1}{T}\int_0^T x(t)x^*(t-\tau)dt = \bar{R}(T,\tau) \tag{3.239}$$

## APPENDIX 3.E PROOF OF THEOREM 3.19

For the case where the signal has finite average power on the infinite interval, it follows from Equation 3.118 that $F$ is a piecewise continuous function. It then follows that

$$\int_{-\infty}^{\infty} e^{j2\pi f\tau} dF(f) = \lim_{\Delta f\to 0}\sum_{i=-\infty}^{\infty} e^{j2\pi(i\Delta f)\tau}[F(i\Delta f)-F((i-1)\Delta f)] \tag{3.240}$$

where a simple function, based on samples of the complex exponential function with a resolution of $\Delta f$, has been used. Using the definition for $F$, as given by Equation 3.118, it follows that the right-hand side of this equation can be written as

$$\lim_{\Delta f\to 0}\sum_{i=-\infty}^{\infty} e^{j2\pi(i\Delta f)\tau}\left[\lim_{T\to\infty}\int_{-\infty}^{i\Delta f} G(T,f)df - \lim_{T\to\infty}\int_{-\infty}^{(i-1)\Delta f} G(T,f)df\right] \tag{3.241}$$

The finiteness of the average power guarantees that both limits defined by $T\to\infty$ are finite. It then follows that

$$\begin{aligned}\int_{-\infty}^{\infty} e^{j2\pi f\tau} dF(f) &= \lim_{\Delta f\to 0}\sum_{i=-\infty}^{\infty} e^{j2\pi(i\Delta f)\tau}\left[\lim_{T\to\infty}\int_{(i-1)\Delta f}^{i\Delta f} G(T,f)\,df\right]\\ &= \lim_{\Delta f\to 0}\sum_{i=-\infty}^{\infty}\left[\lim_{T\to\infty}\int_{(i-1)\Delta f}^{i\Delta f} G(T,f)e^{j2\pi(i\Delta f)\tau}\,df\right]\end{aligned} \tag{3.242}$$

To prove the required result, it is necessary to interchange the order of the limit of $T\to\infty$ and the summation. To this end, the following definitions are made:

$$s_i(T) = \int_{(i-1)\Delta f}^{i\Delta f} G(T,f)e^{j2\pi(i\Delta f)\tau}df \quad S_i(T) = \int_{(i-1)\Delta f}^{i\Delta f} G(T,f)df \tag{3.243}$$

$$S_i = \lim_{T\to\infty} S_i(T) = \lim_{T\to\infty}\int_{(i-1)\Delta f}^{i\Delta f} G(T,f)df \tag{3.244}$$

It immediately follows that $|s_i(T)| \le S_i(T)$ for $i \in \mathbf{Z}$. Further,

$$\sum_{i=-\infty}^{\infty} S_i(T) = \int_{-\infty}^{\infty} G(T,f)\,df = \bar{P}(T) \qquad \sum_{i=-\infty}^{\infty} S_i = \lim_{T\to\infty} \int_{-\infty}^{\infty} G(T,f)\,df = \bar{P}_\infty \tag{3.245}$$

and $\lim_{T\to\infty} \bar{P}(T) = \bar{P}_\infty$, which is finite, that is,

$$\lim_{T\to\infty} \sum_{i=-\infty}^{\infty} S_i(T) = \sum_{i=-\infty}^{\infty} S_i = \sum_{i=-\infty}^{\infty} \lim_{T\to\infty} S_i(T) \tag{3.246}$$

and the interchange of the order of the limit and integral operations is valid. As $|s_i(T)| \le S_i(T)$ for all $i \in \mathbf{Z}$, it follows, from dominated convergence, that

$$\lim_{T\to\infty} \sum_{i=-\infty}^{\infty} s_i(T) = \sum_{i=-\infty}^{\infty} \lim_{T\to\infty} s_i(T) \tag{3.247}$$

Hence, after letting $\Delta f \to 0$,

$$\int_{-\infty}^{\infty} e^{j2\pi f\tau}\,dF(f) = \lim_{T\to\infty} \int_{-\infty}^{\infty} G(T,f)e^{j2\pi f\tau}\,df = \lim_{T\to\infty} \bar{R}(T,\tau) = R_\infty(\tau) \tag{3.248}$$

which is the required result.

## APPENDIX 3.F PROOF OF THEOREM 3.23

Consider the second form for the Cramer transform given in Equation 3.135:

$$X_C(f) = \lim_{T\to\infty} f \int_{-T}^{T} e^{-j\pi f\lambda} \cdot \text{sinc}(f\lambda) \cdot x(\lambda)\,d\lambda \tag{3.249}$$

For any fixed value of $f, f \neq 0$, consider a fixed value of $\lambda_o > 0$ and the result

$$\left|e^{-j\pi f\lambda} \cdot \text{sinc}(f\lambda) \cdot x(\lambda)\right| \le \frac{|x(\lambda)|}{|\pi f\lambda|} < \frac{1+\lambda_o}{\pi f\lambda_o} \cdot \frac{|x(\lambda)|}{1+|\lambda|} \qquad |\lambda| > \lambda_o \tag{3.250}$$

which, for the case of $\lambda > \lambda_o > 0$, follows as

$$\begin{aligned} &\lambda(1+\lambda_o)=\lambda+\lambda\lambda_o>\lambda_o(1+\lambda)=\lambda_o+\lambda\lambda_o \\ \Rightarrow \quad &\frac{\lambda}{1+\lambda}>\frac{\lambda_o}{1+\lambda_o} \\ \Rightarrow \quad &\lambda=\frac{\lambda}{1+\lambda}\cdot(1+\lambda)>\frac{\lambda_o}{1+\lambda_o}\cdot(1+\lambda) \\ \Rightarrow \quad &\frac{1}{\lambda}<\frac{1+\lambda_o}{\lambda_o}\cdot\frac{1}{1+\lambda} \end{aligned} \tag{3.251}$$

Hence,

$$\begin{aligned} |X_C(f)|<f\int_{-\lambda_o}^{\lambda_o}\left|e^{-j\pi f\lambda}\cdot\text{sinc}(f\lambda)\cdot x(\lambda)\right|d\lambda+\lim_{T\to\infty}\frac{1+\lambda_o}{\pi\lambda_o}\cdot\int_{\lambda_o}^{T}\frac{|x(\lambda)|}{1+|\lambda|}d\lambda \\ +\lim_{T\to\infty}\frac{1+\lambda_o}{\pi\lambda_o}\cdot\int_{-T}^{-\lambda_o}\frac{|x(\lambda)|}{1+|\lambda|}d\lambda \end{aligned} \tag{3.252}$$

and $\dfrac{x(t)}{1+|t|}\in \boldsymbol{L}$ guarantees the existence of $X_C(f)$.

## APPENDIX 3.G PROOF OF THEOREM 3.24

Assume $\dfrac{x(t)}{1+|t|}\in \boldsymbol{L}$ and, consistent with Theorem 3.23, assume that $X_C(f)$ is finite for all values of $f$. Using the definition for the Lebesgue–Stieltjes integral, and the definition for $X_C(f)$, it then follows that

$$\begin{aligned} I(t)&=\int_{\boldsymbol{R}}e^{j2\pi ft}dX_C(f)=\lim_{\Delta f\to 0}\sum_{i=-\infty}^{\infty}e^{j2\pi i\Delta ft}[X_C((i+1)\Delta f)-X_C(i\Delta f)] \\ &=\lim_{\Delta f\to 0}\sum_{i=-\infty}^{\infty}e^{j2\pi i\Delta ft}\cdot\left[\frac{j}{2\pi}\int_{-\infty}^{\infty}\frac{e^{-j2\pi(i+1)\Delta f\lambda}-1}{\lambda}\cdot x(\lambda)\,d\lambda-\frac{j}{2\pi}\int_{-\infty}^{\infty}\frac{e^{-j2\pi i\Delta f\lambda}-1}{\lambda}\cdot x(\lambda)\,d\lambda\right] \end{aligned} \tag{3.253}$$

The finiteness of $X_C(f)$, for all values of $f$, leads to

$$I(t)=\lim_{\Delta f\to 0}\sum_{i=-\infty}^{\infty}e^{j2\pi i\Delta ft}\cdot\left[\frac{j}{2\pi}\int_{-\infty}^{\infty}\frac{e^{-j2\pi(i+1)\Delta f\lambda}-e^{-j2\pi i\Delta f\lambda}}{\lambda}\cdot x(\lambda)\,d\lambda\right] \tag{3.254}$$

Consider

$$\frac{e^{-j2\pi(i+1)\Delta f\lambda}-e^{-j2\pi i\Delta f\lambda}}{\Delta f} = \frac{d}{df}\left[e^{-j2\pi f\lambda}\right]\Bigg|_{f=i\Delta f} + \varepsilon(\Delta f, i\Delta f, \lambda)$$
$$= -j2\pi\lambda e^{-j2\pi i\Delta f\lambda} + \varepsilon(\Delta f, i\Delta f, \lambda) \tag{3.255}$$

where

$$\varepsilon(\Delta f, i\Delta f, \lambda) = \frac{e^{-j2\pi(i+1)\Delta f\lambda}-e^{-j2\pi i\Delta f\lambda}}{\Delta f} + j2\pi\lambda e^{-j2\pi i\Delta f\lambda}$$
$$= \frac{e^{-j2\pi i\Delta f\lambda}}{\Delta f}\left[e^{-j2\pi\Delta f\lambda} - 1 + j2\pi\lambda\Delta f\right] \approx -2\pi^2\Delta f\lambda^2 \cdot e^{-j2\pi i\Delta f\lambda} \tag{3.256}$$

assuming $e^{-j2\pi\Delta f\lambda} \approx 1 - j2\pi\Delta f\lambda - 2\pi^2\Delta f^2\lambda^2$, which is valid provided $|\Delta f\lambda| << 1$. Thus,

$$I(t) \approx \lim_{\Delta f\to 0}\sum_{i=-\infty}^{\infty} e^{j2\pi i\Delta ft}\cdot\left[\int_{-\infty}^{\infty} e^{-j2\pi i\Delta f\lambda}[1-j\pi\Delta f\lambda]\cdot x(\lambda)\,d\lambda\right]\Delta f \tag{3.257}$$

Consider

$$I_1(t,\Delta f) = \sum_{i=-\infty}^{\infty} e^{j2\pi i\Delta ft}\cdot\left[\int_{-k_o/\Delta f}^{k_o/\Delta f} e^{-j2\pi i\Delta f\lambda}[1-j\pi\Delta f\lambda]\cdot x(\lambda)\,d\lambda\right]\Delta f \tag{3.258}$$

For $k_o$ fixed the magnitude of the term $j\pi\Delta f\lambda$ can be made arbitrarily small on the interval $[-k_o/\Delta f, k_o/\Delta f]$ by choosing $k_o << 1$. By choosing $k_o << 1$, the approximation leading to the result stated in Equation 3.256 is valid on the interval $[-k_o/\Delta f, k_o/\Delta f]$.

Thus,

$$I_1(t,\Delta f) \approx \sum_{i=-\infty}^{\infty} e^{j2\pi i\Delta ft}\cdot\left[\int_{-k_o/\Delta f}^{k_o/\Delta f} e^{-j2\pi i\Delta f\lambda}\cdot x(\lambda)\,d\lambda\right]\Delta f$$
$$= \lim_{I_o\to\infty}\sum_{i=-I_o}^{I_o} e^{j2\pi i\Delta ft}\cdot\left[\int_{-k_o/\Delta f}^{k_o/\Delta f} e^{-j2\pi i\Delta f\lambda}\cdot x(\lambda)\,d\lambda\right]\Delta f \tag{3.259}$$

Interchanging the order of integration and summation (which is valid as $x$ is locally integrable and the summation is finite) yields

$$I_1(t,\Delta f) \approx \lim_{I_o \to \infty} \int_{-k_o/\Delta f}^{k_o/\Delta f} \left[ \sum_{i=-I_o}^{I_o} e^{j2\pi i \Delta f(t-\lambda)} \Delta f \right] x(\lambda)\, d\lambda \tag{3.260}$$

For $t$ fixed, consider the sequence of numbers defined by the integral and summation in $I_1(t, \Delta f)$ as $\Delta f$ decreases, $I_o$ increases and with $I_o \Delta f = f_o$. The limit is

$$I_2(t,f_o) = \lim_{\Delta f \to 0} I_1(t,\Delta f)\Big|_{I_o \Delta f = f_o} = \int_{-\infty}^{\infty} x(\lambda) \left[ \int_{-f_o}^{f_o} e^{j2\pi(t-\lambda)f}\, df \right] d\lambda \tag{3.261}$$

Evaluating the inner integral yields

$$\begin{aligned} I_2(t,f_o) &= \int_{-\infty}^{\infty} x(\lambda) \left[ \frac{e^{j2\pi f(t-\lambda)}}{j2\pi(t-\lambda)} \Bigg|_{-f_o}^{f_o} \right] d\lambda = \int_{-\infty}^{\infty} x(\lambda) \left[ \frac{e^{j2\pi f_o(t-\lambda)} - e^{-j2\pi f_o(t-\lambda)}}{j2\pi(t-\lambda)} \right] d\lambda \\ &= \int_{-\infty}^{\infty} x(\lambda) \left[ \frac{\sin[2\pi f_o(t-\lambda)]}{\pi(t-\lambda)} \right] d\lambda = \frac{x(t^+) + x(t^-)}{2} \Bigg|_{f_o \to \infty} \end{aligned} \tag{3.262}$$

The last result follows from Theorem 3.6 and the required result is proved as $f_o$ can be made arbitrarily large.

## APPENDIX 3.H PROOF OF THEOREM 3.25

Consider

$$X_C(f) = \lim_{T \to \infty} X_T(f) \tag{3.263}$$

where

$$\begin{aligned} X_T(f) &= \frac{1}{\pi} \int_{-T}^{T} e^{-j\pi f \lambda} \cdot \frac{\sin(\pi f \lambda)}{\lambda} \cdot x(\lambda)\, d\lambda \\ &= \frac{1}{\pi} \int_{-T}^{T} e^{-j\pi f \lambda} \cdot \frac{\sin(\pi f \lambda)}{\lambda} \cdot \left[ \lim_{I_o \to \infty} \sum_{i=-I_o}^{I_o} c_i(T) e^{j2\pi i f_o \lambda} \right] d\lambda \end{aligned} \tag{3.264}$$

assuming a Fourier series decomposition for $x$ is valid almost everywhere on $[-T,T]$ and $f_o = 1/2T$. Define

$$F(I_o,\lambda)=e^{-j\pi f\lambda}\cdot\frac{\sin(\pi f\lambda)}{\lambda}\cdot\sum_{i=-I_o}^{I_o}c_i(T)e^{j2\pi i f_o\lambda} \tag{3.265}$$

and it follows that $\lim_{I_o\to\infty} F(I_o,\lambda)=\pi f\cdot\text{sinc}\,(f\lambda)\cdot e^{-j\pi f\lambda}\cdot x(\lambda)$ almost everywhere. Further, for a fixed value of $f$, the sequence of functions $|F(I_o,\lambda)|$, as $I_o$ varies, is bounded above on $[-T,T]$ by $|\pi f\cdot x(\lambda)|$ as $I_o\to\infty$. As $x$ is integrable on $[-T,T]$, it follows, from the dominated convergence theorem (Theorem 2.19), that

$$\lim_{I_o\to\infty}\int_{-T}^{T}F(I_o,\lambda)d\lambda=\int_{-T}^{T}\lim_{I_o\to\infty}F(I_o,\lambda)\,d\lambda \tag{3.266}$$

and, hence,

$$X_T(f)=\lim_{I_o\to\infty}\frac{1}{\pi}\int_{-T}^{T}\sum_{i=-I_o}^{I_o}c_i(T)\cdot e^{-j\pi(f-2if_o)\lambda}\cdot\frac{\sin(\pi f\lambda)}{\lambda}d\lambda \tag{3.267}$$

The order of integration and summation can then be interchanged by the Fubini–Tonelli theorem (Theorem 2.16) (bounded function on a finite interval and a finite sum) to yield

$$\begin{aligned}X_T(f)&=\sum_{-\infty}^{\infty}\frac{c_i(T)}{\pi}\int_{-T}^{T}e^{-j\pi(f-i2f_o)\lambda}\cdot\frac{\sin(\pi f\lambda)}{\lambda}d\lambda\\&=\sum_{i=-\infty}^{\infty}\frac{2c_i(T)}{\pi}\int_{0}^{T}\cos[\pi(f-2if_o)\lambda]\cdot\frac{\sin(\pi f\lambda)}{\lambda}d\lambda\end{aligned} \tag{3.268}$$

Using the relationship $\sin(A)\cos(B)=0.5(\sin(A-B)+\sin(A+B))$, it follows that

$$X_T(f)=\sum_{i=-\infty}^{\infty}\frac{c_i(T)}{\pi}\left[\int_0^T\frac{\sin(2\pi if_o\lambda)}{\lambda}d\lambda+\int_0^T\frac{\sin[2\pi(f-if_o)\lambda]}{\lambda}d\lambda\right] \tag{3.269}$$

A change of variable $\xi=2\pi k\lambda$, where $k=if_o$ and then $k=f-if_o$, yields, as $d\lambda/\lambda=d\xi/\xi$,

$$X_T(f)=\sum_{i=-\infty}^{\infty}\frac{c_i(T)}{\pi}\left[\int_0^{2\pi if_oT}\frac{\sin(\xi)}{\xi}d\xi+\int_0^{2\pi(f-if_o)T}\frac{\sin(\xi)}{\xi}d\xi\right] \tag{3.270}$$

With the definition $\text{Si}(x) = \int_0^x \frac{\sin(\lambda)}{\lambda} d\lambda$ and as $f_o T = 1/2$, $\text{Si}(-x) = -\text{Si}(x)$, it follows that

$$X_T(f) = \sum_{i=-\infty}^{\infty} \frac{c_i(T)}{\pi} [\text{Si}(i\pi) + \text{Si}(2\pi f T - i\pi)] \tag{3.271}$$

which is the required result.

## APPENDIX 3.I PROOF OF THEOREM 3.26

Consider

$$X_C(f) = \lim_{T \to \infty} \sum_{i=-\infty}^{\infty} \frac{c_i(T)}{\pi} [\text{Si}(i\pi) + \text{Si}(2\pi f T - i\pi)] \tag{3.272}$$

As the sine integral function is bounded with $\text{Si}(\infty) = \pi/2$, $\text{Si}(-\infty) = -\pi/2$, it follows that a sufficient condition for convergence of the summation, for any fixed value of $T$, is

$$\sum_{i=-\infty}^{\infty} |c_i(T)| < \infty \tag{3.273}$$

To establish a better sufficient condition for convergence, consider the graph of $\text{Si}(i\pi) + \text{Si}(2\pi f T - i\pi)$ shown in Figure 3.37 for the case of $f > 0$ and $fT = 100$. For this case, $\text{Si}(i\pi) + \text{Si}(2\pi f T - i\pi)$ is close to a positive constant when $i$ is in the range $I_1$ to $I_2$, where $I_1 \approx 0$ and $I_2 \approx 2fT$. Consider, first, the case where $f$ and $T$ are fixed and write Equation 3.272 for this case as

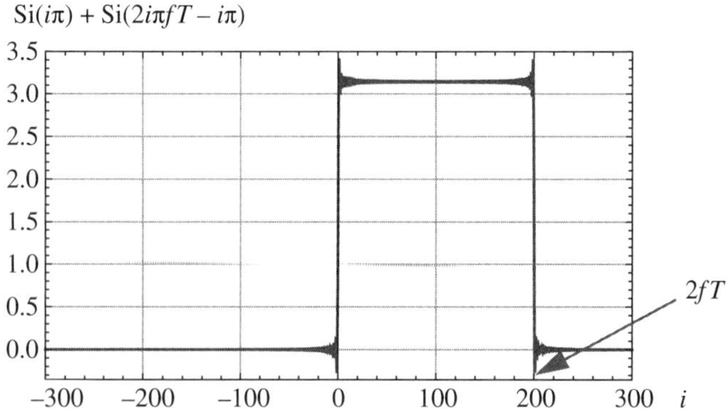


**FIGURE 3.37** Graph of $\text{Si}(i\pi) + \text{Si}(2\pi f T - i\pi)$ for the case of $fT = 100$.

$$X_T(f) = \frac{1}{\pi}\sum_{i=-\infty}^{I_1-1} c_i(T)\ [\mathrm{Si}(i\pi)+\mathrm{Si}(2\pi fT-i\pi)]$$
$$+\frac{1}{\pi}\sum_{i=I_1}^{I_2-1} c_i(T)\ [\mathrm{Si}(i\pi)+\mathrm{Si}(2\pi fT-i\pi)] \tag{3.274}$$
$$+\frac{1}{\pi}\sum_{i=I_2}^{\infty} c_i(T)\ [\mathrm{Si}(i\pi)+\mathrm{Si}(2\pi fT-i\pi)]$$

With the boundedness of $\mathrm{Si}(i\pi)+\mathrm{Si}(2\pi fT-i\pi)$, convergence is determined by the first and last summations for fixed values of $I_1$ and $I_2$. Consider the last summation with $I_2$ such that $\pi I_2 - 2\pi fT > 0$:

$$S_3 = \frac{1}{\pi}\sum_{i=I_2}^{\infty} c_i(T)\ [\mathrm{Si}(i\pi)-\mathrm{Si}(i\pi-2\pi fT)] \qquad \mathrm{Si}(-x) = -\mathrm{Si}(x)$$
$$=\sum_{i=I_2}^{\infty} c_i(T)\left[\int_0^{\pi i}\frac{\sin(\xi)}{\xi}d\xi - \int_0^{i\pi-2\pi fT}\frac{\sin(\xi)}{\xi}d\xi\right] \tag{3.275}$$
$$=\sum_{i=I_2}^{\infty} c_i(T)\left[\int_{i\pi-2\pi fT}^{\pi i}\frac{\sin(\xi)}{\xi}d\xi\right]$$

The magnitude of the integral has the bound

$$\left|\int_{\pi i-2\pi fT}^{\pi i}\frac{\sin(\xi)}{\xi}d\xi\right| \le \int_{\pi i-2\pi fT}^{\pi i}\left|\frac{\sin(\xi)}{\xi}\right|d\xi \le \int_{\pi i-2\pi fT}^{\pi i}\frac{d\xi}{\xi} \le \int_{\pi i-2\pi fT}^{\pi i}\frac{d\xi}{\pi i-2\pi fT}$$
$$= \frac{2\pi fT}{\pi i-2\pi fT} = \frac{2fT}{i(1-2fT/i)} \le \frac{2fT}{i}\cdot\frac{1}{1-2fT/I_2} \tag{3.276}$$

Hence,

$$|S_3| \le \sum_{i=I_2}^{\infty}|c_i(T)|\cdot\frac{2fT}{i}\cdot\frac{1}{1-2fT/I_2} \tag{3.277}$$

A similar argument can be applied for the first summation in Equation 3.274. Hence, boundedness of $X_T(f)$ for fixed values of $f, T$ is guaranteed if $\sum_{i=1}^{\infty}\frac{|c_i(T)|}{i}$ converges.

For the infinite interval case and for a fixed value of $f$, convergence is required for all values of $T$. The above argument applies but account has to be taken of the middle summation in Equation 3.274 as the limit $I_2$ increases linearly with $T$ for a fixed value of $f$. Convergence of

$$\frac{1}{\pi}\sum_{i=I_1}^{I_2-1} c_i(T)[\mathrm{Si}(i\pi)+\mathrm{Si}(2\pi f T - i\pi)] \tag{3.278}$$

when $I_1$ is constant and $I_2$ increases linearly with $T$ is satisfied if $\sum_{i=1}^{\infty} c_i(T)$ is finite as $\mathrm{Si}(i\pi)+\mathrm{Si}(2\pi f T - i\pi)$ is positive and bounded above. Thus, for the case of $T \to \infty$, existence of $X_C(f)$ is guaranteed if both

$$\lim_{T\to\infty}\sum_{i=1}^{\infty} c_i(T) \qquad \lim_{T\to\infty}\sum_{i=1}^{\infty}\frac{|c_i(T)|}{i} \tag{3.279}$$

are finite.

## APPENDIX 3.J CRAMER TRANSFORM OF UNIT STEP FUNCTION

The Cramer transform of the unit step function $u(t-t_o)$, on the interval $[-T,T]$ assuming $|t_o|<T$, is

$$\begin{aligned} X_C(f) &= \frac{1}{\pi}\int_{t_o}^{T} e^{-j\pi f\lambda}\cdot\frac{\sin(\pi f\lambda)}{\lambda}\cdot d\lambda \\ &= \frac{1}{\pi}\left[\int_{t_o}^{T}\cos(\pi f\lambda)\cdot\frac{\sin(\pi f\lambda)}{\lambda}d\lambda - j\int_{t_o}^{T}\sin(\pi f\lambda)\cdot\frac{\sin(\pi f\lambda)}{\lambda}d\lambda\right] \end{aligned} \tag{3.280}$$

As $\sin(A)\cos(A)=0.5\sin(2A)$ and $\sin(A)\sin(A)=0.5[1-\cos(2A)]$, it follows, as the integrand in the first integral is an even function and the integrand in the second integral is an odd function, that

$$X_C(f)=\begin{cases} \dfrac{1}{2\pi}\left[2\displaystyle\int_{0}^{|t_o|}\frac{\sin(2\pi f\lambda)}{\lambda}d\lambda+\int_{|t_o|}^{T}\frac{\sin(2\pi f\lambda)}{\lambda}d\lambda\right]- \\ \dfrac{j}{2\pi}\displaystyle\int_{|t_o|}^{T}\frac{1-\cos(2\pi f\lambda)}{\lambda}d\lambda & t_o<0 \\ \dfrac{1}{2\pi}\left[\displaystyle\int_{t_o}^{T}\frac{\sin(2\pi f\lambda)}{\lambda}d\lambda - j\int_{t_o}^{T}\frac{1-\cos(2\pi f\lambda)}{\lambda}d\lambda\right] & t_o\ge 0 \end{cases} \tag{3.281}$$

Thus,

$$X_C(f)=\frac{1}{2\pi}\left[2u(-t_o)\int_0^{|t_o|}\frac{\sin(2\pi f\lambda)}{\lambda}d\lambda+\int_{|t_o|}^{T}\frac{\sin(2\pi f\lambda)}{\lambda}d\lambda\right]-\frac{j}{2\pi}\int_{|t_o|}^{T}\frac{1-\cos(2\pi f\lambda)}{\lambda}d\lambda \tag{3.282}$$

and, hence,

$$X_C(f)=\frac{\text{sgn}(f)}{2\pi}\left[2u(-t_o)\int_0^{|t_o|}\frac{\sin(2\pi|f|\lambda)}{\lambda}d\lambda+\int_{|t_o|}^{T}\frac{\sin(2\pi|f|\lambda)}{\lambda}d\lambda\right]-\frac{j}{2\pi}\int_{|t_o|}^{T}\frac{1-\cos(2\pi|f|\lambda)}{\lambda}d\lambda \tag{3.283}$$

A change of variable $\xi=2\pi|f|\lambda$ yields, as $d\lambda/\lambda=d\xi/\xi$,

$$X_C(f)=\frac{\text{sgn}(f)}{2\pi}\left[2u(-t_o)\int_0^{2\pi|f||t_o|}\frac{\sin(\xi)}{\xi}d\xi+\int_{2\pi|f||t_o|}^{2\pi fT}\frac{\sin(\xi)}{\xi}d\xi\right]+\frac{j}{2\pi}\int_{2\pi|f||t_o|}^{2\pi|f|T}\frac{\cos(\xi)-1}{\xi}d\xi \tag{3.284}$$

Thus,

$$X_C(f)=\frac{\text{sgn}(f)}{2\pi}\left[[2u(-t_o)-1]\int_0^{2\pi|f||t_o|}\frac{\sin(\xi)}{\xi}d\xi+\int_0^{2\pi|f|T}\frac{\sin(\xi)}{\xi}d\xi\right]+\frac{j}{2\pi}\left[\int_0^{2\pi|f|T}\frac{\cos(\xi)-1}{\xi}d\xi-\int_0^{2\pi|f||t_o|}\frac{\cos(\xi)-1}{\xi}d\xi\right] \tag{3.285}$$

Using the definitions

$$\text{Si}(x)=\int_0^x\frac{\sin(\lambda)}{\lambda}d\lambda \quad \text{Ci}(x)=-\int_x^{\infty}\frac{\cos(\lambda)}{\lambda}d\lambda=\gamma+\ln(x)+\int_0^x\frac{\cos(\lambda)-1}{\lambda}d\lambda \tag{3.286}$$

assuming $x > 0$, it follows that

$$\begin{aligned} X_C(f) &= \frac{\mathrm{sgn}(f)}{2\pi}[[2u(-t_o)-1]\ \mathrm{Si}(2\pi|f||t_o|)+\mathrm{Si}(2\pi|f|T)] \\ &+\frac{j}{2\pi}[\mathrm{Ci}(2\pi|f|T)-\gamma-\ln(2\pi|f|T)-[\mathrm{Ci}(2\pi|f||t_o|)-\gamma-\ln(2\pi|f||t_o|)]] \\ &=\frac{\mathrm{sgn}(f)}{2\pi}[\mathrm{Si}(2\pi|f|T)+[2u(-t_o)-1]\mathrm{Si}(2\pi|f||t_o|)] \\ &+\frac{j}{2\pi}\left[\mathrm{Ci}(2\pi|f|T)-\mathrm{Ci}(2\pi|f||t_o|)+\ln\left(\frac{|t_o|}{T}\right)\right] \end{aligned} \tag{3.287}$$

For the case of $t_o = 0$, it follows from Equations 3.285 and 3.286 that

$$X_C(f)=\frac{1}{2\pi}[\mathrm{sgn}(f)\ \mathrm{Si}(2\pi|f|T)+j[\mathrm{Ci}(2\pi|f|T)-\gamma-\ln(2\pi|f|T)]] \tag{3.288}$$

## APPENDIX 3.K CRAMER TRANSFORM FOR SINUSOIDAL SIGNALS

To establish the Cramer transform of the sinusoidal signals $x(t)=\sin(2\pi f_o t)$ and $x(t)=\cos(2\pi f_o t)$, consider

$$\begin{aligned} X_C(f) &= \frac{1}{\pi}\int_{-\infty}^{\infty} e^{-j\pi f\lambda}\cdot\frac{\sin(\pi f\lambda)}{\lambda}\cdot\begin{Bmatrix}\sin(2\pi f_o\lambda)\\ \cos(2\pi f_o\lambda)\end{Bmatrix}d\lambda \\ &=\frac{1}{\pi}\int_{-\infty}^{\infty}[\cos(\pi f\lambda)-j\sin(\pi f\lambda)]\cdot\frac{\sin(\pi f\lambda)}{\lambda}\cdot\begin{Bmatrix}\sin(2\pi f_o\lambda)\\ \cos(2\pi f_o\lambda)\end{Bmatrix}d\lambda \\ &=\frac{2}{\pi}\int_0^{\infty}\begin{Bmatrix}-j\sin(\pi f\lambda)\sin(2\pi f_o\lambda)\\ \cos(\pi f\lambda)\cos(2\pi f_o\lambda)\end{Bmatrix}\cdot\frac{\sin(\pi f\lambda)}{\lambda}d\lambda \end{aligned} \tag{3.289}$$

where the odd/even properties of the sine and cosine functions have been used. As

$$\begin{aligned} \sin(a)\sin(b) &= 0.5[\cos(a-b)-\cos(a+b)] \\ \cos(a)\cos(b) &= 0.5[\cos(a-b)+\cos(a+b)] \end{aligned} \tag{3.290}$$

it then follows that

$$X_C(f)=\frac{1}{\pi}\int_0^{\infty}\begin{Bmatrix}-j\\ 1\end{Bmatrix}\cdot\begin{Bmatrix}\cos[\pi(f-2f_o)\lambda]-\cos[\pi(f+2f_o)\lambda]\\ \cos[\pi(f-2f_o)\lambda]+\cos[\pi(f+2f_o)\lambda]\end{Bmatrix}\cdot\frac{\sin(\pi f\lambda)}{\lambda}d\lambda \tag{3.291}$$

As $\sin(a)\cos(b)=0.5[\sin(a-b)+\sin(a+b)]$, it follows that

$$\frac{1}{\pi}\int_0^{\infty}\cos[\pi(f\mp 2f_o)\lambda]\cdot\frac{\sin(\pi f\lambda)}{\lambda}d\lambda=\frac{1}{2\pi}\int_0^{\infty}\frac{\sin[2\pi(\pm f_o)\lambda]}{\lambda}+\frac{\sin[2\pi(f\mp f_o)\lambda]}{\lambda}d\lambda$$
$$=\frac{\operatorname{sgn}(\pm f_o)+\operatorname{sgn}(f\mp f_o)}{4} \qquad (3.292)$$

where the result

$$\frac{1}{\pi}\int_0^{\infty}\frac{\sin(p\lambda)}{\lambda}d\lambda=0.5\operatorname{sgn}(p) \qquad (3.293)$$

has been used. Hence, for the sine case,

$$X_C(f)=-j\left[\frac{\operatorname{sgn}(f_o)+\operatorname{sgn}(f-f_o)}{4}-\frac{\operatorname{sgn}(-f_o)+\operatorname{sgn}(f+f_o)}{4}\right]$$
$$=j\left[\frac{\operatorname{sgn}(f+f_o)-\operatorname{sgn}(f-f_o)-2\operatorname{sgn}(f_o)}{4}\right] \qquad (3.294)$$

The last result follows as $\operatorname{sgn}(-f_o)=-\operatorname{sgn}(f_o)$. For the cosine case,

$$X_C(f)=\frac{\operatorname{sgn}(f_o)+\operatorname{sgn}(f-f_o)}{4}+\frac{\operatorname{sgn}(-f_o)+\operatorname{sgn}(f+f_o)}{4}$$
$$=\frac{\operatorname{sgn}(f+f_o)+\operatorname{sgn}(f-f_o)}{4} \qquad (3.295)$$

### 3.K.1 Complex Exponential Case

For the complex exponential case $x(t)=\exp(j2\pi f_o t)$, on the infinite interval, the transform is

$$X_C(f)=\frac{1}{\pi}\int_{-\infty}^{\infty}e^{-j\pi f\lambda}\cdot\frac{\sin(\pi f\lambda)}{\lambda}\cdot e^{j2\pi f_o\lambda}d\lambda=\frac{1}{\pi}\int_{-\infty}^{\infty}e^{-j\pi(f-2f_o)\lambda}\cdot\frac{\sin(\pi f\lambda)}{\lambda}d\lambda$$
$$=\frac{1}{\pi}\int_{-\infty}^{\infty}[\cos(\pi(f-2f_o)\lambda)-j\sin(\pi(f-2f_o)\lambda)]\cdot\frac{\sin(\pi f\lambda)}{\lambda}d\lambda \qquad (3.296)$$
$$=\frac{2}{\pi}\int_0^{\infty}\cos(\pi(f-2f_o)\lambda)\cdot\frac{\sin(\pi f\lambda)}{\lambda}d\lambda$$

where the odd/even properties of the sine and cosine functions have been used. As $\sin(a)\cos(b)=0.5[\sin(a-b)+\sin(a+b)]$, it follows, using the result stated in Equation 3.293, that

$$X_C(f)=\frac{1}{\pi}\int_0^{\infty}\frac{\sin(2\pi f_o\lambda)}{\lambda}+\frac{\sin[2\pi(f-f_o)\lambda]}{\lambda}d\lambda \\ =0.5\,\mathrm{sgn}(f_o)+0.5\,\mathrm{sgn}(f-f_o) \tag{3.297}$$

For the interval $[-T,T]$, a change of limits in Equation 3.297, and the changes of variables $u=2\pi f_o\lambda$ and $v=2\pi(f-f_o)\lambda$ that imply $du/u=d\lambda/\lambda$ and $dv/v=d\lambda/\lambda$, yields

$$X_C(f)=\frac{1}{\pi}\int_0^{2\pi f_o T}\frac{\sin(v)}{v}dv+\frac{1}{\pi}\int_0^{2\pi(f-f_o)T}\frac{\sin(u)}{u}du \tag{3.298}$$

The definition of the sine integral (Eq. 3.286) yields the required result:

$$X_C(f)=\frac{1}{\pi}[\mathrm{Si}(2\pi f_o T)+\mathrm{Si}(2\pi(f-f_o)T)] \tag{3.299}$$

### 3.K.2 Sine and Cosine on [−*T*,*T*]

For the interval $[-T,T]$, the integral in Equation 3.292 becomes

$$\frac{1}{T}\int_0^T\cos[\pi(f\mp 2f_o)\lambda]\cdot\frac{\sin(\pi f\lambda)}{\lambda}d\lambda=\frac{1}{2\pi}\int_0^T\frac{\sin[2\pi(\pm f_o)\lambda]}{\lambda}+\frac{\sin[2\pi(f\mp f_o)\lambda]}{\lambda}d\lambda \\ =\frac{1}{2\pi}\int_0^{2\pi(\pm f_o)T}\frac{\sin(u)}{u}du+\frac{1}{2\pi}\int_0^{2\pi(f\mp f_o)T}\frac{\sin(v)}{v}dv \\ =\frac{\mathrm{Si}(2\pi(\pm f_o)T)}{2\pi}+\frac{\mathrm{Si}(2\pi(f\mp f_o)T)}{2\pi} \tag{3.300}$$

where the changes of variables $u=2\pi(\pm f_o)\lambda$ and $v=2\pi(f\mp f_o)\lambda$, which imply $du/u=d\lambda/\lambda$ and $dv/v=d\lambda/\lambda$, respectively, have been used along with the definition of the sine integral. Thus, from Equation 3.291, it follows that

$$X_C(f) = \begin{cases} \dfrac{-j}{2\pi}[\mathrm{Si}(2\pi f_o T) + \mathrm{Si}(2\pi(f-f_o)T) - [\mathrm{Si}(2\pi(-f_o)T) + \mathrm{Si}(2\pi(f+f_o)T)]] \\ \dfrac{1}{2\pi}[\mathrm{Si}(2\pi f_o T) + \mathrm{Si}(2\pi(f-f_o)T) + [\mathrm{Si}(2\pi(-f_o)T) + \mathrm{Si}(2\pi(f+f_o)T)]] \end{cases}$$
$$= \begin{cases} \dfrac{-j}{2\pi}[2\mathrm{Si}(2\pi f_o T) + \mathrm{Si}(2\pi(f-f_o)T) - \mathrm{Si}(2\pi(f+f_o)T)] & \text{sine} \\ \dfrac{1}{2\pi}[\mathrm{Si}(2\pi(f-f_o)T) + \mathrm{Si}(2\pi(f+f_o)T)] & \text{cosine} \end{cases} \tag{3.301}$$

where the property $\mathrm{Si}(-x) = -\mathrm{Si}(x)$ has been used.

## APPENDIX 3.L PROOF OF THEOREM 3.30

From the definition of the Cramer transform, as given by Equation 3.134, it follows that

$$X_C\left(f+\frac{\Delta f}{2}\right) - X_C\left(f-\frac{\Delta f}{2}\right) = \frac{j}{2\pi}\int_{-\infty}^{\infty} \frac{e^{-j2\pi(f+\Delta f/2)\lambda}-1}{\lambda}\cdot x(\lambda)\,d\lambda - \frac{j}{2\pi}\int_{-\infty}^{\infty} \frac{e^{-j2\pi(f-\Delta f/2)\lambda}-1}{\lambda}\cdot x(\lambda)\,d\lambda \tag{3.302}$$

Assuming $\frac{x(t)}{1+|t|} \in \boldsymbol{L}$, consistent with the Cramer transform being finite (Theorem 3.23), it then follows that

$$X_C\left(f+\frac{\Delta f}{2}\right) - X_C\left(f-\frac{\Delta f}{2}\right) = \frac{j}{2\pi}\int_{-\infty}^{\infty} \frac{e^{-j2\pi(f+\Delta f/2)\lambda}-e^{-j2\pi(f-\Delta f/2)\lambda}}{\lambda}\cdot x(\lambda)\,d\lambda$$
$$= \int_{-\infty}^{\infty} e^{-j2\pi f\lambda}\cdot\frac{e^{j\pi\Delta f\lambda}-e^{-j\pi\Delta f\lambda}}{2j\pi\lambda}\cdot x(\lambda)\,d\lambda \tag{3.303}$$
$$= \int_{-\infty}^{\infty} e^{-j2\pi f\lambda}\cdot\Delta f\cdot\frac{\sin(\pi\Delta f\lambda)}{\pi\Delta f\lambda}\cdot x(\lambda)\,d\lambda = \Delta f\int_{-\infty}^{\infty} e^{-j2\pi f\lambda}\cdot\mathrm{sinc}(\Delta f\lambda)\cdot x(\lambda)\,d\lambda$$

which is the required result.

## APPENDIX 3.M PROOF OF THEOREM 3.31

First, the impulse response of an ideal one-sided band-pass filter, as illustrated in Figure 3.31, with a center frequency of $f_o$ Hz and bandwidth of $\Delta f$ Hz, is

$$\begin{aligned} h(t) &= \int_{f_o-\Delta f/2}^{f_o+\Delta f/2} e^{j2\pi ft}\,dt = \left.\frac{e^{j2\pi ft}}{j2\pi t}\right|_{f_o-\Delta f/2}^{f_o+\Delta f/2} = \frac{e^{j2\pi(f_o+\Delta f/2)t} - e^{j2\pi(f_o-\Delta f/2)t}}{j2\pi t} \\ &= \frac{e^{j2\pi f_o t}}{\pi t}\cdot\frac{e^{j\pi\Delta ft} - e^{-j\pi\Delta ft}}{2j} \\ &= e^{j2\pi f_o t}\cdot\frac{\sin(\pi\Delta ft)}{\pi t} = e^{j2\pi f_o t}\Delta f\,\mathrm{sinc}(\Delta ft) \end{aligned} \tag{3.304}$$

Second, the output of a one-sided ideal band-pass filter to an input signal $x$ is

$$y(t) = \int_{-\infty}^{\infty} x(\lambda)h(t-\lambda)\,d\lambda = \Delta f \int_{-\infty}^{\infty} e^{j2\pi f_o(t-\lambda)}\mathrm{sinc}[\Delta f(t-\lambda)]x(\lambda)\,d\lambda \tag{3.305}$$

and it then follows that

$$y(0) = \Delta f \int_{-\infty}^{\infty} e^{-j2\pi f_o\lambda}\mathrm{sinc}(\Delta f\lambda)x(\lambda)\,d\lambda \tag{3.306}$$

This equals $X_C\left(f_o+\frac{\Delta f}{2}\right) - X_C\left(f_o-\frac{\Delta f}{2}\right)$ as given by Theorem 3.30.

## APPENDIX 3.N PROOF OF THEOREM 3.32

Assume $x \in \boldsymbol{L}[0,T]$ and consider

$$\begin{aligned} X_C(f_2) - X_C(f_1) &= \frac{j}{2\pi}\int_0^T \frac{e^{-j2\pi f_2 t} - 1}{t}\cdot x(t)\,dt - \frac{j}{2\pi}\int_0^T \frac{e^{-j2\pi f_1 t} - 1}{t}\cdot x(t)\,dt \\ &= \int_0^T \frac{e^{-j2\pi f_2 t} - e^{-j2\pi f_1 t}}{-j2\pi t}\cdot x(t)\,dt \end{aligned} \tag{3.307}$$

As

$$\int_{f_1}^{f_2} e^{-j2\pi ft}\,df = \left.\frac{e^{-j2\pi ft}}{-j2\pi t}\right|_{f_1}^{f_2} = \frac{e^{-j2\pi f_2 t} - e^{-j2\pi f_1 t}}{-j2\pi t} \tag{3.308}$$

it follows that

$$X_C(f_2)-X_C(f_1)=\int_0^T\left[\int_{f_1}^{f_2}e^{-j2\pi ft}df\right]x(t)\,dt \tag{3.309}$$

Interchanging the order of integration, valid for $x\in \boldsymbol{L}[0,T]$, yields

$$X_C(f_2)-X_C(f_1)=\int_{f_1}^{f_2}\left[\int_0^T x(t)e^{-j2\pi ft}dt\right]df=\int_{f_1}^{f_2}X(T,f)df \tag{3.310}$$

as required.

## APPENDIX 3.O PROOF OF THEOREM 3.33

First, the Cramer representation of the input signal is

$$x(t)=\int_{\boldsymbol{R}}e^{j2\pi ft}dX_C(f)\Rightarrow x(t-\lambda)=\int_{\boldsymbol{R}}e^{j2\pi f(t-\lambda)}dX_C(f) \tag{3.311}$$

and the integral defining the output signal can be written as

$$\begin{aligned}y(t)&=\int_{-\infty}^{t}x(\lambda)h(t-\lambda)d\lambda-\int_0^{\infty}x(t-\lambda)h(\lambda)d\lambda\\&=\int_0^{\infty}\lim_{F_o\to\infty}\left[\int_{-F_o}^{F_o}e^{j2\pi f(t-\lambda)}dX_C(f)\right]h(\lambda)d\lambda\end{aligned} \tag{3.312}$$

Define

$$G(F_o,t,\lambda)=\left[\int_{-F_o}^{F_o}e^{j2\pi f(t-\lambda)}dX_C(f)\right]h(\lambda) \tag{3.313}$$

and it then follows that

$$\lim_{F_o\to\infty}G(F_o,t,\lambda)=x(t-\lambda)h(\lambda) \tag{3.314}$$

The assumption of $x(t-\lambda)h(\lambda) \in \boldsymbol{L}[0,\infty]$, for all $t \in [0,\infty]$, implies, from the dominated convergence theorem (Theorem 2.19), that

$$\int_0^\infty \lim_{F_o \to \infty} G(F_o, t, \lambda)\, d\lambda = \lim_{F_o \to \infty} \int_0^\infty G(F_o, t, \lambda)\, d\lambda \tag{3.315}$$

Hence,

$$y(t) = \lim_{F_o \to \infty} \int_0^\infty \left[ \int_{-F_o}^{F_o} e^{j2\pi f(t-\lambda)} dX_C(f) \right] h(\lambda)\, d\lambda \tag{3.316}$$

As $\left|e^{j2\pi f(t-\lambda)} dX_C(f) h(\lambda)\right|$ is integrable on $[-F_o, F_o] \times [0,\infty]$, it follows, from the Fubini–Tonelli theorem (Theorem 2.16), that the order of integration can be interchanged to yield

$$y(t) = \lim_{F_o \to \infty} \int_{-F_o}^{F_o} e^{j2\pi ft} \left[ \int_0^\infty h(\lambda) e^{-j2\pi f\lambda} d\lambda \right] dX_C(f) \tag{3.317}$$

The inner integral is the Fourier transform of the impulse response and, hence,

$$y(t) = \int_{-\infty}^{\infty} e^{j2\pi ft} H(f) dX_C(f) = \int_{-\infty}^{\infty} e^{j2\pi ft} dY_C(f) \tag{3.318}$$

where

$$dY_C(f) = H(f) dX_C(f) \tag{3.319}$$

This completes the proof.

# 4

# BACKGROUND: PROBABILITY AND RANDOM VARIABLE THEORY

## 4.1 INTRODUCTION

The characterization of a random process requires an understanding of random variable theory, and this has its basis in probability theory. This chapter provides an overview of the key concepts from probability theory that underpin random variable theory. This is followed by an introduction to random variable theory and an overview of discrete and continuous random variables. The concept of expectation of a random variable is fundamental and a brief discussion is included. As part of this discussion, the characteristic function of a random variable is defined. Random variable theory is extended by the generalization to pairs of random variables and a vector of random variables. Associated concepts for this case include the joint probability mass and density functions, marginal distribution and density functions, conditional probability mass and density functions, and covariance and correlation functions. The chapter concludes with a discussion of Stirling's formula and the important DeMoivre–Laplace and Poisson approximations to the binomial probability mass function. Useful references for probability and random variable theory include Grimmett and Stirzaker (1992) and Loeve (1977).

## 4.2 BASIC CONCEPTS: EXPERIMENTS-PROBABILITY THEORY

The following definitions are fundamental and define the basic concepts that underpin probability and random process theory.

*A Signal Theoretic Introduction to Random Processes*, First Edition. Roy M. Howard.

### 4.2.1 Experiments and Sample Spaces

**Definition: Experiment** An experiment is a procedure that can, theoretically, be repeated an infinite number of times. Associated with the running of an experiment, denoted a trial, is a well-defined outcome. The set of possible outcomes associated with an experiment is assumed to be fixed.

**Definition: Trial of Experiment** The independent running of an experiment, leading to a well-defined outcome, is called a trial of the experiment.

**Definition: Outcome** The result, or outcome, of a single trial of an experiment is called the experimental outcome, or outcome for short. An outcome of an experiment is denoted $\omega$.

**Definition: Sample Space** The sample space associated with an experiment, and denoted $S$ or $\Omega$, is the set of all possible, and distinct, experimental outcomes. Thus,

$$S = \{\omega : \text{as defined by distinct trials of experiment}\} \tag{4.1}$$

There are three important ways to detail the sample space of an experiment:

a. List all possible outcomes.
b. List a small number of outcomes to illustrate the structure of the sample space, for example,

$$S = \{T, HT, HHT, HHHT, \ldots\} \tag{4.2}$$

c. Give a mathematical formula that outcomes must satisfy, for example,

$$S = \left\{(a, b, c) : b^2 - 4ac \geq 0\right\} \tag{4.3}$$

The following is a common notation, respectively, for the countable and uncountable cases:

$$S = \{\omega_i : i \in \boldsymbol{Z}^+\} \quad S = \{\omega : \omega \in \boldsymbol{R}\} \tag{4.4}$$

### 4.2.2 Events

**Definition: Event** An event is a subset of the sample space.
As events are sets, it follows that intersections, unions, etc. of events can be defined.

**Definition: Occurrence of an Event** An event is said to have occurred when the outcome of a trial of an experiment is an element of the subset of the sample space defining the event.

**Definition: Elementary Event** An elementary event is an event consisting of a distinct experimental outcome.

**Definition: Impossible Event** The empty set $\emptyset$ is called the impossible event.

**Definition: Certain Event** The set $S$ or $\Omega$ is called the certain event.

**Definition: Mutually Exclusive, or Disjoint, Events** The events $A$ and $B$ are mutually exclusive, or disjoint, if and only if $A \cap B = \emptyset$, that is, the occurrence of $A$ excludes the occurrence of $B$ and vice versa.

**Definition: Complementary or Contrary Event** Consider an event $A$. The complementary, or contrary, event denoted $A^C$ occurs if and only if the event $A$ does not occur.

**Definition: Implied Event** An event $A$ implies an event $B$ if when $A$ occurs, the event $B$ has also occurred. A sufficient condition for this is for $A \subseteq B$.

**Definition: Equivalent Events** If the event $A$ implies the event $B$, and the event $B$ implies the event $A$, then $A$ and $B$ are equivalent events.

**Definition: Combination of Events** The event of $A$ and $B$, denoted $A \cap B$ or $AB$, is said to occur if both the event $A$ and the event $B$ have occurred.
The event of $A$ or $B$, denoted $A \cup B$, is said to occur if at least one of the events from the set of events $\{A, B\}$ has occurred.

**Definition: Sigma Algebra of Events** The sigma algebra of events, denoted $\boldsymbol{F}$, is the set of all possible events defined on the sample space $S$.

### 4.2.3 Probability of an Event

The terminology $P[A]$ is used to denote the *probability of the event A*.

**Definition: Probability Function** The probability function $P$, or probability measure, is a mapping from the sigma algebra of events defined by the sample space $S$ to the interval [0,1], that is, $P: \boldsymbol{F} \to [0,1]$, which satisfies the following axioms:

1. For any event $A$ defined on $S$, it is the case that $P[A] \geq 0$.
2. $P[S] = 1$.
3. If $A$ and $B$ are mutually exclusive events, then

$$P[A \cup B] = P[A] + P[B] \tag{4.5}$$

If the sample space contains an infinite number of elements, then the following axiom is also required:

4. For all events $A_1, \ldots$ defined on $S$ satisfying the condition $A_i \cap A_j = \emptyset$, for $i \neq j$, the following relationship holds:

$$P\left[\bigcup_{i=1}^{\infty} A_i\right] = \sum_{i=1}^{\infty} P[A_i] \tag{4.6}$$

From these axioms, the following properties can be proved.

**Theorem 4.1 Properties of the Probability Function**
For all events $A$ and $B$, which are subsets of the samples space $S$, it is the case that:

1. $P[\emptyset]=0$.
2. $P[A]\leq 1$.
3. $P[A^C]=1-P[A]$.
4. If $A\subseteq B$, then $P[A]\leq P[B]$.
5. $P[A\cup B]=P[A]+P[B]-P[A\cap B]$.
   This relationship implies $P[A\cup B]\leq P[A]+P[B]$.
6. For $A_1, \ldots, A_N$ belonging to the sample space $S$, it is the case that

$$P\left[\bigcup_{i=1}^{N} A_i\right]=\sum_{i=1}^{N} P[A_i]-\sum_{i,j=1,\, i<j}^{N} P[A_i\cap A_j]+\cdots + (-1)^{N+1}P[A_1\cap\ldots\cap A_N] \tag{4.7}$$

**Proof**
These results follow from the axioms in a straightforward manner.

**Definition: Probability Space** The set of $\{S, \boldsymbol{F}, P\}$ is called the probability space.

***4.2.3.1 Defining Probability of Experimental Outcomes*** By definition, $S$ is the set of possible experimental outcomes defined by an experiment, and each experimental outcome has a defined probability. The following two definitions are fundamental to the analysis related to random variables and apply, respectively, to the cases where the sample space comprises a countable and an uncountable number of outcomes.

**Definition: Probability Mass Function** For the countable case, a probability mass function, $p_\Omega$, defines the probability of specific outcomes:

$$p_\Omega(\omega_i)=P[\text{experimental outcomes is } \omega_i] \tag{4.8}$$

**Definition: Probability Density Function** For the uncountable case, a probability density function, $f_\Omega$, defines the probability that an outcome, assumed to be a real number, is in a small interval:

$$f_\Omega(\omega)d\omega\approx P[\text{experimental outcome is on the interval } [\omega,\omega+d\omega)] \tag{4.9}$$

assuming $d\omega$ is suitably small.

The probability mass, and probability density, functions are considered in detail later in the context of the discussion on random variables.

***4.2.3.2 Notation*** The probability operator is defined on subsets of the sample space $S$. For the case of a countable number of outcomes, the sample space can be defined as

$$S=\{\omega_1,\ldots,\omega_i,\ldots\} \tag{4.10}$$

and $P[\{\omega_1\}]$, $P[\{\omega_2\}]$, etc. are well defined. For notational convenience, it is usual to write $P[\omega_1]$ rather than $P[\{\omega_1\}]$.

### 4.2.4 Conditional Probability

Consider an experiment that is repeated $N$ times and performed under conditions denoted hypothesis $H$. Assume:

Event $A$ occurs $N_A$ times.
Event $B$ occurs $N_B$ times.
Event $A \cap B$ occurs $N_{AB}$ times.

The relative frequency of the event $A \cap B$ is $N_{AB}/N$ and can be written as

$$\frac{N_{AB}}{N} = \frac{N_{AB}}{N_B} \cdot \frac{N_B}{N} \qquad N_B \neq 0 \tag{4.11}$$

As $N \to \infty$ and with convergent ratios, it is reasonable to infer and define

$$P[A \cap B/H] = P[A/(B \text{ and } H)] \cdot P[B/H] \tag{4.12}$$

where $P[A/(B \text{ and } H)]$ is the probability of $A$ assuming conditions, as specified by $B$ and $H$, have occurred. As the conditions specified by $H$ are common to all probabilities, it is usual to omit them and to write

$$P[A \cap B] = P[A/B] \cdot P[B] \tag{4.13}$$

It is common to use the notation $P[AB]$ for $P[A \cap B]$.

**Definition: Conditional Probability** The probability of the event $A$ occurring assuming occurrence of event $B$, that is, the conditional probability of event $A$ assuming the occurrence of event $B$, is denoted $P[A/B]$. Rearranging $P[A \cap B] = P[A/B] \cdot P[B]$ yields

$$P[A/B] = \frac{P[A \cap B]}{P[B]} \qquad P[B] \neq 0 \tag{4.14}$$

A conditional probability is a *standard* probability but is defined under modified conditions.

**Definition: Multiplication Rule for Probabilities** The result $P[A \cap B] = P[A/B] \cdot P[B]$ is called the *multiplication rule* for probabilities.

The multiplication rule for probabilities cannot be proved—it is true by definition. However, it has the empirical justification noted earlier.

#### *4.2.4.1 Theorem of Total Probability*

**Theorem 4.2 Theorem of Total Probability**
If the set of events $\{B_1, B_2, \ldots\}$ define a partition of the sample space of an experiment, then the probability of an event $A$ is given by

$$P[A] = \sum_{i=1}^{\infty} P[A/B_i]P[B_i] \tag{4.15}$$

**Proof**
As $\{B_1, B_2, \ldots\}$ form a partition of the sample space, it follows that the set $A$ can be written as a union of disjoint events

$$A = [A \cap B_1] \cup [A \cap B_2] \cup \ldots \tag{4.16}$$

The stated result follows from the fact that the probabilities of disjoint events are additive.

***4.2.4.2 Bayes' Theorem*** Bayes' theorem finds widespread application.

**Theorem 4.3 Bayes' Theorem**
The probability of the conditional event $B_i/A$ can be written in the form

$$P[B_i/A] = \frac{P[A/B_i]P[B_i]}{\sum_{k=1}^{N} P[A/B_k]P[B_k]} \tag{4.17}$$

assuming the set of events $\{B_1, \ldots, B_N\}$ form a partition of the sample space.

**Proof**
The proof follows from conditional probability theory and the Theorem of Total Probability.

### 4.2.5 Independent Events

It is common for experiments to comprise of unrelated stages or parts. Such stages, or parts, are *independent* and result in independence being an outcome of the structure of the experiment. For example, consider an experiment of throwing a dice twice. The outcome of the second throw is *independent* of the outcome of the first throw.

Alternatively, consider a two-part experiment where the outcomes of the second part are *independent* of the first part. If $A$ denotes an event based solely on the first part, and $B$ denotes an event based solely on the second part, then

$$P[B/A] = P[B] \tag{4.18}$$

Thus, knowing that $A$ has occurred in the first part of the experiment does not affect the probability of the event $B$ occurring in the second part of the experiment. Similarly, $P[A/B] = P[A]$. The conditional probability result

$$P[AB] = P[B/A]P[A] \tag{4.19}$$

then implies

$$P[AB] = P[A]P[B] \tag{4.20}$$

**Definition: Independent Events** Two events, $A$ and $B$, are statistically independent, or independent for short, if

$$P[AB] = P[A]P[B] \tag{4.21}$$

#### *4.2.5.1 Generalization*

**Definition: Independence: General Case** The events $A_1, \ldots, A_N$ are independent if all subcollections of events

$$A_{i_1} \cap A_{i_2} \cap \cdots \cap A_{i_k} \quad i_1, \ldots, i_k \in \{1, \ldots, N\} \tag{4.22}$$

are independent and arise with a probability

$$P[A_{i_1} \cap A_{i_2} \cap \cdots \cap A_{i_k}] = P[A_{i_1}] \cdot \ldots \cdot P[A_{i_k}] \tag{4.23}$$

***4.2.5.2 Independence of Elementary Events*** Because of its importance, the following is stated as being axiomatic.

**Theorem 4.4 Elementary Events Are Independent**
A specific elementary event is independent of all other elementary events.

**Proof**
This is taken as being axiomatic. A rationale is found in the independence of trials of an experiment. For example, in two consecutive trials of an experiment, the probability of the elementary events of $\omega_1$ and $\omega_2$ occurring in sequence is

$$P[\omega_1 \text{ followed by } \omega_2] = P[\omega_1]P[\omega_2] \tag{4.24}$$

### 4.2.6 Countable and Uncountable Sample Spaces

The nature of the experiment determines the nature of the sample space. The following distinctions readily arise:

a. Discrete and finite sample space:

$$S = \{\omega_1, \ldots, \omega_N\} \text{ where } P[\omega_i] = p_i \text{ and } \sum_{i=1}^{N} p_i = 1 \tag{4.25}$$

b. Discrete and infinite, but countable, sample space:

$$S=\{\omega_1,\ldots,\omega_i,\ldots\} \text{ where } P[\omega_i]=p_i \text{ and } \sum_{i=1}^{\infty} p_i=1 \tag{4.26}$$

c. Continuous and uncountable sample space: A continuous sample space arises when the possible outcomes from an experiment are uncountable and form a continuum. For example,

$$S=\{\omega_i:\ \omega_i \in \boldsymbol{R}\}\quad S=\{\omega_i:\ \omega_i \in [0,1]\} \tag{4.27}$$

With an infinite number of outcomes, the probability of an individual *point* outcome (an outcome consistent with a set with zero measure) is zero. Finite probabilities are associated with intervals or collections of outcomes with non-zero length or *measure*. For example,

$$P[\omega_i=0.1]=0 \quad P[\omega_i \in [0.1,0.2]]=p_i \quad p_i>0 \tag{4.28}$$

## 4.3 THE RANDOM VARIABLE

The outcome of an experiment, in general, is not a number and, for example, the sample space could consist of the outcomes $S=\{\text{success, failure}\}$. For such cases, it is useful to associate a number with each outcome, for example, success $=1$, failure $=0$, and such an association defines a random variable.

**Definition: Random Variable** When a number is associated with each experimental outcome $\omega$ arising from an experiment, a random variable $X$ is defined. Formally,

$$X: S \rightarrow S_X \tag{4.29}$$

where $S$ is the sample space of experimental outcomes and $S_X$ is the sample space of outcomes of the random variable $X$, that is, the range of $X$. By definition,

$$S_X=\{X(\omega): \omega \in S\} \tag{4.30}$$

For the countable case, and using the notation $x_i=X(\omega_i)$, the sample space $S_X$ can be written as

$$S_X=\{x_i: x_i=X(\omega_i), \omega_i \in S\} \tag{4.31}$$

The mapping defined by a random variable $X$ is illustrated in Figure 4.1.

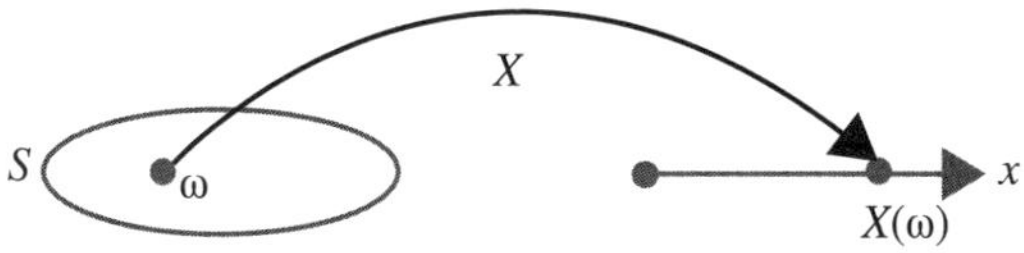


**FIGURE 4.1** Illustration of the mapping defined by a random variable.

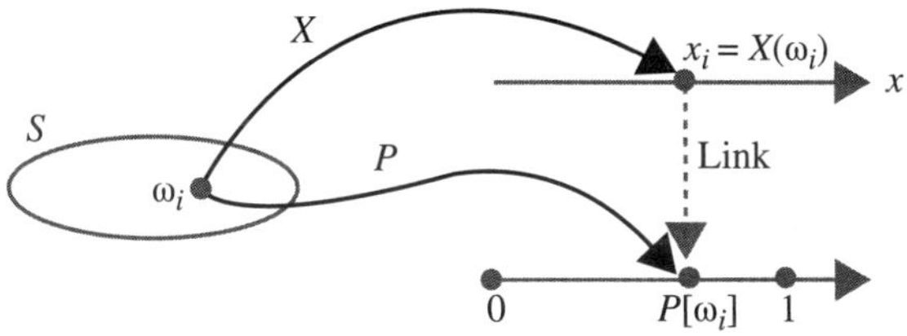


**FIGURE 4.2** Illustration of the mapping associated with a random variable for the case where each experimental outcome is associated with a distinct real number.

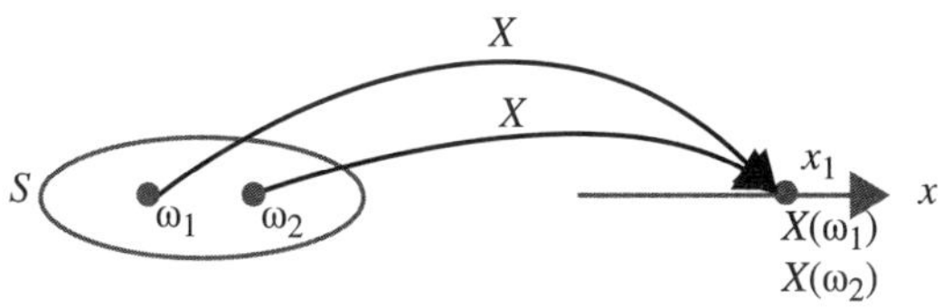


**FIGURE 4.3** Illustration of the mapping associated with a random variable for the general case where two experimental outcomes map to the same number.

### 4.3.1 Notes

For the specific case where the random variable associates a distinct value $X(\omega)$ with each outcome $\omega$ in the sample space, the probability of the outcome $X(\omega)$ equals the probability of the outcome $\omega$. This case is illustrated in Figure 4.2. Mathematically,

$$P[X(\omega_i) = x_i] = P[\omega_i] \tag{4.32}$$

In general, a random variable $X$ assigns the same number to two or more experimental outcomes as illustrated in Figure 4.3. For the situation illustrated in Figure 4.3, it is the case that

$$P[X(\omega) = x_1] = P[\omega_1] + P[\omega_2] \tag{4.33}$$

It is clear that the probability of occurrence of a specific outcome for a random variable is dependent on the probability mass function, $p_\Omega$, or probability density function, $f_\Omega$, associated with the experimental outcomes.

**TABLE 4.1 Forms for the sample space of a random variable**

| $S$ | Sample space $S_X$ |
|---|---|
| $\{\omega_1, \omega_2, \ldots\}$ | $\{X(\omega_1), X(\omega_2), \ldots\}$ |
| $\left\{\begin{array}{l}(\omega_{11},\ldots,\omega_{1N}),\\ (\omega_{21},\ldots,\omega_{2N}),\ldots\end{array}\right\}$ | $\left\{\begin{array}{l}X((\omega_{11},\ldots,\omega_{1N})),\\ X((\omega_{21},\ldots,\omega_{2N})),\ldots\end{array}\right\}$ |
| $\left\{\begin{array}{l}((\omega_{111},\ldots,\omega_{11M}),\ldots,\\ (\omega_{1N1},\ldots,\omega_{1NM})),\ldots\end{array}\right\}$ | $\left\{\begin{array}{l}X(((\omega_{111},\ldots,\omega_{11M}),\ldots,\\ (\omega_{1N1},\ldots,\omega_{1NM}))),\ldots\end{array}\right\}$ |

### 4.3.2 Sample Spaces for Random Variables

Consider a random variable $X$:

$$X : S \rightarrow S_X \quad S_X = \{X(\omega) : \omega \in S\} \tag{4.34}$$

For the case of a 1 : 1 structural mapping between experimental outcomes and numerical values of the associated random variable, the sample spaces, as detailed in Table 4.1, are indicative of those that arise.

### 4.3.3 Random Variable Based on Experimental Outcomes

For the case where the experimental outcomes are numerical values, a random variable $X : S \rightarrow S_X$ can be defined based directly on the distinct experimental outcomes according to $X(\omega) = \omega$. For this case, the sample space $S_X$ is

$$S_X = S = \{\omega_1, \omega_2, \ldots\} \tag{4.35}$$

and similarly for cases where the experiment yields a vector or matrix of numerical values for each trial.

For this case, the probability mass function, or density function, of the random variable $X$ is identical to that of the probability mass function, $p_\Omega$, or the probability density function, $f_\Omega$, associated with the experiment.

## 4.4 DISCRETE AND CONTINUOUS RANDOM VARIABLES

Random variables can be classified as being discrete, continuous, or a mixture of continuous and discrete. The former two types are the most common and dominate random variable theory.

### 4.4.1 Discrete Random Variables

**Definition: Discrete Random Variable** A discrete random variable is a random variable defined on a countable sample space.

Consistent with a countable sample space, a discrete random variable takes on a countable number of values.

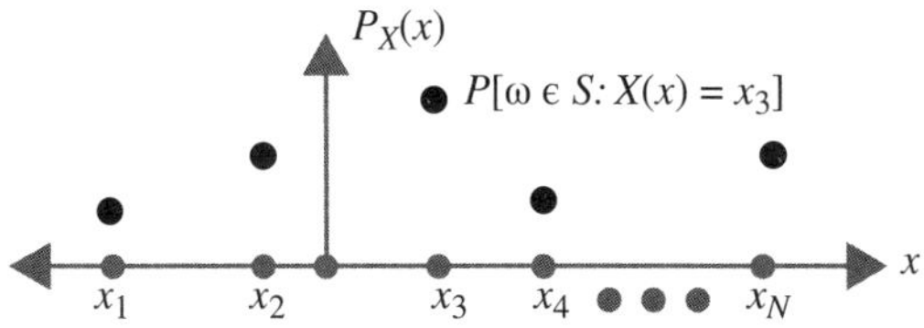


**FIGURE 4.4** Illustration of the probability mass function of a discrete random variable for the case of a finite sample space: $S_X = \{x_1, \ldots, x_N\}$.

**Definition: Probability Mass Function (PMF)** The probability mass function of a discrete random variable $X$, denoted $p_X$, is a mapping from the sample space of the random variable to the interval [0,1] according to

$$p_X : S_X \to [0,1] \quad p_X(x) = P[\omega \in S : X(\omega) = x] \tag{4.36}$$

The probability $P[\omega \in S:\ X(\omega) = x]$ is written, for notational convenience, as $P[X = x]$, that is, $p_X(x) = P[X = x]$.

For the case of a sample space with a finite number of values, the probability mass function has the form illustrated in Figure 4.4.

***4.4.1.1 Properties of Probability Mass Function*** For the case where the random variable $X$ defines the numbers $\{x_1, \ldots, x_i, \ldots\}$, the following properties hold:

1. $p_X(x) \geq 0 \quad x \in \{x_1, \ldots, x_i, \ldots\}$
2. $\sum_{i=1}^{\infty} p_X(x_i) = 1$

### 4.4.2 Continuous Random Variables

**Definition: Continuous Random Variable** A random variable $X$ is of continuous type, that is, a continuous random variable, if it is based on an experiment with an uncountable number of outcomes, its range is a subset of $\boldsymbol{R}$, and the probability of a specific value is zero, that is, $P[X = x] = 0$.

For a continuous random variable $X$, the probability that $X$ takes on a value in the interval $(x - dx/2, x + dx/2)$ is finite and is generally nonzero. This probability, as $dx$ becomes increasingly small, is defined as $f_X(x)dx$ where $f_X$ is the probability density function of $X$.

**Definition: Probability Density Function (PDF)** The probability density function of a continuous random variable $X$ is a mapping from the sample space $S_X$ to the positive real line with the definition

$$f_X : S_X \to \boldsymbol{R}^+ \quad f_X(x) = \lim_{dx \to 0} \frac{P[X \in (x - dx/2, x + dx/2)]}{dx} \tag{4.37}$$

For $dx$ suitably small,

$$f_X(x)dx \approx P[x-dx/2 < X < x+dx/2] \tag{4.38}$$

It is assumed that $f_X$ has, at most, a finite number of discontinuities and is not impulsive for any values in its domain. These properties are consistent with $P[X=x]=0$ for all $x \in \boldsymbol{R}$.

***4.4.2.1 Frequency Interpretation of PDF*** If, in $n$ trials of an experiment, the number of outcomes of a random variable $X$ that fall in the interval $(x-dx/2, x+dx/2)$ equals $n_x$, then $f_X(x)dx \approx n_x/n$ for $n$ suitably large.

### 4.4.3 Cumulative Distribution Function

The cumulative distribution function specifies the probability that a random variable takes on a value in the interval $(-\infty, x]$.

**Definition: Cumulative Distribution Function (CDF)** The cumulative distribution function of a random variable $X$, denoted $F_X$, is a mapping from the sample space of $X$ to the interval [0,1] and is defined according to

$$F_X : S_X \to [0,1] \quad F_X(x) = P[\omega \in S : X(\omega) \le x] \tag{4.39}$$

For notational convenience, the cumulative distribution function is written as

$$F_X(x) = P[X \le x] \tag{4.40}$$

For a discrete random variable,

$$F_X(x) = \sum_{y:y \le x} P[X=y] = \sum_{y:y \le x} p_X(y) \tag{4.41}$$

For a continuous random variable,

$$F_X(x) = P[X \le x] = \int_{-\infty}^{x} f_X(\lambda)d\lambda \tag{4.42}$$

which follows as

$$P[X \in (x-dx/2, x+dx/2)]\big|_{dx \to 0} = f_X(x)dx \tag{4.43}$$

***4.4.3.1 Properties of Cumulative Distribution Function*** The following properties hold for the cumulative distribution function:

1. $0 \le F_X(x) \le 1$.
2. $F_X(-\infty) = 0$ and $F_X(\infty) = 1$.

3. $P[x_1 < x \le x_2] = F_X(x_2) - F_X(x_1)$.
4. $F_X$ is a monotonically increasing function, that is,

$$x_1 < x_2 \Rightarrow F_X(x_1) < F_X(x_2) \tag{4.44}$$

For the continuous random variable case, $F_X$ is a continuous function (this follows from the assumption that $f_X$ has at most a finite number of discontinuities and is not impulsive for any values in its domain) and

$$F_X(x) = \int_{-\infty}^{x} f_X(\lambda)d\lambda \quad f_X(x) = \frac{d}{dx}F_X(x) \tag{4.45}$$

The latter result holds at all points where $F_X$ is differentiable.

## 4.5 STANDARD RANDOM VARIABLES

The number of commonly used random variables is large. The discrete uniform, the Bernoulli, the Binomial, the Poisson, and the geometric are common discrete random variables. Common continuous random variables include the uniform, the Gaussian, the exponential, and the Rayleigh. The random variables that are used in subsequent sections, and chapters, are detailed in the reference section at the end of the book.

## 4.6 FUNCTIONS OF A RANDOM VARIABLE

A random variable $X$ associates a number with each experimental outcome $\omega$ of an experiment, and the values $X$ takes on define the sample space, $S_X$, for the random variable. If there is a function, $g$, which maps each number in the sample space of the random variable $X$ to a new subset of the real line, denoted $S_Y$, then a new random variable $Y$ has been defined according to $Y(\omega) = g(X(\omega))$. These mappings are illustrated in Figure 4.5.

Consistent with the illustration in Figure 4.5, the random variable $Y$ could have been defined directly on the set $S$ of experimental outcomes. However, in many

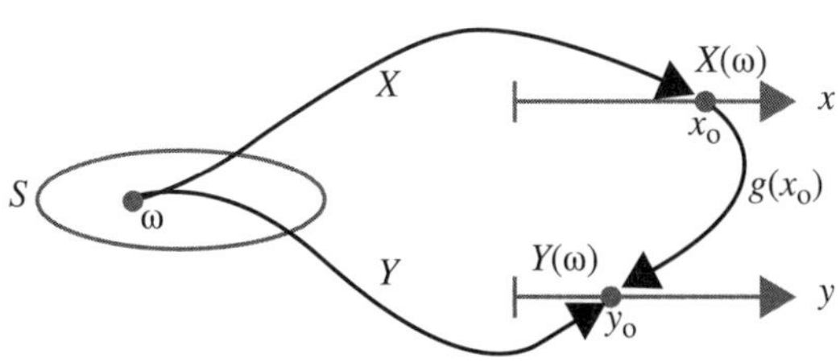


**FIGURE 4.5** Illustration of mappings associated with a random variable that is a function of another random variable.

instances, the outcomes of a random variable, $X$, are observed, and then algebraic operations on the outcomes are undertaken consistent with a mapping as defined by a function $g$. Accordingly, it is natural to consider functions of a random variable.

**Definition: Function of a Random Variable** When a function $g$ operates on each outcome $X(\omega)$ of a random variable $X$, then a function of the random variable is defined. Formally,

$$g : S_X \rightarrow S_Y \tag{4.46}$$

where $S_Y$ is the sample space of values defined by the function.

For notational convenience, $Y = g(X)$ is used instead of $Y(\omega) = g(X(\omega))$.

## 4.7 EXPECTATION

The expectation operator allows the precise definition of various statistical properties of a random variable including the mean and variance.

**Definition: Expectation** The expectation of a random variable $X$, denoted $\boldsymbol{E}[X]$, is a weighted summation over all possible outcomes of the random variable with the weighting being given by the probability of the outcome. For a discrete random variable, with a sample space $S_X = \{x_1, x_2, \ldots\}$, the expectation is defined as

$$\boldsymbol{E}[X] = \sum_{i=1}^{\infty} x_i P[X = x_i] \tag{4.47}$$

For a continuous random variable, with outcomes defined on a set $S_X$, the expectation is defined according to

$$\boldsymbol{E}[X] = \int_{S_X} x f_X(x) dx \tag{4.48}$$

For the continuous random variable case, the expectation definition is a weighted summation definition, and this can be seen by considering the definition of $f_X(x)dx$, consistent with Equation 4.38, and the Riemann sum

$$\boldsymbol{E}[X] = \lim_{dx \to 0} \sum_{i=\lfloor x_1/dx \rfloor}^{\lfloor x_2/dx \rfloor} idx \cdot P[idx \leq X < (i+1)dx] \tag{4.49}$$

assuming $S_X = [x_1, x_2]$.

### 4.7.1 Mean, Variance, and Moments of a Random Variable

The mean and variance are first-order measures of the nature of a random variable.

**Definition: Mean of a Random Variable** By definition, the mean of a random variable $X: S \rightarrow S_X$ is the first moment, is denoted $\mu_X$, and is defined according to

$$\mu_X = \boldsymbol{E}[X] = \begin{cases} \sum_{i:x_i \in S_X} x_i P[X = x_i] & \text{discrete r.v.} \\ \int_{S_X} x f_X(x) dx & \text{continuous r.v.} \end{cases} \tag{4.50}$$

**Definition: Variance of a Random Variable** By definition, the variance of a random variable $X: S \rightarrow S_X$ is the second central moment, is denoted $\sigma_X^2$ or var($X$), and is defined according to

$$\text{var}(X) = \boldsymbol{E}\left[(X - \boldsymbol{E}[X])^2\right] = \begin{cases} \sum_{i:x_i \in S_X} (x_i - \boldsymbol{E}[X])^2 P[X = x_i] & \text{discrete r.v.} \\ \int_{S_X} (x - \boldsymbol{E}[X])^2 f_X(x) dx & \text{continuous r.v.} \end{cases} \tag{4.51}$$

It can readily be shown that the variance is equal to

$$\sigma_X^2 = \boldsymbol{E}\left[X^2\right] - \mu_X^2 \tag{4.52}$$

The mean and variance are the first measures of the sequences of moments and central moments that can be defined.

**Definition: Moment and Central Moment** By definition, the $k$th moment of a random variable $X: S \rightarrow S_X$ is

$$\boldsymbol{E}\left[X^k\right] = \begin{cases} \sum_{i:x_i \in S_X} x_i^k P[X = x_i] & \text{discrete r.v.} \\ \int_{S_X} x^k f_X(x) dx & \text{continuous r.v.} \end{cases} \tag{4.53}$$

By definition, the $k$th central moment of a random variable $X: S \rightarrow S_X$ is

$$\boldsymbol{E}\left[(X - \boldsymbol{E}[X])^k\right] = \begin{cases} \sum_{i:x_i \in S_X} (x_i - \boldsymbol{E}[X])^k P[X = x_i] & \text{discrete r.v.} \\ \int_{S_X} (x - \boldsymbol{E}[X])^k f_X(x) dx & \text{continuous r.v.} \end{cases} \tag{4.54}$$

***4.7.1.1 Notation*** To simplify notation in the aforementioned definitions, the limits of $i = 1$ to $i = \infty$ are often used for the summation, while the limits of $-\infty$ and $\infty$ are often used for the integral.

### 4.7.2 Expectation of a Function of a Random Variable

**Theorem 4.5 Expectation of a Function of a Random Variable**
If $Y = g(X)$, then the expected value of $Y$ is

$$\boldsymbol{E}[Y] = \begin{cases} \sum_{i=1}^{\infty} g(x_i) p_X(x_i) & \text{discrete r.v.} \\ \int_{-\infty}^{\infty} g(x) f_X(x) dx & \text{continuous r.v.} \end{cases} \tag{4.55}$$

**Proof**
The proof follows from the definition of the expectation operator as the summation over each value that the random variable can take on multiplied by the probability that it takes on that value.

### 4.7.3 Characteristic Function

The characteristic function of a random variable facilitates analysis, for example, when determining the probability density function of a summation of identical random variables.

**Definition: Characteristic Function** By definition, the characteristic function of a random variable $X$ is

$$\Phi_X(t) = \boldsymbol{E}\left[e^{jtX}\right] = \begin{cases} \sum_{i=1}^{\infty} e^{jtx_i} p_X(x_i) & \text{discrete r.v.} \\ \int_{-\infty}^{\infty} e^{jtx} f_X(x) dx & \text{continuous r.v.} \end{cases} \tag{4.56}$$

**Definition: Alternative Definition for Characteristic Function** An alternative definition for the characteristic function of a random variable $X$, for the continuous random variable case, is

$$\Phi_X(f) = \boldsymbol{E}\left[e^{j2\pi fX}\right] = \int_{-\infty}^{\infty} e^{j2\pi fX} f_X(x) dx \tag{4.57}$$

which is the inverse Fourier transform of the probability density function.

#### 4.7.3.1 Characteristic Function of a Gaussian PDF

**Theorem 4.6 Characteristic Function of a Gaussian PDF**
The characteristic function of a random variable $X$ with a Gaussian probability density function

$$f_X(x) = \frac{e^{-(x-\mu)^2/2\sigma^2}}{\sqrt{2\pi}\sigma} \tag{4.58}$$

is

$$\Phi_X(f) = \boldsymbol{E}\left[e^{j2\pi fX}\right] = e^{j2\pi f\mu - 2\pi^2 f^2\sigma^2} \tag{4.59}$$

Also,

$$\boldsymbol{E}\left[e^{kX}\right] = e^{k\mu + k^2\sigma^2/2} \tag{4.60}$$

**Proof**
The proof is detailed in Appendix 4.A.

#### 4.7.3.2 Characteristic Function after Linear Transformation

**Theorem 4.7 Characteristic Function after Linear Transformation**
If $Y = aX + b$, then

$$\Phi_Y(f) = e^{j2\pi fb}\Phi_X[af] \tag{4.61}$$

**Proof**
By definition,

$$\begin{aligned}\Phi_Y(f) &= \boldsymbol{E}[\exp(j2\pi fY)] = \boldsymbol{E}[\exp(j2\pi f[aX+b])] \\ &= e^{j2\pi fb}\boldsymbol{E}[\exp(j2\pi faX)] = e^{j2\pi fb}\Phi_X[af]\end{aligned} \tag{4.62}$$

#### 4.7.3.3 Characteristic Function of a Sum of Random Variables

**Theorem 4.8 Characteristic Function of a Weighted Sum of RVs**
The characteristic function of a weighted sum, $S_N$, of independent random variables $X_1, \ldots, X_N$ and defined according to

$$Z = S_N = \sum_{i=1}^{N} Z_i = \sum_{i=1}^{N} w_i X_i \qquad Z_i = w_i X_i, w_i \in \boldsymbol{R} \tag{4.63}$$

is

$$\Phi_{S_N}(f) = \prod_{i=1}^{N} \Phi_{X_i}(w_i f) \tag{4.64}$$

where $\Phi_{X_i}$ is the characteristic function of $X_i$.

**Proof**
The proof is detailed in Appendix 4.B.

**Theorem 4.9 Characteristic Function of a Random Sum of RVs**
Consider the random sum of independent random variables defined according to

$$S_M = \sum_{i=1}^{M} X_i \tag{4.65}$$

where $M$ is a discrete random variable taking on positive values $m_1, m_2, \ldots$, with probabilities $p_1, p_2, \ldots$, and $X_1, X_2, \ldots$ is a sequence of independent random variables, all of which are independent of $M$. The characteristic function of $S_M$ is

$$\Phi_{S_M}(f) = \sum_{k=1}^{\infty} p_k \prod_{i=1}^{m_k} \Phi_{X_i}(f) \tag{4.66}$$

where $\Phi_{X_i}$ is the characteristic function of $X_i$. For the case of identical and independent random variables,

$$\Phi_{S_M}(f) = \sum_{k=1}^{\infty} p_k \Phi_X^{m_k}(f) \tag{4.67}$$

**Proof**
The proof is detailed in Appendix 4.C.

***4.7.3.4 PDF of a Random Sum of Gaussian Random Variables*** Consider the random sum of Gaussian random variables defined according to

$$S_M = \sum_{i=1}^{M} X_i \tag{4.68}$$

where $M$ is a discrete random variable taking on the positive values $m_1, m_2, \ldots$, with probabilities $p_1, p_2, \ldots, X_1, X_2, \ldots$ is a sequence of independent Gaussian random variables and the probability density function of $X_i$ is

$$f_{X_i}(x) = \frac{1}{\sqrt{2\pi\sigma_i^2}} \cdot e^{-(x-\mu_i)^2/2\sigma_i^2} \tag{4.69}$$

**Theorem 4.10 PDF of a Random Sum of Independent Gaussians**
The characteristic function of $S_M$ is

$$\Phi_{S_M}(f) = \sum_{k=1}^{\infty} p_k e^{j2\pi f \,\mathrm{mean}_k - 2\pi^2 f^2 \mathrm{var}_k} \tag{4.70}$$

where

$$\text{mean}_k = \sum_{i=1}^{m_k} \mu_i \quad \text{var}_k = \sum_{i=1}^{m_k} \sigma_i^2 \tag{4.71}$$

The probability density function of $S_M$ is

$$f_{S_M}(x) = \sum_{k=1}^{\infty} p_k \frac{1}{\sqrt{2\pi \text{var}_k}} \cdot e^{-(x-\text{mean}_k)^2/2\text{var}_k} \tag{4.72}$$

Thus, the probability density function is a weighted sum of Gaussian probability density functions.

**Proof**
From Theorem 4.9, the characteristic function of $S_M$ is

$$\begin{aligned} \Phi_{S_M}(f) &= \sum_{k=1}^{\infty} p_k \prod_{i=1}^{m_k} \Phi_{X_i}(f) \qquad \Phi_{X_i}(f) = e^{j2\pi f \mu_i - 2\pi^2 f^2 \sigma_i^2} \\ &= \sum_{k=1}^{\infty} p_k e^{j2\pi f \text{mean}_k - 2\pi^2 f^2 \text{var}_k} \end{aligned} \tag{4.73}$$

The probability density function then follows from Theorem 4.6.

***4.7.3.5 PDF of a Sum of Gaussian Random Variables*** The important result, which is a special case of the general result stated in Theorem 4.11, then follows.

**Theorem 4.11 Sum of Independent Gaussian Random Variables**
The probability density function of a sum of independent Gaussian random variables

$$S_N = \sum_{i=1}^{N} X_i \quad f_{X_i}(x) = \frac{1}{\sqrt{2\pi\sigma_i^2}} \cdot e^{-(x-\mu_i)^2/2\sigma_i^2} \tag{4.74}$$

is a Gaussian probability density function

$$f_{S_N}(x) = \frac{1}{\sqrt{2\pi \text{var}}} \cdot e^{-(x-\text{mean})^2/2\text{var}} \tag{4.75}$$

where

$$\text{mean} = \sum_{i=1}^{N} \mu_i \quad \text{var} = \sum_{i=1}^{N} \sigma_i^2 \tag{4.76}$$

***4.7.3.6 Example*** Consider the random sum of independent Gaussian random variables

$$Z = \sum_{i=1}^{M} X_i \tag{4.77}$$

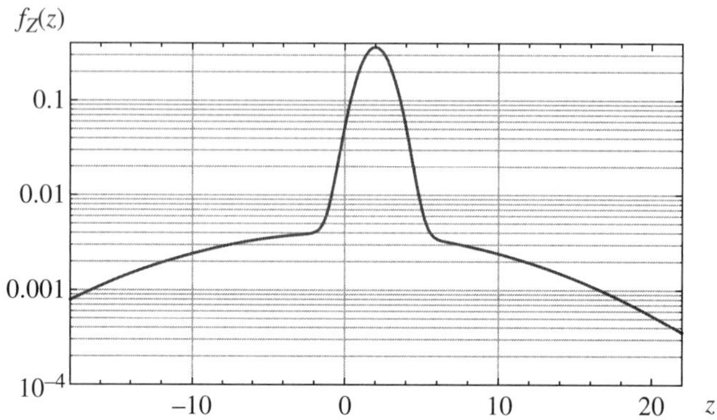


**FIGURE 4.6** Probability density function of a random sum of two independent random variables as specified in the text.

where $M \in \{1,2\}$, $P[M=1]=p_1=1-\varepsilon$, $P[M=2]=p_2=\varepsilon$, $\varepsilon=0.1$, $\mu_1=2$, $\mu_2=-2$, $\sigma_1^2=1$, and $\sigma_2^2=99$. Consistent with Theorem 4.10, the probability density function is

$$f_Z(z)=\sum_{k=1}^{2} p_k \cdot \frac{e^{-(z-\text{mean}_k)^2/2\,\text{var}_k}}{\sqrt{2\pi\,\text{var}_k}} \tag{4.78}$$

where $\text{mean}_1=\mu_1=2$, $\text{mean}_2=\mu_1+\mu_2=0$, $\text{var}_1=\sigma_1^2=1$, and $\text{var}_2=\sigma_1^2+\sigma_2^2=100$. This probability density function is shown in Figure 4.6.

## 4.8 GENERATION OF DATA CONSISTENT WITH DEFINED PDF

Modern mathematical software packages can generate random data consistent with many standard distributions. For non-standard distributions, a method is required to generate data consistent with a defined probability density function. The usual starting point is to generate a number, $u_o$, at random from the interval [0,1].

**Theorem 4.12 Transformation for Data Generation**
For a number $u_o$ from [0, 1] to be consistent with a random variable, with a cumulative distribution function $F_X$, the transformation

$$x_o = F_X^{-1}(u_o) \tag{4.79}$$

is required.

**Proof**
Consider the experiment of choosing a number, at random, from the interval [0, 1]. The sample space of this experiment is

$$S=\{\omega : \omega \in [0,1]\} \tag{4.80}$$

A random variable, $U$, with a uniform distribution, can be defined on $S$ according to $U(\omega)=\omega$. For a random variable $X$ defined on $S$, with a distribution function $F_X$, the following applies:

$$\begin{aligned} F_X(x_o) &= P[\omega \in S : X(\omega) \le x_o] \\ &= P[\omega \in S : U(\omega) \le u_o] \\ &= F_U(u_o) = u_o \end{aligned} \tag{4.81}$$

for some $u_o \in [0,1]$. The monotonicity of a cumulative distribution function guarantees the existence of $F_X^{-1}$ such that $x_o = F_X^{-1}(u_o)$.

### 4.8.1 Example

Consider a random variable with a probability density function, and a cumulative distribution function, defined according to

$$\begin{aligned} f_X(x) &= \frac{\pi}{2x_M} \cdot \sin\left[\frac{\pi x}{x_M}\right] \quad 0 \le x \le x_M \\ F_X(x) &= \frac{1}{2}\left[1 - \cos\left[\frac{\pi x}{x_M}\right]\right] \quad 0 \le x \le x_M \end{aligned} \tag{4.82}$$

The transformation required is

$$x_o = F_X^{-1}(u_o) = \frac{x_M}{\pi} \cdot \operatorname{acos}(1 - 2u_o) \quad u_o \in [0,1] \tag{4.83}$$

## 4.9 VECTOR RANDOM VARIABLES

The experiments, whether real or hypothesized, that underpin random phenomena lead to a variety of sample spaces $S$:

$$S = \{\omega_i : \text{as defined by distinct trials of experiment}\} \tag{4.84}$$

The $i$th trial of an experiment will result, depending on the nature of the random phenomenon being modelled, in a single outcome $\omega_i$, a vector of outcomes $\omega_i = (\omega_{i1}, \ldots, \omega_{iN})$, a matrix of outcomes $\omega_i = ((\omega_{i11}, \ldots, \omega_{i1M}), \ldots, (\omega_{iN1}, \ldots, \omega_{iNM}))$, etc. A vector of outcomes $\omega_i = (\omega_{i1}, \ldots, \omega_{iN})$ may arise from $N$ repeated trials of a subexperiment; a matrix of outcomes $\omega_i = ((\omega_{i11}, \ldots, \omega_{i1M}), \ldots, (\omega_{iN1}, \ldots, \omega_{iNM}))$ may arise from $N$ trials of a subexperiment where each subexperiment is a result of $M$ sub-subexperiments. The following are typical forms for sample spaces:

$$\begin{aligned} &S = \{\omega_i : i \in \mathbf{Z}^+\} \quad S = \{\omega_i : \omega_i = (\omega_{i1}, \ldots, \omega_{iN}), i \in \mathbf{Z}^+\} \\ &S = \{\omega_i : \omega_i = ((\omega_{i11}, \ldots, \omega_{i1M}), \ldots, (\omega_{iN1}, \ldots, \omega_{iNM})), i \in \mathbf{Z}^+\} \end{aligned} \tag{4.85}$$

When a number $x_i$, a vector of numbers $x_i = (x_{i1},\ldots,x_{iN})$, or a matrix of numbers $x_i = ((x_{i11},\ldots,x_{i1M}),\ldots,(x_{iN1},\ldots,x_{iNM}))$, as appropriate, is associated with each experimental outcome, a random variable $X : S \to S_X$ with $X(\omega_i) = x_i$ is defined in the usual manner. The corresponding sample spaces of the random variables are

$$\begin{aligned} &S_X = \{x_i : i \in \mathbf{Z}^+\} \quad S_X = \{x_i : x_i = (x_{i1},\ldots,x_{iN}), i \in \mathbf{Z}^+\} \\ &S_X = \{x_i : x_i = ((x_{i11},\ldots,x_{i1M}),\ldots,(x_{iN1},\ldots,x_{iNM})), i \in \mathbf{Z}^+\} \end{aligned} \tag{4.86}$$

### 4.9.1 Random Variable Defined Based on Experimental Outcomes

For the case where the experimental outcomes are numerical values, a random variable $X : S \to S_X$ can be defined based directly on the experimental outcomes according to $X(\omega) = \omega$. For this case, and with the notation $x_i = X(\omega_i)$, the sample space $S_X$ is $S_X = S = \{\omega_i : \omega_i \in S\}$. For such a case, it is usual to use the probability mass function, $p_\Omega$, or the probability density function, $f_\Omega$, of the experimental outcomes and not the corresponding function, $p_X$ or $f_X$, of the random variable $X$.

### 4.9.2 Vector Random Variables

Consider a random variable $X : S \to S_X$, $S = \{\omega_i : i \in \mathbf{Z}^+\}$, where $S_X$ comprises of vectors in $N$ dimensional space consistent with $X(\omega_i) = (x_{i1},\ldots,x_{iN})$. For this case, the random variable can be interpreted as a vector random variable.

**Definition: Vector Random Variable** A vector random variable $X$ of dimension $N$ is a vector of $N$ one-dimensional random variables

$$X = (X_1, X_2, \ldots, X_N) \tag{4.87}$$

where

$$X(\omega) = (X_1(\omega), X_2(\omega), \ldots, X_N(\omega)) \quad \omega \in S \tag{4.88}$$

The $k$th random variable is defined according to $X_k(\omega_i) = x_{ik}$.

A vector random variable, along with its component random variables, is illustrated in Figure 4.7 for the three-dimensional case.

It is useful to highlight the difficulty that often arises in determining the probability mass function, or the probability density function, for a vector random variable. For the three-dimensional case illustrated in Figure 4.7, the probability $P[X = (x_{i1}, x_{i2}, x_{i3})]$ equals $P[\omega_i]$ when $\omega_i$ is the only outcome that maps to $(x_{i1}, x_{i2}, x_{i3})$. However, in general, there are many outcomes in the sample space that map to the point $(x_{i1}, x_{i2}, x_{i3})$. For an arbitrary point $(x_1, x_2, x_3)$ and for the countable case,

$$P[\omega_i : X(\omega_i) = (x_1, x_2, x_3)] = \sum_{i : X(\omega_i) = (x_1, x_2, x_3)} P[\omega_i] \tag{4.89}$$

Evaluation of such probabilities is often difficult.

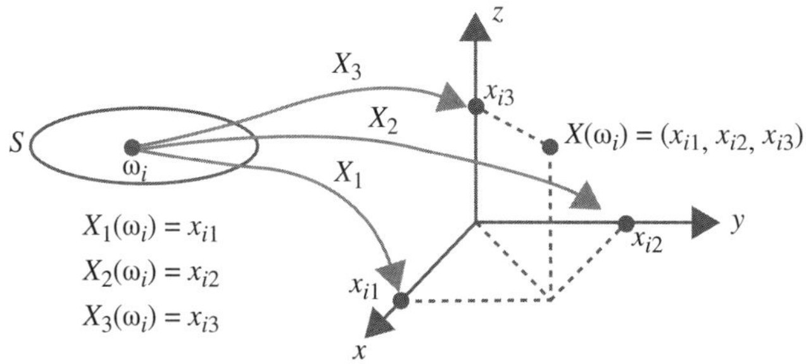


**FIGURE 4.7** Illustration of the component random variables of a vector random variable for the three-dimensional case.

### 4.9.3 Sample Space for Vector Random Variable

Often, the values defined by a vector random variable are utilized directly, and the experiment and sample space for the random phenomena underpinning the vector random variable are left implicit. For this case, the sample space $S_X$ of the random vector is defined directly and in the following manner:

$$S_X = \{x_i = (x_{i1}, x_{i2}, \ldots, x_{iN}) : \text{conditions}\} \tag{4.90}$$

The probability $P[X = x_i]$ associated with a specific outcome $x_i$ is usually clear.

## 4.10 PAIRS OF RANDOM VARIABLES

To establish the properties of multiple random variables, the starting point is to establish the properties for the vector random variable case of two dimensions, that is, properties for the case of $X = (X_1, X_2)$. Of importance is the generalization of the ways of characterizing a random variable for the one-dimensional case: the generalization of the cumulative distribution function, the probability mass function, and the probability density function.

### 4.10.1 Notation

For notational convenience, the vector pair of random variables $(X_1, X_2)$ is written as $(X, Y)$, and the notation $Z = (X, Y)$ is used.

The sample space for $Z = (X, Y)$ has the following general forms:

$$S_Z = \begin{cases} \{(x_i, y_j) : (x_i, y_j) \in R_{XY} \subseteq \mathbf{R}^2\} & \text{discrete r.v.} \\ \{(x, y) : (x, y) \in R_{XY} \subseteq \mathbf{R}^2\} & \text{continuous r.v.} \end{cases} \tag{4.91}$$

where the set $R_{XY}$ consists of a countable and an uncountable number of ordered pairs for, respectively, the discrete and continuous random variable cases.

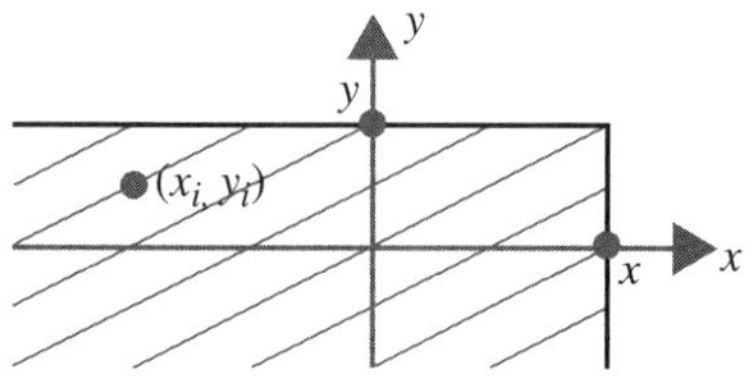


**FIGURE 4.8** Illustration of the region associated with $F_{XY}(x, y)$.

### 4.10.2 Joint Cumulative Distribution Function (Joint CDF)

Consider a pair of random variables $Z = (X, Y)$ defined on a sample space $S$. The generalization of the cumulative distribution function for the one-dimensional case is the joint cumulative distribution function.

**Definition: Joint Cumulative Distribution Function** The joint cumulative distribution function, $F_{XY}$, of two random variables $X$ and $Y$ is defined according to

$$F_{XY}(x, y) = P[\omega \in S : X(\omega) \le x, Y(\omega) \le y] \tag{4.92}$$

For notational convenience, the joint cumulative distribution function is written as

$$F_{XY}(x, y) = P[X \le x, Y \le y] \tag{4.93}$$

The interpretation of the joint cumulative distribution function is as follows: $F_{XY}(x, y)$ is the probability that the ordered pair $(x_i, y_i)$, arising from $i$th outcome of the experiment that defines $Z = (X, Y)$, is in the shaded region illustrated in Figure 4.8.

***4.10.2.1 Properties of Cumulative Distribution Function*** The cumulative distribution function has the following properties:

a. $F_{XY}(x, -\infty) = F_{XY}(-\infty, y) = 0.$

$$F_{XY}(\infty, \infty) = 1 \quad 0 \le F_{XY}(x, y) \le 1.$$

b. If $x_1 \le x_2$ and $y_1 \le y_2$, then $F_{XY}(x_1, y_1) \le F_{XY}(x_2, y_2)$.
c. $F_X(x) = F_{XY}(x, \infty)$ and $F_Y(y) = F_{XY}(\infty, y)$.
d. For the case of $x_1 \le x_2$ and $y_1 \le y_2$:

$$\begin{aligned} P[x_1 < X \le x_2, y_1 < Y \le y_2] = F_{XY}(x_2, y_2) - F_{XY}(x_1, y_2) \\ - F_{XY}(x_2, y_1) + F_{XY}(x_1, y_1) \end{aligned} \tag{4.94}$$

### 4.10.3 Joint Probability Mass Function

**Definition: Joint Probability Mass Function** For the discrete random variable case, the joint probability mass function $p_{XY}$ of $Z = (X, Y)$ is defined as

$$\begin{aligned} p_{XY}(x_i, y_j) &= P[\omega \in S : Z(\omega) = (x_i, y_j)] \\ &= P[\omega \in S : X(\omega) = x_i, Y(\omega) = y_j] \end{aligned} \tag{4.95}$$

and is the probability of occurrence of the set of experimental outcomes that result in $(x_i, y_j)$.

For notational convenience, $P[\omega \in S : X(\omega) = x_i, Y(\omega) = y_j]$ is written as $P[X = x_i, Y = y_j]$, and thus,

$$p_{XY}(x_i, y_j) = P[X = x_i, Y = y_j] \tag{4.96}$$

The joint probability mass function has the property

$$\sum_{i=1}^{\infty}\sum_{j=1}^{\infty} p_{XY}(x_i, y_j) = 1 \tag{4.97}$$

and for any subset $A$ of the sample space of $Z$,

$$P[A] = P[(x_i, y_j) : (x_i, y_j) \in A] = \sum_{i,j:(x_i, y_j) \in A}\sum p_{XY}(x_i, y_j) \tag{4.98}$$

### 4.10.4 Marginal Probability Mass Function

For the joint random variable case, the individual probability mass functions are still of interest, and it is useful to be able to determine these from the joint probability mass function.

**Definition: Marginal Probability Mass Function** For a random variable $Z = (X, Y)$, the probability mass functions for the random variables $X$ and $Y$, termed the marginal probability mass functions, are defined according to

$$p_X(x_i) = \sum_{j=1}^{\infty} p_{XY}(x_i, y_j) \quad p_Y(y_j) = \sum_{i=1}^{\infty} p_{XY}(x_i, y_j) \tag{4.99}$$

To determine $p_X(x_i)$, for example, the requirement is to sum the probabilities of all outcomes for $Y$ consistent with $X = x_i$.

### 4.10.5 Joint Probability Density Function

For the case of two continuous random variables $(X, Y)$ and consistent with the one-dimensional case, the probability of a precise value, $X = x$ and $Y = y$, equals zero for all values of $x$ and $y$. Nonzero probabilities are only associated with intervals or sets. The joint probability density function of two continuous random variables $(X, Y)$ can be defined in an analogous manner to the one-dimensional case.

**Definition: Joint Probability Density Function** The joint probability density function of the random variable pair $(X, Y)$, denoted $f_{XY}$, is defined according to

$$f_{XY}(x, y) = \lim_{dx \to 0, dy \to 0} \frac{P[X \in (x, x + dx), Y \in (y, y + dy)]}{dxdy} \tag{4.100}$$

and for $dx$ and $dy$ suitably small,

$$f_{XY}(x,y)dxdy \approx P[X \in (x, x+dx), Y \in (y, y+dy)] \tag{4.101}$$

***4.10.5.1 Relationship between PDF and CDF*** If $X$ and $Y$ are continuous random variables with a joint probability density function, $f_{XY}$, and a differentiable joint cumulative distribution function $F_{XY}$, then

$$f_{XY}(x,y) = \frac{\partial^2}{\partial y \partial x} F_{XY}(x,y) \quad F_{XY}(x,y) = \int_{-\infty}^{y} \int_{-\infty}^{x} f_{XY}(u,v)dudv \tag{4.102}$$

***4.10.5.2 Properties of Joint Probability Density Function*** The following properties hold for the joint probability density function:

$$f_{XY}(x,y) \geq 0 \quad x, y \in R \quad \int_{-\infty}^{\infty} \int_{-\infty}^{\infty} f_{XY}(x,y)dxdy = 1 \tag{4.103}$$

For any subset $A \subseteq \boldsymbol{R}^2$, it is the case that

$$P[(X,Y) \in A] = \iint_A f_{XY}(x,y)dxdy \tag{4.104}$$

### 4.10.6 Marginal Distribution and Density Functions

Consider a pair of continuous random variables $(X, Y)$. The marginal cumulative distribution function, $F_X$ and $F_Y$, can be determined from the joint cumulative distribution function according to

$$F_X(x) = F_{XY}(x, \infty) \quad F_Y(y) = F_{XY}(\infty, y) \tag{4.105}$$

The marginal probability density function, $f_X$ and $f_Y$, can be determined from the joint probability density function according to

$$f_X(x) = \int_{-\infty}^{\infty} f_{XY}(x,y)dy \quad f_Y(y) = \int_{-\infty}^{\infty} f_{XY}(x,y)dx \tag{4.106}$$

### 4.10.7 Linearity of Expectation Operator

The definition of the joint and marginal probability mass function, and the joint and marginal probability density function, allows the important linearity property of the expectation operator to be proved.

**Theorem 4.13 Linearity of Expectation Operator**
The expectation operator is a linear operator, that is,

$$\boldsymbol{E}[\alpha X + \beta Y] = \alpha \boldsymbol{E}[X] + \beta \boldsymbol{E}[Y] \quad \alpha, \beta \in \boldsymbol{R} \tag{4.107}$$

for any two arbitrary random variables $X$ and $Y$.

**Proof**

Consider the continuous random variable case and a random variable defined according to $Z = \alpha X + \beta Y$. By definition of the expectation operator,

$$\begin{aligned}
\boldsymbol{E}[Z] &= \boldsymbol{E}[\alpha X + \beta Y] = \int_{-\infty}^{\infty}\int_{-\infty}^{\infty} (\alpha x + \beta y) f_{XY}(x, y)\,dx\,dy \\
&= \alpha \int_{-\infty}^{\infty} x \left[ \int_{-\infty}^{\infty} f_{XY}(x, y)\,dy \right] dx + \beta \int_{-\infty}^{\infty} y \left[ \int_{-\infty}^{\infty} f_{XY}(x, y)\,dx \right] dy \\
&= \alpha \int_{-\infty}^{\infty} x f_X(x)\,dx + \beta \int_{-\infty}^{\infty} y f_Y(y)\,dy = \alpha \boldsymbol{E}[X] + \beta \boldsymbol{E}[Y]
\end{aligned} \tag{4.108}$$

### 4.10.8 Conditional Mass and Density Functions

In general, the random variable $Z : S \to S_Z$ where

$$Z = (X, Y) \qquad S_Z = \left\{ (x_i, y_j) : (x_i, y_j) \in R_{XY} \subseteq \boldsymbol{R}^2 \right\} \tag{4.109}$$

is such that the random variables $X$ and $Y$ are not independent, that is, the occurrence of $X = x_i$ contains information about the occurrence, or non-occurrence, of $Y = y_j$ and vice versa.

Consider a communication channel where one trial of an experiment is to send one bit of information, denoted $x$, and to recover this information, denoted $y$, at a receiver. The outcome of the trial is the ordered pair $(x, y)$ where $x, y \in \{0,1\}$ for binary data communication. For a good communication channel, it is expected that $y = 0$ is consistent with $x = 0$ and that $y = 1$ is consistent with $x = 1$, that is, reliable communication is consistent with the output data containing information about the sent, or input, data.

Characterization of the dependence between two random variables is important and conditional probability mass functions, and conditional probability density functions, are widely used. Understanding these functions is facilitated by a discussion of conditioning.

***4.10.8.1 Case 1: Conditioning on Subsets of Joint Sample Space*** Consider $Z = (X, Y)$, where the sample space of $Z$ is defined by Equation 4.109, and consider events that lead to the subsets $A, B \subseteq S_Z$. It follows from the conditional probability result $P[A/B] = P[AB]/P[B],\ \ P[B] \neq 0$, that

$$P[Z \in A / Z \in B] = \frac{P[Z \in A, Z \in B]}{P[Z \in B]} \qquad P[Z \in B] \neq 0 \tag{4.110}$$

The following notation is used:

$$P_Z(A/B) = P[Z \in A / Z \in B] \tag{4.111}$$

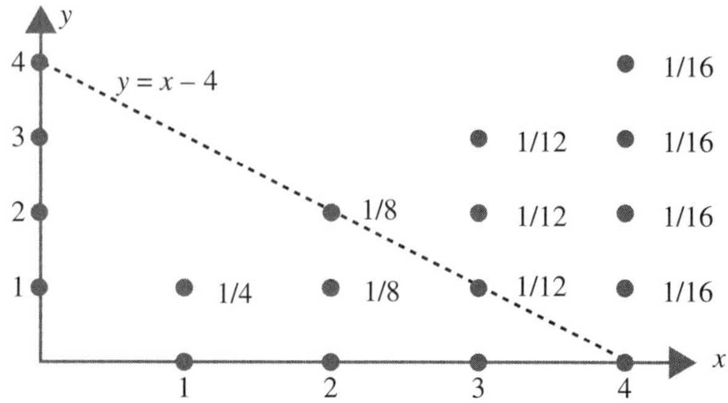


**FIGURE 4.9** Probability of outcomes of the defined random variable $Z$.

***4.10.8.2 Example*** Consider the case of a random variable $Z = (X, Y)$, $Z: S \to S_Z$, which defines the sample space

$$S_Z = \{(1,1),(2,1),(3,1),(4,1),(2,2),(3,2),(4,2),(3,3),(4,3),(4,4)\} \tag{4.112}$$

and where the probabilities of the outcomes are defined in Figure 4.9.

Consider the case of $B = \{(x,y) \in S_Z : x + y \le 4\}$, $A_1 = \{(1,1)\}$, and $A_2 = \{(2,2),(3,2)\}$ and the goal of determining $P_Z[A_1/B]$ and $P_Z[A_2/B]$. First,

$$B = \{(x,y) \in S_Z : x + y \le 4\} = \{(1,1),(2,1),(3,1),(2,2)\} \tag{4.113}$$

and, as the set consists of elementary events, the probability of $B$ is

$$\begin{aligned} P[B] &= P[\{(1,1),(2,1),(3,1),(2,2)\}] \\ &= P[(1,1)] + P[(2,1)] + P[(3,1)] + P[(2,2)] = 7/12 \end{aligned} \tag{4.114}$$

It then follows from Equation 4.110 that

$$P_Z[A_1/B] = \frac{P[A_1, B]}{P[B]} = \frac{P[(1,1)]}{P[B]} = \frac{1/4}{7/12} = 3/7 \tag{4.115}$$

$$P_Z[A_2/B] = \frac{P[\{(2,2),(3,2)\}, B]}{P[B]} = \frac{P[(2,2)]}{P[B]} = \frac{1/8}{7/12} = 3/14 \tag{4.116}$$

***4.10.8.3 Case 2: Conditioning on Subsets of Individual Sample Spaces*** Consider the random variable $Z: S \to S_Z$, $Z = (X, Y)$ with the sample space defined by Equation 4.109. The two component random variables $X$ and $Y$ are mappings:

$$X: S \to S_X \quad Y: S \to S_Y \tag{4.117}$$

The intersection of the events $A \subseteq S_X$, $B \subseteq S_Y$ defines a subset $S_{AB} \subseteq S_Z$:

$$S_{AB} = \{(x, y) \in S_Z : x \in A, y \in B\} \tag{4.118}$$

The probability of occurrence of the joint events $X \in A, Y \in B$ is

$$P[X \in A, Y \in B] = P[S_{AB} \subseteq S_Z] = P[\omega \in S : Z(\omega) \in S_{AB}] \tag{4.119}$$

From conditional probability theory,

$$P[X \in A, Y \in B] = P[X \in A / Y \in B] \cdot P[Y \in B] \qquad P[Y \in B] \neq 0 \tag{4.120}$$

and the following definition can be made.

**Definition: Probability of Conditional Outcome** For $A \subseteq S_X$ and $B \subseteq S_Y$, the conditional probability of $X \in A$, given $Y \in B$ has occurred, is defined as

$$P[X \in A / Y \in B] = \frac{P[X \in A, Y \in B]}{P[Y \in B]} \qquad P[Y \in B] \neq 0 \tag{4.121}$$

and $P[X \in A / Y \in B]$ is written as $p_{X/Y \in B}(A)$. Thus,

$$p_{X/Y \in B}(A) = \frac{P[X \in A, Y \in B]}{P[Y \in B]} \qquad P[Y \in B] \neq 0 \tag{4.122}$$

***4.10.8.4 Conditional Joint Probability Mass Function*** For $Z = (X, Y)$ and with the sample space defined by Equation 4.109, the joint conditional probability mass function can be defined for the case where $X$ and $Y$ are discrete random variables.

**Definition: Joint Conditional Probability Mass Function** For an arbitrary set $A \subseteq S_Z$, the joint conditional probability mass function $p_{XY/A}(x, y)$ is the conditional probability

$$p_{XY/A}(x, y) = P[(X = x, Y = y)/A] \tag{4.123}$$

**Theorem 4.14 Joint Conditional Probability Mass Function**
For an arbitrary set $A \subseteq S_Z$, the joint conditional probability mass function is

$$p_{XY/A}(x, y) = \begin{cases} \dfrac{p_{XY}(x, y)}{P[A]} & (x, y) \in A, P[A] > 0 \\ 0 & (x, y) \notin A, P[A] > 0 \end{cases} \tag{4.124}$$

**Proof**
Consider the case where $(x, y) \in A$. From conditional probability theory,

$$P[(X=x, Y=y)/A] = \frac{P[X=x, Y=y, A]}{P[A]} = \frac{P[X=x, Y=y]}{P[A]} \tag{4.125}$$

as required. For the case where $(x, y) \notin A$, it is the case that $P[X=x, Y=y, A]=0$ and the second result follows.

*4.10.8.5 Joint Conditional Probability Density Function* For $Z=(X, Y)$ and with the sample space defined by Equation 4.109, the joint conditional probability density function can be defined for the case where $X$ and $Y$ are continuous random variables.

**Definition: Joint Conditional Probability Density Function** The joint conditional probability density function, denoted $f_{XY/A}(x, y)$, $A \subseteq S_Z$, is defined as the limit of the conditional probability:

$$f_{XY/A}(x, y) = \lim_{dy\to 0, dx\to 0} \frac{P[(X \in (x, x+dx), Y \in (y, y+dy))/A]}{dxdy} \tag{4.126}$$

For $dx$ and $dy$ sufficiently small, it is the case that

$$f_{XY/A}(x, y)dxdy \approx P[(X \in (x, x+dx), Y \in (y, y+dy))/A] \tag{4.127}$$

**Theorem 4.15 Conditional Joint Probability Density Function**
The conditional joint probability density function, $f_{XY/A}$, can be written in terms of the joint probability density function, $f_{XY}$, according to

$$f_{XY/A}(x, y) = \begin{cases} \dfrac{f_{XY}(x, y)}{P[A]} & (x, y) \in A, P[A] > 0 \\ 0 & (x, y) \notin A, P[A] > 0 \end{cases} \tag{4.128}$$

**Proof**
Consider the case where $(x, y) \in A$. From conditional probability theory,

$$\begin{aligned} P[(X \in (x, x+dx), Y \in (y, y+dy))/A] &= \frac{P[X \in (x, x+dx), Y \in (y, y+dy), A]}{P[A]} \\ &= \frac{P[X \in (x, x+dx), Y \in (y, y+dy)]}{P[A]} \end{aligned} \tag{4.129}$$

and thus, for $dx$, $dy$ sufficiently small,

$$f_{XY/A}(x, y)dxdy \approx \frac{f_{XY}(x, y)dxdy}{P[A]} \Rightarrow f_{XY/A}(x, y) = \frac{f_{XY}(x, y)}{P[A]} \tag{4.130}$$

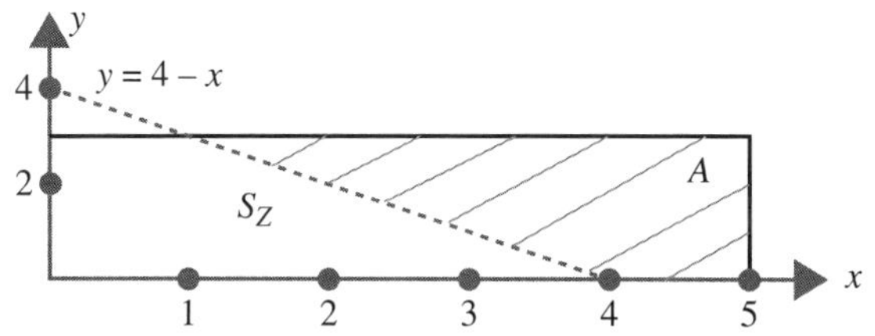


**FIGURE 4.10** Possible outcomes of the random variable $Z$ and the outcomes consistent with the set $A$.

***4.10.8.6 Example*** Consider the determination of the conditional joint probability density function, $f_{XY/A}$, for the case of $Z = (X, Y)$:

$$S_Z = \{(x, y) : 0 \le x \le 5, 0 \le y \le 3\} \tag{4.131}$$

$$A = \{(x, y) : x + y \ge 4, 0 \le x \le 5, 0 \le y \le 3\} \tag{4.132}$$

where $f_{XY}$ is defined according to

$$f_{XY}(x, y) = \begin{cases} 1/15 & 0 \le x \le 5, 0 \le y \le 3 \\ 0 & \text{elsewhere} \end{cases} \tag{4.133}$$

First, Figure 4.10 illustrates the possible outcomes of the random variable $Z$ and the outcomes consistent with the event $A$. Second, the probability of the event $A$ can be determined from the joint probability mass function:

$$P[A] = \int_0^3 \left( \int_{4-y}^5 \frac{1}{15} dx \right) dy = \frac{1}{15} \int_0^3 [5 - (4 - y)] dy = \frac{1}{15} \int_0^3 (1 + y) dy = \frac{1}{2} \tag{4.134}$$

Finally, using the result stated in Theorem 4.15, the required result follows:

$$f_{XY/A}(x, y) = \begin{cases} 2/15 & (x, y) \in A \\ 0 & (x, y) \notin A \end{cases} \tag{4.135}$$

***4.10.8.7 Conditioning Based on Elementary Outcomes of One RV*** The following general results have been established: first, for discrete random variables,

$$p_{XY/A}(x, y) = \begin{cases} \dfrac{p_{XY}(x, y)}{P[A]} & (x, y) \in A, P[A] > 0 \\ 0 & (x, y) \notin A, P[A] > 0 \end{cases} \tag{4.136}$$

Second, for continuous random variables,

$$f_{XY/A}(x,y) = \begin{cases} \dfrac{f_{XY}(x,y)}{P[A]} & (x,y) \in A, P[A] > 0 \\ 0 & (x,y) \notin A, P[A] > 0 \end{cases} \tag{4.137}$$

A specific subcase of these results, that of conditioning on a single outcome of one of the random variables, is widely used and of importance. The following definitions clarify notation.

**Definition: Conditional PMF and PDF—Single Outcome Case** For the discrete random variable case, the notation $p_{Y/X=x}(y)$ has the definition

$$p_{Y/X=x}(y) = P[Y = y/X = x] \tag{4.138}$$

For the continuous random variable case, the notation $f_{Y/X=x}(y)$ means

$$f_{Y/X=x}(y) = \lim_{dy \to 0} \left[ \lim_{dx \to 0} \frac{P[Y \in (y, y+dy)/X \in (x, x+dx)]}{dy} \right] \tag{4.139}$$

and for $dy$ appropriately small

$$f_{Y/X=x}(y)dy \approx \lim_{dx \to 0} P[Y \in (y, y+dy)/X \in (x, x+dx)] \tag{4.140}$$

The notation $X = x$ in $f_{Y/X=x}(y)$ is shorthand notation for $X \in (x, x+dx)$ where $dx$ is arbitrarily small.

The following can be proved.

**Theorem 4.16 PMF and PDF: Conditioning on Single Outcome**
For the case of conditioning on a single outcome $X = x$, the conditional probability mass function is

$$p_{Y/X=x}(y) = \begin{cases} \dfrac{p_{XY}(x,y)}{p_X(x)} & p_X(x) > 0 \\ \text{not defined} & p_X(x) = 0 \end{cases} \tag{4.141}$$

For the case of conditioning on $X \in (x, x+dx)$, the conditional probability density function is

$$f_{Y/X=x}(y) = \begin{cases} \dfrac{f_{XY}(x,y)}{f_X(x)} & f_X(x) > 0 \\ \text{not defined} & f_X(x) = 0 \end{cases} \tag{4.142}$$

Hence,

$$\begin{aligned} p_{XY}(x,y) &= p_{Y/X=x}(y)p_X(x) \quad p_X(x) > 0, \text{ discrete r.v.} \\ f_{XY}(x,y) &= f_{Y/X=x}(y)f_X(x) \quad f_X(x) > 0, \text{ continuous r.v.} \end{aligned} \tag{4.143}$$

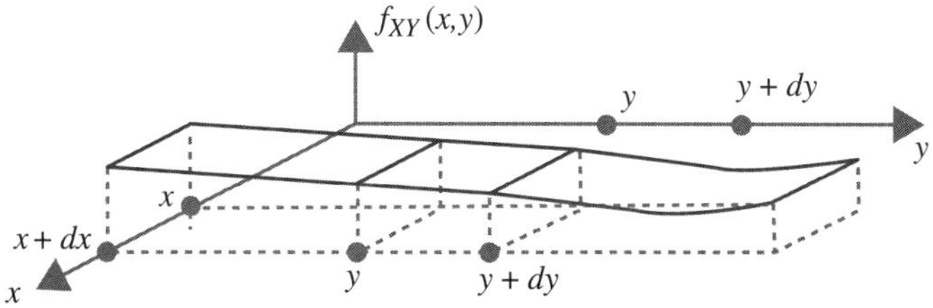


**FIGURE 4.11** Graph illustrating the nature of the joint probability density function for $X \in (x, x + dx)$.

**Proof**

The result for the discrete case follows directly from conditional probability. For the continuous random variable case, consider

$$
\begin{aligned}
&P[Y \in (y, y+dy)/X \in (x, x+dx)] \\
&= \frac{P[X \in (x, x+dx), Y \in (y, y+dy)]}{P[X \in (x, x+dx)]} \\
&\approx \frac{f_{XY}(x,y)\,dxdy}{f_X(x)\,dx} = \frac{f_{XY}(x,y)\,dy}{f_X(x)}
\end{aligned}
\tag{4.144}
$$

for $dx$, $dy$ suitably small. Hence,

$$
\frac{f_{XY}(x,y)}{f_X(x)} = \lim_{dy \to 0}\left[\lim_{dx \to 0} \frac{P[Y \in (y, y+dy)/X \in (x, x+dx)]}{dy}\right] \tag{4.145}
$$

is well defined and the result $f_{Y/X=x}(y) = \dfrac{f_{XY}(x,y)}{f_X(x)}$ follows. Further justification of the notation $f_{Y/X=x}$ can be found by considering the graph of the joint probability density function illustrated in Figure 4.11. The probability of $Y \in (y, y+dy)$, given $X = x$, is close to the probability of $Y \in (y, y+dy)$ given $X = x + dx$ for $dx$ sufficiently small.

#### *4.10.8.8 Independent Case*

**Theorem 4.17 Conditioning on Single Outcome: Independent Case**

For the case where the random variables $X$ and $Y$ are independent, it follows that

$$
p_{Y/X=x}(y) = \begin{cases} p_Y(y) & p_X(x) > 0 \\ \text{not defined} & p_X(x) = 0 \end{cases} \tag{4.146}
$$

$$
f_{Y/X=x}(y) = \begin{cases} f_Y(y) & f_X(x) > 0 \\ \text{not defined} & f_X(x) = 0 \end{cases} \tag{4.147}
$$

**Proof**

These results follow from the assumption of independence which implies $p_{XY}(x, y) = p_X(x)p_Y(y)$ and $f_{XY}(x, y) = f_X(x)f_Y(y)$.

## 4.11 COVARIANCE AND CORRELATION

The covariance function applies to two random variables and is a generalization of the variance function for a single random variable. A related function is the correlation function.

**Definition: Covariance** The covariance of two random variables *X* and *Y*, denoted cov(*X*, *Y*), is defined according to

$$\text{cov}(X, Y) = \boldsymbol{E}[(X - \mu_X)(Y - \mu_Y)] \tag{4.148}$$

where $\mu_X = \boldsymbol{E}[X]$ and $\mu_Y = \boldsymbol{E}[Y]$.

**Definition: Correlation** The correlation of two random variables *X* and *Y*, denoted corr(*X*, *Y*), is defined according to

$$\text{corr}(X, Y) = \boldsymbol{E}[XY] \tag{4.149}$$

**Theorem 4.18 Alternative Form for Covariance**

An alternative form for the covariance function is

$$\begin{aligned} \text{cov}(X, Y) &= \boldsymbol{E}[XY] - \mu_X \mu_Y \\ &= \text{corr}(X, Y) - \mu_X \mu_Y \end{aligned} \tag{4.150}$$

From this form, it directly follows, if $\mu_X = 0$ or $\mu_Y = 0$, that $\text{cov} = (X, Y) = \text{corr}(X, Y) = \boldsymbol{E}[XY]$.

**Proof**

The proof follows from the linearity of the expectation operator (Theorem 4.13) according to

$$\begin{aligned} \text{cov}(X, Y) &= \boldsymbol{E}[(X - \mu_X)\ (Y - \mu_Y)] - \boldsymbol{E}[XY - \mu_X Y \quad \mu_Y X + \mu_X \mu_Y] \\ &= \boldsymbol{E}[XY] - \mu_X \boldsymbol{E}[Y] - \mu_Y \boldsymbol{E}[X] + \mu_X \mu_Y = \boldsymbol{E}[XY] - \mu_X \mu_Y \end{aligned} \tag{4.151}$$

### 4.11.1 Understanding Covariance

Consider two subexperiments with countable samples spaces

$$S_1 = \{\omega_{1i} :\ i \in \boldsymbol{Z}^+\} \quad S_2 = \{\omega_{2i} :\ i \in \boldsymbol{Z}^+\} \tag{4.152}$$

**TABLE 4.2 Outcomes of the random variables Z, X, and Y**

| ω | Z(ω) | X(ω) | Y(ω) |
|---|---|---|---|
| $\omega_1$ | $z_1 = (X(\omega_{11}), Y(\omega_{21}))$ | $x_1 = X(\omega_{11})$ | $y_1 = Y(\omega_{21})$ |
| $\omega_2$ | $z_2 = (X(\omega_{12}), Y(\omega_{22}))$ | $x_2 = X(\omega_{12})$ | $y_2 = Y(\omega_{22})$ |
| $\omega_3$ | $z_3 = (X(\omega_{13}), Y(\omega_{23}))$ | $x_3 = X(\omega_{13})$ | $y_3 = Y(\omega_{23})$ |
| ... | | | |

which underpin an experiment with a sample space $S$:

$$S = \{\omega_i : \ \omega_i = (\omega_{1i}, \omega_{2i}), \ \ i \in \mathbf{Z}^+\} \tag{4.153}$$

The probability of individual outcomes is specified according to

$$\begin{aligned} P[\omega_{1i}] &= p_{\Omega_1}(\omega_{1i}) \quad P[\omega_{2i}] = p_{\Omega_2}(\omega_{2i}) \\ P[\omega_i] &= (\omega_{1i}, \omega_{2i}) = p_{\Omega_1\Omega_2}(\omega_{1i}, \omega_{2i}) \end{aligned} \tag{4.154}$$

Consider the case where a random variable $Z : S \rightarrow S_Z$ is defined on $S$ according to $Z(\omega_i) = (X(\omega_{1i}), Y(\omega_{2i}))$ where $X : S_1 \rightarrow S_X$ and $Y : S_2 \rightarrow S_Y$ define random variables on the two subexperiments in a manner such that distinct outcomes lead to distinct values for the random variables. Notation for the outcomes of the random variables is detailed in Table 4.2.

The joint probability of $(x_i, y_j)$ is $p_{\Omega_1\Omega_2}(x_i, y_j)$, and for the case where $X$ is independent of $Y$, it is the case that $p_{\Omega_1\Omega_2}(x_i, y_j) = p_{\Omega_1}(x_i)p_{\Omega_2}(y_j)$.

The following measures can be proposed for the dependence of the random variables $X$ and $Y$ for the specific case of the outcome $(x_i, y_j)$:

$$\begin{aligned} & p_{\Omega_1\Omega_2}(x_i, y_j) - p_{\Omega_1}(x_i)p_{\Omega_2}(y_j) \qquad \left|p_{\Omega_1\Omega_2}(x_i, y_j) - p_{\Omega_1}(x_i)p_{\Omega_2}(y_j)\right| \\ & \left[p_{\Omega_1\Omega_2}(x_i, y_j) - p_{\Omega_1}(x_i)p_{\Omega_2}(y_j)\right]^2 \ \ \ldots \end{aligned} \tag{4.155}$$

Consider the first measure. When this measure is weighted by the values of the random variables according to

$$\left[p_{\Omega_1\Omega_2}(x_i, y_j) - p_{\Omega_1}(x_i)p_{\Omega_2}(y_j)\right]x_i y_j \tag{4.156}$$

and summed over all possible outcomes, a *mean* measure for the dependence of the random variables $X$ and $Y$ is obtained according to

$$\sum_{i=1}^{\infty}\sum_{j=1}^{\infty}\left[p_{XY}(x_i, y_j) - p_X(x_i)p_Y(y_j)\right]x_i y_j \tag{4.157}$$

This measure is the covariance function of the two random variables $X$ and $Y$ as the following analysis shows:

$$\begin{aligned}&\sum_{i=1}^{\infty}\sum_{j=1}^{\infty}x_i y_j p_{XY}(x_i, y_j) - \sum_{i=1}^{\infty}x_i p_X(x_i)\sum_{j=1}^{\infty}y_j p_y(y_j)\\ &= \boldsymbol{E}[XY] - \boldsymbol{E}[X]\boldsymbol{E}[Y] = \mathrm{cov}(X, Y)\end{aligned} \tag{4.158}$$

The covariance function utilizes the simplest measure for the dependence between two random variables.

### 4.11.2 Uncorrelatedness

The concept of uncorrelatedness for two random variables arises from the definition of the covariance.

**Definition: Uncorrelated Random Variables** Two random variables $X$ and $Y$ are uncorrelated if

$$\mathrm{cov}(X, Y) = 0 \tag{4.159}$$

**Definition: Orthogonal Random Variables** Two random variables $X$ and $Y$ are orthogonal if $\boldsymbol{E}[XY] = 0$.

**Theorem 4.19 Condition for Uncorrelatedness**
The random variables $X$ and $Y$ are uncorrelated, that is, $\mathrm{cov}(X, Y) = 0$, if and only if

$$\boldsymbol{E}[XY] = \boldsymbol{E}[X]\boldsymbol{E}[Y] \tag{4.160}$$

**Proof**
Proof $\Rightarrow$: Assume $X$ and $Y$ are uncorrelated. This implies

$$\mathrm{cov}(X, Y) = \boldsymbol{E}[XY] - \mu_X \mu_Y = 0 \tag{4.161}$$

and, hence, that $\boldsymbol{E}[XY] = \boldsymbol{E}[X]\boldsymbol{E}[Y]$ as required.

Proof $\Leftarrow$: Assume $\boldsymbol{E}[XY] = \boldsymbol{E}[X]\boldsymbol{E}[Y]$. The required result follows as $\mathrm{cov}(X, Y) = \boldsymbol{E}[XY] - \boldsymbol{E}[X]\boldsymbol{E}[Y]$.

#### *4.11.2.1 Independence Implies Uncorrelatedness*

**Theorem 4.20 Independence Implies Uncorrelatedness**
If two random variables $X$ and $Y$ are independent, then they are uncorrelated. The converse is not necessarily true, that is, if $\mathrm{cov}(X, Y) = 0$, then it is not necessarily the case that $p_{XY}(x, y) = p_X(x)p_Y(y)$ or $f_{XY}(x, y) = f_X(x)f_Y(y)$.

For the case of random variables with only positive outcomes and with $f_{XY}(x, y)$ nonzero over the same region where $f_X(x)f_Y(y)$ is nonzero, it is the case that uncorrelatedness implies independence with $f_{XY}(x, y) = f_X(x)f_Y(y)$ except, possibly, on a set of zero measure.

**Proof**

Consider the continuous random variable case and $f_{XY}(x,y) = f_X(x)f_Y(y)$. It then follows that

$$\boldsymbol{E}[XY] = \int_{-\infty}^{\infty}\int_{-\infty}^{\infty} xyf_{XY}(x,y)dxdy = \int_{-\infty}^{\infty} xf_X(x)dx \int_{-\infty}^{\infty} yf_Y(y)dy = \boldsymbol{E}[X]\boldsymbol{E}[Y] \quad (4.162)$$

which implies that $\text{cov}(X,Y) = 0$.

To prove that uncorrelatedness does not necessarily imply independence, assume that $X$ and $Y$ are uncorrelated, that is, assume $\boldsymbol{E}[XY] = \boldsymbol{E}[X]\boldsymbol{E}[Y]$. For the continuous random variable case, this assumption implies:

$$\int_{-\infty}^{\infty}\int_{-\infty}^{\infty} xyf_{XY}(x,y)dxdy = \int_{-\infty}^{\infty} xf_X(x)dx \int_{-\infty}^{\infty} yf_Y(y)dy \quad (4.163)$$

or

$$\int_{-\infty}^{\infty}\int_{-\infty}^{\infty} xy[f_{XY}(x,y) - f_X(x)f_Y(y)]dxdy = 0 \quad (4.164)$$

It is possible to find a joint probability density function, $f_{XY}$, with marginal probability density functions $f_X(x)$ and $f_Y(y)$, such that this relationship is satisfied but

$$f_{XY}(x,y) \neq f_X(x)f_Y(y) \quad (4.165)$$

For the case of random variables with only positive outcomes, the double integral specified in Equation 4.164 is only zero when $f_{XY}(x,y) - f_X(x)f_Y(y) = 0$ except, possibly, on a set of zero measure. The further assumption here is that the nonzero regions of integration are the same.

#### *4.11.2.2 Example: Uncorrelated and Dependent Pair of Random Variables*

Consider the case of a random variable $Z: S \to S_Z$, $Z = (X,Y)$ with outcomes of the form

$$S_Z = \left\{ \begin{array}{l} (x_1, x_1 \pm \Delta), (x_1, -x_1 \pm \Delta), (x_1, x_1 \pm 2\Delta), (x_1, -x_1 \pm 2\Delta), \ldots \\ (x_2, x_2 \pm \Delta), (x_2, -x_2 \pm \Delta), (x_2, x_2 \pm 2\Delta), (x_2, -x_2 \pm 2\Delta), \ldots \\ \ldots \end{array} \right\} \quad (4.166)$$

and with

$$\begin{aligned} f_X(x) &= \frac{1}{\sqrt{2\pi}\sigma} \cdot \exp\left[\frac{-x^2}{2\sigma^2}\right] \\ f_{Y/X=x}(y)dy &\approx P\left[y \in (y, y+dy)\Big|_{x \in (x, x+dx)}\right] \approx \left[\frac{f_X(y-x)}{2} + \frac{f_X(y+x)}{2}\right] dy \end{aligned} \quad (4.167)$$

It then follows that

$$
\begin{aligned}
f_{XY}(x,y) &= f_{Y/X=x}(y)f_X(x) = \frac{f_X(y-x)f_X(x)}{2} + \frac{f_X(y+x)f_X(x)}{2} \\
&= \frac{1}{4\pi\sigma^2}\cdot\left[\exp\left[\frac{-x^2-(y-x)^2}{2\sigma^2}\right] + \exp\left[\frac{-x^2-(y+x)^2}{2\sigma^2}\right]\right]
\end{aligned} \tag{4.168}
$$

$$
\begin{aligned}
f_Y(y) = \int_{-\infty}^{\infty} f_{XY}(x,y)dx = \frac{1}{4\pi\sigma^2}\cdot\int_{-\infty}^{\infty}\left[\exp\left[\frac{-x^2-(y-x)^2}{2\sigma^2}\right]\right. \\
\left. + \exp\left[\frac{-x^2-(y+x)^2}{2\sigma^2}\right]\right]dx = \frac{1}{2\sqrt{\pi}\sigma}\cdot\exp\left[\frac{-y^2}{4\sigma^2}\right]
\end{aligned} \tag{4.169}
$$

$$
f_X(x)f_Y(y) = \frac{1}{2\sqrt{2}\pi\sigma^2}\cdot\exp\left[\frac{-x^2}{2\sigma^2}\right]\cdot\exp\left[\frac{-y^2}{4\sigma^2}\right] \tag{4.170}
$$

It then follows that $f_{XY}(x,y) \neq f_X(x)f_Y(y)$, which implies dependence between $X$ and $Y$. The covariance between $X$ and $Y$ is zero as

$$
\begin{aligned}
&\operatorname{cov}(X,Y) = \boldsymbol{E}[XY] - \boldsymbol{E}[X]\boldsymbol{E}[Y] \\
&\boldsymbol{E}[X] = \boldsymbol{E}[Y] = 0
\end{aligned} \tag{4.171}
$$

$$
\boldsymbol{E}[XY] = \int_{-\infty}^{\infty}\int_{-\infty}^{\infty} xyf_{XY}(x,y)dx = 0 \tag{4.172}
$$

Thus, the random variables are uncorrelated but dependent. The joint probability density function is shown in Figure 4.12 for the case of $\sigma^2 = 1$.

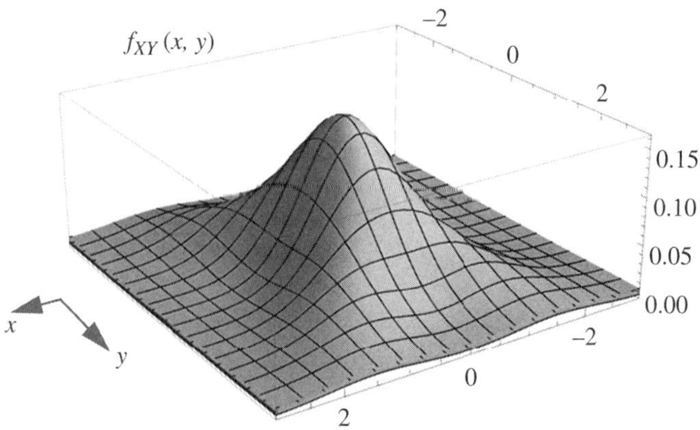


**FIGURE 4.12** Graph of the joint probability density function for the case of $\sigma^2 = 1$.

### 4.11.3 The Correlation Coefficient

The correlation coefficient is widely used as it gives a normalized measure of the correlation between two random variables.

**Definition: Correlation Coefficient** The correlation coefficient of two random variables $X$ and $Y$ is defined according to

$$\rho_{XY} = \frac{\text{cov}(X,Y)}{\sqrt{\text{var}(X)\text{var}(Y)}} = \frac{\boldsymbol{E}[XY]-\boldsymbol{E}[X]\boldsymbol{E}[Y]}{\sqrt{\text{var}(X)\text{var}(Y)}} \tag{4.173}$$

For the zero mean case,

$$\rho_{XY} = \frac{\boldsymbol{E}[XY]}{\sqrt{\text{var}(X)\text{var}(Y)}} \tag{4.174}$$

**Theorem 4.21 Bounds on the Correlation Coefficient**
The correlation coefficient is bounded according to

$$-1 \leq \rho_{XY} \leq 1 \tag{4.175}$$

**Proof**
The proof is detailed in Appendix 4.D.

## 4.12 SUMS OF RANDOM VARIABLES

Consider the random variable $Z$ that is defined as the weighted summation of $N$ random variables $X_1, \ldots, X_N$ according to

$$Z = \sum_{i=1}^{N} \alpha_i X_i \quad \alpha_i \in \boldsymbol{R} \tag{4.176}$$

The following important results hold.

**Theorem 4.22 Mean and Variance of a Sum of Random Variables**
The mean and variance of a sum of random variables is given by

$$\mu_Z = \boldsymbol{E}[Z] = \sum_{i=1}^{N} \alpha_i \boldsymbol{E}[X_i] \tag{4.177}$$

$$\begin{aligned} \sigma_Z^2 &= \text{var}(Z) = \sum_{i=1}^{N}\sum_{j=1}^{N} \alpha_i\alpha_j \text{cov}(X_i, X_j) \\ &= \sum_{i=1}^{N} \alpha_i^2 \text{var}(X_i) + \sum_{i=1}^{N}\sum_{j=1, j\neq 1}^{N} \alpha_i\alpha_j \text{cov}(X_i, X_j) \end{aligned} \tag{4.178}$$

When the random variables are uncorrelated, the variance of the weighted sum equals the sum of the weighted variances, that is,

$$\mathrm{var}(Z) = \sum_{i=1}^{N} \alpha_i^2 \,\mathrm{var}(X_i) \tag{4.179}$$

**Proof**

The result for the mean, as given by Equation 4.177, follows from the linearity of the expectation operator (Theorem 4.13), which justifies interchanging the order of summation and expectation operators. The variance result follows from the definition of the variance in the form $\mathrm{var}(Z) = \boldsymbol{E}(Z-\mu_Z)^2$:

$$\begin{aligned} \mathrm{var}(Z) &= \boldsymbol{E}\left[\sum_{i=1}^{N}\alpha_i X_i - \sum_{i=1}^{N}\alpha_i \boldsymbol{E}[X_i]\right]^2 = \boldsymbol{E}\left[\sum_{i=1}^{N}\alpha_i (X_i - \boldsymbol{E}[X_i])\right]^2 \\ &= \sum_{i=1}^{N}\sum_{j=1}^{N}\alpha_i\alpha_j \boldsymbol{E}\left[(X_i - \boldsymbol{E}[X_i])\ (X_j - \boldsymbol{E}[X_j])\right] = \sum_{i=1}^{N}\sum_{j=1}^{N}\alpha_i\alpha_j \,\mathrm{cov}(X_i, X_j) \end{aligned} \tag{4.180}$$

**Theorem 4.23 PDF of a Sum of Random Variables**

The probability density function, $f_Z$, of the weighted sum of $N$ random variables, as specified in Equation 4.176, is

$$f_Z(z) = \frac{1}{|\alpha_N|}\int_{-\infty}^{\infty}\cdots\int_{-\infty}^{\infty} f_{X_1\ldots X_N}\left(x_1,\ldots,x_{N-1},\frac{z}{\alpha_N}-\frac{\alpha_1}{\alpha_N}x_1-\cdots-\frac{\alpha_{N-1}}{\alpha_N}x_{N-1}\right) dx_1\ldots dx_{N-1} \tag{4.181}$$

For the independent random variable case,

$$\begin{aligned} f_Z(z) = \frac{1}{|\alpha_N|}\int_{-\infty}^{\infty}\cdots\int_{-\infty}^{\infty} & f_{X_1}(x_1) f_{X_2}(x_2)\ldots f_{X_{N-1}}(x_{N-1}) \cdot \\ & f_{X_N}\left(\frac{z}{\alpha_N}-\frac{\alpha_1}{\alpha_N}x_1-\cdots-\frac{\alpha_{N-1}}{\alpha_N}x_{N-1}\right) dx_1\ldots dx_{N-1} \end{aligned} \tag{4.182}$$

**Proof**

With the definition of $U = \frac{\alpha_1}{\alpha_N}X_1 + \cdots + \frac{\alpha_{N-1}}{\alpha_N}X_{N-1}$, the proof is as follows:

$$\begin{aligned} f_Z(z)dz &\approx P[Z \in [z, z+dz]] = P[\alpha_1 X_1 + \cdots + \alpha_N X_N \in [z, z+dz]] \\ &= P\left[X_1\ldots X_{N-1} \in \boldsymbol{R}, X_N \in \left[\frac{z}{\alpha_N} - U, \frac{z+dz}{\alpha_N} - U\right]\right] \\ &\approx \frac{dz}{|\alpha_N|}\int_{-\infty}^{\infty}\cdots\int_{-\infty}^{\infty} f_{X_1\ldots X_N}\left(x_1,\ldots,x_{N-1},\frac{z}{\alpha_N}-\frac{\alpha_1}{\alpha_N}x_1-\cdots-\frac{\alpha_{N-1}}{\alpha_N}x_{N-1}\right) dx_1\ldots dx_{N-1} \end{aligned} \tag{4.183}$$

### 4.12.1 Sum of Gaussian Random Variables

The probability density function for a sum of independent Gaussian random variables has been detailed in Theorem 4.11.

### 4.12.2 Difficulty in Determining the PDF of a Sum of Random Variables

In general and consistent with the $N-1$-fold integral expression specified in Equation 4.181, determining the probability density function of a sum of random variables is difficult.

## 4.13 JOINTLY GAUSSIAN RANDOM VARIABLES

The number of useful joint probability mass functions and joint probability density functions is large. The most widely used joint probability density function is the bivariate Gaussian.

**Definition: Jointly Gaussian, or Bivariate Gaussian, Random Variable** The random variables $X_1$ and $X_2$ are jointly Gaussian, with parameters $\mu_1$, $\mu_2$, $\sigma_1$, $\sigma_2$, and $\rho$, if $\sigma_1 > 0$, $\sigma_2 > 0$, $-1 < \rho < 1$, and

$$f_{X_1X_2}(x_1,x_2) = \frac{1}{2\pi\sigma_1\sigma_2\sqrt{1-\rho^2}} \cdot \exp\left(-\left[\frac{(x_1-\mu_1)^2}{2\sigma_1^2(1-\rho^2)} - \frac{\rho(x_1-\mu_1)\,(x_2-\mu_2)}{\sigma_1\sigma_2(1-\rho^2)} + \frac{(x_2-\mu_2)^2}{2\sigma_2^2(1-\rho^2)}\right]\right) \tag{4.184}$$

This probability density function is shown in Figure 4.13 for the case of $\mu_1 = \mu_2 = 0$, $\sigma_1 = 1$, $\sigma_2 = 2$, and $\rho = 0.5$.

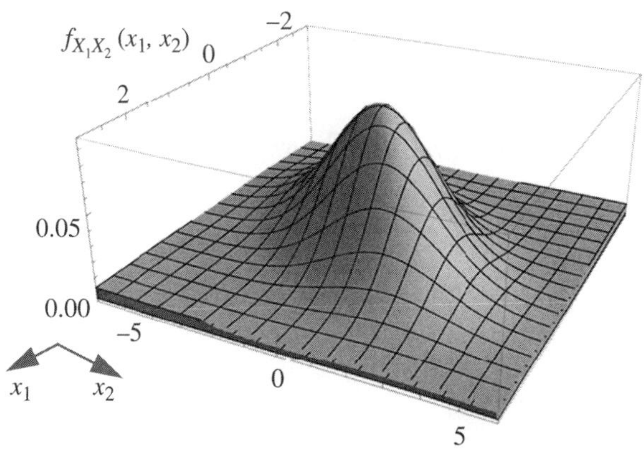


**FIGURE 4.13** Graph of the bivariate Gaussian probability density function for the case of $\mu_1 = \mu_2 = 0$, $\sigma_1 = 1$, $\sigma_2 = 2$, and $\rho = 0.5$.

## 4.14 STIRLING'S FORMULA AND APPROXIMATIONS TO BINOMIAL

The outcomes of complex phenomena, in many instances, can be modelled based on the outcomes of repetitions of a simple experiment. The simplest experiment is the Bernoulli experiment with two outcomes: success or failure. The probability of a specific outcome arising from a large number of repetitions of such an experiment is specified by the binomial probability mass function. This probability mass function is in terms of factorials and powers and computational complexity is reduced if suitable approximations can be defined. The two important approximations are the DeMoivre–Laplace approximation and the Poisson approximation. Both approximations are based on Stirling's formula, which states an approximation to a factorial number.

### 4.14.1 Binomial Probability Mass Function

**Definition: Binomial PMF** The binomial probability mass function, denoted $b(k, N, p)$, is the probability of $k$ successes in $N$ independent trials of a Bernoulli experiment where the probability of success in a given trial is $p$.

**Theorem 4.24 Binomial Probability Mass Function**
The probability of $k$ successes in $N$ independent trials of a Bernoulli experiment, where the probability of success in a given trial is $p$, is given by

$$b(k,N,p) = \frac{N!}{(N-k)!k!} \cdot p^k (1-p)^{N-k} \tag{4.185}$$

**Proof**
The proof readily follows from basic probability and combination theory.

**Theorem 4.25 Mean and Variance of a Binomial Random Variable**
The mean, mean squared value, and variance of a random variable $X$, with a binomial probability mass function, are

$$\begin{aligned} &\boldsymbol{E}[X] = Np \quad \boldsymbol{E}[X^2] = Np(1-p) + N^2p^2 \\ &\text{var}[X] = Np(1-p) \end{aligned} \tag{4.186}$$

**Proof**
These results follow by application of the expectation operator.

### 4.14.2 Stirling's Formula

Stirling's formula (Stirling, 1730; Feller, 1957, p. 50f; Parzen, 1960, p. 242) provides an accurate approximation to the factorial of a number.

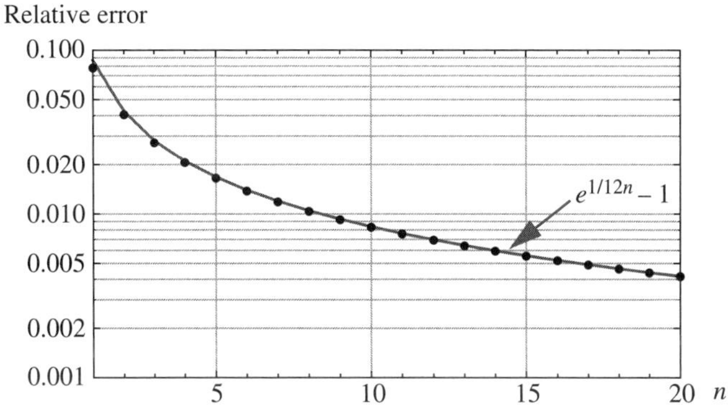


**FIGURE 4.14** Relative error (dots), and relative error bound (line), associated with the approximation of the factorial of a number.

**Theorem 4.26 Stirling's Formula**
The following limit holds:

$$\lim_{n\to\infty} \frac{n!}{\sqrt{2\pi}\sqrt{n}n^n e^{-n}} = 1 \tag{4.187}$$

and an approximation to the factorial of a number is

$$n! \approx \sqrt{2\pi}\sqrt{n}n^n e^{-n} \tag{4.188}$$

The relative error, and a relative error bound, in this approximation is

$$\varepsilon_r = 1 - \frac{\sqrt{2\pi}\sqrt{n}n^n e^{-n}}{n!} \qquad |\varepsilon_r| < e^{1/12n} - 1 \approx \frac{1}{12n} \tag{4.189}$$

The magnitude of the relative error, along with the relative error bound of $e^{1/12n} - 1$, are shown in Figure 4.14.

**Proof**
The proof is detailed in Appendix 4.E.

### 4.14.3 DeMoivre–Laplace Theorem

The first important approximation to the binomial probability mass function is the DeMoivre–Laplace approximation (Feller, 1945; Feller, 1957, p. 168f; Papoulis and Pillai, 2002, p. 105f ).

**Theorem 4.27 DeMoivre–Laplace Theorem**

With the definitions $\sigma=\sqrt{Np(1-p)}$ and $\mu=Np$, the binomial probability mass function, associated with the probability of $k$ successes in $N$ trials of a Bernoulli experiment, can be approximated according to

$$\begin{aligned} b(k,N,p) &\approx \frac{1}{\sqrt{2\pi}\sqrt{k}\sqrt{1-k/N}}\cdot\left(\frac{1}{k/Np}\right)^{k}\left(\frac{1-p}{1-k/N}\right)^{N-k} \\ &= \frac{1}{\sqrt{2\pi}\sqrt{k}\sqrt{1-k/N}}\cdot\exp\left[-k\ln\left(\frac{k}{Np}\right)\right]\cdot\exp\left[(N-k)\ln\left(\frac{1-p}{1-k/N}\right)\right] \\ b(k,N,p) &\approx \frac{e^{-(k-\mu)^2/2\sigma^2}}{\sqrt{2\pi}\sigma} \qquad \max\{0,2Np-N\}<k<\min\{N,2Np\} \end{aligned} \tag{4.190}$$

The first two forms arise from direct use of Stirling's approximation, and the third form, named after DeMoivre and Laplace, has the following approximate relative error for the case of $k\neq\mu$:

$$\text{r.e.}\approx 1-\exp\left[\frac{-(k-\mu)^2}{2\sigma^2}+(k+1/2)\ln\left(\frac{k}{\mu}\right)+(N-k+1/2)\ln\left(1-\frac{k-\mu}{N-\mu}\right)\right] \tag{4.191}$$

For the case of $\frac{2\sigma|\text{r.e.}|}{|1-2p|}>10$, a reasonable approximation for the range of validity for the DeMoivre–Laplace approximation, with a set relative error bound of |r. e.|, is

$$\mu-\sigma^{4/3}\left[\frac{6|\text{r.e.}|}{|1-2p|}\right]^{1/3}<k<\mu+\sigma^{4/3}\left[\frac{6|\text{r.e.}|}{|1-2p|}\right]^{1/3} \tag{4.192}$$

**Proof**

The proof is detailed in Appendix 4.F.

***4.14.3.1 Notes*** For the case of $k=\mu$, the relative error is usually small. For a given $N$ and $\mu$ and a set bound on the relative error, a nonlinear root solving algorithm can be used on Equation 4.191 to solve for the range of $k$ for which the DeMoivre–Laplace approximation is valid.

***4.14.3.2 Example*** The binomial probability mass function, along with the relative error and the approximation to the relative error as given by Equation 4.191, is shown in Figure 4.15 for the case of $N=100$ and $p=0.4$. The DeMoivre–Laplace approximation is valid, with a relative error less than 0.1, for $27\leq k\leq 55$ assuming the relative error expression specified in Equation 4.191. The approximate range for $k$, as given by Equation 4.192, is $29\leq k\leq 52$, which is slightly conservative. This arises as $\frac{2\sigma|\text{r.e.}|}{|1-2p|}=4.9$.

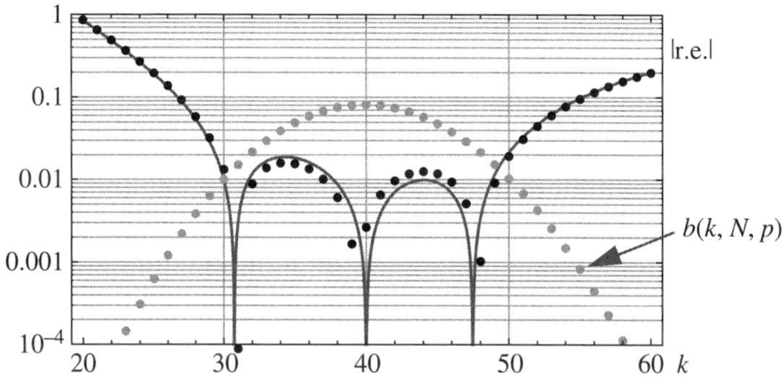


**FIGURE 4.15** Relative error (dots) and relative error approximation (line) for the DeMoivre–Laplace approximation to the binomial probability mass function for the case of $N = 100$ and $p = 0.4$.

***4.14.3.3 Generalization*** The following theorem specifies a useful extension of the DeMoivre–Laplace theorem.

**Theorem 4.28 Generalization of DeMoivre–Laplace Theorem**

For the case of a fixed relative error, $\dfrac{2\sigma|\text{r.e.}|}{|1-2p|} > 10$ and

$$\mu-\sigma^{4/3}\left[\frac{6|\text{r.e.}|}{|1-2p|}\right]^{1/3} < k_1, k_2 < \mu+\sigma^{4/3}\left[\frac{6|\text{r.e.}|}{|1-2p|}\right]^{1/3} \tag{4.193}$$

the following approximation is valid:

$$\begin{aligned} &P[k \text{ successes in } N \text{ trials where } k_1 \le k \le k_2] \\ &= \sum_{k=k_1}^{k_2} b(k,N,p) \approx Z\left[\frac{k_2-Np+0.5}{\sqrt{Np(1-p)}}\right] - Z\left[\frac{k_1-Np-0.5}{\sqrt{Np(1-p)}}\right] \end{aligned} \tag{4.194}$$

where

$$Z(x) = \int_{-\infty}^{x} \frac{1}{\sqrt{2\pi}} \exp\left[\frac{-z^2}{2}\right] dz \tag{4.195}$$

is the normalized Gaussian, or normal, distribution function.

**Proof**

With $q = 1-p$ and the approximation of

$$b(k,N,p) \approx \frac{1}{\sqrt{2\pi}\sqrt{Npq}} \cdot \exp\left[\frac{-(k-Np)^2}{2Npq}\right] \tag{4.196}$$

it follows that

$$\sum_{k=k_1}^{k_2} b(k,N,p) \approx \sum_{k=k_1}^{k_2} \frac{1}{\sqrt{2\pi}\sqrt{Npq}} \cdot \exp\left[\frac{-(k-Np)^2}{2Npq}\right]$$
$$\approx \int_{k_1-0.5}^{k_2+0.5} \frac{1}{\sqrt{2\pi}\sqrt{Npq}} \exp\left[\frac{-(k-Np)^2}{2Npq}\right] dk = \int_{\frac{k_1-Np-0.5}{\sqrt{Npq}}}^{\frac{k_2-Np+0.5}{\sqrt{Npq}}} \frac{1}{\sqrt{2\pi}} \exp\left[\frac{-z^2}{2}\right] dz \tag{4.197}$$

The required result then follows from the definition of $Z(x)$.

### 4.14.4 Poisson Approximation to Binomial

For the case when $p << 1$ and $k << n$, the Poisson approximation to the binomial probability mass function is appropriate (Feller, 1957, p. 142f ).

**Theorem 4.29 Poisson Approximation to Binomial**
If $N >> 1$, $p << 1$, and $Np^2/2 << 1$, then the following approximation is valid:

$$b(k,N,p) \approx \frac{(Np)^k e^{-Np}}{k!} \tag{4.198}$$

for $0 \le k << \min\{\sqrt{2N}, 1/p\}$. The relative error in the approximation is

$$\text{r.e.} = 1 - \frac{(N-k)!N^k e^{-Np}}{N!(1-p)^{N-k}}$$
$$\approx -\left[\frac{k}{2N}(k-1) + p\left(\frac{Np}{2} - k\right)\right] \tag{4.199}$$

**Proof**
The proof is detailed in Appendix 4.G.

***4.14.4.1 Example*** The binomial probability mass function, along with the relative error and relative error approximation as given by the two expressions in Equation 4.199, are shown in Figure 4.16 for the case of $N = 100$ and $p = 0.015$. The Poisson approximation is valid as $N >> 1$, $p << 1$, $Np^2/2 = 0.01125$, and the region of validity, as given by $\min\{\sqrt{2N}, 1/p\}$, is $0 \le k << 14$.

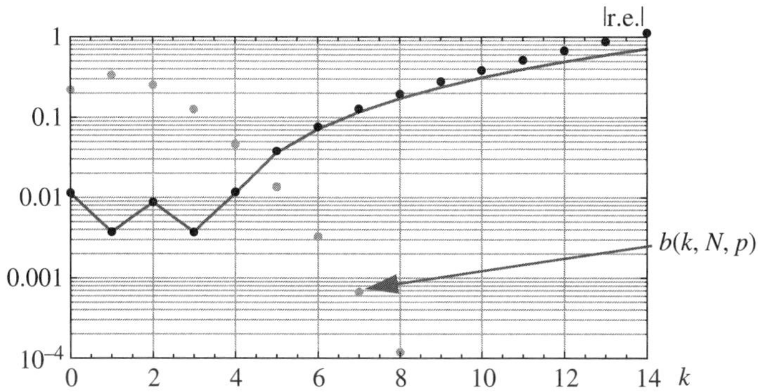


**FIGURE 4.16** Binomial probability mass function for the case of $N = 100$ and $p = 0.015$ and the relative error (dots), and relative error approximation (line), in the Poisson approximation to this probability mass function.

## 4.15 PROBLEMS

**4.1** From the three axioms defining the probability measure, show that:

$$P[\varnothing] = 0.$$

$$\text{If } A \subseteq B, \text{ then } P[A] \leq P[B].$$

$$P[A \cup B] = P[A] + P[B] - P[A \cap B].$$

**4.2** Show that the probability operator is a valid measure.

**4.3** If the events $A$ and $B$ are independent, then show that the events $A$ and $B^C$ are independent.

**4.4** From conditional probability theory, and the Theorem of Total Probability, proves Bayes' theorem:

$$P[B_i/A] = \frac{P[A/B_i]P[B_i]}{\sum_{k=1}^{N} P[A/B_k]P[B_k]} \qquad P[A] \neq 0 \tag{4.200}$$

assuming the set of events $\{B_1, \ldots, B_N\}$ is a partition of the sample space.

**4.5** Binary communication from a transmitter to a receiver is widely used. The sample space for communication of one bit of information in such a communication system is

$$S = \{(0,0),(0,1),(1,0),(1,1)\} \tag{4.201}$$

where the first element in the ordered pair represents the sent information and the second element represents the received information.

a. Specify an expression for the probability of error in the transmission of one bit assuming the following definitions:

$$\begin{aligned} P_0 &= P[0 \text{ sent}] & P_1 &= P[1 \text{ sent}] \\ P_{E0} &= P[1 \text{ received}/0 \text{ sent}] & P_{E1} &= P[0 \text{ received}/1 \text{ sent}] \end{aligned} \quad (4.202)$$

b. Use Bayes' theorem to determine an expression for $P$[1 sent/1 received], that is, the probability of a logic one was sent given a logic one was received.
c. Determine $P$[1 sent/1 received], and the probability of error, for the case of $P_0 = P_1 = 0,5,\ P_{E0} = 10^{-6}$ and $P_{E1} = 2.10^{-6}$.

**4.6** Specify the sample space of the experiment where each trial consists of:

a. Choosing a number at random from the interval $[0,1)$.
b. Choosing a second number, at random, between the first one and one.

**4.7** A trial of an experiment comprises of the running of $N$ independent subexperiments. Each subexperiment yields outcomes from the set $\{A, B, C\}$. Specify the sample space for this experiment.

**4.8** A degenerate random variable is defined on an experiment with only one outcome, that is, $S = \{\omega_1\}$. If a degenerate random variable is defined according to $X(\omega_1) = x_1$, then specify, and graph, the probability mass function, and cumulative distribution function, for this random variable.

**4.9** An indicator random variable is defined according to

$$X(\omega) = \begin{cases} 0 & \omega \in A^C \\ 1 & \omega \in A \end{cases} \quad (4.203)$$

where $A$ is a subset of $S$. Define, and graph, the probability mass function, and the cumulative distribution function, of $X$.

**4.10** Consider an experiment based on a very large number of independent subexperiments with outcomes of success or failure. The probability of success in one of the subexperiments is $p$. Define the random variable $X$ as the number of trials of the subexperiment before the first success. This random variable is called the geometric random variable.

a. Specify the probability mass function for $X$.
b. Graph the probability mass function for the case of $p = 0.1$.
c. Show that

$$P[X > k_o + k_1 / X > k_o] = P[X > k_1] \quad k_o, k_1 \geq 1 \quad (4.204)$$

This result states a memoryless property: if there have been no successes up until the $k_o$th subexperiment, then the probability of the first success occurring after a further $k_1$ subexperiments is independent of $k_o$.

**4.11** Consider a fixed positive integer $r$ and an experiment based on a very large number of independent subexperiments with outcomes of success or failure. The probability of success in one of the subexperiments is $p$. Define the random variable $X$ as the number of failures prior to the $r$th success (the waiting time for the $r$th success).

a. Specify the probability mass function for $X$.
b. Graph the probability mass function for the case of $p=0.1$ and $r=1$.
c. Graph the probability mass function for the case of $p=0.1$ and $r=2$.

**4.12** Photons are incident on a photodetector at random times and such that

$$P[k \text{ photons arrive in } [0,T]] = \frac{(\lambda T)^k e^{-\lambda T}}{k!} \quad k \in \{0,1,\ldots\} \tag{4.205}$$

where $\lambda$ is the mean arrival rate. The probability mass function is Poisson with a parameter $\lambda T$. By considering an infinite number of such detectors, and the experiment of choosing one photodetector at random, an experimental sample space can be defined based on the photon arrival times.

a. For a fixed interval $[0, T]$, a random variable $X$ is defined as the number of photons that have arrived. Graph the probability mass function of this random variable for the case of $\lambda T=0.1$, $\lambda T=1$ and $\lambda T=10$.
b. A random variable $Y$ is defined as the time the first photon arrives. Determine the cumulative distribution function, and the probability density function, of this random variable.

**4.13** A particle is emitted from a source and impacts, at random, on a hemisphere of radius $r$. The point of impact is defined by $\omega=(\theta,\phi)$ where $\theta\in[0,\pi/2]$ is the latitude and $\phi\in[0,2\pi)$ is the longitude. If the emission of one particle from the source is considered as one trial of an experiment, then the set of possible points of impact define an experimental sample space. A random variable is defined on this sample space according to $X(\omega)=\theta$. Determine the cumulative distribution function, and the probability density function, of $X$.

**4.14** Consider an experiment that yields points on the plane and the experimental sample space $S=\{\omega=(x,y): x,y\in \boldsymbol{R}\}$. The points are such that two random variables $X$, $Y$, defined according to $X(\omega)=x$ and $Y(\omega)=y$, are independent and have zero mean Gaussian density functions, that is,

$$f_X(x) = \frac{1}{\sqrt{2\pi}\sigma}\cdot e^{-x^2/2\sigma^2} \quad f_Y(y) = \frac{1}{\sqrt{2\pi}\sigma}\cdot e^{-y^2/2\sigma^2} \tag{4.206}$$

A new random variable, $Z$, is defined according to

$$Z(\omega)=\sqrt{X^2(\omega)+Y^2(\omega)}=\sqrt{x^2+y^2} \quad \omega=(x,y) \tag{4.207}$$

which is the magnitude of the vector defined by each point. Note that

$$P[Z\in[z,z+dz]]=\iint_R f_X(x)f_Y(y)\,dxdy \tag{4.208}$$

where $R=\left\{(x,y):\ \sqrt{x^2+y^2}\ \in[z,z+dz]\right\}$. Determine the probability density function, and cumulative distribution function, of $Z$. This probability density function finds widespread application in communications.

**4.15** Find the expected value, the expected squared value, and the variance of a random variable $X$ that has a Poisson probability mass function with a parameter $\lambda$:

$$p_X(k)=\frac{\lambda^k e^{-\lambda}}{k!} \quad k\in\{0,1,\ldots\} \tag{4.209}$$

**4.16** A random variable $X$ with a Rayleigh distribution has a probability density function given by

$$f_X(x)=\begin{cases}0 & x<0\\ \dfrac{x}{\sigma^2}\cdot e^{-x^2/2\sigma^2} & x\geq 0\end{cases} \tag{4.210}$$

Determine the mean and variance of $X$.

**4.17** A random variable $Y$ is defined on a continuous random variable $X$ according to

$$Y=aX+b \quad a,b\in\boldsymbol{R},a\neq 0 \tag{4.211}$$

Determine the probability density function of $Y$ in terms of the probability density function of $X$.

**4.18** A discrete random variable $Y$ is defined on a continuous random variable $X$ according to $Y=u[X]$. Determine the probability mass function of $X$.

**4.19** Determine the characteristic function of a random variable that has a Poisson distribution with a parameter $\lambda$.

**4.20** A random variable $Z$ is defined as the sum of two independent random variables $X_1$ and $X_2$ where $X_1$ has a Poisson distribution with parameter $\lambda_1$ and $X_2$ has a Poisson distribution with parameter $\lambda_2$.

a. Determine the characteristic function of $Z$.

b. Determine the probability mass function of $Z$.

**4.21** Consider the experiment consistent with a ternary communication signalling protocol, which is defined as follows:

i. A number is chosen, at random, from the set $\{-1,0,1\}$.
ii. A second number is chosen from the set $\{-1,0,1\}$ subject to the constraint that if the number is nonzero, then it cannot be the same as the previous number.
iii. A third number is chosen in a manner consistent with the second number and this procedure is repeated $N$ times.
   a. Specify the set of experimental outcomes for the case of $N=3$.
   b. Specify the probability of each outcome for the case of $N=3$ assuming $P[0]=1/2$, $P[-1]=P[1]=1/4$, $P[-1/\text{prior symbol is } 1]=1/2$, and $P[1/\text{prior symbol is} -1]=1/2$.

**4.22** Consider the experiment defined as follows:

i. A number is chosen at random from the set $\{0, 1\}$. A second number is chosen at random from the interval $[-\Delta,\Delta]$.
ii. The procedure specified in (i) is repeated $N$ times.

Specify the set of experimental outcomes.

**4.23** Consider an experiment based on throwing two die and noting the pair of numbers. An experimental outcome is denoted $\omega=(\omega_1,\omega_2)$. A pair of random variables $Z=(X,Y)$ is defined on this experiment according to

$$X=\min(\omega_1,\omega_2) \quad Y=\max(\omega_1,\omega_2) \tag{4.212}$$

a. Specify the sample space for $Z$.
b. Specify the joint probability mass function of $X$ and $Y$.
c. Determine the marginal probability mass function for $X$.
d. Determine the marginal probability mass function for $Y$.

**4.24** The joint cumulative distribution function of two continuous random variables $X$ and $Y$ is

$$F_{XY}(x,y)=\begin{cases} k\left(1-e^{-\alpha x}\right)\left(1-e^{-\beta y}\right) & x,y\geq 0 \\ 0 & \text{elsewhere} \end{cases} \tag{4.213}$$

a. Determine $k$.
b. Determine $P[X\leq 1, Y\leq 1]$.
c. Determine $P[1<X\leq 2, 2<Y\leq 5]$.
d. Determine the joint probability density function of $X$ and $Y$.
e. Determine the marginal probability density functions of $X$ and $Y$.

**4.25** Two random variables have a joint probability density function

$$F_{XY}(x,y)=\begin{cases} ke^{-2x-y} & 0\le y\le x \\ 0 & \text{otherwise}\end{cases} \tag{4.214}$$

a. Determine $k$.
b. Determine the marginal probability density functions of $X$ and $Y$.
c. Are $X$ and $Y$ independent?
d. Determine the probability of $X+Y\le 2$.

**4.26** An experiment yields the following experimental sample space:

$$S=\{\omega=(\omega_1,\omega_2):\ \omega_1,\omega_2\in\{0,1,\ldots,N-1\}\} \tag{4.215}$$

and the probability of occurrence of a given outcome is $P_\Omega(\omega)=1/N^2$. Two random variables, and an event $A$, are defined on this sample space according to

$$X(\omega)=\omega_1 \quad Y(\omega)=\omega_2 \tag{4.216}$$

$$A=\{(\omega_1,\omega_2):\ \omega_1+\omega_2\le 4\} \tag{4.217}$$

Determine the conditional probability mass function $p_{XY/A}(x,y)$ for the case of $N=4$ and for the general case where $N>4$.

**4.27** The random variables $X$ and $Y$ have a joint probability density function

$$f_{XY}(x,y)=\begin{cases} kx^2 & 0\le x\le 2, 0\le y\le x^2 \\ 0 & \text{otherwise}\end{cases} \tag{4.218}$$

a. Determine the constant $k$.
b. Find the conditional probability density function $f_{XY/A}(x,y)$ for the case of

$$A=\left\{(x,y):\ 0\le x\le 1/\sqrt{2}, 0\le y\le x^2\right\} \tag{4.219}$$

**4.28** The joint probability density function of two random variables $X$ and $Y$ is

$$f_{XY}(x,y)=\begin{cases} 2e^{-x-y} & 0\le y\le x \\ 0 & \text{otherwise}\end{cases} \tag{4.220}$$

a. Determine $f_{Y/X=x}(y) \quad 0\le y\le x, x>0$.
b. Determine $f_{X/Y=y}(x) \quad y>0, y\le x$.

**4.29** In a communication system and at a set time, a signal is detected, which has the form

$$Z = A\sin(X) + Y \tag{4.221}$$

where $X$ and $Y$ are independent random variables with probability density functions:

$$f_X(x) = \frac{1}{2\pi},\ 0 \le x < 2\pi \quad f_Y(y) = \frac{\exp(-y^2/2\sigma^2)}{\sqrt{2\pi}\sigma},\ y \in \boldsymbol{R} \tag{4.222}$$

a. Determine $\boldsymbol{E}[Z]$.

b. Determine $\boldsymbol{E}[Z^2]$ and the variance of $Z$.

c. If the random variable $Y$ represents a noise signal, specify a general condition for the recovery of the amplitude $A$ from $Z^2$.

**4.30** Establish the variance of $Z = aX + bY,\ a, b \in \boldsymbol{R}$.

**4.31** If $X = \cos(\Theta)$ and $Y = \sin(\Theta)$, where $\Theta$ is a random variable with a uniform distribution on $[0, 2\pi)$, then determine the mean and variance of $X$ and $Y$ and the correlation coefficient $\rho_{XY}$.

**4.32** Prove the result that the probability of $k$ successes in $N$ independent trials of an experiment is

$$\frac{N!}{(N-k)!k!} \cdot p^k (1-p)^{N-k} \tag{4.223}$$

when the probability of success in an independent trial is $p$.

**4.33** Determine the mean and variance of a random variable $X$ with a binomial probability mass function.

**4.34** If $X$ and $Y$ are random variables with a bivariate Gaussian joint density function, then determine the marginal density functions of $X$ and $Y$.

**4.35** Consider the DeMoivre–Laplace approximation to the binomial probability mass function for the case of $N = 1000$ and $p = 0.3$. Graph the relative error and the approximation to the relative error as given by

$$\text{r.e.} \approx 1 - \exp\left[\frac{-(k-\mu)^2}{2\sigma^2} + (k+1/2)\ln\left(\frac{k}{\mu}\right) + (N-k+1/2)\ln\left(1 - \frac{k-\mu}{N-\mu}\right)\right] \tag{4.224}$$

Determine the range of validity of the DeMoivre–Laplace approximation for a relative error of less than 0.1 and compare these values with the range

$$\mu - \sigma^{4/3}\left[\frac{6|\text{r.e.}|}{|1-2p|}\right]^{1/3} < k < \mu + \sigma^{4/3}\left[\frac{6|\text{r.e.}|}{|1-2p|}\right]^{1/3} \tag{4.225}$$

**4.36** Check the error bound

$$|\varepsilon_r| \leq \left| \frac{k}{2n}(k-1) + p\left(\frac{np}{2} - k\right) \right| \tag{4.226}$$

for the Poisson approximation to the binomial probability mass function by considering the following values: (i) $n=50$, $p=0.01$, $0 \leq k < \sqrt{n}$, (ii) $n=500$, $p=0.01$, $0 \leq k < \sqrt{n}$, and (iii) $n=500$, $p=0.05$, $0 \leq k < \sqrt{n}$.

**4.37** The probability that a voter in an election is conservative equals 0.4. Consider 1000 voters, chosen at random. Determine the probability, among the 1000 voters, that the number of conservative voters is between, and including the bounds, 370 and 395.

**4.38** In the industrial sector of a country, the probability of a worker having an accident that requires surgery on a single working day is $10^{-6}$.

a. What is the probability that a worker does not have an accident requiring surgery in his career spanning $10^4$ working days.
b. In a working year of 300 days and in a factory with 1000 workers, what is the probability of zero, one, or two accidents requiring surgery? What is the probability of one or more accidents?

## APPENDIX 4.A PROOF OF THEOREM 4.6

By definition

$$\begin{aligned}
\Phi_X(f) = \boldsymbol{E}\left[e^{j2\pi fX}\right] &= \frac{1}{\sqrt{2\pi}\sigma}\int_{-\infty}^{\infty} e^{j2\pi fx} e^{-(x-\mu)^2/2\sigma^2}\,dx \\
&= \frac{1}{\sqrt{2\pi}\sigma}\int_{-\infty}^{\infty} \exp\left[\frac{-\left(x^2 - x\left(2\mu + j4\pi f\sigma^2\right) + \mu^2\right)}{2\sigma^2}\right] dx \\
&= \frac{1}{\sqrt{2\pi}\sigma}\int_{-\infty}^{\infty} \exp\left[\frac{-\left(x-\left(\mu + j2\pi f\sigma^2\right)\right)^2}{2\sigma^2} - \frac{\mu^2}{2\sigma^2} + \frac{\left(\mu + j2\pi f\sigma^2\right)^2}{2\sigma^2}\right] dx \\
&= \exp\left[\frac{j4\pi f\sigma^2\mu - 4\pi^2 f^2\sigma^4}{2\sigma^2}\right] \int_{-\infty}^{\infty} \frac{1}{\sqrt{2\pi}\sigma}\cdot\exp\left[\frac{-\left(x-\left(\mu + j2\pi f\sigma^2\right)\right)^2}{2\sigma^2}\right] dx
\end{aligned} \tag{4.227}$$

The integral is that of a Gaussian probability density function with a mean of $\mu + j2\pi f\sigma^2$ and a variance of $\sigma^2$, and accordingly, the integral is unity. The required result of

$$\Phi_X(f) = \exp\left[j2\pi f\mu - 2\pi^2 f^2\sigma^2\right] \tag{4.228}$$

then follows. The final result arises by using the association $k = j2\pi f$.

## APPENDIX 4.B PROOF OF THEOREM 4.8

By definition

$$\Phi_{S_N}(f) = \boldsymbol{E}[\exp(j2\pi fZ)] = \int_{-\infty}^{\infty} \exp(j2\pi fz) f_{S_N}(z)\, dz \tag{4.229}$$

From Theorem 4.23,

$$\Phi_{S_N}(f) = \int_{-\infty}^{\infty} \exp(j2\pi fz) \cdot \left[\int_{-\infty}^{\infty} \cdots \int_{-\infty}^{\infty} f_{Z_1 \ldots Z_N}\left(z_1, \ldots, z_{N-1}, z - z_1 - \ldots - z_{N-1}\right) dz_1 \ldots dz_{N-1}\right] dz \tag{4.230}$$

Interchanging the order of integration yields

$$\Phi_{S_N}(f) = \int_{-\infty}^{\infty} \cdots \int_{-\infty}^{\infty} \int_{-\infty}^{\infty} \exp(j2\pi fz) \cdot f_{Z_1 \ldots Z_N}\left(z_1, \ldots, z_{N-1}, z - z_1 - \ldots - z_{N-1}\right) dz\, dz_1 \ldots dz_{N-1} \tag{4.231}$$

A change of variable $q = z - z_1 - \cdots - z_{N-1}$ in the inner integral results in

$$\Phi_{S_N}(f) = \int_{-\infty}^{\infty} \cdots \int_{-\infty}^{\infty} \exp\left[j2\pi f \sum_{i=1}^{N-1} z_i\right] \cdot \int_{-\infty}^{\infty} \exp(j2\pi fq) \cdot f_{Z_1 \ldots Z_N}(z_1, \ldots, z_{N-1}, q)\, dq\, dz_1 \ldots dz_{N-1} \tag{4.232}$$

The assumption of independence yields the required result:

$$\Phi_{S_N}(f) = \int_{-\infty}^{\infty} \exp(j2\pi fz_1) f_{Z_1}(z_1)\, dz_1 \cdot \ldots \cdot \int_{-\infty}^{\infty} \exp(j2\pi fz_{N-1}) f_{Z_{N-1}}(z_{N-1})\, dz_{N-1} \cdot \int_{-\infty}^{\infty} \exp(j2\pi fq) f_{Z_N}(q)\, dq = \prod_{i=1}^{N} \Phi_{Z_i}(f) \tag{4.233}$$

The result for a weighted sum follows from the relationship given in Theorem 4.7: if $Y = aX + b$, then $\Phi_Y(f) = e^{j2\pi fb} \Phi_X[af]$.

## APPENDIX 4.C PROOF OF THEOREM 4.9

Consider the disjoint outcomes of $S_M$ and their associated probabilities:

$$\begin{aligned} P\left[M=m_1, \sum_{i=1}^{m_1} X_i \in [x, x+dx]\right] &= p_1 f_{M_1}(x)\,dx \ \ldots \\ P\left[M=m_k, \sum_{i=1}^{m_k} X_i \in [x, x+dx]\right] &= p_k f_{M_k}(x)\,dx \ \ldots \end{aligned} \tag{4.234}$$

where $f_{M_k}$ is the probability density function of $X_1 + \cdots + X_{M_k}$. It then follows that

$$\begin{aligned} \Phi_{S_M}(f) &= \boldsymbol{E}[\exp(j2\pi f S_M)] = \int_{-\infty}^{\infty} \exp(j2\pi fx) p_1 f_{M_1}(x)\,dx + \cdots + \\ &\int_{-\infty}^{\infty} \exp(j2\pi fx) p_k f_{M_k}(x)\,dx + \ldots = \sum_{k=1}^{\infty} p_k \int_{-\infty}^{\infty} \exp(j2\pi fx) f_{M_k}(x)\,dx \\ &= \sum_{k=1}^{\infty} p_k \Phi_{S_{m_k}}(f) \qquad S_{m_k} = \sum_{i=1}^{m_k} X_i \end{aligned} \tag{4.235}$$

The assumption of independent random variables, and the result for the characteristic function of a sum of independent random variables stated in Theorem 4.8, yields the required result:

$$\Phi_{S_M}(f) = \sum_{k=1}^{\infty} p_k \prod_{i=1}^{m_k} \Phi_{X_i}(f) \tag{4.236}$$

## APPENDIX 4.D PROOF OF THEOREM 4.21

Consider a random variable $W = X - \alpha Y$ for some constant $\alpha$. Consider

$$\begin{aligned} \mathrm{var}(W) &= \boldsymbol{E}[W^2] - \mu_W^2 = \boldsymbol{E}\left[(X-\alpha Y)^2\right] - (\mu_X - \alpha\mu_Y)^2 \\ &= \boldsymbol{E}[X^2] + \alpha^2 \boldsymbol{E}[Y^2] - 2\alpha \boldsymbol{E}[XY] - \left(\mu_X^2 - 2\alpha\mu_X\mu_Y + \alpha^2\mu_Y^2\right) \\ &= \mathrm{var}(X) - 2\alpha\,\mathrm{cov}(X, Y) + \alpha^2\,\mathrm{var}(Y) \end{aligned} \tag{4.237}$$

Since $\mathrm{var}(W) \geq 0$, it follows that

$$\mathrm{var}(X) - 2\alpha\,\mathrm{cov}(X, Y) + \alpha^2\,\mathrm{var}(Y) \geq 0 \tag{4.238}$$

and

$$2\alpha \text{cov}(X, Y) \le \text{var}(X) + \alpha^2 \text{var}(Y) \tag{4.239}$$

for all values of α. With $\alpha = \sqrt{\text{var}(X)}/\sqrt{\text{var}(Y)}$, the upper bound for the correlation coefficient results according to

$$\begin{aligned} &2\frac{\sqrt{\text{var}(X)}}{\sqrt{\text{var}(Y)}}\text{cov}(X, Y) \le 2\,\text{var}(X) \;\Rightarrow \\ &\text{cov}(X, Y) \le \sqrt{\text{var}(Y)}\,\sqrt{\text{var}(X)} \;\Rightarrow\; \frac{\text{cov}(X, Y)}{\sqrt{\text{var}(Y)}\,\sqrt{\text{var}(X)}} \le 1 \end{aligned} \tag{4.240}$$

The lower bound for the correlation coefficient arises by choosing $\alpha = -\sqrt{\text{var}(X)}/\sqrt{\text{var}(Y)}$ and proceeding in a similar manner.

## APPENDIX 4.E PROOF OF STIRLING'S FORMULA

Stirling's formula can be proved by considering upper and lower bounds on the summation

$$S_n = \ln(n!) = \sum_{i=1}^{n} \ln(i) = \sum_{i=2}^{n} \ln(i) \tag{4.241}$$

To determine an upper bound on $S_n$, consider the integral $\int_1^n \ln(x)\,dx$ and the integral approximation illustrated in Figure 4.17. Consistent with the areas defined in this figure, it follows that

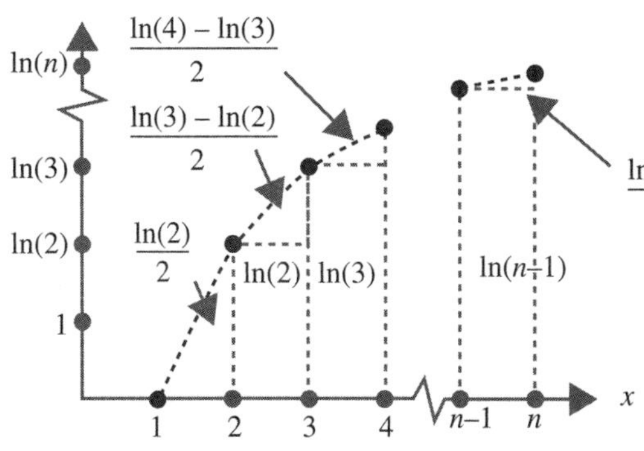


**FIGURE 4.17** Areas defined by an affine approximation to the integral ln(x) on the interval [1, n].

$$\begin{aligned}\int_1^n \ln(x)\,dx &> \sum_{i=1}^{n-1}\left[\ln(i) + \frac{\ln(i+1)-\ln(i)}{2}\right] = \frac{1}{2}\sum_{i=1}^{n-1}[\ln(i+1)+\ln(i)] \\ &= \frac{1}{2}\left[\sum_{i=2}^{n}\ln(i) + \sum_{i=1}^{n-1}\ln(i)\right] = \frac{1}{2}\left[\sum_{i=1}^{n}\ln(i) + \sum_{i=1}^{n}\ln(i) - \ln(n)\right] \\ &= S_n - \frac{\ln(n)}{2}\end{aligned} \tag{4.242}$$

Hence, an upper bound for $S_n$ is

$$S_n < \int_1^n \ln(x)\,dx + \frac{\ln(n)}{2} \tag{4.243}$$

To determine a lower bound for $S_n$, consider an affine, and tangent, approximation to $\ln(x)$ around the point $x = i$ as illustrated in Figure 4.18. Consistent with the tangent approximation illustrated in this Figure, it follows that

$$S_n = \sum_{i=2}^{n}\ln(i) > \int_{3/2}^{n+1/2}\ln(x)\,dx = \int_{3/2}^{n}\ln(x)\,dx + \int_{n}^{n+1/2}\ln(x)\,dx \tag{4.244}$$

As $\ln(x)$ is monotonically increasing, it follows that

$$\int_{n}^{n+1/2}\ln(x)\,dx > \frac{\ln(n)}{2} \tag{4.245}$$

and, hence,

$$S_n > \int_{3/2}^{n}\ln(x)\,dx + \frac{\ln(n)}{2} \tag{4.246}$$

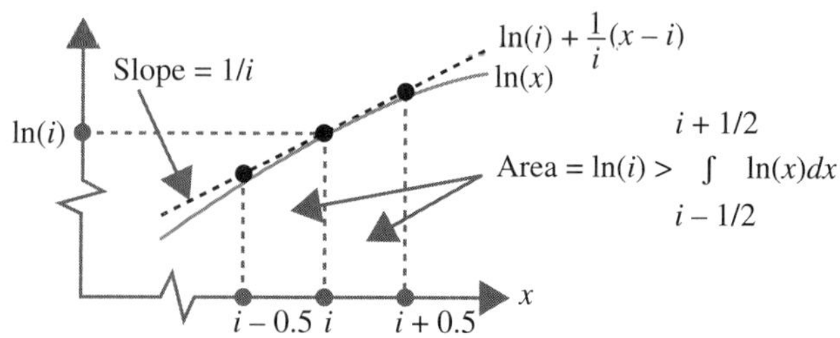


**FIGURE 4.18** An affine, and tangent, approximation to $\ln(x)$ around the point $x = i$.

Combining the upper and lower bounds for $S_n$, as given by Equations 4.243 and 4.246, yields

$$\int_{3/2}^{n} \ln(x)\,dx + \frac{\ln(n)}{2} < S_n < \int_{1}^{n} \ln(x)\,dx + \frac{\ln(n)}{2} \tag{4.247}$$

Evaluation of the integrals using the result

$$\int_{\alpha}^{n} \ln(x)\,dx = x\ln(x) - x\big|_{\alpha}^{n} = n\ln(n) - \alpha\ln(\alpha) - (n-\alpha) \tag{4.248}$$

and substitution of the definition $S_n = \ln(n!)$ yields

$$n\ln(n) - \frac{3}{2}\ln\left(\frac{3}{2}\right) - \left(n - \frac{3}{2}\right) + \frac{\ln(n)}{2} < \ln(n!) < n\ln(n) - (n-1) + \frac{\ln(n)}{2} \tag{4.249}$$

Hence,

$$\left(n + \frac{1}{2}\right)\ln(n) + \frac{3}{2}\left(1 - \ln\left(\frac{3}{2}\right)\right) - n < \ln(n!) < \left(n + \frac{1}{2}\right)\ln(n) - n + 1 \tag{4.250}$$

As the exponential function is a monotonically increasing function, it follows that

$$\begin{gathered} \exp\left[\left(n + \frac{1}{2}\right)\ln(n)\right]\exp\left[\frac{3}{2}\left(1 - \ln\left(\frac{3}{2}\right)\right)\right]\exp(-n) < n! < \\ \exp\left[\left(n + \frac{1}{2}\right)\ln(n)\right]\exp(-n)\exp(1) \end{gathered} \tag{4.251}$$

and thus,

$$\exp\left[\frac{3}{2}\left(1 - \ln\left(\frac{3}{2}\right)\right)\right] = 2.43952 < n!e^n n^{-(n+1/2)} < \exp[1] = 2.718282 \tag{4.252}$$

It can be shown that the limit of the sequence $\{n!e^n n^{-(n+1/2)}\}_{n=1}^{\infty}$ is $\sqrt{2\pi} = 2.5066283$. It is the case that $n!e^n n^{-(n+1/2)}$ equals 2.527597, 2.508718, 2.506837, respectively, for the case of $n = 10, 100, 1000$. Hence,

$$\lim_{n\to\infty} \frac{n!}{\sqrt{2\pi}\sqrt{n}n^n e^{-n}} = 1 \tag{4.253}$$

The relative error bound in the approximation $n! \approx \sqrt{2\pi}\sqrt{n}n^n e^{-n}$ is

$$\varepsilon_r = 1 - \frac{\sqrt{2\pi}\sqrt{n}n^n e^{-n}}{n!} \tag{4.254}$$

## APPENDIX 4.F PROOF OF THEOREM 4.27

Consider the use of Stirling's formula $n! \approx \sqrt{2\pi}\sqrt{n}n^n e^{-n}$ on the result for $k$ successes in $N$ trials, that is,

$$\begin{aligned} b(k,N,p) &= \frac{N!}{(N-k)!k!} \cdot p^k (1-p)^{N-k} \\ &\approx \frac{\sqrt{2\pi}\sqrt{N}N^N e^{-N}}{\left[\sqrt{2\pi}\sqrt{N-k}(N-k)^{N-k} e^{-(N-k)}\right]\left[\sqrt{2\pi}\sqrt{k}k^k e^{-k}\right]} \cdot p^k (1-p)^{N-k} \\ &= \frac{\sqrt{N}}{\sqrt{2\pi}\sqrt{N-k}\sqrt{k}} \cdot \left(\frac{Np}{k}\right)^k \left[\frac{N(1-p)}{N-k}\right]^{N-k} \\ &= \frac{1}{\sqrt{2\pi}\sqrt{k}\sqrt{1-k/N}} \cdot \left(\frac{1}{k/Np}\right)^k \left(\frac{1-p}{1-k/N}\right)^{N-k} \end{aligned} \tag{4.255}$$

which is the first required result. Using the relationship $A^b = \exp[b\ln(A)]$, $A>0$ yields the second required form

$$b(k,N,p) \approx \frac{1}{\sqrt{2\pi}\sqrt{k}\sqrt{1-k/N}} \cdot \exp\left[-k\ln\left(\frac{k}{Np}\right)\right] \cdot \exp\left[(N-k)\ln\left(\frac{1-p}{1-k/N}\right)\right] \tag{4.256}$$

To establish the third result, define the difference between $k$ and the mean value of $k$, as given by $Np$, as $\Delta_k$, that is, $\Delta_k = k - Np$. The following results then hold

$$\begin{aligned} &k = \Delta_k + Np \qquad \frac{k}{Np} = 1 + \frac{\Delta_k}{Np} \qquad 1 - \frac{k}{N} = 1 - p - \frac{\Delta_k}{N} \\ &N - k = N - (\Delta_k + Np) = N(1-p) - \Delta_k \end{aligned} \tag{4.257}$$

Substitution of these results into Equation 4.255 yields

$$\begin{aligned} &b(k,N,p) \approx \\ &\approx \frac{1}{\sqrt{2\pi}\sqrt{\Delta_k + Np}\sqrt{1-p-\frac{\Delta_k}{N}}} \cdot \frac{1}{\left(1+\frac{\Delta_k}{Np}\right)^{\Delta_k + Np}} \left(\frac{1-p}{1-p-\frac{\Delta_k}{N}}\right)^{N(1-p)-\Delta_k} \end{aligned} \tag{4.258}$$

With the manipulations

$$\sqrt{\Delta_k + Np}\sqrt{1-p-\frac{\Delta_k}{N}} = \sqrt{Np}\sqrt{1+\frac{\Delta_k}{Np}} \cdot \sqrt{1-p}\sqrt{1-\frac{\Delta_k}{N(1-p)}} \tag{4.259}$$

and $(1-p)/(1-p-\Delta_k/N) = 1/(1-\Delta_k/[N(1-p)])$, it follows that

$$b(k,N,p) \approx \frac{1}{\sqrt{2\pi}\sqrt{Np(1-p)}} \cdot \left(1+\frac{\Delta_k}{Np}\right)^{-\Delta_k - Np - 1/2} \left(1-\frac{\Delta_k}{N(1-p)}\right)^{-N(1-p)+\Delta_k - 1/2} \tag{4.260}$$

Consider the expression

$$T = \left(1+\frac{\Delta_k}{Np}\right)^{-\Delta_k - Np - 1/2} \left(1-\frac{\Delta_k}{N(1-p)}\right)^{-N(1-p)+\Delta_k - 1/2} \tag{4.261}$$

Taking the natural logarithm yields

$$\ln(T) = \left(-Np - \Delta_k - \frac{1}{2}\right) \ln\left(1+\frac{\Delta_k}{Np}\right) + \left[-N(1-p)+\Delta_k - \frac{1}{2}\right] \ln\left(1-\frac{\Delta_k}{N(1-p)}\right) \tag{4.262}$$

Consider the expansion $\ln(1+x) = x - \frac{x^2}{2} + \frac{x^3}{3} - \frac{x^4}{4} + \cdots$, $x \in (-1,1)$, which can be applied provided

$$\begin{aligned} & -1 < \frac{\Delta_k}{Np} = \frac{k-Np}{Np} < 1 \qquad -1 < \frac{-\Delta_k}{N(1-p)} = \frac{-(k-Np)}{N(1-p)} < 1 \\ & \Rightarrow \; 0 < k < 2Np \qquad 2Np - N < k < N \\ & \Rightarrow \; \max\{0, 2Np-N\} < k < \min\{N, 2Np\} \end{aligned} \tag{4.263}$$

Use of this expansion leads to

$$\begin{aligned} \ln(T) = & \left[-Np - \Delta_k - \frac{1}{2}\right] \left[\frac{\Delta_k}{Np} - \frac{\Delta_k^2}{2N^2p^2} + \frac{\Delta_k^3}{3N^3p^3} - \frac{\Delta_k^4}{4N^4p^4} + \ldots\right] \\ & + \left[-N(1-p)+\Delta_k - \frac{1}{2}\right] \left[\frac{-\Delta_k}{N(1-p)} - \frac{\Delta_k^2}{2N^2(1-p)^2} - \frac{\Delta_k^3}{3N^3(1-p)^3} - \frac{\Delta_k^4}{4N^4(1-p)^4} - \ldots\right] \end{aligned} \tag{4.264}$$

Expanding and collecting terms yields the following result:

$$\ln(T) = \frac{-\Delta_k^2}{2Np(1-p)} + e(N,p,\Delta_k) \tag{4.265}$$

where

$$e(N,p,\Delta_k)=\frac{-\Delta_k(1-2p)}{2Np(1-p)}+\frac{\Delta_k^2\left((1-p)^2+p^2\right)}{4N^2p^2(1-p)^2}$$

$$+\Delta_k^3\left[\frac{1-2p}{6N^2p^2(1-p)^2}+\frac{-(1-p)^3+p^3}{6N^3p^3(1-p)^3}\right]+\Delta_k^4\left[\frac{-[(1-p)^3+p^3]}{12N^3p^3(1-p)^3}+\frac{(1-p)^4+p^4}{8N^4p^4(1-p)^4}\right]+\dots$$

$$+\frac{\Delta_k^i}{i}\left[\frac{\left[(-1)^{i-1}(1-p)^{i-1}-p^{i-1}\right]}{(i-1)(Np(1-p))^{i-1}}+\frac{[(-1)^i(1-p)^i+p^i]}{2(Np(1-p))^i}\right]+\dots \tag{4.266}$$

It then follows that

$$b(k,N,p)\approx\frac{1}{\sqrt{2\pi}\sqrt{Np(1-p)}}\cdot\exp\left[\frac{-\Delta_K^2}{2Np(1-p)}\right]\cdot\exp[e(N,p,\Delta_k)] \tag{4.267}$$

When $e(N,p,\Delta_k)\ll 1$, the DeMoivre–Laplace approximation follows from the definitions $\Delta_k=k-\mu,\mu=Np$ and $\sigma^2=Np(1-p)$, that is,

$$b(k,N,p)\approx\frac{1}{\sqrt{2\pi}\sigma}\cdot\exp\left[\frac{-(k-\mu)^2}{2\sigma^2}\right] \tag{4.268}$$

### 4.F.1 Relative Error

The relative error in the DeMoivre–Laplace approximation is

$$\text{r.e.}=1-\frac{\dfrac{1}{\sqrt{2\pi}\sigma}\cdot\exp\left[\dfrac{-(k-\mu)^2}{2\sigma^2}\right]}{\dfrac{N!}{(N-k)!k!}\cdot p^k(1-p)^{N-k}} \tag{4.269}$$

The DeMoivre–Laplace approximation is based on using Stirling's formula, and the relative error in this approximation is small for $N$ moderate to large. Accordingly, using Equation 4.260, a reasonable approximation to the relative error is

$$\text{r.e.} \approx 1 - \frac{\exp\left[\dfrac{-\Delta_k^2}{2Np(1-p)}\right]}{\left(1+\dfrac{\Delta_k}{Np}\right)^{-\Delta_k - Np - 1/2}\left(1+\dfrac{\Delta_k}{N(1-p)}\right)^{-N(1-p)+\Delta_k - 1/2}}$$

$$= 1 - \frac{e^{-(k-\mu)^2/2\sigma^2}}{\left(\dfrac{k}{\mu}\right)^{-k-1/2}\left(1-\dfrac{k-\mu}{N-\mu}\right)^{-N+k-1/2}} \tag{4.270}$$

$$= 1 - \frac{e^{-(k-\mu)^2/2\sigma^2}}{\exp\left[-(k+1/2)\ln\left(\dfrac{k}{\mu}\right)\right]\cdot\exp\left[-(N-k+1/2)\ln\left(1-\dfrac{k-\mu}{N-\mu}\right)\right]}$$

To obtain the last equation, the result $\alpha^\beta = \exp[\beta\ln(\alpha)]$ has been used and this is the stated relative error expression.

An alternative approximation to the relative error can be obtained by considering Equation 4.267:

$$\text{r.e.} \approx 1 - \exp[-e(N,p,\Delta_k)] \approx e(N,p,\Delta_k) \tag{4.271}$$

assuming $|e(N,p,\Delta_k)| << 1$. For the case of $Np >> 1, \Delta_k << Np$ and $p \neq 1/2$, a simplified relative error approximation can be obtained from Equation 4.266 according to

$$\text{r.e.} \approx \frac{-\Delta_k(1-2p)}{2Np(1-p)} + \frac{\Delta_k^2\left((1-p)^2+p^2\right)}{4N^2p^2(1-p)^2} + \frac{\Delta_k^3(1-2p)}{6N^2p^2(1-p)^2} \tag{4.272}$$

The first and third terms typically dominate the second term and with $\Delta_k = k-\mu,\ \sigma^2 = Np(1-p)$:

$$\text{r.e.} \approx \frac{-(1-2p)}{2\sigma}\left[x - \frac{x^3}{3}\right] \quad x = \frac{k-\mu}{\sigma} \tag{4.273}$$

To determine the range of values of $k$, where the approximation is within a set relative error, the equation $\dfrac{-(1-2p)}{2\sigma}\left[x-\dfrac{x^3}{3}\right] = \text{r.e.}$ can be solved for $x$ and hence $k$. To this end, consider the function

$$f(x) = \frac{x^3}{3} - x = \frac{2\sigma(r.e.)}{1-2p} \tag{4.274}$$

For the case of $\frac{2\sigma(\text{r.e.})}{1-2p} \geq 10$, the approximation $f(x) \approx \frac{x^3}{3}$ yields a relative error of less than 10% in the solution of $f(x) = \frac{2\sigma(\text{r.e.})}{1-2p}$. With such an approximation, it follows that values of $k$ in the range

$$-\left[\frac{6\sigma|\text{r.e.}|}{|1-2p|}\right]^{1/3} < \frac{k-\mu}{\sigma} < \left[\frac{6\sigma|\text{r.e.}|}{|1-2p|}\right]^{1/3} \tag{4.275}$$

satisfy the relative error criterion. Hence, for a set relative error,

$$\mu-\sigma^{4/3}\left[\frac{6|\text{r.e.}|}{|1-2p|}\right]^{1/3} < k < \mu+\sigma^{4/3}\left[\frac{6|\text{r.e.}|}{|1-2p|}\right]^{1/3} \tag{4.276}$$

## APPENDIX 4.G PROOF OF THEOREM 4.29

As $N!/(N-k)! = N(N-1)\ldots(N-k+1)$, it follows that

$$\begin{aligned} b(k,N,p) &= \frac{N(N-1)\ldots(N-k+1)}{k!} \cdot p^k(1-p)^{N-k} \\ &= \frac{N^k\left(1-\frac{1}{N}\right)\ldots\left(1-\frac{k-1}{N}\right)}{k!} \cdot p^k(1-p)^{N-k} \end{aligned} \tag{4.277}$$

Consider $T_1 = (1-p)^{N-k}$, which can be written as

$$\begin{aligned} T_1 &= \exp[(N-k)\ln(1-p)] \\ &= \exp\left[N\left(1-\frac{k}{N}\right)\left(-p-\frac{p^2}{2}-\frac{p^3}{3}-\frac{p^4}{4}+\ldots\right)\right] \end{aligned} \tag{4.278}$$

using the expansion $\ln(1+x) = x - \frac{x^2}{2} + \frac{x^3}{3} - \frac{x^4}{4} + \ldots$, $x \in (-1,1)$. It then follows, assuming $k << N$ and $p << 1$, that $T_1 \approx e^{-Np}$. With the additional assumption of $N >> 1$, the required approximation follows, that is,

$$b(k,N,p) \approx \frac{(Np)^k e^{-Np}}{k!} \quad N >> 1, k << N, p << 1 \tag{4.279}$$

The relative error in this approximation is

$$\text{r.e.} = 1 - \frac{(Np)^k e^{-Np}/k!}{\frac{N!}{(N-k)!k!} \cdot p^k (1-p)^{N-k}} = 1 - \frac{(N-k)! N^k e^{-Np}}{N!(1-p)^{N-k}} \tag{4.280}$$

Stirling's formula $n! \approx \sqrt{2\pi}\sqrt{n} n^n e^{-n}$ yields

$$\begin{aligned}
\text{r.e.} &\approx 1 - \frac{N^k e^{-Np}}{(1-p)^{N-k}} \cdot \frac{\sqrt{2\pi}\sqrt{N-k}(N-k)^{N-k} e^{-(N-k)}}{\sqrt{2\pi}\sqrt{N} N^N e^{-N}} \\
&= 1 - \frac{e^{k-Np}}{(1-p)^{N-k}} \cdot \frac{(N-k)^{N-k+1/2}}{N^{N-k+1/2}} = 1 - \frac{e^{k-Np}}{(1-p)^{N-k}} \cdot \left(1 - \frac{k}{N}\right)^{N-k+1/2} \\
&= 1 - \exp\left[k - Np + (N-k+1/2)\ln\left(1 - \frac{k}{N}\right) - (N-k)\ln(1-p)\right]
\end{aligned} \tag{4.281}$$

The relative error is small when the magnitude of the argument of the exponential term is much less than one such that the approximation

$$\text{r.e.} \approx -\left[k - Np + (N-k+1/2)\ln\left(1 - \frac{k}{N}\right) - (N-k)\ln(1-p)\right] \tag{4.282}$$

is valid. To determine when the relative error is small, consider the approximation noted earlier for $\ln(1+x)$, which yields

$$\begin{aligned}
\text{r.e.} \approx -\Bigg[&k - Np + (N-k+1/2)\left(\frac{-k}{N} - \frac{k^2}{2N^2} - \frac{k^3}{3N^3} - \frac{k^4}{4N^4} - \ldots\right) - \\
&(N-k)\left(-p - \frac{p^2}{2} - \frac{p^3}{3} - \frac{p^4}{4} - \ldots\right)\Bigg]
\end{aligned} \tag{4.283}$$

Expanding out and collecting terms yields

$$\begin{aligned}
\text{r.e.} \approx -\Bigg[&\frac{-k}{2N}\left[1 + \frac{k}{2N} + \frac{k^2}{3N^2} + \ldots\right] + \frac{k^2}{2N}\left[1 + \frac{k}{3N} + \frac{k^2}{6N^2} + \ldots\right] + \\
&\frac{Np^2}{2}\left[1 + \frac{2p}{3} + \frac{p^2}{2} + \ldots\right] - pk\left[1 + \frac{p}{2} + \frac{p^2}{3} + \ldots\right]\Bigg]
\end{aligned} \tag{4.284}$$

With the existing assumptions of $N >> 1, k << N, p << 1$, the further assumptions of $k^2/2N << 1, Np^2/2 << 1$ and $pk << 1$ are required for the relative error to be small. The restrictions on $k$ then are

$$k << N, k << \sqrt{2N}, k << 1/p \tag{4.285}$$

The assumption of $N >> 1$ implies that $k << \min\{\sqrt{2N}, 1/p\}$. In summary, the constraints for the validity of the Poisson approximation are $N >> 1$, $p << 1, Np^2/2 << 1$, $k << \min\{\sqrt{2N}, 1/p\}$. When these constraints hold, the relative error can be approximated according to

$$\text{r.e.} \approx -\left[\frac{-k}{2N} + \frac{k^2}{2N} + \frac{Np^2}{2} - pk\right] = -\left[\frac{-k}{2N}(k-1) + p\left(\frac{Np}{2} - k\right)\right] \tag{4.286}$$

which is the required result.

# 5

# INTRODUCTION TO RANDOM PROCESSES

## 5.1 RANDOM PROCESSES

The random variation with time, of measures associated with phenomena in the physical, technical, and social worlds, underpins significant interest in random processes. This chapter defines a random process and establishes notation that is used in subsequent chapters. Underpinning a random process is an experiment, and the two prototypical experiments, the Bernoulli and the Poisson, are defined along with a discussion related to the specification of experimental outcomes. Finally, a preliminary classification of random processes is given.

## 5.2 DEFINITION OF A RANDOM PROCESS

**Definition: random process** When a function, whose range is a subset of the real line or complex plane, is associated with each experimental outcome of an experiment, a random process $X$ is defined. Formally,

$$X : S \rightarrow S_X \tag{5.1}$$

where $S$ is the sample space of experimental outcomes and the set $S_X$ is a set of functions, denoted the signal sample space, defined by the random process.

*A Signal Theoretic Introduction to Random Processes*, First Edition. Roy M. Howard.

**Definition: Signal Sample Space** The signal sample space, $S_X$, is the set of functions defined by the random process:

$$S_X = \{X(\omega): \quad X(\omega): \boldsymbol{R} \to \boldsymbol{C}, \quad \omega \in S\} \tag{5.2}$$

### 5.2.1 Notation

A random process is a mapping:

$$X: S \to S_X \quad X: \omega \to X(\omega) \tag{5.3}$$

and the set of signals defined by the random process is

$$S_X = \{X(\omega): \boldsymbol{R} \to \boldsymbol{C}, \quad \omega \in S\} \tag{5.4}$$

The following options exist for specifying an individual signal defined by a random process: first,

$$X(\omega): t \to \dots \quad \text{or} \quad X(\omega, t) = \dots \tag{5.5}$$

Second, with the use of the notation $x_\omega = X(\omega)$,

$$x_\omega : t \to \dots \quad \text{or} \quad x_\omega(t) = \dots \tag{5.6}$$

Third, the reference to $\omega$ can be dropped and the notation then is

$$X: t \to \dots \quad \text{or} \quad X(t) = \dots \tag{5.7}$$

The first option in the form

$$X(\omega, t) = \dots \quad S_X = \{X(\omega, t)\} = \dots : \quad \omega \in S\} \tag{5.8}$$

provides good clarity and will be used.

The notation $X(\Omega, t)$ will be used to denote a random process with a specific signal being defined as $X(\omega, t)$, $\omega \in S$. Here, $\Omega$ represents the random variable whose outcomes are $\{\omega : \omega \in S\}$.

#### 5.2.1.1 *Notes*

For each experimental outcome, $\omega \in S$, a random process $X$ assigns a real or complex function to $\omega$ according to $X(\omega)$. The mapping is illustrated in Figure 5.1.

In many instances, there is a one-to-one mapping between an experimental outcome and a signal defined by a random process. For such a case, $P[X(\omega)] = P[\omega]$. In general, the probability of occurrence of a specific outcome (signal) of a random process is dependent on the probability mass function, $p_\Omega$, or probability density function, $f_\Omega$, defining the associated experimental outcome(s).

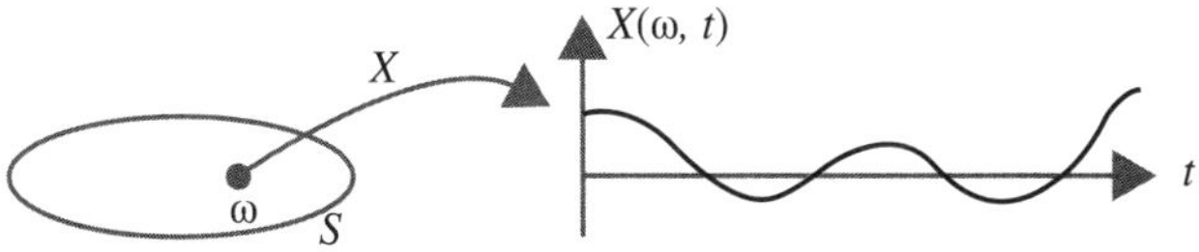


**FIGURE 5.1** Illustration of mapping associated with a random process.

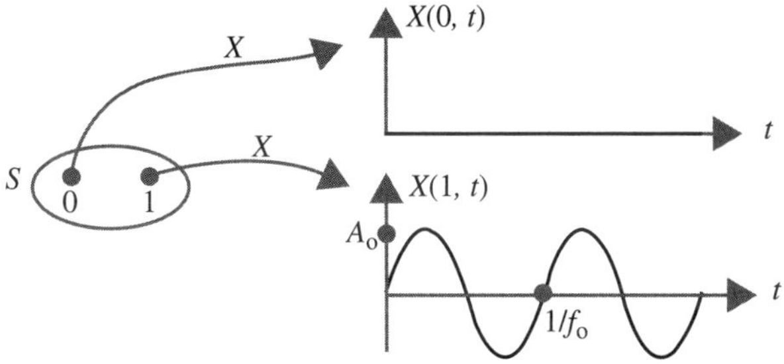


**FIGURE 5.2** Illustration of the two signals defined by the random process specified in Equation 5.9 for the case of $\phi_o = 0$.

With many random phenomena in science and engineering, a random signal is observed. The experiment underpinning the observed signal may be readily known or can be inferred or can be hypothesized.

## 5.3 EXAMPLES OF RANDOM PROCESSES

The following examples illustrate the nature of random processes and the notation associated with their use. For the defined random processes, there is a one-to-one mapping between an experimental outcome and a signal.

### 5.3.1 On–Off Sinusoid

Consider the random process $X : S \to S_X$ defined according to

$$\begin{aligned} &S = \{\omega : \quad \omega \in \{0,1\}\} \quad p_\Omega(\omega) = \begin{cases} 1-p & \omega = 0 \\ p & \omega = 1 \end{cases} \\ &X(\omega,t) = \omega A_o \sin(2\pi f_o t + \phi_o) u(t) \\ &S_X = \{X(\omega,t) = \omega A_o \sin(2\pi f_o t + \phi_o) u(t) : \quad \omega \in S\} \end{aligned} \tag{5.9}$$

The random process defines two signals that are shown in Figure 5.2. For notational convenience, the random process is written as

$$X(\Omega,t) = \Omega A_o \sin(2\pi f_o t + \phi_o) u(t) \tag{5.10}$$

where $\Omega$ is a random variable with a probability mass function $p_\Omega$.

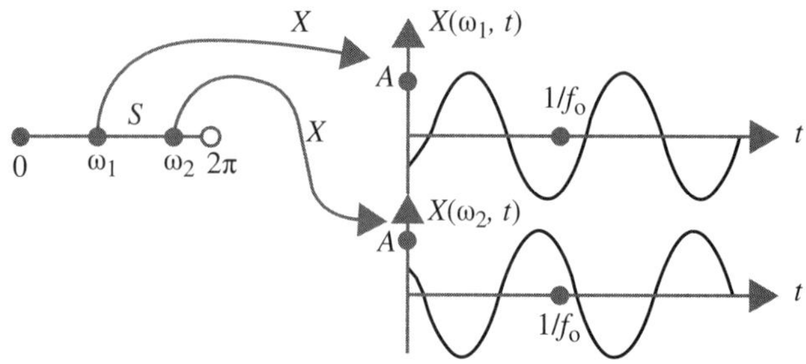


**FIGURE 5.3** Illustration of the signals defined by a random process whose signals are sinusoidal but have a random phase.

### 5.3.2 Sinusoid with Random Phase

Consider the random process $X : S \rightarrow S_X$ defined according to

$$\begin{aligned} &S = \{\omega : \quad \omega \in [0, 2\pi)\} \quad f_\Omega(\omega) = \begin{cases} 1/2\pi & 0 \le \omega < 2\pi \\ 0 & \text{elsewhere} \end{cases} \\ &X(\omega, t) = A\sin(2\pi f_o t + \omega)u(t) \\ &S_X = \{X(\omega, t) = A\sin(2\pi f_o t + \omega)u(t) : \quad \omega \in S\} \end{aligned} \tag{5.11}$$

Such a random process defines a set of sinusoids with random phase as illustrated in Figure 5.3.

For notational convenience, the random process is written as

$$X(\Omega, t) = A\sin(2\pi f_o t + \Omega)u(t) \tag{5.12}$$

where $\Omega$ is a random variable taking on values from the interval $[0, 2\pi)$ with a probability density function $f_\Omega(\omega) = 1/2\pi$.

### 5.3.3 Sinusoid with Random Amplitude

Consider the random process $X : S \rightarrow S_X$ defined according to

$$\begin{aligned} &S = \{\omega : \omega \in \boldsymbol{R}\} \quad f_\Omega(\omega) = \frac{1}{\sqrt{2\pi}\sigma} \cdot \exp\left[\frac{-\omega^2}{2\sigma^2}\right] \\ &X(\omega, t) = \omega\sin(2\pi f_o t + \phi_o)u(t) \\ &S_X = \{X(\omega, t) = \omega\sin(2\pi f_o t + \phi_o)u(t) : \quad \omega \in S\} \end{aligned} \tag{5.13}$$

This random process defines sinusoidal signals, as illustrated in Figure 5.4, with random amplitudes and where the amplitudes are consistent with a Gaussian probability density function with a variance $\sigma^2$.

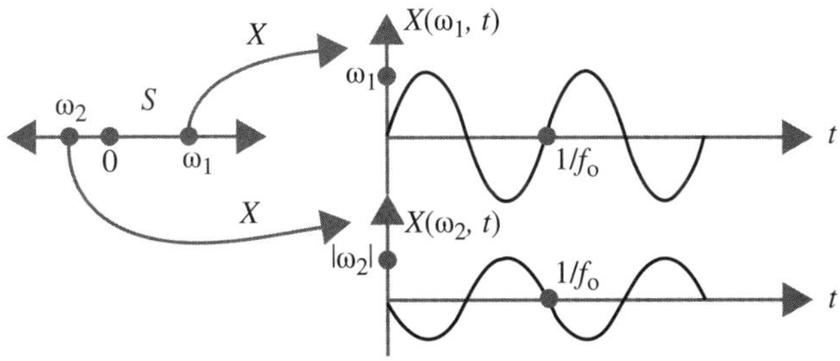


**FIGURE 5.4** Illustration of the signals defined by a random process whose signals are sinusoidal but with random amplitude. The phase $\phi_o$ is assumed to be zero.

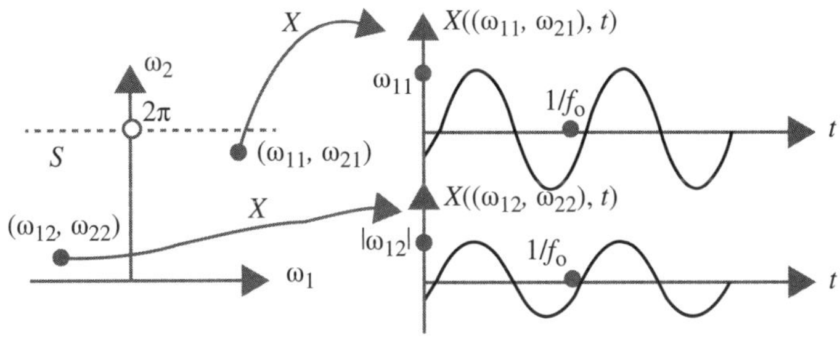


**FIGURE 5.5** Illustration of the signals defined by a random process whose signals are sinusoidal with random amplitude and random phase.

For notational convenience, the random process is written as

$$X(\Omega,t) = \Omega \sin(2\pi f_o t + \phi_o)u(t) \tag{5.14}$$

where $\Omega$ is a random variable with a probability density function $f_\Omega$.

### 5.3.4 Amplitude and Frequency Modulation

Consider the random process $X: S \to S_X$ defined according to

$$\begin{aligned}
&S = \{\omega:\ \omega = (\omega_1, \omega_2), \omega_1 \in \boldsymbol{R}, \omega_2 \in [0, 2\pi)\}\\
&f_\Omega(\omega) = f_{\Omega_1\Omega_2}(\omega_1,\omega_2) = \begin{cases} \dfrac{1}{2\pi}\cdot\dfrac{1}{\sqrt{2\pi}\sigma}\cdot\exp\left[\dfrac{-\omega_1^2}{2\sigma^2}\right] & 0 \le \omega_2 < 2\pi \\ 0 & \text{otherwise} \end{cases}\\
&X((\omega_1,\omega_2),t) = \omega_1 \sin(2\pi f_o t + \omega_2)u(t)\\
&S_X = \{X((\omega_1,\omega_2),t) = \omega_1 \sin(2\pi f_o t + \omega_2)u(t) : (\omega_1,\omega_2) \in S\}
\end{aligned} \tag{5.15}$$

The random process defines sinusoidal signals, illustrated in Figure 5.5, with random amplitude and random phase.

For notational convenience, the random process is written as

$$X(\Omega,t)=\Omega_1 \sin(2\pi f_o t+\Omega_2)u(t) \tag{5.16}$$

where $\Omega=(\Omega_1,\Omega_2)$. Here, $\Omega_1$ and $\Omega_2$ are independent random variables with, respectively, Gaussian and uniform probability density functions as can be inferred from the joint probability mass function $f_{\Omega_1\Omega_2}$.

### 5.3.5 Binary Digital Communication Signalling

Consider an experiment yielding a vector of $N$-independent outcomes and defining a sample space with $2^N$ elements:

$$S=\{\omega=(\omega_1,\ldots,\omega_N): \quad \omega_k\in\{0,1\}, k\in\{1,\ldots,N\}\} \tag{5.17}$$

where the probability of a specific outcome, comprising of $r$ ones and $N-r$ zeroes in a specific order, is given by

$$p_\Omega((\omega_1,\ldots,\omega_N))=p^r(1-p)^{N-r} \tag{5.18}$$

assuming $P[\omega_k=1]=p$ and $P[\omega_k=0]=1-p$.

A binary digital communication random process for the interval $[0,ND]$ can be defined on $S$ according to

$$\begin{aligned} &X: S\to S_X \\ &X((\omega_1,\ldots,\omega_N),t)=\sum_{k=1}^{N}\omega_k p(t-(k-1)D) \end{aligned} \tag{5.19}$$

where $p:\boldsymbol{R}\to\boldsymbol{R}$ is the signalling waveform and the signalling duration is $D$ consistent with a signalling rate of $r=1/D$. The signal sample space comprises $2^N$ waveforms:

$$S_X=\left\{X((\omega_1,\ldots,\omega_N),t)=\sum_{k=1}^{N}\omega_k p(t-(k-1)D):\begin{cases}\omega_k\in\{0,1\}\\(\omega_1,\ldots,\omega_N)\in S\end{cases}\right\} \tag{5.20}$$

For notational convenience, the random process is written as

$$X(\Omega,t)=\sum_{k=1}^{N}\Omega_k p(t-(k-1)D) \tag{5.21}$$

where $\Omega=(\Omega_1,\ldots,\Omega_N)$ is a vector of independent and discrete random variables taking on values from the set $\{0,1\}$ with probabilities, respectively, given by $1-p$ and $p$.

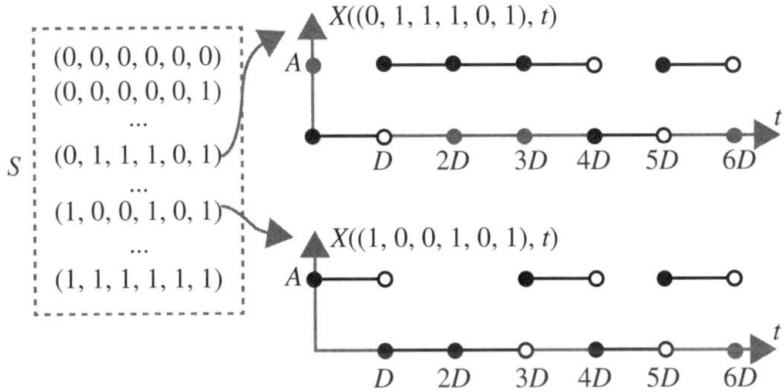


**FIGURE 5.6** Two waveforms defined by a binary digital communication random process for the interval $[0, 6D]$ and for the case where the signalling waveform is a rectangular pulse of duration $D$.

Two waveforms from the signal sample space are illustrated in Figure 5.6, for the prototypical but inefficient case (inefficient in terms of energy per bit of information) of a rectangular pulse signalling waveform:

$$p(t) = \begin{cases} A & 0 \leq t < D \\ 0 & \text{elsewhere} \end{cases} \tag{5.22}$$

## 5.4 EXPERIMENTS AND EXPERIMENTAL OUTCOMES

The aforementioned examples illustrate the influence that the nature of the experiment underpinning a random process has on the nature of the random process. Of importance is whether the experiment has possibly an uncountable, or a countable, number of outcomes and whether there are a single outcome, a vector of outcomes, a matrix of outcomes, etc. associated with each trial of the experiment.

### 5.4.1 Experiments and Subexperiments

When a trial of an experiment leads to a vector of outcomes, for example, $\omega = (\omega_1, \ldots, \omega_N)$, the experiment, in many instances, can be considered to consist of $N$ individual subexperiments. The following updated definition for an experiment, along with the definition of a subexperiment, is useful in the context of defining random processes for the usual case where there is a one-to-one relationship between a distinct experimental outcome and a signal defined by the random process.

**Definition: Experiment for Random Processes** An experiment underpinning an random process is a procedure that can, theoretically, be repeated an infinite number of times, and each trial of the experiment leads to a specific signal as defined by the random process. The set of possible outcomes is assumed not to change with the running of the experiment.

**Definition: Subexperiment** A subexperiment is a procedure that, by itself, can be treated as an experiment but is used with other subexperiments to form one experimental outcome for the sample space underpinning a random process.

It is possible to define sub-sub-experiments, etc. in an obvious manner.

**Definition: Algorithmic Subexperiment** The subexperiments comprising an experiment are algorithmic if the subexperimental outcomes occur sequentially and, at least in part, the outcomes arise according to a set algorithm with specific outcomes depending on prior outcomes.

As an example, consider the experiment defined by choosing two numbers according to the two-step algorithm: first, a number, denoted $x$, is chosen, at random, from the interval $[0, 1]$. Second, a second number, denoted $y$, is chosen, at random, from the interval $[0, x]$. The experimental outcome is the ordered pair $(x, y)$.

#### 5.4.1.1 *Correlated or Uncorrelated Experimental Outcomes*

Consistent with the trials of an experiment being independent of one another, all distinct experimental outcomes, forming the sample space underpinning a random process, are independent. However, the outcomes of subexperiments, in general, will be correlated. Such correlatedness, in general, leads to the random process being correlated at different times. Similarly, algorithmic subexperiments lead, in general, to correlated subexperimental outcomes.

### 5.4.2 Specifying Sample Spaces: Single–Vector–Matrix Cases

The $i$th trial of an experiment will, in general, lead to a single outcome $\omega_i$, a vector of outcomes $\omega_i = (\omega_{i1}, \ldots, \omega_{iN})$, a matrix of outcomes $\omega_i = ((\omega_{i11}, \ldots, \omega_{i1M}), \ldots, (\omega_{iN1}, \ldots, \omega_{iNM}))$, etc. A vector of outcomes $\omega_i = (\omega_{i1}, \ldots, \omega_{iN})$ may arise from $N$ repeated trials of a subexperiment or $N$ trials of individual subexperiments; a matrix of outcomes $\omega_i = ((\omega_{i11}, \ldots, \omega_{i1M}), \ldots, (\omega_{iN1}, \ldots, \omega_{iNM}))$ may arise from $N$ trials of a subexperiment where each subexperiment is a result of $M$ sub-sub-experiments. In summary, the following are typical forms for sample spaces:

$$
\begin{aligned}
S &= \{\omega_1, \omega_2, \ldots\} \\
S &= \{(\omega_{11}, \ldots, \omega_{1N}), (\omega_{21}, \ldots, \omega_{2N}), \ldots\} \\
S &= \left\{ \begin{array}{l} ((\omega_{111}, \ldots, \omega_{11M}), \ldots, (\omega_{1N1}, \ldots, \omega_{1NM})), \\ ((\omega_{211}, \ldots, \omega_{21M}), \ldots, (\omega_{2N1}, \ldots, \omega_{2NM})), \ldots \end{array} \right\}
\end{aligned}
\tag{5.23}
$$

assuming a countable number of experimental outcomes. The notation $\omega_{1NM}$ is shorthand notation for $\omega_{1,N,M}$, etc. The interpretation of subscript notation is usually clear from the context.

#### 5.4.2.1 *Probability of Experimental Outcomes*

Using the usual notation $p_\Omega$ and $f_\Omega$, respectively, for a probability mass function and a probability density function and assuming the experimental outcomes are numbers, the probabilities associated with the single, vector, and matrix outcomes of a trial of an experiment are as follows:

1. If the sample space comprises of individual numbers, consistent with $S=\{\omega_1,\omega_2,\ldots\}$, then a random variable $\Omega$ is defined. The probability of the $i$th experimental outcome is

$$\begin{cases} P[\omega_i]=p_\Omega(\omega_i) & \text{countable case} \\ P[\omega_i\in[\omega,\omega+d\omega]]\approx f_\Omega(\omega)d\omega & \text{uncountable case} \end{cases} \tag{5.24}$$

2. If the sample space comprises of vectors of numbers, consistent with $S=\{(\omega_{11},\ldots,\omega_{1N}),(\omega_{21},\ldots,\omega_{2N}),\ldots\}$, then a vector of random variables $(\Omega_1, \ldots, \Omega_N)$ is defined. The probability of the $i$th outcome is

$$\begin{cases} P[(\omega_{i1},\ldots,\omega_{iN})]=p_{\Omega_1\ldots\Omega_N}(\omega_{i1},\ldots,\omega_{iN}) & \text{countable case} \\ P[\omega_{i1}\in[\omega_1,\omega_1+d\omega_1],\ldots,\omega_{iN}\in[\omega_N,\omega_N+d\omega_N]]\approx & \\ \quad f_{\Omega_1\ldots\Omega_N}(\omega_1,\ldots,\omega_N)d\omega_1\ldots d\omega_N & \text{uncountable case} \end{cases} \tag{5.25}$$

3. If the sample space comprises of matrices of numbers consistent with the $i$th experimental outcome having the form $((\omega_{i11}, \ldots, \omega_{i1M}), \ldots, (\omega_{iN1}, \ldots, \omega_{iNM}))$, then a matrix of random variables $((\Omega_{11}, \ldots, \Omega_{1M}), \ldots, (\Omega_{N1}, \ldots, \Omega_{NM}))$ is defined. The probability of the $i$th outcome, assuming the individual vectors are independent, is

$$\begin{cases} P[((\omega_{i11},\ldots,\omega_{i1M}),\ldots,(\omega_{iN1},\ldots,\omega_{iNM}))]= & \\ \quad P_{\Omega_{11}\ldots\Omega_{1M}}(\omega_{i11},\ldots,\omega_{i1M})\cdot\ldots\cdot P_{\Omega_{N1}\ldots\Omega_{NM}}(\omega_{iN1},\ldots,\omega_{iNM}) & \text{countable case} \\ & \\ P[\omega_{i11}\in[\omega_{11},\omega_{11}+d\omega_{11}],\ldots,\omega_{iNM}\in[\omega_{NM},\omega_{NM}+d\omega_{NM}]]\approx & \\ \quad f_{\Omega_{11}\ldots\Omega_{1M}}(\omega_{11},\ldots,\omega_{1M})d\omega_{11}\ldots d\omega_{1M}\cdot\ldots\cdot & \\ \quad f_{\Omega_{N1}\ldots\Omega_{NM}}(\omega_{N1},\ldots,\omega_{NM})d\omega_{N1}\ldots d\omega_{NM} & \text{uncountable case} \end{cases} \tag{5.26}$$

***5.4.2.2 Specifying Random Processes*** Consistent with the common forms for the sample space of experimental outcomes, as detailed in Equation 5.23, the forms for the signal and the signal sample space of the associated random processes (assuming one-dimensional signals) are detailed in Table 5.1. For a general signal, the following notation, consistent with that of Section 5.2.1, is used:

$$\begin{aligned} &X(\omega,t) \quad \text{experiment with individual outcomes} \\ &X((\omega_1,\ldots,\omega_N),t) \quad \text{experiment with vector of outcomes} \\ &X(((\omega_{11},\ldots,\omega_{1M}),(\omega_{N1},\ldots,\omega_{NM})),t) \quad \text{matrix of outcomes} \end{aligned} \tag{5.27}$$

**TABLE 5.1 Common forms for the sample space and signal sample space associated with a random process**

| Sample space $S$ | Notation for $i$th signal defined by random process and the signal sample space $S_X$ |
|---|---|
| $\{\omega_1, \omega_2, \ldots\}$ | $X(\omega_i, t)$ <br> $S_X = \{X(\omega_i, t) : \omega_i \in S\}$ |
| $\left\{\begin{matrix}(\omega_{11},\ldots,\omega_{1N}), \\ (\omega_{21},\ldots,\omega_{2N}),\ldots\end{matrix}\right\}$ | $X((\omega_{i1}, \ldots, \omega_{iN}), t)$ <br> $S_X = \left\{\begin{matrix}X((\omega_{i1},\ldots,\omega_{iN}),t): \\ (\omega_{i1},\ldots,\omega_{iN}) \in S\end{matrix}\right\}$ |
| $\{((\omega_{111},\ldots,\omega_{11M}),\ldots,$ <br> $(\omega_{1N1},\ldots,\omega_{1NM})),\ldots\}$ | $X(((\omega_{i11},\ldots,\omega_{i1M}),\ldots,$ <br> $(\omega_{iN1},\ldots,\omega_{iNM})),t)$ <br> $S_X = \left\{\begin{matrix}X((\omega_{i11},\ldots,\omega_{i1M}),\ldots, \\ ((\omega_{iN1},\ldots,\omega_{iNM})),t): \\ ((\omega_{i11},\ldots,\omega_{i1M}),\ldots, \\ (\omega_{iN1},\ldots,\omega_{iNM})) \in S\end{matrix}\right\}$ |

## 5.5 PROTOTYPICAL EXPERIMENTS

The Bernoulli and Poisson experiments underpin a wide range of random phenomena.

### 5.5.1 Bernoulli Experiment

**Definition: Bernoulli Experiment** A Bernoulli experiment is an experiment with two possible outcomes consistent with the sample space

$$S = \{\text{success, failure}\} \tag{5.28}$$

where $P[\text{success}] = p,\ P[\text{failure}] = 1-p$.

For notational convenience, it is usual to associate the number zero with *failure* and the number one with *success* leading to the sample space

$$S = \{0,1\} \tag{5.29}$$

where $P[1] = p$ and $P[0] = 1-p$. A random variable $\Omega$ is defined by this association, and its probability mass function is defined according to $p_\Omega[0] = 1-p$ and $p_\Omega[1] = p$.

**Definition: Generalized Bernoulli Experiment** A generalized Bernoulli experiment, of order $N$, comprises $N$ independent trials of a Bernoulli experiment as $N$ subexperiments. The resulting sample space $S$ consists of $2^N$ distinct vectors of length $N$:

$$S = \left\{(\omega_{i1},\ldots,\omega_{iN}) : \quad \omega_{ik} \in \{0,1\}, k \in \{1,\ldots,N\}, i \in \{1,\ldots,2^N\}\right\} \tag{5.30}$$

which can be written in the simpler form

$$S = \{(\omega_1,\ldots,\omega_N) : \quad \omega_k \in \{0,1\}, k \in \{1,\ldots,N\}\} \tag{5.31}$$

The associated vector random variable $\Omega = (\Omega_1 \ldots \Omega_N)$ has a probability mass function given by

$$p_\Omega(\omega_{i1},\ldots,\omega_{iN}) = p_{\Omega_1\ldots\Omega_N}(\omega_{i1},\ldots,\omega_{iN}) = p_{\Omega_1}(\omega_{i1})\ldots p_{\Omega_N}(\omega_{iN})$$
$$\omega_{ik} \in \{0,1\}, p_{\Omega_k}(0) = 1-p, p_{\Omega_k}(1) = p \tag{5.32}$$

or using the simpler notation

$$p_\Omega(\omega_1,\ldots,\omega_N) = p_{\Omega_1\ldots\Omega_N}(\omega_1,\ldots,\omega_N) = p_{\Omega_1}(\omega_1)\ldots p_{\Omega_N}(\omega_N)$$
$$\omega_k \in \{0,1\}, p_{\Omega_k}(0) = 1-p, p_{\Omega_k}(1) = p \tag{5.33}$$

An experimental outcome $(\omega_1, \ldots, \omega_N)$ consists of $k$ *successes* and $N-k$ *failures* and, consistent with Equation 5.32, occurs with a probability of $p^k(1-p)^{N-k}$. Here, the *successes* and *failures* occur in a specific order, and the experimental outcome is often called an *ordered outcome*.

The probability of an outcome comprising of $k$ *successes* and $N-k$ *failures*, when order is not considered, is given by the binomial probability mass function, that is,

$$P[k \text{ successes in } N \text{ trials}] = b(k,N,p) = \frac{N!}{k!(N-k)!} \cdot p^k(1-p)^{N-k} \tag{5.34}$$

### 5.5.2 Poisson Experiment

Underpinning the Poisson experiment is the experiment of choosing a number at random from a set interval, and the interval $[0,T)$ is convenient.

**Definition: Random Experiment for Interval** The experiment of selecting a number, at random, from the interval $[0,T)$ defines a sample space

$$S = \{\omega: \ \omega \in [0,T)\} \tag{5.35}$$

and an associated random variable $\Omega$ with a probability density function

$$f_\Omega(\omega) = \begin{cases} 1/T & 0 \le \omega < T \\ 0 & \text{elsewhere} \end{cases} \tag{5.36}$$

**Definition: Pseudo-Poisson Experiment for Interval $[0,T)$** Consider the experiment defined by ordering the outcomes of $N$ independent subexperiments where each subexperiment is a random experiment for $[0,T)$. The sample space is

$$S = \{(\gamma_1,\ldots,\gamma_N): \ \gamma_k \in [0,T), k \in \{1,\ldots,N\}, \gamma_1 \le \ldots \le \gamma_N\} \tag{5.37}$$

Such an experiment yields an average density of points of $\lambda = N/T$ points per unit interval and is a *pseudo-Poisson* experiment. This term is used because if it is known that there are $N-k$ points in the interval $[0, T-\tau]$, then it follows that there are $k$ points in the interval $(T-\tau, T)$. For this case, there is dependence between the number of points in the two disjoint intervals, and such a property is not consistent with the goal of a set of independent points as given by a Poisson experiment.

The associated vector random variable $\Omega = (\Gamma_1, \ldots, \Gamma_N)$ has a probability mass function given by

$$f_{\Gamma_1 \cdots \Gamma_N}(\gamma_1, \ldots, \gamma_N) = N! f_{\Gamma_1}(\gamma_1) \ldots f_{\Gamma_N}(\gamma_N) = \frac{N!}{T^N} \tag{5.38}$$

The $N!$ here arises from the ordering of the outcomes, and this result is proved in Chapter 11.

#### 5.5.2.1 *Poisson Experiment*

Consider the sequence of experiments $\{E_i\}_{i=1}^{\infty}$, defined by pseudo-Poisson experiments, on the intervals:

$$\begin{array}{ll} [0, T_o) & N_o \text{ points with rate } \lambda = N_o/T_o \\ [0, 2T_o) & 2N_o \text{ points with rate } \lambda = N_o/T_o \\ \ldots & \\ [0, iT_o) & iN_o \text{ points with rate } \lambda = N_o/T_o \\ \ldots & \end{array} \tag{5.39}$$

The corresponding sequence of sample spaces $\{S_i\}_{i=1}^{\infty}$ can be defined:

$$\begin{array}{ll} S_1 = \{(\gamma_1, \ldots, \gamma_{N_o}): & \gamma_k \in [0, T_o), \gamma_1 \leq \ldots \leq \gamma_{N_o}\} \\ S_2 = \{(\gamma_1, \ldots, \gamma_{2N_o}): & \gamma_k \in [0, 2T_o), \gamma_1 \leq \ldots \leq \gamma_{2N_o}\} \\ \ldots & \\ S_i = \{(\gamma_1, \ldots, \gamma_{iN_o}): & \gamma_k \in [0, iT_o), \gamma_1 \leq \ldots \leq \gamma_{iN_o}\} \\ \ldots & \end{array} \tag{5.40}$$

Such a sequence of sample spaces underpins the definition of a Poisson experiment:

**Definition: Poisson Experiment** A Poisson experiment with a rate $\lambda$, and for an interval $[0, T)$, is any pseudo-Poisson experiment, with an associated sample space from $\{S_i\}_{i=1}^{\infty}$, which is such that for any maximum number of points $N_{max}$ to be considered, the following approximation, as proved in Theorem 5.1, is valid:

$$P[k \text{ points on } [0, T)] \approx \frac{(\lambda T)^k e^{-\lambda T}}{k!} \qquad k \in \{0, 1, \ldots, N_{max}\} \tag{5.41}$$

A Poisson experiment is one where, by choosing points arbitrarily on an interval $[0,iT)$, the points, and number of points, appear randomly on the interval $[0,T)$ consistent with an *infinite* interval being chosen.

**Theorem 5.1 PMF for the Number of Points Defined by Poisson Experiment**
The number of points in an interval $[0,T)$, as defined by a Poisson experiment with a rate $\lambda$, is a random variable with the following probability mass function:

$$P[k \text{ points on } [0,T)] \approx \frac{(\lambda T)^k e^{-\lambda T}}{k!} \qquad k \in \{0,1,\ldots\} \tag{5.42}$$

**Proof**
Consider the pseudo-Poisson experiment defined on the interval $[0,iT_o)$ where the number of points is $iN_o$ and the rate is $\lambda = N_o/T$. The $iN_o$ points arise from repeating a random experiment on the interval $[0,iT_o)$ $iN_o$ times. By considering *success* in a trial of this random experiment as a point in the interval $[0,T)$, which occurs with a probability of $1/i$, it follows that the probability of $k$ *successes* in $iN_o$ trials, consistent with $k$ points on the interval $[0,T)$, is given by the binomial probability mass function:

$$P[k \text{ points on } [0,T)] = b(k, iN_o, 1/i) \approx \frac{(iN_o/i)^k e^{-iN_o/i}}{k!} = \frac{(\lambda T)^k e^{-\lambda T}}{k!} \tag{5.43}$$

as $N_o = \lambda T$. The approximation arises from the Poisson approximation to the binomial defined in Theorem 4.29. The approximation can be made arbitrarily close by choosing $i$ sufficiently large.

**Definition: Poisson Sample Space** The sample space associated with a Poisson experiment for an interval $[0,T)$, with a rate $\lambda > 0$, is defined as a set of infinite length vectors (meaning a vector of sufficiently long length to cover the interval of interest such that Eq. 5.41 is valid) and is defined according to

$$S = \{(\gamma_1,\ldots,\gamma_k,\ldots): \ \gamma_k \in \boldsymbol{R}^+, \gamma_1 \le \ldots \le \gamma_k \le \ldots\} \tag{5.44}$$

### *5.5.2.2 Union of Outcomes from Two Poisson Experiments*

**Theorem 5.2 Union of Outcomes from Two Poisson Experiments**
Consider two independent Poisson experiments which, on the $i$th trial, result in the outcomes

$$\begin{aligned} &(\omega_{i1},\ldots,\omega_{ik},\ldots): \omega_{ik} \in \boldsymbol{R}^+, \omega_{i1} \le \ldots \le \omega_{ik} \le \ldots \quad \text{rate} = \lambda_1 \\ &(\psi_{i1},\ldots,\psi_{ik},\ldots): \psi_{ik} \in \boldsymbol{R}^+, \psi_{i1} \le \ldots \le \psi_{ik} \le \ldots \quad \text{rate} = \lambda_2 \end{aligned} \tag{5.45}$$

If these outcomes are combined to form a new experimental outcome

$$(\gamma_{i1},\ldots,\gamma_{ik},\ldots): \ \gamma_{ik} \in \boldsymbol{R}^+, \gamma_{i1} \le \ldots \le \gamma_{ik} \le \ldots \tag{5.46}$$

then this outcome is consistent with one experimental outcome from a Poisson experiment with a rate $\lambda_1 + \lambda_2$.

**Proof**

This result arises from considering two pseudo-Poisson experiments with $N_1 = \lambda_1 T_o$ and $N_2 = \lambda_2 T_o$ outcomes, chosen at random on the interval $[0, T_o)$ for $T_o$ sufficiently large. Combining the outcomes yields $N_1 + N_2$ outcomes on the interval $[0, T_o)$ at an average rate of $(N_1 + N_2)/T_o = \lambda_1 + \lambda_2$.

### 5.5.3 Experiments with an Infinite Number of Subexperiments

In many instances, and consistent with the Poisson experiment defined previously, the number of suboutcomes associated with a single trial of a given experiment is unbounded. Often, this situation arises by repeating a subexperiment an unbounded number of times—something that is clearly not possible. Accordingly, the following definition is made.

**Definition: Experiment Based on an Infinite Number of Subexperiments** An experiment with an infinite number of suboutcomes in a single trial is one where the number of subexperiments has been repeated a sufficiently large number of times, as required, to ensure the integrity of the definition of the random process over the time interval of interest.

## 5.6 RANDOM VARIABLES DEFINED BY A RANDOM PROCESS

Consider the case where an random process $X: S \to S_X$ defines the sample space of one-dimensional functions:

$$S_X = \{X(\omega, t): \quad \omega \in S\} \tag{5.47}$$

as illustrated in Figure 5.7. For this case, the values defined by the functions at a specific time $t_0$ define a random variable, denoted $X_{t_0}$, with a sample space $S_{t_0}$:

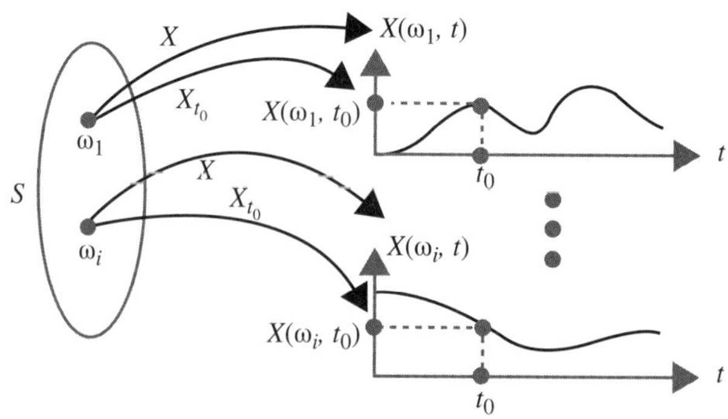


**FIGURE 5.7** Illustration of the values of a random variable defined by a random process at a fixed time.

$$
\begin{aligned}
&X_{t_0} : S \rightarrow S_{t_0} \quad X_{t_0}(\omega) = X(\omega, t_0) \\
&S_{t_0} = \{X_{t_0}(\omega) : \ X_{t_0}(\omega) = X(\omega, t_0), \ \ \omega \in S\}
\end{aligned}
\tag{5.48}
$$

By varying the time from $t_0$ to $t_1$ to $t_2$, etc., an infinite sequence of random variables can be defined. In some contexts, it is appropriate to define a random process as a set of random variables.

## 5.7 CLASSIFICATION OF RANDOM PROCESSES

Classification is fundamental to most areas of science and is of importance to random process theory. A starting point for classification is the demarcation between one-, two-, and higher-dimensional random processes.

### 5.7.1 One-Dimensional Random Processes

The usual case encountered in the physical sciences is of a time-varying signal associated with a set point in space, for example, the signal received at the antenna in a wireless communication system or the signal at a test point in a system, and where the observed signal can be associated with a trial of an experiment and, hence, a random process. Such a signal is one-dimensional and the set of possible signals defines a one-dimensional random process.

As an example, consider the following signal arising from a random process:

$$X(((a_1,\gamma_1),\ldots,(a_k,\gamma_k),\ldots),t) = \sum_{k=1}^{\infty} a_k p(t-\gamma_k) \qquad t \geq 0 \tag{5.49}$$

where $a_k$ and $\gamma_k$ are outcomes, respectively, of random variables $A_k$ and $\Gamma_k$ and $p : \boldsymbol{R} \rightarrow \boldsymbol{R}$ is a pulse function. The random process underpinning the outcome can be written, for notational convenience, as

$$X(\Omega,t) = \sum_{k=1}^{\infty} A_k p(t-\Gamma_k) \tag{5.50}$$

where $\Omega = ((A_1,\Gamma_1),(A_2,\Gamma_2),\ldots)$. One signal, $X(((a_1, \gamma_1), \ldots, (a_k, \gamma_k), \ldots), t)$, defined by this random process is graphed in Figure 5.8 for the case of independent amplitudes taken from a Gaussian distribution with a mean of zero and a variance of unity, that is, $A_k \sim N[0,1]$, and with the times $\gamma_1, \ldots, \gamma_k, \ldots$ being consistent with outcomes of a Poisson experiment with a rate of $\lambda = 10$ per unit time interval. A pulse function $p$ with unity energy and defined according to

$$p(t) = \frac{2t^2}{\sqrt{3}\tau_p^{5/2}} e^{-t/\tau_p} u(t) \qquad \tau_p = 1 \tag{5.51}$$

has been used.

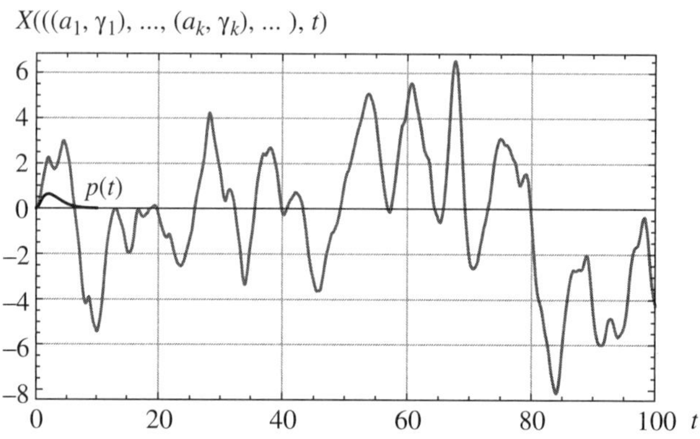


**FIGURE 5.8** One signal defined by the random process specified by Equations 5.50 and 5.51 for the case of $\tau_p = 1$.

### 5.7.2 Two-Dimensional Random Processes

Naturally, there are phenomena that yield, at each point on a line or along a path, a time-varying parameter such that a two-dimensional signal is defined. Examples include the electric field, or electrostatic potential, along the edge of a surface, the wind speed along a path, etc. When the observed signal can be associated with a trial of an experiment and, hence, a random process, a two-dimensional random process is defined.

As an example of such a random process, consider the signal

$$X(((a_1,x_1,\gamma_1),\ldots,(a_k,x_k,\gamma_k),\ldots),x,t) = \sum_{k=1}^{\infty} a_k p(x-x_k,t-\gamma_k) \quad x \in [0,x_o], t \geq 0 \tag{5.52}$$

where $a_k$, $x_k$, and $\gamma_k$ are outcomes, respectively, of random variables $A_k$, $X_k$, and $\Gamma_k$ and $p: \boldsymbol{R}^2 \to \boldsymbol{R}$ is a two-dimensional pulse function. The random process underpinning the outcome can be written, for notational convenience, as

$$X(\Omega,x,t) = \sum_{k=1}^{\infty} A_k p(x-X_k,t-\Gamma_k), \tag{5.53}$$

where $\Omega = ((A_1,X_1,\Gamma_1),(A_2,X_2,\Gamma_2),\ldots)$. One signal from the random process is shown in Figure 5.9 for the case of $A_1, \ldots$ being independent zero-mean Gaussian random variables with unit variances, $X_1, \ldots, X_k, \ldots$ and $\Gamma_1, \ldots, \Gamma_k, \ldots$ being random variables defined by the outcomes of independent Poisson experiments with a rate of 10 per unit interval. The pulse function used is defined according to

$$p(x,t) = \frac{4x^2t^2}{3\tau_x^{5/2}\tau_p^{5/2}} e^{-x/\tau_x} e^{-t/\tau_p} u(x)u(t) \qquad \tau_x = \tau_p = 1 \tag{5.54}$$

and has unit energy. The graph of this function is shown in Figure 5.10.

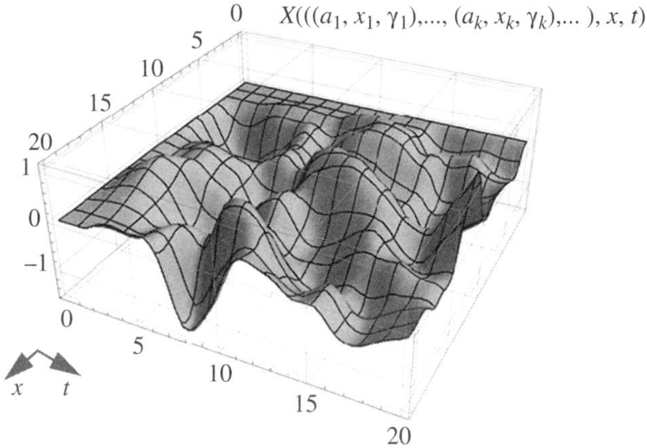


**FIGURE 5.9** A two-dimensional signal defined by the random process specified by Equation 5.53 and based on the pulse function defined by Equation 5.54.

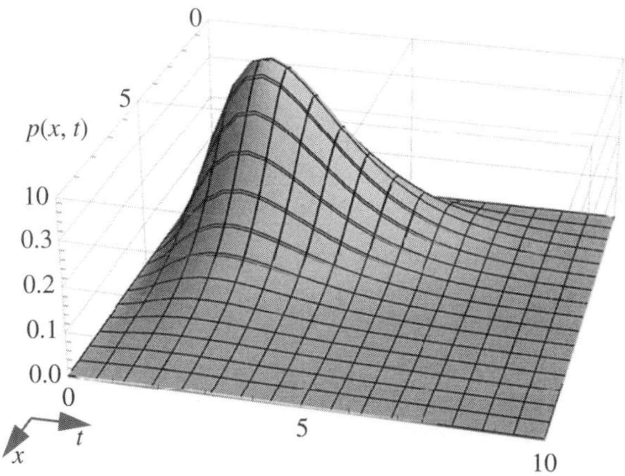


**FIGURE 5.10** Illustration of the two-dimensional pulse function defined by Equation 5.54 for the case of $\tau_x = \tau_p = 1$. The pulse energy is unity.

### 5.7.3 Higher-Dimensional Random Processes

Random processes based on appropriate experiments, which define higher-dimensional functions can be defined in a similar manner. For example, for the three-dimensional case, and using simplified notation, one random process can be defined according to

$$X(\Omega,x,y,t) = \sum_{k=1}^{\infty} A_k p(x - X_k, y - Y_k, t - \Gamma_k) \qquad \begin{cases} x \in [0,x_o], y \in [0,y_o] \\ t \geq 0 \end{cases} \tag{5.55}$$

where $\Omega = ((A_1, X_1, Y_1, \Gamma_1), (A_2, X_2, Y_2, \Gamma_2), \ldots)$ and $p$ is an appropriate pulse function. This random process defines signals that specify variation, with time, on a surface $[0,x_o] \times [0,y_o]$.

## 5.8 CLASSIFICATION: ONE-DIMENSIONAL RPs

The classification of one-dimensional (and similarly for higher-dimensional) random processes can be done in many ways. A useful classification, assuming distinct experimental outcomes lead to distinct signals in the signal sample space, is based on the following distinctions:

1. Whether the sample space $S$ comprises of a countable, or uncountable, number of elements
2. Whether the signals in the signal sample space are discrete time, or continuous time, signals
3. Whether the range of the signals in the signal sample space is consistent with a countable, or an uncountable, set

These distinctions along with the following definitions provide a basis for classifying one-dimensional random processes.

**Definition: Countable and Uncountable Random Process** A random process is countable if the number of elements in the sample space is countable; otherwise, it is uncountable.

**Definition: State** A state of a random process is a set value that a signal defined by the random process takes on at a specific time.

**Definition: State Space** The set of possible states defined by a random process is called the state space.

With these definitions, the classification of random processes, as detailed in Table 5.2, can be made.

The three different types of random processes are illustrated in Figure 5.11 for the countable sample space case assuming $S=\{0,1\}$.

**TABLE 5.2 Classification of one-dimensional random processes: $\boldsymbol{R}_C$ denotes a countable set of $\boldsymbol{R}$, while $\boldsymbol{R}_U$ denotes an uncountable set of $\boldsymbol{R}$**

| Sample space: $S$ | $i$th signal: $X(\omega_i)$ | Classification |
|---|---|---|
| Countable case: $S=\{\omega_i : i\in \boldsymbol{Z}^+\}$ | $X(\omega_i):\boldsymbol{R}_C \to \boldsymbol{R}_C$ | Discrete time–discrete state |
| | $X(\omega_i):\boldsymbol{R}_C \to \boldsymbol{R}_U$ | Not possible |
| | $X(\omega_i):\boldsymbol{R}_U \to \boldsymbol{R}_C$ | Continuous time–discrete state |
| | $X(\omega_i):\boldsymbol{R}_U \to \boldsymbol{R}_U$ | Continuous time–continuous state |
| Uncountable case: $S=(\omega :\ \ \omega\in \boldsymbol{R})$ | $X(\omega):\boldsymbol{R}_C \to \boldsymbol{R}_C$ | Discrete time–discrete state |
| | $X(\omega):\boldsymbol{R}_C \to \boldsymbol{R}_U$ | Discrete time–continuous state |
| | $X(\omega):\boldsymbol{R}_U \to \boldsymbol{R}_C$ | Continuous time–discrete state |
| | $X(\omega):\boldsymbol{R}_U \to \boldsymbol{R}_U$ | Continuous time–continuous state |

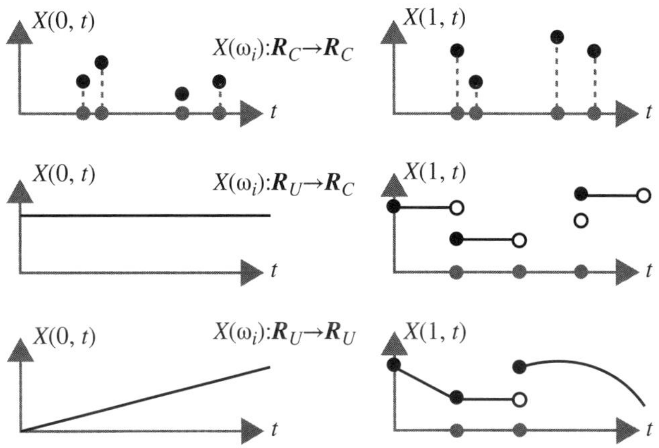


**FIGURE 5.11** Three types of random processes for the countable sample space with $S = \{0,1\}$. Top: signals from a discrete time–discrete state random process. Middle: signals from a continuous time–discrete state random process. Bottom: signals from a continuous time–continuous state random process.

**TABLE 5.3 State space demarcation of random processes**

| | State space | |
|---|---|---|
| Countability of domain of random process | Countable # states: discrete state random process | Uncountable # states: continuous state random process |
| Countable: discrete time random process | Discrete time–discrete state random process | Discrete time–continuous state random process |
| Uncountable: continuous time random process | Continuous time–discrete state random process | Continuous time–continuous state random process |

## 5.8.1 Classification According to State

The signals defined by a random process can be defined at a countable, or an uncountable, number of points. The range of the signals defined by a random process, similarly, can be either a countable or uncountable set of numbers. Such distinctions lead to the classification of random processes according to their state space as detailed in Table 5.3. A discrete time–continuous state random process can arise, for example, from a random process with a signal sample space comprising of an uncountable number of signals and where the signals, while only defined at discrete times, can take on values from a subset of the real line. Signals from the different types of random processes detailed in Table 5.3 are illustrated in Figures 5.12 and 5.13.

***5.8.1.1 Quantization*** In the conversion of information, from an analogue form to a digital form, quantization is widely used. In general, both time and amplitude can be quantized as illustrated in Figure 5.14. The change in classification of a random process arising from quantization is detailed in Table 5.4.

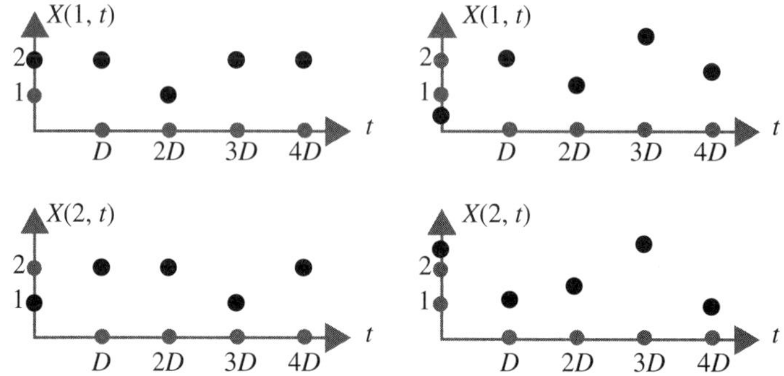


**FIGURE 5.12** Illustration of two waveforms from different discrete time random processes. Left: waveforms from a discrete time–discrete state random process. Right: waveforms from a discrete time–continuous state random process.

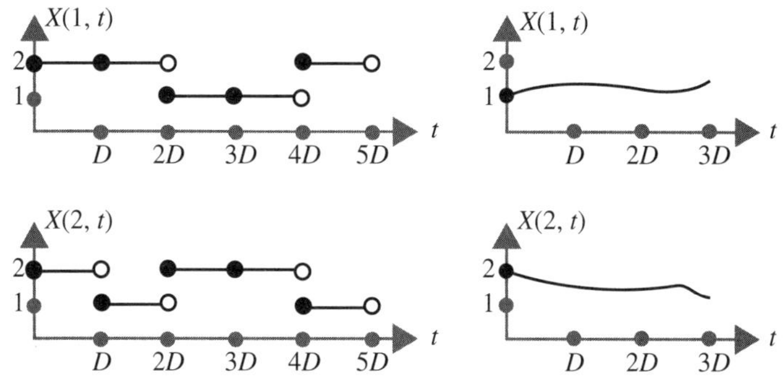


**FIGURE 5.13** Illustration of two waveforms from continuous time random processes. Left: waveforms from a continuous time–discrete state random process. Right: waveforms from a continuous time–continuous state random process.

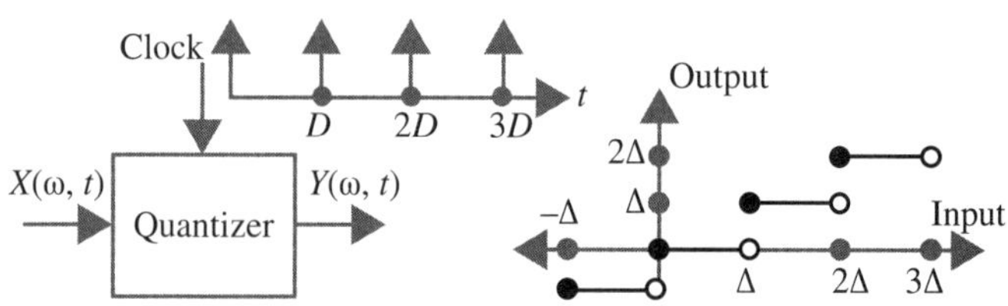


**FIGURE 5.14** Time and amplitude quantization of signals from a random process. The quantization of the amplitude is illustrated in the right-hand graph.

**TABLE 5.4 Change in the classification of a random process arising from quantization**

| | After quantization | | |
|---|---|---|---|
| Before quantization | Time quantization | Amplitude quantization | Time and amplitude quantization |
| Discrete time–discrete state random process | | | |
| Discrete time–continuous state random process | | Discrete time–discrete state random process | |
| Continuous time–discrete state random process | Discrete time–discrete state random process | | |
| Continuous time–continuous state random process | Discrete time–discrete state random process (countable random process)<br>Discrete time–continuous state random process (uncountable random process) | Continuous time–discrete state random process | Discrete time–discrete state random process |

## 5.9 SUMS OF RANDOM PROCESSES

Consider two one-dimensional random processes $X_1$ and $X_2$ based on the respective experimental sample spaces $S_1$ and $S_2$, which may be correlated or independent etc. The respective signal sample spaces are denoted $S_{X_1}$ and $S_{X_2}$. A new random process $X_3$ can be defined on the Cartesian product of the individual experimental sample spaces. One example is the sum of the individual random processes with $X_3$ being defined according to

$$\begin{aligned} &X_3 : S_1 \times S_2 \rightarrow S_{X_3} \\ &X_3((\omega_1,\omega_2),t) = X_1(\omega_1,t) + X_2(\omega_2,t) \\ &S_{X_3} = \{X_1(\omega_1,t) + X_2(\omega_2,t) : \ \omega_1 \in S_1, \omega_2 \in S_2\} \end{aligned} \tag{5.56}$$

For notational convenience, $X_3$ is written as

$$X_3(\Omega,t) = X_1(\Omega_1,t) + X_2(\Omega_2,t), \tag{5.57}$$

where $\Omega = (\Omega_1,\Omega_2)$.

## 5.10 PROBLEMS

**5.1** Consider the experiment where one trial is defined by:

i. Selecting a number at random from the interval $[-\Delta/2,\ \Delta/2]$
ii. Repeating (i) $N$ times

a. Specify the experimental sample space.
b. A random process is defined on this experiment by associating the numbers with consecutive time intervals of duration $D$ commencing at $t=0$. Define the random process and draw representative signals defined by the random process. This random process provides a first-order model for quantization noise arising from a digital to analogue converter.

**5.2** Consider the experiment where one trial is defined according to:

i. Selecting a number from the real line and consistent with a Gaussian probability density function with zero mean and a variance $\sigma^2$
ii. Repeating (i) $N$ times

a. Specify the experimental sample space.
b. A random process is defined on this experiment by associating the numbers with the jitter of the rising and falling edges of a binary clock signal commencing at $t=0$, with amplitudes of zero and $A$, and with a period of $D$ seconds. Define the random process and draw representative signals for the case where $\sigma=0.1D$.

**5.3** Consider an experiment where one trial is defined according to:

i. Selecting a number, at random, from the interval $[0, T]$
ii. Selecting a second number, at random, from the interval $[0, \Delta]$
iii. Pairing the numbers from (i) and (ii) and repeating the selection $N$ times
iv. Ordering the pairs of numbers based on the first number

a. Specify the experimental sample space.
b. A random process is defined on this experiment by associating the first number in each pair with the commencement of a causal pulse function $p_1$ and the second number in each pair with the delay from the first pulse for the commencement of a causal pulse function $p_2$. Define the random process and draw representative signals for the case where $\Delta < T/N$ and the duration of both pulse functions is much less than $T/N$. Assume the energy in $p_2$ is less than the energy in $p_1$. Such a random process is an example of a random process where one large event triggers a second event.

**5.4** Consider a random process defined on a Poisson experiment with a rate $\lambda$ in a manner consistent with a random walk: at each point defined by a trial of the experiment, the level moves up $\Delta$ with a probability of $p$ and moves down $\Delta$ with a probability of $1-p$.

a. Specify the experiment underpinning the random process.
b. Mathematically specify the random process.
c. Draw representative signals from the random process.

**5.5** Consider a two-dimensional random walk random process where, at each unit time interval, the two components of the spatial coordinates of the path are incremented, at random, from the set $\{-1, 1\}$. At $t=0$, the random walk is at the origin.

a. Specify the experiment underpinning the random process.
b. Mathematically specify the random process as a set of trajectories in three dimensions (two spatial dimensions plus time).
c. Draw representative trajectories from the random process.

**5.6** Consider an experiment with a sample space defined according to

$$S=\{\omega_i: \ \omega_i \in \{0,1,\ldots\}\} \tag{5.58}$$

and with probabilities of outcomes defined by a Poisson probability mass function, that is, $p_\Omega(\omega_i)=\dfrac{\lambda^{\omega_i}e^{-\lambda}}{\omega_i!}$, $\lambda>0$. Give examples of random processes based on this experiment that are:

a. Discrete time–discrete state random processes
b. Continuous time–discrete state random processes
c. Continuous time–continuous state random processes

For each random process, specify the state space.

**5.7** Consider an experiment with a sample space defined according to

$$S=\{\omega_i: \ \omega_i \in [0,1]\} \tag{5.59}$$

and with probabilities of outcomes defined by a uniform distribution, that is, $f_\Omega(\omega_i)=1$, $\omega_i \in [0,1]$. Give examples of random processes based on this experiment that are:

a. Discrete time–discrete state random processes
b. Discrete time–continuous state random processes
c. Continuous time–discrete state random processes
d. Continuous time–continuous state random processes

For each random process, specify the state space.

# 6

# PROTOTYPICAL RANDOM PROCESSES

## 6.1 INTRODUCTION

In this chapter, the prototypical, and important, random processes encountered in the physical sciences are defined. The Bernoulli, generalized Bernoulli, the random walk, Poisson, the random telegraph signal, shot noise, clustered, signalling, jitter, white noise, $1/f$ noise, birth–death, and orthogonal increment random processes are defined. The experiments, and notation for random processes, defined in the previous chapter are used. The simple as possible random process is the Bernoulli random process.

## 6.2 BERNOULLI RANDOM PROCESSES

**Definition: Bernoulli Random Processes** A Bernoulli random process is defined on a Bernoulli experiment according to

$$\begin{aligned}
&X : S \to S_X \\
&S = \{\text{failure, success}\} \quad \text{or} \quad S = \{0,1\} \\
&p_\Omega(\omega) = \begin{cases} 1-p & \omega = 0 \\ p & \omega = 1 \end{cases} \\
&S_X = \{X(\omega,t) : \omega \in S\} = \{X(0,t), X(1,t)\}
\end{aligned} \tag{6.1}$$

*A Signal Theoretic Introduction to Random Processes*, First Edition. Roy M. Howard.

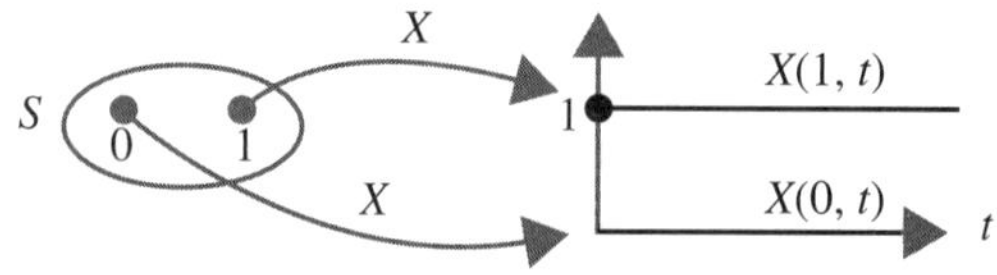


**FIGURE 6.1** Two signals arising from the Bernoulli random process defined in the text.

Clearly,

$$\begin{aligned} P[X(0,t)] &= P[\omega = 0] = 1 - p \\ P[X(1,t)] &= P[\omega = 1] = p \end{aligned} \tag{6.2}$$

One example of a Bernoulli random process is defined by the signal sample space

$$S_X = \{0, u(t)\} \qquad \begin{cases} X(0,t) = 0 \\ X(1,t) = u(t) \end{cases} \tag{6.3}$$

This random process, and the defined signals, is illustrated in Figure 6.1.

### 6.2.1 Generalized Bernoulli Random Processes

**Definition: Generalized Bernoulli Random Process** A generalized Bernoulli random process, of order $N$, is defined on a generalized Bernoulli experiment according to

$$\begin{aligned} & X : S \to S_X \\ & S = \{(\omega_1, \ldots, \omega_N) : \ \omega_k \in \{0,1\}, k \in \{1, \ldots, N\}\} \\ & p_{\Omega_1 \ldots \Omega_N}(\omega_1, \ldots, \omega_N) = p_{\Omega_1}(\omega_1) \ldots p_{\Omega_N}(\omega_N) \\ & \quad p_{\Omega_k}(0) = 1 - p, p_{\Omega_k}(1) = p \\ & S_X = \{X((\omega_1, \ldots, \omega_N), t) : \ (\omega_1, \ldots, \omega_N) \in S\} \end{aligned} \tag{6.4}$$

One example of a generalized Bernoulli random process is that of binary digital communication signalling as detailed in Section 5.3.5. Another important case is that of the random walk.

### 6.2.2 Random Walk

The quintessential nonstationary random process is the random walk random process, or random walk for short. The limit of a random walk, in terms of an increasingly small step size and step interval, assuming appropriate scaling, yields the Wiener process or one-dimensional Brownian motion.

A random walk of order $N$ is a generalized Bernoulli random process of order $N$ and is defined according to

$$X:S\rightarrow S_X \qquad X((\omega_1,\ldots,\omega_N),t)=\sum_{k=1}^{N}[-1+2\omega_k]u(t-(k-1))$$
$$S_X=\left\{\begin{array}{c}X((\omega_1,\ldots,\omega_N),t)=\sum_{k=1}^{N}[-1+2\omega_k]u(t-(k-1)):\\ (\omega_1,\ldots\omega_N)\in S\end{array}\right\} \tag{6.5}$$

where

$$-1+2\omega_k=\begin{cases}-1 & \omega_k=0\\ 1 & \omega_k=1\end{cases} \tag{6.6}$$

For notational convenience, the random walk random process is written as

$$X(\Omega,t)=\sum_{k=1}^{N}A_k u(t-(k-1)) \tag{6.7}$$

where $\Omega=(A_1,\ldots A_N)$ is a vector of identical and independent random variables. The sample space of $A_k$ is $\{-1,1\}$ and $P[A_k=1]=p,\ P[A_k=0]=1-p$. Several waveforms from the signal sample space are illustrated in Figures 6.2 and 6.3 for the case of $p=0.5$.

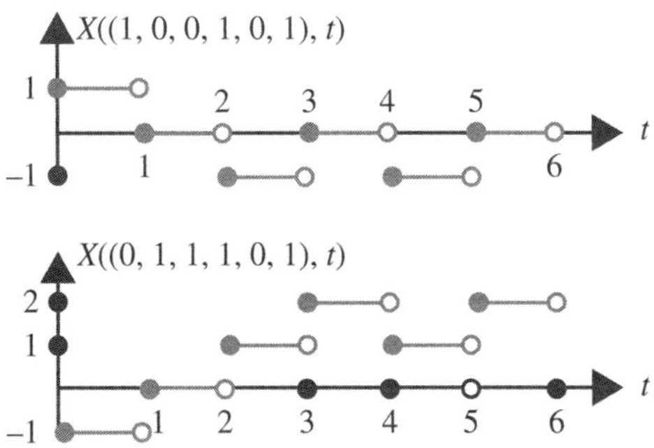


**FIGURE 6.2** Two waveforms defined by the random walk random process.

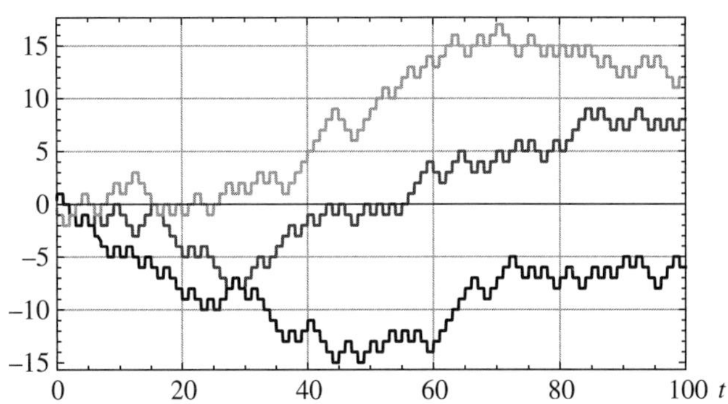


**FIGURE 6.3** Three waveforms defined by the random walk random process.

## 6.3 POISSON RANDOM PROCESSES

The Poisson experiment defined in the previous chapter underpins the definition of many random phenomena. Some common examples are detailed in the following subsections and assume the Poisson experiment sample space:

$$S = \{(\gamma_1, \ldots, \gamma_k, \ldots) : \gamma_k \in \boldsymbol{R}^+,\ \gamma_1 \le \ldots \le \gamma_k \le \ldots\} \tag{6.8}$$

where the point rate is $\lambda$ and

$$P[k \text{ points on } [0,T)] \approx \frac{(\lambda T)^k e^{-\lambda T}}{k!} \qquad k \in \{0,1,\ldots\} \tag{6.9}$$

The fundamental Poisson random process is the Poisson point process.

### 6.3.1 Poisson Point Process

**Definition: Poisson Point Process** A Poisson point process $X$ is a random process based on a Poisson experiment and defined according to

$$\begin{aligned} &X : S \to S_X \qquad X((\gamma_1,\ldots,\gamma_k,\ldots),t) = \sum_{k=1}^{\infty} \delta_K(t-\gamma_k) \\ &S_X = \left\{ X((\gamma_1,\ldots,\gamma_k,\ldots),t) = \sum_{k=1}^{\infty} \delta_K(t-\gamma_k) :\ (\gamma_1,\ldots,\gamma_k,\ldots) \in S \right\} \end{aligned} \tag{6.10}$$

where $\delta_K$ is the Kronecker delta. The Kronecker delta can also be replaced by the Dirac delta $\delta$ (which is to imply an infinite sequence of functions consistent with the Dirac delta and where the functions in this sequence have the properties as required by the application). The Dirac delta form is useful for modelling inputs into a system with time domain filtering.

For notational convenience, a Poisson point random process is written as

$$X(\Omega,t) = \sum_{k=1}^{\infty} \delta_K(t-\Gamma_k) \quad \text{or} \quad X(\Omega,t) = \sum_{k=1}^{\infty} \delta(t-\Gamma_k) \tag{6.11}$$

where $\Omega = (\Gamma_1, \Gamma_2, \ldots)$ is a vector of random variables consistent with the outcomes of a Poisson experiment. One outcome of a Poisson point process is illustrated in Figure 6.4 for the Kronecker delta case.

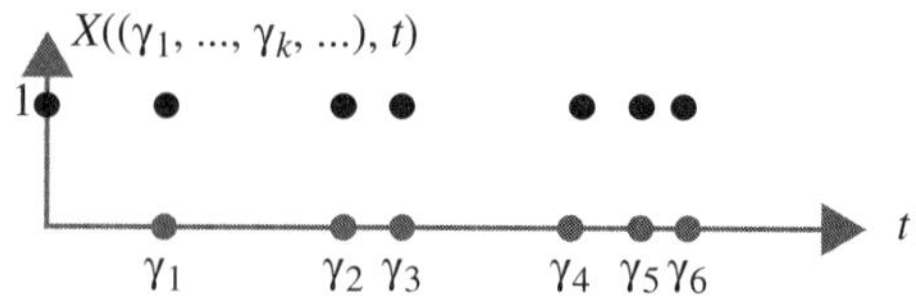


**FIGURE 6.4** One waveform defined by the Poisson point process assuming the Kronecker delta definition.

*6.3.1.1 Arrival Times and Interarrival Times* A Poisson point random process defines a vector of random variables consistent with the arrival times, and interarrival times, of the points.

**Definition: Vector Random Variable of Arrival Times** The arrival times of a Poisson point random process define a vector random variable, $X: S \to S_X$, according to

$$\begin{aligned} &X((\gamma_1,\ldots,\gamma_k,\ldots)) = (\gamma_1,\ldots,\gamma_k,\ldots) \\ &S_X = S \end{aligned} \tag{6.12}$$

**Definition: *K*th Arrival Time** The *k*th arrival time is defined by the random variable $X_k: S \to S_{X_k}$ according to

$$\begin{aligned} &X_k((\gamma_1,\ldots,\gamma_k,\ldots)) = \gamma_k \qquad k \geq 1 \\ &S_{X_k} = \{\gamma_k: \ (\gamma_1,\ldots,\gamma_k,\ldots) \in S\} \end{aligned} \tag{6.13}$$

**Definition: Vector Random Variable of Interarrival Times** The interarrival times define a vector random variable, $X_\Delta: S \to S_\Delta$, according to

$$\begin{aligned} &X_\Delta((\gamma_1,\ldots,\gamma_k,\ldots)) = (\gamma_1 - 0, \gamma_2 - \gamma_1,\ldots,\gamma_k - \gamma_{k-1},\ldots) \\ &S_\Delta = \left\{ \begin{matrix} (\gamma_1, \gamma_2 - \gamma_1,\ldots,\gamma_k - \gamma_{k-1},\ldots): \\ (\gamma_1,\ldots,\gamma_k,\ldots) \in S \end{matrix} \right\} \end{aligned} \tag{6.14}$$

**Definition: *k*th Interarrival Time** The *k*th interarrival time is defined by the random variable $X_{\Delta,k}: S \to S_{\Delta_k}$ according to

$$\begin{aligned} &X_{\Delta,k}((\gamma_1,\ldots,\gamma_k,\ldots)) = \gamma_k - \gamma_{k-1} \qquad \gamma_0 = 0, k \geq 1 \\ &S_{\Delta_k} = \{\gamma_k - \gamma_{k-1}: \ (\gamma_1,\ldots,\gamma_{k-1},\gamma_k,\ldots) \in S\} \end{aligned} \tag{6.15}$$

Equivalently, $X_{\Delta,1} = X_1$ and $X_{\Delta,k} = X_k - X_{k-1}$ for $k \geq 2$.

### 6.3.2 Poisson Counting Process

**Definition: Poisson Counting Process** A Poisson counting process $X$ is a random process based on a Poisson experiment and defined according to

$$\begin{aligned} &X: S \to S_X \qquad X((\gamma_1,\ldots,\gamma_k,\ldots),t) = \sum_{k=1}^{\infty} u(t-\gamma_k) \\ &S_X = \left\{ X((\gamma_1,\ldots,\gamma_k,\ldots),t) = \sum_{k=1}^{\infty} u(t-\gamma_k): \ (\gamma_1,\ldots,\gamma_k,\ldots) \in S \right\} \end{aligned} \tag{6.16}$$

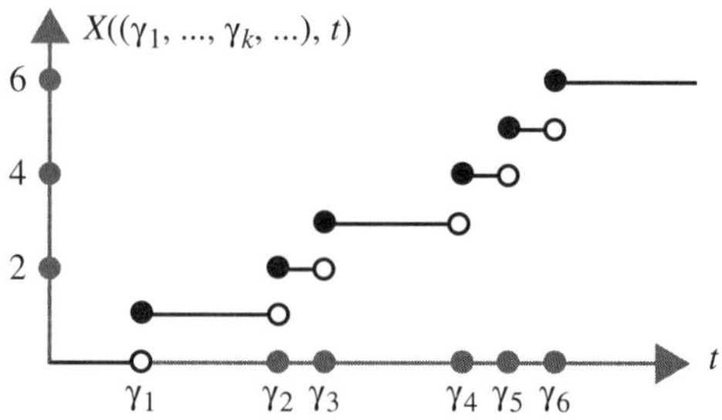


**FIGURE 6.5** One signal defined by a Poisson counting process.

For notational convenience, a Poisson counting random process is written as

$$X(\Omega, t) = \sum_{k=1}^{\infty} u(t - \Gamma_k) \tag{6.17}$$

where $\Omega = (\Gamma_1, \Gamma_2, \ldots)$ is a vector of random variables consistent with the outcomes of a Poisson experiment. One signal defined by a Poisson counting process is illustrated in Figure 6.5.

At a time $t_o$, the level of a signal from a Poisson counting process equals the number of outcomes of the subexperiments defining the Poisson experiment that are in the interval $[0, t_o]$, that is, the number of Poisson points in the interval $[0, t_o]$.

The following theorem specifies the probability mass function of the random variable defined by the level of a Poisson counting process at the time $t_o$.

**Theorem 6.1 PMF for RV Defined by Poisson Counting Process**
The random variable $X_{t_o} : S \to S_{X_{t_o}}$ defined by the level of a Poisson counting process, with a rate $\lambda$ and at the time $t_o$, that is,

$$X_{t_o}((\gamma_1, \ldots, \gamma_k, \ldots)) = \sum_{k=1}^{\infty} u(t_o - \gamma_k), \qquad (\gamma_1, \ldots, \gamma_k, \ldots) \in S \tag{6.18}$$

has the probability mass function

$$p_{X_{t_o}}(k) = \frac{(\lambda t_o)^k e^{-\lambda t_o}}{k!} \qquad t_o > 0,\ k \in \mathbf{Z}^+ \tag{6.19}$$

**Proof**

The probability mass function, by definition, is $p_{X_{t_o}}(k) = P[X_{t_o} = k]$, and the level $k$ is consistent with $k$ points of the underlying Poisson experiment being in the interval $[0, t_o]$. The proof of the stated result is detailed in the proof of Theorem 5.1.

### 6.3.3 Shot Noise Random Process

A shot noise random process arises when a signal $p: \mathbf{R} \rightarrow \mathbf{R}$ with finite energy, that is, $p \in \mathbf{L}^2$, is associated with each outcome of a Poisson experiment.

**Definition: Shot Noise Random Process** A shot noise random process $X$ is based on a Poisson experiment and defined according to

$$
\begin{aligned}
&X: S \rightarrow S_X \qquad X((\gamma_1,\ldots,\gamma_k,\ldots),t) = \sum_{k=1}^{\infty} p(t-\gamma_k) \\
&S_X = \left\{ X((\gamma_1,\ldots,\gamma_k,\ldots),t) = \sum_{k=1}^{\infty} p(t-\gamma_k) : \ (\gamma_1,\ldots,\gamma_k,\ldots) \in S \right\}
\end{aligned}
\tag{6.20}
$$

For notational convenience, a shot noise random process is written as

$$X(\Omega,t) = \sum_{k=1}^{\infty} p(t-\Gamma_k) \tag{6.21}$$

where $\Omega = (\Gamma_1, \Gamma_2, \ldots)$ is a sequence of random variables defined by the outcomes of a Poisson experiment.

Clearly, the pulse rate of a shot noise random process equals the point rate $\lambda$ of the underlying Poisson experiment. The components of one signal of a shot noise random process are illustrated in Figure 6.6. A shot noise random process arises, equivalently, by the filtering of a Dirac delta-based Poisson point process as illustrated in Figure 6.7. Two signals from two distinct shot noise random processes are shown in Figure 6.8 for the case of $p(t) = t\exp(-t/\tau)/\tau^2$ (a pulse with unity area), $\tau = 1$ and with rates of $\lambda = 0.1$ and $\lambda = 1$.

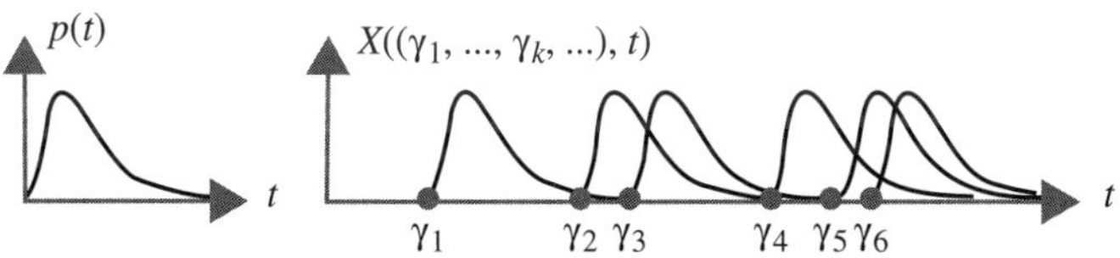


**FIGURE 6.6** Individual components of one signal defined by a shot noise random process.

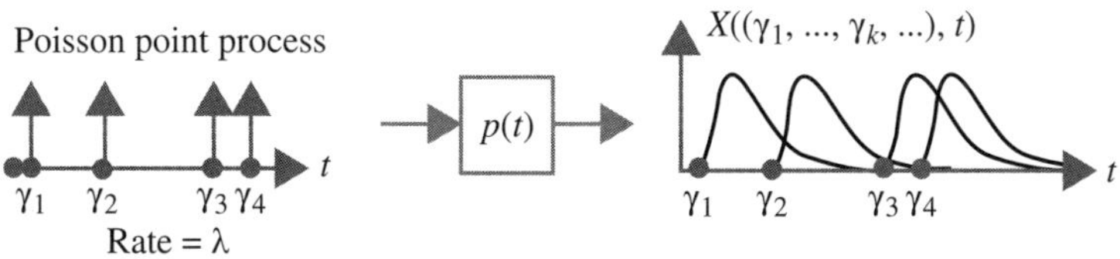


**FIGURE 6.7** Illustration of shot noise as linear filtering of a Dirac delta based Poisson point process.

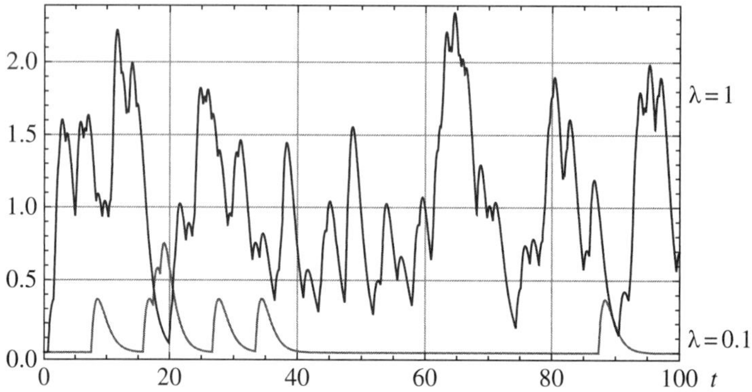


**FIGURE 6.8** Signals from shot noise random processes with rates of $\lambda = 0.1$ and $\lambda = 1$.

### 6.3.4 Generalized Shot Noise

A generalized shot noise random process arises when a signal $p \in \boldsymbol{L}^2$ is associated with each outcome of a Poisson experiment and when the amplitude of each signal is chosen in a random manner. The experiment underpinning such a random process yields the following matrix of outcomes:

$$S = \left\{ \begin{array}{l} ((a_1,\gamma_1),\ldots,(a_k,\gamma_k),\ldots): \;\; a_k \in S_{A_k}, \\ (\gamma_1,\ldots,\gamma_k,\ldots) \in \text{ Poisson sample space} \end{array} \right\} \tag{6.22}$$

where $S_{A_k}$ is the sample space of a random variable $A_k$. The Poisson subexperiment is assumed to independent of the subexperiment defining the vector of random variables $(A_1, \ldots, A_k, \ldots)$.

**Definition: Generalized Shot Noise Random Process** A generalized shot noise random process $X$ is based on the sample space defined by Equation 6.22 and is defined according to

$$\begin{array}{ll} X : S \to S_X & X(((a_1,\gamma_1),\ldots,(a_k,\gamma_k),\ldots),t) = \sum_{k=1}^{\infty} a_k p(t-\gamma_k) \\ S_X = \left\{ \begin{array}{l} X(((a_1,\gamma_1),\ldots,(a_k,\gamma_k),\ldots),t) = \sum_{k=1}^{\infty} a_k p(t-\gamma_k): \\ ((a_1,\gamma_1),\ldots,(a_k,\gamma_k),\ldots) \in S \end{array} \right\} & \end{array} \tag{6.23}$$

For notational convenience, a generalized shot noise random process is written as

$$X(\Omega,t) = \sum_{k=1}^{\infty} A_k p(t-\Gamma_k) \tag{6.24}$$

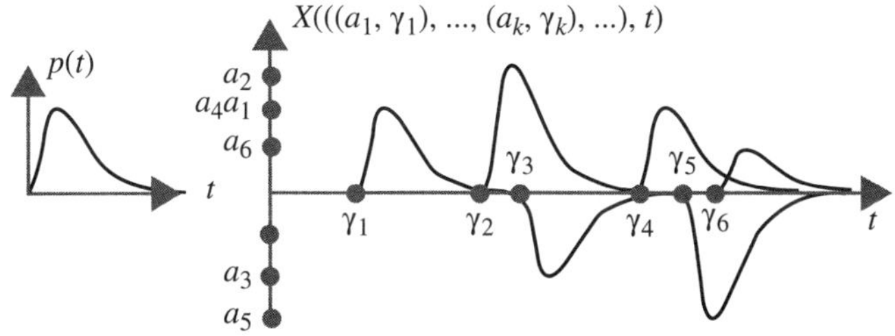


**FIGURE 6.9** Individual components of a signal from a generalized shot noise random process.

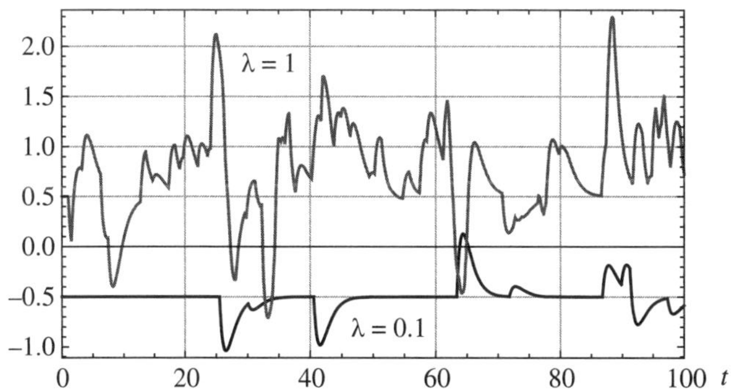


**FIGURE 6.10** Signals from generalized shot noise random processes with rates of $\lambda = 0.1$ and $\lambda = 1$. The signals are offset with levels of $\pm 0.5$ for clarity.

where $\Omega = ((A_1, \Gamma_1), \ldots, (A_k, \Gamma_k), \ldots)$, $\Gamma_1, \Gamma_2, \ldots$ is a sequence of random variables defined by the outcomes of a Poisson experiment and $(A_1, A_2, \ldots)$ is a vector of random variables that define the pulse amplitudes. The components of one signal from a generalized shot noise process are illustrated in Figure 6.9. Two signals from a generalized shot noise random process are shown in Figure 6.10 for the case of $p(t) = t\exp(-t/\tau)/\tau^2$ (a pulse with unity area), $\tau = 1$, and for the amplitudes $A_1, \ldots$ being independent Gaussian random variables with zero mean and unity variance.

### 6.3.5 Causal Random Telegraph Signal

A causal random telegraph random process, usually called a random telegraph signal (or dichotomous random process), is defined by a change of state, from 1 to $-1$ or from $-1$ to 1, at each point defined by a Poisson experiment.

**Definition: Causal Random Telegraph Random Process** A causal random telegraph random process, $X : S \rightarrow S_X$, is defined on a Poisson experiment according to

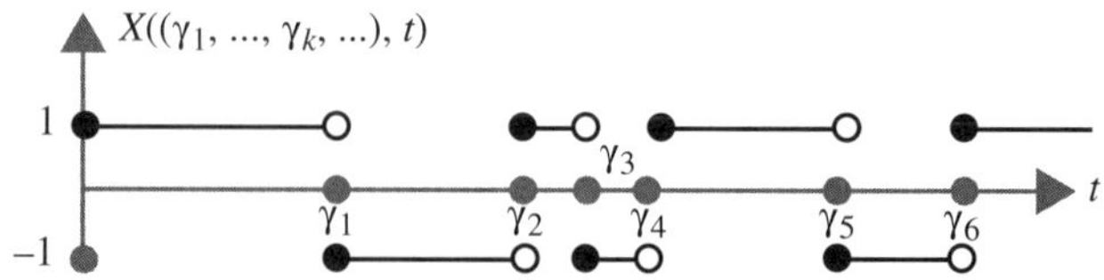


**FIGURE 6.11** One waveform from a causal random telegraph random process.

$$X((\gamma_1,\ldots,\gamma_k,\ldots),t) = u(t) + 2\sum_{k=1}^{\infty}(-1)^k u(t-\gamma_k)$$
$$= \sum_{k=0}^{\infty} a_k u(t-\gamma_k) \qquad \begin{cases} a_0 = 1,\ \gamma_0 = 0 \\ a_1, a_3, \ldots = -2 \\ a_2, a_4, \ldots = 2 \end{cases} \tag{6.25}$$

$$S_X = \left\{ \begin{array}{l} X((\gamma_1,\ldots,\gamma_k,\ldots),t) = \sum_{k=0}^{\infty} a_k u(t-\gamma_k): \\ \gamma_0 = 0, (\gamma_1,\ldots,\gamma_k,\ldots) \in S, a_0 = 1, a_k = (-1)^k \cdot 2 \end{array} \right\} \tag{6.26}$$

For notational convenience, a causal random telegraph random process is written as

$$X(\Omega,t) = \sum_{k=0}^{\infty} a_k u(t-\Gamma_k) \tag{6.27}$$

where $\Gamma_0 = 0$ and $\Omega = (\Gamma_1, \Gamma_2, \ldots)$ is a vector of random variables consistent with the outcomes of a Poisson experiment. One signal from a causal random telegraph random process is illustrated in Figure 6.11.

### 6.3.6 Random Telegraph Signal: Type II

A causal random telegraph signal can be modified by removing the deterministic nature of the change at each time specified by the outcome of a Poisson experiment and allowing a probabilistic change at such times.

To this end, consider the experiment of associating a level, from the set $\{-1, 1\}$, with each point defined by an outcome of a trial of a Poisson experiment such that the following sample space results:

$$S = \left\{ \begin{array}{l} ((a_1,\gamma_1),\ldots,(a_k,\gamma_k),\ldots) : a_k \in \{-1,1\}, \\ (\gamma_1,\ldots,\gamma_k,\ldots) \in \text{Poisson sample space} \end{array} \right\} \tag{6.28}$$

This experiment defines the matrix of random variables $\Omega = ((A_1,\Gamma_1),\ldots,(A_k,\Gamma_k),\ldots)$. With

$$P[A_k = 1, A_{k+1} = -1] = P[A_k = -1, A_{k+1} = 1] = p/2 \tag{6.29}$$

it follows that the probability of a change of level at a time specified by the outcome of the Poisson experiment is $p$.

Such an experiment underpins the definition of a random telegraph random process of type II according to the following.

**Definition: Random Telegraph Signal, Type II** A random telegraph random process of type II, $X: S \to S_X$, is defined on the sample space $S$ according to

$$X(((a_1,\gamma_1),\ldots,(a_k,\gamma_k),\ldots),t)=\sum_{k=0}^{\infty}a_k[u(t-\gamma_k)-u(t-\gamma_{k+1})]$$
$$a_0=1,\gamma_0=0,a_k\in\{-1,1\} \tag{6.30}$$

$$S_X=\left\{\begin{array}{l}X(((a_1,\gamma_1),\ldots,(a_k,\gamma_k),\ldots),t)=\sum_{k=0}^{\infty}a_k[u(t-\gamma_k)-u(t-\gamma_{k+1})]:\\ a_0=1,\gamma_0=0,((a_1,\gamma_1),\ldots,(a_k,\gamma_k),\ldots)\in S\end{array}\right\} \tag{6.31}$$

One waveform from such a random process is illustrated in Figure 6.12.

For notational convenience, a causal random telegraph random process of type II is written as

$$X(\Omega,t)=\sum_{k=0}^{\infty}A_k[u(t-\Gamma_k)-u(t-\Gamma_{k+1})] \qquad A_0=1,\ \Gamma_0=0 \tag{6.32}$$

where $\Omega=((A_1,\Gamma_1),\ldots,(A_k,\Gamma_k),\ldots),(A_1,A_2,\ldots)$ is a vector of dependent random variables with individual sample spaces of $\{-1,1\}$ and $\Gamma_1,\Gamma_2,\ldots$ are a sequence of random variables as defined by a Poisson experiment.

The correlation between consecutive random variables in the sequence $A_1, A_2, \ldots$ is defined according to

$$\begin{aligned}P[A_k=-1,A_{k+1}=-1]&=P[A_k=1,\ A_{k+1}=1]=1/2-p/2\\ P[A_k=-1,A_{k+1}=1]&=P[A_k=1,\ A_{k+1}=-1]=p/2\end{aligned} \tag{6.33}$$

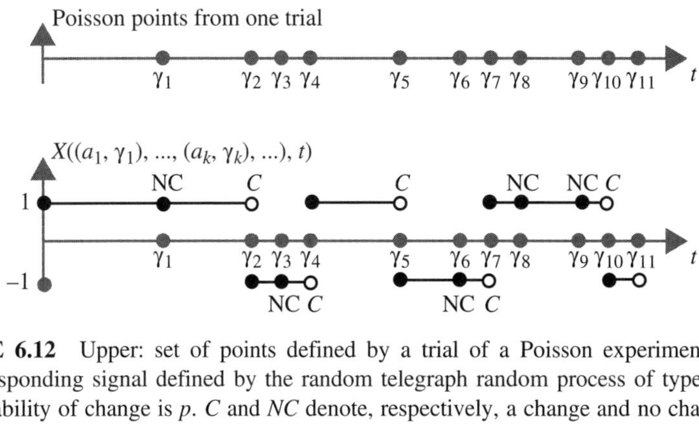


**FIGURE 6.12** Upper: set of points defined by a trial of a Poisson experiment. Lower: the corresponding signal defined by the random telegraph random process of type II where the probability of change is $p$. $C$ and $NC$ denote, respectively, a change and no change.

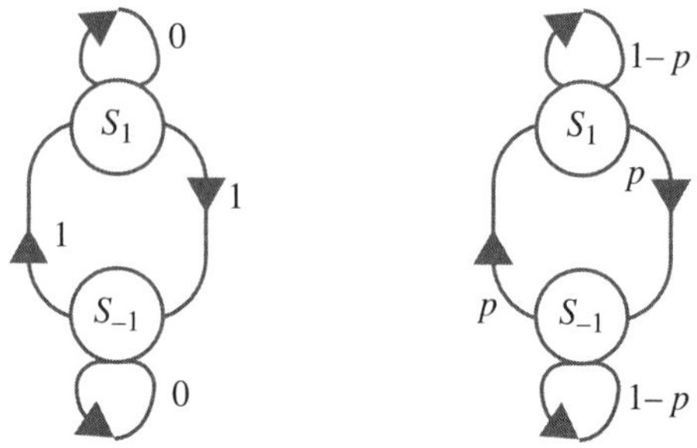


**FIGURE 6.13** Left: state diagram for a random telegraph signal. Right: state diagram for a random telegraph random process of type II where the probability of change is $p$. Implicit in the state diagrams is a clock as defined by the associated Poisson point random process.

consistent with

$$\begin{aligned} &P[\text{change}] = P[A_k = -1, A_{k+1} = 1] + P[A_k = 1,\ A_{k+1} = -1] = p \\ &P[\text{no change in level}] = 1 - p \end{aligned} \tag{6.34}$$

State diagrams for both types of random telegraph random processes are illustrated in Figure 6.13.

***6.3.6.1 Applications*** Random telegraph signals arise in a number of physical situations. First, consider a submicron MOSFET with a small number of trap sites. When an electron is trapped at such a site, it, potentially, leads to an increase in the scattering of other carriers comprising the current flow. This leads to a consequent increase in the resistance of the channel. The current flow then exhibits a random telegraph form around a mean value.

Second, consider the case of a potential across an ion channel in a cell membrane. The membrane can change, at random times, between the situation where the channel is open to ion flow and closed to ion flow. Accordingly, the potential across the ion channel will exhibit a random telegraph form around a mean level (Liebovitch, 1988).

Third, atomic or molecular systems may operate between two energy states when subject to a randomly varying input. When this is the case, the energy of the system exhibits a random telegraph form around a mean level.

Finally, random processes with a form consistent with a random telegraph signal can arise when two similar random processes $X_1$ and $X_2$ are compared according to

$$X(\Omega, t) = \text{sgn}[X_1(\Omega_1, t) - X_2(\Omega_2, t)] \tag{6.35}$$

A positive level indicates the level of the first random process is greater than the second random process and a negative level is the converse case. In general, the zero crossing statistics of $X_1 - X_2$ will determine whether or not $X$ is a random telegraph signal.

## 6.4 CLUSTERED RANDOM PROCESSES

A clustered random process arises when one major event triggers a number of minor events and when the major events occur at random and at a rate much slower than secondary or minor events. The prototypical case is that of major earthquakes which cause, in turn, a series of later minor earth tremors. Clustered random processes can also potentially arise with the arrival of electrons and photons. For example, the neuronal response to light input, in specific instances, has a clustered form (Saleh and Teich, 1983).

The following clustered random process is related to the Neyman–Scott model (Neyman and Scott, 1958, 1972).

### 6.4.1 Experiment for a Clustered Random Process

Consider an experiment where a trial yields the following sets of times:

$$\omega=\left(\left(t_1,\gamma_{11},\ldots,\gamma_{1m_1}\right),\ldots,\left(t_k,\gamma_{k1},\ldots,\gamma_{km_k}\right),\ldots\right) \tag{6.36}$$

which arise from the following subexperiments:

1. A trial of a Poisson experiment, with a rate of $\lambda$ per second, yields the times $(t_1, \ldots, t_k, \ldots)$ that define the commencement of the major events.
2. Each major event is associated with a number of secondary or minor events. This number is random and is determined from a trial of an experiment underpinning a discrete random variable with a probability mass function $p_M(m)$ with a sample space $S_M$. Running this experiment repeatedly yields the numbers $m_1, \ldots, m_k, \ldots$ with $m_k$ being the number of minor events associated with the major event occurring at time $t_k$.
3. Finally, the times of the minor events are established by running, repeatedly, an experiment underpinning a positive random variable with a probability density function $f_\Gamma(\gamma)$ and a sample space of $S_\Gamma$. The times associated with the $m_k$ minor events define the vector $\left(\gamma_{k1},\ldots,\gamma_{km_k}\right)$.

The sample space defined by the experiment is

$$S=\left\{\begin{array}{l}\omega=\left(\left(t_1,\gamma_{11},\ldots,\gamma_{1m_1}\right),\ldots,\left(t_k,\gamma_{k1},\ldots,\gamma_{km_k}\right),\ldots\right):m_k\in S_M,\\ \\ \gamma_{km_k}\in S_\Gamma,(t_1,\ldots,t_k,\ldots)\in \text{ Poisson sample space}\end{array}\right\} \tag{6.37}$$

### 6.4.2 Clustered Point Random Process

**Definition: Clustered Point Random Process** A clustered point random process $X:S\rightarrow S_X$, based on the aforementioned experiment, is defined according to

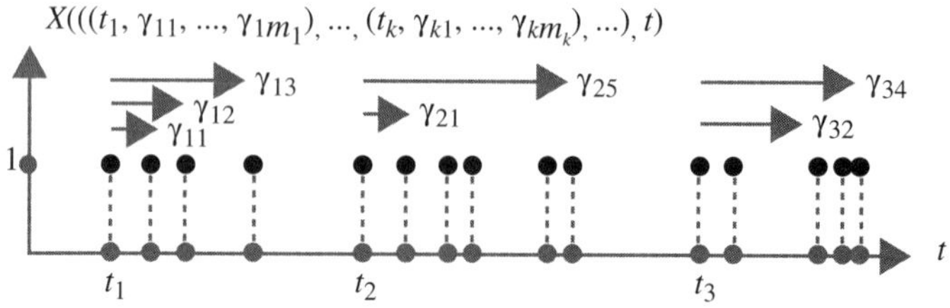


**FIGURE 6.14** One outcome of a clustered point random process where $m_1 = 3$, $m_2 = 5$, $m_3 = 4$, ….

$$X\left(\left(\left(t_1,\gamma_{11},\ldots,\gamma_{1m_1}\right),\ldots,\left(t_k,\gamma_{k1},\ldots,\gamma_{km_k}\right),\ldots\right),t\right) = \sum_{k=1}^{\infty}\left[\delta_K(t-t_k) + \sum_{r=1}^{m_k}\delta_K(t-t_k-\gamma_{kr})\right] \tag{6.38}$$

$$S_X = \left\{ \begin{array}{l} X\left(\left(\left(t_1,\gamma_{11},\ldots,\gamma_{1m_1}\right),\ldots,\left(t_k,\gamma_{k1},\ldots,\gamma_{km_k}\right),\ldots\right),t\right) \\ = \displaystyle\sum_{k=1}^{\infty}\left[\delta_K(t-t_k) + \sum_{r=1}^{m_k}\delta_K(t-t_k-\gamma_{kr})\right] : \\ \left(\left(t_1,\gamma_{11},\ldots,\gamma_{1m_1}\right),\ldots,\left(t_k,\gamma_{k1},\ldots,\gamma_{km_k}\right),\ldots\right) \in S \end{array} \right\} \tag{6.39}$$

For notational convenience, a clustered point random process is written as

$$X(\Omega,t) = \sum_{k=1}^{\infty}\left[\delta_K(t-T_k) + \sum_{r=1}^{M_k}\delta_K(t-T_k-\Gamma_{kr})\right] \tag{6.40}$$

where $\Omega = ((T_1,\Gamma_{11},\ldots,\Gamma_{1M_1}),\ldots,(T_k,\Gamma_{k1},\ldots,\Gamma_{kM_k}),\ldots)$, $T_1, T_2, \ldots$ are a sequence of random variables consistent with the outcomes of a Poisson experiment with a rate of $\lambda$ per second, $M_1, \ldots, M_k, \ldots$ are independent random variables with a probability mass function $p_M(m)$, and $(\Gamma_{11},\ldots,\Gamma_{1M_1}),\ldots,(\Gamma_{k1},\ldots,\Gamma_{kM_k})\ldots$ are vectors of independent random variables with each random variable having a probability density functions $f_\Gamma$.

***6.4.2.1 Example*** One outcome of a clustered point random process is illustrated in Figure 6.14.

### 6.4.3 Clustered Random Process

By associating a pulse function $p : \boldsymbol{R}^+ \to \boldsymbol{R}$ with each point of a clustered point random process, a clustered random process is defined.

**Definition: Clustered Random Process** A clustered random process is defined according to

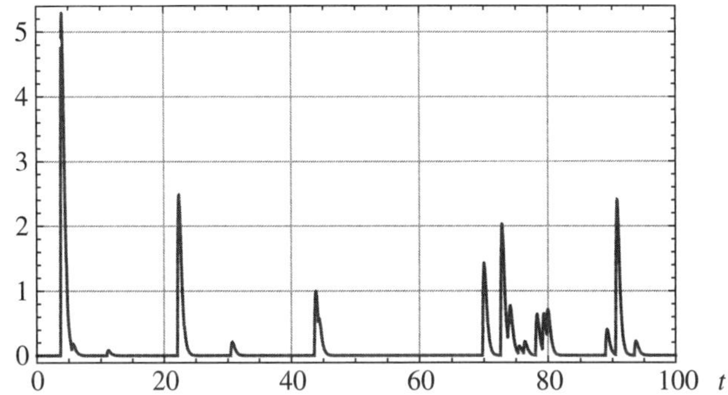


**FIGURE 6.15** One signal from a generalized clustered random process.

$$X(\Omega,t)=\sum_{k=1}^{\infty}\left[p(t-T_k)+\sum_{r=1}^{M_k}p(t-T_k-\Gamma_{kr})\right] \tag{6.41}$$

When the amplitudes of the pulse function are randomly chosen, a generalized clustered random process is defined.

**Definition: Generalized Clustered Random Process** A generalized clustered random process is defined according to

$$X(\Omega,t)=\sum_{k=1}^{\infty}\left[A_k p(t-T_k)+\sum_{r=1}^{M_k}A_{kr}p(t-T_k-\Gamma_{kr})\right] \tag{6.42}$$

where $A_1,\ldots$ and $A_{11},\ldots,A_{1M_1},A_{21},\ldots,A_{2M_2},\ldots$ are a sequence of random variables.

One signal from a generalized clustered random processes is shown in Figure 6.15 for the case of $p(t)=t\exp(-t/\tau)/\tau^2$, $\tau=0.25$, (a pulse with unity area), the rate for major events being 0.1 per second and the rate for minor events being 0.5 per second. The random variables $A_1,\ldots$ defining the major amplitudes have been assumed to be independent random variables with an exponential distribution with parameter $\lambda=1$ (mean amplitude of 1), that is, $f_A(a)=\lambda\exp(-\lambda a)$. The random variables $A_{11},\ldots,A_{1M_1},A_{21},\ldots,A_{2M_2},\ldots$ defining the minor amplitudes have been assumed to be independent with an exponential distribution with a parameter $\lambda=10$ (mean amplitude of 0.1), and the random variables $M_1,\ldots,M_k,\ldots$ defining the number of minor events have been assumed to be independent with a Poisson distribution with parameter $\lambda=3$ (mean number of 3), that is, $f_M(m)=\lambda^m\exp(-\lambda)/m!$.

## 6.5 SIGNALLING RANDOM PROCESSES

The communication of information underpins an intelligent world, is based on agreed protocols, and is based on the ability to send and recover information. Technological communication is usually based on encoding information in waveforms from a signalling set to generate information signals.

**Definition: Signalling Set** A signalling set $S_\Phi$ is a set of waveforms with well-defined properties:

$$S_\Phi = \left\{ \phi(i) : \boldsymbol{R} \to \boldsymbol{R}, \int_0^\infty |\phi(i,t)|\, dt < \infty, i \in \{1, \ldots, N_\Phi\}, \text{properties} \right\} \tag{6.43}$$

The set $\{1, \ldots, N_\Phi\}$ is the index set. In a communication context, the waveforms in the signalling set are chosen to have properties that facilitate efficient information transmission and information recovery at the receiver.

**Definition: Information Signal** An information signal is one where a waveform from a signalling set is associated with each signalling interval of $D$ seconds. The waveform rate is $r = 1/D$.

The set of all possible information signals underpins a signalling random process.

### 6.5.1 Signalling Random Processes

***6.5.1.1 Experiment Underpinning a Signalling Random Process*** A signalling random process for the interval $[0, ND]$ is based on the experiment where a trial is defined according to:

1. A number, $\omega_1$, is chosen at random, from the index set $\{1, \ldots, N_\Phi\}$.
2. A second number, $\omega_2$, is chosen from the index set $\{1, \ldots, N_\Phi\}$ taking into account any dependency with the first chosen number as required by the signalling protocol.

...

$N$. The $N$th number, $\omega_N$, is chosen from the index set $\{1, \ldots, N_\Phi\}$ taking into account any dependencies with the previous $N-1$ numbers as required by the signalling protocol.

The experimental sample space is

$$S = \{(\omega_1, \ldots, \omega_N) :\ \omega_k \in \{1, \ldots, N_\Phi\}\} \tag{6.44}$$

and, in general, $S$ comprises of $N_\Phi^N$ vectors. A probability mass function

$$p_\Omega((\omega_1, \ldots, \omega_N)) \tag{6.45}$$

where $\Omega = (\Omega_1, \ldots, \Omega_N)$ is a vector of random variables with outcomes $(\omega_1, \ldots, \omega_N) \in S$, defines the probability of a specific experimental outcome.

***6.5.1.2 Signalling Random Process*** A signalling random process is defined on the sample space $S$ according to the following.

**Definition: Signalling Random Process** A signalling random process $X$, for the interval $[0, ND]$, and based on the signalling set $S_\Phi$, is defined according to

$$X: S \to S_X \qquad X((\omega_1,\ldots,\omega_N),t) = \sum_{k=1}^{N} \phi(\omega_k, t-(k-1)D)$$
$$S_X = \left\{ \begin{array}{l} X((\omega_1,\ldots,\omega_N),t) = \sum_{k=1}^{N} \phi(\omega_k, t-(k-1)D) : \\ (\omega_1,\ldots,\omega_N) \in S, \phi \in S_\Phi \end{array} \right\} \tag{6.46}$$

For notational convenience, a signalling random process is written as

$$X(\Omega,t) = \sum_{k=1}^{N} \phi(\Omega_k, t-(k-1)D) \tag{6.47}$$

***6.5.1.3 Examples*** The component waveforms from a signalling random process are illustrated in Figure 6.16 for the case of $S_\Phi = \{\phi(1,t), \phi(2,t)\}$.

Consider the case where the signalling set comprises of three signals:

$$S_\Phi = \{\phi(-1,t), \phi(0,t), \phi(1,t)\} \tag{6.48}$$

$$\phi(0,t) = 0$$
$$\phi(\pm 1,t) = \begin{cases} \pm A\sin(2\pi f_C t)\sin(\pi t/D) & t \in [0,D] \\ 0 & \text{elsewhere} \end{cases} \tag{6.49}$$

One signal from a signalling random process based on this signalling set is shown in Figure 6.17.

### 6.5.2 Generalized Signalling Random Process

A signalling random process can be generalized by varying the duration of the signalling intervals in a random manner. This can be achieved, for example, when the times

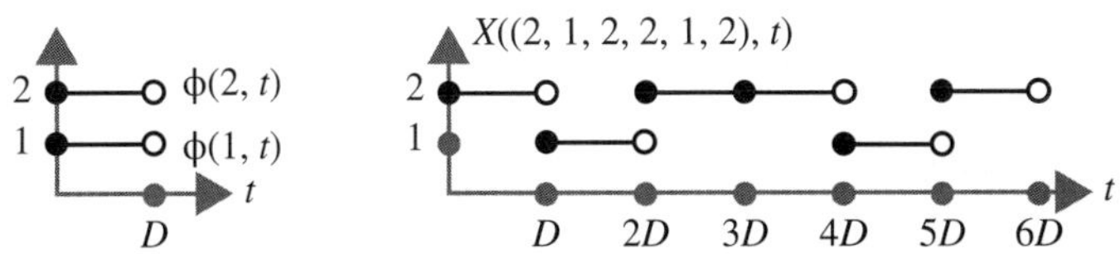


**FIGURE 6.16** Left: graph of signals in signalling set. Right: one waveform from a signalling random process based on the signalling set that comprises of the two signals defined in the graph on the left.

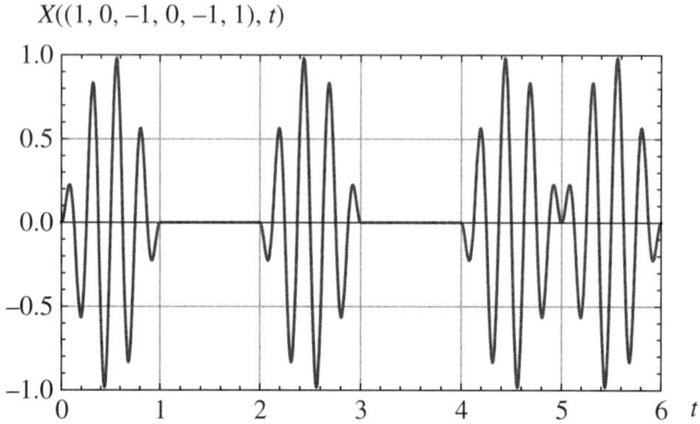


**FIGURE 6.17** One signal from the defined signalling random process with the following parameters: $A = 1$, $D = 1$, and $f_C = 4$.

demarcating the signalling intervals are defined by the outcome of a Poisson experiment. Such an approach leads to a generalized signalling random process based on the following experiment:

### *6.5.2.1 Experiment Underpinning a Generalized Signalling Random Process*

Consider an experiment that is based on two subexperiments:

1. A trial of a Poisson experiment with a rate $\lambda$ per unit interval yields the vector of times $(\gamma_1, \ldots, \gamma_N, \ldots)$.
2. A trial of the experiment underpinning a signalling random process yields the $N$ numbers $(\omega_1, \ldots, \omega_N)$ from the index set $(1, \ldots, N_\Phi)$.
3. Pairing these outcomes yield the matrix of numbers: $((\omega_1, \gamma_1), \ldots (\omega_N, \gamma_N))$.

The sample space for this experiment is

$$S = \left\{ \begin{array}{l} ((\omega_1,\gamma_1),\ldots,(\omega_N,\gamma_N)):\ (\gamma_1,\ldots,\gamma_N,\ldots) \in \text{ Poisson sample space} \\ \omega_1,\ldots,\omega_N \in \{1,\ldots,N_\Phi\} \end{array} \right\} \qquad (6.50)$$

**Definition: Generalized Signalling Random Process** A generalized signalling random process $X : S \rightarrow S_X$ is based on associating one signalling waveform from a signalling set, $S_\Phi$, with times as specified by the outcomes of a Poisson point random process with a signalling rate of $\lambda$:

$$X(((\omega_1,\gamma_1),\ldots,(\omega_N,\gamma_N)),t) = \sum_{k=1}^{N} \phi(\omega_k, t-\gamma_k) \qquad \phi \in S_\Phi$$

$$S_X = \left\{ \begin{array}{l} X(((\omega_1,\gamma_1),\ldots,(\omega_N,\gamma_N)),t) = \sum_{k=1}^{N} \phi(\omega_k, t-\gamma_k): \\ ((\omega_1,\gamma_1),\ldots,(\omega_N,\gamma_N)) \in S, \phi \in S_\Phi \end{array} \right\} \qquad (6.51)$$

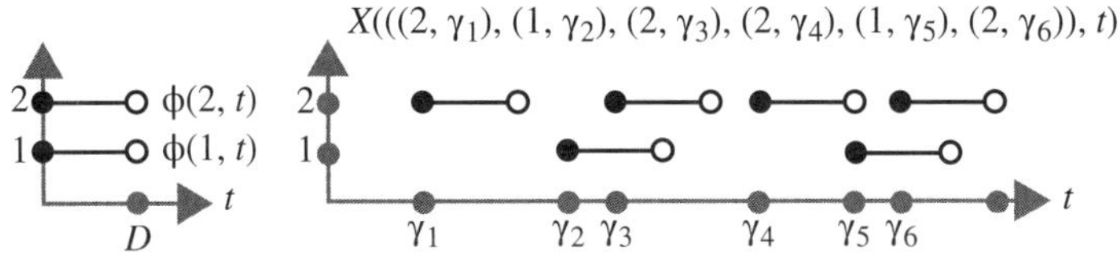


**FIGURE 6.18** Left: graph of signals in signalling set. Right: components of one signal of a generalized signalling random process based on the signalling set $S_\Phi = \{\phi(1,t), \phi(2,t)\}$.

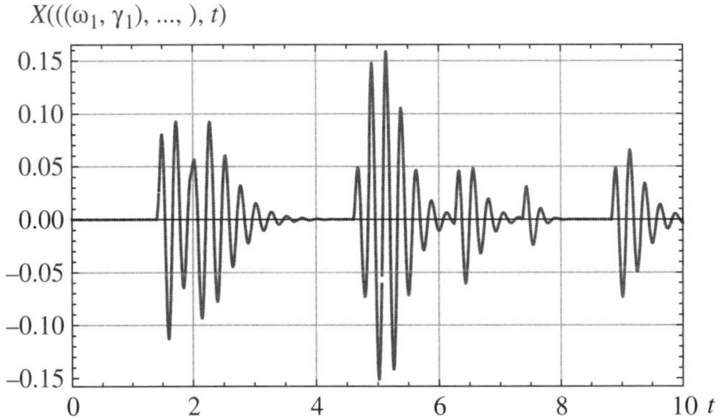


**FIGURE 6.19** One signal from the defined generalized signalling random process.

For notational convenience, a signalling random process is written as

$$X(\Omega, t) = \sum_{k=1}^{N} \phi(\Omega_k, t - \Gamma_k) \tag{6.52}$$

where $\Omega = ((\Omega_1, \Gamma_1), \ldots, (\Omega_N, \Gamma_N)), \Gamma_1, \Gamma_2, \ldots$ are a sequence of random variables defined by the outcomes of a Poisson experiment and $(\Omega_1, \ldots, \Omega_N)$ is a vector of random variables with outcomes $(\omega_1, \ldots, \omega_N)$, $\omega_k \in \{1, \ldots, N_\Phi\}$.

The number of signalling waveforms, $N$, can be made sufficiently large to ensure the statistical integrity of the random process on a defined interval.

***6.5.2.2 Examples*** The component waveforms of one signal from a generalized signalling random process are illustrated in Figure 6.18 for the case of $S_\Phi = \{\phi(1,t), \phi(2,t)\}$ and where the signalling waveforms are pulses.

Consider the case where the signalling set comprises of signals:

$$S_\Phi = \left\{ \phi(i,t) = \sin(2\pi f_C t) \cdot \frac{t}{\tau^2} \cdot \exp[-(i+1)t/\tau] u(t) : \; i \in \{0, 1, \ldots\} \right\} \tag{6.53}$$

One signal from a generalized signalling random process based on this signalling set is shown in Figure 6.19 for the case of $\lambda = 1$, $f_C = 4$, $\tau = 1$ and where signals are chosen from the index set independently and consistent with a Poisson distribution, that is,

$$P[\Omega_k = i] = \frac{\lambda_o^i}{i!}\exp(-\lambda_o) \qquad i \in \{0,1,\ldots\}, \lambda_o = 4 \tag{6.54}$$

## 6.6 JITTER

The noise inherent in electronic devices, for example, results in the situation where the processing of a deterministic signal is generally imperfect. A common case is where additive noise on a signal introduces variations, in comparison to the zero noise case, in the instants a signal crosses a set threshold. Such variations are called jitter. Jitter arises in many practical applications including synchronization of signals, hard limiting of signals, and digital circuitry. The archetypical jitter case is that of a jittered binary signal as illustrated in Figure 6.20. The jittered output signal is illustrated in Figure 6.21.

Assume the experiment, which underpins the jitter process for the interval $[0, ND]$, defines the sample space:

$$S = \{ (j_1, \ldots, j_N) : j_k \in S_k \} \tag{6.55}$$

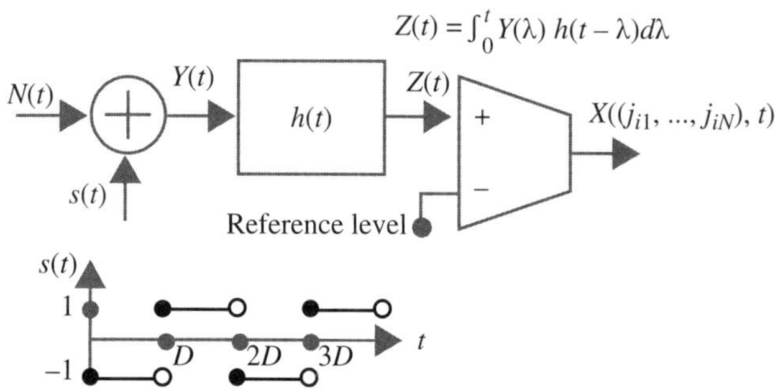


**FIGURE 6.20** Prototypical jitter situation resulting from the filtering of a noise signal $N(t)$, and a binary signal $s(t)$, followed by a circuit that responds to the first passage time across the reference level.

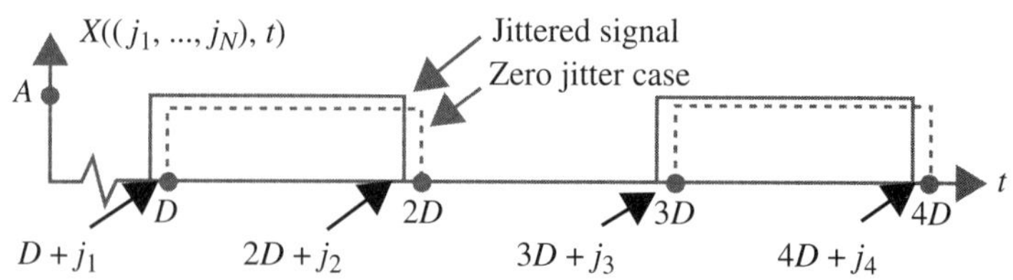


**FIGURE 6.21** Illustration of a jittered output signal.

where $S_k$ is the set of possible outcomes for the jitter associated with the $k$th level crossing, which defines the outcomes of the jitter random variable $J_k$. In general, the jitter values are correlated and

$$P[J_1 \in [j_1, j_1 + dj_1], \ldots, J_N \in [j_N, j_N + dj_N]] \approx f_\Omega(j_1, \ldots, j_N) dj_1 \ldots dj_N \tag{6.56}$$

where $\Omega = (J_1, \ldots, J_N)$ and $f_\Omega$ is the joint probability density function of the jitter random variables.

**Definition: Jittered Binary Random Process** The random process $X$ defining the possible jittered output signals for the interval $[0, ND]$ is defined according to

$$\begin{aligned} &X : S \rightarrow S_X \\ &X((j_1, \ldots, j_N), t) = A \sum_{k=1}^{N} (-1)^{k+1} u(t - kD - j_k) \\ &S_X = \left\{ \begin{aligned} &X((j_1, \ldots, j_N), t) = A \sum_{k=1}^{N} (-1)^{k+1} u(t - kD - j_k) : \\ &(j_1, \ldots, j_N) \in S \end{aligned} \right\} \end{aligned} \tag{6.57}$$

For notational convenience, a jittered binary random process is written as

$$X(\Omega, t) = A \sum_{k=1}^{N} (-1)^{k+1} u(t - kD - J_k) \tag{6.58}$$

where $\Omega = (J_1, \ldots, J_N)$ is a vector of random variables with $J_k$ specifying the $k$th jitter times.

An alternative form for a jittered binary signal is

$$X((j_1, \ldots, j_N), t) = A \sum_{k=1}^{N/2} p\left( \frac{t - (2k-1)D - j_{2k-1}}{D + j_{2k} - j_{2k-1}} \right) \tag{6.59}$$

where the pulse function has the form defined in Figure 6.22.

### 6.6.1 Jittered Pulse Train

The generalization of a rectangular pulse waveform subject to jitter is a pulse train with both amplitude and timing jitter as illustrated in Figure 6.23.

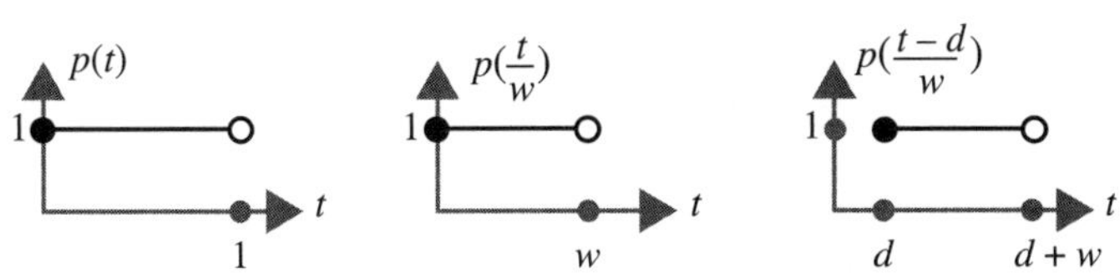


**FIGURE 6.22** Definition of the rectangular pulse function $p$.

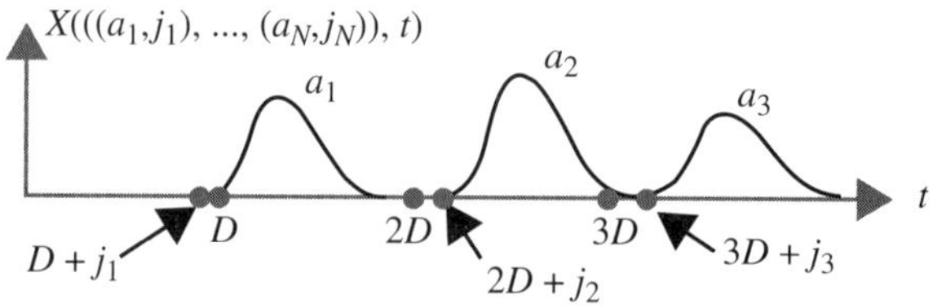


**FIGURE 6.23** Illustration of a signal from a jittered pulse train.

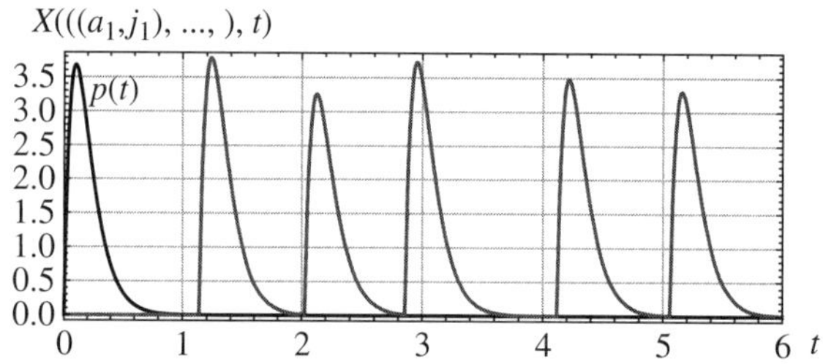


**FIGURE 6.24** Graph of the pulse function $p$ and one signal from a jittered pulse train with parameters defined in text.

A random process that is consistent with such a pulse train can be defined, using simplified notation, according to

$$X(\Omega,t)=\sum_{k=1}^{N} A_k p(t-kD-J_k) \qquad \Omega=((A_1,J_1),\ldots,(A_N,J_N)) \tag{6.60}$$

where the underlying experimental sample space is

$$S=\{((a_1,j_1),\ldots,(a_N,j_N)): \quad a_k\in S_{A_k}, j_k\in S_{J_k}\} \tag{6.61}$$

Here, $A_k$ and $J_k$, respectively, are the random variables defining the amplitude and timing jitter of the $k$th pulse and which have the respective sample spaces of $S_{A_k}$ and $S_{J_k}$.

***6.6.1.1 Example*** One signal from a jittered pulse train, based on an exponentially decaying pulse with unit area

$$p(t)=\frac{t}{\tau^2}\cdot\exp(-t/\tau)u(t) \tag{6.62}$$

is shown in Figure 6.24 for the case of $\tau=0.1$, $D=1$, and with both the amplitude and jitter values being independently taken from a Gaussian distribution with zero mean and a variance of 0.01.

## 6.7 WHITE NOISE

The most widely type of noise signals encountered in electronics and communication systems is Gaussian white noise, and such signals arise from a Gaussian white noise random process. Here, the term *Gaussian* refers to the amplitude distribution at any set time (which defines a random variable), while the term *white* refers to the spectrum, or power spectral density, being constant over a set frequency band—usually a band much greater than the signal bandwidth of the system in which the noise occurs. There is no necessity for white noise to be Gaussian. The following pragmatic definition for a white noise random process will be used.

**Definition: White Noise Random Process** A white noise random process is one where the power spectral density, as given by the expected value of the individual signal power spectral densities, is constant over the dominant frequency band of interest.

### 6.7.1 White Noise: Approach I

A white noise random process can be defined in many different ways and underpinned by different experiments. The following experiment, and resulting white noise random process, provides a good intuitive introduction to white noise (Rice, 1944, p. 287).

***6.7.1.1 Experiment for a White Noise Random Process*** Consider the experiment defined by repeating the following subexperiment $N$ times:

1. Choose a number, at random, from the interval $[0, 2\pi)$. Denote this number $\phi_k$.
2. Generate two numbers, $\alpha_k$ and $\beta_k$, according to

$$\alpha_k = \sin(\phi_k) \qquad \beta_k = \cos(\phi_k) \tag{6.63}$$

The experiment leads to a sequence of pairs of random numbers ($\alpha_k$, $\beta_k$) where $\alpha_k^2 + \beta_k^2 = 1$. The resulting sample space is

$$S = \{((\alpha_1, \beta_1), \ldots, (\alpha_N, \beta_N)): \quad \alpha_k = \sin(\phi_k), \beta_k = \cos(\phi_k), \phi_k \in [0, 2\pi)\} \tag{6.64}$$

#### *6.7.1.2 White Noise Random Process*

**Definition: White Noise Random Process** A white noise random process, $X$, for the interval $[0, T_o]$ with a fundamental frequency of $f_o, f_o = 1/T_o$, a flat spectrum on the interval $[0, f_{max}]$, and an average power of $A_o^2$, is defined on $S$ according to

$$X: S \rightarrow S_X$$
$$X(((\alpha_1, \beta_1), \ldots, (\alpha_N, \beta_N)), t) = A_o \sum_{k=1}^{N} \frac{\sqrt{2}\alpha_k}{\sqrt{N}} \cos(2\pi k f_o t) + \frac{\sqrt{2}\beta_k}{\sqrt{N}} \sin(2\pi k f_o t) \tag{6.65}$$

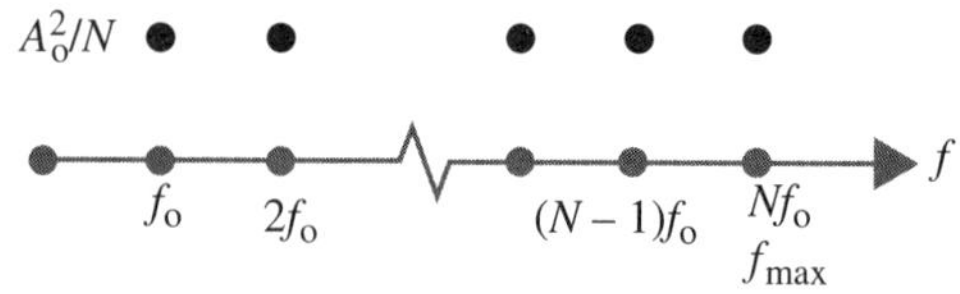


**FIGURE 6.25** Frequencies, and powers, of the sinusoidal components of a signal from a white noise random process defined for the interval $[0, T_o]$ and with a flat spectrum over the interval $[0, f_{max}]$.

$$S_X = \left\{ \begin{array}{l} X(((\alpha_1, \beta_1), \ldots, (\alpha_N, \beta_N)), t) = \\ A_o \sum_{k=1}^{N} \frac{\sqrt{2}\alpha_k}{\sqrt{N}} \cos(2\pi k f_o t) + \frac{\sqrt{2}\beta_k}{\sqrt{N}} \sin(2\pi k f_o t) : \\ ((\alpha_1, \beta_1), \ldots, (\alpha_N, \beta_N)) \in S \end{array} \right\} \quad (6.66)$$

where $N = \lfloor f_{max}/f_o \rfloor$. Each signal defined by the random process has $N$ component signals, with individual powers of $A_o^2/N$, which, collectively, have frequencies that define the set $\{f_o, 2f_o, \ldots, Nf_o\}$. These frequencies, along with the associated powers, are illustrated in Figure 6.25.

For notational convenience, a white noise random process can be written as

$$X(\Omega, t) = A_o \sum_{k=1}^{N} \frac{\sqrt{2}A_k}{\sqrt{N}} \cos(2\pi k f_o t) + \frac{\sqrt{2}B_k}{\sqrt{N}} \sin(2\pi k f_o t) \quad (6.67)$$

where $\Omega = ((A_1, B_1), \ldots, (A_N, B_N))$, $A_k = \sin(\Phi_k)$, $B_k = \cos(\Phi_k)$, and $\Phi_1, \ldots, \Phi_N$ are identical and independent random variables with a probability density function

$$f_{\Phi_k}(\phi) = \begin{cases} 1/2\pi & 0 \le \phi < 2\pi \\ 0 & \text{elsewhere} \end{cases} \qquad k \in \{1, \ldots, N\} \quad (6.68)$$

Examples of signals from a white noise random process are shown in Figures 6.26 and 6.27.

### 6.7.2 White Noise: Approach II

The white noise random process defined by Equations 6.65 and 6.66 is based on set frequencies. A more general approach is to use frequencies randomly chosen, at an average rate, from an interval $[0, f_{max}]$. With a minimum frequency of $f_o$, the average rate is $\lambda = 1/f_o$ and with, on average, $N = f_{max}/f_o$, frequency components for each signal. Such a white noise random process is underpinned by the following experiment:

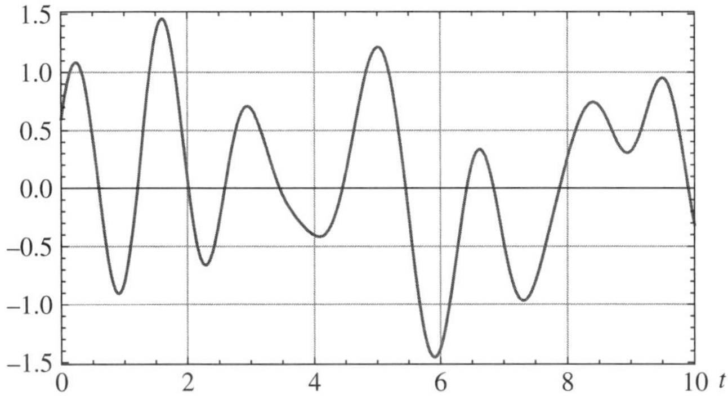


**FIGURE 6.26** Graph of a signal from a white noise random process defined on the interval [0, 100] and with the following parameters: $A_o = 1, f_o = 1/100, f_{max} = 1$, and $N = 100$.

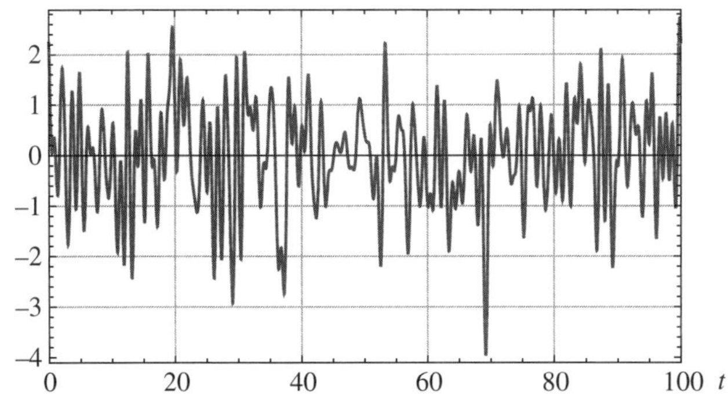


**FIGURE 6.27** Graph of a signal from a white noise random process defined on the interval [0, 100] and with the following parameters: $A_o = 1, f_o = 1/100, f_{max} = 1$, and $N = 100$.

***6.7.2.1 Experiment*** Consider the experiment where for each trial:

1. A trial of a Poisson experiment, with a rate $\lambda = 1/f_o$, yields $n$ values in the interval $[0, f_{max}]$ which defines a vector $(f_1, \ldots, f_n)$.
2. A subexperiment of taking a number at random from the interval $[0, 2\pi)$ is repeated $n$ times to create the vector $(\omega_1, \ldots, \omega_n)$.

The sample space of experimental outcomes is

$$S = \left\{ \begin{array}{r} ((\omega_1, f_1), \ldots, (\omega_n, f_n)) : \; (f_1, \ldots, f_n) \text{ defined by Poisson exp.,} \\ \omega_k \in [0, 2\pi), k \in \{1, \ldots, n\} \end{array} \right\} \tag{6.69}$$

***6.7.2.2 White Noise Random Process*** A white noise random process, based on this experiment, can be defined according to the following.

**Definition: White Noise Random Process** A white noise random process, $X$, for the interval $[0, kT_o]$, $k \in \mathbf{Z}^+$, with a fundamental frequency of $f_o, f_o = 1/T_o$, with frequencies chosen at random from the interval $[0, f_{max}]$ at a rate $\lambda = 1/f_o$, and with an average power of $A_o^2$, is defined according to

$$X : S \rightarrow S_X$$
$$X(((\omega_1, f_1), \ldots, (\omega_n, f_n)), t) = \sum_{k=1}^{n} \frac{\sqrt{2}A_o}{\sqrt{n}} \sin(2\pi f_k t + \omega_k)$$
$$S_X = \left\{ \begin{array}{l} X(((\omega_1, f_1), \ldots, (\omega_n, f_n)), t) = \sum_{k=1}^{n} \frac{\sqrt{2}A_o}{\sqrt{n}} \sin(2\pi f_k t + \omega_k) : \\ \qquad ((\omega_1, f_1), \ldots, (\omega_n, f_n)) \in S \end{array} \right\} \tag{6.70}$$

For notational convenience, this white noise random process is written as

$$X(\Omega, t) = \sum_{k=1}^{N} \frac{\sqrt{2}A_o}{\sqrt{N}} \sin(2\pi F_k t + \Omega_k) \tag{6.71}$$

where $\Omega = \{N, (\Omega_1, F_1), \ldots, (\Omega_N, F_N)\}$ is a set of random variables as defined by the experiment $S$. Here, $N$ is a random variable defined by the number of Poisson points in the interval $[0, f_{max}]$.

#### *6.7.2.3 Link to the Cramer Representation*

Consider a Poisson counting process $Z(\Omega, f)$ defined as

$$Z(\Omega, f) = \sum_{k=1}^{N} u(f - \Gamma_k) \tag{6.72}$$

where $\Omega = (N, (\Gamma_1, \Gamma_2, \ldots, \Gamma_N))$, $(\Gamma_1, \Gamma_2, \ldots, \Gamma_N)$ is a vector of random variables consistent with the outcomes of a Poisson experiment with a rate $\lambda = 1/f_o$ and $N$ is a random variable defined by the number of Poisson points in the interval $[0, f_{max}]$. One outcome of such a random process is illustrated in Figure 6.28.

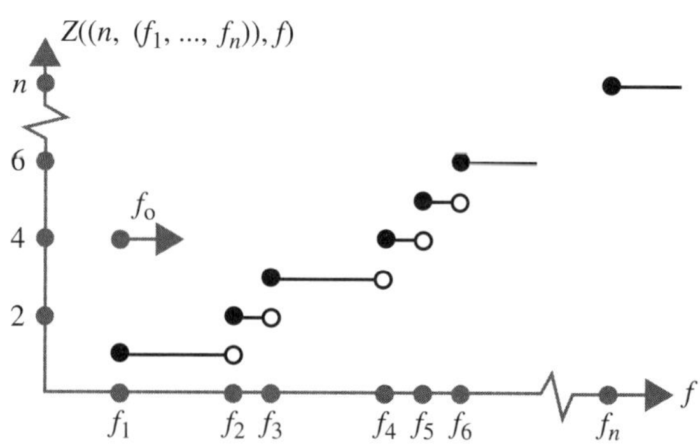


**FIGURE 6.28** One signal defined by the Poisson counting process.

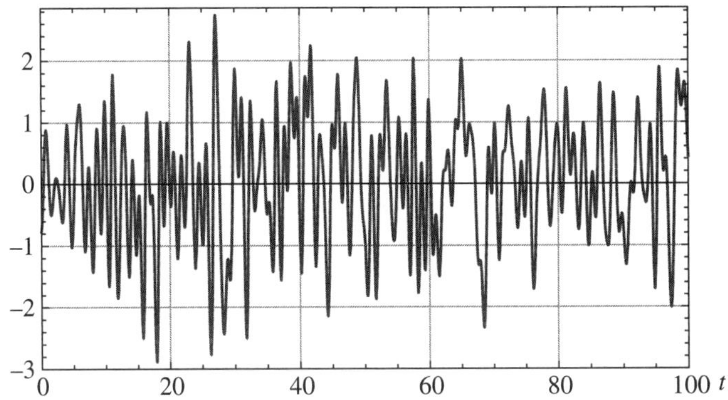


**FIGURE 6.29** Graph of one signal from a white noise random process defined on the interval [0, 100] and comprising, on average, of 100 sinusoids.

A white noise random process, for the interval $[0,\infty]$ and with an average power of $A_o^2$, can be defined according to

$$X(\Omega,t) = \int_R \frac{\sqrt{2}A_o}{\sqrt{N}}\sin(2\pi ft)dZ(\Omega,f) \tag{6.73}$$

where the integral is a Lebesgue–Stieltjes integral. The association of the term Cramer arises from the similarity of the definition with the Cramer representation of a signal in terms of an integrator function. The signals from such a random process will have a transient component as all the sinusoids are in phase at $t=0$.

***6.7.2.4 Example*** Consider a white noise random process, defined by Equation 6.70, on the interval [0, 100] with $A_o = 1$, $f_{max} = 1$ and, on average, comprising of 100 sinusoids with frequencies chosen consistent with a Poisson point process with a rate of $\lambda = 100$. One signal from this random process is shown in Figure 6.29. The power spectral density of this random process, obtained by averaging the individual power spectral densities of 100 signals, is illustrated in Figure 6.30.

### 6.7.3 Gaussian White Noise

As the number of sinusoids increases, in the aforementioned definitions for a white noise random process, it follows, from the central limit theorem, that the random variable defined by the white noise random process at any set time will have, approximately, a Gaussian probability density function. Hence, the aforementioned definitions are consistent with Gaussian white noise when the number of sinusoids is large.

***6.7.3.1 Example*** Consider the white noise random process, defined by Equation 6.71, on the interval [0, 100] with $A_o = 1, f_{max} = 1$ and, on average, comprising of 100 sinusoids with frequencies chosen consistent with a Poisson point process

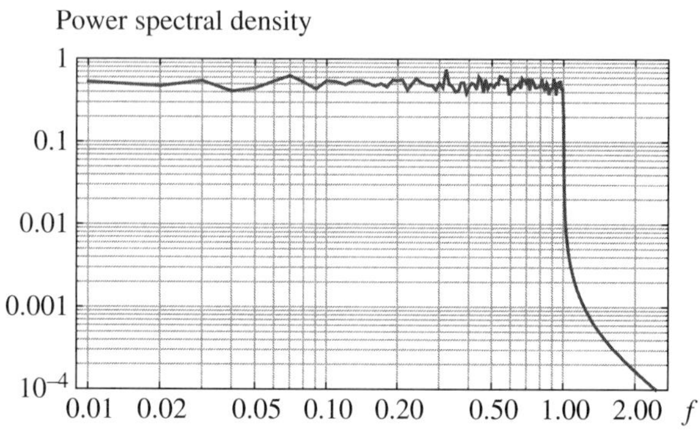


**FIGURE 6.30** Graph of the power spectral density of the defined white noise random process with $f_{max} = 1$ and obtained by averaging the individual power spectral densities of 100 signals.

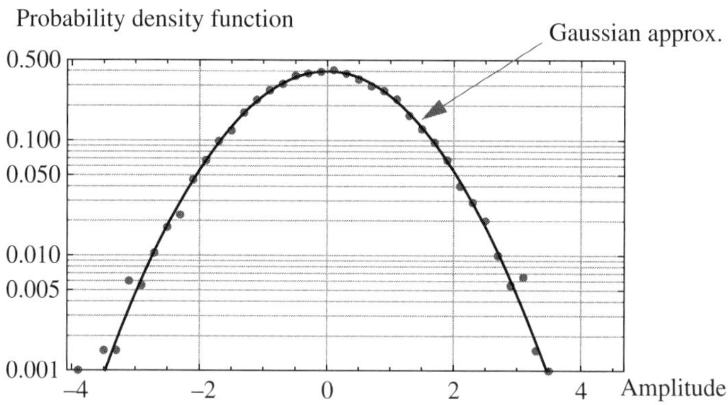


**FIGURE 6.31** Histogram-based approximation to the probability density function of the amplitude, at $t = 50$, and based on 10,000 signals from a white noise random process defined on the interval [0, 100] and comprising, on average, of 100 sinusoids.

with a rate of $\lambda = 100$. The probability density function of the amplitude, estimated from 10,000 signals, is shown in Figure 6.31 along with a Gaussian approximation based on the sample mean and sample variance.

### 6.7.4 Filtered White Noise

In many contexts, the noise at the input of a system has a flat spectrum over the system bandwidth and the system significantly filters the noise leading to filtered white noise. A model for such noise is of interest. Consider a causal linear filter with an impulse response $h$ whose Laplace transform is denoted $H$. For such a system, a filtered white noise random process can be defined by modifying Equation 6.71 according to

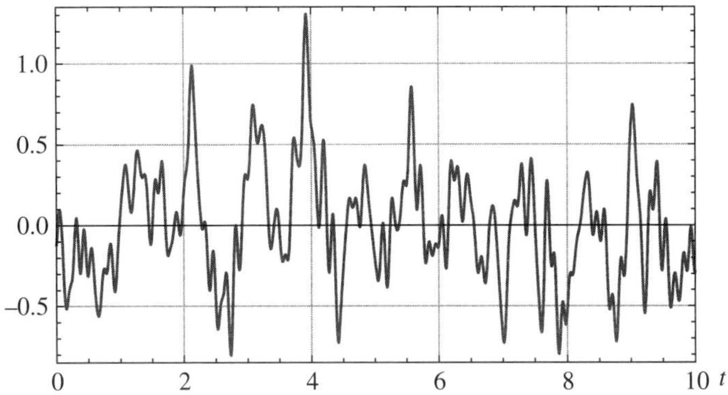


**FIGURE 6.32** Graph of one signal from a filtered white noise random process defined on the interval [0, 10] and comprising, on average, of 100 sinusoids.

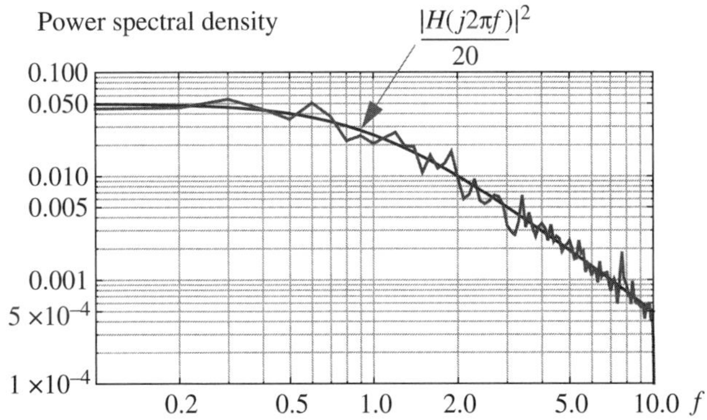


**FIGURE 6.33** Graph of the power spectral density of the defined filtered white noise random process with $f_{max} = 10$ and obtained by averaging the individual power spectral densities of 100 signals.

$$X(\Omega,t) = \sum_{k=1}^{N} \frac{\sqrt{2}A_o A(F_k)}{\sqrt{N}} \sin(2\pi F_k t + \Phi(F_k) + \Omega_k) \tag{6.74}$$

where

$$H(j2\pi F_k) = A(F_k)e^{j\Phi(F_k)} \tag{6.75}$$

The assumption here is that the bandwidth of the linear system is much less than $f_{max}$.

***6.7.4.1 Example*** Consider a white noise random process, defined by Equation 6.74, and with linear filtering by a filter with a transfer function

$$H(s) = \frac{1}{1 + \frac{s}{2\pi f_p}} \quad \Leftrightarrow \quad h(t) = \frac{1}{\tau} \cdot e^{-t/\tau} u(t) \qquad \tau = \frac{1}{2\pi f_p} \tag{6.76}$$

One signal from a filtered white noise random process is shown in Figure 6.32 for the case of $f_p = 1$, $f_{\max} = 10$, $A_o = 1$, $T = 10$, and $f_o = 1/10$. On average, each signal comprises of 100 sinusoids with frequencies chosen consistent with a Poisson point process with a rate of $\lambda = 10$. The power spectral density of this random process, obtained by averaging the individual power spectral densities of 100 signals, is illustrated in Figure 6.33.

## 6.8 1/*f* NOISE

In general, there are many influences, often independent of one another, having a random nature and different intrinsic time constants, which affect a given output of a complex system. Further, it is often the case that the output signal of such systems exhibit a power spectral density of the form $1/f^{\alpha}$, where $\alpha$ is close to unity. Such randomness is called 1/*f* noise.

Systems that exhibit 1/*f* noise are ubiquitous and such noise is often a signature of complexity. 1/*f* noise has been found in graphene, in electronic devices, in music signals, in measures of the human response to stimuli, in traffic flow, and in many other phenomena in the physical and social worlds. Such noise has been the subject of detailed investigation and modelling. Milotti (2002) provides a good review of 1/*f* noise, and Grigolini et al. (2009) provides a good overview of systems that exhibit such noise.

**Definition: 1/*f* Noise** Random processes that exhibit a power spectral density of the form 1/*f* over a finite frequency range—usually several decades of frequency—are called 1/*f* noise random processes or 1/*f* noise for short.

### 6.8.1 Model for 1/*f* Noise

One model for a random process that exhibits a 1/*f* power spectral density over a defined frequency range is that of a summation of $N$ equal power generalized shot noise random processes with rates that form a geometric series with a ratio of 2 and with pulse functions that have a measure of duration that is inversely proportional to the pulse rate. The model is

$$X(\Omega, t) = \sum_{i=1}^{N} X_i(\Omega_i, t) \qquad X_i(\Omega_i, t) = \sum_{k=1}^{\infty} A_{ik} p_i(t - \Gamma_{ik}) \tag{6.77}$$

where $X_i(\Omega_i, t)$ is a generalized shot noise random process, consistent with Equation 6.24. This leads to the following model for 1/*f* noise:

**Definition: Model for 1/f Noise** One model for a 1/f noise random process, with a fundamental pulse rate of $\lambda_o$, is

$$X(\Omega,t)=\sum_{i=1}^{N}\sum_{k=1}^{\infty}A_{ik}p_i(t-\Gamma_{ik}) \tag{6.78}$$

where

$$\Omega=(((A_{11},\Gamma_{11}),\ldots,(A_{1r},\Gamma_{1r}),\ldots),\ldots,((A_{N1},\Gamma_{N1}),\ldots,(A_{Nr},\Gamma_{Nr}),\ldots)) \tag{6.79}$$

is a matrix of random variables. For the $i$th generalized signalling random process, $i\in\{1,\ldots,N\}$, the pulse rate is $\lambda_i=\lambda_o/2^{i-1}$, and the pulse times are defined by a sequence of random variables $\Gamma_{i1},\Gamma_{i2},\ldots$ consistent with a Poisson experiment with the same rate. The sequence of random variables $A_{i1},A_{i2},\ldots$ defining the pulse amplitudes are assumed to be independent, with an identical probability density function and with zero mean and a variance $\sigma_i^2$. The pulse functions are required to be such that a measure of their duration scales according to the inverse of the pulse rate. Finally, the pulse energy, pulse rate, and amplitude variance must satisfy the restriction (see Theorem 9.13), for the case where the mean of the random variables $A_{i1},A_{i2},\ldots$ is zero, of

$$\int_0^{\infty}p_i^2(t)dt=\frac{P_o/N}{\sigma_i^2\lambda_i}\qquad \lambda_i=\lambda_o/2^{i-1} \tag{6.80}$$

Here, $P_o/N$ is the average power, in steady state, of the individual generalized signalling random processes; the overall average power of the random process being $P_o$.

### 6.8.2 Example

An example of a signal from a 1/f noise random process is shown in Figures 6.34 and 6.35 for the case of the summation of $N=14$ generalized shot noise random processes with overall unit power, that is, $P_o=1$. The power spectral density, obtained from 100 1/f noise signals, is shown in Figure 6.36. For the $i$th generalized signalling random process, $i\in\{1,\ldots,N\}$, the pulse function is a raised cosine pulse, with unity energy, and defined according to

$$p_i(t)=\frac{1}{\sqrt{3}\sqrt{2^{i-2}}}\left[1+\cos\left(2\pi\cdot\frac{1}{2^{i-1}D_o}\cdot\left(t-2^{i-2}D_o\right)\right)\right]\cdot \\ \left[u(t)-u\left(t-2^{i-1}D_o\right)\right]\quad i\in\{1,\ldots,14\} \tag{6.81}$$

Here, $D_o$ is the duration of the fundamental pulse associated with the first generalized shot noise random process and which is set to unity. The fundamental pulse rate is

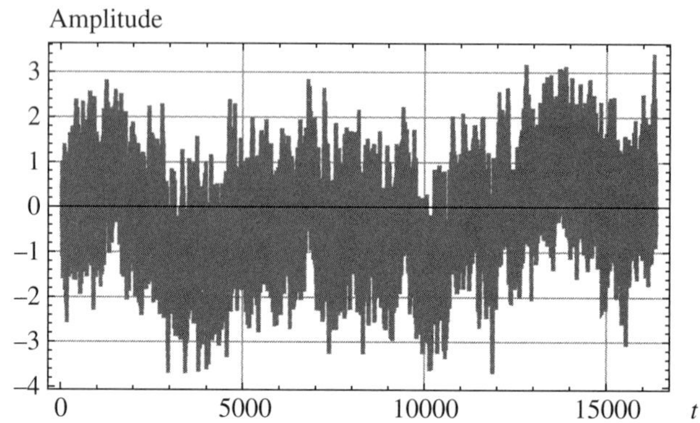


**FIGURE 6.34** Graph of a signal, over the interval $[0, 2^{14}]$, from the $1/f$ noise random process defined by Equation 6.78 with $N = 14$ and based on pulse functions with unit energy as defined by Equation 6.81. The following parameters have been used: $D_o = 1$, $\lambda_o = 10$, $\lambda_i = \lambda_o/2^{i-1}$, $P_o = 1$, and $\sigma_i^2$ set consistent with Equation 6.80.

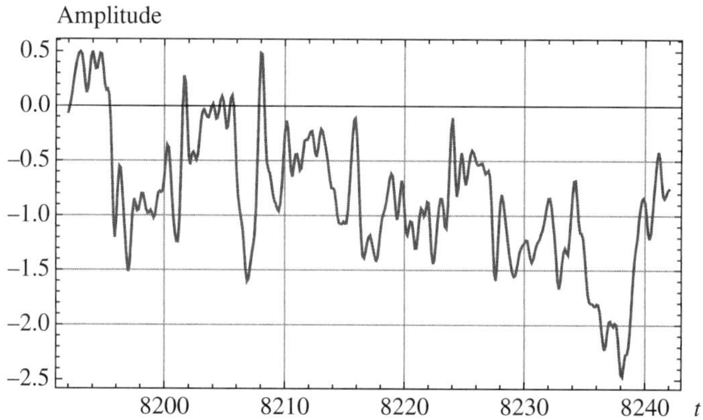


**FIGURE 6.35** Expanded view of the $1/f$ signal shown in Figure 6.34.

$\lambda_o = 10$ and $\lambda_i = \lambda_o/2^{i-1}$. The amplitudes of the pulses are assumed to be consistent with a Gaussian distribution with zero mean and with variances $\sigma_i^2$ set consistent with Equation 6.80. The pulse functions $p_1(t)$ to $p_5(t)$ are shown in Figure 6.37.

The form of the pulse function is relatively unimportant, in terms of generating a $1/f$ noise spectrum over a defined frequency range, provided a measure of the pulse duration scales according to the reciprocal of the pulse rate. The low-frequency variations of the noise signal, as evident in Figure 6.34, are typical of $1/f$ noise. Further details about the modelling of $1/f$ noise, consistent with Equation 6.78, can be found in Howard and Raffel (2003). Consistent with the diverse manifestations of $1/f$ noise, many models for $1/f$ noise exist, for example, Montroll and Shlesinger (1982).

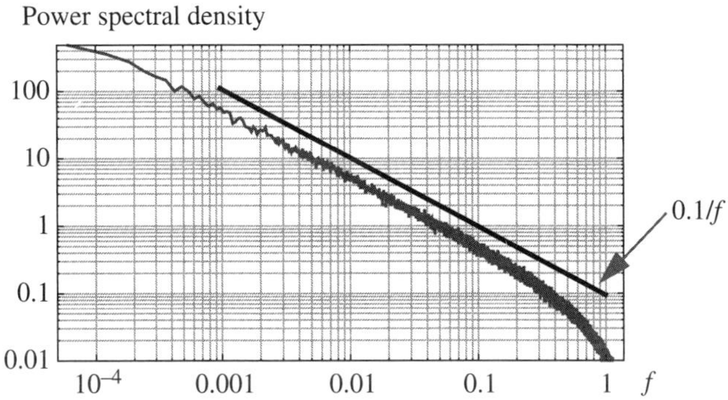


**FIGURE 6.36** The power spectral density obtained by averaging the individual power spectral densities from 100 $1/f$ noise signals.

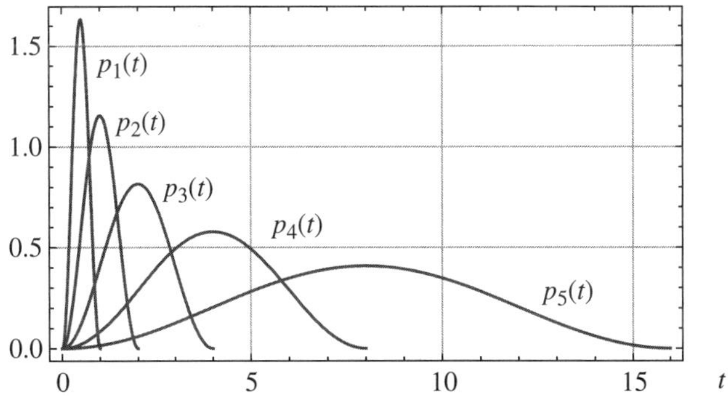


**FIGURE 6.37** Pulse functions used for the first to fifth generalized shot noise random processes that underpin the $1/f$ noise random process defined in the text.

## 6.9 BIRTH–DEATH RANDOM PROCESSES

Many random phenomena can be characterized in a state manner and by the associated changes of states. For the case where the states are specified by integer numbers, a state represents the population of a set of entities, and the population changes in an integer manner according to *births* and *deaths*, then a birth–death random process is defined. Birth–death random processes are useful, for example, in modelling the throughput in queueing systems, modelling population dynamics of species, etc.

### 6.9.1 Experiment Underpinning Birth–Death Random Process

The following algorithmic experiment, where a state parameter $k$ represents the population and the initial population at $t = t_0$ is $k_0$, $k_0 > 0$, underpins a birth–death random process:

0. At the commencement time of $t_0$ the state parameter, $k$ is set to $k_0$.
1. Two Poisson experiments are performed that define points on the interval $[t_0, \infty)$: the first, with rate $\lambda_1$, defines points for a birth process; the second, with rate $\mu_1$, defines points for a death process. The earliest point time arising from these two experiments is taken and denoted $t_1$. If the earliest time is associated with a birth, then the state parameter is incremented, that is, $k_1 = k_0 + 1$; if the earliest time is associated with a death, then the state parameter is decremented, that is, $k_1 = k_0 - 1$.

...

*j*. Two Poisson experiments are performed that define points on the interval $[t_{j-1}, \infty)$: the first, with rate $\lambda_j$, defines points for a birth process; the second, with rate $\mu_j$, defines points for a death process. The earliest point time arising from these two experiments is taken and denoted $t_j$. If the earliest time is associated with a birth, then the state parameter is incremented, that is, $k_j = k_{j-1} + 1$; if the earliest time is associated with a death, then the state parameter is decremented, that is, $k_j = k_{j-1} - 1$.

...

If at any time the population drops to zero, that is, $\exists j$ such that $k_j = 0$, then the algorithm terminates and all future population values are zero.

The outcome of a trial of this experiment is a sequence of ordered pairs specifying the birth–death times and the resulting population:

$$\left((t_0, k_0), (t_1, k_1), \ldots, (t_j, k_j), \ldots\right) \tag{6.82}$$

The sample space associated with the experiment is

$$S = \left\{ \begin{array}{l} \left((t_0, k_0), (t_1, k_1), \ldots, (t_j, k_j), \ldots\right) : \\ (t_j, k_j) \text{ as defined by algorithm} \end{array} \right\} \tag{6.83}$$

### 6.9.2 Birth–Death Random Processes

A birth–death random process arises from the experiment defined in Section 6.9.1 by associating a level, as given by the state, between the state transition times.

**Definition: Birth–Death Random Process** A birth–death random process $X$ is defined on the sample space $S$, as specified in Equation 6.83, according to

$$\begin{aligned} &X : S \to S_X \\ &X\left(\left((t_0, k_0), (t_1, k_1), \ldots, (t_j, k_j), \ldots\right), t\right) = k_0 u(t - t_0) \\ &\quad + \sum_{j=1}^{\infty} (k_j - k_{j-1}) u(t - t_j) \qquad t \geq t_0 \end{aligned} \tag{6.84}$$

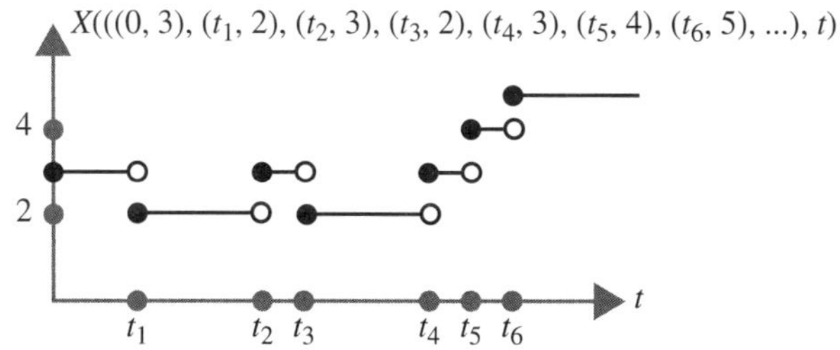


**FIGURE 6.38** One waveform defined by a birth–death random process for the case of an initial population of $k_0 = 3$. The sequence is consistent with a death, a birth, a death, a birth, a birth, a birth, ….

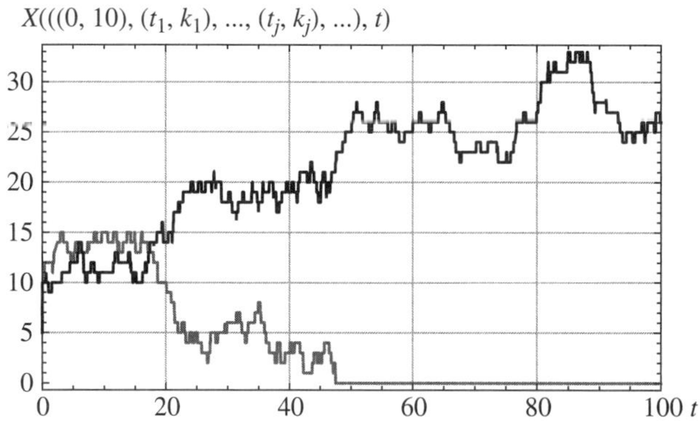


**FIGURE 6.39** Two signals from a birth–death random process with unity birth and death rates and an initial population of 10.

$$S_X = \left\{ \begin{array}{r} X\left(\left((t_0,k_0),(t_1,k_1),\ldots,(t_j,k_j),\ldots\right),t\right) = k_0 u(t-t_0) + \\ \sum_{j=1}^{\infty} \left(k_j - k_{j-1}\right) u\left(t - t_j\right) : \\ t \geq t_0, \left((t_0,k_0),(t_1,k_1),\ldots,(t_j,k_j),\ldots\right) \in S \end{array} \right\} \tag{6.85}$$

For notational convenience, a birth–death random process is written as

$$X(\Omega,t) = k_0 u(t-t_0) + \sum_{j=1}^{\infty} \left(K_j - K_{j-1}\right) u\left(t - \Gamma_j\right) \qquad K_0 = k_0, t \geq t_0 \tag{6.86}$$

where $\Omega = \left((\Gamma_1,K_1),(\Gamma_2,K_2),\ldots,(\Gamma_j,K_j),\ldots\right)$, $\Gamma_1, \Gamma_2, \ldots$ is a sequence of random variables consistent with the outcomes of the sequences of two Poisson experiments in the algorithm and $K_1, K_2, \ldots$ is a sequence of random variables defining the population levels at the transition times.

*6.9.2.1 Examples* One signal defined by a birth–death random process is illustrated in Figure 6.38. Two signals from a birth–death random process with $\lambda_1 = \mu_1 = 1$, $k_0 = 10$, and $t_0 = 0$ are shown in Figure 6.39.

## 6.10 ORTHOGONAL INCREMENT RANDOM PROCESSES

An orthogonal increment random process underpins the definition of the Cramer representation of stationary random processes and the use of the Lebesgue–Stieltjes integral for stochastic calculus. An orthogonal increment process is best understood by first defining an *increment function*.

### 6.10.1 Increment Function

**Definition: Increment Function** A function $f: \boldsymbol{R} \to \boldsymbol{R}$ is an increment function if, on all finite intervals, it can be written in the form of a sum of unit step functions, that is, if it can be written as

$$f(t) = \sum_{i=1}^{N} \alpha_i u(t - t_i) \qquad \alpha_i \in \boldsymbol{R} \tag{6.87}$$

An increment function is a right continuous function. One example of an increment function is illustrated in Figure 6.40.

**Theorem 6.2 Change in an Increment Function**
The change in an increment function $f: \boldsymbol{R} \to \boldsymbol{R}$, as defined by $f(t_\beta) - f(t_\alpha)$, assuming $t_\beta > t_\alpha$, is

$$f(t_\beta) - f(t_\alpha) = \sum_{i=1}^{N} \alpha_i \Delta_i \tag{6.88}$$

where

$$\Delta_i = u(t_\beta - t_i) - u(t_\alpha - t_i) = \begin{cases} 0 & t_\alpha, t_\beta < t_i; t_\alpha, t_\beta \geq t_i \\ 1 & t_\alpha < t_i, t_\beta \geq t_i \end{cases} \tag{6.89}$$

**Proof**
By definition

$$f(t_\beta) - f(t_\alpha) = \sum_{i=1}^{N} \alpha_i \left[ u(t_\beta - t_i) - u(t_\alpha - t_i) \right] = \sum_{i=1}^{N} \alpha_i \Delta_i \tag{6.90}$$

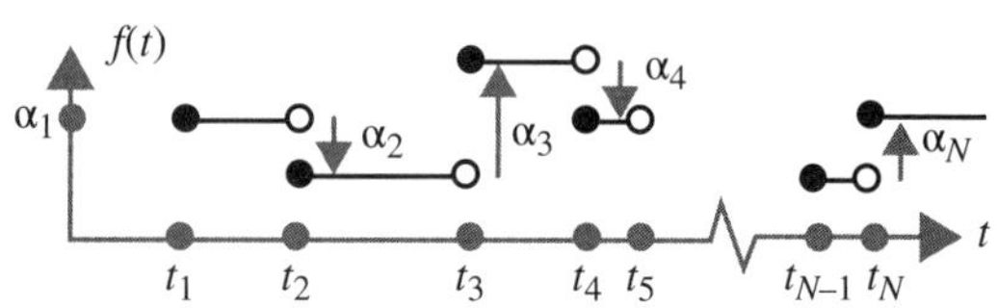


**FIGURE 6.40** Graph of an increment function.

where

$$\Delta_i = u(t_\beta - t_i) - u(t_\alpha - t_i) \tag{6.91}$$

The required result follows as

$$\Delta_i = \begin{cases} 0 & t_\alpha, t_\beta < t_i; \; t_\alpha, t_\beta \geq t_i \\ 1 & t_\alpha < t_i, t_\beta \geq t_i \end{cases} \tag{6.92}$$

***6.10.1.1 Important Result*** The following result proved in Chapter 2, as Theorem 2.10, is fundamental to orthogonal increment random processes.

**Theorem 6.3 Approximating Right Cont. Function by an Increment Function** On a finite interval [α, β], a right continuous function $f: \boldsymbol{R} \to \boldsymbol{R}$, with a finite number of discontinuities on any given finite interval, can be approximated, arbitrarily closely, by an increment function, that is, $\forall \varepsilon > 0$, there exist a positive integer $N$ and pairs of numbers $(\alpha_1, t_1), \ldots, (\alpha_N, t_N)$, such that

$$\left| f(t) - \sum_{i=1}^{N} \alpha_i u(t - t_i) \right| < \varepsilon \qquad t, t_i \in [\alpha, \beta] \tag{6.93}$$

### 6.10.2 Definition: Orthogonal Increment Random Process

**Definition: Orthogonal Increment Random Process** A random process $X: \Omega \to S_X$ is an orthogonal increment random process if, for all arbitrary times $t, t_1, t_2, t_3, t_4$, it is the case that

$$\boldsymbol{E}[X(\Omega, t)] = 0 \qquad \boldsymbol{E}[X(\Omega, t) X^*(\Omega, t)] < \infty \qquad t \in \boldsymbol{R} \tag{6.94}$$

and

$$\boldsymbol{E}[(X(\Omega, t_4) - X(\Omega, t_3))(X(\Omega, t_2) - X(\Omega, t_1))^*] = 0 \tag{6.95}$$

when

$$(t_1, t_2] \cap (t_3, t_4] = \varnothing, \qquad t_1 < t_2 \leq t_3 < t_4 \tag{6.96}$$

Physically, an orthogonal increment random process is one where the difference in the signal level between two points multiplied by the difference in the signal level between another two points, when the intervals defined by the pairs of points are disjoint, on average, is zero. This is illustrated in Figure 6.41.

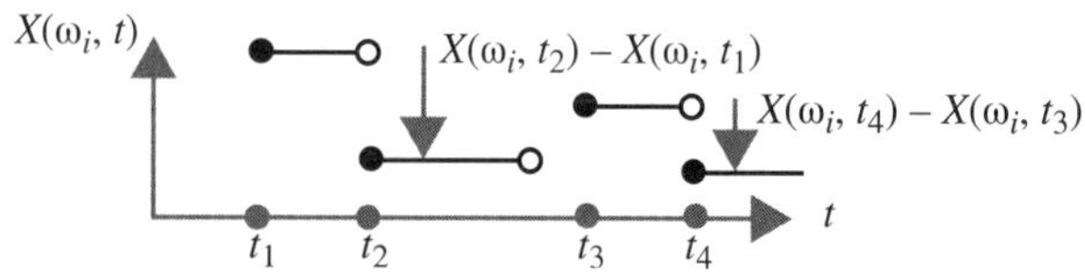


**FIGURE 6.41** Illustration of the changes in the $i$th signal from an orthogonal increment random process. On average, the change in the signal level between $t_2$ and $t_1$ multiplied by the change in the signal level between $t_4$ and $t_3$ is zero.

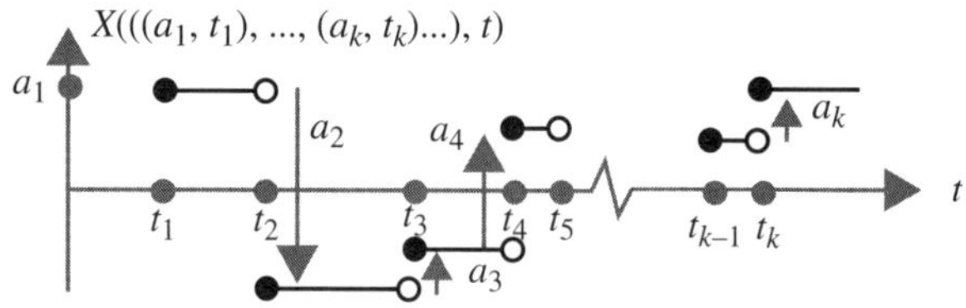


**FIGURE 6.42** Graph of one signal from the Poisson point based random process defined in Equation 6.97.

### 6.10.3 Examples of Orthogonal Increment Random Processes

#### *6.10.3.1 Poisson Point-Based Random Process*

**Theorem 6.4 Poisson Point-Based Random Process**
A random process based on a Poisson experiment and defined according to

$$
\begin{aligned}
X(\Omega,t) &= \sum_{k=1}^{\infty} A_k u(t-\Gamma_k) \qquad t\geq 0 \\
\Omega &= ((A_1,\Gamma_1),\ldots,(A_k,\Gamma_k),\ldots)
\end{aligned}
\tag{6.97}
$$

is an orthogonal increment random process on the interval $[0,\infty]$ when $A_1, \ldots, A_k, \ldots$ are zero mean independent random variables with respective probability density functions $f_{A_1},\ldots,f_{A_k},\ldots$ and $\Gamma_1, \ldots, \Gamma_k, \ldots$ are a sequence of random variables consistent with the outcomes of a Poisson experiment.

**Proof**
The proof is detailed in Appendix 6.A.

One signal defined by the random process specified in Equation 6.97 is illustrated in Figure 6.42.

#### *6.10.3.2 Random Walk*

**Theorem 6.5 Random Walk: Orthogonal Increment Random Process**
A random walk with a step size of $A_o$, a step spacing of $\Delta t$, and defined according to

$$X(\Omega,t)=\sum_{k=1}^{N}A_k u(t-k\Delta t) \quad t\geq 0 \tag{6.98}$$

where $\Omega=(A_1,A_2,\ldots)$ is a vector of independent and identical random variables with a sample space $S_A=\{-A_o,A_o\}$, is an orthogonal increment random process on the interval $[0, N\Delta t]$.

**Proof**

The follows directly from Theorem 6.4.

***6.10.3.3 Brownian Motion*** Brownian motion can be defined as the limit of a sequence of random walk random processes.

**Definition: Brownian Motion** One-dimensional Brownian motion can be defined on the interval $[0, T]$ by the sequence of random processes $\{X_i(\Omega_i,t)\}_{i=1}^{\infty}$ where

$$X_i(\Omega_i,t)=\sum_{k=1}^{i^2}A_k u(t-k\Delta t_i) \qquad \Delta t_i=\frac{T}{i^2}, t\geq 0 \tag{6.99}$$

Here, $\Omega_i=(A_1,A_2,\ldots)$ is a vector of identical and independent random variables with a sample space $S_A=\{-A_o/i,A_o/i\}$ where $A_o$ is a fixed constant and $P[A_k=-A_o/i]=P[A_k=A_o/i]=1/2$.

***6.10.3.4 Notes*** A rationale for the scaling of the amplitude and time steps according to $A_o/i$ and $T/i^2$ is given in Chapter 8 (Section 8.6).

Examples of waveforms from the sequence of random processes defining Brownian motion are shown in Figures 6.43 and 6.44.

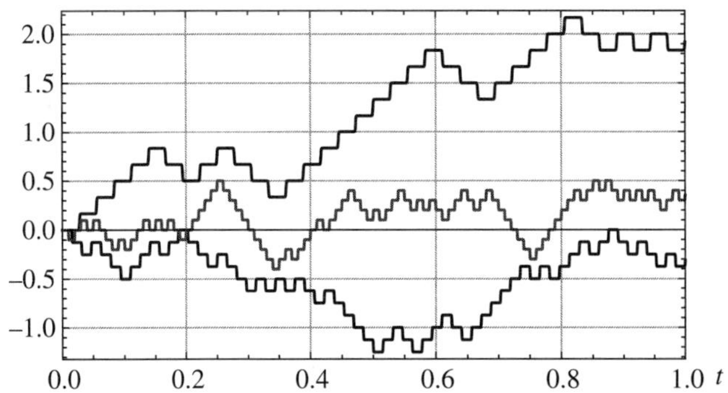


**FIGURE 6.43** Examples of signals from the 6th (upper waveform), 8th (lower waveform), and 10th (middle waveform) random processes defining Brownian motion for the case of $A_o=T=1$.

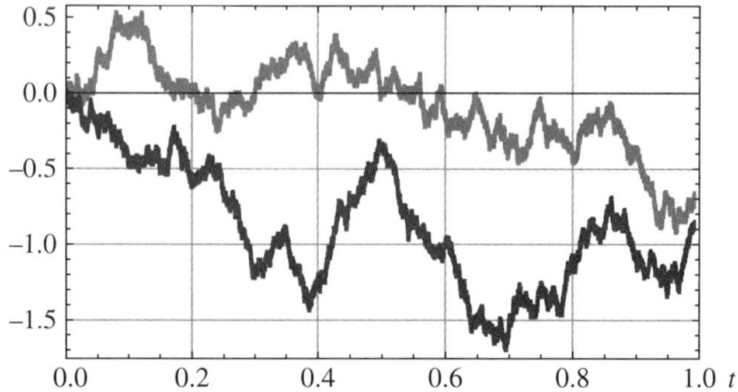


**FIGURE 6.44** Two signals from the 100th random process defining Brownian motion for the case of $A_o = T = 1$.

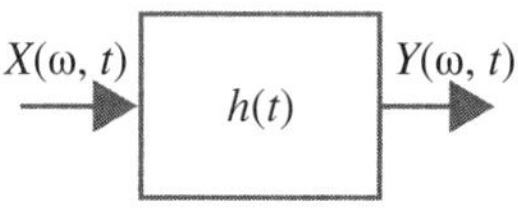


**FIGURE 6.45** Linear filtering by a causal system with an impulse response $h$.

**Theorem 6.6 Brownian Motion Is an Orthogonal Increment Random Process** Brownian motion, as defined by the sequence of random processes consistent with Equation 6.99, is an orthogonal increment random process.

**Proof**

All random processes in the sequence, consistent with Theorem 6.4, are orthogonal increment random processes. Accordingly, Brownian motion is an orthogonal increment random process.

## 6.11 LINEAR FILTERING OF RANDOM PROCESSES

Consider the case of a random process $X: S \to S_X$ with a signal sample space defined according to

$$S_X = \{X(\omega, t): \quad \omega \in S\} \tag{6.100}$$

For the case of linear filtering, as illustrated in Figure 6.45, there is a one to one correspondence between each signal $X(\omega, t)$ in the input sample signal sample space $S_X$ and an output signal $Y(\omega, t)$:

$$Y(\omega,t) = \int_0^t X(\omega,\lambda)h(t-\lambda)d\lambda \tag{6.101}$$

where $h$ is the system impulse response. Accordingly, the random process $Y$, defined by the output signals of the linear system, can be defined according to

$$Y : S \to S_Y$$
$$S_Y = \left\{ Y(\omega, t): \quad Y(\omega, t) = \int_0^t X(\omega, \lambda) h(t-\lambda) d\lambda, \omega \in S \right\} \tag{6.102}$$

## 6.12 SUMMARY OF RANDOM PROCESSES

A summary of the random processes defined in this chapter is given in Table 6.1. In this table the following notation has been used: $\Gamma_1, \Gamma_2, \ldots$ are a sequence of random variables consistent with the outcomes of a Poisson experiment with a rate of $\lambda$ per second, $A_1, A_2, \ldots$ are a sequence of random variables defining pulse amplitudes, $S_A$ is the sample space of $A_1, A_2, \ldots$ assuming they are identical random variables, and $p : [0, \infty) \to \boldsymbol{R}$ is a pulse function with finite energy.

**TABLE 6.1 Summary of random processes. In general, the random processes are tabulated according to increasing levels of randomness based on amplitude and then time variation**

| Random process | Definition of random process | |
|---|---|---|
| Random walk | $X(\Omega, t) = \sum_{k=1}^{N} A_k u(t-(k-1))$ | $S_A = \{-1, 1\}$ |
| Poisson point process | $X(\Omega, t) = \sum_{k=1}^{\infty} \delta_K(t-\Gamma_k)$ or $X(\Omega, t) = \sum_{k=1}^{\infty} \delta(t-\Gamma_k)$ | |
| Poisson counting process | $X(\Omega, t) = \sum_{k=1}^{\infty} u(t-\Gamma_k)$ | |
| Shot noise | $X(\Omega, t) = \sum_{k=1}^{\infty} p(t-\Gamma_k)$ | |
| Generalized shot noise | $X(\Omega, t) = \sum_{k=1}^{\infty} A_k p(t-\Gamma_k)$ | |
| Random telegraph signal | $X(\Omega, t) = \sum_{k=0}^{\infty} a_k u(t-\Gamma_k)$ | $\begin{cases} a_0 = 1, \ \Gamma_0 = 0 \\ a_1, a_3, \ldots = -2 \\ a_2, a_4, \ldots = 2 \end{cases}$ |
| Random telegraph signal—type II | $X(\Omega, t) = \sum_{k=0}^{\infty} A_k [u(t-\Gamma_k) - u(t-\Gamma_{k+1})]$ | $A_0 = 1, \ \Gamma_0 = 0$<br>$A_k, A_{k+1}$ correlated<br>$S_A = \{-1, 1\}$ |

(*continued*)

**TABLE 6.1** *(continued)*

| Random process | Definition of random process |
|---|---|
| Clustered Point random process | $X(\Omega,t)=\sum_{k=1}^{\infty}\left[\delta_K(t-T_k)+\sum_{r=1}^{M_k}\delta_K(t-T_k-\Gamma_{kr})\right]$ <br> $T_1,\ldots$Poisson points, $M_k$ : sample space $S_M$, $\Gamma_{kr}$ : sample space $S_\Gamma$ |
| Clustered random process | $X(\Omega,t)=\sum_{k=1}^{\infty}\left[p(t-T_k)+\sum_{r=1}^{M_k}p(t-T_k-\Gamma_{kr})\right]$ <br> $T_1,\ldots$Poisson points, $M_k$ : sample space $S_M$, $\Gamma_{kr}$ : sample space $S_\Gamma$ |
| Jittered binary random process | $X(\Omega,t)=A\sum_{k=1}^{\infty}(-1)^{k+1}u(t-kD-J_k)$ $J_k$- sample space $S_J$ |
| Jittered pulse train | $X(\Omega,t)=\sum_{k=1}^{\infty}A_kp(t-kD-J_k)$ $J_k$- sample space $S_J$ |
| White noise: flat spectrum on interval $(-f_{\max},f_{\max})$ | $X(\Omega,t)=A_o\sum_{k=1}^{N}\frac{\sqrt{2}A_k}{\sqrt{N}}\cos(2\pi kf_ot)+\frac{\sqrt{2}B_k}{\sqrt{N}}\sin(2\pi kf_ot)$ <br> $A_k=\sin(\Phi_k)$, $B_k=\cos(\Phi_k)$, $S_{\Phi_k}=[0,2\pi)$, $f_o=1/T$, $f_{\max}=Nf_o$ <br> or <br> $X(\Omega,t)=\sum_{k=1}^{N}\frac{\sqrt{2}A_o}{\sqrt{N}}\sin(2\pi F_kt+\Omega_k)$ <br> $F_1,F_2,\ldots$ defined by Poisson experiment : $\lambda=1/f_o$ <br> $N$- number of points in $[0,f_{\max}]$, $S_{\Omega_k}=[0,2\pi)$ |
| Signalling random process | $X(\Omega,t)=\sum_{k=1}^{\infty}\phi(\Omega_k,t-(k-1)D)$ signalling rate $=1/D$, $\phi\in S_\Phi$ |
| Signalling random process: generalized | $X(\Omega,t)=\sum_{k=1}^{\infty}\phi(\Omega_k,t-\Gamma_k)$ $\phi\in S_\Phi$ |
| $1/f$ noise | $X(\Omega,t)=\sum_{i=1}^{N}\sum_{k=1}^{\infty}A_{ik}p_i(t-\Gamma_{ik})$ $A_{ik}$ : sample space $S_i$ <br> $\Gamma_{ik}$ defined by Poisson experiment : $\lambda_i=\lambda_o/2^{i-1}$ |
| Birth–death random process | $X(\Omega,t)=k_0u(t-t_0)+\sum_{j=1}^{\infty}(K_j-K_{j-1})u[t-\Gamma_j]$ <br> $K_0=k_0,\ t\geq t_0,\ K_j=K_{j-1}\pm 1$ |
| Brownian Motion | $X(\Omega,t)=\{X_i(\Omega_i,t)\}_{i=1}^{\infty}$ $X_i(\Omega_i,t)=\sum_{k=1}^{i^2}A_ku(t-k\Delta t_i)$ <br> $A_k\in\{-A_o/i,A_o/i\}$, $\Delta t_i=T/i^2$ |

## 6.13 PROBLEMS

**6.1** Draw graphs of signals from a random walk random process where the probability of an outcome of the underlying Bernoulli experiment is $p = P[X = 1] = 0.55$ and $p = 0.6$. For each case, specify the expected value after $N$ steps.

**6.2** Consider the sampling of a random process, at random times, and at a rate of $\lambda$ per second.

a. Specify the experiment underpinning the sampling and define the generated random process.
b. Draw representative signals from the random process for the case where the samples form a sample space $\{-1, 0, 1\}$ and the sampling rate is unity.

**6.3** Consider two independent Poisson counting random processes $X_1(\Omega_1, t)$ and $X_2(\Omega_2, t)$ that are based on Poisson experiments with a rate $\lambda$. New random processes are defined according to

$$\begin{aligned} Z_1(\Omega, t) &= X_1(\Omega_1, t) - X_2(\Omega_2, t) \\ Z_2(\Omega, t) &= \text{sgn}[X_1(\Omega_1, t) - X_2(\Omega_2, t)] \end{aligned} \tag{6.103}$$

a. Give a physical interpretation for $Z_1$ and $Z_2$.
b. Specify the experimental sample space, underpinning $Z_1$ and $Z_2$. Specify $\Omega$ and define $Z_1$ and $Z_2$ in terms of experimental outcomes.
c. For the case of $\lambda = 1$, generate signals from $X_1, X_2, Z_1$, and $Z_2$.

**6.4** Consider a shot noise random process based on a Gaussian pulse

$$p(t) = \frac{1}{\sqrt{2\pi}\sigma} \cdot \exp\left(-t^2/2\sigma^2\right) \qquad \sigma^2 = t_{\text{fwhm}}^2/8\ln(2) \tag{6.104}$$

which has unity area and a full width half maximum of $t_{\text{fwhm}}$.

a. Graph signals from the random process with rates of $\lambda = 0.1$ and $\lambda = 1$. Assume $t_{\text{fwhm}} = 1$.
b. For the case where each pulse amplitude is randomly selected from the set $\{-1, 0, 1\}$, graph signals from the shot noise random process for the case of $\lambda = 0.1$ and $\lambda = 1$. Assume $t_{\text{fwhm}} = 1$.

**6.5** Consider the case of photons arriving at random times, at a rate $\lambda$, and randomly on a circular plate of radius $r_o$. Behind the plate is a spherical lens of radius $r_o$ and a photodetector. The configuration is such that the detected photon energy is modified by $\cos(\pi r/2r_o), 0 \le r \le r_o$, for a photon arriving on the circular plate at a point $(r, \phi)$ assuming polar coordinates.

a. Define the experiment underpinning the photodetector signal.
b. Define the random process that is consistent with the photodetector current assuming the photodetector response to a photon is a pulse function $p$.
c. Graph signals from the random process assuming the pulse function has the form $p(t) = u(t) - u(t-\tau)$. Assume $r_o = 1$, $\lambda = 1$, and $\tau = 0.1$.

**6.6** Consider a random telegraph signal of type II that is based on a Poisson experiment with a point rate of $\lambda$. Graph signals from such a random process, along with the underlying Poisson points, for the case of $\lambda = 1$ and $p = 1$, $p = 0.5$, and $p = 0.1$.

**6.7** Consider a clustered random process based on traffic collisions on freeways where an initial accident causes subsequent collisions. Consider the case of the initial accident occurring, at random, and at a normalized rate of $\lambda_1 = 1$. The number of secondary collisions is given by a Poisson distribution

$$P[\text{number of secondary collisions} = k] = \frac{\lambda_S^k \exp(-\lambda_S)}{k!} \tag{6.105}$$

where $\lambda_S = 1/2$. Secondary collisions occur at a uniform rate of $\lambda_2 = 10$.

a. Specify the experiment underpinning the points associated with the timing of the collisions. Specify the sample space for the experiment.
b. Specify the clustered point random process that defines the commencement times of collisions.
c. A pulse function that gives a measure (such as damage cost) of the damage with time, of a specific collision commencing at $t = 0$, is

$$p(t) = \frac{2t}{\sigma_p^2} \cdot \exp\left[\frac{-t^2}{\sigma_p^2}\right] u(t) \qquad \sigma_p^2 = 0.01 \tag{6.106}$$

A random variable $A$ defines the level of damage in a specific collision and has a probability density function as specified by a Pareto distribution of type II:

$$f_A(a) = \frac{\alpha}{k\left(1 + \frac{a}{k}\right)^{1+\alpha}} \tag{6.107}$$

with parameters $\alpha = 3$ and $k = 2$. The mean and variance of $A$ are $\mu_A = \dfrac{k}{\alpha - 1}$ $(\alpha > 1)$ and $\sigma^2 = \dfrac{k^2\alpha}{(\alpha-2)(\alpha-1)^2}$ $(\alpha > 2)$.

Define the clustered random process that models the collision damage with time. Graph signals from this random process and graph the component signals for the damage arising from the initial and secondary collisions.

**6.8** For quadrature phase shift keying (QPSK), the signalling set is

$$S_{\Phi} = \left\{ \begin{matrix} A\cos\left[2\pi f_C t + (i-1)\frac{\pi}{2} + \frac{\pi}{4}\right][u(t) - u(t-D)] : \\ i \in \{1,2,3,4\} \end{matrix} \right\} \tag{6.108}$$

A signalling random process is based on this signalling set with a signalling period of $D$ seconds. The signals for each time interval are chosen independent of one another.

a. Specify the experiment underpinning this random process for the time interval $[0, ND]$.
b. Specify the experimental sample space.
c. Specify the random process.
d. Graph signals from the random process for the case of $A = 1,\ D = 1,\ f_C = 4$.

**6.9** Consider the second definition for a white noise random process. Consider the time interval [0, 1] and frequencies chosen, consistent with the outcomes of a Poisson point process, with a rate of 100 and in the range $[0, f_{\max}]$ where $f_{\max} = 100$.

a. Generate a time signal from this random process.
b. From one thousand signals, take time samples at a specific time, for example, $t = 0.5$. From these samples, generate a normalized histogram and compare the histogram values with a Gaussian probability density function with a mean and variance as determined by the sample mean and sample variance. Compare the sample variance with the expected average power of $A_o^2$.

**6.10** Consider two birth–death random processes that both have an initial population of $k_0 = 10$ at $t_0 = 0$ and a constant birth rate of $\lambda = 1$. The first has a constant death rate of $\mu = 0.9$, while the second has a constant death rate of $\mu = 1.1$. Graph signals from these random processes.

**6.11** Consider a sequence of random walks $\{X_i(\Omega, t)\}_{i=1}^{\infty}$ defined on the interval [0, 1] and the following cases:

i. Scaling of the time step by $1/i^2$ and scaling of the amplitude by $1/\sqrt{i}$
ii. Scaling of the time step by $1/i^2$ and scaling of the amplitude by $1/i$
iii. Scaling of the time step by $1/i^2$ and scaling of the amplitude by $1/i^{3/2}$

a. For each of the three cases, generate signals for $i = 10$ and $i = 100$.
b. For a fixed value of time and for the three cases, specify the variance of the random walks. Note which scaling leads to a constant variance as $i$ increases.

## APPENDIX 6.A PROOF OF THEOREM 6.4

For all times the random process has zero mean and finite mean squared value. Consider the expectation $\boldsymbol{E}[(X(\Omega,t_4)-X(\Omega,t_3))\ (X(\Omega,t_2)-X(\Omega,t_1))]$:

$$\begin{aligned}
&\boldsymbol{E}[(X(\Omega,t_4)-X(\Omega,t_3))\ (X(\Omega,t_2)-X(\Omega,t_1))] = \\
&\boldsymbol{E}\left[\sum_{i=1}^{N}\sum_{j=1}^{N} A_i[u(t_2-\Gamma_i)-u(t_1-\Gamma_i)]A_j\left[u(t_4-\Gamma_j)-u(t_3-\Gamma_j)\right]\right] \\
&= \sum_{i=1}^{N}\sum_{j=1}^{N} \boldsymbol{E}\left[A_iA_j\right]\boldsymbol{E}\left[[u(t_2-\Gamma_i)-u(t_1-\Gamma_i)]\ \left[u(t_4-\Gamma_j)-u(t_3-\Gamma_j)\right]\right] \\
&= \sum_{i=1}^{N} \boldsymbol{E}\left[A_i^2\right]\boldsymbol{E}[[u(t_2-\Gamma_i)-u(t_1-\Gamma_i)]\ [u(t_4-\Gamma_i)-u(t_3-\Gamma_i)]]
\end{aligned} \tag{6.109}$$

as $\boldsymbol{E}\left[A_iA_j\right]=0$ for $i \neq j$ by assumption.

Consider the random variables $u(t_2-\Gamma_i)-u(t_1-\Gamma_i)$ and $u(t_4-\Gamma_i)-u(t_3-\Gamma_i)$ for the case where $(t_1,t_2]\cap(t_3,t_4]=\varnothing$ and $t_1<t_2\le t_3<t_4$. Each outcome of the experiment underpinning the random process yields one value for

$$[u(t_2-\Gamma_i)-u(t_1-\Gamma_i)]\ [u(t_4-\Gamma_i)-u(t_3-\Gamma_i)] \tag{6.110}$$

Possible cases are illustrated in Figure 6.46, and the following results can arise:

$$\begin{aligned}
u(t_2-t_i)-u(t_1-t_i)&=0 \quad \text{and} \quad u(t_4-t_i)-u(t_3-t_i)=0 \\
u(t_2-t_i)-u(t_1-t_i)&=1 \quad \text{and} \quad u(t_4-t_i)-u(t_3-t_i)=0 \\
u(t_2-t_i)-u(t_1-t_i)&=0 \quad \text{and} \quad u(t_4-t_i)-u(t_3-t_i)=1
\end{aligned} \tag{6.111}$$

assuming $(t_1,t_2]\cap(t_3,t_4]=\varnothing,\ t_1<t_2\le t_3<t_4$.

Thus, it is the case that the random variables $u(t_2-\Gamma_i)-u(t_1-\Gamma_i)$ and $u(t_4-\Gamma_i)-u(t_3-\Gamma_i)$ are either zero or one, but it is not possible for both of them to be one when $(t_1,t_2]\cap(t_3,t_4]=\varnothing,\ t_1<t_2\le t_3<t_4$. Hence, the expectation is zero as required.

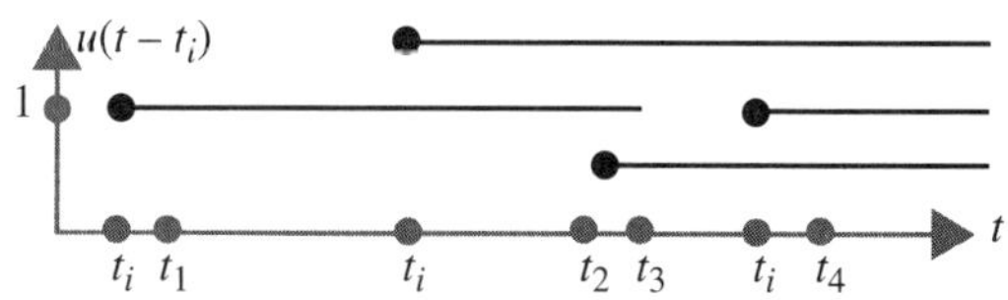


**FIGURE 6.46** Graphs of $u(t-t_i)$ (offset) for various values of $t_i$.

# 7

# CHARACTERIZING RANDOM PROCESSES

## 7.1 INTRODUCTION

There are many ways a random process can be characterized, and the characterization is usually linked to the application or situation in which it arises and the information required. The most common ways of characterizing random processes is via the evolution with time of the probability mass/density function, an autocorrelation function, and a power spectral density function. This chapter provides an introduction to such characterization, and this is followed by associated material including correlation, the average power in a random process, stationarity, Cramer's representation of random processes, and the state space characterization of random processes. One-dimensional random processes are assumed.

### 7.1.1 Notation for One-Dimensional Random Processes

Consistent with the notation introduced in Chapter 5 and used in Chapter 6, the following notation is used for a random process $X$:

$$\begin{aligned} &X : S \rightarrow S_X \\ &S = \{\omega : \text{as defined by experiment}\} \\ &S_X = \{X(\omega, t) : \quad \omega \in S\} \end{aligned} \tag{7.1}$$

For notational convenience, a random process is written as $X(\Omega, t)$ with the interpretation

*A Signal Theoretic Introduction to Random Processes*, First Edition. Roy M. Howard.

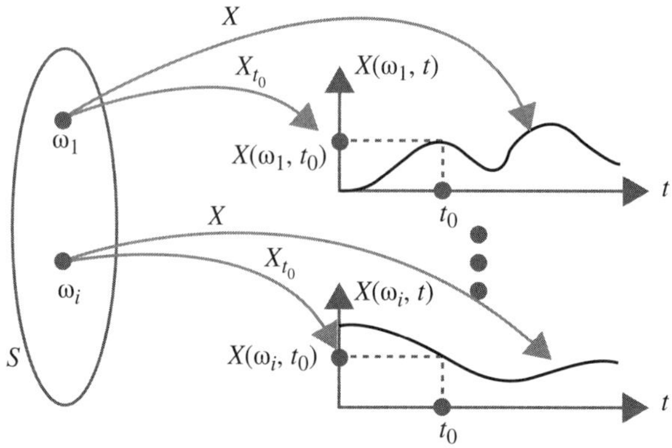


**FIGURE 7.1** Illustration of signals defined by a random process for the one-dimensional case and the values determined by the random variable defined by the random process at the time $t_0$.

$$\begin{array}{ll} X(\omega,t) & \text{individual signal defined by random process} \\ X(\Omega,t) & \text{set of signals defined by random process, i.e., } S_X \end{array} \tag{7.2}$$

The probability of a specific experimental outcome is governed by a probability mass function or a probability density function:

$$\begin{array}{l} P[\omega=\omega_i]=p_\Omega(\omega_i) \quad \text{countable case} \\ P[\omega\in[\omega_i,\omega_i+d\omega]]=f_\Omega(\omega_i)d\omega \quad \text{uncountable case} \end{array} \tag{7.3}$$

In this chapter, the signals are assumed to be one-dimensional as illustrated in Figure 7.1.

*7.1.1.1 Random Variables Defined by a Random Process* For a random process that defines one-dimensional signals, the values defined by the signals of the random process, at a specific time $t_0$, define a random variable $X_{t_0}$:

$$\begin{array}{l} X_{t_0}:S\to S_{t_0} \\ X_{t_0}(\omega)=X(\omega,t_0) \\ S_{t_0}=\{X_{t_0}(\omega): \quad \omega\in S\} \end{array} \tag{7.4}$$

By varying the time from $t_0$ to $t_1$ to $t_2$, etc., an infinite sequence of random variables can be defined.

### 7.1.2 Associated Random Processes

Consider a one-dimensional random process:

$$\begin{array}{l} X:S\to S_X \\ S=\{\omega: \quad \text{as defined by experiment}\} \\ S_X=\{X(\omega,t): \quad \omega\in S\} \end{array} \tag{7.5}$$

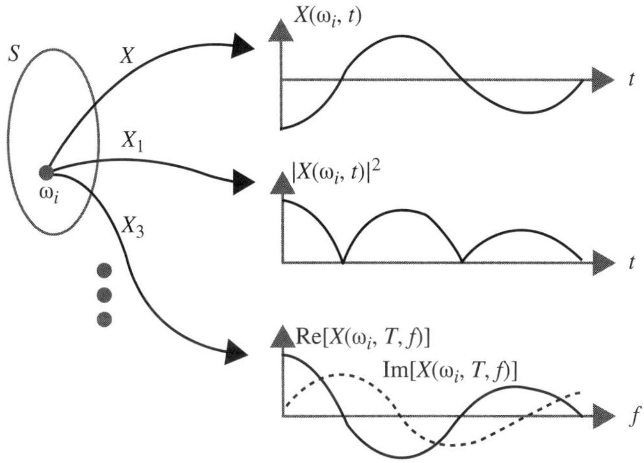


**FIGURE 7.2** Illustration of the $i$th signals defined by a random process $X$ and the associated random processes $X_1$ and $X_3$.

The following associated random processes, for example, can be defined:

$$\begin{aligned}
&X_1 : S \rightarrow S_1 \quad S_1 = \left\{ |X(\omega,t)|^2 : \ \omega \in S \right\} \\
&X_2 : S \rightarrow S_2 \quad S_2 = \{X(\omega,t)X^*(\omega,t-\tau) : \ \omega \in S\} \\
&X_3 : S \rightarrow S_3 \quad S_3 = \left\{ X(\omega,T,f) = \int_0^T X(\omega,t)e^{-j2\pi ft}dt : \ \omega \in S \right\} \\
&X_4 : S \rightarrow S_4 \quad S_4 = \left\{ \frac{|X(\omega,T,f)|^2}{T} : \ \omega \in S \right\}
\end{aligned} \tag{7.6}$$

Here, $X_1$ is a random process defined by the magnitude squared of the signals arising from $X$. $X_2$ is a random process, for $\tau$ fixed, defined by the autocorrelation function of the signals arising from $X$. $X_3$ is a random process defined by the Fourier transform of the signals arising from $X$, and $X_4$ is a random process defined by the power spectral density of the signals arising from $X$. Signals from $X$, $X_1$, and $X_3$ are illustrated in Figure 7.2.

## 7.2 TIME EVOLUTION OF PMF OR PDF

For all values of time, where the signals defined by a random process are valid, a random variable is defined that has a probability mass function, a probability density function, or outcomes that form both a countable set and an uncountable set (the mixed random variable case). The former two cases dominate the latter case, and these are considered. The probability mass function, or probability density function, changes with time to result in an evolving function. The evolution with time of the probability mass function, or probability density function, provides useful information about a random process.

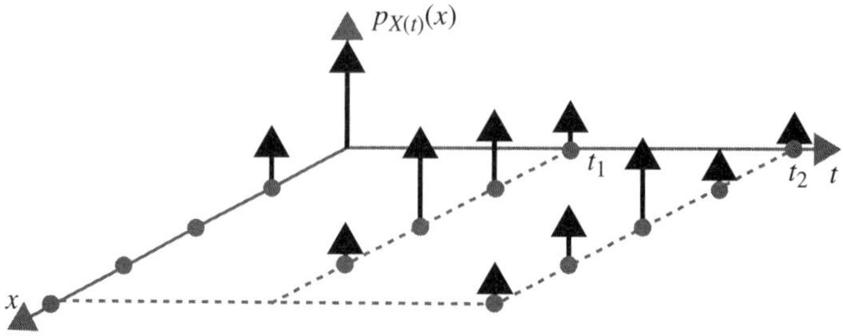


**FIGURE 7.3** Illustration of the evolution of probability mass function for a discrete time–discrete state random process.

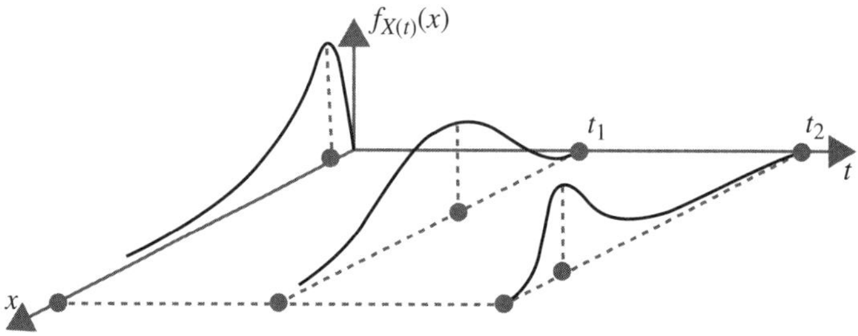


**FIGURE 7.4** Illustration of the evolution of the probability density function for a continuous time–continuous state random process.

For a discrete state random process, the time-evolving probability mass function is denoted $p_{X(t)}(x)$ and is defined according to

$$p_{X(t)}(x) = P[X_t = x] \tag{7.7}$$

where, for a fixed time $t$, the random variable defined, consistent with the notation above, is denoted $X_t$.

For a continuous state random process, the time-evolving probability density function is denoted $f_{X(t)}(x)$ and is defined according to

$$f_{X(t)}(x)dx \approx P[X_t \in (x, x + dx)] \tag{7.8}$$

for $dx$ sufficiently small.

This evolution is illustrated in Figure 7.3 for the case of a discrete time–discrete state random process and in Figure 7.4 for the case of a continuous time–continuous state random process.

## 7.3 FIRST-, SECOND-, AND HIGHER-ORDER CHARACTERIZATION

As a random process, at a set time, defines a random variable, it follows that random variable theory finds widespread application in characterizing random processes. The following first-, second-, and higher-order characterizations are fundamental.

### 7.3.1 First-Order Characterization

Consider a random variable $X_{t_o}$ defined by a random process at a fixed time $t_o$. The following parameters and functions can be defined: first, the mean and variance:

$$\begin{aligned}\mu_{t_o} &= \boldsymbol{E}[X_{t_o}] \\ \text{var}(X_{t_o}) &= \boldsymbol{E}\left[(X_{t_o}-\mu_{t_o})^2\right] = \boldsymbol{E}\left[X_{t_o}^2\right]-\mu_{t_o}^2\end{aligned} \tag{7.9}$$

Second, the cumulative distribution function:

$$F_{t_o}(x) = P[X_{t_o} \le x] \tag{7.10}$$

Third, the probability mass function for the case where $X_{t_o}$ is a discrete random variable and the probability density function for the case where $X_{t_o}$ is a continuous random variable:

$$\begin{aligned}&p_{t_o}(x) = P[X_{t_o} = x] \\ &f_{t_o}(x)dx \approx P[X_{t_o} \in [x, x+dx]] \quad f_{t_o}(x) = \frac{\partial}{\partial x}F_{t_o}(x)\end{aligned} \tag{7.11}$$

Here, $dx$ is assumed to be sufficiently small.

### 7.3.2 Second-Order Characterization

Consider the two random variables $X_{t_1}$ and $X_{t_2}$ defined by the fixed times $t_1$ and $t_2$. Apart from the mean, variance, cumulative distribution function, probability mass function, or probability density function, of the individual random variables, the following functions can be defined. First, the joint cumulative distribution function:

$$F_{t_1 t_2}(x_1, x_2) = P[X_{t_1} \le x_1, X_{t_2} \le x_2] \tag{7.12}$$

Second, the joint probability mass function for the case where $X_{t_1}$ and $X_{t_2}$ are discrete random variables and the joint probability density function for the case where $X_{t_1}$ and $X_{t_2}$ are continuous random variables:

$$\begin{aligned}&P_{t_1 t_2}(x_1, x_2) = P[X_{t_1} = x_1, X_{t_2} = x_2] \\ &f_{t_1 t_2}(x_1, x_2)dx_1 dx_2 \approx P[X_{t_1} \in [x_1, x_1 + dx_1], X_{t_2} \in [x_2, x_2 + dx_2]] \\ &f_{t_1 t_2}(x_1, x_2) = \frac{\partial}{\partial x_2}\frac{\partial}{\partial x_1}F_{t_1 t_2}(x_1, x_2)\end{aligned} \tag{7.13}$$

Here, $dx_1$ and $dx_2$ are assumed to be sufficiently small.

### 7.3.3 *N*th-Order Characterization

Consider the $N$ random variables $X_{t_1}, \ldots, X_{t_N}$ defined by the fixed times $t_1, \ldots, t_N$. Apart from the mean, variance, cumulative distribution function, probability mass/density

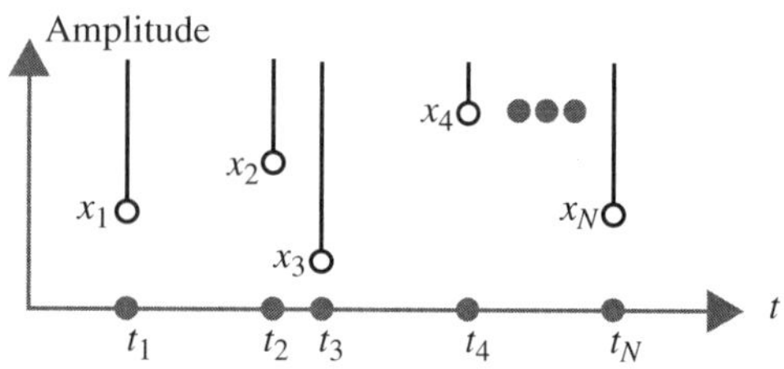


**FIGURE 7.5** Illustration of the constraints on a random process as defined by the joint cumulative distribution function.

function of the individual random variables or the joint cumulative distribution and joint probability mass/density function of pairs of random variables, the following functions can be defined. First, the joint cumulative distribution function of the $N$ random variables:

$$F_{t_1 \dots t_N}(x_1,\dots,x_N) = P[X_{t_1} \le x_1,\dots,X_{t_N} \le x_N] \tag{7.14}$$

The constraints, on a random process, defined by the joint cumulative distribution function are illustrated in Figure 7.5.

Second, the joint probability mass function for the case where $X_{t_1},\dots,X_{t_N}$ are discrete random variables and the joint probability density function for the case where $X_{t_1},\dots,X_{t_N}$ are continuous random variables, respectively, are defined according to

$$p_{t_1 \dots t_N}(x_1,\dots,x_N) = P[X_{t_1} = x_1,\dots,X_{t_N} = x_N] \tag{7.15}$$

$$\begin{aligned} &f_{t_1 \dots t_N}(x_1,\dots,x_N)dx_1\dots dx_N \approx P[X_{t_1} \in [x_1, x_1 + dx_1],\dots,X_{t_N} \in [x_N, x_N + dx_N]] \\ &f_{t_1 \dots t_N}(x_1,\dots,x_N) = \frac{\partial}{\partial x_N}\cdots\frac{\partial}{\partial x_1} F_{t_1 \dots t_N}(x_1,\dots,x_N) \end{aligned} \tag{7.16}$$

Here, $dx_1 \dots dx_N$ are assumed to be sufficiently small.

As a random process defined on the real line defines an uncountable number of random variables, it follows that a complete specification of such a random process, in terms of a joint probability density function, is not possible.

### 7.3.4 Mean and Variance and Average Power

Based on a first-order characterization, the following functions give useful information about a random process $X$.

**Definition: Mean** The mean $\mu_X : \boldsymbol{R} \to \boldsymbol{R}$ of a random process $X : S \to S_X$ is defined according to

$$\mu_X(t) = \boldsymbol{E}[X_t] = \boldsymbol{E}[X(\Omega,t)] = \begin{cases} \sum\limits_{\omega_i \in S} p_\Omega(\omega_i) X(\omega_i,t) & \text{countable case} \\ \int\limits_{-\infty}^{\infty} X(\omega,t) f_\Omega(\omega)d\omega & \text{uncountable case} \end{cases} \tag{7.17}$$

**Definition: Mean Squared Value** The mean squared value of a random process $X : S \to S_X$ is defined according to

$$E\left[|X_t|^2\right] = E\left[|X(\Omega,t)|^2\right] = \begin{cases} \displaystyle\sum_{\omega_i \in S} P_\Omega(\omega_i)|X(\omega_i,t)|^2 & \text{countable case} \\ \displaystyle\int_{-\infty}^{\infty} |X(\omega,t)|^2 f_\Omega(\omega)d\omega & \text{uncountable case} \end{cases} \tag{7.18}$$

**Definition: Variance** The variance $\sigma_X^2 : \boldsymbol{R} \to \boldsymbol{R}$ of a random process $X : S \to S_X$ is defined according to

$$\begin{aligned} \sigma_X^2(t) &= E\left[|X_t - \mu_X(t)|^2\right] = E\left[|X_t|^2\right] - |\mu_X(t)|^2 \\ &= E\left[|X(\Omega,t) - \mu_X(t)|^2\right] = E\left[|X(\Omega,t)|^2\right] - |\mu_X(t)|^2 \\ &= \begin{cases} \displaystyle\sum_{\omega_i \in S} p_\Omega(\omega_i)|X(\omega_i,t)|^2 - |\mu_X(t)|^2 & \text{countable case} \\ \displaystyle\int_{-\infty}^{\infty} |X(\omega,t)|^2 f_\Omega(\omega)d\omega - |\mu_X(t)|^2 & \text{uncountable case} \end{cases} \end{aligned} \tag{7.19}$$

**Definition: Average Power** The average power $\bar{P}(T)$ of a random process $X : S \to S_X$, on the interval $[0, T]$, is the weighted sum of the individual signal powers, that is,

$$\bar{P}(T) = \begin{cases} \displaystyle\sum_{\omega_i \in S} p_\Omega(\omega_i)\frac{1}{T}\int_0^T |X(\omega_i,t)|^2 dt & \text{countable case} \\ \displaystyle\int_{-\infty}^{\infty} \left[\frac{1}{T}\int_0^T |X(\omega,t)|^2 dt\right] f_\Omega(\omega)d\omega & \text{uncountable case} \end{cases} \tag{7.20}$$

Interchanging the order of summation and integration, or the order of integration, yields

$$\bar{P}(T) = \frac{1}{T}\int_0^T E\left[|X(\Omega,t)|^2\right] dt \tag{7.21}$$

*7.3.4.1 Instantaneous Power* The average power, as defined by Equation 7.21, is consistent with the following definition of instantaneous power.

**Definition: Instantaneous Power** The instantaneous power $P(t)$ of a random process $X: S \rightarrow S_X$ at a time $t$ is defined as

$$P(t) = \boldsymbol{E}\left[|X(\Omega,t)|^2\right] \tag{7.22}$$

*7.3.4.2 Example* Consider a random process defined as the sum of two random processes:

$$X(\Omega,t) = X_1(\Omega,t) + X_2(\Omega,t) \tag{7.23}$$

The instantaneous power of $X$ is

$$\begin{aligned} P(t) &= \boldsymbol{E}\left[|X(\Omega,t)|^2\right] = \boldsymbol{E}\left[[X_1(\Omega,t) + X_2(\Omega,t)] \cdot \left[X_1^*(\Omega,t) + X_2^*(\Omega,t)\right]\right] \\ &= \boldsymbol{E}\left[|X_1(\Omega,t)|^2\right] + \boldsymbol{E}\left[|X_2(\Omega,t)|^2\right] + 2\text{Re}\left[\boldsymbol{E}\left[X_1(\Omega,t)X_2^*(\Omega,t)\right]\right] \\ &= P_1(t) + P_2(t) + 2\text{Re}\left[\boldsymbol{E}\left[X_1(\Omega,t)X_2^*(\Omega,t)\right]\right] \end{aligned} \tag{7.24}$$

where $P_i(t) = \boldsymbol{E}|X_i(\Omega,t)|^2$ for $i \in \{1,2\}$ is the instantaneous power in $X_i$. Thus, uncorrelatedness implies addition of individual powers; correlation implies that the addition of individual powers does not hold.

Consider the case of

$$X_1(\Omega_1,t) = A_1 \sin(2\pi f_1 t + \Phi_1) \tag{7.25}$$

$$X_2(\Omega_2,t) = A_2 \sin(2\pi f_2 t + \Phi_2) \tag{7.26}$$

where $\Omega_1 = (A_1, \Phi_1)$ and $\Omega_2 = (A_2, \Phi_2)$. It then follows that

$$P(t) = P_1(t) + P_2(t) + 2\boldsymbol{E}[A_1A_2]\boldsymbol{E}[\sin(2\pi f_1 t + \Phi_1)\sin(2\pi f_2 t + \Phi_2)] \tag{7.27}$$

assuming $A_1$ and $A_2$ are independent of $\Phi_1$ and $\Phi_2$. For the case where one or both of $\boldsymbol{E}[A_1A_2]$ and $\boldsymbol{E}[\sin(2\pi f_1 t + \Phi_1)\sin(2\pi f_2 t + \Phi_2)]$ are zero, the instantaneous power of $X$ equals the sum of the instantaneous powers of $X_1$ and $X_2$.

### 7.3.5 Transient, Steady-State, Periodic, and Aperiodic Random Processes

The definitions for the mean, and mean squared value, of a random process allows the following classification of random processes.

**Definition: Transient Random Process** A random process $X$ is a transient random process if $\lim_{t \to \infty} \boldsymbol{E}\left[|X(\Omega,t)|^2\right] = 0$, and there exists a time $t_o$ such that $\boldsymbol{E}\left[|X(\Omega,t_o)|^2\right] > 0$.

**Definition: Steady-State Random Process with a Transient Component** A random process $X$ reaches steady state, after a transient period, if $\lim_{t\to\infty} \mu_X(t) = k_1$ and $\lim_{t\to\infty} \boldsymbol{E}\left[|X(\Omega,t)|^2\right] = k_2$ for positive constants $k_1, k_2$ and there exists a time $t_o$ such that $\boldsymbol{E}\left[|X(\Omega,t_o)|^2\right] \neq k_2$.

**Definition: Periodic Random Process with Transient Component** A random process $X$ is a periodic random process, with a transient component, if there exists a positive constant $T$ such that $\lim_{t\to\infty} \mu_X(t+T) = \lim_{t\to\infty} \mu_X(t)$ and there exists a time $t_o$ such that $\mu_X(t_o + T) \neq \mu_X(t_o)$.

**Definition: Aperiodic Random Process** A random process $X$ is an aperiodic random process if for all $T > 0$ it is not the case that $\lim_{t\to\infty} \mu_X(t+T) = \lim_{t\to\infty} \mu_X(t)$.

## 7.4 AUTOCORRELATION AND POWER SPECTRAL DENSITY

The two most widely used approaches for characterizing a random process is via an autocorrelation function and via a power spectral density function. The transition from the signal definitions for the autocorrelation function and the power spectral density function to equivalent definitions for a random process is via the expectation operator. A review of the signal definitions for these functions is useful.

### 7.4.1 Definitions for Individual Signals

Consider a random process $X: S \to S_X$. Each signal in the signal sample space $S_X$ has an autocorrelation, a time-averaged autocorrelation, and a power spectral density consistent with the definitions in Chapter 3. For the ωth signal and the interval [0, $T$], these definitions are

$$R(\omega,T,t_1,t_2) = X(\omega,t_1)X^*(\omega,t_2) \qquad t_1,t_2 \in [0,T] \tag{7.28}$$

$$\begin{aligned} R(\omega,T,t,\tau) &= X(\omega,t)X^*(\omega,t-\tau) \qquad t,t-\tau \in [0,T] \\ R(\omega,T,t,\tau) &= X(\omega,t+\tau)X^*(\omega,t) \qquad t,t+\tau \in [0,T] \end{aligned} \tag{7.29}$$

$$\bar{R}(\omega,T,\tau) = \begin{cases} \dfrac{1}{T}\displaystyle\int_0^{T+\tau} R(\omega,T,t,\tau)dt = \dfrac{1}{T}\displaystyle\int_0^{T+\tau} X(\omega,t)X^*(\omega,t-\tau)dt & \tau < 0 \\ \dfrac{1}{T}\displaystyle\int_\tau^{T} R(\omega,T,t,\tau)dt = \dfrac{1}{T}\displaystyle\int_\tau^{T} X(\omega,t)X^*(\omega,t-\tau)dt & \tau > 0 \end{cases} \tag{7.30}$$

$$G(\omega,T,f) = \frac{|X(\omega,T,f)|^2}{T} \tag{7.31}$$

where $X(\omega, T, f)$ is the Fourier transform of $X(\omega, t)$ evaluated over the interval [0, $T$].

The time-averaged autocorrelation–power spectral density function relationships are

$$G(\omega,T,f)=\int_{-T}^{T}\bar{R}(\omega,T,\tau)e^{-j2\pi f\tau}d\tau \qquad \bar{R}(\omega,T,\tau)=\int_{-\infty}^{\infty}G(\omega,T,f)e^{j2\pi f\tau}df \tag{7.32}$$

Sufficient conditions for the existence of the individual power spectral density and time-averaged autocorrelation functions have been detailed in Chapter 3, and the relationships between $G(\omega, T, f)$ and $\bar{R}(\omega,T,\tau)$ are valid when $X(\omega,t)\in \boldsymbol{L}^2[0,T]$ and $X(\omega, t)$ is piecewise differentiable on $[0, T]$.

***7.4.1.1 Associated Autocorrelation and PSD Random Processes*** The collection of autocorrelation and power spectral density functions defined by a random process $X$ define the following associated random processes:

$$\begin{aligned}&R:S\to S_R\\&S=\{\omega:\ \text{as defined by experiment}\}\\&S_R=\{R(\omega,T,t_1,t_2):\ \omega\in S\}\end{aligned} \tag{7.33}$$

$$\begin{aligned}&R:S\to S_R\\&S=\{\omega:\ \text{as defined by experiment}\}\\&S_R=\{R(\omega,T,t,\tau):\ \omega\in S\}\end{aligned} \tag{7.34}$$

$$\begin{aligned}&\bar{R}:S\to S_{\bar{R}}\\&S=\{\omega:\ \text{as defined by experiment}\}\\&S_{\bar{R}}=\{\bar{R}(\omega,T,\tau):\ \omega\in S\}\end{aligned} \tag{7.35}$$

$$\begin{aligned}&G:S\to S_G\\&S=\{\omega:\ \text{as defined by experiment}\}\\&S_G=\{G(\omega,T,f):\ \omega\in S\}\end{aligned} \tag{7.36}$$

Convenient notation for these associated random processes are $R(\Omega, T, t_1, t_2)$, $R(\Omega, T, t, \tau)$, $\bar{R}(\Omega,T,\tau)$, and $G(\Omega, T, f)$.

### 7.4.2 Definitions for Autocorrelation and PSD

The following definitions can be made by considering the expectation over the appropriate signal sample space.

**Definition: Autocorrelation Function** The autocorrelation function of a random process $X:S\to S_X$ is defined as the weighted average of the individual autocorrelation functions according to

$$R(T,t_1,t_2)=\boldsymbol{E}[R(\Omega,T,t_1,t_2)]=\boldsymbol{E}\left[X_{t_1}X_{t_2}^*\right]=\boldsymbol{E}[X(\Omega,t_1)X^*(\Omega,t_2)]$$
$$=\begin{cases}\sum\limits_{\omega_i\in S}p_\Omega(\omega_i)R(\omega_i,T,t_1,t_2)=\sum\limits_{\omega_i\in S}p_\Omega(\omega_i)X(\omega_i,t_1)X^*(\omega_i,t_2)\\ \int\limits_{-\infty}^{\infty}R(\omega,T,t_1,t_2)f_\Omega(\omega)d\omega=\int\limits_{-\infty}^{\infty}X(\omega,t_1)X^*(\omega,t_2)f_\Omega(\omega)d\omega\end{cases}\tag{7.37}$$

for $t_1,t_2\in[0,T]$. The upper result is for the countable case; the lower result is for the uncountable case. For $t,t-\tau\in[0,T]$:

$$R(T,t,t-\tau)=\boldsymbol{E}[R(\Omega,T,t,t-\tau)]=\boldsymbol{E}\left[X_tX_{t-\tau}^*\right]=\boldsymbol{E}[X(\Omega,t)X^*(\Omega,t-\tau)]$$
$$=\begin{cases}\sum\limits_{\omega_i\in S}p_\Omega(\omega_i)X(\omega_i,t)X^*(\omega_i,t-\tau) & \text{countable case}\\ \int\limits_{-\infty}^{\infty}X(\omega,t)X^*(\omega,t-\tau)f_\Omega(\omega)d\omega & \text{uncountable case}\end{cases}\tag{7.38}$$

For $t,t+\tau\in[0,T]$:

$$R(t+\tau,t)=\boldsymbol{E}[R(\Omega,T,t+\tau,t)]=\boldsymbol{E}\left[X_{t+\tau}X_t^*\right]=\boldsymbol{E}[X(\Omega,t+\tau)X^*(\Omega,t)]$$
$$=\begin{cases}\sum\limits_{\omega_i\in S}p_\Omega(\omega_i)X(\omega_i,t+\tau)X^*(\omega_i,t) & \text{countable case}\\ \int\limits_{-\infty}^{\infty}X(\omega,t+\tau)X^*(\omega,t)f_\Omega(\omega)d\omega & \text{uncountable case}\end{cases}\tag{7.39}$$

**Definition: Time-Averaged Autocorrelation Function** The time-averaged autocorrelation function of a random process $X:S\to S_X$ is defined as the weighted average of the individual time-averaged autocorrelation functions according to

$$\bar{R}(T,\tau)=\boldsymbol{E}[\bar{R}(\Omega,T,\tau)]=\boldsymbol{E}\left[\frac{1}{T}\int_{t_{\min}}^{t_{\max}}[X(\Omega,t+\tau)X^*(\Omega,t)]dt\right]$$
$$=\begin{cases}\sum\limits_{\omega_i\in S}p_\Omega(\omega_i)\bar{R}(\omega_i,T,\tau)=\sum\limits_{\omega_i\in S}p_\Omega(\omega_i)\frac{1}{T}\int_{t_{\min}}^{t_{\max}}[X(\omega_i,t+\tau)X^*(\omega_i,t)]dt\\ \int\limits_{-\infty}^{\infty}\bar{R}(\omega,T,\tau)f_\Omega(\omega)d\omega=\int\limits_{-\infty}^{\infty}\left[\frac{1}{T}\int_{t_{\min}}^{t_{\max}}[X(\omega,t+\tau)X^*(\omega,t)]dt\right]f_\Omega(\omega)d\omega\end{cases}\tag{7.40}$$

where $t_{max} = T,\ t_{min} = -\tau$ when $t < 0$ and $t_{max} = T - \tau,\ t_{min} = 0$ when $\tau > 0$. The upper result is for the countable case; the lower result is for the uncountable case.

**Definition: Power Spectral Density** The power spectral density of a random process $X : S \rightarrow S_X$ is defined according to

$$G(T,f) = \boldsymbol{E}[G(\Omega, T, f)] = \boldsymbol{E}\left[\frac{|X(\Omega, T, f)|^2}{T}\right]$$
$$= \begin{cases} \sum_{\omega_i \in S} p_\Omega(\omega_i) G(\omega_i, T, f) = \sum_{\omega_i \in S} p_\Omega(\omega_i) \frac{|X(\omega_i, T, f)|^2}{T} & \text{c.c.} \\ \int_{-\infty}^{\infty} G(\omega, T, f) f_\Omega(\omega) d\omega = \int_{-\infty}^{\infty} \frac{|X(\omega, T, f)|^2}{T} \cdot f_\Omega(\omega) d\omega & \text{u.c.} \end{cases} \tag{7.41}$$

### 7.4.3 Simplified Notation: Countable Signal Sample Space Case

For the case of a countable space of experimental outcomes

$$S = \{\omega_i : i \in \mathbf{Z}^+\} \tag{7.42}$$

where the probability of the $i$th outcome is $P[\Omega = \omega_i] = p_i$, the notation for a random process $X$, as follows, is useful:

$$\begin{aligned} &X : S \rightarrow S_X \quad X(\omega_i, t) = x_i(t) \\ &S_X = \{x_i(t) : \ i \in \mathbf{Z}^+\} \quad P[X(\Omega, t) = x_i(t)] = p_i \end{aligned} \tag{7.43}$$

The following simplified notation holds for the autocorrelation and power spectral density functions: for the $i$th signal,

$$R_i(T, t_1, t_2) = x_i(t_1) x_i^*(t_2) \tag{7.44}$$

$$R_i(T, t, \tau) = x_i(t) x_i^*(t - \tau) \quad \text{or} \quad R_i(T, t, \tau) = x_i(t + \tau) x_i^*(t) \tag{7.45}$$

$$\bar{R}_i(T, \tau) = \begin{cases} \frac{1}{T} \int_0^{T+\tau} R_i(T, t, \tau) dt = \frac{1}{T} \int_0^{T+\tau} x_i(t) x_i^*(t - \tau) dt & \tau < 0 \\ \frac{1}{T} \int_\tau^T R_i(T, t, \tau) dt = \frac{1}{T} \int_\tau^T x_i(t) x_i^*(t - \tau) dt & \tau > 0 \end{cases} \tag{7.46}$$

$$G_i(T, f) = \frac{|X_i(T, f)|^2}{T} \tag{7.47}$$

and the time-averaged autocorrelation–power spectral density function relationships are

$$G_i(T,f) = \int_{-T}^{T} \bar{R}_i(T,\tau)e^{-j2\pi f\tau}d\tau \qquad \bar{R}_i(T,\tau) = \int_{-\infty}^{\infty} G_i(T,f)e^{j2\pi f\tau}df \tag{7.48}$$

***7.4.3.1 Definitions*** The following definitions for functions that characterize a random process can then be made.

**Definition: Mean**

$$\mu_X(t) = \boldsymbol{E}[X_t] = \sum_{i=1}^{\infty} p_i x_i(t) \tag{7.49}$$

**Definition: Variance**

$$\sigma_X^2(t) = \sum_{i=1}^{\infty} p_i |x_i(t)|^2 - \left| \sum_{i=1}^{\infty} p_i x_i(t) \right|^2 \tag{7.50}$$

**Definition: Average Power**

$$\bar{P}(T) = \sum_{i=1}^{\infty} p_i \left[ \frac{1}{T} \int_0^T |x_i(t)^2 dt| \right] = \frac{1}{T} \int_0^T \sum_{i=1}^{\infty} p_i |x_i(t)|^2 dt \tag{7.51}$$

**Definition: Autocorrelation Function**

$$R(T,t_1,t_2) = \boldsymbol{E}\left[X_{t_1} X_{t_2}^*\right] = \sum_{i=1}^{\infty} p_i x_i(t_1) x_i^*(t_2) \quad t_1,t_2 \in [0,T] \tag{7.52}$$

$$R(T,t,t-\tau) = \boldsymbol{E}\left[X_t X_{t-\tau}^*\right] = \sum_{i=1}^{\infty} p_i x_i(t) x_i^*(t-\tau) \quad t,t-\tau \in [0,T] \tag{7.53}$$

or

$$R(t+\tau,t) = \boldsymbol{E}\left[X_{t+\tau} X_t^*\right] = \sum_{i=1}^{\infty} p_i x_i(t+\tau) x_i^*(t) \quad t,t+\tau \in [0,T] \tag{7.54}$$

**Definition: Time-Averaged Autocorrelation Function**

$$\bar{R}(T,\tau)=\begin{cases}\dfrac{1}{T}\displaystyle\int_{0}^{T+\tau}\sum_{i=1}^{\infty}p_i x_i(t)x_i^*(t-\tau)dt & \tau<0\\ \dfrac{1}{T}\displaystyle\int_{\tau}^{T}\sum_{i=1}^{\infty}p_i x_i(t)x_i^*(t-\tau)dt & \tau>0\end{cases} \tag{7.55}$$

**Definition: Power Spectral Density**

$$G(T,f)=\sum_{i=1}^{\infty}p_i G_i(T,f)=\frac{1}{T}\sum_{i=1}^{\infty}p_i|X_i(T,f)|^2 \tag{7.56}$$

### 7.4.4 Notation: Vector Case

In many cases, the experiment underpinning the random process $X:S\to S_X$ defines $N$ random variables $\Omega=(\Omega_1,\ldots,\Omega_N)$ where

$$\begin{aligned}&P[\Omega_1=\omega_1,\ldots,\Omega_N=\omega_N]=p_{\Omega_1\ldots\Omega_N}(\omega_1,\ldots,\omega_N)\\&P[\Omega_1\in(\omega_1+d\omega_1),\ldots,\Omega_N\in(\omega_N+d\omega_N)]\approx f_{\Omega_1\ldots\Omega_N}(\omega_1,\ldots,\omega_N)d\omega_1\ldots d\omega_N\end{aligned} \tag{7.57}$$

The first definition is for the discrete random variable case; the second definition is for the continuous random variable case. For the continuous random variable case, the following definitions apply.

**Definition: Autocorrelation Function**

$$\begin{aligned}&R(T,t,t-\tau)=\boldsymbol{E}[X(\Omega,t)X(\Omega,t-\tau)^*]=\\&\int_{-\infty}^{\infty}\cdots\int_{-\infty}^{\infty}X((\omega_1,\ldots,\omega_N),t)X((\omega_1,\ldots,\omega_N),t-\tau)^*f_{\Omega_1\ldots\Omega_N}(\omega_1,\ldots,\omega_N)d\omega_1\ldots d\omega_N\end{aligned} \tag{7.58}$$

**Definition: Power Spectral Density**

$$\begin{aligned}&G(T,f)=\boldsymbol{E}\left[\frac{|X(\Omega,T,f)|^2}{T}\right]=\\&\frac{1}{T}\int_{-\infty}^{\infty}\cdots\int_{-\infty}^{\infty}|X((\omega_1,\ldots,\omega_N),T,f)|^2 f_{\Omega_1\ldots\Omega_N}(\omega_1,\ldots,\omega_N)d\omega_1\ldots d\omega_N\end{aligned} \tag{7.59}$$

Analogous definitions hold for the countable signal sample space case.

### 7.4.5 Infinite Interval Case

The time-averaged autocorrelation function, and the power spectral density function, of individual signals from a random process can be defined on the infinite interval $[0, \infty]$ as a limit according to

$$R_\infty(\omega, \tau) = \lim_{T \to \infty} \bar{R}(\omega, T, \tau) \quad \omega \in S \tag{7.60}$$

$$G_\infty(\omega, f) = \lim_{T \to \infty} G(\omega, T, f) \quad \omega \in S \tag{7.61}$$

The expectation over the ensemble of signals leads to the following definitions.

**Definition: Time-Averaged Autocorrelation Function—Infinite Interval** The time-averaged autocorrelation function for the infinite interval is defined according to

$$R_\infty(\tau) = \lim_{T \to \infty} \bar{R}(T, \tau) \tag{7.62}$$

where $\bar{R}(T, \tau)$ is the expectation defined by Equation 7.40.

**Definition: Power Spectral Density—Infinite Interval** The power spectral density for the infinite interval is defined according to

$$G_\infty(f) = \lim_{T \to \infty} G(T, f) \tag{7.63}$$

where $G(T, f)$ is the expectation defined by Equation 7.41.

### 7.4.6 Existence: Finite Interval

**Theorem 7.1 Average Power**
If all signals in the signal sample space have finite energy on $[0, T]$, that is, $X(\omega, t) \in \boldsymbol{L}^2[0, T]$, $\omega \in S$, and with an upper bound

$$\sup\left\{\frac{1}{T}\int_0^T |X(\omega, t)|^2 dt : \ \omega \in S\right\} < \infty \tag{7.64}$$

then the average power of the random process

$$\bar{P}(T) = \sum_{\omega_i \in S} p_\Omega(\omega_i)\left[\frac{1}{T}\int_0^T |X(\omega_i, t)|^2 dt\right] \tag{7.65}$$

is finite.

**Proof**

The proof follows from the assumptions and the definition of the average power.

**Theorem 7.2 Existence of Autocorrelation and Power Spectral Density**

If all signals in the signal sample space have finite energy on $[0, T]$, that is, $X(\omega,t) \in \boldsymbol{L}^2[0,T]$, $\omega \in S$, and the average power of the random process is finite, then the time-averaged autocorrelation function specified in Equation 7.40 is valid, and the power spectral density function defined in Equation 7.41 is valid, that is, the following are finite:

$$\bar{R}(T,\tau) = \begin{cases} \sum_{i=1}^{\infty} p_{\Omega}(\omega_i)\bar{R}(\omega_i,T,\tau) & \text{countable case} \quad -T<\tau<T \\ \int_{-\infty}^{\infty} \bar{R}(\omega,T,\tau)f_{\Omega}(\omega)d\omega & \text{uncountable case} \quad -T<\tau<T \end{cases} \tag{7.66}$$

$$G(T,f) = \begin{cases} \sum_{\omega_i \in S} p_{\Omega}(\omega_i)G(\omega_i,T,f) & \text{countable case,} \quad f \in \boldsymbol{R} \\ \int_{-\infty}^{\infty} G(\omega,T,f)f_{\Omega}(\omega)d\omega & \text{uncountable case,} \quad f \in \boldsymbol{R} \end{cases} \tag{7.67}$$

Further, $\bar{R}(T,\tau) \in \boldsymbol{L}[-T,T]$ and $G(T,f) \in \boldsymbol{L}$. The latter result implies that the power spectral density is finite for all values of $f$ except at most at a countable number of values. Finally, the average power is the integral of the power spectral density, that is,

$$\bar{P}(T) = \int_{-\infty}^{\infty} G(T,f)df \tag{7.68}$$

**Proof**

The proof is detailed in Appendix 7.A.

### 7.4.7 Existence: Infinite Interval

**Theorem 7.3 Existence of Time-Averaged Autocorrelation Function**

The time-averaged autocorrelation function for the infinite interval, defined according to

$$R_{\infty}(\tau) = \lim_{T\to\infty} \bar{R}(T,\tau) \tag{7.69}$$

is finite if $\sup\{|\bar{R}(\omega,T,\tau)| : \; T,\tau \in \boldsymbol{R}^+, \omega \in S\} < \infty$ whereupon

$$R_\infty(\tau) = \begin{cases} \sum\limits_{\omega_i \in S} p_\Omega(\omega_i) R_\infty(\omega_i, \tau) & \text{countable case} \\ \int\limits_{-\infty}^{\infty} R_\infty(\omega, \tau) f_\Omega(\omega) d\omega & \text{uncountable case} \end{cases} \tag{7.70}$$

Here, $R_\infty(\omega_i, \tau) = \lim\limits_{T \to \infty} \bar{R}(\omega_i, T, \tau), \omega_i \in S$. A sufficient condition for the assumption to be valid is for $X(\omega, t) \in \boldsymbol{L}^2[0, T]$, for all $T \in \boldsymbol{R}^+$, $\omega \in S$, and an upper bound on the signal powers over the interval $[0, \infty)$, that is,

$$\sup\left\{ \frac{1}{T} \int\limits_0^T |X(\omega, t)|^2 dt : \ T \in \boldsymbol{R}^+, \omega \in S \right\} < \infty \tag{7.71}$$

**Proof**

The proof is detailed in Appendix 7.B.

### *7.4.7.1 Power Spectral Density: Infinite Interval*

**Theorem 7.4 Existence of Power Spectral Density: Infinite Interval**

The power spectral density for the infinite interval, defined according to

$$G_\infty(f) = \lim_{T \to \infty} G(T, f) \tag{7.72}$$

exists if $\sup\{G(\omega, T, f) : \ T \in \boldsymbol{R}^+, \omega \in S\}$ is finite whereupon

$$G_\infty(f) = \begin{cases} \sum\limits_{\omega_i \in S} p_\Omega(\omega_i) G_\infty(\omega_i, f) & \text{countable case} \\ \int\limits_{-\infty}^{\infty} G_\infty(\omega, f) f_\Omega(\omega) d\omega & \text{uncountable case} \end{cases} \tag{7.73}$$

Sufficient conditions for $G(\omega, T, f)$ to be bounded as $T \to \infty$ are for $X(\omega, t) \in \boldsymbol{L}[0, \infty]$ or for $R_\infty(\omega, \tau) \in \boldsymbol{L}$.

**Proof**

The proof is detailed in Appendix 7.B.

## 7.4.8 Power Spectral Density–Autocorrelation Relationship

***7.4.8.1 Finite Interval*** The relationship between the power spectral density function and the time-averaged autocorrelation function is via the Fourier transform according to the following theorem.

**Theorem 7.5 Power Spectral Density–Autocorrelation Relationship**
If all signals are piecewise differentiable and have finite energy on [0, *T*], that is, $X(\omega,t) \in \boldsymbol{L}^2[0,T], \omega \in S$, and the average power of the random process is finite on [0, *T*], then the power spectral density–time-averaged autocorrelation function relationships for the random process are

$$G(T,f) = \int_{-T}^{T} \bar{R}(T,\tau)e^{-j2\pi f\tau}d\tau \qquad \bar{R}(T,\tau) = \int_{-\infty}^{\infty} G(T,f)e^{j2\pi f\tau}df \tag{7.74}$$

Further, the average power is the integral of the power spectral density, that is,

$$\bar{P}(T) = \bar{R}(T,0) = \int_{-\infty}^{\infty} G(T,f)df \tag{7.75}$$

**Proof**
The proof is detailed in Appendix 7.C.

***7.4.8.2 Infinite Interval Case: Wiener–Khintchine Relationships*** Additional restrictions are required for the power spectral density–autocorrelation function relationships to hold on the infinite interval.

**Theorem 7.6 Power Spectral Density–Autocorrelation: Infinite Interval**
Assume the restrictions for the finite case hold for all $T \in \boldsymbol{R}^+$, that is, assume $X(\omega,t) \in \boldsymbol{L}^2[0,T]$, $\omega \in S$, finite average power on [0, *T*], and $X(\omega, t)$ is piecewise differentiable on [0, *T*], $\omega \in S$. Further, assume $G(T,f)$ is nonimpulsive such that there exists a function $B_G \in \boldsymbol{L}$ consistent with $G(T,f) \le B_G(f)$ for all *T*. Assume there exists a function $B_R \in \boldsymbol{L}$ such that $\bar{R}(T,\tau) \le B_R(\tau)$ for all *T*. With these assumptions, it follows that the Fourier transform relationships

$$R_\infty(\tau) = \int_{-\infty}^{\infty} G_\infty(f)e^{j2\pi f\tau}df \qquad G_\infty(f) = \int_{-\infty}^{\infty} R_\infty(\tau)e^{-j2\pi f\tau}d\tau \tag{7.76}$$

are valid. Further, the average power is the integral of the power spectral density, that is,

$$\bar{P}_\infty = R_\infty(0) = \int_{-\infty}^{\infty} G_\infty(f)df \tag{7.77}$$

Commonly, these Fourier transform relationships are called the Wiener–Khintchine relations (Champeney, 1987, p. 104; Parzen, 1962a, p. 110).

**Proof**

The proof is detailed in Appendix 7.D.

*7.4.8.3 Impulsive Case: Formal Definition of Wiener–Khintchine* When the signals defined by a random process contain periodic components, the power spectral density $G(T,f)$ becomes impulsive, at specific values of $f$, as $T \to \infty$, and it is not possible to interchange the order of limit and integration in the following equation:

$$R_\infty(\tau) = \lim_{T\to\infty} \bar{R}(T,\tau) = \lim_{T\to\infty} \int_{-\infty}^{\infty} G(T,f)e^{j2\pi f\tau}df \tag{7.78}$$

For this case, a spectral distribution function is defined.

**Definition: Spectral Distribution Function** The spectral distribution function is defined according to

$$F(f) = \lim_{T\to\infty} \int_{-\infty}^{f} G(T,\lambda)d\lambda \tag{7.79}$$

Note that $F(-\infty)=0$, $\bar{P}_\infty = F(\infty)$, and $F$ is a monotonically increasing function.

**Theorem 7.7 Wiener–Khintchine Theorem: Formal Definition**

The spectral theorem for the autocorrelation function is (Champeney, 1987, p. 108; Grimmett and Stirzaker, 1992, p. 353; Parzen, 1962a, p. 110)

$$R_\infty(\tau) = \int_{-\infty}^{\infty} e^{j2\pi f\tau} dF(f) \tag{7.80}$$

This result is a formal statement of the Wiener–Khintchine theorem (Priestley, 1981, p. 218).

**Proof**

The proof is the same as that given for Theorem 3.19.

### 7.4.9 Notes on Spectral Characterization

The most widely used approach for characterizing random phenomena in engineering and science is through the use of the power spectral density function. Further, it is common to characterize stationary random processes according to the nature of their spectrum and without reference to the underlying time domain signals or underlying experiments. Examples include white noise with a power spectral density specified as

$$G_\infty(f) = \eta \tag{7.81}$$

or $1/f$ noise random processes with a power spectral density specified as

$$G_\infty(f) = \frac{k}{f} \tag{7.82}$$

Both definitions, however, are problematic as they imply infinite power random processes.

## 7.5 CORRELATION

For notational convenience, the argument $T$ is dropped from the autocorrelation functions and a subscript of $XX$ for a random process $X$ is added, that is, $R(T, t_1, t_2)$ is written as $R_{XX}(t_1, t_2)$.

### 7.5.1 Correlation Coefficient

A correlation coefficient can be defined, as in Section 4.11.3, for any two random variables. Accordingly, the following definition can be made.

**Definition: Correlation Coefficient** The correlation coefficient of a random process $X : S \to S_X$ at the times $t+\tau$ and $t$ is defined according to

$$\begin{aligned}\rho_{XX}(t+\tau,t) &= \frac{E[X(\Omega,t+\tau)X^*(\Omega,t)] - E[X(\Omega,t+\tau)]E[X^*(\Omega,t)]}{\sqrt{\sigma_X^2(t+\tau)\sigma_X^2(t)}} \\ &= \frac{R_{XX}(t+\tau,t) - \mu_X(t+\tau)\mu_X(t)}{\sqrt{\sigma_X^2(t+\tau)\sigma_X^2(t)}} = \frac{C_{XX}(t+\tau,t)}{\sqrt{\sigma_X^2(t+\tau)\sigma_X^2(t)}}\end{aligned} \tag{7.83}$$

Here, $C_{XX}(t+\tau,t)$ is the covariance function between $X(\Omega,t+\tau)$ and $X(\Omega, t)$:

$$C_{XX}(t+\tau,t) = E[X(\Omega,t+\tau)X^*(\Omega,t)] - E[X(\Omega,t+\tau)]E[X^*(\Omega,t)] \tag{7.84}$$

For the zero-mean case,

$$\rho_{XX}(t+\tau,t) = \frac{E[X(\Omega,t+\tau)X^*(\Omega,t)]}{\sqrt{\sigma_X^2(t+\tau)\sigma_X^2(t)}} = \frac{R_{XX}(t+\tau,t)}{\sqrt{\sigma_X^2(t+\tau)\sigma_X^2(t)}} \tag{7.85}$$

Note that $\rho_{XX}(t,t) = 1$.

***7.5.1.1 Correlation Time*** It can be useful to have a measure of the time over which a random process is correlated. One approach is to determine the time $\tau$ for the correlation coefficient to change from unity at the times $(t, t)$ to a predefined level at the times $(t+\tau,t)$.

**Definition: Effective Correlation Time** The effective correlation time function $C_t : \boldsymbol{R} \to \boldsymbol{R}$ of a random process $X$, at a time $t$, is defined according to

$$C_t(t) = \tau \qquad \tau : \rho_{XX}(t+\tau, t) = \rho_o, \tau > 0 \tag{7.86}$$

where $\rho_o$ is a set value between 0 and 1. A useful value for $\rho_o$ is 0.1.

As the correlation coefficient gives a measure of the correlation of the random process at the times $t$ and $t+\tau$, a level of 0.1 for $\rho_o$, for example, is consistent with a high level of independence. This justifies the use of the phrase "*effective correlation time*."

### 7.5.2 Expected Change in a Specified Interval

Another useful characteristic of a random process is its expected change over a set interval.

**Definition: Change Random Process** The change in a random process $X : S \to S_X$ between the time $t$ and $t + \Delta t$ defines an associated two-dimensional change random process:

$$\begin{aligned} &\Delta X : S \to S_{\Delta X} \\ &\Delta X(\Omega, t, \Delta t) = X(\Omega, t+\Delta t) - X(\Omega, t) \\ &S_{\Delta X} = \{X(\omega, t+\Delta t) - X(\omega, t) : \quad \omega \in S, t \geq 0, \Delta t \geq 0\} \end{aligned} \tag{7.87}$$

**Theorem 7.8 Expected Mean Squared Change**

For a random process $X : S \to S_X$, the expected mean squared change in the interval $[t, t+\Delta t]$ is

$$\begin{aligned} \boldsymbol{E}\left[|\Delta X(\Omega, t, \Delta t)|^2\right] &= \boldsymbol{E}\left[|X(\Omega, t+\Delta t) - X(\Omega, t)|^2\right] \\ &= \boldsymbol{E}\left[|X(\Omega, t+\Delta t)|^2\right] + \boldsymbol{E}\left[|X(\Omega, t)|^2\right] - 2\mathrm{Re}[R_{XX}(t+\Delta t, t)] \end{aligned} \tag{7.88}$$

For a zero-mean random process,

$$\boldsymbol{E}\left[|\Delta X(\Omega, t, \Delta t)|^2\right] = \sigma_X^2(t+\Delta t) + \sigma_X^2(t) - 2\mathrm{Re}[R_{XX}(t+\Delta t, t)] \tag{7.89}$$

**Proof**

Consider

$$\begin{aligned} &\boldsymbol{E}\left[[X(\Omega, t+\Delta t) - X(\Omega, t)][X(\Omega, t+\Delta t) - X(\Omega, t)]^*\right] = \boldsymbol{E}\left[|X(\Omega, t+\Delta t)|^2\right] \\ &+ \boldsymbol{E}\left[|X(\Omega, t)|^2\right] - \boldsymbol{E}[X(\Omega, t+\Delta t)X(\Omega, t)^*] - \boldsymbol{E}[X(\Omega, t+\Delta t)^* X(\Omega, t)] \\ &= \boldsymbol{E}\left[|X(\Omega, t+\Delta t)|^2\right] + \boldsymbol{E}\left[|X(\Omega, t)|^2\right] - 2\mathrm{Re}[R_{XX}(t+\Delta t, t)] \end{aligned} \tag{7.90}$$

For a zero-mean random process, $\sigma_X^2(t+\Delta t) = \boldsymbol{E}\left[|X(\Omega, t+\Delta t)|^2\right]$, and the required result follows.

*7.5.2.1 Application* For a set resolution in amplitude $\Delta x$ and for a given random process, it is possible to find the maximum time resolution $\Delta t$ such that

$$\boldsymbol{E}\left[|X(\Omega, t+\Delta t) - X(\Omega, t)|^2\right] = \Delta x^2 \tag{7.91}$$

This information can be used to set the sampling time required to ascertain, for example, the first passage time or maximum level, consistent with an amplitude resolution of $\Delta x$.

## 7.6 NOTES ON AVERAGE POWER AND AVERAGE ENERGY

When an orthonormal basis set is used for signal decomposition, the average power, or average energy, in a random process is given by the weighted average of the powers, or energies, in the individual component signals.

**Theorem 7.9 Average Power or Energy: Decomposition Form**
Consider an orthonormal basis set $\{b_0, b_1, \ldots\}$ for the interval $[\alpha, \beta]$ and a random process $X: S \to S_X$ defined on this interval. Each signal in the signal sample space $S_X$ has the decomposition

$$X(\omega, t) = \sum_{k=0}^{\infty} c_k(\omega) b_k(t) \quad t \in [\alpha, \beta], c_k(\omega) = \int_{\alpha}^{\beta} X(\omega, t) b_k^*(t) dt \tag{7.92}$$

and the average power, or energy, in the random process is

$$\bar{P} = \frac{1}{\beta - \alpha} \sum_{k=0}^{\infty} \boldsymbol{E}\left[|C_k|^2\right] \quad \bar{E} = \sum_{k=0}^{\infty} \boldsymbol{E}\left[|C_k|^2\right] \tag{7.93}$$

where $C_k$ is the random variable defined by $\{c_k(\omega): \ \omega \in S\}$. The average power, or average energy, associated with the $k$th component signal (for the countable case) is

$$\begin{aligned}\bar{P}_k &= \frac{1}{\beta - \alpha} \sum_{\omega_i \in S} p_\Omega(\omega_i) \int_{\alpha}^{\beta} |c_k(\omega_i) b_k(t)|^2 dt = \frac{\boldsymbol{E}\left[|C_k|^2\right]}{\beta - \alpha} \\ \bar{E}_k &= \sum_{\omega_i \in S} p_\Omega(\omega_i) \int_{\alpha}^{\beta} |c_k(\omega_i) b_k(t)|^2 dt = \boldsymbol{E}\left[|C_k|^2\right]\end{aligned} \tag{7.94}$$

Thus, the average power, or average energy, of the random process is the sum of the average powers, or energies, in the component signals.

**Proof**

Consider the countable case and the average power. Consistent with Equation 7.21, the average power on the interval is

$$\begin{aligned}\bar{P} &= \frac{1}{\beta-\alpha}\int_{\alpha}^{\beta} \boldsymbol{E}\left[|X(\Omega,t)|^2\right]dt = \frac{1}{\beta-\alpha}\int_{\alpha}^{\beta}\left[\sum_{\omega_i\in S} p_{\Omega}(\omega_i)\left|\sum_{k=0}^{\infty} c_k(\omega_i)b_k(t)\right|^2\right]dt \\ &= \frac{1}{\beta-\alpha}\int_{\alpha}^{\beta}\left[\sum_{\omega_i\in S} p_{\Omega}(\omega_i)\sum_{u=0}^{\infty}\sum_{v=0}^{\infty} c_u(\omega_i)c_v^*(\omega_i)b_u(t)b_v^*(t)\right]dt \\ &= \frac{1}{\beta-\alpha}\sum_{u=0}^{\infty}\sum_{v=0}^{\infty}\left[\sum_{\omega_i\in S} p_{\Omega}(\omega_i)c_u(\omega_i)c_v^*(\omega_i)\right]\int_{\alpha}^{\beta} b_u(t)b_v^*(t)dt \\ &= \frac{1}{\beta-\alpha}\sum_{u=0}^{\infty}\sum_{v=0}^{\infty}\boldsymbol{E}\left[C_uC_v^*\right]\int_{\alpha}^{\beta} b_u(t)b_v^*(t)dt\end{aligned} \tag{7.95}$$

Using the orthonormality property of the basis functions, the required result follows:

$$P = \frac{1}{\beta-\alpha}\sum_{k=0}^{\infty}\boldsymbol{E}\left[|C_k|^2\right] \tag{7.96}$$

### 7.6.1 Power and Energy: Uncorrelated Coefficients

Orthonormality guarantees that the average power, or average energy, on an interval is the summation of the powers in the individual signal components. It is of interest if this result also holds for the case where the waveforms in a decomposition are not necessarily orthonormal or orthogonal.

**Theorem 7.10 Power and Energy for Uncorrelated Coefficient Case**

Consider a set of signals $\{b_0, b_1, \ldots\}$, which are not necessarily orthogonal, defined on $[\alpha, \beta]$ and an random process $X: S \to S_X$ defined on this interval. Assume each signal in the signal sample space $S_X$ has the decomposition

$$X(\omega,t) = \sum_{k=0}^{\infty} c_k(\omega)b_k(t) \qquad t\in[\alpha,\beta] \tag{7.97}$$

For the case where the random variables $C_i$ and $C_k$ defined, respectively, by $\{c_i(\omega):\ \omega\in S\}$ and $\{c_k(\omega):\ \omega\in S\}$, $i\neq k$, are uncorrelated, that is, $\boldsymbol{E}\left[C_iC_k^*\right]=0$, it is the case that the average power/energy is the weighted sum of the powers/energies in the component signals according to

$$\bar{P} = \frac{1}{\beta-\alpha}\sum_{k=0}^{\infty} \boldsymbol{E}\left[|C_k|^2\right]\int_{\alpha}^{\beta}|b_k(t)|^2 dt$$
$$\bar{E} = \sum_{k=0}^{\infty} \boldsymbol{E}\left[|C_k|^2\right]\int_{\alpha}^{\beta}|b_k(t)|^2 dt \tag{7.98}$$

When the signals $b_0, b_1, \ldots$ have unity energy, it follows that

$$\bar{P} = \frac{1}{\beta-\alpha}\sum_{k=0}^{\infty} \boldsymbol{E}\left[|C_k|^2\right] \quad \bar{E} = \sum_{k=0}^{\infty} \boldsymbol{E}\left[|C_k|^2\right] \tag{7.99}$$

Thus, uncorrelated coefficients are a sufficient condition to ensure that the average power equals the weighted sum of the average powers of the constituent signal components.

**Proof**

The proof follows in an analogous manner to that of the proof of Theorem 7.9. Specifically, from Equation 7.95:

$$\bar{P} = \frac{1}{\beta-\alpha}\sum_{i=0}^{\infty}\sum_{k=0}^{\infty} \boldsymbol{E}\left[C_i C_k^*\right]\int_{\alpha}^{\beta} b_i(t) b_k^*(t)dt = \frac{1}{\beta-\alpha}\sum_{k=0}^{\infty} \boldsymbol{E}\left[|C_k|^2\right]\int_{\alpha}^{\beta}|b_k(t)|^2 dt \tag{7.100}$$

### 7.6.2 Sufficient Conditions for Uncorrelated Coefficients

Consider a sinusoidal basis set $\{e^{j2\pi i f_o t} : f_o = 1/T, i \in \boldsymbol{Z}\}$ for the interval [0, $T$]. Assume a random process $X : S \rightarrow S_X$ which is such that each defined signal has a Fourier series decomposition on the interval [0, $T$]:

$$X(\omega, t) = \sum_{k=-\infty}^{\infty} c_k(\omega) e^{j2\pi k f_o t} \quad c_k(\omega) = \frac{1}{T}\int_{0}^{T} X(\omega, t) e^{-j2\pi k f_o t} dt \tag{7.101}$$

The $k$th coefficients define a random variable $C_k$ with outcomes $\{c_k(\omega): \ \omega \in S\}$. The uncorrelatedness of the coefficients $C_i$ and $C_k$ for $i \neq k$, that is, $\boldsymbol{E}\left[C_i C_k^*\right] = 0$, as $T \rightarrow \infty$ depends on the autocorrelation function, $R_{XX}(t+\tau, t)$, and on the interval [0, $T$].

**Theorem 7.11 Sufficient Conditions for Uncorrelated Coefficients**

If the autocorrelation function $R_{XX}(t+\tau, t)$ is integrable with respect to $\tau$ for all $t \in [0, T]$, and this integral is bounded as $T \rightarrow \infty$, that is,

$$\sup\left\{\int_{-t}^{T-t} |R_{XX}(t+\tau,t)|d\tau: \quad t\in[0,T], T\in(0,\infty)\right\}<\infty \tag{7.102}$$

then $\boldsymbol{E}\left[C_i C_k^*\right]=0$ for $i\neq k$, on the interval [0, $T$] as $T\rightarrow\infty$.

**Proof**

The proof is detailed in Appendix 7.E.

***7.6.2.1 Notes*** For $t$ fixed the region of integration for the integral in Equation 7.102 is illustrated in Figure 7.6.

The random variables $C_i$ and $C_k$, in general, vary with $T$. However, the result of the convergence of $\boldsymbol{E}\left[C_i C_k^*\right]$ to zero, $i\neq k$, is guaranteed when the autocorrelation function is integrable on the infinite interval such that Equation 7.102 holds.

Writing the autocorrelation function in terms of the correlation function (see Eq. 7.83) according to

$$R_{XX}(t+\tau,t)=\mu_X(t+\tau)\mu_X(t)+\rho_{XX}(t+\tau,t)\sqrt{\sigma_X^2(t+\tau)\sigma_X^2(t)} \tag{7.103}$$

it follows that sufficient conditions for Equation 7.102 to hold are as follows: $\mu_X(t)=0$ for all $t$, boundedness of the variance, that is,

$$\sup\left\{\sigma_X^2(t): \quad t\in[0,\infty]\right\}<\sigma_{\max}^2 \tag{7.104}$$

and for the random process to be increasingly uncorrelated at the times $t+\tau$ and $t$, as $t\rightarrow\infty$, consistent with the existence of constants $\tau_o,\delta>0$, $\tau_o$ independent of $t$, such that for all $|\tau|>\tau_o$ (see Fig. 7.6), it is the case that

$$|\rho_{XX}(t+\tau,t)|<\frac{1}{|\tau|^{1+\delta}} \qquad t>0, |\tau|>\tau_o \tag{7.105}$$

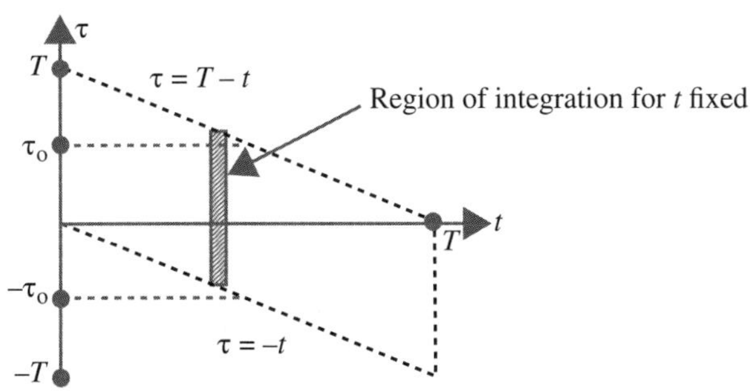


**FIGURE 7.6** Region where the autocorrelation function $R_{XX}(t+\tau, t)$ is nonzero assuming the interval [0,$T$].

The condition specified by Equation 7.102 may not hold, for example, when $\mu_X(t+\tau)$ is periodic with $\tau$ for $t$ fixed.

Stationarity allows a precise statement, for the finite interval and for the sinusoidal basis set case, for when the coefficients $C_i$ and $C_k$ are uncorrelated for $i \neq k$.

***7.6.2.2 Conditions for Correlated Coefficient to be Negligible*** Consider a basis set $\{b_1, b_2, \ldots\}$ for the interval $[\alpha, \beta]$, where the basis signals do not necessarily form an orthogonal set. With such a basis set, each signal in the signal sample space of a random process $X: S \to S_X$ has the decomposition

$$X(\omega,t) = \sum_{k=1}^{\infty} c_k(\omega) b_k(t) \qquad t \in [\alpha,\beta], \omega \in S \tag{7.106}$$

and the $k$th coefficients define a random variable $C_k$ with outcomes $\{c_k(\omega): \ \omega \in S\}$. Consistent with Equation 7.95, the average power, or average energy, in the random process is

$$\begin{aligned} \bar{P} &= \frac{1}{\beta-\alpha} \sum_{i=1}^{\infty} \sum_{k=1}^{\infty} \boldsymbol{E}\left[C_i C_k^*\right] \int_{\alpha}^{\beta} b_i(t) b_k^*(t) dt \\ \bar{E} &= \sum_{i=1}^{\infty} \sum_{k=1}^{\infty} \boldsymbol{E}\left[C_i C_k^*\right] \int_{\alpha}^{\beta} b_i(t) b_k^*(t) dt \end{aligned} \tag{7.107}$$

The average power, or average energy, depends on the nature of $\boldsymbol{E}\left[C_i C_k^*\right]$, $i,k \in \boldsymbol{Z}^+$, and it is useful to define the following correlation matrix:

$$\begin{bmatrix} \boldsymbol{E}\left[|C_1|^2\right] & \boldsymbol{E}\left[|C_1 C_2^*|\right] & \boldsymbol{E}\left[|C_1 C_3^*|\right] & \ldots & \boldsymbol{E}\left[|C_1 C_i^*|\right] & \ldots \\ \boldsymbol{E}\left[|C_2 C_1^*|\right] & \boldsymbol{E}\left[|C_2|^2\right] & \boldsymbol{E}\left[|C_2 C_3^*|\right] & \ldots & \boldsymbol{E}\left[|C_2 C_i^*|\right] & \\ \boldsymbol{E}\left[|C_3 C_1^*|\right] & \boldsymbol{E}\left[|C_3 C_2^*|\right] & \boldsymbol{E}\left[|C_3|^2\right] & \ldots & \boldsymbol{E}\left[|C_3 C_i^*|\right] & \\ \ldots & & & & & \\ \boldsymbol{E}\left[|C_i C_1^*|\right] & \boldsymbol{E}\left[|C_i C_2^*|\right] & \boldsymbol{E}\left[|C_i C_3^*|\right] & \ldots & \boldsymbol{E}\left[|C_i|^2\right] & \\ \ldots & & & & & \end{bmatrix} \tag{7.108}$$

**Theorem 7.12 Conditions for Correlated Coefficients To Be Negligible**
The average power, or average energy, of a random process is approximated by the summation associated with the diagonal terms in the correlation matrix, that is,

$$\bar{P} \approx \frac{1}{\beta-\alpha}\sum_{k=1}^{\infty} \boldsymbol{E}\left[|C_k|^2\right]\int_{\alpha}^{\beta}|b_k(t)|^2 dt \qquad \bar{E} \approx \sum_{k=1}^{\infty} \boldsymbol{E}\left[|C_k|^2\right]\int_{\alpha}^{\beta}|b_k(t)|^2 dt \tag{7.109}$$

when

$$\left|\sum_{k=1}^{i-1} 2\mathrm{Re}\left[\boldsymbol{E}\left[C_i C_k^*\right]\int_{\alpha}^{\beta} b_i(t) b_k^*(t)dt\right]\right| \ll \boldsymbol{E}\left[|C_i|^2\right]\int_{\alpha}^{\beta}|b_i(t)|^2 dt \quad i \in \{2,3,\ldots\} \tag{7.110}$$

**Proof**
The proof is detailed in Appendix 7.F.

***7.6.2.3 Orthonormal Basis Set and Uncorrelated Coefficients*** It has been shown (Theorems 7.9 and 7.10) that orthogonality of the basis signals and uncorrelatedness of the basis signal coefficients are sufficient conditions for the average power of a random process to equal the summation of the average power of the individual signal components. It is of interest if orthogonality of the basis signals is linked to the uncorrelatedness of the basis signal coefficients.

Consider an arbitrary orthonormal basis set $\{b_1, \ldots\}$ and the decomposition for the $i$th signal defined by the random process $X: S \to S_X$, on the interval $[0, T]$, according to

$$X(\omega_i, t) = \sum_{k=1}^{\infty} c_k(\omega_i) b_k(t) \qquad c_k(\omega_i) = \int_0^T X(\omega_i, t) b_k^*(t)dt \tag{7.111}$$

The following theorem states sufficient conditions on the basis functions for uncorrelated coefficients.

**Theorem 7.13 Orthonormal Basis Set and Uncorrelated Coefficients**
The random variables $C_i$ and $C_k$ defined by the $i$th and $k$th coefficients, and the sample spaces $\{c_i(\omega): \ \omega \in S\}$ and $\{c_k(\omega): \ \omega \in S\}$, are uncorrelated if

$$\int_0^T\int_0^T b_k(t) R_{XX}(t,\lambda) b_i^*(\lambda)d\lambda dt = \begin{cases} 0 & i \neq k \\ \boldsymbol{E}\left[|C_i|^2\right] & i = k \end{cases} \tag{7.112}$$

**Proof**
For the countable case, the expected value of $C_i$ and $C_k$ is

$$
\begin{aligned}
\boldsymbol{E}\left[C_i C_k^*\right] &= \sum_{\omega_r \in S} p_\Omega(\omega_r) \left[\int_0^T X(\omega_r,\lambda) b_i^*(\lambda) d\lambda\right] \left[\int_0^T X^*(\omega_r,t) b_k(t) dt\right] \\
&= \int_0^T\int_0^T \left[\sum_{\omega_r \in S} p_\Omega(\omega_r) X(\omega_r,t) X^*(\omega_r,\lambda)\right] b_i^*(\lambda) b_k(t) d\lambda dt \\
&= \int_0^T\int_0^T b_k(t) R_{XX}(t,\lambda) b_i^*(\lambda) d\lambda dt
\end{aligned}
\qquad (7.113)
$$

***7.6.2.4 Karhunen–Loeve Decomposition*** For a set random process, with $R_{XX}(t, \lambda)$ known, solving Equation 7.112 for the basis functions results is a Karhunen–Loeve basis set.

## 7.7 CLASSIFICATION: STATIONARITY VS NON-STATIONARITY

A widely used classification of random phenomena is based on the concept of stationarity, and two definitions are important: strict-sense stationarity and wide-sense stationarity. The latter is widely used.

### 7.7.1 Definitions

To define strict-sense stationarity, first, consider two random processes defined on a sample space $S$ and on the infinite interval $(-\infty, \infty)$:

$$
\begin{array}{ll}
X : S \to S_X & S_X = \{X(\omega,t): \ \omega \in S\} \\
V : S \to S_V & S_V = \{V(\omega,t): \ V(\omega,t) = X(\omega,t+\tau), \omega \in S\}
\end{array}
\qquad (7.114)
$$

Each signal defined by $V$ is a time-shifted version of the corresponding signal from $X$.

**Definition: Strict-Sense Stationarity** A random process $X : S \to S_X$ is strict-sense stationary if its statistical properties are the same as the random process $V : S \to S_X$ for all possible values of the time shift $\tau$, that is, $X$ is strict-sense stationary if for all times $t_1, t_2, \ldots, t_N$ and for all values of $\tau$, it is the case that

$$
\mu_X(t) = \mu_V(t) \qquad \sigma_X^2(t) = \sigma_V^2(t) \qquad t \in \boldsymbol{R} \qquad (7.115)
$$

$$
F_{X_{t_1} \ldots X_{t_N}}(x_1, \ldots, x_N) = F_{V_{t_1} \ldots V_{t_N}}(x_1, \ldots, x_N) \qquad x_1, \ldots, x_N \in \boldsymbol{R} \qquad (7.116)
$$

etc. An equivalent requirement is: for all times $t_1, t_2, \ldots, t_N$, and for all times $\tau > 0$, it is the case that

$$F_{X_{t_1}\ldots X_{t_N}}(x_1,\ldots,x_N)=F_{X_{t_1}+\tau,\ldots,X_{t_N}+\tau}(x_1,\ldots,x_N),\quad x_1,\ldots,x_N\in\boldsymbol{R} \tag{7.117}$$

**Definition: Wide-Sense Stationarity** A random process $X:S\to S_X$ is wide-sense stationary if its mean does not vary with time and if its autocorrelation function depends only on the difference between the two times being considered, that is, $X$ is wide-sense stationary if

$$\begin{aligned}&\mu_X(t)=\mu \quad t\in\boldsymbol{R}\\&R_{XX}(t_1+\tau,t_1)=R_{XX}(t_2+\tau,t_2)\quad t_1,t_2,\tau\in\boldsymbol{R}\end{aligned} \tag{7.118}$$

A sufficient, but not a necessary, condition for wide-sense stationarity is for

$$F_{X_{t_1},X_{t_2}}(\lambda_1,\lambda_2)=F_{X_{t_1+\tau},X_{t_2+\tau}}(\lambda_1,\lambda_2)\quad \lambda_1,\lambda_2\in\boldsymbol{R} \tag{7.119}$$

for all times $t_1$, $t_2$ and for values of $\tau>0$.

***7.7.1.1 Finite Interval*** Wide-sense stationarity for the interval $[0, T]$ implies

$$\begin{aligned}&\boldsymbol{E}[X(\Omega,t)]=\mu \quad 0\le t<T\\&R_{XX}(t_1+\tau,t_1)=\boldsymbol{E}[X(\Omega,t_1+\tau)X^*(\Omega,t_1)]=R_{XX}(\tau)\quad \begin{cases}0\le t_1<T\\0\le t_1+\tau<T\end{cases}\end{aligned} \tag{7.120}$$

***7.7.1.2 Stationary and Nonstationary Random Processes*** Most random phenomena commencing at a set time, which exhibit an initial transient response, are nonstationary. Many random phenomena, which exhibit steady-state behavior after the transient period, will exhibit characteristics consistent with stationarity. A random walk, for example, is clearly nonstationary.

### 7.7.2 Examples of Stationary Random Processes

The following subsections detail random processes that are wide-sense stationary:

***7.7.2.1 Example 1*** Consider an random process $X:S\to S_X$ defined by a sinusoid with random phase:

$$X(\Omega,t)=A\sin(2\pi f_o t+\Omega) \tag{7.121}$$

$$f_\Omega(\omega)=\begin{cases}0 & \text{elsewhere}\\ \dfrac{1}{2\pi} & 0\le\omega<2\pi\end{cases} \tag{7.122}$$

which has the following properties:

$$\mu_X(t) = \boldsymbol{E}[X(\Omega,t)] = \int_0^{2\pi} A\sin(2\pi f_o t + \omega)\cdot\frac{1}{2\pi}\cdot d\omega = 0 \tag{7.123}$$

$$\begin{aligned} R_{XX}(t+\tau,t) &= \boldsymbol{E}[X(\Omega,t+\tau)X(\Omega,t)] \\ &= \int_0^{2\pi} A\sin(2\pi f_o(t+\tau)+\omega)\cdot A\sin(2\pi f_o t+\omega)\cdot\frac{1}{2\pi}\cdot d\omega \\ &= \frac{A^2}{2}\cdot\cos(2\pi f_o \tau) \end{aligned} \tag{7.124}$$

and is wide-sense stationary.

*7.7.2.2* ***Example 2*** Consider the random process defined by the sum of two sinusoids of the same frequency but with random amplitudes:

$$X(\Omega,t) = \Omega_1\cos(2\pi f_o t) + \Omega_2\sin(2\pi f_o t) \tag{7.125}$$

where $\Omega = (\Omega_1, \Omega_2)$ is a pair of random variables whose outcomes are governed by the joint probability density function $f_{\Omega_1\Omega_2}$.

The mean and autocorrelation functions are

$$\begin{aligned} \mu_X(t) &= \boldsymbol{E}[X(\Omega,t)] = \int_{-\infty}^{\infty}\int_{-\infty}^{\infty}[\omega_1\cos(2\pi f_o t) + \omega_2\sin(2\pi f_o t)]f_{\Omega_1\Omega_2}(\omega_1,\omega_2)d\omega_1 d\omega_2 \\ &= \cos(2\pi f_o t)\int_{-\infty}^{\infty}\omega_1 f_{\Omega_1}(\omega_1)d\omega_1 + \sin(2\pi f_o t)\int_{-\infty}^{\infty}\omega_2 f_{\Omega_2}(\omega_2)d\omega_2 \\ &= \boldsymbol{E}[\Omega_1]\cos(2\pi f_o t) + \boldsymbol{E}[\Omega_2]\sin(2\pi f_o t) \end{aligned} \tag{7.126}$$

$$\begin{aligned} R_{XX}(t+\tau,t) &= \boldsymbol{E}[X(\Omega,t+\tau)X^*(\Omega,t)] \\ &= \int_{-\infty}^{\infty}\int_{-\infty}^{\infty}[\omega_1\cos(2\pi f_o(t+\tau)) + \omega_2\sin(2\pi f_o(t+\tau))]\cdot \\ &\quad [\omega_1\cos(2\pi f_o t) + \omega_2\sin(2\pi f_o t)]f_{\Omega_1\Omega_2}(\omega_1,\omega_2)d\omega_1 d\omega_2 \end{aligned} \tag{7.127}$$

Multiplying terms yields

$$R_{XX}(t+\tau,t) = \cos(2\pi f_o(t+\tau))\cos(2\pi f_o t)\int_{-\infty}^{\infty}\omega_1^2 f_{\Omega_1}(\omega_1)d\omega_1$$
$$+\sin(2\pi f_o(t+\tau))\sin(2\pi f_o t)\int_{-\infty}^{\infty}\omega_2^2 f_{\Omega_2}(\omega_2)d\omega_2$$
$$+\sin[2\pi f_o(2t+\tau)]\int_{-\infty}^{\infty}\int_{-\infty}^{\infty}\omega_1\omega_2 f_{\Omega_1\Omega_2}(\omega_1,\omega_2)d\omega_1 d\omega_2 \quad (7.128)$$

as

$$\cos(2\pi f_o(t+\tau))\sin(2\pi f_o t)+\sin(2\pi f_o(t+\tau))\cos(2\pi f_o t)=\sin[2\pi f_o(2t+\tau)] \quad (7.129)$$

Thus,

$$R_{XX}(t+\tau,t) = \cos(2\pi f_o(t+\tau))\cos(2\pi f_o t)\boldsymbol{E}\left[\Omega_1^2\right]$$
$$+\sin(2\pi f_o(t+\tau))\sin(2\pi f_o t)\boldsymbol{E}\left[\Omega_2^2\right]$$
$$+\sin[2\pi f_o(2t+\tau)]\boldsymbol{E}[\Omega_1\Omega_2] \quad (7.130)$$

and it then follows that the random process is wide-sense stationary if $\Omega_1$ and $\Omega_2$ have zero mean, have identical variances, and are uncorrelated, that is, $\boldsymbol{E}[\Omega_1]=\boldsymbol{E}[\Omega_2]=0$, $\boldsymbol{E}\left[\Omega_1^2\right]=\boldsymbol{E}\left[\Omega_2^2\right]$, and $\boldsymbol{E}[\Omega_1\Omega_2]=0$. Here, the result

$$\cos(2\pi f_o(t+\tau))\cos(2\pi f_o t)+\sin(2\pi f_o(t+\tau))\sin(2\pi f_o t)=\cos(2\pi f_o t) \quad (7.131)$$

has been used. With the stated assumptions,

$$\mu_X(t)=0 \qquad R_{XX}(t+\tau,t)=R_{XX}(\tau)=\cos(2\pi f_o\tau)\boldsymbol{E}\left[\Omega_1^2\right] \quad (7.132)$$

***7.7.2.3 Example 3*** Consider the random process defined by a sinusoid with random frequency variations around a set frequency and with random phase:

$$X(\Omega,t)=A\cos(2\pi(f_o+\Omega_1)t+\Omega_2) \quad (7.133)$$

where $\Omega=(\Omega_1,\Omega_2)$ is a pair of independent random variables with respective density functions $f_{\Omega_1}$ and $f_{\Omega_2}$.

The mean of this random process is

$$\mu_X(t) = \boldsymbol{E}[X(\Omega,t)] = \int_{-\infty}^{\infty}\int_{-\infty}^{\infty} A\cos[2\pi(f_o+\omega_1)t+\omega_2]f_{\Omega_1}(\omega_1)f_{\Omega_2}(\omega_2)d\omega_1 d\omega_2$$

$$= A\int_{-\infty}^{\infty}\cos[2\pi(f_o+\omega_1)t]f_{\Omega_1}(\omega_1)d\omega_1\int_{-\infty}^{\infty}\cos(\omega_2)f_{\Omega_2}(\omega_2)d\omega_2$$

$$-A\int_{-\infty}^{\infty}\sin[2\pi(f_o+\omega_1)t]f_{\Omega_1}(\omega_1)d\omega_1\int_{-\infty}^{\infty}\sin(\omega_2)f_{\Omega_2}(\omega_2)d\omega_2 \tag{7.134}$$

For the case where $f_{\Omega_2}(\omega_2) = \begin{cases} 1/2\pi & -\pi \le \omega_2 < \pi \\ 0 & \text{elsewhere} \end{cases}$, it follows that $\mu_X(t) = 0$ as required by stationarity.

The autocorrelation function, by definition, is

$$R_{XX}(t+\tau,t) = \boldsymbol{E}[X(\Omega,t+\tau)X^*(\Omega,t)]$$

$$= \int_{-\infty}^{\infty}\int_{-\infty}^{\infty}[A\cos(2\pi(f_o+\omega_1)(t+\tau)+\omega_2)][A\cos(2\pi(f_o+\omega_1)t+\omega_2)]$$

$$f_{\Omega_1}(\omega_1)f_{\Omega_2}(\omega_2)d\omega_1 d\omega_2 \tag{7.135}$$

As $\cos(a)\cos(b) = 0.5[\cos(a-b)+\cos(a+b)]$, it follows that

$$R_{XX}(t+\tau,t) = \frac{A^2}{2}\int_{-\infty}^{\infty}\int_{-\infty}^{\infty}[\cos(2\pi(f_o+\omega_1)\tau)+\cos(2\pi(f_o+\omega_1)(2t+\tau)+2\omega_2)]$$

$$f_{\Omega_1}(\omega_1)f_{\Omega_2}(\omega_2)d\omega_1 d\omega_2 \tag{7.136}$$

For the case where $\Omega_2$ has a uniform distribution on $[-\pi,\pi)$, it follows that

$$R_{XX}(t+\tau,t) = \frac{A^2}{2}\boldsymbol{E}[\cos(2\pi(f_o+\Omega_1)\tau)] \tag{7.137}$$

which is consistent with wide-sense stationarity.

### 7.7.3 Implications of Stationarity

Several important results can be stated for stationary random processes.

**Theorem 7.14 Stationarity Implies Hermitian Symmetry for Autocorrelation**
A wide-sense stationary random process $X: S \to S_X$ is such that its autocorrelation function

$$R_{XX}(\tau) = \boldsymbol{E}[X(\Omega, t_1 + \tau) X^*(\Omega, t_1)] \qquad 0 \le t_1, t_1 + \tau < T \tag{7.138}$$

has Hermitian symmetry, that is,

$$R^*_{XX}(-\tau) = R_{XX}(\tau) \tag{7.139}$$

**Proof**
By definition

$$R_{XX^*}(-\tau) = \boldsymbol{E}[X^*(\Omega, t_1 - \tau) X(\Omega, t_1)] \qquad 0 \le t_1, t_1 - \tau < T \tag{7.140}$$

Choose $t_1 = t_2 + \tau$ and with the constraints $0 \le t_2, t_2 + \tau < T$, it is the case that

$$\begin{aligned} R_{XX^*}(-\tau) &= \boldsymbol{E}[X^*(\Omega, t_2) X(\Omega, t_2 + \tau)] \\ &= \boldsymbol{E}[X(\Omega, t_2 + \tau) X^*(\Omega, t_2)] = R_{XX}(\tau) \end{aligned} \tag{7.141}$$

as required.

### 7.7.4 Wide-Sense Stationarity and Correlation of Coefficients

Consider a random process $X: S \to S_X$ defined on the interval [α, β] and the Fourier series decomposition of each signal:

$$X(\omega, t) = \sum_{k=-\infty}^{\infty} c_k(\omega) e^{j2\pi k f_o t} \qquad f_o = \frac{1}{\beta - \alpha}, \omega \in S \tag{7.142}$$

The coefficients define a set of random variables $C_k, k \in \boldsymbol{Z}$ where the sample space of $C_k$ is $\{c_k(\omega): \ \omega \in S\}$. The following theorem states sufficient conditions for wide-sense stationarity (Koopmans, 1974/1995, p. 40):

**Theorem 7.15 Uncorrelated Coefficients Imply Stationarity**
If

$$\boldsymbol{E}[C_k] = \begin{cases} \mu_X & k = 0 \\ 0 & k \neq 0 \end{cases} \tag{7.143}$$

$$\boldsymbol{E}\left[C_u C_v^*\right] = \begin{cases} \boldsymbol{E}\left[|C_u|^2\right] & u = v \\ 0 & u \neq v \end{cases} \tag{7.144}$$

then the random process $X$ is wide-sense stationary on [α, β].

**Proof**

Consider the countable case. First, consider the mean:

$$E[X(\Omega,t)] = \sum_{\omega_i \in S} p_\Omega(\omega_i) \sum_{k=-\infty}^{\infty} c_k(\omega_i) e^{j2\pi k f_o t} = \sum_{\omega_i \in S} p_\Omega(\omega_i) c_0(\omega_i)$$
$$+ \sum_{\substack{k=-\infty \\ k\neq 0}}^{\infty} \sum_{\omega_i \in S} p_\Omega(\omega_i) c_k(\omega_i) e^{j2\pi k f_o t} = E[C_0] + \sum_{k=-\infty, k\neq 0}^{\infty} E[C_k] e^{j2\pi k f_o t} \tag{7.145}$$

The mean can only be constant if $E[C_k] = \begin{cases} \mu_X & k=0 \\ 0 & k\neq 0 \end{cases}$ which is the required result. Second, consider the autocorrelation function of $X$:

$$E[X(\Omega,t+\tau)X^*(\Omega,t)]$$
$$= \sum_{\omega_i \in S} p_\Omega(\omega_i) \left[ \sum_{u=-\infty}^{\infty} c_u(\omega_i) e^{j2\pi u f_o (t+\tau)} \right] \left[ \sum_{v=-\infty}^{\infty} c_v(\omega_i)^* e^{-j2\pi v f_o t} \right]$$
$$= \sum_{u=-\infty}^{\infty} \sum_{v=-\infty}^{\infty} \left[ \sum_{\omega_i \in S} p_\Omega(\omega_i) [c_u(\omega_i) c_v(\omega_i)^*] \right] e^{j2\pi u f_o \tau} e^{j2\pi(u-v) f_o t} \tag{7.146}$$
$$= \sum_{u=-\infty}^{\infty} E\left[|C_u|^2\right] e^{j2\pi u f_o \tau} + \sum_{u=-\infty}^{\infty} \sum_{\substack{v=-\infty \\ v\neq u}}^{\infty} E\left[C_u C_v^*\right] e^{j2\pi u f_o \tau} e^{j2\pi(u-v) f_o t}$$

For the autocorrelation function not to vary with $t$, the requirement is for $E\left[C_u C_v^*\right] = 0$, when $u \neq v$, which is the required result.

#### 7.7.4.1 *Stationarity Implies Uncorrelated Fourier Coefficients*

Theorem 7.15 states that if the coefficients defined by a Fourier series decomposition of the signals in a random process are uncorrelated and have zero mean apart from the zeroth-order coefficient, then the random process is wide-sense stationary. The converse is also true (Kawata, 1969; Papoulis, 1965, pp. 367, 461; Yaglom, 1962, p. 36).

**Theorem 7.16 Stationarity Implies Uncorrelated Fourier Coefficients**

If a random process $X : S \to S_X$ is wide-sense stationary on an interval $[\alpha, \beta]$ and each signal has a Fourier series decomposition as specified by Equation 7.142 on this interval, then the random variables defined by the Fourier coefficients are uncorrelated, that is, $E\left[C_u C_v^*\right] = 0$ for $u \neq v$.

**Proof**

The proof is detailed in Appendix 7.G.

## 7.8 CRAMER'S REPRESENTATION

Consider a random process $X : S \rightarrow S_X$ defined on the interval $[0, T]$ with $S_X = \{X(\omega,t): \ \omega \in S\}$. Consistent with the discussion in Chapter 3, the Cramer representation of the signal $X(\omega, t)$, denoted $X_C(\omega, f)$, by definition, is

$$X_C(\omega,f) = \frac{j}{2\pi}\int_0^T \frac{e^{-j2\pi f\lambda}-1}{\lambda}\cdot X(\omega,\lambda)d\lambda \qquad X(\omega,t) = \int_R e^{j2\pi ft} dX_C(\omega,f) \tag{7.147}$$

The set of signals $\{X_C(\omega,f): \ \omega \in S\}$ defines the associated random process $X_C$:

$$\begin{aligned} &X_C : S \rightarrow S_{X_C} \\ &S = \{\omega: \ \text{as defined by experiment}\} \\ &S_{X_C} = \{X_C(\omega,f): \ \omega \in S\} \end{aligned} \tag{7.148}$$

The shorthand notation

$$X_C(\Omega,f) = \frac{j}{2\pi}\int_0^T \frac{e^{-j2\pi f\lambda}-1}{\lambda}\cdot X(\Omega,\lambda)d\lambda \tag{7.149}$$

is useful.

Using the signal definitions in Chapter 3, the following integrated spectrum, spectrum, and power spectrum definitions can be made.

**Definition: Integrated Spectrum** The integrated spectrum, denoted $X_C(f)$, is the weighted average of the individual integrated spectrums:

$$X_C(f) = \boldsymbol{E}[X_C(\Omega,f)] = \begin{cases} \displaystyle\sum_{\omega_i \in S} p_\Omega(\omega_i) X_C(\omega_i,f) & \text{countable case} \\ \displaystyle\int_{-\infty}^{\infty} X_C(\omega,f) f_\Omega(\omega) d\omega & \text{uncountable case} \end{cases} \tag{7.150}$$

**Definition: Spectrum, Power Spectrum, and Integrated Power Spectrum** The spectrum, $dX_C(f)$, power spectrum, $|dX_C(f)|^2$, and integrated power spectrum, $G_I(f)$, of the random process $X$ are defined, respectively, according to the weighted summations:

$$\begin{aligned} dX_C(f) &= \lim_{df \to 0} X_C(f + df/2) - X_C(f - df/2) \\ &= \lim_{df \to 0} \boldsymbol{E}[X_C(\Omega, f + df/2) - X_C(\Omega, f - df/2)] \\ &= \begin{cases} \displaystyle\lim_{df \to 0} \sum_{\omega_i \in S} p_\Omega(\omega_i)[X_C(\omega_i, f + df/2) - X_C(\omega_i, f - df/2)] & \text{c.c.} \\ \displaystyle\lim_{df \to 0} \int_{-\infty}^{\infty} [X_C(\omega, f + df/2) - X_C(\omega, f - df/2)]\, f_\Omega(\omega) d\omega & \text{u.c.} \end{cases} \end{aligned} \tag{7.151}$$

$$
\begin{aligned}
|dX_C(f)|^2 &= \lim_{df\to 0} \boldsymbol{E}\left[|X_C(\Omega,f+df/2)-X_C(\Omega,f-df/2)|^2\right] \\
&= \begin{cases} \displaystyle\lim_{df\to 0}\sum_{\omega_i\in S} p_\Omega(\omega_i)|X_C(\omega_i,f+df/2)-X_C(\omega_i,f-df/2)|^2 & \text{c.c.} \\ \displaystyle\lim_{df\to 0}\int_{-\infty}^{\infty} |X_C(\omega,f+df/2)-X_C(\omega,f-df/2)|^2 f_\Omega(\omega)d\omega & \text{u.c.} \end{cases}
\end{aligned} \tag{7.152}
$$

$$
\begin{aligned}
G_I(f) &= \boldsymbol{E}\left[|X_C(\Omega,f)|^2\right] \\
&= \begin{cases} \displaystyle\sum_{\omega_i\in S} p_\Omega(\omega_i)|X_C(\omega_i,f)|^2 & \text{countable case} \\ \displaystyle\int_{-\infty}^{\infty} |X_C(\omega,f)|^2 f_\Omega(\omega)d\omega & \text{uncountable case} \end{cases}
\end{aligned} \tag{7.153}
$$

**Definition: Power Spectral Density** For the case where the integrated power spectrum is continuous, the power spectral density can be defined according to

$$
\begin{aligned}
G(f) &= \lim_{df\to 0} \frac{G_I(f+df/2)-G_I(f-df/2)}{df} \\
&= \lim_{df\to 0} \frac{\boldsymbol{E}\left[|X_C(\Omega,f+df/2)|^2\right]-\boldsymbol{E}\left[|X_C(\Omega,f-df/2)|^2\right]}{df}
\end{aligned} \tag{7.154}
$$

### 7.8.1 Fundamental Results

The following theorems state fundamental results.

**Theorem 7.17 Integrated Power Spectrum**
Consider a wide-sense stationary random process $X : S \to S_X$, defined on an interval $[-T,T]$, which is such that each signal in the signal sample space can be written as a Fourier series according to

$$
X(\omega,t) = \sum_{i=-\infty}^{\infty} c_i(\omega)e^{j2\pi i f_o t} \qquad f_o = \frac{1}{2T}, \omega \in S \tag{7.155}
$$

For this case, the integrated power spectrum is

$$G_I(f)=\boldsymbol{E}\left[|X_C(\Omega,f)|^2\right]=\sum_{i=-\infty}^{\infty}\frac{\boldsymbol{E}\left[|C_i|^2\right]}{\pi^2}\cdot[\mathrm{Si}(i\pi)+\mathrm{Si}(2\pi fT-i\pi)]^2$$

$$\approx\begin{cases}\sum_{i=0}^{\lfloor f/f_o\rfloor}\frac{\boldsymbol{E}\left[|C_i|^2\right]}{4}[\mathrm{sgn}(i)+\mathrm{sgn}(f-if_o)]^2 & f>0\\ 0 & f=0\\ \sum_{i=-\lfloor|f|/f_o\rfloor}^{0}\frac{\boldsymbol{E}\left[|C_i|^2\right]}{4}[\mathrm{sgn}(i)+\mathrm{sgn}(f-if_o)]^2 & f<0\end{cases} \tag{7.156}$$

where the sample space of the random variable $C_k$ is $\{c_k(\omega):\ \omega\in S\}$.

It then follows that

$$G_I(f_2)-G_I(f_1)=\boldsymbol{E}\left[|X_C(\Omega,f_2)|^2\right]-\boldsymbol{E}\left[|X_C(\Omega,f_1)|^2\right]=$$
$$\sum_{i=-\infty}^{\infty}\frac{\boldsymbol{E}\left[|C_i|^2\right]}{\pi^2}\cdot\left[[\mathrm{Si}(i\pi)+\mathrm{Si}(2\pi f_2T-i\pi)]^2-[\mathrm{Si}(i\pi)+\mathrm{Si}(2\pi f_1T-i\pi)]^2\right] \tag{7.157}$$

and for the case of $f_2>f_1>0$:

$$G_I(f_2)-G_I(f_1)\approx$$
$$\sum_{i=\lfloor f_1/f_o\rfloor}^{\lfloor f_2/f_o\rfloor}\frac{\boldsymbol{E}\left[|C_i|^2\right]}{4}\cdot\left[[\mathrm{sgn}(i)+\mathrm{sgn}(f_2-if_o)]^2-[\mathrm{sgn}(i)+\mathrm{sgn}(f_1-if_o)]^2\right] \tag{7.158}$$

**Proof**

The proof is detailed in Appendix 7.H.

The graph of $G_I(f)$ is illustrated in Figure 7.7 assuming the approximate form defined in Equation 7.156.

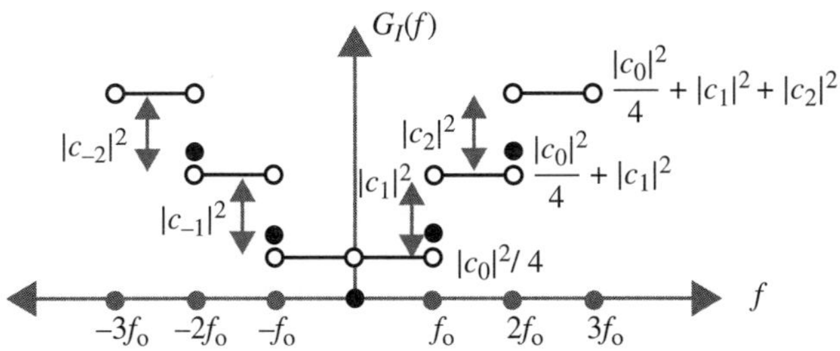


**FIGURE 7.7** Graph of $G_I(f)=\boldsymbol{E}\left[|X_C(\Omega,f)|^2\right]$. The notation $|c_i|^2$ has been used for $\boldsymbol{E}[|c_i|^2]$.

**Theorem 7.18 Cramer: Orthogonal Increment Random Process**
On the interval $[-T,T]$ and for the case of $T \to \infty$, it follows, for a wide-sense stationary random process $X: S \to S_X$, that

$$E\left[[X_C(\Omega,f_2)-X_C(\Omega,f_1)]\,[X_C(\Omega,f_4)-X_C(\Omega,f_3)]^*\right]=0 \tag{7.159}$$

when $f_1<f_2\leq f_3<f_4$ and $(f_1,f_2]\cap(f_3,f_4]=\varnothing$.

Further, for $0<f_1<f_2$ and for $T$ sufficiently large,

$$\begin{aligned} &E\left[|X_C(\Omega,f_2)-X_C(\Omega,f_1)|^2\right]\approx \\ &\left|E\left[|X_C(\Omega,f_2)|^2\right]-E\left[|X_C(\Omega,f_1)|^2\right]\right|=|G_I(f_2)-G_I(f_1)|\approx\sum_{i=\lfloor f_1/f_o\rfloor}^{\lfloor f_2/f_o\rfloor}E\left[|C_i|^2\right] \end{aligned} \tag{7.160}$$

where $f_o=1/2T$ and $C_i$ is the random variable defined by the $i$th Fourier coefficients.

**Proof**
The proof is detailed in Appendix 7.I.

***7.8.1.1 Notes*** The first result of Theorem 7.18 implies that the integrated spectrum, $X_C(\Omega,f)$, is an orthogonal increment random process on the infinite interval. This result does not hold for the interval $[0, T]$ as $T \to \infty$.

The second result of Theorem 7.18 states that the expected magnitude squared value of the difference in the integrated spectrum between $f_1$ and $f_2$ equals the power in the sinusoidal components with frequencies between the two values. This result justifies the definition of the power spectrum $|dX_C(f)|^2$, as given by Equation 7.152, as a valid power spectrum with a resolution of $f_o=1/2T$ for the interval $[-T,T]$. In summary:

**Theorem 7.19 Power Spectrum–Integrated Spectrum Relationship**
For a wide-sense stationary random process defined on $[-T,T]$, $T\to\infty$, it follows, for $0<f_1<f_2$, that

$$\begin{aligned} E\left[|X_C(\Omega,f_2)-X_C(\Omega,f_1)|^2\right]&=\left|E\left[|X_C(\Omega,f_2)|^2\right]-E\left[|X_C(\Omega,f_1)|^2\right]\right| \\ &=|G_I(f_2)-G_I(f_1)| \end{aligned} \tag{7.161}$$

and hence,

$$\begin{aligned} |dX_C(f)|^2 &= \lim_{df\to 0}E\left[|X_C(\Omega,f+df/2)-X_C(\Omega,f-df/2)|^2\right] \\ &\lim_{df\to 0}\left|E\left[|X_C(\Omega,f+df/2)|^2\right]-E\left[|X_C(\Omega,f-df/2)|^2\right]\right| \\ &=\lim_{df\to 0}\left[|G_I(f+df/2)-G_I(f-df/2)|\right] \end{aligned} \tag{7.162}$$

**Proof**
The proof follows from Theorem 7.18 and the definitions for $G_I(f)$ and $dX_C(f)$.

***7.8.1.2 Notes*** Theorem 7.19 states that the expected mean squared change of $X_C(\Omega, f)$, which is an orthogonal increment function, between $f_2$ and $f_1$ equals the change in the integrated power spectrum, that is, the power, between $f_2$ and $f_1$.

To determine the power spectrum, as given by $|dX_C(f)|^2$, it is sufficient to determine the integrated power spectrum as given by $G_I(f) = \boldsymbol{E}\left[|X_C(\Omega,f)|^2\right]$.

For the case where the integrated power spectrum is continuous, the power spectral density can be defined according to

$$G(f) = \lim_{df \to 0} \frac{G_I(f + df/2) - G_I(f - df/2)}{df} \tag{7.163}$$

***7.8.1.3 Power***

**Theorem 7.20 Total Power**
For a stationary random process defined on $[-T, T]$, $T \to \infty$, it is the case that the average power is given by the Lebesgue–Stieltjes integral of the expected value of the mean squared integrated spectrum:

$$\bar{P}_\infty = \int_{-\infty}^{\infty} \left| d\boldsymbol{E}\left[|X_C(\Omega,f)|^2\right] \right| = \int_{-\infty}^{\infty} |dG_I(f)| \tag{7.164}$$

**Proof**
The proof is detailed in Appendix 7.J. This result is consistent with the illustration in Figure 7.7.

### 7.8.2 Continuous Spectrum

Consider any arbitrary interval of the form $[-T, T]$ and a random process $X : S \to S_X$, which is such that all signals in $S_X$ can be decomposed, using a Fourier series, into the form

$$X(\omega, t) = \sum_{i=-\infty}^{\infty} c_i(\omega) e^{-j2\pi i f_o t} \qquad f_o = \frac{1}{2T}, \omega \in S \tag{7.165}$$

Such a decomposition leads to a discrete line power spectrum of the form shown in Figure 7.8. It is of interest if a random process exists, which has a spectrum with spectral components that are spaced arbitrarily closely, such that a continuous spectrum results.

***7.8.2.1 Example of a Random Process with a Continuous Spectrum*** Consider a sequence of random processes $\{X_i\}_{i=1}^{\infty}$ where the $i$th random process is defined according to

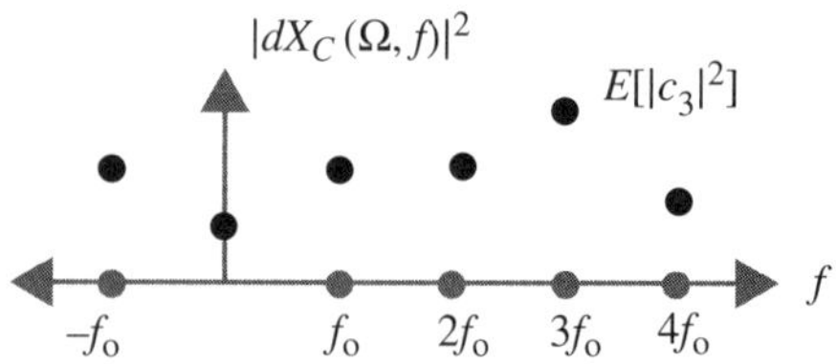


**FIGURE 7.8** Power spectrum as defined by a Fourier series.

$$X_i(K,\Theta,t) = A\sin\left[2\pi\left(f_C + \frac{KB}{2^i}\right)t + \Theta\right] \qquad t\in[-T,T], f_C>0 \tag{7.166}$$

Here, $B$ is a constant with the restriction $0<B<f_C$, $\Theta$ is a continuous random variable with a uniform distribution on $[0,2\pi)$, and $K$ is a discrete uniform random variable with a sample space $\{0,1,\ldots,2^i-1\}$ and with $P[K=0]=1/2^i,\ldots,$ $P[K=2^i-1]=1/2^i$. The two random variables are assumed to be independent. For the $i$th random process, the waveforms are sinusoidal with frequencies randomly selected from the set $\{f_C+kB/2^i:\ k\in\{0,1,\ldots,2^i-1\}\}$, and associated with each waveform is random phase.

For the $i$th random process, the expected power in the frequency range $\left[f_C+\frac{kB}{2^i}, f_C+\frac{(k+1)B}{2^i}\right)$ is determined by the power of a sinusoid with a frequency $f_C+kB/2^i$, which occurs with a probability $1/2^i$. The expected power is $(A^2/2)\cdot(1/2^i)$. The expected power in the interval $[f_C, f_C+B]$ equals the average power, which is $A^2/2$.

The Cramer transform of one outcome of the $i$th random process as specified by

$$\begin{aligned} X_i(k,\theta_k,t) &= A\sin\left[2\pi\left(f_C+\frac{kB}{2^i}\right)t+\theta_k\right] \\ &= \frac{A}{2j}\left[e^{j\theta_k}\exp\left[j2\pi\left(f_C+\frac{kB}{2^i}\right)t\right] - e^{-j\theta_k}\exp\left[-j2\pi\left(f_C+\frac{kB}{2^i}\right)t\right]\right] \end{aligned} \tag{7.167}$$

is given, for the infinite interval $(-\infty,\infty)$, by (see Table 3.1)

$$\begin{aligned} X_{C_i}(k,\theta_k,f) = {} & \frac{Ae^{j\theta_k}}{4j}\cdot\left[\operatorname{sgn}\left(f_C+\frac{kB}{2^i}\right)+\operatorname{sgn}\left[f-\left(f_C+\frac{kB}{2^i}\right)\right]\right] \\ & -\frac{Ae^{-j\theta_k}}{4j}\cdot\left[\operatorname{sgn}\left(-f_C-\frac{kB}{2^i}\right)+\operatorname{sgn}\left[f+\left(f_C+\frac{kB}{2^i}\right)\right]\right] \end{aligned} \tag{7.168}$$

Thus,

$$X_{C_i}(k,\theta_k,f)=\begin{cases}\dfrac{Ae^{-j\theta_k}}{2j} & f<-f_C-kB/2^i\\ 0 & -f_C-kB/2^i<f<f_C+kB/2^i\\ \dfrac{Ae^{j\theta_k}}{2j} & f>f_C+kB/2^i\end{cases} \tag{7.169}$$

and

$$|dX_{C_i}(k,\theta_k,f)|^2=\frac{A^2}{4}\cdot\delta_K\left[f-\left(f_C+\frac{kB}{2^i}\right)\right]+\frac{A^2}{4}\cdot\delta_K\left[f+\left(f_C+\frac{kB}{2^i}\right)\right] \tag{7.170}$$

These functions are illustrated in Figure 7.9.

The probability of each signal in the $i$th random process is $1/2^i$. It then follows that the power spectrum, $\boldsymbol{E}\left[|dX_{C_i}(K,\Theta,f)|^2\right]$, of the $i$th random process has the form illustrated in Figure 7.10.

***7.8.2.2 Meaning of a Continuous Spectrum*** For such a sequence of random processes, it is the case, for any fixed finite resolution in $f$ of $B/2^i$, that the random processes $X_1, \ldots, X_i$ have a power spectrum with a discrete uniform distribution as illustrated in Figure 7.10. The power spectrum for the random processes $X_{i+1}, \ldots$ will appear increasingly uniform and continuous when viewed with the same resolution.

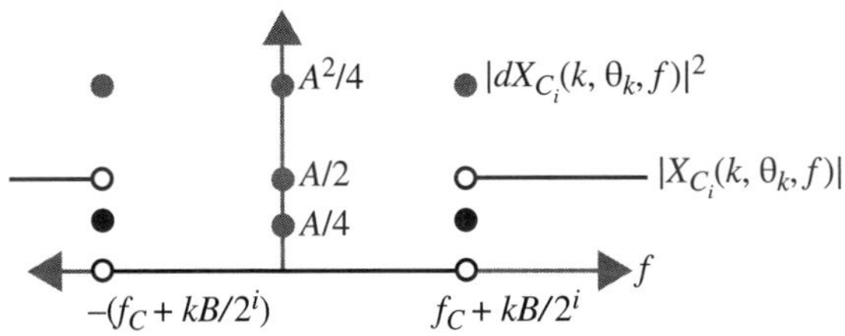


**FIGURE 7.9** Graph of the magnitude of the Cramer transform and the power spectrum associated with $X_i$ $(k, \theta_k, t)$.

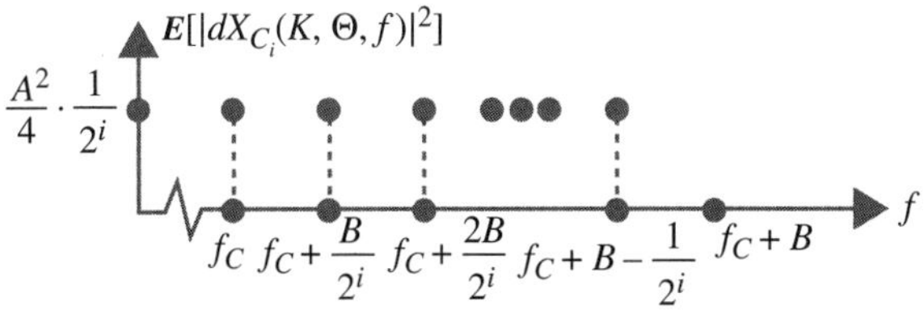


**FIGURE 7.10** Graph of the power spectrum for the $i$th random process.

### 7.8.3 Example: Binary Communication Random Process

Consider a random process that defines binary communication signals on the interval $[-T-D/2,\ T+D/2]$, $T=ND$, according to

$$X(\Omega,t)=\sum_{i=-N}^{N}A_i[u(t-(i-0.5)D)-u(t-(i+0.5)D)] \tag{7.171}$$

where $\Omega=(A_{-N},\ldots,A_N), A_{-N},\ldots,A_N$, are identical and independent random variables with zero mean and variance $\sigma_A^2$.

As detailed in Chapter 3 (Eq. 3.152), the Cramer transform of a unit step function $u(t-t_o)$ on the interval $[-\infty,T]$, assuming $t_o\neq 0$ and $f\neq 0$, is

$$\begin{aligned}X_C(f)=&\frac{\operatorname{sgn}(f)}{2\pi}[\operatorname{Si}(2\pi|f|T)+[2u(-t_o)-1]\operatorname{Si}(2\pi|f||t_o|)]\\&+\frac{j}{2\pi}\left[\operatorname{Ci}(2\pi|f|T)-\operatorname{Ci}(2\pi|f||t_o|)+\ln\left(\frac{|t_o|}{T}\right)\right]\end{aligned} \tag{7.172}$$

Hence,

$$\begin{aligned}X_C(\Omega,f)=&\frac{1}{2\pi}\sum_{i=-N}^{N}A_i\Big[\operatorname{sgn}(f)[2u(0.5-i)-1]\operatorname{Si}(2\pi|f||(i-0.5)D|)\\&-\operatorname{sgn}(f)[2u(-i-0.5)-1]\operatorname{Si}(2\pi|f||(i+0.5)D|)\\&+j\left[\operatorname{Ci}(2\pi|f||(i+0.5)D|)-\operatorname{Ci}(2\pi|f||(i-0.5)D|)+\ln\left[\frac{i-0.5}{i+0.5}\right]\right]\Big]\end{aligned} \tag{7.173}$$

Define

$$\begin{aligned}B(i,f)=&\operatorname{sgn}(f)[2u(0.5-i)-1]\operatorname{Si}(2\pi|(i-0.5)f|D)\\&-\operatorname{sgn}(f)[2u(-i-0.5)-1]\operatorname{Si}(2\pi|(i+0.5)f|D)\\&+j\left[\ln\left|\frac{i-0.5}{i+0.5}\right|-\operatorname{Ci}(2\pi|(i-0.5)f|D)+\operatorname{Ci}(2\pi|(i+0.5)f|D)\right]\end{aligned} \tag{7.174}$$

and it follows that

$$|X_C(\Omega,f)|^2=\frac{1}{4\pi^2}\sum_{i=-N}^{N}\sum_{k=-N}^{N}A_iA_kB(i,f)B^*(k,f) \tag{7.175}$$

and the integrated power spectrum is

$$\begin{aligned}G_I(f)=\boldsymbol{E}|X_C(\Omega,f)|^2&=\frac{1}{4\pi^2}\sum_{i=-N}^{N}\sum_{k=-N}^{N}\boldsymbol{E}[A_iA_k]B(i,f)B^*(k,f)\\&=\frac{\sigma_A^2}{4\pi^2}\sum_{i=-N}^{N}|B(i,f)|^2\end{aligned} \tag{7.176}$$

as $\boldsymbol{E}[A_iA_k] = 0$ when $i \neq k$. Here,

$$|B(i,f)|^2 = [[2u(0.5-i)-1]\mathrm{Si}(2\pi|(i-0.5)f|D) - [2u(-i-0.5)-1]\mathrm{Si}(2\pi|(i+0.5)f|D)]^2$$
$$+\left[\ln\left|\frac{i-0.5}{i+0.5}\right| - \mathrm{Ci}(2\pi|(i-0.5)f|D) + \mathrm{Ci}(2\pi|(i+0.5)f|D)\right]^2 \tag{7.177}$$

The power spectral density can then be evaluated according to

$$G(f) = \lim_{df\to 0}\frac{|G_I(f+df/2) - G_I(f-df/2)|}{df} \tag{7.178}$$

The integrated power spectrum is shown in Figure 7.11 for the case of $N = 16$, $D = 1$, amplitudes from the set $\{-1, 1\}$ and $\sigma_A^2 = 1$. The corresponding power spectral density is shown in Figure 7.12.

For reference, the power spectral density obtained according to

$$G(T,f) = \frac{\boldsymbol{E}|X(\Omega,T,f)|^2}{T} \tag{7.179}$$

is as $T \to \infty$,

$$G_\infty(f) = r\sigma_A^2|P(f)|^2 \tag{7.180}$$

where $P(f)$ is the Fourier transform of the signalling pulse $u(t+0.5D) - u(t-0.5D)$, that is,

$$P(f) = D\,\mathrm{sinc}(fD) \tag{7.181}$$

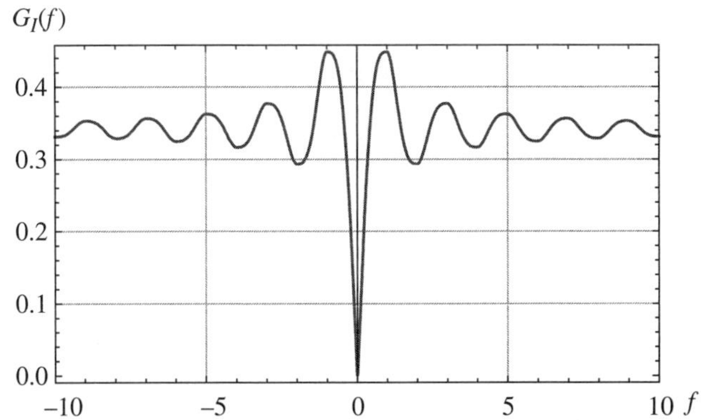


**FIGURE 7.11** Integrated power spectrum of a binary communication signal defined on the interval $[-ND - D/2, ND + D/2]$ for the case of $N = 16$, $D = 1$, amplitudes from the set $\{-1,1\}$, and $\sigma_A^2 = 1$.

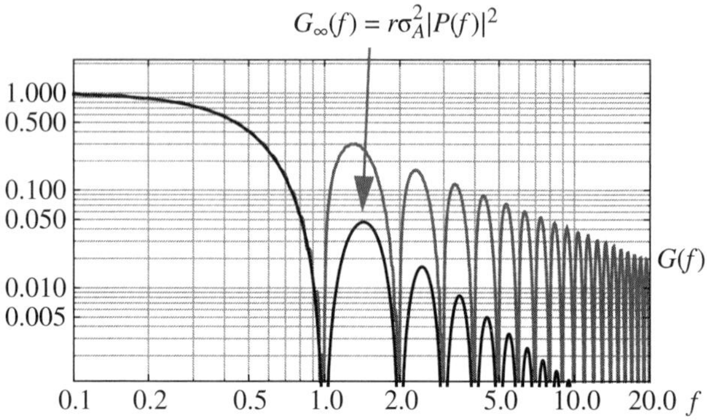


**FIGURE 7.12** Power spectral density of a binary communication signal defined on the interval $[-ND-D/2,\ ND+D/2]$ for the case of $N=16$, $D=1$, amplitudes from the set $\{-1,1\}$, and $\sigma_A^2=1$. $G(f)$ is the power spectral density defined by the Cramer transform and evaluated with a frequency resolution of $df=1/2(ND+D/2)$. $G_\infty(f)$ is the power spectral density assuming a sinusoidal basis set.

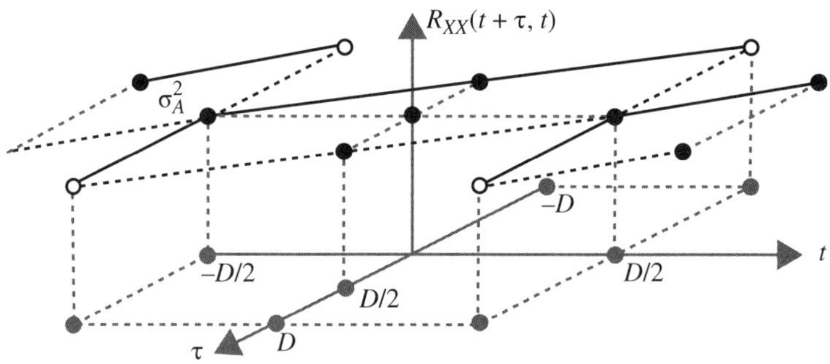


**FIGURE 7.13** Autocorrelation function of the defined binary communication random process.

This power spectral density is shown in Figure 7.12.

***7.8.3.1 Notes*** The random process is not wide-sense stationary as is evident from the autocorrelation function

$$R_{XX}(t+\tau,t)=\begin{cases}\sigma_A^2 & t,t+\tau\in[(0.5-i)D,(0.5+i)D], i\in\{-N,N\}\\ 0 & \text{elsewhere}\end{cases} \tag{7.182}$$

which is illustrated in Figure 7.13. This nonstationarity results in the power spectral density, as given by Equation 7.178, not being a valid power spectral density in terms of the usual requirement of the sum of the power in the individual components being equal to the total power. The next section details a stationary random process (white noise) whose power spectral density defined by the Cramer transform is a power spectral density that satisfies such a requirement.

### 7.8.4 White Noise: Approach II

Consider a white noise random process $X:S\to S_X$ defined according to

$$X(\Omega,t)=\sum_{i=1}^{N}\frac{\sqrt{2}A_o}{\sqrt{N}}\sin(2\pi F_i t+\Phi_i) \tag{7.183}$$

where $\Omega=((F_1,\Phi_1),\ldots,(F_N,\Phi_N))$, $F_1,\ldots,F_N$ are independent random variables with a uniform distribution on $[0,f_{max}]$, $\Phi_1,\ldots,\Phi_N$ are independent random variables with a uniform distribution on the interval $[0,2\pi)$, and $N$ is fixed according to $N=f_{max}/f_o$ where $f_o$ is the nominal frequency resolution.

**Theorem 7.21 Integrated Power Spectrum and Power Spectral Density**
On the interval $[-T,T]$ and with $T>>1/f_o$, the integrated power spectrum and power spectral density of the white noise random process defined by Equation 7.183, respectively, are

$$\begin{aligned}G_I(f)&=\frac{A_o^2}{4\pi^2 f_{max}}\cdot\int_0^{f_{max}}\Big[[\mathrm{Si}(2\pi(f+f_i)T)-\mathrm{Si}(2\pi(f-f_i)T)-2\mathrm{Si}(2\pi f_i T)]^2\\&\quad+[\mathrm{Si}(2\pi(f+f_i)T)+\mathrm{Si}(2\pi(f-f_i)T)]^2\Big]df_i\\&\approx\frac{A_o^2}{2}\cdot\frac{\min\{f,f_{max}\}}{f_{max}}\qquad f>0, G_I(-f)=G_i(f)\end{aligned} \tag{7.184}$$

$$G(f)\approx\begin{cases}\dfrac{A_o^2}{2f_{max}} & 0<f<f_{max}, -f_{max}<f<0\\[2ex] \dfrac{A_o^2}{4f_{max}} & f=\pm f_{max}\\[2ex] 0 & f=0, |f|>f_{max}\end{cases} \tag{7.185}$$

The integrated power spectrum and power spectral density are illustrated in Figure 7.14 for the infinite interval case. The average power as given by $2\lim_{f\to\infty}G_I(f)$ and the integral of the power spectral density both equal $A_o^2$ as required.

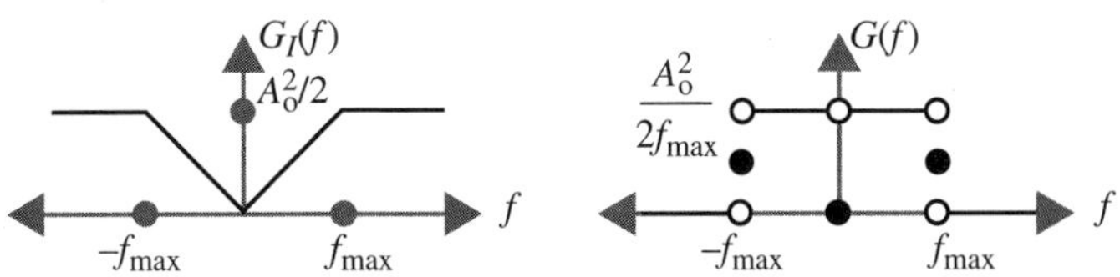


**FIGURE 7.14** Left: graph of the integrated power spectrum. Right: graph of the power spectral density.

**Proof**

The proof is detailed in Appendix 7.K.

***7.8.4.1 Example*** In Figures 7.15 and 7.16, the integrated power spectrum as given by Equation 7.184 and the associated power spectral density as given by

$$G(f) = \lim_{df \to 0} \frac{|G_I(f + df/2) - G_I(f - df/2)|}{df} \tag{7.186}$$

are shown for the finite interval case with $T = 10$, $f_{\max} = 1$, and $A_o = 1$ and an average signal power of $A_o^2 = 1$.

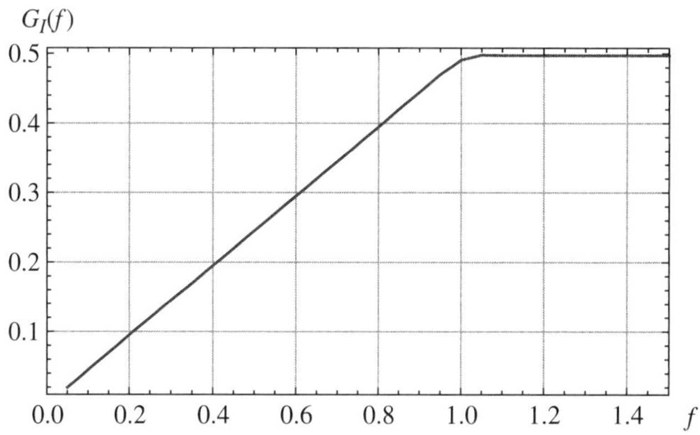


**FIGURE 7.15** Integrated power spectrum for a white noise random process defined on the interval $[-T, T]$ for $T = 10$, $A_o = 1$, and $f_{\max} = 1$.

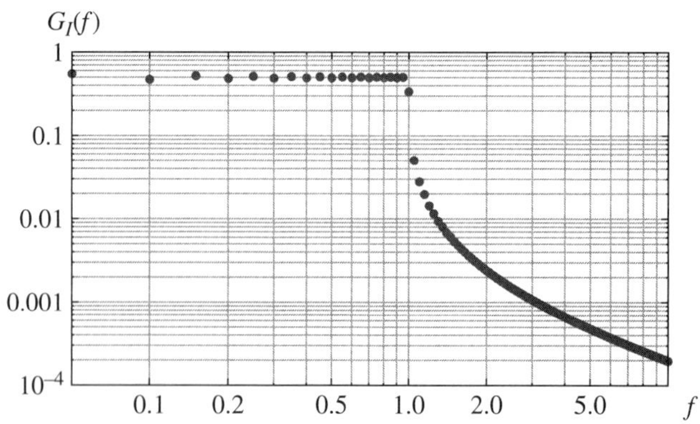


**FIGURE 7.16** Power spectral density for a white noise random process defined on the interval $[-T, T]$ for $T = 10$, $A_o = 1$, and $f_{\max} = 1$. A resolution of $df = 1/2T$ has been used.

## 7.9 STATE SPACE CHARACTERIZATION OF RANDOM PROCESSES

In some instances, the time nature of a random process is less important than the states that the random process takes on, and for this case, Markov theory is often appropriate (Allen, 1978, p. 129f; Grimmett and Stirzaker, 1992, p. 156f ).

### 7.9.1 Markov Processes

Markov processes are characterized by a lack of memory: the future is dependent on the current state of the random process and not on its past. The theory of Markov processes can be applied to queueing systems, population dynamics, diffusion processes, probability of error calculations in communication systems, etc.

***7.9.1.1 Classification of Markov Processes*** As noted in Chapter 5, one-dimensional random processes define signals that are either discrete time–discrete state, discrete time–continuous state, continuous time–discrete state, or continuous time–continuous state. Consistent with this demarcation, Markov processes can be classified as detailed in Table 7.1.

#### *7.9.1.2 Definition of a Markov Random Process*

**Definition: Markov Random Process** A random process, $X : S \to S_X$, is a Markov random process if future values can be predicted, with a specified probability, from present values and future values are independent of past values. Mathematically, and for the discrete time and discrete state case, $X$ is a Markov random process if

$$P[X_{t_{k+1}} = x_{k+1}/X_{t_k} = x_k, X_{t_{k-1}} = x_{k-1}, \ldots, X_{t_1} = x_1]$$
$$= P[X_{t_{k+1}} = x_{k+1}/X_{t_k} = x_k] \tag{7.187}$$

for all possible sets of ordered times $\{t_1, t_2, \ldots, t_k, t_{k+1}, \ldots\}$.

For other cases, more complex definitions are required, and the analysis of such random processes is more difficult.

**TABLE 7.1 Demarcation of Markov random processes**

| | State space | |
|---|---|---|
| Discrete time–continuous time | Countable # states: Markov chain | Uncountable # states: Markov process |
| Countable → discrete time Markov process | Discrete time Markov chain | Discrete time Markov process |
| Uncountable → continuous time Markov process | Continuous time Markov chain | Continuous time Markov process |

**Definition: Non-Markovian Random Process** A random process that is not a Markov random process is called a non-Markovian random process.

### 7.9.2 Discrete Time–Discrete State Random Processes

#### *7.9.2.1 State Diagrams for Discrete Time–Discrete State Random Processes*

Consider a discrete time–discrete state random process $X$ defined on a sample space $S$ according to

$$\begin{aligned} &X: S \rightarrow S_X \\ &S_X = \{X(\omega): S_T \rightarrow S_S, \quad S_T, S_S \subseteq \boldsymbol{R}, \omega \in S\} \end{aligned} \tag{7.188}$$

where the set $S_T = \{t_1, t_2, \ldots, t_N, \ldots\}$ specifies the times at which signals from the random process are defined, and at these times the signals take on values from the state space $S_S = \{s_1, s_2, \ldots, s_N, \ldots\}$. The possible states are illustrated in Figure 7.17.

At each possible time, a discrete random variable is defined. For the $k$th time, the random variable $X_{t_k}: S \rightarrow S_S$ is defined along with the associated probability mass function:

$$p_{X_{t_k}}(s_i) = P[X_{t_k} = s_i] \quad i \in \{1, \ldots, N, \ldots\} \tag{7.189}$$

At the time $t_k$, the transitions, as illustrated in Figure 7.17, are possible. For the time $t_1$, the probabilities associated with these transitions form a matrix according to

$$P^{12} = \begin{bmatrix} p_{11}^{12} & p_{12}^{12} & \cdots & p_{1N}^{12} & \cdots \\ p_{21}^{12} & p_{22}^{12} & \cdots & p_{2N}^{12} & \cdots \\ \cdots & & & & \\ p_{N1}^{12} & p_{N2}^{12} & \cdots & p_{NN}^{12} & \cdots \\ \cdots & & & & \end{bmatrix} \qquad p_{ij}^{12} = P\left[X_{t_2} = s_j / X_{t_1} = s_i\right] \tag{7.190}$$

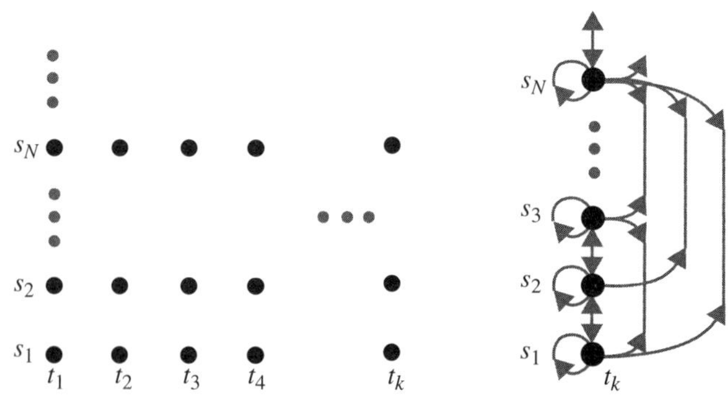


**FIGURE 7.17** Left: illustration of the possible states for a discrete time–discrete state random process. Right: illustration of the possible transitions at the time $t_k$.

where the superscript of 12 indicates the transition as time changes from $t_1$ to $t_2$. At the time $t_2$, the same transitions as at time $t_1$ are possible but, in general, with different probabilities. For the special memoryless case—the Markov case—where the probabilities associated with these transitions are dependent only on the state level and not on the prior transition, the following transition probability matrix can be defined:

$$P^{23} = \begin{bmatrix} p_{11}^{23} & p_{12}^{23} & \cdots & p_{1N}^{23} & \cdots \\ p_{21}^{23} & p_{22}^{23} & \cdots & p_{2N}^{23} & \cdots \\ \cdots & & & & \\ p_{N1}^{23} & p_{N2}^{23} & \cdots & p_{NN}^{23} & \cdots \\ \cdots & & & & \end{bmatrix} \quad p_{ij}^{23} = P\left[X_{t_3} = s_j / X_{t_2} = s_i\right] \tag{7.191}$$

For the nonmemoryless case—the non-Markov case—the transition probabilities have the form

$$p_{ijk}^{23} = P\left[X_{t_3} = s_k / X_{t_2} = s_j, X_{t_1} = s_i\right] \tag{7.192}$$

etc.

#### *7.9.2.2 Characterizing a Trajectory of a Signal from a Random Process*

The probability of a given trajectory for a random process over $N$ possible transitions is

$$\begin{aligned} &P[X_1 = s_1, X_2 = s_2, X_3 = s_3, \ldots, X_N = s_N] \\ &\quad = P[X_N = s_N / X_1 = s_1, \ldots, X_{N-1} = s_{N-1}] P[X_1 = s_1, \ldots, X_{N-1} = s_{N-1}] \\ &\quad = P[X_N = s_N / X_{N-1} = s_{N-1}, \ldots, X_1 = s_1] \cdot \\ &\qquad P[X_{N-1} = s_{N-1} / X_{N-2} = s_{N-2}, \ldots, X_1 = s_1] \cdot P[X_{N-2} = s_{N-2}, \ldots, X_1 = s_1] \end{aligned} \tag{7.193}$$

Expanding yields

$$\begin{aligned} &P[X_1 = s_1, X_2 = s_2, X_3 = s_3, \ldots, X_N = s_N] \\ &\quad = P[X_N = s_N / X_{N-1} = s_{N-1}, \ldots, X_1 = s_1] \cdot \\ &\qquad P[X_{N-1} = s_{N-1} / X_{N-2} = s_{N-2}, \ldots, X_1 = s_1] \cdot \\ &\quad \cdots \\ &\qquad P[X_2 = s_2 / X_1 = s_1] P[X_1 = s_1] \end{aligned} \tag{7.194}$$

For a Markov process, it then follows that

$$\begin{aligned} &P[X_1 = s_1, X_2 = s_2, X_3 = s_3, \ldots, X_N = s_N] \\ &\quad = P[X_N = s_N / X_{N-1} = s_{N-1}] \cdot P[X_{N-1} - s_{N-1} / X_{N-2} = s_{N-2}] \cdot \\ &\quad \cdots \\ &\qquad P[X_2 = s_2 / X_1 = s_1] P[X_1 = s_1] \end{aligned} \tag{7.195}$$

and a two-state transition matrix is sufficient.

### 7.9.3 Homogenous Markov Chain

**Definition: Homogenous Markov Chain** A discrete Markov process is homogenous if the transition probabilities do not vary with time, that is,

$$p_{ij} = P\left[X_{t_{k+1}} = s_j / X_{t_k} = s_i\right] = P\left[X_{t_{r+1}} = s_j / X_{t_r} = s_i\right] \quad k, r \in \mathbf{Z}^+ \tag{7.196}$$

**Definition: Transition Probability Matrix** The transition probability matrix for a homogenous Markov chain is denoted $P$ and is defined as

$$P = \begin{bmatrix} X_{t_{k+1}} = s_1 / X_{t_k} = s_1 & X_{t_{k+1}} = s_2 / X_{t_k} = s_1 & \ldots & X_{t_{k+1}} = s_N / X_{t_k} = s_1 & \ldots \\ X_{t_{k+1}} = s_1 / X_{t_k} = s_2 & X_{t_{k+1}} = s_2 / X_{t_k} = s_2 & \ldots & X_{t_{k+1}} = s_N / X_{t_k} = s_2 & \ldots \\ \ldots & & & & \\ X_{t_{k+1}} = s_1 / X_{t_k} = s_N & X_{t_{k+1}} = s_2 / X_{t_k} = s_N & \ldots & X_{t_{k+1}} = s_N / X_{t_k} = s_N & \ldots \\ \ldots & & & & \end{bmatrix} \tag{7.197}$$

Using the notation of Equation 7.196, $P$ can be written as

$$P = \left[p_{ij}\right] = \begin{bmatrix} p_{11} & p_{12} & \cdots & p_{1N} & \cdots \\ p_{21} & p_{22} & \cdots & p_{2N} & \cdots \\ \ldots & & & & \\ p_{N1} & p_{N2} & \cdots & p_{NN} & \cdots \\ \ldots & & & & \end{bmatrix} \tag{7.198}$$

#### *7.9.3.1 Example: Random Walk*

A random walk, $X$, with states $\{0, \pm 1, \pm 2, \ldots\}$ and with transition probabilities

$$p_{ij} = P[X_{t_{k+1}} = j / X_{t_k} = i] = \begin{cases} p & j = i+1 \\ 1-p & j = i-1 \\ 0 & \text{elsewhere} \end{cases} \quad k \in \{0,1,2,\ldots\} \tag{7.199}$$

is a homogenous Markov chain. The transition probability matrix is

$$P = \begin{bmatrix} \cdots & & \cdots & & \cdots & \\ \cdots & p_{-1,-2} & p_{-1,-1} & p_{-1,0} & p_{-1,1} & \cdots \\ \cdots & p_{0,-2} & p_{0,-1} & p_{0,0} & p_{0,1} & \cdots \\ \cdots & p_{1,-2} & p_{1,-1} & p_{1,0} & p_{1,1} & p_{1,2} \\ \cdots & & & & & \end{bmatrix} = \begin{bmatrix} \cdots & & \cdots & & \cdots & & \\ \cdots & 1-p & 0 & p & \cdots & & \\ \cdots & 0 & 1-p & 0 & p & \cdots & \\ \cdots & 0 & 0 & 1-p & 0 & p & \\ \cdots & & & & & & \end{bmatrix} \tag{7.200}$$

***7.9.3.2 Example: Ehrenfest Model*** Consider $m$ particles, as illustrated in Figure 7.18, that are distributed on either side of a membrane and that diffuse from one side to the other. When the probability of a particle diffusing from one side to the other side is proportional to the number of particles on the first side, the Ehrenfest model is defined. Consider a random process $X$ defined as the number of particles on the right side of the membrane. An increase in $X$ from a level $i$ is consistent with one of the $m-i$ particles moving from the left side to the right side and a decrease in $X$ from a level $i$ is consistent with one of the $i$ particles moving from the right side to the left side. With the Ehrenfest model, the transition probabilities, for $i,j \in \{0,1,\ldots,m\}$ are

$$p_{ij} = P[X_{t_{k+1}} = j / X_{t_k} = i] = \begin{cases} \dfrac{m-i}{m} = 1 - \dfrac{i}{m} & j = i+1 \\ \dfrac{i}{m} & j = i-1 \\ 0 & \text{otherwise} \end{cases} \quad (7.201)$$

The random process $X$ is a homogenous Markov chain with a transition probability matrix:

$$P = \begin{bmatrix} p_{00} & p_{01} & p_{02} & \cdots & p_{0m} \\ p_{10} & p_{11} & p_{12} & \cdots & p_{1m} \\ p_{20} & p_{21} & p_{22} & \cdots & p_{2m} \\ p_{m-1,0} & & \cdots & p_{m-1,m-1} & p_{m-1,m} \\ p_{m0} & & \cdots & p_{m,m-1} & p_{mm} \end{bmatrix} = \begin{bmatrix} 0 & 1 & 0 & \cdots & 0 \\ 1/m & 0 & 1-1/m & \cdots & 0 \\ 0 & 2/m & 0 & \cdots & 0 \\ 0 & & \cdots & 0 & 1/m \\ 0 & & \cdots & 1 & 0 \end{bmatrix}$$

(7.202)

### 7.9.4 N-Step Transition Probabilities for Homogenous Case

For a homogenous Markov chain, the following definitions can be made.

**Definition: N-Step Transition Probability** The $n$-step transition probability, denoted $p_{ij}(n)$, is

$$P_{ij}(n) = P\left[X_{t_{k+n}} = s_j / X_{t_k} = s_i\right] \quad k,n \in Z^+ \quad (7.203)$$

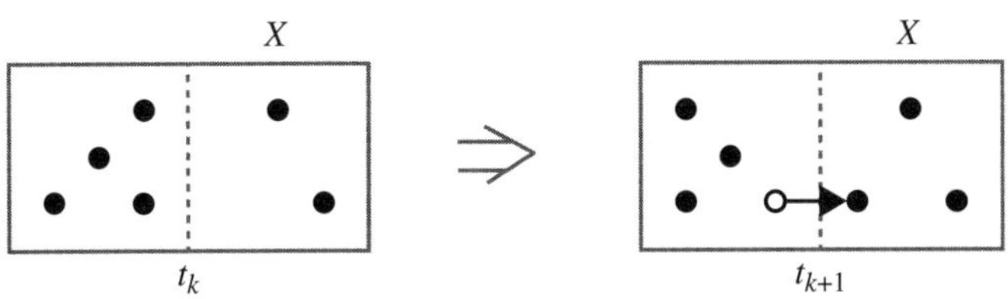


**FIGURE 7.18** Diffusion of particles from one side of a membrane to the other side.

**Definition: N-Step Transition Probability Matrix** The $n$-step transition probability matrix, denoted $P(n)$, is

$$P(n) = [p_{ij}(n)] \tag{7.204}$$

The following results then arise.

**Theorem 7.22 Basic Properties**
For a homogenous Markov chain with $M$ states, it is the case that $p_{ij}(n) \geq 0$ and

$$\sum_{j=1}^{M} p_{ij}(n) = 1 \quad i \in \mathbf{Z}^{+} \tag{7.205}$$

**Proof**
By definition of a probability, $p_{ij}(n) \geq 0$. The second result follows because the summation is over all possible paths, in $n$ steps, from the $i$th state. Formally,

$$\begin{aligned}\sum_{j=1}^{M} p_{ij}(n) &= \sum_{j=1}^{M} P[X_{t_{k+n}} = s_j / X_{t_k} = s_i] = \sum_{j=1}^{M} \frac{P[X_{t_{k+n}} = s_j, X_{t_k} = s_i]}{P[X_{t_k} = s_i]} \\ &= \frac{1}{P[X_{t_k} = s_i]} \sum_{j=1}^{M} P[X_{t_k} = s_i / X_{t_{k+n}} = s_j] P[X_{t_{k+n}} = s_j] = \frac{P[X_{t_k} = s_i]}{P[X_{t_k} = s_i]} = 1\end{aligned} \tag{7.206}$$

The second last equality arises from the Theorem of Total Probability and as the states $s_1, s_2, \ldots$ form a partition of the sample space for the $(k+n)$th sample time.

*7.9.4.1 Chapman–Kolmogorov Equation* Consider the possible paths between the $i$th state at time $t_k$ and the $j$th state at time $t_{k+m+n}$ as illustrated in Figure 7.19.

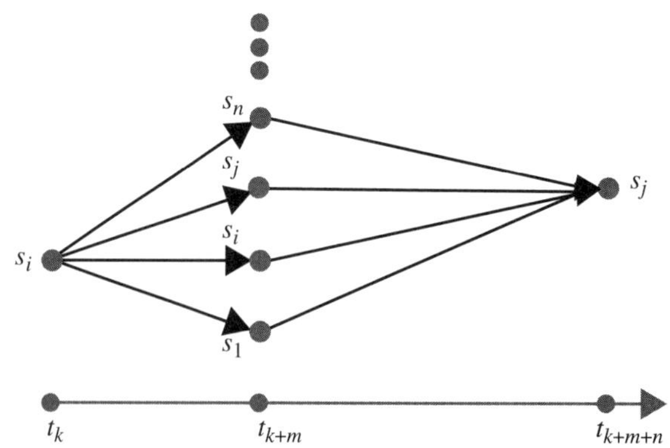


**FIGURE 7.19** Possible paths for a discrete time–discrete state random process between the $i$th state at time $t_k$ time and the $j$th state at time $t_{k+m+n}$.

**Theorem 7.23 Chapman–Kolmogorov Equation**
For a discrete time–discrete state random process with $M$ states, and which satisfies the Markov property of the transition probabilities being dependent on the current state and not prior states, it is the case that

$$P\left[X_{t_{k+m+n}}=s_j/X_{t_k}=s_i\right]=\sum_{q=1}^{M}P\left[X_{t_{k+m}}=s_q/X_{t_k}=s_i\right]P\left[X_{t_{k+m+n}}=s_j/X_{t_{k+m}}=s_q\right] \quad (7.207)$$

For the homogenous Markov chain case and with simplified notation,

$$p_{ij}(m+n)=\sum_{q=1}^{M}p_{iq}(m)p_{qj}(n) \quad (7.208)$$

**Proof**
The proof is detailed in Appendix 7.L.

### *7.9.4.2 N-Step Transition Probability Matrix*

**Theorem 7.24 N-Step Transition Probability Matrix**
For a homogenous Markov chain, the $n$-step transition probability matrix can be evaluated according to

$$P(n)=P^n(1) \quad (7.209)$$

**Proof**
The proof relies on the Chapman–Kolmogorov equation and is detailed in Appendix 7.M.

***7.9.4.3 Example*** Consider an $N$-stage communication system, as illustrated in Figure 7.20, with identical stages and with binary data being transmitted. The probabilities characterizing each stage are

$$\begin{array}{ll} p_{00}=P[0\text{ received}/0\text{ sent}] & p_{01}=P[1\text{ received}/0\text{ sent}] \\ p_{10}=P[0\text{ received}/1\text{ sent}] & p_{11}=P[1\text{ received}/1\text{ sent}] \end{array} \quad (7.210)$$

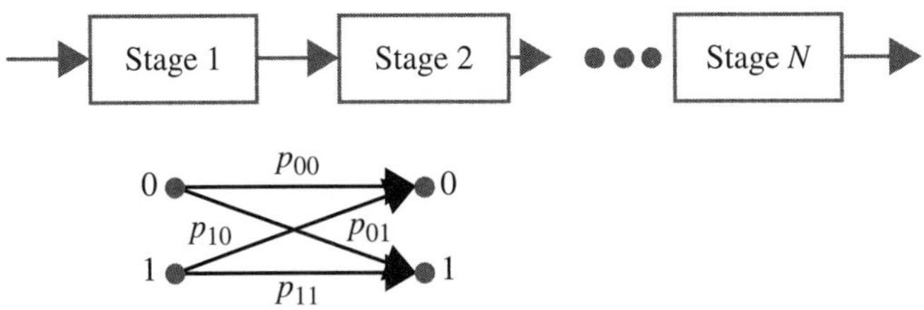


**FIGURE 7.20** Top: $N$-stage communication system. Bottom: transmission matrix for each stage.

and define the transmission matrix illustrated in Figure 7.20. The input data is characterized according to

$$P[0 \text{ input to stage } 1] = p_0 \quad P[1 \text{ input to stage } 1] = p_1 \tag{7.211}$$

with $p_0 + p_1 = 1$.

An important measure of the communication system is the probability of error in transmission after $N$ stages, which is defined as

$$\begin{aligned} P_E = &P[1 \text{ received after } N \text{ stages}/0 \text{ sent}]p_0 \\ &+ P[0 \text{ received after } N \text{ stages}/1 \text{ sent}]p_1 \end{aligned} \tag{7.212}$$

A simple approach to finding the probability of error is to note that the communication link defines an $N$-stage homogenous Markov chain where each stage has two possible states with a transition probability matrix of

$$P(1) = \begin{bmatrix} p_{00} & p_{01} \\ p_{10} & p_{11} \end{bmatrix} \tag{7.213}$$

It then follows, from Theorem 7.24, that the transmission probability matrix for $N$ stages is

$$P(N) = \begin{bmatrix} p_{00}(N) & p_{01}(N) \\ p_{10}(N) & p_{11}(N) \end{bmatrix} = \begin{bmatrix} p_{00} & p_{01} \\ p_{10} & p_{11} \end{bmatrix}^N \tag{7.214}$$

and the probability of error is

$$P_E = p_{01}(N)p_0 + p_{10}(N)p_1 \tag{7.215}$$

For the case of $p_0 = p_1 = 0.5$, $p_{01} = 10^{-5}$, and $p_{10} = 10^{-6}$, the probability of error is shown in Figure 7.21.

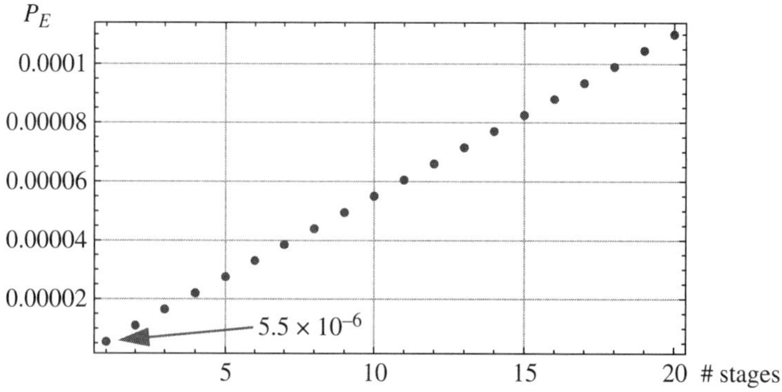


**FIGURE 7.21** Probability of error in a binary communication system as the number of stages varies.

### 7.9.5 PMF Evolution for Homogenous Markov Chain

For a homogenous Markov chain, the $n$-step transition probability matrix $P(n)=P^n(1)$ facilitates establishing the evolution of the probability mass function once the initial state probabilities are known. The initial states, associated with the time $t_0$, define a random variable $X_{t_0}$ with outcomes $s_1, \ldots, s_i, \ldots$ and a probability mass function

$$P[X_{t_0}=s_i]=p_{X_{t_0}}(s_i) \tag{7.216}$$

The $n$th transition time, $t_n$, defines a random variable $X_{t_n}$ with outcomes $s_1, \ldots, s_i, \ldots$ and a probability mass function

$$P[X_{t_n}=s_i]=p_{X_{t_n}}(s_i) \tag{7.217}$$

This probability mass function is specified in the following theorem.

**Theorem 7.25 PMF Evolution of a Homogenous Markov Chain**
The probability mass function evolution of the states in a homogenous Markov chain is specified by the vector of probabilities

$$P_n=\left[p_{X_{t_n}}(s_i)\right]=\left(p_{X_{t_n}}(s_1),p_{X_{t_n}}(s_2),\ldots,p_{X_{t_n}}(s_M)\right) \tag{7.218}$$

as $n$ varies assuming $M$ states. This vector of probabilities is given by

$$P_n=P_0P(n)=P_0P^n(1) \tag{7.219}$$

where

$$P_0=\left(p_{X_{t_0}}(s_1),p_{X_{t_0}}(s_2),\ldots,p_{X_{t_0}}(s_M)\right) \tag{7.220}$$

**Proof**
From the Theorem of Total Probability,

$$\begin{aligned} p_{X_{t_n}}(s_i) &= P[X_{t_n}=s_i]=\sum_{k=1}^{M}P[X_{t_n}=s_i/X_{t_0}=s_k]P[X_{t_0}=s_k] \\ &= \sum_{k=1}^{M}P[X_{t_0}=s_k]p_{ki}(n) \end{aligned} \tag{7.221}$$

which is equal to the $i$th term arising from the matrix multiplication $P_0P(n)$.

***7.9.5.1 Example*** Consider a one-step processing system, as illustrated in Figure 7.22, with one level of redundancy such that the state diagram, as detailed in Figure 7.23, is appropriate. The state diagram arises from the state assignment detailed in Table 7.2. It is assumed that there is a time between when a unit fails and when it is taken for repair, and this time varies depending on the nature of the fault. It is also assumed that the repair time varies with the nature of the fault.

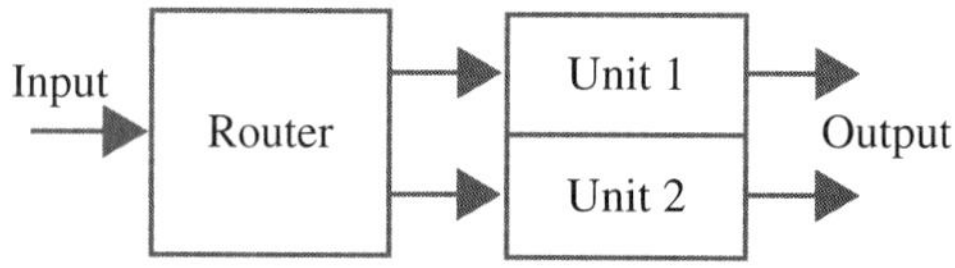


**FIGURE 7.22** A one-step processing system with one level of redundancy. The router directs the input to unit one when it is operational and to unit 2 when unit 1 has failed.

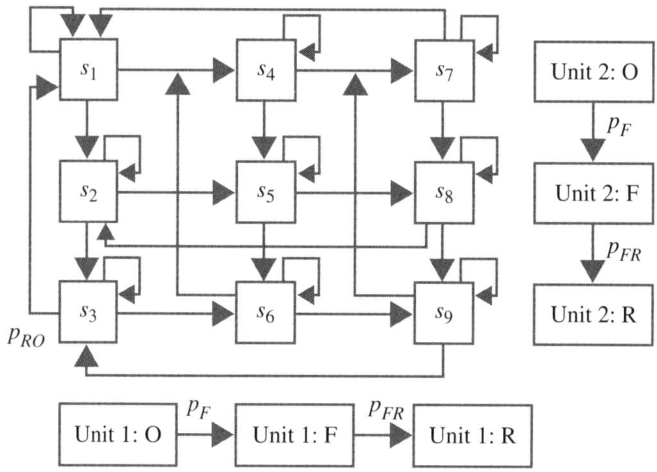


**FIGURE 7.23** State diagram for system. O, F, and R stand, respectively, for operational, fail, and repair.

**TABLE 7.2 State assignment for the system of Figure 7.22**

| State | Unit 1 | Unit 2 | System status |
|---|---|---|---|
| $s_1$ | Operational, in use | Operational, standby | Operational |
| $s_2$ | | Failed | Operational |
| $s_3$ | | Being repaired | Operational |
| $s_4$ | Failed | Operational, in use | Operational |
| $s_5$ | | Failed | Nonoperational |
| $s_6$ | | Being repaired | Nonoperational |
| $s_7$ | Being repaired | Operational, in use | Operational |
| $s_8$ | | Failed | Nonoperational |
| $s_9$ | | Being repaired | Nonoperational |

For the case where the system state is updated at set rate of $1/\Delta t$, the transition probability matrix is

$$P(1)=\begin{bmatrix} p_{11} & p_{12} & 0 & p_{14} & 0 & 0 & 0 & 0 & 0 \\ 0 & p_{22} & p_{23} & 0 & p_{25} & 0 & 0 & 0 & 0 \\ p_{31} & 0 & p_{33} & 0 & 0 & p_{36} & 0 & 0 & 0 \\ 0 & 0 & 0 & p_{44} & p_{45} & 0 & p_{47} & 0 & 0 \\ 0 & 0 & 0 & 0 & p_{55} & p_{56} & 0 & p_{58} & 0 \\ 0 & 0 & 0 & p_{64} & 0 & p_{66} & 0 & 0 & p_{69} \\ p_{71} & 0 & 0 & 0 & 0 & 0 & p_{77} & p_{78} & 0 \\ 0 & p_{82} & 0 & 0 & 0 & 0 & 0 & p_{88} & p_{89} \\ 0 & 0 & p_{93} & 0 & 0 & 0 & p_{97} & 0 & p_{99} \end{bmatrix} \quad (7.222)$$

The following definitions, associated with a time interval of duration $\Delta t$ seconds, apply: $p_F$ is the fault probability given initial operation status, $p_{FR}$ is the probability of moving to a repair state given an initial failed status state, and $p_{RO}$ is the probability of moving to an operational state given an initial repair state. With these definitions, it then follows that

$$p_{14}=p_{25}=p_{36}=p_{12}=p_{45}=p_{78}=p_F \quad (7.223)$$

$$p_{23}=p_{56}=p_{89}=p_{47}=p_{58}=p_{69}=p_{FR} \quad (7.224)$$

$$p_{31}=p_{64}=p_{97}=p_{71}=p_{82}=p_{93}=p_{RO} \quad (7.225)$$

$$p_{11}=1-2p_F, p_{22}=p_{44}=1-p_F-p_{FR}, p_{33}=p_{77}=1-p_F-p_{RO} \quad (7.226)$$

$$p_{55}=1-2p_{FR}, p_{66}=p_{88}=1-p_{FR}-p_{RO}, p_{99}=1-2p_{RO} \quad (7.227)$$

The evolution of the state probabilities with time, as given by $P_0P^n(1)$, is shown in Figure 7.24 for the case of $p_F=0.1$ (an unrealistically high value but a value that demonstrates the nature of the state probability evolution), $p_{FR}=0.4$, $p_{RO}=0.333$, and $P_0=(1,0,0,0,0,0,0,0,0)$.

Of interest is the probability that the system is operational. At the time $t_n=n\Delta t$, this is given by (see Table 7.2)

$$p_{\text{operational}}(t_n)=p_{X_{t_n}}(s_1)+p_{X_{t_n}}(s_2)+p_{X_{t_n}}(s_3)+p_{X_{t_n}}(s_4)+p_{X_{t_n}}(s_7) \quad (7.228)$$

The evolution with time of the probability of the system being nonoperational is shown in Figure 7.25 for the case of $p_F=0.0001$, $p_{FR}=0.4$, $p_{RO}=0.33$, and $P_0=(1,0,0,0,0,0,0,0,0)$.

### 7.9.6 First Passage Time for Homogenous Markov Chain

The follow definitions underpin the determination of characteristics related to a state being reached for the first time.

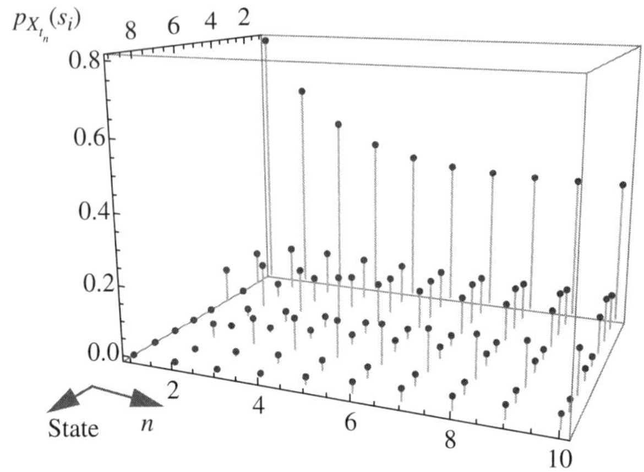


**FIGURE 7.24** Probability mass function evolution with the number of time intervals of duration $\Delta t$. Here, $t_n = n\Delta t$.

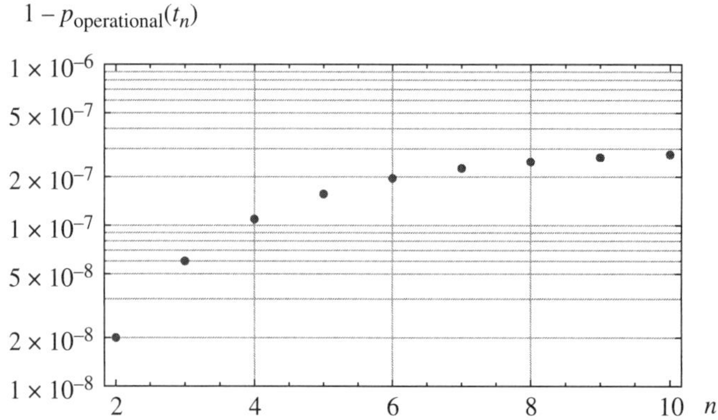


**FIGURE 7.25** Probability of system being nonoperational for the case of $p_F = 0.0001$, $p_{FR} = 0.4$, $p_{RO} = 0.33$, and $t_n = n\Delta t$.

**Definition: First Passage Time to State from Specified State** The probability that a homogenous Markov chain reaches the state $s_j$ from the state $s_i$ for the first time after $n$ steps is denoted $p_{ij}^F(n)$. This probability is given by

$$p_{ij}^F(n) = P\left[X_{t_n} = s_j, X_{t_{n-1}} \neq s_j, \ldots, X_{t_1} \neq s_j / X_{t_0} = s_i\right] \tag{7.229}$$

**Definition: First Passage Time to State** The probability that a homogenous Markov chain reaches the state $s_j$ from the state $s_i$ at some future time is denoted $p_{ij}^F$. This probability is given by

$$p_{ij}^F = \sum_{n=1}^{\infty} p_{ij}^F(n) \tag{7.230}$$

**Definition: Time of First Visit** For a given trial of a Markov chain, the time of the first visit to state $s_j$ from the state $s_i$ is denoted $t_{ij}$ and is given by

$$t_{ij} = t_n - t_0: \quad n = \min\left\{k: \; X_{t_k} = s_j / X_{t_0} = s_i\right\} \tag{7.231}$$

## 7.10 TIME SERIES CHARACTERIZATION

Time series characterization is an important area, and AR, MA, and ARIMA models find widespread application. Brillinger (2001), for example, provides a good introduction.

## 7.11 PROBLEMS

**7.1** Consider an experiment with a sample space $S = \{\omega : \omega \in \{1,2,\ldots\}\}$ and where the probability of an outcome is specified by the probability mass function $p_\Omega(\omega) = 1/2^\omega$. A random process is defined on the experiment according to

$$X : S \to S_X$$

$$X(\omega,t) = \begin{cases} A_o/\omega & t \in [0,T] \\ 0 & \text{elsewhere} \end{cases} \tag{7.232}$$

a. Explicitly define $S_X$.
b. Define the following associated random processes:
   i. The random process defined by the autocorrelation function
   ii. The random process defined by the time-averaged autocorrelation function
   iii. The random process defined by the Fourier transform
   iv. The random process defined by the power spectral density function
c. Determine the time-averaged autocorrelation function and the power spectral density function for the random process.
d. Show that the Fourier transform of the time-averaged autocorrelation function equals the power spectral density.
e. Determine the average power of the random process. Show that the average power equals the value of the time-averaged autocorrelation function with an argument of zero and is also equal to the integral of the power spectral density function.

Note that $\sum_{k=1}^{\infty} \frac{1}{k^2 2^k} = \frac{\pi^2}{12} - \frac{1}{2} \cdot [\ln(2)]^2$ (Gradsteyn and Ryzhik, 1980, eq. 0.241.)

**7.2** Consider a system, where on power down, a transient signal component is evident. This is modelled by the random process $X:S\rightarrow S_X$ according to

$$S=\{\omega:\omega\in \boldsymbol{R}^+\} \quad f_\Omega(\omega)=\frac{\omega}{\sigma_\omega^2}e^{-\omega^2/2\sigma_\omega^2}, \omega>0$$
$$X(\omega,t)=\omega p(t) \tag{7.233}$$

for a causal pulse function $p$.

a. Determine the mean and variance of the random process.
b. Determine an expression for the probability density function evolution with time of the random process.
c. For the case of $p(t)=\sin(2\pi f_C t)e^{-t/\tau}u(t), f_C=2$, $\tau=1$, and $\sigma_\omega^2=1$, graph the mean, variance, and probability density function evolution with time.

**7.3** Consider a sinusoidal signal with a phase component that is randomly varying and that can be approximated by a random walk. The set of possible sinusoids defines a random process according to

$$X(\Omega,t)=A_o\sin\left[2\pi f_o t+\sum_{k=1}^{\infty}\Omega_k u(t-k\Delta t)\right]u(t) \tag{7.234}$$

where $\Omega=(\Omega_1,\Omega_2,\ldots)$ and $\Omega_1$, $\Omega_2$, … are independent and identical random variables with a sample space $\{-\phi_o,\phi_o\}$ and with $P[\Omega_k=-\phi_o]=P[\Omega_k=\phi_o]=0.5$. The following result is useful: for a random walk with a step size of $\Delta$x, after $N$ steps, the possible levels are $-N\Delta x,(-N+2)\Delta x,\ldots,(N-2)\Delta x,N\Delta$ and the probability of the level $k\Delta x$ is

$$P[\text{level}=k\Delta x]=\frac{N!}{\left(\frac{N+k}{2}\right)!\left(\frac{N-k}{2}\right)!}\cdot\frac{1}{2^N} \quad k\in\{-N,-N+2,\ldots,N\}$$

a. Determine analytical expressions for the mean, mean squared value, and variance of the random process.
b. Determine an analytical expression for the autocorrelation function of the random process. To determine the joint probability mass function of the random walk at the times $t$ and $t+\tau$, it is useful to use the result

$$P[X_1=x_1,X_2=x_2]=P[X_2=x_2/X_1=x_1]P[X_1=x_1] \tag{7.235}$$

c. Note that the expressions for the mean, variance, and autocorrelation function allow the correlation coefficient of the random process at the times $t$ and $t+\tau$ to be determined.

d. For the case of $f_o = 1$, $A_o = 1$, $\Delta t = 1/10f_o$, and $\phi_o = \pi/10$, graph examples of the phase variation with time, and graph the variation with time of the mean and variance. Graph the autocorrelation and correlation coefficient with respect to time and the delay $\tau$. By varying $\Delta t$ and $\phi_o$, the duration of the transient period, before the random process becomes approximately stationary, can be ascertained.

**7.4** Consider the case of a sinusoidal signal that is corrupted by additive noise such that a random process $X : S \to S_X$ results:

$$\begin{gathered} S = \{(\omega_1, \omega_2, \omega_3, \omega_4) : \omega_k \in \{-1, 1\}\} \\ P[\omega_k = -1] = 1 - p, P[\omega_k = 1] = p \end{gathered} \tag{7.236}$$

$$X((\omega_1, \omega_2, \omega_3, \omega_4), t) = A_o \sin(2\pi f_C t) + \sum_{k=1}^{4} A_N \sin\left(2\pi 2^{k-1} f_o t + \omega_k \phi_o\right)$$

a. For the case of $A_o = 1$, $A_N = 0.1$, $f_o = 1$, $\phi_o = \pi/4$, and $f_C = 8f_o$, graph some of the possible signals, and the noise components, from the random process.
b. Determine the average power of the signals in the random process and the average power of the random process. Assume an interval $T$ that is an integer multiple of $1/f_o$.
c. Establish expressions for the coefficients in the decomposition

$$X((\omega_1, \omega_2, \omega_3, \omega_4), t) = \sum_{i=-\infty}^{\infty} c_i(\omega_1, \omega_2, \omega_3, \omega_4) e^{j2\pi i f_o t} \tag{7.237}$$

on the interval $[0, 1/f_o]$ for the case of $f_C = 8f_o$. Hence, determine the average power of the random process as a summation over the expected value of the magnitude squared of the coefficients.

**7.5** Consider a random process $X : S \to S_X$ defined according to

$$\begin{aligned} &S = \{\omega_1, \omega_2, \omega_3\} \\ &P[\omega_1] = p_1, P[\omega_2] = p_2, P[\omega_3] = p_3 \\ &X(\omega, t) = C_1(\omega) b_1(t) + C_2(\omega) b_2(t) \quad t \geq 0 \end{aligned} \tag{7.238}$$

where $C_1$ and $C_2$ are random variables defined on $S$.

a. Determine general expressions for the mean and autocorrelation function of the random process.
b. If $C_1$ and $C_2$ are uncorrelated, then specify the autocorrelation function.

c. Consider the case of $p_1 = 1/2$, $p_2 = 1/4$, and $p_3 = 1/4$ and where the sample spaces of the random variables $C_1$ and $C_2$, respectively, are $\{1, -3, 1\}$ and $\{1, 0, -2\}$. Show that $C_1$ and $C_2$ are uncorrelated. Determine expressions for the mean and autocorrelation functions for this case. Is the random process stationary? What nonzero functions $b_1$ and $b_2$ would ensure stationarity?

d. Using the above specified parameters and assuming

$$b_1(t) = \sin(2\pi f_1 t)u(t) \quad b_2(t) = u(t) \tag{7.239}$$

graph the autocorrelation function for the case of $f_1 = 1$.

**7.6** Consider an experiment with a sample space

$$S\{(\omega_1, \omega_2): \ \omega_1 \in \boldsymbol{R}^+, \omega_2 \in \boldsymbol{R}^+\} \tag{7.240}$$

and with outcomes governed by a joint probability density function

$$f_{\Omega_1\Omega_2}(\omega_1, \omega_2) = f_{\Omega_1}(\omega_1) f_{\Omega_2}(\omega_2) \tag{7.241}$$

A random process $X: S \to S_X$ is defined on this sample space according to

$$X((\omega_1, \omega_2), t) = g_1(\omega_1) g_2(\omega_2, t) \tag{7.242}$$

for defined functions $g_1$ and $g_2$.

a. Determine general expressions for the mean, variance, and autocorrelation function of the random process.
   For the following parts, consider the specific definitions:

$$g_1(\omega_1) = \omega_1 \quad g_2(\omega_2, t) = t\sin(2\pi\omega_2 t)\exp(-t/\tau_2) \tag{7.243}$$

$$\begin{aligned} f_{\Omega_1}(\omega_1) &= \lambda_1 \exp(-\lambda_1 \omega_1) \quad \omega_1 > 0 \\ f_{\Omega_2}(\omega_2) &= \frac{\pi}{2f_o} \sin\left(\frac{\pi\omega_2}{f_o}\right) \quad 0 \le \omega_2 < f_o \end{aligned} \tag{7.244}$$

b. Determine expressions for the mean, variance, and autocorrelation function.

c. For the case of $\lambda_1 = 1, f_o = 1$, and $\tau_2 = 1$, graph signals from the random process.

d. For the above defined parameters, graph the mean, variance, autocorrelation function, and correlation coefficient.

e. For the above defined parameters, graph the effective correlation time for $\rho_o = 0.1$.

f. For the above defined parameters, graph the expected mean squared change.

**7.7** A random process $X$ is based on an experiment with outcomes

$$S = \{(\omega_1, \omega_2): \ \omega_1 \in [-\Delta f, \Delta f), \omega_2 \in [0, 2\pi)\} \tag{7.245}$$

and where the outcomes are governed by the probability density function $f_\Omega(\omega_1,\omega_2) = f_{\Omega_1}(\omega_1)\cdot f_{\Omega_2}(\omega_2)$, $\Omega = (\Omega_1,\Omega_2)$, $f_{\Omega_1}(\omega_1) = 1/2\Delta f$, and $f_{\Omega_2}(\omega_2) = 1/2\pi$. The random process is defined according to

$$X((\omega_1,\omega_2),t) = A_o \sin(2\pi(f_1+\omega_1)t+\omega_2) \qquad t\in[-T,T] \tag{7.246}$$

a. Determine the mean and autocorrelation function of this random process, and confirm that it is a wide-sense stationary random process.
b. Using the Cramer transform, determine the integrated spectrum and integrated power spectrum for the random process. For the latter, leave your result in an integral form.
c. For the case of $T=10$, $A_o=1$, $f_1=1$, and $\Delta f=1/2$, use numerical integration to graph the integrated power spectrum and the associated power spectral density.

**7.8** Consider the case of a progressive disease in a specific species with no known cure but with several effective treatments. The initial treatment is available when the disease is diagnosed and classed as being in stage 1. When the disease is classed as being in stage 2, a second treatment is available. When the disease is classed as being in stage 3, only new experimental treatments are available. The disease is such that the initial treatment is not effective a second time. The state diagram, as detailed in Figure 7.26, is appropriate. The state diagram arises from the state assignment detailed in Table 7.3. The transition probabilities listed in Figure 7.26 are for a specified time interval.

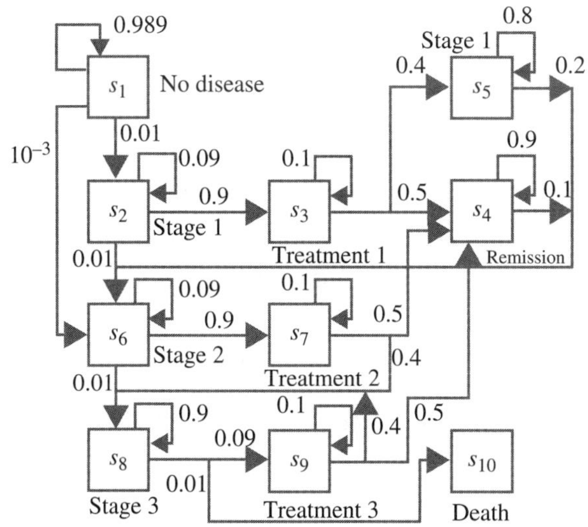


**FIGURE 7.26** State diagram for disease progression and state transition probabilities.

**TABLE 7.3 State assignment for states defined in Figure 7.26**

| State | Definition |
|---|---|
| $s_1$ | Prior to initial disease diagnosis |
| $s_2$ | Stage 1 of disease |
| $s_3$ | First treatment |
| $s_4$ | Remission phase |
| $s_5$ | First treatment unsuccessful |
| $s_6$ | Stage 2 of disease |
| $s_7$ | Second treatment |
| $s_8$ | Stage 3 of disease |
| $s_9$ | Experimental treatment |
| $s_{10}$ | Death |

a. Specify the transition probability matrix.
b. Graph the time evolution of the state probabilities assuming state 1 at $t = 0$.
c. Determine the probability that an entity from the species, chosen at random, has not been diagnosed with the disease, is in stage three of the disease, or has died after 10, 20, and 100 time intervals.

## APPENDIX 7.A PROOF OF THEOREM 7.2

The proof is given for the countable case; the proof for the uncountable case follows in an analogous manner. First, $X(\omega_i,t) \in \boldsymbol{L}^2[0,T]$ guarantees the existence of $\bar{R}(\omega_i,T,\tau)$. As $X(\omega_i,t) \in \boldsymbol{L}^2[0,T]$ implies $X(\omega_i,t) \in \boldsymbol{L}[0,T]$ (Theorem 2.13), it follows that the Fourier transform of $X(\omega_i, t)$, denoted $X(\omega_i, T, f)$, exists, and thus, $G(\omega_i, T, f)$ is well defined.

For the time-averaged autocorrelation function, consider the case of $\tau < 0$:

$$\begin{aligned}\bar{R}(T,\tau) &= \sum_{\omega_i \in S} p_\Omega(\omega_i)\left[\frac{1}{T}\int_0^{T+\tau} X(\omega_i,t)X^*(\omega_i,t-\tau)dt\right] \\ &\leq \sum_{\omega_i \in S} p_\Omega(\omega_i)\sqrt{\frac{1}{T}\int_0^{T+\tau}|X(\omega_i,t)|^2dt}\cdot\sqrt{\frac{1}{T}\int_0^{T+\tau}|X^*(\omega_i,t-\tau)|^2dt} \\ &\leq \sum_{\omega_i \in S} p_\Omega(\omega_i)\sqrt{\frac{1}{T}\int_0^{T}|X(\omega_i,t)|^2dt}\cdot\sqrt{\frac{1}{T}\int_0^{T}|X(\omega_i,t)|^2dt} = \bar{P}(T)\end{aligned} \tag{7.247}$$

where Schwarz's inequality (Theorem 2.15) has been used. Thus, the assumption of finite average power implies that $\bar{R}(T,\tau)$ is finite for $-T \leq \tau \leq T$, which implies $\bar{R}(T,\tau) \in \boldsymbol{L}[-T,T]$.

For the power spectral density, the assumption of finite average power implies:

$$\bar{P}(T) = \sum_{\omega_i \in S} p_\Omega(\omega_i)\left[\frac{1}{T}\int_0^T |X(\omega_i,t)|^2 dt\right] = \sum_{\omega_i \in S} p_\Omega(\omega_i)\left[\frac{1}{T}\int_{-\infty}^{\infty} |X(\omega_i,T,f)|^2 df\right]$$
$$= \sum_{\omega_i \in S} p_\Omega(\omega_i)\left[\int_{-\infty}^{\infty} G(\omega_i,T,f)df\right] = \int_{-\infty}^{\infty} \sum_{\omega_i \in S} p_\Omega(\omega_i) G(\omega_i,T,f)df \tag{7.248}$$

Here, Parseval's relationship (Theorem 3.9) has been used. The interchange of the summation and integral is valid, from the Fubini–Tonelli theorem (Theorem 2.16), as all terms are positive and the summation of the integral exists by the assumption of finite average power. Thus, $\bar{P}(T) = \int_{-\infty}^{\infty} G(T,f)df$, which implies $G(T,f) \in \boldsymbol{L}$ and, hence, the existence of $G(T,f)$ for all $f$ except, potentially, at a countable number of points.

## APPENDIX 7.B PROOF OF THEOREMS 7.3 AND 7.4

### 7.B.1 Autocorrelation Function: Theorem 7.3

Consider the countable case: by definition

$$R_\infty(\tau) = \lim_{T\to 0} \bar{R}(T,\tau) = \lim_{T\to 0} \sum_{\omega_i \in S} p_\Omega(\omega_i)\bar{R}(\omega_i,T,\tau) \tag{7.249}$$

Convergence is guaranteed as

$$\sup\{|\bar{R}(\omega_i,T,\tau)| : \ T,\tau \in \boldsymbol{R}^+, \omega_i \in S\} < \infty \tag{7.250}$$

has assumed to be finite. This assumption also ensures that the interchange of the limit and summation, consistent with the dominated convergence theorem (Theorem 2.19), is valid, that is,

$$R_\infty(\tau) = \sum_{\omega_i \in S} p_\Omega(\omega_i) \lim_{T\to 0} \bar{R}(\omega_i,T,\tau) = \sum_{\omega_i \in S} p_\Omega(\omega_i) R_\infty(\omega_i,\tau) \tag{7.251}$$

As shown in Appendix 7.A, $\bar{R}(T,\tau)$ is finite if $X(\omega_i,t) \in \boldsymbol{L}^2[0,T]$, and the average power in the random process $\bar{P}(T)$ is finite. These results extend to the infinite interval when $X(\omega_i,t) \in \boldsymbol{L}^2[0,T]$ for all $T \in \boldsymbol{R}^+$, $\omega_i \in S$, and when there is an upper bound on the signal powers over the interval $[0, \infty)$ as specified by Equation 7.71.

### 7.B.2 Power Spectral Density: Theorem 7.4

Consider the countable case:

$$G_\infty(f) = \lim_{T\to 0} \sum_{\omega_i \in S} p_\Omega(\omega_i) G(\omega_i, T, f) \tag{7.252}$$

Finiteness of $G_\infty(f)$ is guaranteed as

$$\sup\{G(\omega_i, T, f):\ T \in \boldsymbol{R}^+, \omega_i \in S\} < \infty \tag{7.253}$$

has been assumed. This assumption also ensures the validity of the interchange of the limit and summation, according to the dominated convergence theorem (Theorem 2.19), i.e.,

$$G_\infty(f) = \sum_{\omega_i \in S} p_\Omega(\omega_i) \lim_{T\to 0} G(\omega_i, T, f) = \sum_{\omega_i \in S} p_\Omega(\omega_i) G_\infty(\omega_i, f) \tag{7.254}$$

A sufficient condition for $G(\omega_i, T, f)$ to be bounded as $T \to \infty$ is for $X(\omega_i, t) \in \boldsymbol{L}[0, \infty]$. A stronger condition can be found by considering

$$G(\omega_i, T, f) = \int_{-T}^{T} \bar{R}(\omega_i, T, \tau) e^{-j2\pi f \tau} d\tau \le \int_{-T}^{T} |\bar{R}(\omega_i, T, \tau)| d\tau \le \int_{-\infty}^{\infty} |R_\infty(\omega_i, \tau)| d\tau \tag{7.255}$$

Hence, if $R_\infty(\omega_i, \tau) \in \boldsymbol{L}$, then $G(\omega_i, T, f)$ is bounded.

## APPENDIX 7.C PROOF OF THEOREM 7.5

First, consistent with Theorem 3.14, the assumption of finite energy, and piecewise differentiability, on $[0, T]$ for each signal, ensures that

$$G(\omega, T, f) = \int_{-T}^{T} \bar{R}(\omega, T, \tau) e^{-j2\pi f \tau} d\tau \qquad \bar{R}(\omega, T, \tau) = \int_{-\infty}^{\infty} G(\omega, T, f) e^{j2\pi f \tau} df \tag{7.256}$$

for all $\omega \in S$. Consider the countable case: to prove that the power spectral density is the Fourier transform of the time-averaged autocorrelation function, consider the integral

$$I_1(f) = \int_{-T}^{T} \bar{R}(T, \tau) e^{-j2\pi f \tau} d\tau = \int_{-T}^{T} \left[\sum_{\omega_i \in S} p_\Omega(\omega_i) \bar{R}(\omega_i, T, \tau)\right] e^{-j2\pi f \tau} d\tau \tag{7.257}$$

Consistent with Theorem 7.2, the assumption of finite average signal power on [0, *T*] implies that the summation is bounded above, that is, $\bar{R}(T,\tau)$ is finite. It then follows from the Fubini–Tonelli theorem that the order of the summation and the integral operation can be interchanged to yield

$$\begin{aligned} I_1(f) &= \sum_{i=1}^{\infty} p_\Omega(\omega_i) \int_{-T}^{T} \bar{R}(\omega_i, T, \tau) e^{-j2\pi f\tau} d\tau \\ &= \sum_{i=1}^{\infty} p_\Omega(\omega_i) G(\omega_i, T, f) = G(T, f) \end{aligned} \tag{7.258}$$

To prove that the inverse Fourier transform of the power spectral density function equals the time-averaged autocorrelation function, consider

$$I_2(\tau) = \int_{-\infty}^{\infty} G(T,f) e^{j2\pi f\tau} df = \int_{-\infty}^{\infty} \left[ \sum_{\omega_i \in S} p_\Omega(\omega_i) G(\omega_i, T, f) \right] e^{j2\pi f\tau} df \tag{7.259}$$

Consistent with Theorem 7.2, the assumption of finite average signal power on [0, *T*] implies that the summation is bounded above, that is, $G(T,f)$ is finite and, further, $G(T,f) \in \boldsymbol{L}$. It then follows from the Fubini–Tonelli theorem that the order of the summation and the integral operation can be interchanged to yield

$$I_2(\tau) = \sum_{\omega_i \in S} p_\Omega(\omega_i) \int_{-\infty}^{\infty} G(\omega_i, T, f) e^{j2\pi f\tau} df = \sum_{\omega_i \in S} p_\Omega(\omega_i) \bar{R}(\omega_i, T, \tau) = \bar{R}(T, \tau) \tag{7.260}$$

As $\bar{R}(T,\tau)$ is finite for all values of $\tau$, it follows that

$$\bar{R}(T,0) = \int_{-\infty}^{\infty} G(T,f) df = \bar{P}(T) \tag{7.261}$$

## APPENDIX 7.D PROOF OF THEOREM 7.6

First, the assumed conditions, consistent with Theorem 7.5, ensure that the $\bar{R}(T,\tau)$ and $G(T,f)$ exist for all $T \in \boldsymbol{R}^+$ and that they are related via the Fourier and inverse Fourier transforms.

To relate $G_\infty$ to $R_\infty$, consider

$$G_\infty(f) = \lim_{T\to\infty} G(T,f) = \lim_{T\to\infty} \int_{-T}^{T} \bar{R}(T,\tau) e^{-j2\pi f\tau} d\tau \tag{7.262}$$

It is necessary to interchange the limit and integration operations in this equation. To achieve this, first, note, consistent with Theorem 7.2, that $\bar{R} \in \boldsymbol{L}[-T,T]$. Second, assume there exists a function $B_R \in \boldsymbol{L}$ such that $\bar{R}(T,\tau) \le B_R(\tau)$ for all $T$. Then, from the dominated convergence theorem (Theorem 2.19), it follows that the order of limit and integral can be interchanged to yield

$$G_\infty(f) = \int_{-\infty}^{\infty} \lim_{T\to\infty} \bar{R}(T,\tau)e^{-j2\pi f\tau}d\tau = \int_{-\infty}^{\infty} R_\infty(\tau)e^{-j2\pi f\tau}d\tau \tag{7.263}$$

To relate $R_\infty$ to $G_\infty$, consider

$$R_\infty(\tau) = \lim_{T\to\infty} \bar{R}(T,\tau) = \lim_{T\to\infty} \int_{-\infty}^{\infty} G(T,f)e^{j2\pi f\tau}df \tag{7.264}$$

Again, it is necessary to interchange the limit and integration operations. However, and in general, the power spectral density function $G(T,f)$ will have impulsive components, and it is not possible to justify the interchange of the limit and integral operations. With the assumption of the existence of a function $B_G \in \boldsymbol{L}$ such that $G(T,f) \le B_G(f)$ for all $T$, it follows, from the dominated convergence theorem (Theorem 2.19), that the order of limit and integral can be interchanged to yield

$$R_\infty(\tau) = \int_{-\infty}^{\infty} \lim_{T\to\infty} G(T,f)e^{j2\pi f\tau}df = \int_{-\infty}^{\infty} G_\infty(f)e^{j2\pi f\tau}df \tag{7.265}$$

## APPENDIX 7.E PROOF OF THEOREM 7.11

Assume the countable case and consider $\boldsymbol{E}\left[C_u C_v^*\right]$:

$$\begin{aligned}
\boldsymbol{E}\left[C_u C_v^*\right] &= \sum_{\omega_i \in S} p_\Omega(\omega_i)\left[\frac{1}{T}\int_0^T X(\omega_i,\lambda)e^{-j2\pi u f_o \lambda}d\lambda\right]\left[\frac{1}{T}\int_0^T X^*(\omega_i,t)e^{j2\pi v f_o t}dt\right] \\
&= \frac{1}{T^2}\int_0^T\int_0^T \left[\sum_{\omega_i \in S} p_\Omega(\omega_i)X(\omega_i,\lambda)X^*(\omega_i,t)\right]e^{j2\pi(vt-u\lambda)f_o}d\lambda dt \\
&= \frac{1}{T^2}\int_0^T\int_0^T R_{XX}(\lambda,t)e^{j2\pi(vt-u\lambda)f_o}d\lambda dt \qquad R_{XX}(\lambda,t) = \boldsymbol{E}[X(\Omega,\lambda)X^*(\Omega,t)]
\end{aligned} \tag{7.266}$$

A change of variable $\tau = \lambda - t$ for $\lambda$ results in the area for integration as illustrated in Figure 7.6. The expectation can then be written as

$$\boldsymbol{E}\left[C_u C_v^*\right] = \frac{1}{T^2}\int_0^T \left[\int_{-t}^{T-t} R_{XX}(t+\tau,t)e^{j2\pi(vt-u(\tau+t))f_o}d\tau\right]dt$$
$$= \frac{1}{T^2}\int_0^T \left[\int_{-t}^{T-t} R_{XX}(t+\tau,t)e^{-j2\pi u f_o \tau}d\tau\right]e^{j2\pi(v-u)f_o t}dt \tag{7.267}$$

Define $I_R$ as

$$I_R = \sup\left\{\int_{-t}^{T-t} |R_{XX}(t+\tau,t)|d\tau : \quad t \in [0,T], T \in (0,\infty)\right\} \tag{7.268}$$

It then follows that

$$\left|\boldsymbol{E}\left[C_u C_v^*\right]\right| \leq \frac{1}{T^2}\int_0^T \left[\int_{-t}^{T-t} |R_{XX}(t+\tau,t)|d\tau\right]dt < \frac{I_R}{T} \tag{7.269}$$

which clearly converges to zero as $T \to \infty$.

## APPENDIX 7.F PROOF OF THEOREM 7.12

Consider the double summation consistent with the average power:

$$\bar{P} = \frac{1}{\beta-\alpha}\sum_{i=1}^{\infty}\sum_{k=1}^{\infty}\boldsymbol{E}\left[C_i C_k^*\right]\int_\alpha^\beta b_i(t)b_k^*(t)dt \tag{7.270}$$

Expanding the summation out yields

$$\bar{P} = \frac{1}{\beta-\alpha}\cdot\left[\boldsymbol{E}\left[|C_i|^2\right]\int_\alpha^\beta |b_1(t)|^2 dt\right.$$
$$+\boldsymbol{E}\left[|C_2|^2\right]\int_\alpha^\beta |b_2(t)|^2 dt + \boldsymbol{E}\left[C_2 C_1^*\right]\int_\alpha^\beta b_2(t)b_1^*(t)dt + \boldsymbol{E}\left[C_1 C_2^*\right]\int_\alpha^\beta b_1(t)b_2^*(t)dt + \cdots$$
$$+\boldsymbol{E}\left[|C_i|^2\right]\int_\alpha^\beta |b_i(t)|^2 dt + \boldsymbol{E}\left[C_i C_1^*\right]\int_\alpha^\beta b_i(t)b_1^*(t)dt + \cdots + \boldsymbol{E}\left[C_i C_{i-1}^*\right]\int_\alpha^\beta b_i(t)b_{i-1}^*(t)dt$$
$$\left.+\boldsymbol{E}\left[C_{i-1}C_i^*\right]\int_\alpha^\beta b_{i-1}(t)b_i^*(t)dt + \cdots + \boldsymbol{E}\left[C_1 C_i^*\right]\int_\alpha^\beta b_1(t)b_i^*(t)dt + \cdots\right] \tag{7.271}$$

and the following requirement then follows for the off-diagonal terms to be negligible:

$$\left|\sum_{k=1}^{i-1} 2\mathrm{Re}\left[\boldsymbol{E}\left[C_i C_k^*\right]\int_{\alpha}^{\beta} b_i(t) b_k^*(t)dt\right]\right| << \boldsymbol{E}\left[|C_i|^2\right]\int_{\alpha}^{\beta}|b_i(t)|^2 dt \quad i \in \{2,3,\ldots\} \quad (7.272)$$

## APPENDIX 7.G PROOF OF THEOREM 7.16

Consider the countable case:

$$\begin{aligned} R_{XX}(t+\tau,\lambda) &= \boldsymbol{E}[X(\Omega,t+\tau)X^*(\Omega,t)] \\ &= \sum_{\omega_i \in S} p_\Omega(\omega_i)\left[\sum_{u=-\infty}^{\infty} c_u(\omega_i)e^{j2\pi u f_o(t+\tau)}\right]\left[\sum_{v=-\infty}^{\infty} c_v(\omega_i)^* e^{-j2\pi v f_o t}\right] \\ &= \sum_{u=-\infty}^{\infty} \boldsymbol{E}\left[|C_u|^2\right]e^{j2\pi u f_o \tau} + \sum_{u=-\infty}^{\infty}\sum_{\substack{v=-\infty \\ u\neq v}}^{\infty} \boldsymbol{E}\left[C_u C_v^*\right]e^{j2\pi u f_o \tau}e^{j2\pi(u-v)f_o t} \end{aligned} \quad (7.273)$$

The assumption of wide-sense stationarity implies that the second summation is independent of $t$. Due to the harmonic nature of the terms in this summation, this is only possible if

$$\boldsymbol{E}\left[C_u C_v^*\right]e^{j2\pi u f_o \tau}e^{j2\pi(u-v)f_o t} + \boldsymbol{E}\left[C_{-u} C_{-v}^*\right]e^{-j2\pi u f_o \tau}e^{-j2\pi(u-v)f_o t} = 0 \quad (7.274)$$

Expanding the exponential terms out gives the condition

$$\begin{aligned} &\boldsymbol{E}\left[C_u C_v^*\right][\cos(2\pi(u-v)f_o t + k_\tau) + j\sin(2\pi(u-v)f_o t + k_\tau)] + \\ &\boldsymbol{E}\left[C_{-u} C_{-v}^*\right][\cos(2\pi(u-v)f_o t + k_\tau) - j\sin(2\pi(u-v)f_o t + k_\tau)] = 0 \end{aligned} \quad (7.275)$$

where $k_\tau = 2\pi u f_o \tau$. Due to the orthogonality of the sine and cosine functions, the requirement is for

$$\boldsymbol{E}\left[C_u C_v^*\right] + \boldsymbol{E}\left[C_{-u} C_{-v}^*\right] = 0 \quad \boldsymbol{E}\left[C_u C_v^*\right] - \boldsymbol{E}\left[C_{-u} C_{-v}^*\right] = 0 \quad (7.276)$$

and, hence, the condition that $\boldsymbol{E}\left[C_u C_v^*\right] = 0$ for $u \neq v$.

## APPENDIX 7.H PROOF OF THEOREM 7.17

Consider the countable case:

$$G_I(f) = \boldsymbol{E}\left[|X_C(\Omega,f)|^2\right] = \sum_{\omega_r \in S} p_\Omega(\omega_r)|X_C(\omega_r,f)|^2 \quad (7.277)$$

For the interval $[-T,T]$, the Cramer transform of $\sum_{i=-\infty}^{\infty} c_i(\omega_r)e^{j2\pi i f_o t}$ (see Theorem 3.29) is

$$X_C(\omega_r,f)=\sum_{i=-\infty}^{\infty}\frac{c_i(\omega_r)}{\pi}\cdot[\mathrm{Si}(i\pi)+\mathrm{Si}(2\pi fT-i\pi)] \tag{7.278}$$

and the approximation to this equation is

$$X_C(\omega_r,f)\approx\sum_{i=-\infty}^{\infty}\frac{c_i(\omega_r)}{2}[\mathrm{sgn}(i)+\mathrm{sgn}(f-if_o)] \tag{7.279}$$

Thus,

$$\begin{aligned}\boldsymbol{E}\left[|X_C(\Omega,f)|^2\right]&\approx\sum_{i=-\infty}^{\infty}\sum_{k=-\infty}^{\infty}\sum_{\omega_r\in S}p_\Omega(\omega_r)\cdot\frac{c_i(\omega_r)}{2}\cdot\frac{c_k^*(\omega_r)}{2}\cdot\\&\quad[\mathrm{sgn}(i)+\mathrm{sgn}(f-if_o)][\mathrm{sgn}(k)+\mathrm{sgn}(f-kf_o)]\\&=\sum_{i=-\infty}^{\infty}\sum_{k=-\infty}^{\infty}\frac{1}{4}\boldsymbol{E}[C_iC_k^*]\cdot\\&\quad[\mathrm{sgn}(i)+\mathrm{sgn}(f-if_o)][\mathrm{sgn}(k)+\mathrm{sgn}(f-kf_o)]\end{aligned} \tag{7.280}$$

For the stationary case $\boldsymbol{E}[C_iC_k^*]=0$ when $i\neq k$ (see Theorem 7.16) and, thus

$$\boldsymbol{E}\left[|X_C(\Omega,f)|^2\right]\approx\sum_{i=-\infty}^{\infty}\frac{1}{4}\boldsymbol{E}\left[|C_i|^2\right]\cdot[\mathrm{sgn}(i)+\mathrm{sgn}(f-if_o)]^2 \tag{7.281}$$

The exact expression follows in an analogous manner and is

$$\boldsymbol{E}\left[|X_C(\Omega,f)|^2\right]=\sum_{i=-\infty}^{\infty}\frac{\boldsymbol{E}\left[|C_i|^2\right]}{\pi^2}\cdot[\mathrm{Si}(i\pi)+\mathrm{Si}(2\pi fT-i\pi)]^2 \tag{7.282}$$

In Figure 7.27, the graph of $[\mathrm{sgn}(i)+\mathrm{sgn}(f-if_o)]^2$ is shown. It then follows that

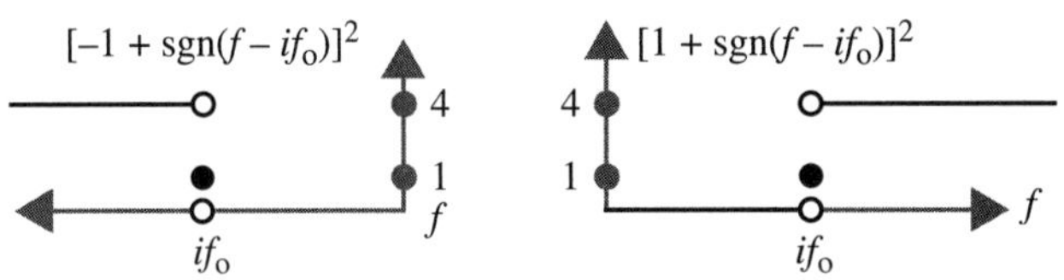


**FIGURE 7.27** Graph of $[\mathrm{sgn}(i)+\mathrm{sgn}(f-if_o)]^2$. Left: $i<0$. Right: $i>0$.

$$E\left[|X_C(\Omega,f)|^2\right] \approx \begin{cases} \sum_{i=0}^{\lfloor f/f_o \rfloor} \frac{E\left[|C_i|^2\right]}{4}[\text{sgn}(i) + \text{sgn}(f - if_o)]^2 & f > 0 \\ 0 & f = 0 \\ \sum_{i=-\lfloor |f|/f_o \rfloor}^{0} \frac{E\left[|C_i|^2\right]}{4}[\text{sgn}(i) + \text{sgn}(f - if_o)]^2 & f < 0 \end{cases} \tag{7.283}$$

## APPENDIX 7.I PROOF OF THEOREM 7.18

Denote $I = E[[X_C(\Omega,f_2) - X_C(\Omega,f_1)][X_C(\Omega,f_4) - X_C(\Omega,f_3)]^*]$ and consider the countable case:

$$I = \sum_{\omega_r \in S} p_\Omega(\omega_r)[X_C(\omega_r,f_2) - X_C(\omega_r,f_1)][X_C(\omega_r,f_4) - X_C(\omega_r,f_3)]^* \tag{7.284}$$

For an interval $[-T,T]$, $f_o = 1/2T$, it follows, from Theorem 3.32 for the case of $0 < f_1 < f_2$, that

$$X_C(\omega_r,f_2) - X_C(\omega_r,f_1) \approx \sum_{i=\lfloor f_1/f_o \rfloor}^{\lfloor f_2/f_o \rfloor} \frac{c_i(\omega_r)}{2}[\text{sgn}(f_2 - if_o) - \text{sgn}(f_1 - if_o)] \tag{7.285}$$

It then follows, for the case of $f_1 < f_2 < f_3 < f_4$, that

$$\begin{aligned} I \approx \sum_{\omega_r \in S} p_\Omega(\omega_r) & \left[\sum_{i=\lfloor f_1/f_o \rfloor}^{\lfloor f_2/f_o \rfloor} \frac{c_i(\omega_r)}{2}[\text{sgn}(f_2 - if_o) - \text{sgn}(f_1 - if_o)]\right] \cdot \\ & \left[\sum_{k=\lfloor f_3/f_o \rfloor}^{\lfloor f_4/f_o \rfloor} \frac{c_k(\omega_r)}{2}[\text{sgn}(f_4 - kf_o) - \text{sgn}(f_3 - kf_o)]\right]^* \end{aligned} \tag{7.286}$$

Interchanging the order of summations yields

$$\begin{aligned} I \approx & \sum_{i=\lfloor f_1/f_o \rfloor}^{\lfloor f_2/f_o \rfloor} \sum_{k=\lfloor f_3/f_o \rfloor}^{\lfloor f_4/f_o \rfloor} \sum_{\omega_r \in S} p_\Omega(\omega_r) \frac{c_i(\omega_r) c_k^*(\omega_r)}{4} \cdot \\ & [\text{sgn}(f_2 - if_o) - \text{sgn}(f_1 - if_o)][\text{sgn}(f_4 - kf_o) - \text{sgn}(f_3 - kf_o)] \\ = & \sum_{i=\lfloor f_1/f_o \rfloor}^{\lfloor f_2/f_o \rfloor} \sum_{k=\lfloor f_3/f_o \rfloor}^{\lfloor f_4/f_o \rfloor} \frac{E[C_i C_k^*]}{4} \cdot [\text{sgn}(f_2 - if_o) - \text{sgn}(f_1 - if_o)] \cdot \\ & [\text{sgn}(f_4 - kf_o) - \text{sgn}(f_3 - kf_o)] \end{aligned} \tag{7.287}$$

Wide-sense stationarity implies, consistent with Theorem 7.16, that $\boldsymbol{E}\left[C_i C_k^*\right]=0$ for $i \neq k$, and it then follows that $I \to 0$ as $T \to \infty, f_o \to 0$, as required.

It also follows from the result $\boldsymbol{E}\left[C_i C_k^*\right]=0$ for $i \neq k$, and for the case of $f_3=f_1, f_4=f_2$, that

$$\begin{aligned} I &\approx \sum_{i=\lfloor f_1/f_o \rfloor}^{\lfloor f_2/f_o \rfloor} \frac{\boldsymbol{E}\left[|C_i|^2\right]}{4}\left[\operatorname{sgn}(f_2 - if_o) - \operatorname{sgn}(f_1 - if_o)\right]^2 \\ &= \frac{\boldsymbol{E}\left[\left|C_{\lfloor f_2/f_o \rfloor}\right|^2\right]}{4} + \sum_{i=\lfloor f_1/f_o \rfloor+1}^{\lfloor f_2/f_o \rfloor-1} \boldsymbol{E}\left[|C_i|^2\right] + \frac{\boldsymbol{E}\left[\left|C_{\lfloor f_1/f_o \rfloor}\right|^2\right]}{4} \end{aligned} \tag{7.288}$$

assuming $f_1$ and $f_2$ are integer multiples of $f_o$. Consider the result from Theorem 7.17 for $f>0$:

$$\boldsymbol{E}\left[|X_C(\Omega,f)|^2\right] \approx \sum_{i=0}^{\lfloor f/f_o \rfloor} \frac{\boldsymbol{E}\left[|C_i|^2\right]}{4}\left[\operatorname{sgn}(i) + \operatorname{sgn}(f - if_o)\right]^2 \tag{7.289}$$

which, for $0<f_1<f_2$, implies

$$\begin{aligned} \boldsymbol{E}\left[|X_C(\Omega,f_2)|^2\right] - \boldsymbol{E}\left[|X_C(\Omega,f_1)|^2\right] \approx &\sum_{i=0}^{\lfloor f_2/f_o \rfloor} \frac{\boldsymbol{E}\left[|C_i|^2\right]}{4}\left[\operatorname{sgn}(i) + \operatorname{sgn}(f_2 - if_o)\right]^2 \\ &- \sum_{i=0}^{\lfloor f_1/f_o \rfloor} \frac{\boldsymbol{E}\left[|C_i|^2\right]}{4}\left[\operatorname{sgn}(i) + \operatorname{sgn}(f_1 - if_o)\right]^2 \end{aligned} \tag{7.290}$$

Thus,

$$\begin{aligned} \boldsymbol{E}\left[|X_C(\Omega,f_2)|^2\right] - \boldsymbol{E}\left[|X_C(\Omega,f_1)|^2\right] \approx &\frac{\boldsymbol{E}\left[\left|C_{\lfloor f_2/f_o \rfloor}\right|^2\right]}{4} + \sum_{i=\lfloor f_1/f_o \rfloor+1}^{\lfloor f_2/f_o \rfloor-1} \boldsymbol{E}\left[|C_i|^2\right] \\ &+ \frac{3\boldsymbol{E}\left[\left|C_{\lfloor f_1/f_o \rfloor}\right|^2\right]}{4} \end{aligned} \tag{7.291}$$

assuming $f_1$ and $f_2$ are integer multiples of $f_o$. As $T \to \infty, f_o \to 0$, and the significance of an individual term in the summation becomes increasingly small, it then follows from Equations 7.288 and 7.291 that

$$\begin{aligned} \boldsymbol{E}\left[|X_C(\Omega,f_2)|^2\right] - \boldsymbol{E}\left[|X_C(\Omega,f_1)|^2\right] &\approx \boldsymbol{E}\left[|X_C(\Omega,f_2) - X_C(\Omega,f_1)|^2\right] \\ &\approx \sum_{i=\lfloor f_1/f_o \rfloor}^{\lfloor f_2/f_o \rfloor} \boldsymbol{E}\left[|C_i|^2\right] \end{aligned} \tag{7.292}$$

as required.

## APPENDIX 7.J PROOF OF THEOREM 7.20

Consider the countable case and the definition of power:

$$\begin{aligned}\bar{P} &= \sum_{\omega_i \in S} p_\Omega(\omega_i) \left[ \frac{1}{2T} \int_{-T}^{T} |X(\omega_i, t)|^2 dt \right] \\ &= \sum_{\omega_i \in S} p_\Omega(\omega_i) \frac{1}{2T} \int_{-T}^{T} \left[ \left[ \int_{-\infty}^{\infty} e^{j2\pi ft} dX_C(\omega_i, f) \right] \right] \left[ \int_{-\infty}^{\infty} e^{-j2\pi\zeta t} dX_C^*(\omega_i, \zeta) \right] dt \\ &= \sum_{\omega_i \in S} p_\Omega(\omega_i) \frac{1}{2T} \int_{-T}^{T} \left[ \left[ \lim_{df \to 0} \sum_{k=-\infty}^{\infty} e^{j2\pi kdft} \left[ X_C\left( \omega_i, kdf + \frac{df}{2} \right) - X_C\left( \omega_i, kdf - \frac{df}{2} \right) \right] \right] \cdot \right. \\ &\quad \left. \left[ \lim_{d\zeta \to 0} \sum_{l=-\infty}^{\infty} e^{-j2\pi ld\zeta t} \left[ X_C^*\left( \omega_i, ld\zeta + \frac{d\zeta}{2} \right) - X_C^*\left( \omega_i, id\zeta - \frac{d\zeta}{2} \right) \right] \right] \right] dt \end{aligned} \tag{7.293}$$

Interchanging the order of summations, it follows that

$$\begin{aligned}\bar{P} &= \frac{1}{2T} \int_{-T}^{T} \left[ \lim_{df \to 0} \sum_{k=-\infty}^{\infty} \lim_{d\zeta \to 0} \sum_{l=-\infty}^{\infty} e^{j2\pi(kdf - ld\zeta)t} \sum_{\omega_i \in S} p_\Omega(\omega_i) \cdot \right. \\ &\quad \left. \left[ X_C\left( \omega_i, kdf + \frac{df}{2} \right) - X_C\left( \omega_i, kdf - \frac{df}{2} \right) \right] \left[ X_C^*\left( \omega_i, ld\zeta + \frac{d\zeta}{2} \right) - X_C^*\left( \omega_i, ld\zeta - \frac{d\zeta}{2} \right) \right] \right] dt \\ &= \frac{1}{2T} \int_{-T}^{T} \left[ \lim_{df \to 0} \sum_{k=-\infty}^{\infty} \lim_{d\zeta \to 0} \sum_{l=-\infty}^{\infty} e^{j2\pi(kdf - ld\zeta)t} \cdot \right. \\ &\quad \left. \boldsymbol{E}\left[ \left[ X\left( \Omega, kdf + \frac{df}{2} \right) - X_C\left( \Omega, kdf - \frac{df}{2} \right) \right] \left[ X_C^*\left( \Omega, ld\zeta + \frac{d\zeta}{2} \right) - X_C^*\left( \Omega, ld\zeta - \frac{d\zeta}{2} \right) \right] \right] \right] dt \end{aligned} \tag{7.294}$$

For $k \neq l$ and the same step size for $df$ and $d\zeta$, it is the case, consistent with Theorem 7.18, that

$$\begin{aligned} &\boldsymbol{E}\left[ \left[ X_C\left( \Omega, kdf + \frac{df}{2} \right) - X_C\left( \Omega, kdf - \frac{df}{2} \right) \right] \cdot \right. \\ &\quad \left. \left[ X_C\left( \Omega, ld\zeta + \frac{d\zeta}{2} \right) - X_C\left( \Omega, ld\zeta - \frac{d\zeta}{2} \right) \right]^* \right] = 0 \end{aligned} \tag{7.295}$$

Hence,

$$\bar{P} = \frac{1}{2T}\int_{-T}^{T}\left[\lim_{df\to 0}\sum_{k=-\infty}^{\infty}\boldsymbol{E}\left[\left|X_C\left(\Omega, kdf + \frac{df}{2}\right) - X_C\left(\Omega, kdf - \frac{df}{2}\right)\right|^2\right]\right]dt$$
$$= \lim_{df\to 0}\sum_{k=-\infty}^{\infty}\boldsymbol{E}\left[\left|X_C\left(\Omega, kdf + \frac{df}{2}\right) - X_C\left(\Omega, kdf - \frac{df}{2}\right)\right|^2\right] \quad (7.296)$$

From Theorem 7.18, it then follows that

$$\bar{P} = \lim_{df\to 0}\sum_{k=-\infty}^{\infty}\left|\boldsymbol{E}\left|X_C\left(\Omega, kdf + \frac{df}{2}\right)\right|^2 - \boldsymbol{E}\left|X_C\left(\Omega, kdf - \frac{df}{2}\right)\right|^2\right|$$
$$= \int_{-\infty}^{\infty}\left|d\boldsymbol{E}\left[|X_C(\Omega, f)|^2\right]\right| \quad (7.297)$$

which is the required result.

## APPENDIX 7.K PROOF OF THEOREM 7.21

Consider the alternative trigonometric form for $X(\Omega, t)$:

$$X(\Omega, t) = \sum_{i=1}^{N}\frac{\sqrt{2}A_o}{\sqrt{N}}[\sin(2\pi F_i t)\cos(\Phi_i) + \cos(2\pi F_i t)\sin(\Phi_i)] \quad (7.298)$$

The Cramer transforms, respectively, of $\sin(2\pi f_i t)$ and $\cos(2\pi f_i t)$ for the interval $[-T, T]$ are specified in Theorem 3.28:

$$X_C(f) = \begin{cases} \frac{j}{2\pi}[\mathrm{Si}(2\pi(f+f_i)T) - \mathrm{Si}(2\pi(f-f_i)T) - 2\mathrm{Si}(2\pi f_i T)] & \text{sine} \\ \frac{1}{2\pi}[\mathrm{Si}(2\pi(f+f_i)T) + \mathrm{Si}(2\pi(f-f_i)T)] & \text{cosine} \end{cases} \quad (7.299)$$

and for the infinite interval (Theorem 3.27):

$$X_C(f) = \begin{cases} j\cdot\frac{\mathrm{sgn}(f+f_i) - \mathrm{sgn}(f-f_i) - 2\,\mathrm{sgn}(f_i)}{4} & \text{sine} \\ \frac{\mathrm{sgn}(f+f_i) + \mathrm{sgn}(f-f_i)}{4} & \text{cosine} \end{cases} \quad (7.300)$$

For the case of $T >> 1/2f_i$, the infinite interval result is a good approximation.

The associated random process defined by the Cramer transform of each signal of $X(\Omega, t)$ is

$$X_C(\Omega,f) = \frac{j\sqrt{2}A_o}{2\pi\sqrt{N}}\sum_{i=1}^{N}\cos(\Phi_i)[\mathrm{Si}(2\pi(f+F_i)T)-\mathrm{Si}(2\pi(f-F_i)T)-2\mathrm{Si}(2\pi F_i T)] + \frac{\sqrt{2}A_o}{2\pi\sqrt{N}}\sum_{i=1}^{N}\sin(\Phi_i)[\mathrm{Si}(2\pi(f+F_i)T)+\mathrm{Si}(2\pi(f-F_i)T)] \quad (7.301)$$

and this random process can be approximated according to

$$X_C(\Omega,f) \approx \frac{j\sqrt{2}A_o}{4\sqrt{N}}\sum_{i=1}^{N}\cos(\Phi_i)[\mathrm{sgn}(f+F_i)-\mathrm{sgn}(f-F_i)-2\,\mathrm{sgn}(F_i)] + \frac{\sqrt{2}A_o}{4\sqrt{N}}\sum_{i=1}^{N}\sin(\Phi_i)[\mathrm{sgn}(f+F_i)+\mathrm{sgn}(f-F_i)] \quad (7.302)$$

The associated random process defined by the magnitude squared of the Cramer transform of each signal of $X(\Omega, t)$ is

$$|X_C(\Omega,f)|^2 = \frac{A_o^2}{2\pi^2 N}\sum_{i=1}^{N}\sum_{k=1}^{N}[\cos(\Phi_i)\cos(\Phi_k) \cdot[\mathrm{Si}(2\pi(f+F_i)T)-\mathrm{Si}(2\pi(f-F_i)T)-2\mathrm{Si}(2\pi F_i T)] \cdot[\mathrm{Si}(2\pi(f+F_k)T)-\mathrm{Si}(2\pi(f-F_k)T)-2\mathrm{Si}(2\pi F_k T)] + \sin(\Phi_i)\sin(\Phi_k)[\mathrm{Si}(2\pi(f+F_i)T)+\mathrm{Si}(2\pi(f-F_i)T)] \cdot[\mathrm{Si}(2\pi(f+F_k)T)+\mathrm{Si}(2\pi(f-F_k)T)]] \quad (7.303)$$

and this random process can be approximated according to

$$|X_C(\Omega,f)|^2 \approx \frac{A_o^2}{8N}\sum_{i=1}^{N}\sum_{k=1}^{N}[\cos(\Phi_i)\cos(\Phi_k)\ [\mathrm{sgn}(f+F_i)-\mathrm{sgn}(f-F_i)-2\,\mathrm{sgn}(F_i)] \cdot[\mathrm{sgn}(f+F_k)-\mathrm{sgn}(f-F_k)-2\,\mathrm{sgn}(F_k)] + \sin(\Phi_i)\sin(\Phi_k)[\mathrm{sgn}(f+F_i)+\mathrm{sgn}(f-F_i)]\cdot[\mathrm{sgn}(f+F_k)+\mathrm{sgn}(f-F_k)]] \quad (7.304)$$

The joint density function of $F_1, \ldots, F_N, \Phi_1, \ldots, \Phi_N$ is

$$f_{F_1,\ldots,F_N,\Phi_1,\ldots,\Phi_N}(f_1,\ldots,f_N,\phi_1,\ldots,\phi_N) = \begin{cases} \dfrac{1}{f_{\max}^N}\cdot\dfrac{1}{(2\pi)^N} & \begin{matrix} 0<f_i<f_{\max} \\ 0\le\phi_i<2\pi \end{matrix} \\ 0 & \text{elsewhere} \end{cases} \quad (7.305)$$

and it then follows that

$$G_I(f)=\boldsymbol{E}\left[|X_C(\Omega,f)|^2\right]\approx\frac{A_o^2}{8N}\cdot\frac{1}{f_{\max}^N}\cdot\frac{1}{(2\pi)^N}\cdot\int_0^{f_{\max}}\cdots\int_0^{f_{\max}}\cdots\int_0^{2\pi}\cdots\int_0^{2\pi}\left[\sum_{i=1}^{N}\sum_{k=1}^{N}\right.$$

$$\left[\cos(\phi_i)\cos(\phi_k)[\mathrm{sgn}(f+f_i)-\mathrm{sgn}(f-f_i)-2\mathrm{sgn}(f_i)]\right.$$

$$\cdot[\mathrm{sgn}(f+f_k)-\mathrm{sgn}(f-f_k)-2\mathrm{sgn}(f_k)]$$

$$+\sin(\phi_i)\sin(\phi_k)[\mathrm{sgn}(f+f_i)+\mathrm{sgn}(f-f_i)]$$

$$\left.\left.\cdot[\mathrm{sgn}(f+f_k)+\mathrm{sgn}(f-f_k)]\right]\right]d\phi_1\ldots d\phi_N df_1\ldots df_N \tag{7.306}$$

Interchanging the order of integrations and summations, and splitting the summations into the diagonal and nondiagonal components, yields

$$G_I(f)=G_1(f)+G_2(f) \tag{7.307}$$

where

$$G_1(f)\approx\frac{A_o^2}{8N}\cdot\frac{1}{f_{\max}}\cdot\frac{1}{2\pi}\cdot\sum_{i=1}^{N}\int_0^{f_{\max}}\int_0^{2\pi}\left[\cos(\phi_i)^2[\mathrm{sgn}(f+f_i)-\mathrm{sgn}(f-f_i)-2\mathrm{sgn}(f_i)]^2\right.$$
$$\left.+\sin(\phi_i)^2[\mathrm{sgn}(f+f_i)+\mathrm{sgn}(f-f_i)]^2\right]\mathrm{d}\phi_i df_i \tag{7.308}$$

$$G_2(f)\approx\frac{A_o^2}{8N}\cdot\frac{1}{f_{\max}^2}\cdot\frac{1}{(2\pi)^2}\cdot\sum_{i=1}^{N}\sum_{k=1,k\neq i}^{N}\int_0^{f_{\max}}\int_0^{f_{\max}}\int_0^{2\pi}\int_0^{2\pi}[\cos(\phi_i)\cos(\phi_k)$$

$$[\mathrm{sgn}(f+f_i)-\mathrm{sgn}(f-f_i)-2\mathrm{sgn}(f_i)]\cdot[\mathrm{sgn}(f+f_k)-\mathrm{sgn}(f-f_k)-2\mathrm{sgn}(f_k)]$$

$$+\sin(\phi_i)\sin(\phi_k)[\mathrm{sgn}(f+f_i)+\mathrm{sgn}(f-f_i)]\cdot[\mathrm{sgn}(f+f_k)$$

$$+\mathrm{sgn}(f-f_k)]]d\phi_i d\phi_k df_i df_k \tag{7.309}$$

As the integral of a sinusoid over its period is zero, it is the case that $G_2(f)=0$. Further, as the integral of $\cos(\phi_i)^2$ or $\sin(\phi_i)^2$ over $[0, 2\pi]$ equals $\pi$, it follows that

$$G_I(f)\approx\frac{A_o^2}{16f_{\max}}\cdot\int_0^{f_{\max}}\left[[\mathrm{sgn}(f+f_i)-\mathrm{sgn}(f-f_i)-2\,\mathrm{sgn}(f_i)]^2+[\mathrm{sgn}(f+f_i)+\mathrm{sgn}(f-f_i)]^2\right]df_i \tag{7.310}$$

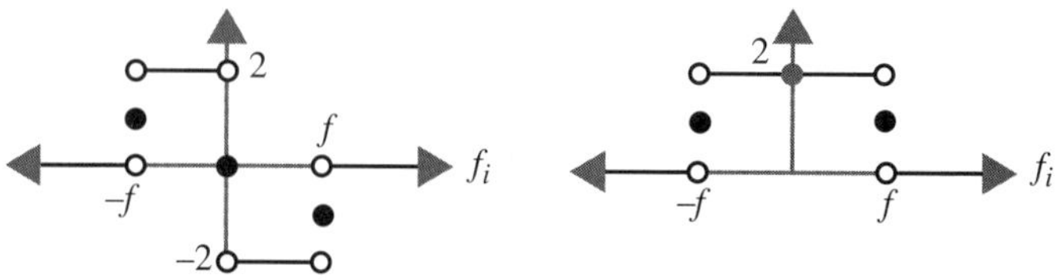


**FIGURE 7.28** Left: graph of sgn $(f+f_i)$ − sgn $(f-f_i)$ −2 sgn $(f_i)$. Right: graph of sgn $(f+f_i)$ + sgn $(f-f_i)$. The case of $f>0$ is assumed.

The same procedure leads to the true form

$$G_I(f) = \frac{A_o^2}{4\pi^2 f_{\max}} \cdot \int_0^{f_{\max}} \Big[[\mathrm{Si}(2\pi(f+f_i)T) - \mathrm{Si}(2\pi(f-f_i)T) - 2\mathrm{Si}(2\pi f_i T)]^2 + [\mathrm{Si}(2\pi(f+f_i)T) + \mathrm{Si}(2\pi(f-f_i)T)]^2\Big]\, df_i \tag{7.311}$$

The functions $\mathrm{sgn}(f+f_i) - \mathrm{sgn}(f-f_i) - 2\,\mathrm{sgn}(f_i)$ and $\mathrm{sgn}(f+f_i) + \mathrm{sgn}(f-f_i)$ have the graphs shown in Figure 7.28, and it then follows, for $f>0$, that

$$G_I(f) \approx \frac{A_o^2}{2f_{\max}} \int_0^{\min\{f, f_{\max}\}} df_i = \frac{A_o^2}{2} \cdot \frac{\min\{f, f_{\max}\}}{f_{\max}} \tag{7.312}$$

and $G_I(-f) = G_I(f)$. The power spectral density then is

$$G(f) = \lim_{df \to 0} \frac{|G_I(f+df/2) - G_I(f-df/2)|}{df} \approx \begin{cases} \dfrac{A_o^2}{2f_{\max}} & 0<f<f_{\max},\ -f_{\max}<f<0 \\ \dfrac{A_o^2}{4f_{\max}} & f = \pm f_{\max} \\ 0 & f=0, |f|>f_{\max} \end{cases} \tag{7.313}$$

## APPENDIX 7.L PROOF OF THEOREM 7.23

Consider the following two results: first, from conditional probability theory,

$$P\left[X_{t_{k+m+n}} = s_j / X_{t_k} = s_i\right] = \frac{P\left[X_{t_{k+m+n}} = s_j, X_{t_k} = s_i\right]}{P\left[X_{t_k} = s_i\right]} \tag{7.314}$$

Second, from the Theorem of Total Probability,

$$
\begin{aligned}
&P\left[X_{t_{k+m+n}}=s_j, X_{t_k}=s_i\right] \\
&=\sum_{q=1}^{M} P\left[X_{t_{k+m+n}}=s_j, X_{t_k}=s_i / X_{t_{k+m}}=s_q\right] P\left[X_{t_{k+m}}=s_q\right] \\
&=\sum_{q=1}^{M} P\left[X_{t_{k+m+n}}=s_j, X_{t_k}=s_i, X_{t_{k+m}}=s_q\right]
\end{aligned}
\tag{7.315}
$$

The substitution of Equations 7.313 into 7.312 yields

$$
\begin{aligned}
P\left[X_{t_{k+m+n}}=s_j / X_{t_k}=s_i\right] &= \sum_{q=1}^{M} \frac{P\left[X_{t_{k+m+n}}=s_j, X_{t_k}=s_i, X_{t_{k+m}}=s_q\right]}{P\left[X_{t_k}=s_i\right]} \\
&= \sum_{q=1}^{M} \frac{P\left[X_{t_{k+m+n}}=s_j / X_{t_k}=s_i, X_{t_{k+m}}=s_q\right] P\left[X_{t_k}=s_i, X_{t_{k+m}}=s_q\right]}{P\left[X_{t_k}=s_i\right]}
\end{aligned}
\tag{7.316}
$$

The Markov property of transition probabilities depending on the present, and not past, values results in

$$
P\left[X_{t_{k+m+n}}=s_j / X_{t_k}=s_i, X_{t_{k+m}}=s_q\right]=P\left[X_{t_{k+m+n}}=s_j / X_{t_{k+m}}=s_q\right] \tag{7.317}
$$

The substitution of this result, and the use of conditional probability theory, yields

$$
\begin{aligned}
&P\left[X_{t_{k+m+n}}=s_j / X_{t_k}=s_i\right] \\
&\sum_{q=1}^{M} P\left[X_{t_{k+m+n}}=s_j / X_{t_{k+m}}=s_q\right] P\left[X_{t_{k+m}}=s_q / X_{t_k}=s_i\right]
\end{aligned}
\tag{7.318}
$$

as required.

## APPENDIX 7.M PROOF OF THEOREM 7.24

The proof is by induction. First, the theorem is true for the case of $n=1$. Second, consider $P(2)=\left[p_{ij}(2)\right]$. From the Chapman–Kolmogorov equation, it is the case for an $M$-state random process that

$$
p_{ij}(2)=\sum_{q=1}^{M} p_{iq}(1) p_{qj}(1)=p_{i1}p_{1j}+p_{i2}p_{2j}+\cdots+p_{ii}p_{ij}+\cdots+p_{iM}p_{Mj} \tag{7.319}
$$

Now,

$$P(1)P(1) = \begin{bmatrix} p_{11} & p_{12} & \cdots & p_{1M} \\ & & \cdots & \\ p_{i1} & p_{i2} & \cdots & p_{iM} \\ & & \cdots & \\ p_{M1} & p_{M2} & \cdots & p_{MM} \end{bmatrix} \begin{bmatrix} p_{11} & \cdots & p_{1j} & \cdots & p_{1M} \\ & & \cdots & & \\ p_{i1} & \cdots & p_{ij} & \cdots & p_{iM} \\ & & \cdots & & \\ p_{M1} & \cdots & p_{Mj} & & p_{MM} \end{bmatrix} \tag{7.320}$$

and it is clear that the $ij$th element of this matrix is consistent with Equation 7.317, and the theorem holds for the case of $n=2$. Third, consider the case of $P(n+1)=[p_{ij}(n+1)]$, $n \in \{1,2,\ldots\}$. From the Chapman–Kolmogorov equation, it follows that

$$p_{ij}(n+1) = \sum_{q=1}^{M} p_{iq}(n)p_{qj}(1) = p_{i1}(n)p_{1j}(1) + p_{i2}(n)p_{2j}(1) + \cdots + p_{iM}(n)p_{Mj}(1) \tag{7.321}$$

It is the case that

$$P(n)P(1) = \begin{bmatrix} p_{11}(n) & p_{12}(n) & \cdots & p_{1M}(n) \\ & & \cdots & \\ p_{i1}(n) & p_{i2}(n) & \cdots & p_{iM}(n) \\ & & \cdots & \\ p_{M1}(n) & p_{M2}(n) & \cdots & p_{MM}(n) \end{bmatrix} \begin{bmatrix} p_{11}(1) & \cdots & p_{1j}(1) & \cdots & p_{1M}(1) \\ & & \cdots & & \\ p_{i1}(1) & \cdots & p_{ij}(1) & \cdots & p_{iM}(1) \\ & & \cdots & & \\ p_{M1}(1) & \cdots & p_{Mj}(1) & & p_{MM}(1) \end{bmatrix} \tag{7.322}$$

and it is clear that the $ij$th element of this matrix is consistent with Equation 7.319, and the result $P(n)=P(n)P(1)$ holds. Finally,

$$P(n+1) = P(n)P(1) = P(n-1)P(1)P(1) = \cdots = P^n(1) \tag{7.323}$$

concludes the proof.

# 8

# PMF AND PDF EVOLUTION

## 8.1 INTRODUCTION

At a set time, a random process defines a random variable that has a probability mass function, or a probability density function, as appropriate. It is of interest to be able to determine this probability mass/density function for all times—resulting in the probability mass/density function evolution with time. However, in general, such evolution is analytically difficult to establish. This evolution can be established for a few important random processes including the random walk, one-dimensional Brownian motion, and the case of random processes defined as a sum of independent random variables at each time instant.

Probability density function evolution is consistent with *probability flow*, and for continuous time Markov random processes, the probability density function evolution leads to the Fokker–Planck equation.

When an analytic solution cannot be obtained, estimation techniques for the probability mass/density function can be used, and this chapter commences with a brief introduction. For the continuous case, a useful approach is to estimate the probability density function at set times $\Delta t, 2\Delta t, 3\Delta t, \ldots$ and use a normalized reconstruction function, $r:\boldsymbol{R}\rightarrow\boldsymbol{R}, r(0)=1, r(k)=0, k\in\{\ldots,-1,1,\ldots\}$, to estimate the probability density function at other times according to

$$\hat{f}_{X(t)}(x)\approx\sum_{k=1}^{\infty}\hat{f}_{X(k\Delta t)}(x)r\left(\frac{t-k\Delta t}{\Delta t}\right)\qquad t>0 \tag{8.1}$$

---

*A Signal Theoretic Introduction to Random Processes*, First Edition. Roy M. Howard.

Here, $\hat{f}_{X(k\Delta t)}(x)$ is an estimate of the probability density function of the random process $X$ at the time $k\Delta t$.

## 8.2 PROBABILITY MASS/DENSITY FUNCTION ESTIMATION

If $x_1,\ldots, x_N$ are independent samples taken from a random variable, then an important issue is to determine an approximation to the probability mass/density function of the random variable. Silverman (1986) is a good reference. The following two definitions are fundamental:

**Definition: Estimate** When data is used to find an approximation to an unknown parameter, an estimate of the parameter is made.

**Definition: Estimator** An approximation to a function, over all values in its domain, is called an estimator of the function.

### 8.2.1 Parametric and Non-parametric Approaches

If a model for a probability mass/density function is proposed, then the issue is to estimate the parameters in the model. For example, if a Gaussian model is proposed, then the mean and variance need to be estimated. Kay (1993) is a useful reference.

A nonparametric approach uses the available data directly, without parameter estimation, to find an estimator for the probability mass/density function. The prototypical case is that of kernel density estimation.

### 8.2.2 Assumption and Notation

It is assumed that $N$ independent samples, $x_1,\ldots, x_N$, from a random variable $X$, are available. For the continuous case, the probability density function of $X$ and an estimator for this function are denoted, respectively, $f_X(x)$ and $\hat{f}_X(x)$. The mean and variance of $X$ are denoted, respectively, $\mu_X$ and $\sigma_X^2$. Analogous definitions follow for the discrete random variable case.

### 8.2.3 Measures of Performance

An estimator, $\hat{f}_X(x)$, of the probability density function $f_X(x)$ yields the following integrated, and integrated square, errors:

$$\int_{-\infty}^{\infty} \left|f(x)-\hat{f}_X(x)\right| dx \quad \text{or} \quad \int_{-\infty}^{\infty} \left[f(x)-\hat{f}_X(x)\right]^2 dx \tag{8.2}$$

Two useful error measures for a probability density function estimator, over the range $-M\Delta x < x < M\Delta x$, are the integrated error

$$\varepsilon_1 = \Delta x \sum_{i=-M}^{M} \left| f_X(i\Delta x) - \hat{f}_X(i\Delta x) \right| \tag{8.3}$$

and the root integrated square error

$$\varepsilon_2 = \sqrt{\Delta x \sum_{i=-M}^{M} \left[ f_X(i\Delta x) - \hat{f}_X(i\Delta x) \right]^2} \tag{8.4}$$

The former is preferable when the approximation of the tail of a probability density function is important as the latter de-emphasizes errors when the probability function takes on small values.

An alternative measure, over the range $-M\Delta x < x < M\Delta x$, is the integrated mean square error:

$$\varepsilon = \Delta x \sum_{i=-M}^{M} \left[ f_X(i\Delta x) - \hat{f}_X(i\Delta x) \right]^2 \tag{8.5}$$

This measure is widely used and is called the MISE for mean square integrated error (Silverman, 1986, p. 35).

***8.2.3.1 Kullback–Leibler Divergence or Relative Entropy*** An alternative error measure is the average log-likelihood measure, or the Kullback–Leibler divergence, or the relative entropy measure:

$$\int_{-\infty}^{\infty} f_X(x) \left[ \ln[f_X(x)] - \ln[\hat{f}_X(x)] \right] dx = \int_{-\infty}^{\infty} f_X(x) \ln\left[ \frac{f_X(x)}{\hat{f}_X(x)} \right] dx \tag{8.6}$$

### 8.2.4 CDF Estimation

Consistent with the step approximation, illustrated in Figure 8.1, to the cumulative distribution function of a random variable $X$, an estimator for the cumulative distribution function, based on $N$ independent samples, $x_1, \ldots x_N$, is

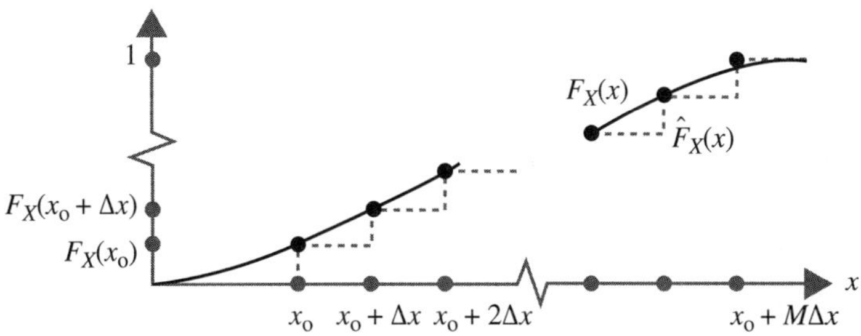


**FIGURE 8.1** Step approximation to the cumulative distribution function.

$$\hat{F}_X(x) = \sum_{i=0}^{M} \hat{c}_i u(x - (x_o + i\Delta x)) \tag{8.7}$$

or

$$\hat{F}_X(x) = \sum_{i=0}^{M} \frac{\hat{c}_i}{2}\left[1 + \tanh\left(\frac{x - (x_o + i\Delta x)}{\Delta}\right)\right] \qquad \Delta << \Delta x \tag{8.8}$$

assuming the transition from zero to one occurs mainly in the interval $(x_o, x_o + M\Delta x)$ and where the coefficients are defined according to

$$\hat{c}_0 = \frac{1}{N}\sum_{k=1}^{N} u(x_o - x_k) \qquad \hat{c}_1 = \frac{1}{N}\sum_{k=1}^{N} u(x_o + \Delta x - x_k) - \hat{c}_0 \ \ldots \tag{8.9}$$

$$\hat{c}_i = \frac{1}{N}\sum_{k=1}^{N} u(x_o + i\Delta x - x_k) - \sum_{k=0}^{i-1} \hat{c}_k \qquad \ldots \qquad \hat{c}_M = 1 - \sum_{k=0}^{M-1} \hat{c}_k \tag{8.10}$$

An alternative estimator is

$$\hat{F}_X(x) = \frac{1}{N}\sum_{i=1}^{N} u(x - x_i) \tag{8.11}$$

## 8.3 NON/SEMI-PARAMETRIC PDF ESTIMATION

The following subsections detail useful nonparametric and semiparametric approaches for estimation of the probability density function of a random variable $X$ based on $N$ independent samples, $x_1, \ldots, x_N$ of the random variable being available.

### 8.3.1 Normalized Histogram

The normalized histogram, with a bin width of $\Delta$, is defined as

$$h(x) = \frac{1}{N}\sum_{i=1}^{N} p\left(\frac{x - \lfloor x_i/\Delta \rfloor \Delta}{\Delta}\right) \qquad p(x) = \begin{cases} 1 & 0 \le x < 1 \\ 0 & \text{elsewhere} \end{cases} \tag{8.12}$$

and normalization leads to the following estimator:

$$\hat{f}_X(x) = \frac{1}{N\Delta} \cdot \sum_{i=1}^{N} p\left(\frac{x - \lfloor x_i/\Delta \rfloor \Delta}{\Delta}\right) \tag{8.13}$$

A normalized histogram estimator may contain large step discontinuities. A continuous estimator is preferable. One important estimator is the kernel density estimator.

### 8.3.2 Kernel Density Estimation

A generalization of the histogram estimator, as given by Equation 8.13, called a kernel density estimator, is defined according to

$$\hat{f}_X(x) = \frac{1}{Nh} \cdot \sum_{i=1}^{N} K\left(\frac{x - x_i}{h}\right) \tag{8.14}$$

where $K$ is the weighing function and $h$ is a scaling parameter that needs to be optimally chosen. The simplest weighting function is the rectangular function:

$$K(x) = \begin{cases} 1/2 & -1 \le x < 1 \\ 0 & \text{elsewhere} \end{cases} \tag{8.15}$$

and this defines the uniform kernel estimator. The issue is to find a kernel such that

$$\int_{-\infty}^{\infty} \left| f(x) - \hat{f}_X(x) \right| dx \quad \text{or} \quad \int_{-\infty}^{\infty} \left[ f(x) - \hat{f}_X(x) \right]^2 dx \tag{8.16}$$

is minimized subject to $\hat{f}_X(x) \ge 0$.

Useful references for kernel density estimation include Parzen (1962b), Silverman (1986), and Thompson and Tapia (1990).

### 8.3.3 Characteristic Function Approach

By definition, the characteristic function of a random variable $X$ is

$$\Phi_X(f) = \boldsymbol{E}\left[e^{j2\pi fX}\right] = \int_{-\infty}^{\infty} e^{j2\pi fX} f_X(x)dx \tag{8.17}$$

which is the inverse Fourier transform of the probability density function. An estimator for the characteristic function is

$$\widehat{\Phi}_X(f) = \frac{1}{N} \sum_{i=1}^{N} e^{j2\pi f x_i} \tag{8.18}$$

and an estimator for $f_X(x)$ can be obtained from $\widehat{\Phi}_X(f)$ by taking its Fourier transform. However, the Fourier transform of Equation 8.18 does not exist (or, using the distribution theory, consists of impulses), and windowing is required according to

$$\hat{f}_X(x) = \int_{-\infty}^{\infty} \widehat{\Phi}_X(f) W(f) e^{-j2\pi fx} df \tag{8.19}$$

where $W$ is the windowing function. This leads to the kernel estimate of the probability density function according to

$$\hat{f}_X(x) = \frac{1}{N}\sum_{i=1}^{N} w(x-x_i) \tag{8.20}$$

where $w$ is the Fourier transform of $W$. The proof of this result is detailed in Appendix Appendix 8.A.

An alternative approach is detailed in the following theorem:

**Theorem 8.1 Characteristic Function Estimator for PDF**
With the use of $2N_f+1$ samples of the characteristic function, with a spacing of $\Delta f$, and a sinc reconstruction function, the characteristic function can be estimated, over the range $-N_f\Delta f \le f \le N_f\Delta f$, according to

$$\widehat{\Phi}_X(f) = \sum_{k=-N_f}^{N_f} \widehat{\Phi}_X(k\Delta f)\operatorname{sinc}\left(\frac{f-k\Delta f}{\Delta f}\right) \qquad \widehat{\Phi}_X(k\Delta f) = \frac{1}{N}\sum_{i=1}^{N} e^{j2\pi k\Delta f x_i} \tag{8.21}$$

Based on $\widehat{\Phi}_X(f)$, an estimator for the probability density function, valid over the range $-1/2\Delta f < x < 1/2\Delta f$, can be defined according to

$$\hat{f}_X(x) = \begin{cases} \Delta f \displaystyle\sum_{k=-N_f}^{N_f} \widehat{\Phi}_X(k\Delta f)\exp(-j2\pi k\Delta f x) & -\dfrac{1}{2\Delta f} < x < \dfrac{1}{2\Delta f} \\ 0 & \text{elsewhere} \end{cases} \tag{8.22}$$

**Proof**
The is detailed in Appendix 8.A.

### 8.3.4 Orthonormal Basis Set Decomposition

Consider an orthonormal basis set $\{b_1,\ldots, b_i,\ldots\}$ defined on the interval where the probability density function of interest is nonzero. With such a basis set, the probability density function of a random variable $X$ can be written as (Schwartz, 1967)

$$f_X(x) = c_1 b_1(x) + c_2 b_2(x) + \cdots \tag{8.23}$$

where the $i$th coefficient is

$$c_i = \boldsymbol{E}[b_i(X)] = \int_{-\infty}^{\infty} b_i(x) f_X(x)\,dx \tag{8.24}$$

The coefficient $c_i$ can be estimated from $N$ independent samples, $x_1,\ldots,x_N$, according to

$$\hat{c}_i = \frac{1}{N}\sum_{i=1}^{N} b_i(x_i) \tag{8.25}$$

and, consequently, an estimator for $f_X(x)$ is

$$\hat{f}_X(x) = \hat{c}_1 b_1(x) + \hat{c}_2 b_2(x) + \cdots \tag{8.26}$$

The estimate of the $i$th coefficient, as given by Equation 8.25, is unbiased and has a variance that decreases according to $1/N$ (Schwartz, 1967).

### 8.3.5 Example

Consider a complex system comprising of independent subsystems and the case where the system output is defined by a random variable $Z$ according to

$$Z = \begin{cases} X_1 & P[Z = X_1] = p_1 \\ X_2 & P[Z = X_2] = p_2 \\ & \ldots \end{cases} \tag{8.27}$$

Here, $X_i$ is the random variable defining the system output due to the $i$th subsystem. From the Theorem of Total Probability, it follows that

$$\begin{aligned} f_Z(z)dz \approx P[Z \in (z, z+dz)] &= \sum_{i=1}^{\infty} P[Z \in (z, z+dz)/Z = X_i]P[Z = X_i] \\ &= \sum_{i=1}^{\infty} P[X_i \in (z, z+dz)]P[Z = X_i] \end{aligned} \tag{8.28}$$

and, hence,

$$f_Z(z) \approx \sum_{i=1}^{\infty} p_i f_{X_i}(z) \tag{8.29}$$

where $f_{X_i}$ is the probability density function of $X_i$.

The probabilities $p_1, p_2,\ldots$ are assumed to be consistent with the number of trials for a success in a sequence of independent Bernoulli trials of an experiment where the probability of success in an individual trial is $p_o$. Such probabilities are given by the probability mass function of a geometric random variable, that is,

$$p_i = (1 - p_o)^{i-1} p_o \qquad p_o \in (0, 1),\ i \in \{1,2,\ldots\} \tag{8.30}$$

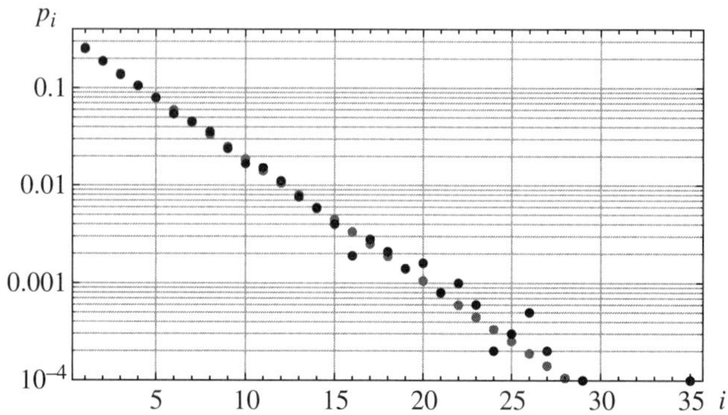


**FIGURE 8.2** Probability mass function, and simulation values based on 10,000 independent trials, for a geometric random variable with $p_o = 0.25$.

The $i$th random variable is assumed to be Gaussian with a mean of $\sqrt{i}$ and a variance of $i$, that is

$$f_{X_i}(x) = \frac{1}{\sqrt{2\pi i}} \cdot \exp\left[\frac{-\left(x-\sqrt{i}\right)^2}{2i}\right] \tag{8.31}$$

It then follows that

$$f_Z(z) = \sum_{i=1}^{\infty} \frac{(1-p_o)^{i-1} p_o}{\sqrt{2\pi i}} \cdot \exp\left[\frac{-\left(x-\sqrt{i}\right)^2}{2i}\right] \tag{8.32}$$

and, using the result stated in Theorem 4.6, the characteristic function of $Z$ is

$$\Phi_Z(f) = p_o \sum_{i=1}^{\infty} (1-p_o)^{i-1} e^{j2\pi f\sqrt{i} - 2\pi^2 f^2 i} \tag{8.33}$$

The probability mass function, as defined by Equation 8.30, and the probability density function, as defined by Equation 8.32, are shown, respectively, in Figures 8.2 and 8.3 for the case of $p_o = 0.25$.

Estimators, based on 10000 independent samples, for the probability density function are shown in Figures 8.3, 8.4, 8.5, and 8.6, respectively, for the histogram estimator defined by Equation 8.13, the kernel density estimator as given by Equation 8.14, the characteristic function estimator as given by Equation 8.22, and the basis set estimator as given by Equation 8.26. For the kernel density estimator, a rectangular kernel function, as defined by Equation 8.15, has been used, along with an optimum value of $h$ of $h = 0.7$ (optimum in the sense of minimizing the integrated error as specified by Eq. 8.3). For the characteristic function estimator, the parameters $N_f = 100$ and $\Delta f = 0.006$ have been used. A 21-term Gaussian-based basis set,

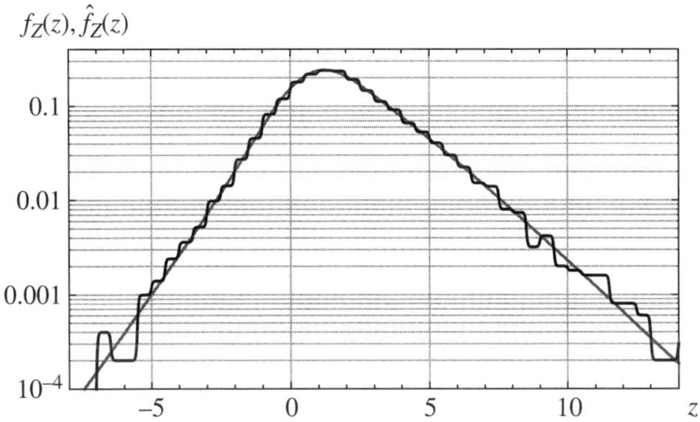


**FIGURE 8.3** True probability density function and an estimator based on histogram values with a bin width of 0.5.

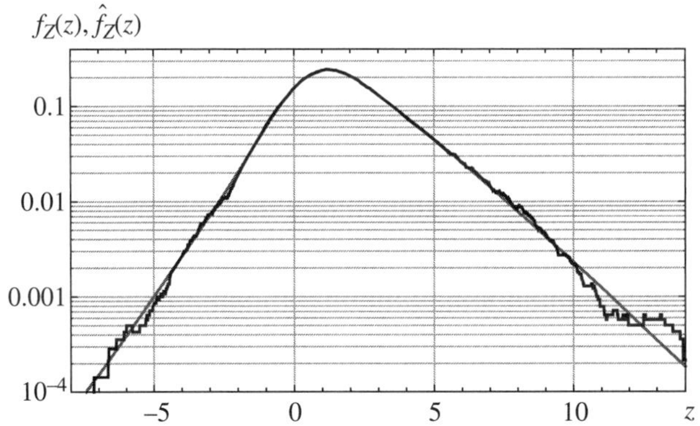


**FIGURE 8.4** True probability density function and the associated kernel density estimator function with an optimum value of $h = 0.7$.

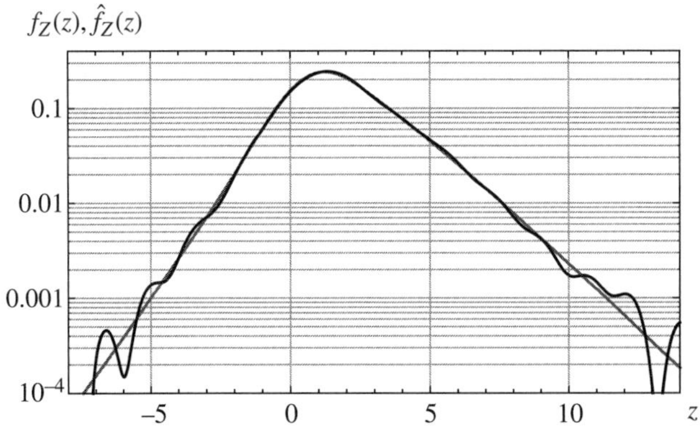


**FIGURE 8.5** True probability density function and the associated characteristic function estimator.

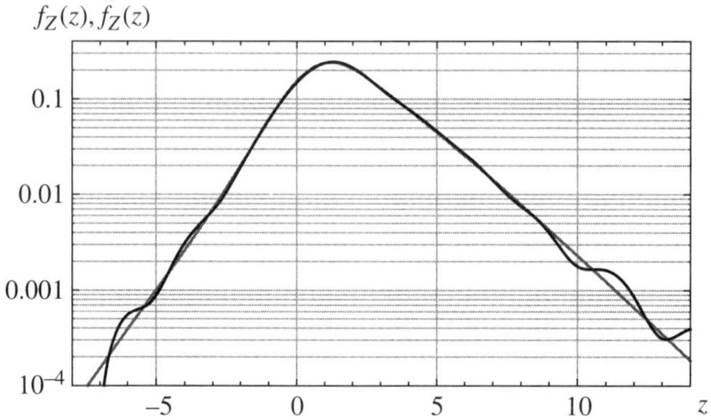


**FIGURE 8.6** True probability density function and the associated estimator function based on a 21-term Gaussian-based basis set.

**TABLE 8.1 Integrated errors**

| PDF estimator | $\varepsilon_1$ | $\varepsilon_2$ |
|---|---|---|
| Normalized histogram | 0.067 | 0.028 |
| Uniform kernel | 0.0145 | 0.0047 |
| Characteristic function | 0.023 | 0.0085 |
| Gaussian-based basis set | 0.022 | 0.0081 |

as defined in Theorem 3.3 and based on the sample mean and sample variance, has been used for the basis set estimator.

The integrated errors, as specified by Equations 8.3 and 8.4, for the different estimators are detailed in Table 8.1 for the case of $\Delta x = 0.022$.

## 8.4 PMF/PDF EVOLUTION: SIGNAL PLUS NOISE

One of the simplest cases to consider for probability mass function evolution, or probability density function evolution, is that of a random process defined by a noise signal added to a deterministic signal. Consider a random process $X{:}S \rightarrow S_X$ defined according to

$$X(\Omega, t) = s(t) + N(\Omega, t) \tag{8.34}$$

where $s$: $\boldsymbol{R} \rightarrow \boldsymbol{R}$ is a deterministic signal and $N{:}S \rightarrow S_N$ is a noise random process.

### 8.4.1 Discrete Case: PMF Evolution

Consider the case where the noise random process, $N(\Omega,t)$, takes on discrete values for all values of time and has a known probability mass function evolution as specified by $p_{N(t)}(n_k)$ and $n_k \in S^{N(t)}$ where $S^{N(t)}$ is the sample space of the random variable defined by $N(\Omega, t)$, for $t$ fixed. The simplest case is where the noise random process is stationary and $p_{N(t)}(n_k)$ is independent of time.

**Theorem 8.2 PMF Evolution of a Signal and Noise**
The probability mass function evolution with time of $X : S \to S_X$ is given by

$$p_{X(t)}(x) = \begin{cases} p_N(x - s(t)) & x - s(t) \in S^{N(t)} \\ 0 & \text{elsewhere} \end{cases} \tag{8.35}$$

For the case where the noise is stationary and $p_{N(t)}(n_k) = p_N(n_k) = p_k$:

$$p_{X(t)}(x) = \begin{cases} p_N(x - s(t)) & x - s(t) \in S^N \\ 0 & \text{elsewhere} \end{cases} \tag{8.36}$$

For this case, the probability mass function evolution with time comprises of trajectories in three dimensions as defined by

$$\{(n_k + s(t), t, p_k) : t \geq 0\} \qquad n_k \in S^N \tag{8.37}$$

**Proof**
The proof follows from the definition of the probability mass function:

$$\begin{aligned} p_{X(t)}(x) &= P[X(\Omega, t) = x] = P[s(t) + N(\Omega, t) = x] \\ &= P[N(\Omega, t) = x - s(t)] \\ &= p_{N(t)}(x - s(t)) \qquad x - s(t) \in S^{N(t)} \end{aligned} \tag{8.38}$$

For the stationary case, where $S^N = \{n_1, n_2, \ldots\}$, the probability mass function is defined for the values: $x - s(t) = n_1, x - s(t) = n_2, \ldots$, and the corresponding trajectories are $\{(n_1 + s(t), t, p_1) : t \geq 0\}, \{(n_2 + s(t), t, p_2) : t \geq 0\}, \ldots$.

***8.4.1.1 Example*** Consider the case of a pulse signal with unity area, $s(t) = t^2 \exp(-t/\tau)/2\tau^3$, which is subject to additive noise from the set $S_N = \{-1/4, 0, 1/4\}$ with respective probabilities 1/4, 1/2, 1/4. The probability mass function evolution with time of $X(\Omega, t) = s(t) + N(\Omega, t)$ then is

$$p_{X(t)}(x) = \begin{cases} p_N(x - s(t)) & x - s(t) \in \{-1/4, 0, 1/4\} \\ 0 & \text{elsewhere} \end{cases} \tag{8.39}$$

and the three trajectories are

$$\begin{aligned} &\{(-1/4 + s(t),\ t,\ 1/4) : t \geq 0\} \\ &\{(s(t),\ t,\ 1/2) : t \geq 0\} \\ &\{(1/4 + s(t),\ t,\ 1/4) : t \geq 0\} \end{aligned} \tag{8.40}$$

The probability mass function evolution with time is shown in Figure 8.7 for the case of $\tau = 1$.

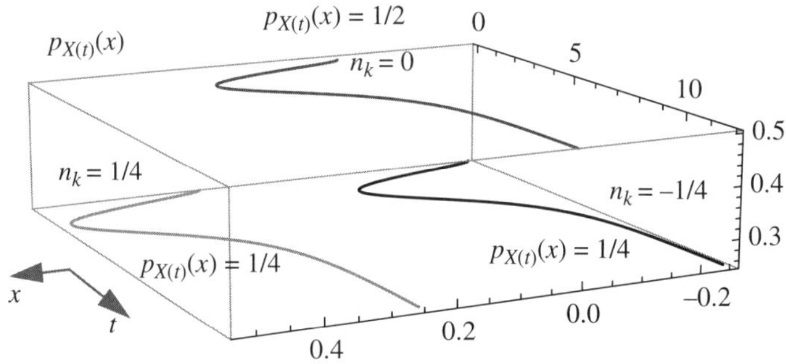


**FIGURE 8.7** Trajectories defining the evolution of the probability mass function with time for a transient pulse plus noise taking on values from the set {−1/4, 0, 1/4}.

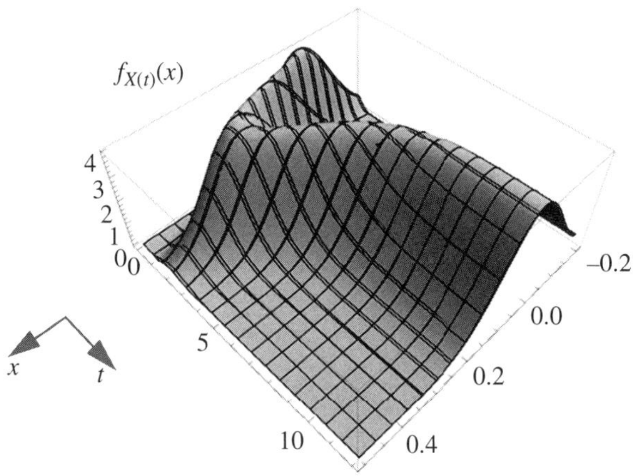


**FIGURE 8.8** Evolution of the probability density function with time for a transient pulse plus additive Gaussian noise.

## 8.4.2 Continuous Case: PDF Evolution

Consider the case where $N(\Omega, t)$ is a noise random process with a probability density function $f_{N(t)}(n)$ that evolves with time in a known manner. In many cases, the noise random process is stationary and $f_{N(t)}(n)$ is independent of time.

**Theorem 8.3 PDF Evolution of a Signal and Noise**
The probability density function evolution with time of $X : S \rightarrow S_X$ is given by

$$f_{X(t)}(x) = f_{N(t)}(x - s(t)) \tag{8.41}$$

For the case where the noise is stationary and $f_{N(t)}(n) = f_N(n)$:

$$f_{X(t)}(x) = f_N(x - s(t)) \tag{8.42}$$

**Proof**

The proof follows from the definition of the probability density function:

$$\begin{aligned} f_{X(t)}(x)dx &\approx P[X(\Omega, t) \in [x, x+dx]] \\ &= P[s(t) + N(\Omega, t) \in [x, x+dx]] \\ &= P[N(\Omega, t) \in [x-s(t), x-s(t)+dx]] \approx f_{N(t)}(x-s(t))dx \end{aligned} \tag{8.43}$$

***8.4.2.1 Example*** Consider the case of a pulse signal with unity area, $s(t) = t^2 \exp(-t/\tau)/2\tau^3$, which is subject to additive Gaussian noise with a probability density function $f_N(n) = \exp(-n^2/2\sigma^2)/\sqrt{2\pi}\sigma$. The probability density function evolution with time of $X(\Omega, t) = s(t) + N(\Omega, t)$ then is

$$f_{X(t)}(x) = \frac{1}{\sqrt{2\pi}\sigma} \cdot \exp\left[\frac{-1}{2\sigma^2}\left[x - \frac{t^2 \exp(-t/\tau)}{2\tau^3}\right]^2\right] \tag{8.44}$$

The probability density function evolution with time is shown in Figure 8.8 for the case of $\tau = 1$ and $\sigma^2 = 0.01$.

## 8.5 PMF EVOLUTION OF A RANDOM WALK

A random walk random process, with a step size of $\Delta x$ and a step duration $\Delta t$ and comprising of $N$ steps, is defined according to

$$X(\Omega, t) = \sum_{i=1}^{N} A_i u(t - i\Delta t) \qquad t \in [0, N\Delta t] \tag{8.45}$$

where $\Omega = (A_1, \ldots, A_N)$, $A_1, \ldots, A_N$ are identical and independent random variables with outcomes from the set $\{-\Delta x, \Delta x\}$ and $u$ is the unit step function. The evolution of the probability mass function of such a random process is specified in the following theorem:

**Theorem 8.4 PMF Evolution of a Random Walk**

The probability mass function of a random walk random process, with a probability of an upward step equal to $p$, evolves according to

$$\begin{aligned} p_{X(t)}(k\Delta x) &= P[X(\Omega, t) = k\Delta x] = b\left(\frac{t/\Delta t + k}{2}, \frac{t}{\Delta t}, p\right) \\ &= \frac{(t/\Delta t)!}{\left(\frac{t/\Delta t + k}{2}\right)! \left(\frac{t/\Delta t - k}{2}\right)!} \cdot p^{(t/\Delta t + k)/2} \cdot (1-p)^{(t/\Delta t - k)/2} \\ & t \in \{\Delta t, 2\Delta t, \ldots\} \end{aligned} \tag{8.46}$$

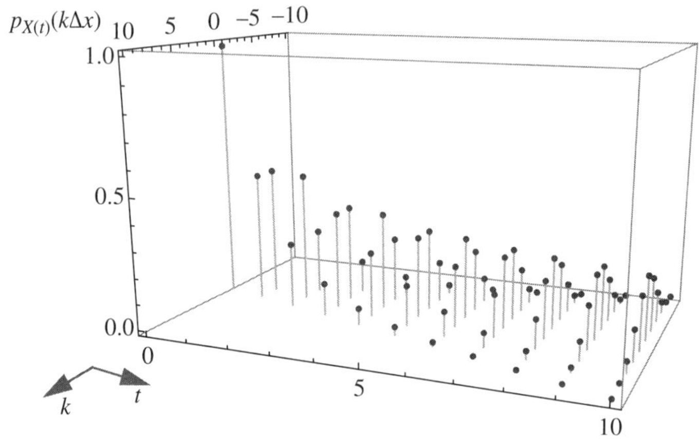


**FIGURE 8.9** Evolution of the probability mass function with time for the case of $\Delta x = \Delta t = 1$ and $p = 0.5$.

with the restrictions: $\frac{-t}{\Delta t} \le k \le \frac{t}{\Delta t}$; $k$ is even if $t/\Delta t$ is even, and $k$ is odd if $t/\Delta t$ is odd.

**Proof**

To reach a level of $k\Delta x$, after $N = t/\Delta t$ steps of a random walk, requires $b$ up steps and $N - b$ down steps where $k = b - (N - b)$. This implies $b = (N + k)/2$. Accordingly, a level of $k\Delta x$ after $N$ steps is consistent with $(N + k)/2$ successes and $(N - k)/2$ failures in $N$ trials. The result then follows from Theorem 4.24.

### 8.5.1 Example

The probability mass function is shown in Figure 8.9 for the first 10 steps of a random walk.

### 8.5.2 Approximation to Probability Mass Function

It is useful to have an approximation to the probability mass function evolution as specified by Equation 8.46 to avoid the calculation of the factorial terms.

**Theorem 8.5 Approximation to Probability Mass Function**

The probability mass function of a random walk with a step size of $\Delta x$, a step duration $\Delta t$, and a probability of an upward step of $p$, can be approximated according to

$$p_{X(t)}(x) = \frac{2\Delta x e^{-\left(\frac{x-(2p-1)Dt}{\Delta x}\right)^2/8p(1-p)Dt}}{\sqrt{2\pi}\sqrt{4p(1-p)Dt}} \qquad D = \frac{\Delta x^2}{\Delta t} \tag{8.47}$$

for the case of $t \in \{\Delta t, 2\Delta t, \ldots\}$ and where

$$\begin{aligned} &x \in \{0, \pm 2\Delta x, \ldots, \pm t\Delta x/\Delta t\} && t/\Delta t \text{ even} \\ &x \in \{\pm \Delta x, \pm 3\Delta x, \ldots, \pm t\Delta x/\Delta t\} && t/\Delta t \text{ odd} \end{aligned} \tag{8.48}$$

The approximation is valid for

$$\frac{(2p-1)Dt}{\Delta x} - \frac{2[p(1-p)Dt]^{2/3}}{\Delta x^{1/3}}\left[\frac{6|\text{r.e.}|}{1-2p}\right]^{1/3} < x < \frac{(2p-1)Dt}{\Delta x} + \frac{2[p(1-p)Dt]^{2/3}}{\Delta x^{1/3}}\left[\frac{6|\text{r.e.}|}{1-2p}\right]^{1/3} \qquad \frac{2\sqrt{tp(1-p)}|\text{r.e.}|}{\sqrt{\Delta t}|1-2p|} > 10 \tag{8.49}$$

provided $p \neq 1/2$ and where r.e. is the relative error. For the case of $p = 1/2$, the approximation is usually good.

**Proof**

The proof is detailed in Appendix 8.B.

*8.5.2.1 Notes* The constant $D$ is termed the diffusion constant.

The mean increase in the level of the random walk in one step is $p\Delta x-(1-p)\Delta x=(2p-1)\Delta x$ and in $N$ steps is $N(2p-1)\Delta x=(2p-1)Dt/\Delta x$, where $t = N\Delta t$.

*8.5.2.2 Symmetrical Case* For the case of $p = 0.5$, the random walk, statistically, has zero mean and has a symmetrical probability mass function.

**Theorem 8.6 PMF for Zero Mean Random Walk**

The probability mass function of a random walk, with a step size of $\Delta x$ and a step duration $\Delta t$, can be approximated, for the case of $p = 0.5$, according to

$$p_{X(t)}(x) \approx \frac{2\Delta x e^{-x^2/2Dt}}{\sqrt{2\pi}\sqrt{Dt}} \qquad D = \frac{\Delta x^2}{\Delta t},\ t \in \{\Delta t, 2\Delta t, \ldots\} \tag{8.50}$$

provided $x \in \{0, \pm 2\Delta x, \ldots, \pm t\Delta x/\Delta t\}$ for the case of $t/\Delta t$ even and $x \in \{\pm \Delta x, \pm 3\Delta x, \ldots, \pm t\Delta x/\Delta t\}$ for the case of $t/\Delta t$ odd.

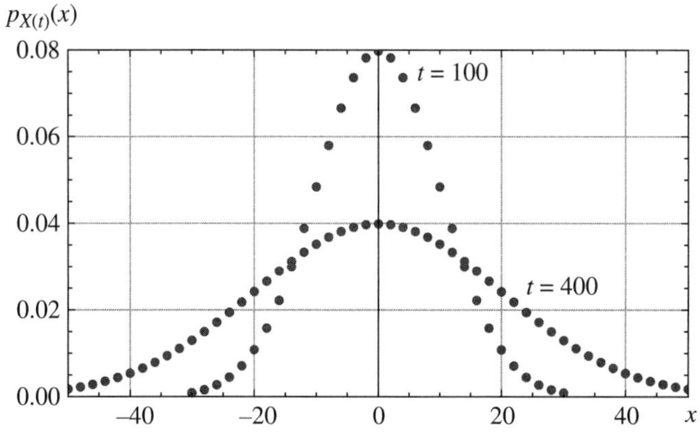


**FIGURE 8.10** Probability mass function for the level of a random walk with $\Delta t = \Delta x = 1$ and $p = 0.5$ for the cases of $t = 100$ and $t = 400$.

**Proof**

The proof follows directly from Theorem 8.5.

*8.5.2.3 Examples* Two examples of the probability mass function, as specified by Theorem 8.6, are shown in Figure 8.10.

## 8.6 PDF EVOLUTION: BROWNIAN MOTION

The following is an alternative, but equivalent, definition for Brownian motion to that given in Chapter 6.

**Definition: Wiener Process or One-Dimensional Brownian Motion** Consider a random process that, when viewed with a resolution of $(\Delta t_o, \Delta x_o)$, is consistent with a zero mean random walk with a variance of $\sigma^2(t)$ at the times $t \in \{\Delta t_o, 2\Delta t_o, \ldots\}$. Further, if when viewed with resolutions

$$\left(\frac{\Delta t_o}{4}, \frac{\Delta x_o}{2}\right), \left(\frac{\Delta t_o}{16}, \frac{\Delta x_o}{4}\right), \ldots, \left(\frac{\Delta t_o}{i^2}, \frac{\Delta x_o}{i}\right), \ldots \tag{8.51}$$

the same variance is maintained at the times $t \in \{\Delta t_o, 2\Delta t_o, \ldots\}$ along with the same probability mass form, that is,

$$\begin{array}{lll} p_{X(t)}(x) \approx \dfrac{2\Delta x e^{-x^2/2Dt}}{\sqrt{2\pi}\sqrt{Dt}} & D = \dfrac{\Delta x^2}{\Delta t}, t \in \{\Delta t, 2\Delta t, \ldots\} & \\ x \in \{0, \pm 2\Delta x, \ldots, \pm t\Delta x/\Delta t\} & t/\Delta t \text{ even} \\ x \in \{\pm \Delta x, \pm 3\Delta x, \ldots, \pm t\Delta x/\Delta t\} & t/\Delta t \text{ odd} \end{array} \tag{8.52}$$

for $(\Delta t, \Delta x) = (\Delta t_o/i^2, \Delta x_o/i)$, $i \in \mathbf{Z}^+$, then the random process is a Wiener process or a one-dimensional Brownian motion process.

### 8.6.1 Rationale for Definition

If a random process with zero mean can be modelled, after assuming a time and level resolution of $(\Delta t, \Delta x)$, by a random walk, then, according to Theorem 8.6, its probability mass function at a time $t$ can be approximated by Equation 8.52. If the random process is viewed with a finer time resolution consistent with $\Delta t \to \Delta t/2$, then the probability mass function at the same time $t$ will have the same form, but scaled, if the diffusion constant $D$ is the same. This constraint implies that the level resolution must scale according to $\Delta x \to \Delta x/\sqrt{2}$. With the finer resolution of $(\Delta t/2, \Delta x/\sqrt{2})$, it is the case that

$$\begin{array}{lll} p_{X(t)}(x) \approx \dfrac{\sqrt{2}\Delta x e^{-x^2/2Dt}}{\sqrt{2\pi}\sqrt{Dt}} & D = \dfrac{\Delta x^2}{\Delta t}, t \in \left\{\dfrac{\Delta t}{2}, \Delta t, \dfrac{3\Delta t}{2}, \ldots\right\} \\ x \in \{0, \pm\sqrt{2}\Delta x, \ldots, \pm\sqrt{2}t\Delta x/\Delta t\} & 2t/\Delta t \text{ even} \\ x \in \{\pm\Delta x/\sqrt{2}, \pm 3\Delta x/\sqrt{2}, \ldots, \pm\sqrt{2}t\Delta x/\Delta t\} & 2t/\Delta t \text{ odd} \end{array} \tag{8.53}$$

An example of this scaling is illustrated in Figure 8.11.

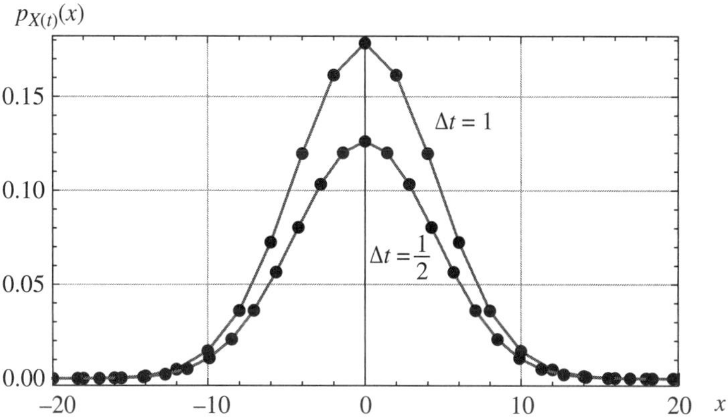


**FIGURE 8.11** Probability mass function at $t = 20$ for the level of a random walk with $p = 0.5$ for the cases of $\Delta t = \Delta x = 1$ and $\Delta t = 1/2$, $\Delta t = 1/\sqrt{2}$.

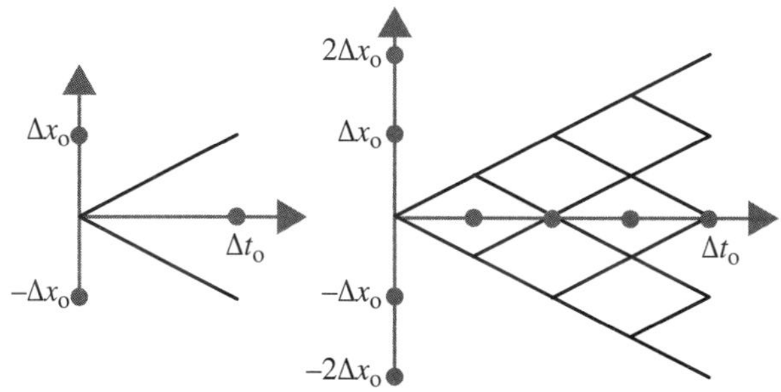


**FIGURE 8.12** Trellis indicating possible paths of a random walk. Left: trellis for the case of $(\Delta t_o, \Delta x_o)$. Right: trellis for the case of $(\Delta t_o/4, \Delta x_o/2)$.

### 8.6.2 Justification of Scaling: Constant Variance

Consider a zero mean random walk with $p = 0.5$ and a quantization level of $(\Delta t_o, \Delta x_o)$. At a time $\Delta t_o$, there are two possible outcomes, $\pm\Delta x_o$, each with probability 0.5 as illustrated in the left diagram of Figure 8.12. The mean squared outcome at $\Delta t_o$ is $(-\Delta x_o)^2/2 + \Delta x_o^2/2 = \Delta x_o^2$. For the case, as illustrated in the right diagram of Figure 8.12, where the quantization level is $(\Delta t_o/4, \Delta x_o/2)$, the possible levels at $\Delta t_o$ are $-2\Delta x_o, -\Delta x_o, 0, \Delta x_o$, and $-2\Delta x_o$. The probability of these levels, respectively, is consistent with zero, one, …, four success in four trials of a Bernoulli experiment where the probability of success equals 0.5. It then follows that the mean squared outcome at $\Delta t_o$ is

$$\frac{(-2\Delta x_o)^2}{16} + \frac{(-\Delta x_o)^2}{4} + \frac{3(0)^2}{8} + \frac{\Delta x_o^2}{4} + \frac{(2\Delta x_o)^2}{16} = \Delta x_o^2 \tag{8.54}$$

This can be repeated for other levels of quantization to show the scaling is correct with maintaining, at a set time, the same variance for the probability mass function.

### 8.6.3 From PMF to PDF

Consider the case of a random process $X$ that exhibits symmetrical random walk behavior for all resolutions, $(\Delta t, \Delta x) = (\Delta t_o / i^2,\ \Delta x_o / i)$, $i \in \mathbf{Z}^+$, that is,

$$p_{X(t)}(x) \approx \frac{2\Delta x e^{-x^2/2Dt}}{\sqrt{2\pi}\sqrt{Dt}} \qquad D = \frac{\Delta x_o^2}{\Delta t_o} = \frac{\Delta x^2}{\Delta t}, \quad t \in \{\Delta t, 2\Delta t, \ldots\}$$
$$x \in \{0, \pm 2\Delta x, \ldots, \pm t\Delta x/\Delta t\} \qquad t/\Delta t \text{ even}$$
$$x \in \{\pm \Delta x, \pm 3\Delta x, \ldots, \pm t\Delta x/\Delta t\} \qquad t/\Delta t \text{ odd} \tag{8.55}$$

Thus,

$$P[X(\Omega, t) \in [x - \Delta x, x + \Delta x)] \approx \frac{2\Delta x e^{-x^2/2Dt}}{\sqrt{2\pi}\sqrt{Dt}} \tag{8.56}$$

For $\Delta x$ sufficiently small, consistent with $\Delta t$ being small, it is possible to treat the random walk as a continuous time random process with a probability density function as defined in the following theorem:

**Theorem 8.7 PDF for Brownian Motion**
The probability density function evolution of a Brownian motion random process $X$ is consistent with

$$f_{X(t)}(x) = \frac{e^{-x^2/2Dt}}{\sqrt{2\pi}\sqrt{Dt}} \tag{8.57}$$

**Proof**
The proof follows from the previous discussion and with the relationship

$$f_{X(t)}(x) 2\Delta x \approx P[X(\Omega, t) \in [x - \Delta x, x + \Delta x)] \tag{8.58}$$

for $\Delta x$ sufficiently small.

#### *8.6.3.1 Extension for Case of Initial Value*

**Theorem 8.8 PDF for Brownian Motion with Initial Condition**
For the case of Brownian motion with an initial condition of $X(\Omega, t_o) = x_o$, the evolution of the probability density function is according to

$$f_{X(t)/(x_o, t_o)}(x) = \frac{e^{-(x-x_o)^2/2D(t-t_o)}}{\sqrt{2\pi}\sqrt{D(t-t_o)}} \qquad t > t_o \tag{8.59}$$

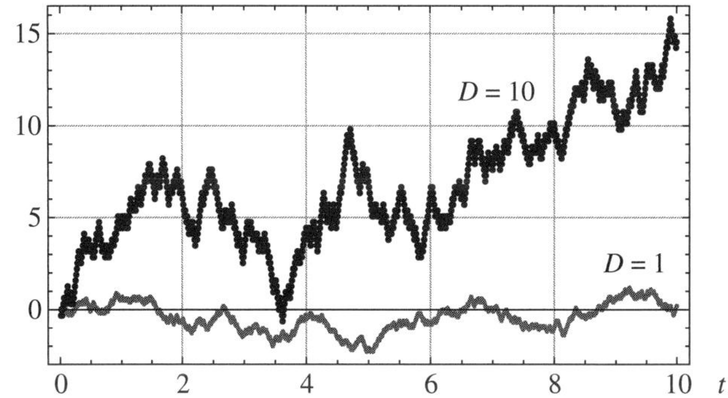


**FIGURE 8.13** Two signals from a Brownian motion random process based on a time resolution of $\Delta t = 0.01$. Lower: $D = 1$, $\Delta x = 0.1$. Upper: $D = 10$, $\Delta x = 1/\sqrt{10}$.

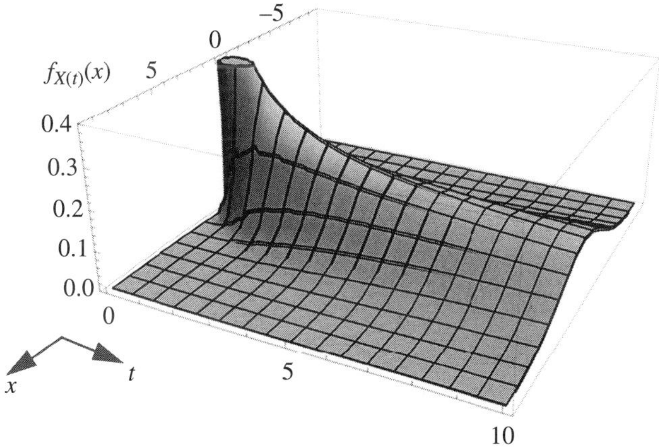


**FIGURE 8.14** Evolution of the probability density function of Brownian motion for the case of $D = 1$.

**Proof**

This follows directly from the previous analysis taking into account the initial condition of $X(\Omega, t_o) = x_o$ and the definition

$$f_{X(t)/(x_o,\, t_o)}(x) 2\Delta x \approx P[X(\Omega, t) \in [x - \Delta x, x + \Delta x)/X(\Omega, t_o) = x_o] \tag{8.60}$$

for $\Delta x$ sufficiently small.

***8.6.3.2 Example*** Two signals from a Brownian motion random process are shown in Figure 8.13.

The evolution, with time, of the probability density function of one-dimensional Brownian motion is illustrated in Figure 8.14 for the case of $D = 1$.

## 8.7 PDF EVOLUTION: SIGNALLING RANDOM PROCESS

Consider a signalling random process, $X$, with amplitude modulation and defined according to

$$X(\Omega,t)=\sum_{k=0}^{\lfloor t/D\rfloor} A_k h(t-kD) \tag{8.61}$$

where $\Omega=\{A_0,\ldots,A_k,\ldots\}$, $A_0,\ldots,A_k,\ldots$ are independent random variables and the probability density function and sample space of $A_k$ are, respectively, $f_{A_k}$ and $S_{A_k}$. The pulse function $h:\boldsymbol{R}\to\boldsymbol{R}$ is assumed to be causal and continuous with $h(0)=0$. The signal sample space is

$$S_X=\left\{X((a_0,\ldots,a_k,\ldots),t)=\sum_{k=0}^{\lfloor t/D\rfloor} a_k h(t-kD):\ a_k\in S_{A_k}\right\} \tag{8.62}$$

Consider a fixed time $t_o$ and $X(\Omega,t_o)=\sum_{k=0}^{\lfloor t_o/D\rfloor} A_k h(t_o-kD)$, which is a weighted sum of $\lfloor t_o/D\rfloor+1$ random variables:

$$A_0h(t_o),A_1h(t_o-D),\ldots,A_{\lfloor t_o/D\rfloor}h(t_o-\lfloor t_o/D\rfloor D) \tag{8.63}$$

Clearly, the determination of the evolution of the probability density function requires the evaluation of the probability density of a sum of random variables. For the case of independent random variables, as has been assumed, an analytic solution can be found as the following example demonstrates. However, in general, establishing the probability density function of a sum is problematic.

### 8.7.1 Example

For the case of independent Gaussian random variables with

$$f_{A_k}(a)=\frac{1}{\sqrt{2\pi}\sigma_k}\cdot\exp\left[\frac{-(a-\mu_k)^2}{2\sigma_k^2}\right]\qquad k\in\{0,1,\ldots\} \tag{8.64}$$

it follows, from Theorem 4.11, that

$$\begin{gathered}f_{X(t)}(x)=\frac{1}{\sqrt{2\pi}\sigma(t)}\cdot\exp\left[\frac{-(x-\mu(t))^2}{2\sigma^2(t)}\right]\\ \mu_X(t)=\sum_{k=0}^{\lfloor t/D\rfloor}\mu_k h(t-kD)\qquad \sigma_X^2(t)=\sum_{k=0}^{\lfloor t/D\rfloor}\sigma_k^2h^2(t-kD)\end{gathered} \tag{8.65}$$

For the case of $D=1$, $\mu_k=0$, $\sigma_k^2=1$, and a pulse function

$$h(t)=\frac{t\exp(-t/\tau)}{\tau^2}\cdot u(t) \tag{8.66}$$

two cases are considered: First, the case of $\tau = 2$ and, second, the case of $\tau = 0.5$. The pulse function is shown in Figure 8.15 for these two cases, and two signals from the random process are shown in Figure 8.16. The evolution, with time, of the variance is shown in Figure 8.17 and the probability density function time evolution is shown for the two cases, respectively, in Figures 8.18 and 8.19.

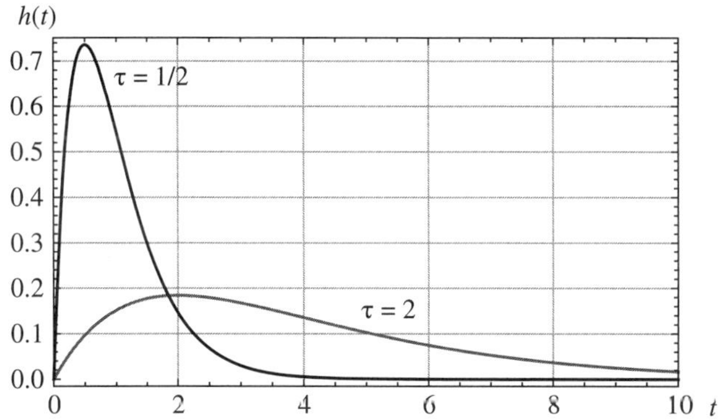


**FIGURE 8.15** Graph of pulse function.

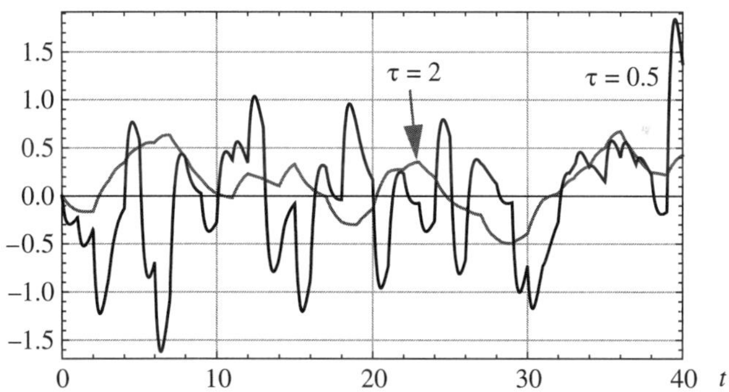


**FIGURE 8.16** Two signals from the amplitude modulated signalling random process.

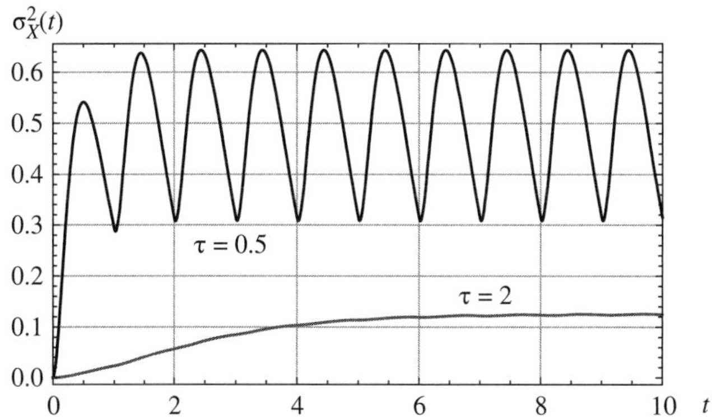


**FIGURE 8.17** Evolution of the variance with time.

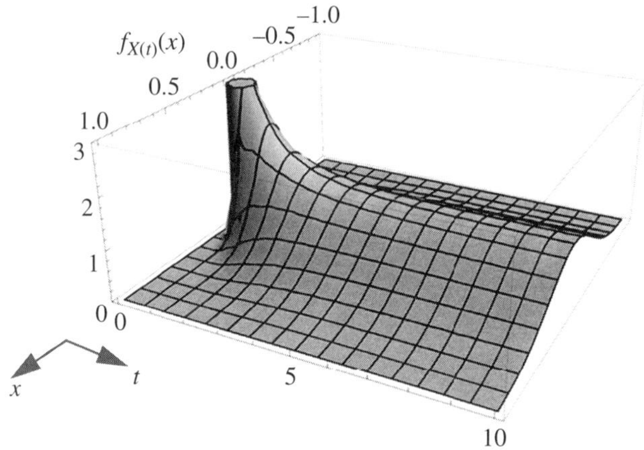


**FIGURE 8.18** Evolution, with time, of the probability density function for the defined signalling random process for the case of $\tau = 2$.

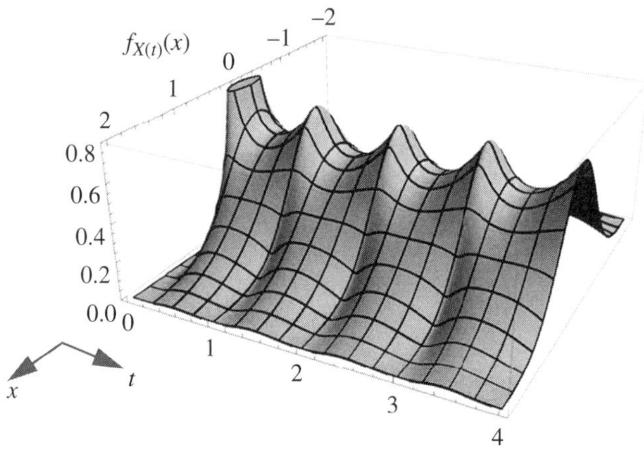


**FIGURE 8.19** Evolution, with time, of the probability density function for the defined signalling random process for the case of $\tau = 1/2$.

## 8.8 PDF EVOLUTION: GENERALIZED SHOT NOISE

Consider a generalized shot noise random process $X : S \to S_X$ defined according to

$$X(\Omega, t) = \sum_{k=1}^{\infty} A_k h(t - \Gamma_k) \tag{8.67}$$

where $\Omega = ((A_1, \Gamma_1), (A_2, \Gamma_2), \ldots)$, $A_1, \ldots, A_k, \ldots$ are independent random variables and the probability density function and sample space of $A_k$ are, respectively, $f_{A_k}$

and $S_{A_k}$. The random variables $\Gamma_1, \ldots, \Gamma_k, \ldots$ are defined by a Poisson experiment with a rate of $\lambda$. The sample space of $\Gamma_k$ is $S_{\Gamma_k}$. The pulse function $h: \boldsymbol{R} \to \boldsymbol{R}$ is assumed to be causal and continuous with $h(0)=0$. The signal sample space is

$$S_X = \left\{ X(((a_1, t_1),\ldots,(a_k, t_k),\ldots), t) = \sum_{k=1}^{\infty} a_k h(t-t_k) : a_k \in S_{A_k}, t_k \in S_{\Gamma_k} \right\} \tag{8.68}$$

Consider the random variable defined by the random process at a set time $t_o$:

$$X(\Omega, t_o) = \sum_{k=1}^{\infty} A_k h(t_o - \Gamma_k) \tag{8.69}$$

With the definitions

$$X_0 = 0,\ X_1 = A_1 h(t_o - \Gamma_1),\ X_2 = A_2 h(t_o - \Gamma_2), \ldots \qquad \Gamma_1 < \Gamma_2 < \ldots \tag{8.70}$$

it follows that $X(\Omega, t_o)$ is a random sum of random variables, that is,

$$X(\Omega, t_o) = \sum_{k=0}^{M(t_o)} X_k \quad P[M(t_o) = m] = \frac{\lambda^m t_o^m}{k!} \cdot e^{-\lambda t_o}, \quad m \in \{0, 1, 2, \ldots\} \tag{8.71}$$

The result for $P[M(t_o) = m]$ arises from a Poisson point process with a fixed rate $\lambda$ (Theorem 12.13). The probability density function evolution can then be stated as follows:

**Theorem 8.9 Probability Density Function Evolution**
The probability density function of the random process $X$ at a time $t_o$ is:

$$\begin{aligned} f_{X(t_o)}(x)dx &\approx P[X(\Omega, t_o) \in [x, x+dx]] \\ &= e^{-\lambda t_o}\delta_k(x) + \lambda t_o e^{-\lambda t_o} \cdot P[A_1 h(t_o - \Gamma_1) \in [x, x+dx]] + \\ &\quad \frac{\lambda^2 t_o^2}{2!} \cdot e^{-\lambda t_o} \cdot P[A_1 h(t_o - \Gamma_1) + A_2 h(t_o - \Gamma_2) \in [x, x+dx]] + \cdots \end{aligned} \tag{8.72}$$

**Proof**
From Equation 8.71, it follows, from the possible disjoint outcomes, that

$$\begin{aligned} P[X(\Omega, t_o) \in [x, x+dx]] &= P[X_0 \in [x, x+dx], 0 \text{ points in } [0, t_o]] \\ &\quad + P[X_0 + X_1 \in [x, x+dx], 1 \text{ point in } [0, t_o]] + \cdots \\ &= P\left[X_0 \in [x, x+dx] \big|_{0 \text{ point in } [0, t_o]}\right] \cdot P[0 \text{ points in } [0, t_o]] \\ &\quad + P\left[X_o + X_1 \in [x, x+dx] \big|_{1 \text{ point in } [0, t_o]}\right] \cdot P[1 \text{ point in } [0, t_o]] + \cdots \end{aligned} \tag{8.73}$$

where

$$P\left[X_0 \in [x, x+dx]\big|_{0 \text{ points in } [0, t_o]}\right] = \begin{cases} 0 & x \neq 0 \\ 1 & x = 0 \end{cases} \tag{8.74}$$

$$P\left[X_0 + X_1 \in [x, x+dx]\big|_{1 \text{ point in } [0, t_o]}\right] = P[A_1 h(t_o - \Gamma_1) \in [x, x+dx]] \tag{8.75}$$

$$\begin{aligned} &P\left[X_0 + X_1 + X_2 \in [x, x+dx]\big|_{2 \text{ points in } [0, t_o]}\right] \\ &= P[A_1 h(t_o - \Gamma_1) + A_2 h(t_o - \Gamma_2) \in [x, x+dx]] \end{aligned} \tag{8.76}$$

etc.

### 8.8.1 Notes

First, the random variables $X_1, X_2, \ldots$ are correlated as the random variables $\Gamma_1, \Gamma_2, \ldots$ are correlated due to the ordering. For example, the joint probability density function of $A_1$, $A_2$, $\Gamma_1$, and $\Gamma_2$ is

$$f_{A_1 A_2 \Gamma_1 \Gamma_2}(a_1, a_2, t_1, t_2) = f_{A_1}(a_1) \cdot f_{A_2}(a_2) \cdot f_{\Gamma_1 \Gamma_2}(t_1, t_2) \tag{8.77}$$

which implies that $X_1$ and $X_2$ are correlated. This can also readily be seen by writing $\Gamma_2 = \Gamma_1 + \Delta\Gamma_1$ where the interarrival time between $\Gamma_1$ and $\Gamma_2$ is $\Delta\Gamma_1$.

It then follows that the sum, as given by Equation 8.71, is a correlated sum. Determination of the probability density function of a correlated sum is problematic. Accordingly, the determination of an analytic expression for the evolution of the probability density function of a shot noise process is problematic.

### 8.8.2 Estimation of the Probability Density Function

A pragmatic approach is to establish an estimator for the evolution of the probability density function. For a fixed time $t_o$, the random process defines a random variable

$$X(\Omega, t_o) = \sum_{k=1}^{M(t_o)} A_k h(t_o - \Gamma_k) \tag{8.78}$$

and an estimator for the probability density function of this random variable can be obtained from samples. By varying the time $t_o$, an estimator for the probability density function evolution can be obtained. The following procedure is appropriate:

a. In order to generate points, consistent with a Poisson point process, consider the interval $[0, 10t_o]$. Take a point at random from this interval and repeat this $10t_o\lambda$ times. Over the interval $[0, t_o]$, such samples are consistent with those generated by a Poisson point process with a rate $\lambda$.
b. Order the samples and determine the number of samples, $m$, on the interval $[0, t_o]$. Denote these samples $t_1, \ldots, t_m$.

c. Determine $m$ independent samples of the amplitudes $A_1,\ldots,A_m$, which are denoted as $a_1,\ldots, a_m$.

d. Determine $N_o$ samples of the resulting signal

$$X(((a_1,t_1),\ldots,(a_m,t_m)),t)=\sum_{k=1}^{m}a_k h(t-t_k) \tag{8.79}$$

at the times $\Delta t$, $2\Delta t$, …, $N_o\Delta t$, where $t_o = N_o\Delta t$.

e. Repeat this procedure $N$ times to obtain $N_o$ samples of $N$ independent signals.

An approximation to the probability density function of $X(\Omega, i\Delta t)$ can then be obtained from the $i$th sample of each of the $N$ signals by using one of the approaches for probability density function estimation detailed in Section 8.3. The following subsection details an approach using a Gaussian-based basis set.

***8.8.2.1 PDF Evolution Using a Gaussian-Based Basis Set*** Consider the Gaussian-based basis set, for a fixed time $t$, based on Gaussian functions with a mean $\mu_G(t)$ and a variance $\sigma_G^2(t)$, as defined in Theorem 3.3:

$$\begin{cases} b_0(x,t)=\sqrt{2}\pi^{1/4}\sqrt{\sigma}H_0\left[\dfrac{x-\mu_G(t)}{\sigma_G(t)}\right]\cdot\dfrac{e^{-(x-\mu_G(t))^2/2\sigma_G^2(t)}}{\sqrt{2\pi}\sigma_G(t)},\ldots, \\ b_i(x,t)=\dfrac{\pi^{1/4}\sqrt{\sigma}H_i[(x-\mu_G(t))/\sigma_G(t)]}{2^{(i-1)/2}\sqrt{i!}}\cdot\dfrac{e^{-(x-\mu_G(t))^2/2\sigma_G^2(t)}}{\sqrt{2\pi}\sigma_G(t)},\ldots\} \end{cases} \tag{8.80}$$

Consistent with the discussion in Section 8.3.2, the probability density function at a specific time $t$ can be modelled according to

$$f_{X(t)}(x)=c_0(t)b_0(x,\,t)+\cdots+c_i(t)b_i(x,\,t)+\cdots \tag{8.81}$$

where $c_i(t)=\boldsymbol{E}[b_i(X(\Omega,t),t)]$.

From values of $c_i(t)$, $\mu(t)$, and $\sigma^2(t)$ evaluated at the times $\Delta t$, $2\Delta t,\ldots$, approximate values for all other times can be established by using an interpolating function $r$, with $r(0)=1$ and $r(k)=0$, $k\in\boldsymbol{Z}$, $k\neq 0$, defined according to

$$\mu(t)\approx\sum_{k=1}^{\infty}\mu(k\Delta t)r\left(\frac{t-k\Delta t}{\Delta t}\right)\qquad \sigma^2(t)\approx\sum_{k=1}^{\infty}\sigma^2(k\Delta t)r\left(\frac{t-k\Delta t}{\Delta t}\right) \tag{8.82}$$

$$c_i(t)\approx\sum_{k=1}^{\infty}c_i(k\Delta t)r\left(\frac{t-k\Delta t}{\Delta t}\right) \tag{8.83}$$

The probability density function evolution with time can then be approximated according to

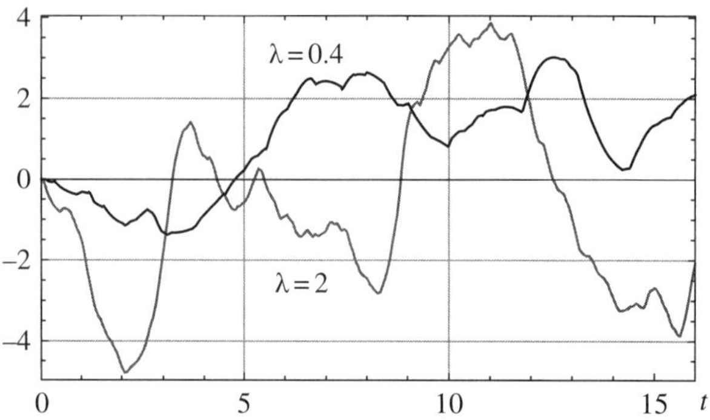


**FIGURE 8.20** Signals from the two defined generalized shot noise random processes.

$$f_{X(t)}(x) \approx \sum_{k=1}^{\infty} r\left(\frac{t-k\Delta t}{\Delta t}\right) \sum_{i=0}^{\infty} c_i(k\Delta t) b_i(x, k\Delta t) \tag{8.84}$$

The usual case is where, for each time $\Delta t$, $2\Delta t, \ldots$, there are $N$ independent sample values available. For the time $k\Delta t$, the samples are denoted as $x_1(k\Delta t), \ldots, x_N(k\Delta t)$. From these samples, the mean, variance, and basis set coefficients can be estimated according to

$$\hat{\mu}(k\Delta t) = \frac{1}{N} \cdot \sum_{j=1}^{N} x_j(k\Delta t) \tag{8.85}$$

$$\hat{\sigma}^2(k\Delta t) = \frac{1}{N} \cdot \sum_{j=1}^{N} x_j^2(k\Delta t) - \left[\frac{1}{N} \cdot \sum_{j=1}^{N} x_j(k\Delta t)\right]^2 \tag{8.86}$$

$$\hat{c}_i(k\Delta t) = \frac{1}{N} \cdot \sum_{j=1}^{N} b\big(x_j(k\Delta t), k\Delta t\big) \tag{8.87}$$

The use of these estimates, for $k \in \{1, 2, \ldots\}$ in Equation 8.84, leads to an estimator for the probability density function evolution.

***8.8.2.2 Example*** Consider two generalized shot noise random processes, one with $\lambda = 2$ and one with $\lambda = 0.4$ and based on a pulse function

$$h(t) = \frac{te^{-t/\tau}}{\tau^2} \cdot u(t) \tag{8.88}$$

where $\tau = 1$. The amplitudes of the pulses are assumed to be independent, with zero mean, unit variance and to have a Gaussian distribution. One signal from each of these random processes is shown in Figure 8.20. To establish an estimator for the probability density function evolution with time, samples from 10,000 signals, with a time

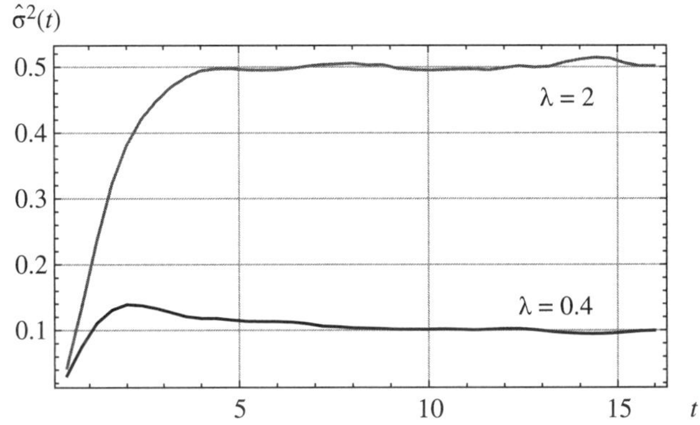


**FIGURE 8.21** Estimation of the evolution of the variance with time for the two defined generalized shot noise random processes.

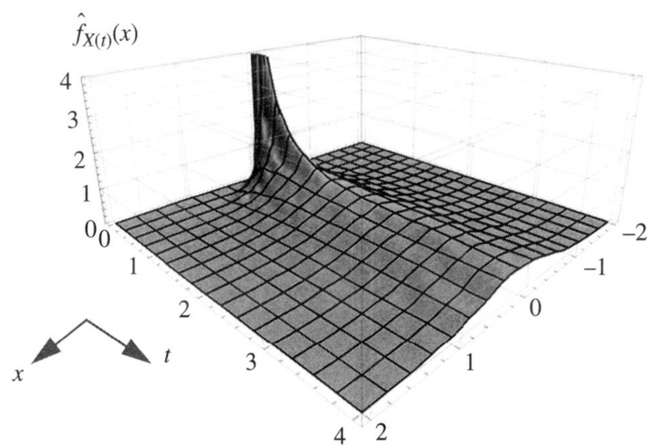


**FIGURE 8.22** Estimator for the probability density function evolution of the defined generalized shot noise random process for the case of $\lambda = 2$.

resolution of 0.1 s when $\lambda = 2$ and 0.2 s when $\lambda = 0.4$, have been used to establish estimates for the mean, variance, and basis function coefficients consistent with Equations 8.85, 8.86, and 8.87. Estimates of the variance are shown in Figure 8.21. The estimator for the probability density function evolution is shown in Figure 8.22 for the case of $\lambda = 2$ and in Figure 8.24 for the case of $\lambda = 0.4$. The results are based on a 21-term Gaussian-based basis set with $\mu_G(k\Delta t) = \hat{\mu}(k\Delta t)$, $\sigma_G^2(k\Delta t) = \hat{\sigma}^2(k\Delta t)/4$, and the triangular interpolation function:

$$r(t) = (1 - |t|)[u(t+1) - u(t-1)] \tag{8.89}$$

The probability mass function associated with the zero outcome is shown in Figure 8.23 for the case of $\lambda = 2$.

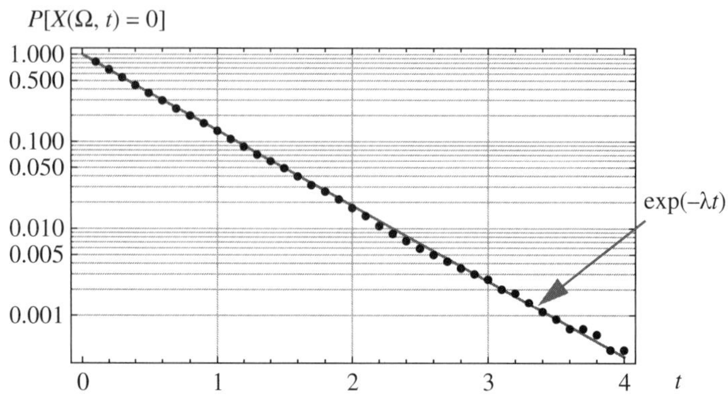


**FIGURE 8.23** Estimator for the probability density function evolution of the defined generalized shot noise random process for the case of $\lambda = 2$.

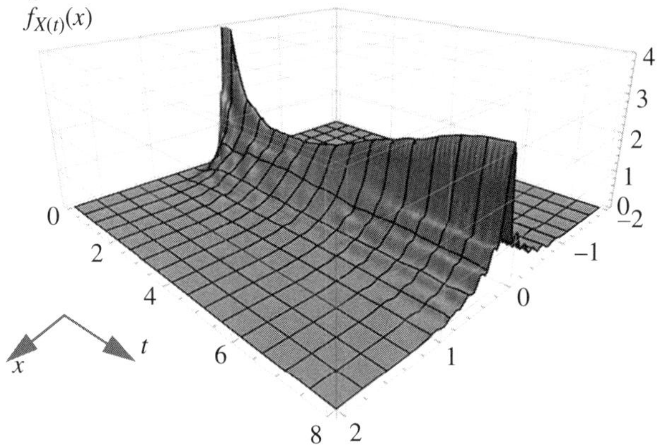


**FIGURE 8.24** Estimated values for the probability mass function for the zero outcome case of the defined generalized shot noise random process when $\lambda = 0.4$.

## 8.9 PDF EVOLUTION: SWITCHING IN A CMOS INVERTER

Consider a CMOS inverter with a capacitive load and its associated model, which are detailed in Figures 8.25 and 8.26. The model assumes, as is typically the case, that the internal MOSFET capacitance are small relative to the load capacitance and can be included in this capacitance. The current sources in the model have, as a first-order approximation, the following definitions:

$$i_N(v_{GS}, v_{DS}, t) = \begin{cases} 0 & \text{cut off}: v_{GS}(t) \le V_T \\ k_n\left[(v_{GS}(t) - V_T)v_{DS}(t) - \dfrac{v_{DS}^2(t)}{2}\right] & \text{triode region}: \\ & v_{DS}(t) \le v_{GS}(t) - V_T \\ \dfrac{k_n}{2}(v_{GS}(t) - V_T)^2 & \text{saturation}: v_{DS}(t) > v_{GS}(t) - V_T \end{cases} \tag{8.90}$$

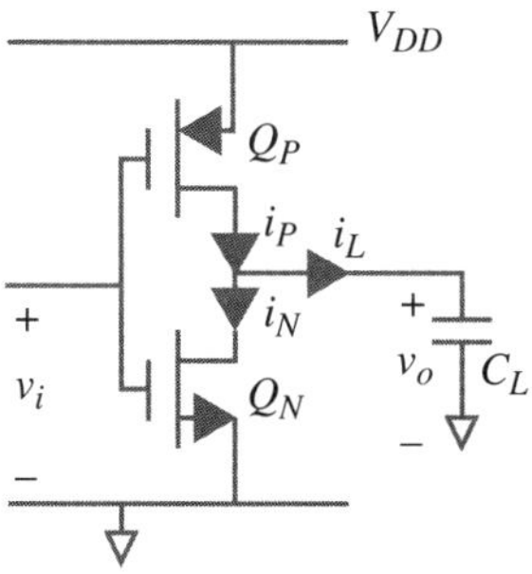


**FIGURE 8.25** Structure of a CMOS inverter.

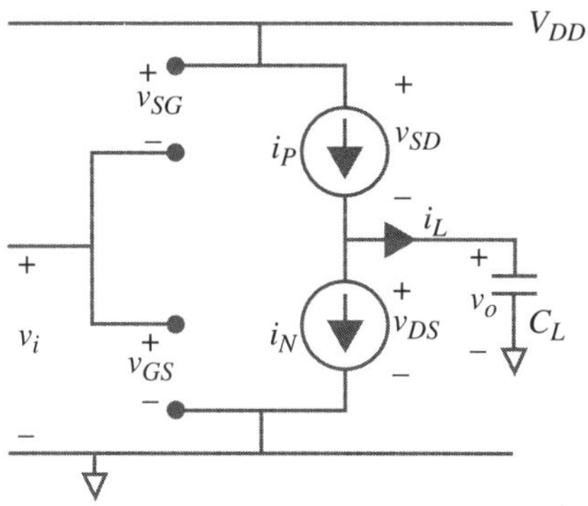


**FIGURE 8.26** Model for CMOS inverter.

$$i_p(v_{SG}, v_{SD}, t) = \begin{cases} 0 & \text{cut off}: v_{SG}(t) \le V_T \\ k_p\left[(v_{SG}(t) - V_T)v_{SD}(t) - \dfrac{v_{SD}^2(t)}{2}\right] & \text{triode region}: \\ & v_{SD}(t) \le v_{SG}(t) - V_T \\ \dfrac{k_p}{2}(v_{SG}(t) - V_T)^2 & \text{saturation}: v_{SD}(t) > v_{SG}(t) - V_T \end{cases} \tag{8.91}$$

where $V_T$ is the threshold voltage and

$$k_n = \frac{\mu_n C_{\text{ox}} W}{L} k_p = \frac{\mu_p C_{\text{ox}} W}{L} C_{\text{ox}} = \frac{\varepsilon_r \varepsilon_{\text{o}}}{t_{\text{ox}}} \tag{8.92}$$

Here, $\mu_n$ and $\mu_p$ are, respectively, the surface mobility of electrons and holes, $\varepsilon_r$ is the relative permittivity of the oxide material, $\varepsilon_{\text{o}}$ is the permittivity of free space, and $t_{\text{ox}}$ is the oxide thickness.

The output voltage of the CMOS inverter is the solution of the first-order nonlinear differential equation

$$C_L \frac{d}{d} v_{\text{o}}(t) = i_P(v_i, v_{\text{o}}, t) - i_N(v_i, v_{\text{o}}, t) \tag{8.93}$$

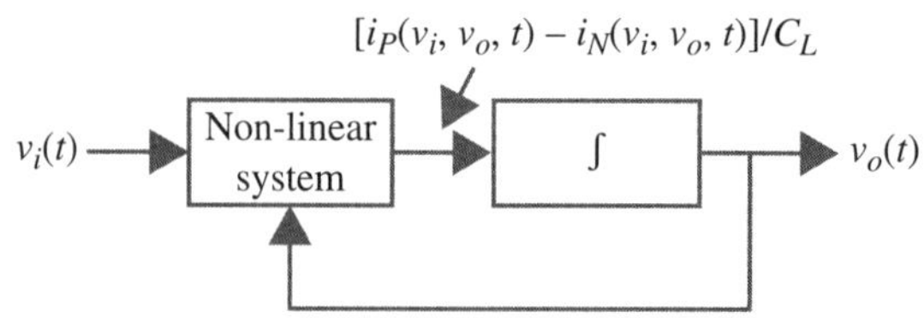


**FIGURE 8.27** Equivalent model: Hammerstein system with feedback.

Integrating yields

$$C_L[v_o(t)-v_o(0)] = \int_0^t [i_P(v_i,v_o,t)-i_N(v_i,v_o,t)]dt \tag{8.94}$$

which is a Hammerstein system with feedback as illustrated in Figure 8.27.

To facilitate numerical solution of the differential equation, the currents can be modelled according to

$$\begin{aligned} i_N(v_{GS},v_{DS},t) &= k_n\left[(v_{GS}(t)-V_T)v_{DS}(t)-\frac{v_{DS}^2(t)}{2}\right]\cdot u[v_{GS}(t)-V_T]\cdot \\ &u[(v_{GS}(t)-V_T)-v_{DS}(t)]+ \\ &\frac{k_n}{2}(V_{GS}(t)-V_T)^2\cdot u[v_{GS}(t)-V_T]\cdot u[v_{DS}(t)-(v_{GS}(t)-V_T)] \end{aligned} \tag{8.95}$$

$$\begin{aligned} i_P(v_{SG},v_{SD},t) &= k_p\left[(v_{SG}(t)-V_T)v_{SD}(t)-\frac{v_{SD}^2(t)}{2}\right]\cdot u[v_{SG}(t)-V_T]\cdot \\ &u[(v_{SG}(t)-V_T)-v_{SD}(t)]+ \\ &\frac{k_p}{2}(V_{SG}(t)-V_T)^2\cdot u[v_{SG}(t)-V_T]\cdot u[v_{SD}(t)-(v_{SG}(t)-V_T)] \end{aligned} \tag{8.96}$$

These equations are nondifferentiable at the transition points between the cutoff, triode, and saturation regions. To avoid the complications this involves (a piecewise solution of the differential equation), a useful approach is to approximate the unit step function by the hyperbolic tangent function according to

$$u(x) \approx \frac{1}{2}\left[1+\tanh\left(\frac{x}{\Delta}\right)\right] \tag{8.97}$$

for a suitably small value of $\Delta$.

The input to the CMOS inverter, for case of switching from high to low, has a finite fall time $t_R$ and is modelled according to

$$v_i(t) = \frac{V_{DD}}{2}[1-\tanh((t-t_R)/\tau)]+n(t) \tag{8.98}$$

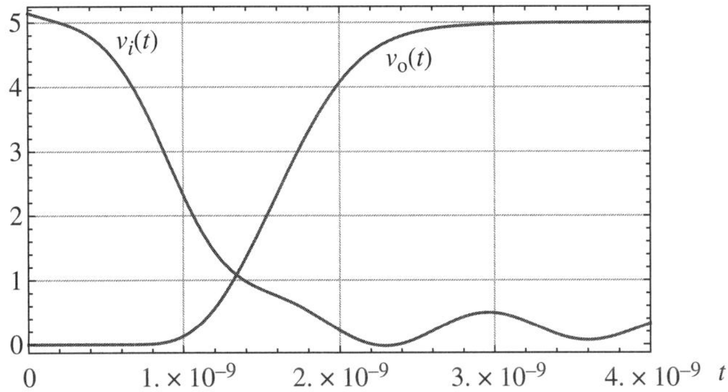


**FIGURE 8.28** Input and output signal.

where $n$ is the input noise signal and the constant $\tau$ is defined according to

$$\tau = \frac{t_R}{2\,\mathrm{atanh}(0.8)} \tag{8.99}$$

The noise signal $n$ is assumed to be from a Gaussian white noise random process $N$, with a flat spectrum on the interval $[0, f_{max}]$, a rms level of $A_o$, and defined consistent with Section 6.7.2.2:

$$N(\Omega, t) = \sum_{k=1}^{N_o} \frac{\sqrt{2}A_o}{\sqrt{N}} \sin(2\pi F_k t + \Phi_k) \tag{8.100}$$

Here, $\Omega = (N_o, (\Omega_1, F_1), \ldots, (\Omega_{N_o}, F_{N_O}))$ is a vector of random variables; the expected value of $N_o$ defines the average number of sinusoids in each signal and the average frequency spacing of the sinusoids is $f_o = f_{max}/\boldsymbol{E}[N_o]$.

Consider the case of a CMOS inverter with the following parameters: $V_{DD} = 5\text{V}$, $k_n = k_p = 0.01$, $V_T = 1\text{V}$, $C_L = 10pF$, $t_R = 1$ ns, $f_{max} = 10^9$ Hz, $\boldsymbol{E}[N_o] = 1000$, $A_o = 0.25$, and $\Delta = 0.1$. One possible input signal and the corresponding output signal are shown in Figure 8.28. Using results from 10000 input signals, the jitter probability density function, associated with timing jitter around the level $V_{DD}/2$, is shown in Figure 8.29, along with the probability density function of the propagation delay. An estimator, for the probability density function evolution of the random process defined by the output signals, is shown in Figure 8.30. The probability density function evolution is based on a 21-term Gaussian basis set as discussed in Section 8.8.2.1 and a triangular reconstruction function as defined by Equation 8.89. The sample resolution used was $\Delta t = 0.1$ ns. More results are detailed in Howard (2013).

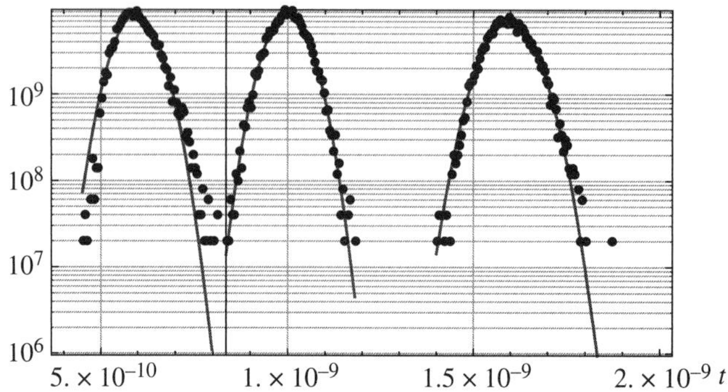


**FIGURE 8.29** Jitter probability density function of input (middle), output (right), and propagation delay (left) associated with the first passage time across the level of $V_{DD}/2$. The smooth curves are a Gaussian approximation.

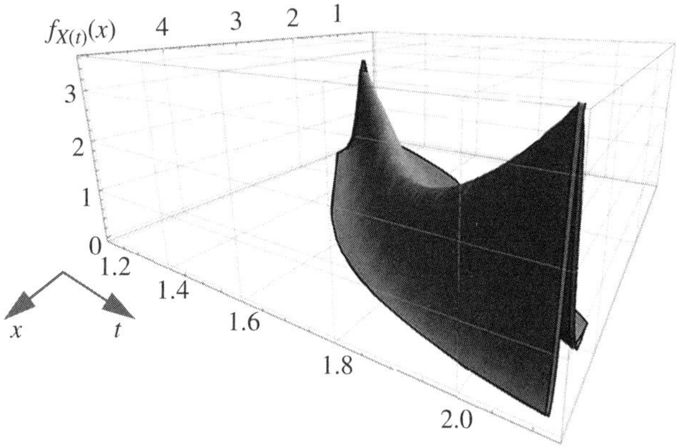


**FIGURE 8.30** PDF evolution of the random process defined by the output signals of a CMOS inverter as the input switches from high to low: 5 V to 0 V. Time is in ns.

## 8.10 PDF EVOLUTION: GENERAL CASE

The previous discussion shows that establishing an analytic solution for the probability density function evolution of a random process is, in general, problematic. One general approach is detailed in Li and Chen (2006, 2008) and Li et al. (2012) but requires the solution of a nonlinear partial differential equation followed by high order integration.

A starting point for considering probability density function evolution is conservation of probability:

### 8.10.1 Conservation of Probability

Consider the probability density function evolution, $f_{X(t)}(x)$, of a random process $X$ as illustrated in Figure 8.31. The total probability in the volume defined by $x \in [-\infty, \infty]$, $t \in [0, t_o]$, and $f_{X(t)}(x)$ is

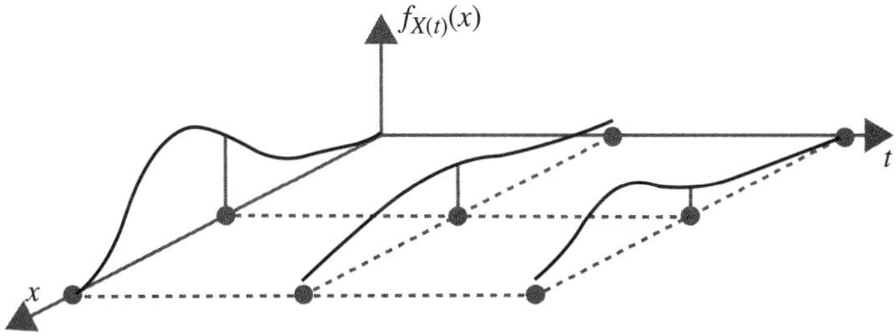


**FIGURE 8.31** Probability function evolution with time.

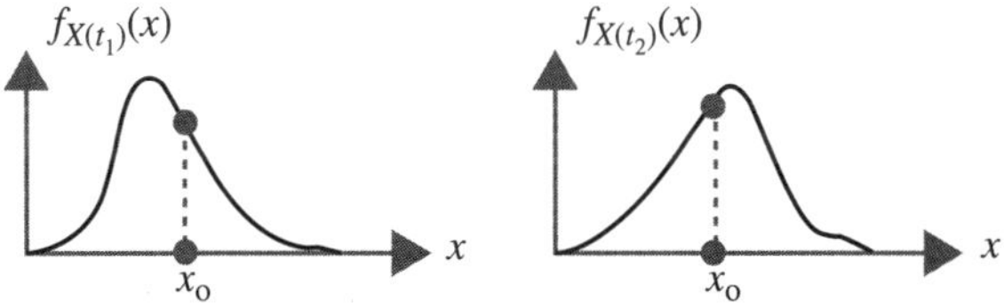


**FIGURE 8.32** Illustration of the change in the probability density function between two times.

$$P_{[0,t_o]} = \int_0^{t_o} \left[ \int_{-\infty}^{\infty} f_{X(t)}(x)dx \right] dt = \int_0^{t_o} dt = t_o \tag{8.101}$$

This result implies that there is conservation of probability within a fixed time interval, and changes in the probability density function evolution must be consistent with this conservation.

***8.10.1.1 Probability Flow*** Consider the change in the probability density function between two times as illustrated in Figure 8.32. Define $F(x_o, t)$ as the flow of probability past the level $x_o$, up until the time $t$, according to

$$F(x_o,t) = \int_{x_o}^{\infty} f_{X(t)}(x)dx - \int_{x_o}^{\infty} f_{X(0)}(x)dx \tag{8.102}$$

This flow must be consistent with the conservation of probability with time. The change in the probability density function in the interval $[x_o, x_o + dx]$ and during the time interval $[t, t + dt]$ is illustrated in Figure 8.33. This change is specified in the following theorem:

**Theorem 8.10 Probability Flow**
For a time-evolving probability density function, $f_{X(t)}(x)$, the probability flow into an interval $[x_o, x_o + dx]$ during the time interval $[t, t + dt]$ equals the increase in probability in the interval $[x_o, x_o + dx]$ and is given by

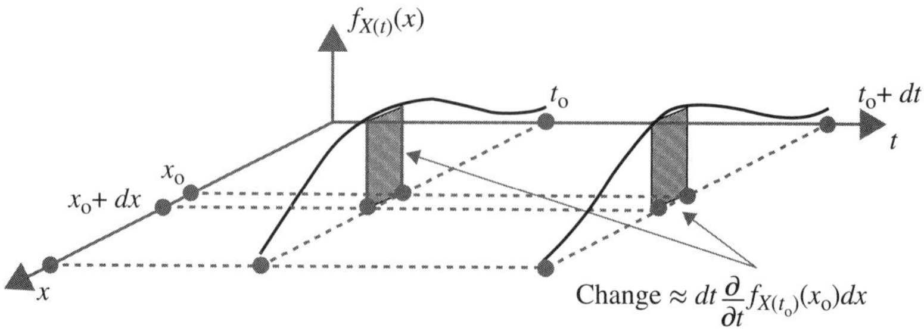


**FIGURE 8.33** Probability function evolution with time.

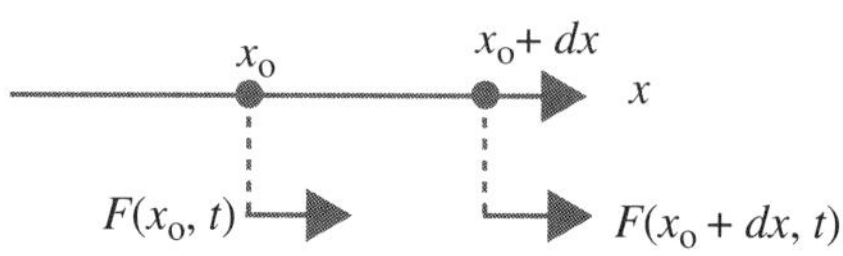


**FIGURE 8.34** Probability flow into and out of the interval $[x_o, x_o + dx]$.

$$dtdx\frac{\partial}{\partial t}f_{X(t)}(x_o) \tag{8.103}$$

Thus, the increase in probability in the interval $[x_o, x_o + dx]$ during $[t, t + dt]$, that is, the net probability flux into the interval, equals $dt$ times the rate of change of probability in the interval $[x_o, x_o + dx]$ with respect to time.

**Proof**

This result can be proved in three ways with each way being informative: First, from the definition of a probability density function, the increase in probability in the interval $[x_o, x_o + dx]$, during the time interval $[t, t + dt]$, approximately is

$$\begin{aligned} \text{probability increase} &\approx f_{X(t+dt)}(x_o)dx - f_{X(t)}(x_o)dx \\ &\approx dtdx\frac{\partial}{\partial t}f_{X(t)}(x_o) \end{aligned} \tag{8.104}$$

Second, consider the definition of probability flow, as specified by Equation 8.102, and an interval $[x_o, x_o + dx]$ as illustrated in Figure 8.34.

The increase in probability in the interval $[x_o, x_o + dx]$ during the time interval $[t, t + dt]$ is due to a net change in the probability flowing into the interval, and this is given by

$$\text{probability increase} = F(x_o, t + dt) - F(x_o, t) - [F(x_o + dx, t + dt) - F(x_o + dx, t)] \tag{8.105}$$

Using the definition of $F(x_o, t)$, the probability increase equals

$$\text{probability increase} = \int_{x_o}^{\infty} f_{X(t+dt)}(x)dx - \int_{x_o}^{\infty} f_{X(t)}(x)dx - \left[\int_{x_o+dx}^{\infty} f_{X(t+dt)}(x)dx - \int_{x_o+dx}^{\infty} f_{X(t)}(x)dx\right] \tag{8.106}$$

Thus,

$$\begin{aligned}\text{probability increase} &= \int_{x_o}^{x_o+dx} f_{X(t+dt)}(x)dx - \int_{x_o}^{x_o+dx} f_{X(t)}(x)dx \\ &= \int_{x_o}^{x_o+dx} \left[f_{X(t+dt)}(x) - f_{X(t)}(x)\right]dx \approx dtdx\frac{\partial}{\partial t}f_{X(t)}(x_o)\end{aligned} \tag{8.107}$$

which is the required result.

Third, the change in probability in a time interval $[t_o, t_o + dt]$ can be established by a first-order Taylor's series:

$$f_{X(t_o+dt)}(x_o + dx) \approx f_{X(t_o)}(x_o) + dt\frac{\partial}{\partial t}f_{X(t_o)}(x_o) + dx\frac{\partial}{\partial x}f_{X(t_o)}(x_o) \tag{8.108}$$

Consider the case of $dx$ and $dt$ sufficiently small so that the first-order Taylor series expansion is valid. For a fixed value of $dt$ and when $\frac{\partial}{\partial t}f_{X(t_o)}(x_o)$ is nonzero, $dx$ can be made arbitrarily small such that

$$f_{X(t_o+dt)}(x_o + dx) \approx f_{X(t_o+dt)}(x_o) \qquad \left|dx\frac{\partial}{\partial x}f_{X(t_o)}(x_o)\right| << \left|dt\frac{\partial}{\partial t}f_{X(t_o)}(x_o)\right| \tag{8.109}$$

and

$$f_{X(t_o+dt)}(x_o) \approx f_{X(t_o)}(x_o) + dt\frac{\partial}{\partial t}f_{X(t_o)}(x_o) \tag{8.110}$$

The change in the probability on the interval $[x_o, x_o + dx]$, as $t$ varies from $t_o$ to $t_o + dt$, then is the required result:

$$\text{prob. increase} = f_{X(t_o+dt)}(x_o)dx - f_{X(t_o)}(x_o)dx \approx dt\frac{\partial}{\partial t}f_{X(t_o)}(x_o)dx \tag{8.111}$$

### 8.10.2 Chapman–Kolmogorov Equation for Continuous Case

For the case of a Markov process with an uncountable number of outcomes, the Chapman–Kolmogorov equation has a similar form to that for the discrete case, as detailed in Theorem 7.23, where the summation is replaced by an integral and probabilities are replaced by values from a probability density function.

**Theorem 8.11 Chapman–Kolmogorov Equation for Continuous Case**
For the case of a Markov process, with an uncountable number of outcomes at times specified by the process, the Chapman–Kolmogorov equation is

$$f_{X(t_3)/X(t_1)=x_1}(x_3) = \int_{-\infty}^{\infty} f_{X(t_2)/X(t_1)=x_1}(x_2) \cdot f_{X(t_3)/X(t_2)=x_2}(x_3)dx_2 \tag{8.112}$$

assuming $t_1 < t_2 < t_3$.

**Proof**
The proof is detailed in Appendix 8.C.

### 8.10.3 Fokker–Planck Equation

For a continuous time Markov process and where the changes are consistent with diffusion (small changes in a small time interval), the probability density function evolution of a random process is given by the Fokker–Planck equation.

**Theorem 8.12 Fokker–Planck Equation**
The probability density function evolution of a continuous time Markov process satisfies the Fokker–Planck equation:

$$\begin{aligned} \frac{\partial}{\partial t}\left[f_{X(t)/X(t_o)=x_0}(x)\right] &= -\frac{\partial}{\partial x}\left[A(x)f_{X(t)/X(t_0)=x_0}(x)\right] \\ &\quad + \frac{1}{2}\cdot\frac{\partial^2}{\partial x^2}\left[B(x)f_{X(t)/X(t_0)=x_0}(x)\right] + \ldots \quad t > t_0 \\ f_{X(t_0)/X(t_0)=x_0}(x) &= \delta(x-x_0) \end{aligned} \tag{8.113}$$

where

$$A(x) = \lim_{\Delta t \to 0} \frac{1}{\Delta t} \int_{-\infty}^{\infty} (x-\lambda) f_{X(t+\Delta t)/X(t_0)=x}(\lambda)d\lambda \tag{8.114}$$

$$B(x) = \lim_{\Delta t \to 0} \frac{1}{\Delta t} \int_{-\infty}^{\infty} (x-\lambda)^2 f_{X(t+\Delta t)/X(t_0)=x}(\lambda)d\lambda \tag{8.115}$$

**Proof**
The proof is detailed in Appendix 8.D.

***8.10.3.1 Notes*** There is significant literature on the Fokker–Planck equation, for example, Risken (1989) and Gardner (1985), and approaches for the non-Markov case, for example Li and Chen (2006, 2008, 2012).

### 8.10.4 Appreciation of the Problem

As the material in this chapter shows, establishing an analytical expression for the probability density function evolution of a stochastic process is nontrivial. To provide an insight into why this is the case, consider a random process defined by the differential equation

$$\frac{\mathrm{d}}{\mathrm{d}t}X(\Omega, t) - F(X(\Omega, t), t) = 0 \tag{8.116}$$

where $\Omega$ is a vector of random variables and where the initial condition $X(\Omega, 0)$ is defined. Integrating yields

$$X(\Omega, t) = X(\Omega, 0) + \int_0^t F(X(\Omega, \lambda), \lambda)d\lambda \tag{8.117}$$

For a suitably small value of $\Delta t$, a Riemann sum approximation to the integral yields

$$\begin{aligned} X(\Omega, i\Delta t) \approx\ & \Delta t F(X(\Omega, (i-1)\Delta t), (i-1)\Delta t) + \Delta t F(X(\Omega, (i-2)\Delta t), (i-2)\Delta t) \\ & + \cdots + \Delta t F(X(\Omega, 0), 0) + X(\Omega, 0) \end{aligned} \tag{8.118}$$

Then

$$\begin{aligned} f_{X(i\Delta t)}(x)dx &= P[X(\Omega, i\Delta t) \in [x, x + dx]] \\ &\approx P[\Delta t F(X(\Omega, (i-1)\Delta t), (i-1)\Delta t) + \Delta t F(X(\Omega, (i-2)\Delta t), (i-2)\Delta t) \\ &\quad + \cdots + \Delta t F(X(\Omega, 0), 0) + X(\Omega, 0) \in [x, x + dx]] \end{aligned} \tag{8.119}$$

which is a sum of functions of random variables that, in general, are correlated. It is this that makes the determination of the probability density function evolution of a random process difficult, and this leads to indirect approaches that result, for example, in the Fokker–Planck equation for continuous time Markov processes, which is a partial differential equation for the probability density function evolution.

## 8.11 PROBLEMS

**8.1** Consider a complex system that exhibits a transient on power down, which is one of two possible forms. For both forms, the energy varies randomly but with a Rayleigh distribution:

$$f_{X_i}(x) = \frac{x}{\sigma_i^2} \cdot \exp\left[\frac{-x^2}{2\sigma_i^2}\right] \qquad i \in \{1,2\} \tag{8.120}$$

The transient energy defines a random variable $Z$. Consider the case where the probability of either form is 1/2 and where $\sigma_1^2 = 1$ and $\sigma_2^2 = 16$.

a. Determine the probability density function of $Z$. Graph this probability density function using the stated parameters.
b. Use 10000 independent samples for $Z$ to determine a histogram estimator, a kernel density estimator, a characteristic function estimator, and a basis set estimator for its probability density function. For the kernel density estimator, use a rectangular kernel function with an optimum value for $h$ of $h = 0.3$. For the characteristic function estimator, use parameters $N_f = 100$ and $\Delta f = 0.01$. For the basis set estimator, use a 21-term Gaussian-based basis set and based on the sample mean and sample variance.
c. For each of the estimators, determine the integrated errors.

**8.2** Consider a reflectometer system where the received signal can be modelled as a delayed rectangular pulse plus additive noise.

a. For the case of a received signal $s(t) = A_o[u(t-D) - u(t-D-\tau)]$ and additive noise with a probability density function $f_N(n) = \exp\left[-(n-\mu)^2/2\sigma^2\right]/\sqrt{2\pi}\sigma$, specify the probability density function evolution with time of the associated random process.
b. Graph the probability density function evolution with time for the case of, $A_o = 1$, $D = 1$, $\tau = D/4$, $\mu = A_o/4$, and $\sigma = A_o/10$.

**8.3** The phase of an oscillator exhibits drift and random variations and is modelled as a random walk with a phase jump of $\Delta x = \pm\pi/100$ every $\Delta t = 0.01$ s and with $P[\Delta x = \pi/100] = 0.6$. Graph the probability mass function evolution of the phase with time.

**8.4** A causal random telegraph signal is based on a Poisson point process with a rate $\lambda$. Determine the probability mass evolution with time of the two states of this random telegraph signal by using the result:

$$P[k \text{ points in } (0,t)] = \frac{\lambda^k t^k e^{-\lambda t}}{k!} \qquad k \in \{0,1,\ldots\} \tag{8.121}$$

Graph the probability mass function evolution with time for the case of $\lambda = 1$.

**8.5** Consider a signalling random process $X$ that models binary communication according to

$$X(\Omega,t) = \sum_{k=0}^{\infty} A_k h(t-kD) \tag{8.122}$$

where $\Omega=\{A_0,\ldots,A_k,\ldots\}$, $A_0,\ldots,A_k,\ldots$ are identical and independent random variables with a sample space $S_A=\{-1,1\}$ and $P[A_k=1]=p$ and $P[A_k=-1]=1-p$.

a. Specify and graph the probability mass function evolution with time for the case of $h(t)=u(t)-u(t-D)$ and $p=1/2$.
b. Specify and graph the probability mass function evolution with time for the case of $p=0.6$ and

$$h(t)=\left[u(t)-u\left(t-\frac{D}{2}\right)\right]-\left[u\left(t-\frac{D}{2}\right)-u(t-D)\right] \tag{8.123}$$

c. Specify and graph the probability mass function evolution with time for the case of $p=0.4$ and

$$h(t)=\sin\left(\frac{\pi t}{D}\right)[u(t)-u(t-D)] \tag{8.124}$$

d. Specify and graph the probability mass function evolution with time for the case of $p=0.5$ and

$$h(t)=\operatorname{sinc}\left(\frac{t-D/2}{D/2}\right)[u(t)-u(t-2D)] \tag{8.125}$$

**8.6** Consider a receiver signal which is modelled as a modulated shot noise random process that is subject to additive noise. The model is

$$X(\Omega,t)=e^{-t/\tau}\sum_{i=1}^{\infty}p(t-\Gamma_i)+N(\Omega_N,t) \tag{8.126}$$

where $\Omega$ = $(\Gamma_1,\ldots,\Omega_N)$, $\Gamma_1,\ldots$ are a sequence of random variables defined by the underlying Poisson experiment that has a rate $\lambda$, $\Omega_N$ is a vector of random variables underpinning the noise random process, and the pulse function $p$ is defined according to $p(t)=u(t)-u(t-\Delta)$ where $\Delta \ll 1/\lambda$. The noise random process is assumed to be independent of the shot noise random process and is assumed to be stationary with a probability density function at a set time of $f_N$.

a. Determine an expression for the probability density function evolution of the random process assuming that, at most, two pulses overlap with non-negligible probability.
b. Graph the probability density function evolution with time assuming $\lambda=1$, $\tau=10$, $\Delta=0.2$, and $f_N$ is a zero mean Gaussian function with a variance of 0.01.

**8.7** The powering up of a complex system is done in sequential stages and one measure of the transient component signal generated at a particular point is defined by a random process according to

$$X(\Omega,t)=\sum_{k=0}^{N}A_k\cdot\frac{(t-kD)^2e^{-(t-kD)/\tau_k}}{2\tau_k^3}\cdot(2+\sin[2\pi f_k(t-kD)])\cdot u(t-kD) \tag{8.127}$$

where $\Omega=(A_0,\ldots,A_N)$.

a. For the case of $A_0, \ldots, A_N$ being independent Gaussian random variables with zero mean and with respective variances $\sigma_0^2,\ldots,\sigma_N^2$, determine the variance of $X$ and the probability density function evolution with time of $X$.

b. For the case of $D=1$, $\tau_0=\ldots=\tau_N=D$, $f_0=\ldots=f_N=2/D$, $\sigma_k^2=\sigma_o^2/(k+1)$, $k\in\{0,\ldots,N\},\sigma_o^2=1$, and $N$ large, graph some signals from the random process, the variance of the random process, and the probability density function evolution of the random process with time.

**8.8** Consider a jittered pulse train defined according to

$$X(\Omega,t)=\sum_{k=1}^{N}A_kp(t-kD-J_k) \qquad \Omega=((A_1,J_1),\ldots,(A_N,J_N)) \qquad t\in[0,ND] \tag{8.128}$$

where

$$p(t)=\frac{1}{\sqrt{2\pi}\sigma}\cdot\exp(-t^2/2\sigma^2) \qquad \sigma^2=t_{\text{fwhm}}^2/8\ln(2) \tag{8.129}$$

and all the random variables are independent. Assume $A_k$ has a Rayleigh distribution with parameter $\sigma_A$ (variance of $\sigma_A^2(4-\pi)/2$) and $J_k$ has a Gaussian distribution with zero mean and a variance $\sigma_J^2$. Assume $D=1$, $t_{\text{fwhm}}=1/2$, $\sigma_A=1$, and $\sigma_J=0.1D$.

a. Graph a few signals from the random process.

b. Using data from 10,000 signals, graph an estimator for the mean and variance of the random process.

c. Using data from 10,000 signals and a Gaussian-based basis set, graph an approximation to the probability density function evolution with time of the random process.

**8.9** Consider a random process comprising of a linear signal plus noise with an exponential distribution such that its probability density function evolves with time according to

$$f_{X(t)}(x)=\lambda e^{-\lambda(x-kt)}u(x-kt) \qquad k>0,t>0 \tag{8.130}$$

a. Graph the probability density function evolution for the case of $\lambda = 1$ and $k = 1$.

b. Determine the probability flow past a level $x_o$ in the interval $[0, t]$ as given by

$$F(x_o, t) = \int_{x_o}^{\infty} f_{X(t)}(x)dx - \int_{x_o}^{\infty} f_{X(0)}(x)dx \tag{8.131}$$

c. For $dx \to 0$ and $dt \to 0$, determine the probability increase in the interval $[x_o, x_o + dx]$ during the interval $[t, t + dt]$ as given by

$$\begin{aligned} \text{probability increase} &= F(x_o, t+dt) - F(x_o, t) \\ &\quad - [F(x_o + dx, t+dt) - F(x_o + dx, t)] \end{aligned} \tag{8.132}$$

Show that this probability increase is consistent with $dtdx\frac{\partial}{\partial t} f_{X(t)}(x_o)$.

## APPENDIX 8.A PROOF OF THEOREM 8.1

Consider

$$\widehat{\Phi}_X(f) = \frac{1}{N}\sum_{i=1}^{N} e^{j2\pi f x_i} \tag{8.133}$$

and

$$\hat{f}_X(x) = \int_{-\infty}^{\infty} \widehat{\Phi}_X(f) W(f) e^{-j2\pi f x} df = \frac{1}{N}\sum_{i=1}^{N} \int_{-\infty}^{\infty} W(f) e^{-j2\pi f (x - x_i)} df \tag{8.134}$$

Consider the Fourier transform of $W$:

$$w(x) = \int_{-\infty}^{\infty} W(f) e^{-j2\pi f x} df \quad \Rightarrow \quad w(x - x_i) = \int_{-\infty}^{\infty} W(f) e^{-j2\pi f (x - x_i)} df \tag{8.135}$$

The required result then follows:

$$\hat{f}_X(x) = \frac{1}{N}\sum_{i=1}^{N} w(x - x_i) \tag{8.136}$$

### 8.A.1 Proof of Theorem 8.1

Consider reconstruction of the characteristic function, from samples taken over the range $-N_f\Delta f \le f \le N_f\Delta f$, according to

$$\widehat{\Phi}_X(f) \approx \sum_{k=-N_f}^{N_f} \phi_k \operatorname{sinc}\left(\frac{f-k\Delta f}{\Delta f}\right) \qquad \phi_k = \widehat{\Phi}_X(k\Delta f) \tag{8.137}$$

where a sinc reconstruction function has been used. As the probability density function is the Fourier transform of the characteristic function, it follows that

$$\begin{aligned} f_X(x) &= \int_{-\infty}^{\infty} \Phi_X(f)\exp(-j2\pi fx)df \approx \int_{-\infty}^{\infty} \left[\sum_{k=-N_f}^{N_f} \phi_k \operatorname{sinc}\left(\frac{f-k\Delta f}{\Delta f}\right)\right] \exp(-j2\pi fx)df \\ &= \sum_{k=-N_f}^{N_f} \phi_k \int_{-\infty}^{\infty} \operatorname{sinc}\left(\frac{f-k\Delta f}{\Delta f}\right)\exp(-j2\pi fx)df \end{aligned} \tag{8.138}$$

Hence, $f_X(x)$ can be approximated by the weighted sum of the Fourier transform of delayed sinc functions. Using the Fourier transform results,

$$\begin{aligned} \operatorname{sinc}\left(\frac{t}{\Delta t}\right) &\quad \leftrightarrow \quad \begin{cases} \Delta t & -1/2\Delta t \le f < 1/2\Delta t \\ 0 & \text{elsewhere} \end{cases} \\ v(t-k\Delta t) &\quad \leftrightarrow \quad V(f)e^{-j2\pi k\Delta tf} \end{aligned} \tag{8.139}$$

it follows that an estimator for $f_X$ is

$$\hat{f}_X(x) = q(x)\sum_{k=-N_f}^{N_f} \phi_k \exp(-j2\pi k\Delta fx) \tag{8.140}$$

where

$$q(x) = \begin{cases} \Delta f & -1/2\Delta f \le x < 1/2\Delta f \\ 0 & \text{elsewhere} \end{cases} \tag{8.141}$$

The required result then follows, namely,

$$\hat{f}_X(x) = \Delta f \sum_{k=-N_f}^{N_f} \phi_k \exp(-j2\pi k\Delta fx) \qquad -\frac{1}{2\Delta f} < x < \frac{1}{2\Delta f} \tag{8.142}$$

## APPENDIX 8.B PROOF OF THEOREM 8.5

Consider the result from Theorem 8.4,

$$P[X(\Omega,t) = k\Delta x] = b\left(\frac{t/\Delta t + k}{2}, \frac{t}{\Delta t}, p\right) \tag{8.143}$$

and the approximation given by the DeMoivre–Laplace theorem (Theorem 4.27)

$$b[k_o, N, p] \approx \frac{e^{-(k_o-\mu)^2/2\sigma^2}}{\sqrt{2\pi}\sigma} \tag{8.144}$$

where $\sigma = \sqrt{Np(1-p)}$ and $\mu = Np$. The region of validity is

$$\mu - \sigma^{4/3}\left[\frac{6|\text{r.e.}|}{1-2p}\right]^{1/3} < k_o < \mu + \sigma^{4/3}\left[\frac{6|\text{r.e.}|}{1-2p}\right]^{1/3} \tag{8.145}$$

when $\dfrac{2\sigma|\text{r.e.}|}{|1-2p|} > 10$. After $N$ steps, at a time $t = N\Delta t$, a random walk is at a level $x = k\Delta x$, which implies

$$k = x/\Delta x \qquad N = t/\Delta t \qquad \mu = pt/\Delta t \qquad \sigma = \sqrt{tp(1-p)/\Delta t} \tag{8.146}$$

Substitution of these definitions into Equation 8.144 yields

$$P[X(\Omega, t) = x] = b\left(\frac{t}{2\Delta t} + \frac{x}{2\Delta x}, \frac{t}{\Delta t}, p\right) \approx \frac{e^{-\left(\frac{x}{2\Delta x}+\frac{t}{2\Delta t}-\frac{tp}{\Delta t}\right)^2/2\left(\frac{t}{\Delta t}\right)p(1-p)}}{\sqrt{2\pi}\sqrt{\left(\frac{t}{\Delta t}\right)p(1-p)}} \tag{8.147}$$

Consistent with Equation 8.145, the region of validity is

$$\begin{gathered}\frac{tp}{\Delta t} - \left[\left(\frac{t}{\Delta t}\right)p(1-p)\right]^{2/3}\left[\frac{6|\text{r.e.}|}{1-2p}\right]^{1/3} < \frac{x}{2\Delta x} + \frac{t}{2\Delta t} < \\ \frac{tp}{\Delta t} + \left[\left(\frac{t}{\Delta t}\right)p(1-p)\right]^{2/3}\left[\frac{6|\text{r.e.}|}{1-2p}\right]^{1/3} \qquad \frac{2\sqrt{tp(1-p)/\Delta t}|\text{r.e.}|}{|1-2p|} > 10\end{gathered} \tag{8.148}$$

Simplifying yields

$$P[X(t) = x] \approx \frac{\Delta x e^{-\left(x+\frac{\Delta xt}{\Delta t}-\frac{2\Delta xtp}{\Delta t}\right)^2/8\left(\frac{\Delta x^2 t}{\Delta t}\right)p(1-p)}}{\sqrt{2\pi}\sqrt{\left(\frac{\Delta x^2 t}{\Delta t}\right)p(1-p)}} \tag{8.149}$$

provided

$$\begin{gathered}\frac{2\Delta xtp}{\Delta t} - 2\left[\left(\frac{\Delta x^{3/2} t}{\Delta t}\right)p(1-p)\right]^{2/3}\left[\frac{6(\text{r.e.})}{1-2p}\right]^{1/3} < x + \frac{\Delta xt}{\Delta t} < \\ \frac{2\Delta xtp}{\Delta t} + 2\left[\left(\frac{\Delta x^{3/2} t}{\Delta t}\right)p(1-p)\right]^{2/3}\left[\frac{6(\text{r.e.})}{1-2p}\right]^{1/3} \qquad \frac{2\sqrt{tp(1-p)}|\text{r.e.}|}{\sqrt{\Delta t}|1-2p|} > 10\end{gathered} \tag{8.150}$$

With the use of the definition $D = \Delta x^2/\Delta t$, the required results follow.

## APPENDIX 8.C PROOF OF THEOREM 8.11

Consider the joint probability density function

$$\begin{aligned}
&f_{X(t_1)X(t_2)X(t_3)}(x_1, x_2, x_3)dx_1dx_2dx_3 \\
&\quad = P[X(\Omega, t_1) \in [x_1, x_1 + dx_1], X(\Omega, t_2) \in [x_2, x_2 + dx_2], X(\Omega, t_3) \in [x_3, x_3 + dx_3]] \\
&\quad = P[X_{t_3} \in [x_3, x_3 + dx_3]/X_{t_1} \in [x_1, x_1 + dx_1], X_{t_2} \in [x_2, x_2 + dx_2]] \cdot \\
&\qquad P[X_{t_1} \in [x_1, x_1 + dx_1], X_{t_2} \in [x_2, x_2 + dx_2]] \\
&\quad = P[X_{t_3} \in [x_3, x_3 + dx_3]/X_{t_1} \in [x_1, x_1 + dx_1], X_{t_2} \in [x_2, x_2 + dx_2]] \cdot \\
&\qquad P[X_{t_2} \in [x_2, x_2 + dx_2]/X_{t_1} \in [x_1, x_1 + dx_1]] \cdot P[X_{t_1} \in [x_1, x_1 + dx_1]]
\end{aligned} \tag{8.151}$$

With the Markov assumption, it then follows that

$$\begin{aligned}
&f_{X(t_1)X(t_2)X(t_3)}(x_1, x_2, x_3)dx_1dx_2dx_3 = f_{X(t_3)/X(t_2)=x_2}(x_3)dx_3 \cdot \\
&f_{X(t_2)/X(t_1)=x_1}(x_2)dx_2 \cdot f_{X(t_1)}(x_1)dx_1
\end{aligned} \tag{8.152}$$

and thus

$$\begin{aligned}
f_{X(t_1)X(t_2)X(t_3)}(x_1,x_2,x_3) = f_{X(t_3)/X(t_2)=x_2}(x_3) \cdot f_{X(t_2)/X(t_1)=x_1}(x_2) \cdot \\
f_{X(t_1)}(x_1)
\end{aligned} \tag{8.153}$$

Consider the general result for a marginal density function

$$\begin{aligned}
\int_{-\infty}^{\infty} f_{X(t_1)X(t_2)X(t_3)}(x_1,x_2,x_3)dx_2 &= f_{X(t_1)X(t_3)}(x_1,x_3) \\
&= f_{X(t_3)/X(t_1)=x_1}(x_3) \cdot f_{X(t_1)}(x_1)
\end{aligned} \tag{8.154}$$

The use of Equation 8.153 then yields

$$\int_{-\infty}^{\infty} f_{X(t_3)/X(t_2)=x_2}(x_3) \cdot f_{X(t_2)/X(t_1)=x_1}(x_2) \cdot f_{X(t_1)}(x_1)dx_2 = f_{X(t_3)/X(t_1)=x_1}(x_3) \cdot f_{X(t_1)}(x_1) \tag{8.155}$$

and cancellation of the term $f_{X(t_1)}(x_1)$ yields the required result.

## APPENDIX 8.D PROOF OF THEOREM 8.12

First, consider the Chapman–Kolmogorov equation (Theorem 8.11) for the case illustrated in Figure 8.35.

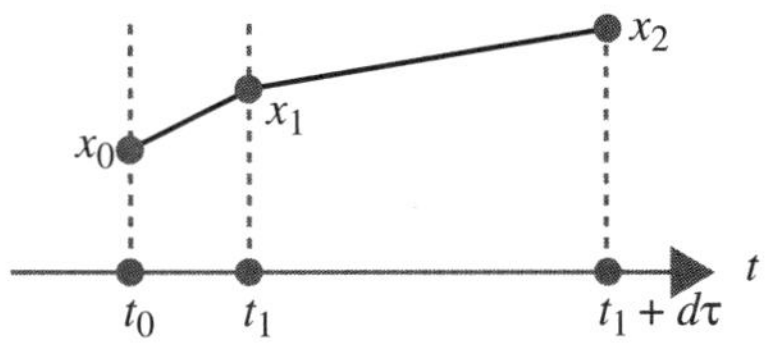


**FIGURE 8.35** Level changes with time.

$$\begin{aligned} f_{X(t_1+d\tau)/X(t_0)=x_0}(x_2) &= \int_{-\infty}^{\infty} f_{X(t_1)/X(t_0)=x_0}(x_1)\cdot f_{X(t_1+d\tau)/X(t_1)=x_1}(x_2)dx_1 \\ &= \int_{-\infty}^{\infty} f_{X(t_1)/X(t_0)=x_0}(x_2-\lambda)\cdot f_{X(t_1+d\tau)/X(t_1)=x_2-\lambda}(x_2)d\lambda \end{aligned} \tag{8.156}$$

The last equation is valid as the summation is positive (a summation of probabilities) and over all values of the integration variable.

Second, a first-order Taylor series expansion for $f_{X(t+d\tau)/X(t_0)=x_0}(x_2)$ with respect to $t$, and with a reference of $t_1$, yields for an appropriately small value of $d\tau$:

$$f_{X(t_1+d\tau)/X(t_0)=x_0}(x_2) \approx f_{X(t_1)/X(t_0)=x_0}(x_2) + d\tau\frac{\partial}{\partial t}\left[f_{X(t)/X(t_0)=x_0}(x_2)\right]|_{t\to t_1} \tag{8.157}$$

Equating Equations 8.156 and 8.157 yields

$$\begin{aligned} d\tau\frac{\partial}{\partial t}\left[f_{X(t)/X(t_0)=x_0}(x_2)\right]|_{t\to t_1} &= -f_{X(t_1)/X(t_0)=x_0}(x_2) \\ &+ \int_{-\infty}^{\infty} f_{X(t_1)/X(t_0)=x_0}(x_2-\lambda)\cdot f_{X(t_1+d\tau)/X(t_1)=x_2-\lambda}(x_2)d\lambda \end{aligned} \tag{8.158}$$

By multiplying both sides by $dx$, it follows that the left-hand side, consistent with Theorem 8.10, is equal to the change in probability, in the interval $[x_2, x_2+dx]$, of the random process $X(\Omega,t)/X(\Omega,t_0)=x_0$ during the time from $t_1$ to $t_1+d\tau$.

Third, a Taylor's series of the integrand in Equation 8.158, with respect to $x_2$ and around a point $x_{20}$, is

$$\begin{aligned} & f_{X(t_1)/X(t_0)=x_0}(x_{20}+\Delta-\lambda)\cdot f_{X(t_1+d\tau)/X(t_1)=x_{20}+\Delta-\lambda}(x_{20}+\Delta) \\ &\approx f_{X(t_1)/X(t_0)=x_0}(x_{20}-\lambda)\cdot f_{X(t_1+d\tau)/X(t_1)=x_{20}-\lambda}(x_{20}) \\ &+\Delta\cdot\frac{\partial}{\partial x_2}\left[f_{X(t_1)/X(t_0)=x_0}(x_2-\lambda)\cdot f_{X(t_1+d\tau)/X(t_1)=x_2-\lambda}(x_2)\right]\Big|_{x_2\to x_{20}} \\ &+\frac{\Delta^2}{2}\cdot\frac{\partial^2}{\partial x_2^2}\left[f_{X(t_1)/X(t_0)=x_0}(x_2-\lambda)\cdot f_{X(t_1+d\tau)/X(t_1)=x_2-\lambda}(x_2)\right]\Big|_{x_2\to x_{20}} + \cdots \end{aligned} \tag{8.159}$$

This equation is valid for any fixed value of $\lambda$ and for $\Delta$ in the region of convergence of the expansion. For a fixed value of $\lambda$ and $x_2$, consider the Taylor series expansion around the point $x_{20} = x_2 + \lambda$:

$$\begin{aligned}
&f_{X(t_1)/X(t_0)=x_0}(x_2+\Delta)\cdot f_{X(t_1+d\tau)/X(t_1)=x_2+\Delta}(x_2+\Delta+\lambda) \\
&\quad \approx f_{X(t_1)/X(t_0)=x_0}(x_2)\cdot f_{X(t_1+d\tau)/X(t_1)=x_2}(x_2+\lambda) \\
&\quad + \Delta\cdot\frac{\partial}{\partial x_2}\left[f_{X(t_1)/X(t_0)=x_0}(x_2)\cdot f_{X(t_1+d\tau)/X(t_1)=x_2}(x_2+\lambda)\right] \\
&\quad + \frac{\Delta^2}{2}\cdot\frac{\partial^2}{\partial x_2^2}\left[f_{X(t_1)/X(t_0)=x_0}(x_2)\cdot f_{X(t_1+d\tau)/X(t_1)=x_2}(x_2+\lambda)\right] + \cdots
\end{aligned} \tag{8.160}$$

For the case of $\Delta = -\lambda$, the expansion is

$$\begin{aligned}
&f_{X(t_1)/X(t_0)=x_0}(x_2-\lambda)\cdot f_{X(t_1+d\tau)/X(t_1)=x_2-\lambda}(x_2) \\
&\quad \approx f_{X(t_1)/X(t_0)=x_0}(x_2)\cdot f_{X(t_1+d\tau)/X(t_1)=x_2}(x_2+\lambda) \\
&\quad + (-\lambda)\cdot\frac{\partial}{\partial x_2}\left[f_{X(t_1)/X(t_0)=x_0}(x_2)\cdot f_{X(t_1+d\tau)/X(t_1)=x_2}(x_2+\lambda)\right] \\
&\quad + \frac{\lambda^2}{2}\cdot\frac{\partial^2}{\partial x_2^2}\left[f_{X(t_1)/X(t_0)=x_0}(x_2)\cdot f_{X(t_1+d\tau)/X(t_1)=x_2}(x_2+\lambda)\right] + \cdots
\end{aligned} \tag{8.161}$$

Substitution of this approximation into Equation 8.158 yields

$$\begin{aligned}
&d\tau\frac{\partial}{\partial t}\left[f_{X(t)/X(t_0)=x_0}(x_2)\right]\Big|_{t\to t_1} = -f_{X(t_1)/X(t_0)=x_0}(x_2) \\
&\quad + f_{X(t_1)/X(t_0)=x_0}(x_2)\int_{-\infty}^{\infty} f_{X(t_1+d\tau)/X(t_1)=x_2}(x_2+\lambda)\,d\lambda \\
&\quad + \frac{\partial}{\partial x_2}\left[f_{X(t_1)/X(t_0)=x_0}(x_2)\int_{-\infty}^{\infty}(-\lambda) f_{X(t_1+d\tau)/X(t_1)=x_2}(x_2+\lambda)\,d\lambda\right] \\
&\quad + \frac{\partial^2}{\partial x_2^2}\left[f_{X(t_1)/X(t_0)=x_0}(x_2)\int_{-\infty}^{\infty}\frac{\lambda^2}{2}\cdot f_{X(t_1+d\tau)/X(t_1)=x_2}(x_2+\lambda)\,d\lambda\right] + \cdots
\end{aligned} \tag{8.162}$$

The first two terms on the right-hand side cancel out leaving

$$d\tau\frac{\partial}{\partial t}\left[f_{X(t)/X(t_0)=x_0}(x_2)\right]\Big|_{t\to t_1} =$$

$$(-1)\frac{\partial}{\partial x_2}\left[f_{X(t_1)/X(t_0)=x_0}(x_2)\int_{-\infty}^{\infty}\lambda f_{X(t_1+d\tau)/X(t_1)=x_2}(x_2+\lambda)d\lambda\right] \tag{8.163}$$

$$+\frac{\partial^2}{\partial x_2^2}\left[f_{X(t_1)/X(t_0)=x_0}(x_2)\int_{-\infty}^{\infty}\frac{\lambda^2}{2}\cdot f_{X(t_1+d\tau)/X(t_1)=x_2}(x_2+\lambda)d\lambda\right]+\cdots$$

A change of variable in the integrals of $\xi = x_2 + \lambda$, along with the definitions $t_1 \to t$ and $x_2 \to x$, yields

$$\frac{\partial}{\partial t}\left[f_{X(t)/X(t_0)=x_0}(x)\right] =$$

$$(-1)\frac{\partial}{\partial x}\left[f_{X(t)/X(t_0)=x_0}(x)\lim_{d\tau\to 0}\frac{1}{d\tau}\left[\int_{-\infty}^{\infty}(\xi-x)f_{X(t+d\tau)/X(t)=x}(\xi)d\xi\right]\right] \tag{8.164}$$

$$+\frac{\partial^2}{\partial x^2}\left[f_{X(t)/X(t_0)=x_0}(x)\lim_{d\tau\to 0}\frac{1}{d\tau}\left[\int_{-\infty}^{\infty}\frac{(\xi-x)^2}{2}\cdot f_{X(t+d\tau)/X(t)=x}(\xi)d\xi\right]\right]+\cdots$$

which is the required result.

# 9

# THE AUTOCORRELATION FUNCTION

## 9.1 INTRODUCTION

As is evident from the previous chapter, a complete statistical specification of a random process, in general, is very difficult to achieve. It is usual to use partial statistical results, such as the autocorrelation function and power spectral density function, to obtain important information about a random process. This chapter is devoted to the autocorrelation function, and the autocorrelation function for the following random processes is detailed: a sinusoid with random phase, the random telegraph signal, generalized shot noise random processes, signalling random processes, jittered random processes, and the random walk.

## 9.2 NOTATION AND DEFINITIONS

The mean, variance, and autocorrelation function of a random process have been defined in Chapter 7. For notational convenience, in this chapter, the argument $T$ is dropped from the autocorrelation functions, and a subscript of $XX$ for a random process $X$ is added, that is, $R(T, t_1, t_2)$ is written as $R_{XX}(t_1, t_2)$. The definitions and notation are as follows.

**Definition: Mean, Variance, and Autocorrelation** For a random process $X:S\rightarrow S_X$ defined on the interval [0, *T*], the mean, variance, and autocorrelation functions are, respectively, defined according to

$$\begin{aligned}\mu_X(t) &= \boldsymbol{E}[X(\Omega,t)] = \boldsymbol{E}[X_t] \\ \sigma_X^2(t) &= \boldsymbol{E}[|X(\Omega,t)|^2] - |\mu_X(t)|^2 = \boldsymbol{E}[|X_t|^2] - |\mu_X(t)|^2 \\ R_{XX}(t_1,t_2) &= \boldsymbol{E}[X(\Omega,t_1)X^*(\Omega,t_2)] = \boldsymbol{E}[X_{t_1}X_{t_2}{}^*]\end{aligned} \tag{9.1}$$

### 9.2.1 Autocorrelation and Power Spectral Density

While the autocorrelation function contains important information in its own right, it is also useful in determining the power spectral density according to Theorem 7.5:

**Theorem 9.1 Power Spectral Density–Autocorrelation Relationship**
The power spectral density of a random process *X* defined on the interval [0,*T*] is given by

$$G(T,f) = \int_{-T}^{T} \bar{R}_{XX}(T,\tau)e^{-j2\pi f\tau}d\tau \tag{9.2}$$

where $\bar{R}_{XX}$ is the time average autocorrelation function on the interval [0, *T*] and defined according to

$$\bar{R}_{XX}(T,\tau) = \frac{1}{T}\int_{-T}^{T} R_{XX}(t+\tau,t)dt = \frac{1}{T}\int_{-T}^{T} \boldsymbol{E}[X(\Omega,t+\tau)X^*(\Omega,t)]dt \tag{9.3}$$

### 9.2.2 Autocovariance Function

A related function to the autocorrelation function is the autocovariance function.

**Definition: Autocovariance Function** The autocovariance function of a random process *X*, denoted $C_{XX}$, is defined according to

$$C_{XX}:\boldsymbol{R}^2\rightarrow\boldsymbol{R} \quad C_{XX}(t_1,t_2) = \boldsymbol{E}[(X_{t_1}-\mu_X(t_1))(X_{t_2}-\mu_X(t_2))^*] \tag{9.4}$$

The autocovariance function is a *generalized* variance definition.

**Theorem 9.2 Autocovariance–Autocorrelation Function Relationship**
The autocovariance and autocorrelation functions are related according to

$$\begin{aligned}C_{XX}(t_1,t_2) &= R_{XX}(t_1,t_2) - \mu_X(t_1)\mu_X{}^*(t_2) \\ R_{XX}(t_1,t_2) &= C_{XX}(t_1,t_2) + \mu_X(t_1)\mu_X{}^*(t_2)\end{aligned} \tag{9.5}$$

**Proof**

These relationships arise from the linearity of the expectation operator according to

$$\begin{aligned} C_{XX}(t_1,t_2) &= \boldsymbol{E}[(X_{t_1}-\mu_X(t_1))(X_{t_2}-\mu_X(t_2))^*] \\ &= \boldsymbol{E}[X_{t_1}X_{t_2}{}^* - \mu_X(t_1)X_{t_2}{}^* - \mu_X{}^*(t_2)X_{t_1} + \mu_X(t_1)\mu_X{}^*(t_2)] \\ &= \boldsymbol{E}[X_{t_1}X_{t_2}{}^*] - \mu_X(t_1)\mu_X{}^*(t_2) = R_{XX}(t_1,t_2) - \mu_X(t_1)\mu_X{}^*(t_2) \end{aligned} \tag{9.6}$$

## 9.3 BASIC RESULTS AND INDEPENDENCE INFORMATION

The following theorems detail basic results for the autocorrelation and autocovariance functions.

**Theorem 9.3 Autocorrelation for Identical Times**

For the case of identical times, that is, $t_2 = t_1$, it is the case that the autocorrelation function equals the variance plus the magnitude squared of a random variable defined at the time $t_1$, that is,

$$R_{XX}(t_1,t_1) = \boldsymbol{E}\left[|X_{t_1}|^2\right] = \sigma_X^2(t_1) + |\mu_X(t_1)|^2 \tag{9.7}$$

**Proof**

This result is simply a statement of the result that the mean squared value of a random variable equals the variance of the random variable plus the magnitude squared of the mean of the random variable. A direct proof of this result is

$$\begin{aligned} R_{XX}(t_1,t_2) &= \boldsymbol{E}\left[|X_{t_1}|^2\right] = \boldsymbol{E}[[X_{t_1}-\mu_X(t_1)+\mu_X(t_1)][X_{t_1}-\mu_X(t_1)+\mu_X(t_1)]^*] \\ &= \boldsymbol{E}[(X_{t_1}-\mu_X(t_1))(X_{t_1}-\mu_X(t_1))^* + \mu_X{}^*(t_1)(X_{t_1}-\mu_X(t_1)) \\ &\quad + \mu_X(t_1)(X_{t_1}-\mu_X(t_1))^* + \mu_X(t_1)\mu_X{}^*(t_1)] \\ &= \boldsymbol{E}\left[|X_{t_1}-\mu_X(t_1)|^2\right] + |\mu_X(t_1)|^2 = \sigma_X^2(t_1) + |\mu_X(t_1)|^2 \end{aligned} \tag{9.8}$$

**Theorem 9.4 Autocorrelation Function: Alternative Form**

If a random process $X: S \to S_X$ is written in terms of a random process $V: S \to S_X$ according to

$$X(\Omega,t) = V(\Omega,t) + \mu_X(t) \tag{9.9}$$

where $V$ has zero mean for all time, that is, $\boldsymbol{E}[V(\Omega,t)] = 0,\ t \in \boldsymbol{R}$, then the autocorrelation function can be written as

$$R_{XX}(t_1,t_2) = R_{VV}(t_1,t_2) + \mu_X(t_1)\mu_X{}^*(t_2) \tag{9.10}$$

**Proof**

The proof follows simply from the definitions of $X$ and $V$:

$$\begin{aligned} R_{XX}(t_1,t_2) &= \boldsymbol{E}[X_{t_1}X_{t_2}^*] = \boldsymbol{E}[(V_{t_1}+\mu_X(t_1))(V_{t_2}+\mu_X(t_2))^*] \\ &= \boldsymbol{E}[V_{t_1}V_{t_2}^* + \mu_X(t_1)V_{t_2}^* + \mu_X^*(t_2)V_{t_1} + \mu_X(t_1)\mu_X^*(t_2)] \\ &= \boldsymbol{E}[V_{t_1}V_{t_2}^*] + \mu_X(t_1)\mu_X^*(t_2) = R_{VV}(t_1,t_2) + \mu_X(t_1)\mu_X^*(t_2) \end{aligned} \tag{9.11}$$

### 9.3.1 Independence Information

An insight into the nature of the autocorrelation function can be obtained by, first, considering the case of $t_2 = t_1$ and, consistent with Theorem 9.3, the result:

$$R_{XX}(t_1,t_2) = \boldsymbol{E}\left[|X_{t_1}|^2\right] = \sigma_X^2(t_1) + |\mu_X(t_1)|^2 \tag{9.12}$$

Second, random processes are usually such that the random variable $X_{t_1}$ becomes increasingly independent of the random variable $X_{t_2}$ as $|t_1 - t_2|$ increases. When independence holds, it is the case that

$$R_{XX}(t_1,t_2) = \boldsymbol{E}[X_{t_1}X_{t_2}^*] = \boldsymbol{E}[X_{t_1}]\boldsymbol{E}[X_{t_2}^*] = \mu_X(t_1)\mu_X^*(t_2) \tag{9.13}$$

Thus, for the case of increasing independence as $|t_1 - t_2|$ increases, it follows that $R_{XX}(t_1, t_2)$ changes from $\sigma_X^2(t_1) + |\mu_X(t_1)|^2$ to $\mu_X(t_1)\mu_X^*(t_2)$. For the case of a random process with a constant mean, the change, as illustrated in Figure 9.1, is from $\sigma_X^2(t_1) + |\mu_X|^2$ to $|\mu_X|^2$. Accordingly, the autocorrelation function provides information about the time interval required, from a set time, for a set level of independence to be achieved.

### 9.3.2 Cross-Correlation between Two Random Processes

A natural generalization of the autocorrelation function is the cross-correlation function.

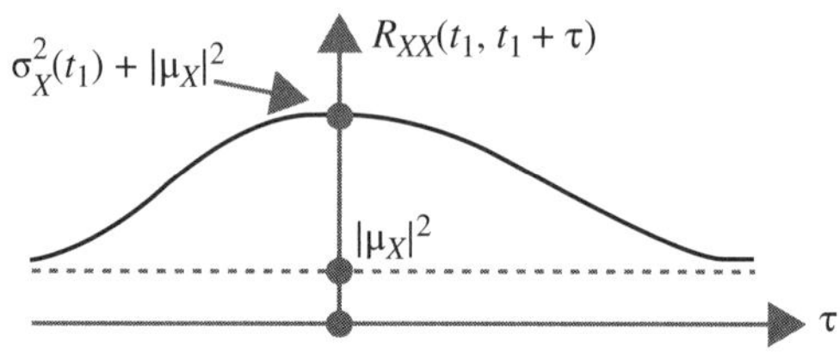


**FIGURE 9.1** Illustration of the change in the autocorrelation function for the case of constant mean and for the case of increasing independence as $|\tau|$ increases where $\tau = t_2 - t_1$.

**Definition: Cross-Correlation between Two Random Processes** The cross-correlation between two random processes $X$ and $Y$ is defined according to

$$R_{XY}(t_1,t_2)=\boldsymbol{E}[X(\Omega,t_1)Y^*(\Omega,t_2)] \tag{9.14}$$

***9.3.2.1 Autocorrelation of the Sum of Two Random Processes*** Consider two random processes $X$ and $Y$ and a random process $Z$ defined according to $Z(\Omega,t)=X(\Omega,t)+Y(\Omega,t)$. The autocorrelation of $Z$ can be written in terms of the autocorrelations of $X$ and $Y$, and the cross-correlation between $X$ and $Y$, according to:

**Theorem 9.5 Autocorrelation: Sum of Two Random Processes**
The autocorrelation of $Z(\Omega,t)=X(\Omega,t)+Y(\Omega,t)$ is

$$R_{ZZ}(t+\tau,t)=R_{XX}(t+\tau,t)+R_{YY}(t+\tau,t)+R_{XY}(t+\tau,t)+R_{YX}(t+\tau,t) \tag{9.15}$$

For the case of independent random processes

$$R_{ZZ}(t+\tau,t)=R_{XX}(t+\tau,t)+R_{YY}(t+\tau,t)+\mu_X(t+\tau)\mu_Y{}^*(t)+\mu_X{}^*(t)\mu_Y(t+\tau) \tag{9.16}$$

where $\mu_X$ and $\mu_Y$ are, respectively, the means of the random processes $X$ and $Y$.

For the independent case and when at least one of the random processes has a zero mean, the autocorrelation function of $Z$ is the sum of the individual autocorrelation functions, that is,

$$R_{ZZ}(t+\tau,t)=R_{XX}(t+\tau,t)+R_{YY}(t+\tau,t) \tag{9.17}$$

**Proof**
The proof follows from the definition of $Z$ and the linearity of the expectation operator:

$$\begin{aligned} R_{ZZ}(t+\tau,t)&=\boldsymbol{E}[Z(\Omega,\,t+\tau)Z^*(\Omega,\,t)]\\ &=\boldsymbol{E}[(X(\Omega,\,t+\tau)+Y(\Omega,\,t+\tau)(X^*(\Omega,\,t)+Y^*(\Omega,\,t))]\\ &=\boldsymbol{E}[X(\Omega,\,t+\tau)X^*(\Omega,\,t)+Y(\Omega,\,t+\tau)Y^*(\Omega,\,t)\\ &\qquad+X(\Omega,\,t+\tau)Y^*(\Omega,\,t)+Y(t+\tau)X^*(\Omega,\,t)]\\ &=R_{XX}(t+\tau,\,t)+R_{YY}(t+\tau,\,t)+R_{XY}(t+\tau,\,t)+R_{YX}(t+\tau,\,t) \end{aligned} \tag{9.18}$$

## 9.4 SINUSOID WITH RANDOM AMPLITUDE AND PHASE

Consider a random process $X$ defined according to

$$X(\Omega,t)=A\sin(2\pi f_C t+\Phi) \qquad t\geq 0 \tag{9.19}$$

where $\Omega = (A, \Phi)$ and the random variables $A$ and $\Phi$ have a joint probability density function $f_{A\Phi}$. For the independent case, $A$ and $\Phi$ have probability density functions, respectively, of $f_A$ and $f_\Phi$.

**Theorem 9.6 Mean, Variance, and Autocorrelation Function**
The mean, variance, and autocorrelation of the random process $X$, for $0 \le t, t+\tau$, respectively, are

$$\mu_X(t) = \int_{-\infty}^{\infty}\int_{-\infty}^{\infty} a\sin(2\pi f_C t + \phi) f_{A\Phi}(a, \phi)\,da\,d\phi \tag{9.20}$$

$$\sigma_X^2(t) = \frac{\boldsymbol{E}[A^2]}{2} - \frac{1}{2}\int_{-\infty}^{\infty}\int_{-\infty}^{\infty} a^2\cos(4\pi f_C t + 2\phi) f_{A\Phi}(a, \phi)\,da\,d\phi - \mu_X^2(t) \tag{9.21}$$

$$R_{XX}(t+\tau, t) = \frac{\boldsymbol{E}[A^2]}{2}\cos(2\pi f_C \tau) - \frac{1}{2}\int_{-\infty}^{\infty}\int_{-\infty}^{\infty} \frac{a^2}{2}[\cos(2\pi f_C(2t+\tau) + 2\phi)] f_{A\Phi}(a, \phi)\,da\,d\phi \tag{9.22}$$

For the independent case and with the definition of $\mu_A = \boldsymbol{E}[A]$,

$$\mu_X(t) = \mu_A \int_{-\infty}^{\infty} \sin(2\pi f_C t + \phi) f_\Phi(\phi)\,d\phi \tag{9.23}$$

$$\sigma_X^2(t) = \frac{\boldsymbol{E}[A^2]}{2} \cdot \left[1 - \int_{-\infty}^{\infty} \cos(4\pi f_C t + 2\phi) f_\Phi(\phi)\,d\phi\right] - \mu_X^2(t) \tag{9.24}$$

$$R_{XX}(t+\tau, t) = \frac{\boldsymbol{E}[A^2]}{2}\left[\cos(2\pi f_C \tau) - \int_{-\infty}^{\infty} \cos(2\pi f_C(2t+\tau) + 2\phi) f_\Phi(\phi)\,d\phi\right] \tag{9.25}$$

**Proof**
The proof is detailed in Appendix 9.A.

### 9.4.1 Example

For the independent case where

$$f_A(a) = \frac{1}{\sqrt{2\pi}\sigma_A}\exp\left[\frac{(x-\mu_A)^2}{2\sigma_A^2}\right] \qquad f_\Phi(\phi) = \begin{cases} 1/2\pi & -\pi \le \phi < \pi \\ 0 & \text{elsewhere} \end{cases} \tag{9.26}$$

it follows that

$$\mu_X(t)=0 \qquad \sigma_X^2(t)=\frac{\sigma_A^2+\mu_A^2}{2} \qquad t\geq 0 \tag{9.27}$$

$$R_{XX}(t+\tau,t)=\frac{\sigma_A^2+\mu_A^2}{2}\cdot\cos(2\pi f_C\tau) \qquad t\geq 0, t+\tau\geq 0 \tag{9.28}$$

The random process is wide-sense stationary.

## 9.5 RANDOM TELEGRAPH SIGNAL

Consider a Poisson experiment with a sample space

$$S=\{(\gamma_1,\ldots,\gamma_k,\ldots):\ \gamma_k\in \boldsymbol{R}^+,\gamma_1\leq\ \ldots\ \leq\gamma_k\leq\ \ldots\} \tag{9.29}$$

where the point rate is $\lambda$ and a causal random telegraph random process $X:S\rightarrow S_X$ defined according to

$$X(\Omega,t)=\sum_{j=0}^{\infty}a_j u(t-\Gamma_j) \qquad \begin{cases} a_0=1 \\ a_1,a_3,\ldots=-2 \\ a_2,a_4,\ldots=2 \end{cases} \tag{9.30}$$

Here, $\Omega=(\Gamma_0,\Gamma_1,\ldots)$, $\Gamma_0=0$ and $\Gamma_1,\Gamma_2,\ldots$ are a sequence of random variables consistent with the outcomes of the Poisson experiment. One signal from a causal random telegraph random process is illustrated in Figure 9.2.

### 9.5.1 Mean, Variance, and Autocorrelation Function

**Theorem 9.7 Mean, Variance, and Autocorrelation Function**
The mean, variance, and autocorrelation of the causal random telegraph random process, respectively, are

$$\mu_X(t)=e^{-2\lambda t} \qquad \sigma_X^2(t)=1-e^{-4\lambda t} \qquad t\geq 0 \tag{9.31}$$

$$R_{XX}(t+\tau,t)=e^{-2\lambda|\tau|} \qquad 0\leq t,t+\tau \tag{9.32}$$

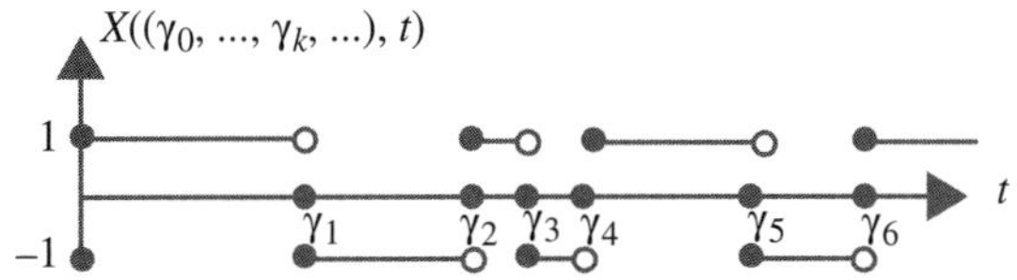


**FIGURE 9.2** One signal from a causal random telegraph random process.

The mean square value, as expected, is

$$E\left[X^2(\Omega, t)\right] = R_{XX}(t, t) = 1 \qquad t \geq 0 \tag{9.33}$$

**Proof**
The proof is detailed in Appendix 9.B.

## 9.6 GENERALIZED SHOT NOISE

Consider a generalized shot noise random process defined according to

$$X(\Omega, t) = \sum_{k=1}^{\infty} A_k p(t - \Gamma_k) \tag{9.34}$$

where $\Omega = ((A_1, \Gamma_1), \ldots), \Gamma_1, \Gamma_2, \ldots$ are a sequence of random variables consistent with the outcomes of a Poisson experiment and $(A_1, A_2, \ldots)$ is a vector of random variables that define the amplitudes. The components of one signal defined by a generalized shot noise process are illustrated in Figure 9.3 assuming the experiment underpinning this signal yields the values $(a_1, t_1), \ldots, (a_k, t_k), \ldots$. The average waveform rate is denoted $r$.

The following assumptions are made: First, the random variables $A_1, A_2, \ldots$ defining the amplitudes are all identical and independent of one another and are taken from a sample space $S_A$ according to a probability density function $f_A$. The random variables $\Gamma_1, \Gamma_2, \ldots, \Gamma_N$ are assumed to be defined by unordered outcomes of independent Poisson experiments on the finite interval of interest of [0, $T$] and to have a uniform distribution, that is, $f_\Gamma(t) = 1/T$. The point rate is $\lambda = N/T$. Finally, $A_i$ is assumed to be independent of $\Gamma_k$ for all positive integer values of $i$, $k$.

### 9.6.1 Mean, Variance, and Autocorrelation Function

**Theorem 9.8 Mean, Variance, and Autocorrelation Function: General Case**
The mean, variance, and autocorrelation functions of the generalized shot noise random process defined above, and for the interval [0, $T$], that is, $0 \leq t \leq T$, $0 \leq t + \tau \leq T$, respectively, are

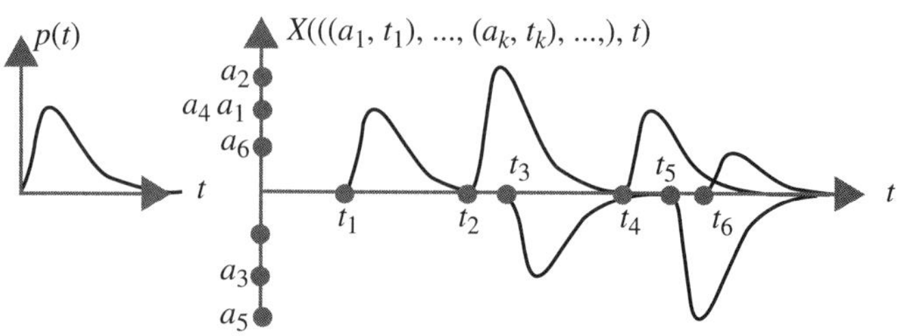


**FIGURE 9.3** Individual components of one signal defined by a generalized shot noise process.

$$\mu_X(t) = r\mu_A \int_0^T p(t-\lambda)d\lambda \tag{9.35}$$

$$\sigma_X^2(t) = r\boldsymbol{E}\left[A^2\right] \int_0^T p^2(t-\lambda)d\lambda - \frac{r^2\mu_A^2}{N}\left[\int_0^T p(t-\lambda)d\lambda\right]^2 \tag{9.36}$$

$$\begin{aligned} R_{XX}(t+\tau,t) &= r\boldsymbol{E}[A^2] \int_0^T p(t+\tau-\lambda)p(t-\lambda)d\lambda \\ &+ r^2\mu_A^2\left(1-\frac{1}{N}\right)\int_0^T\int_0^T p(t+\tau-\lambda_1)p(t-\lambda_2)d\lambda_1 d\lambda_2 \end{aligned} \tag{9.37}$$

where $\mu_A$ and $\boldsymbol{E}[A^2]$, respectively, are the mean and mean squared value of the amplitude random variables and $r = N/T$ is the waveform rate.

For the case where the amplitudes have zero mean, that is, $\mu_A = 0$, the autocorrelation and variance are

$$\sigma_X^2(t) = r\sigma_A^2 \int_0^T p^2(t-\lambda)d\lambda \tag{9.38}$$

$$R_{XX}(t+\tau,\, t) = r\sigma_A^2 \int_0^T p(t+\tau-\lambda)p(t-\lambda)d\lambda \tag{9.39}$$

where $\sigma_A^2$ is the variance of $A$.

**Proof**

The proof is detailed in Appendix 9.C.

**Theorem 9.9 Mean, Variance, and Autocorrelation Function: Causal Case**
For the interval $[0, T]$ and for the case of a causal pulse function, the mean, variance, and autocorrelation function of a generalized shot noise random process simplify to

$$\mu_X(t) = r\mu_A \int_0^t p(\lambda)d\lambda \qquad 0 \le t \le T \tag{9.40}$$

$$\sigma_X^2(t) = r\boldsymbol{E}\left[A^2\right] \int_0^t p^2(\lambda)d\lambda - \frac{r^2\mu_A^2}{N}\left[\int_0^t p(\lambda)d\lambda\right]^2 \qquad 0 \le t \le T \tag{9.41}$$

$$R_{XX}(t+\tau,t) = \begin{cases} rE[A^2]\int_0^t p(\lambda+\tau)p(\lambda)d\lambda + r^2\mu_A^2\left(1-\frac{1}{N}\right)\int_0^{t+\tau} p(\lambda)d\lambda \int_0^t p(\lambda)d\lambda & 0 \le \tau \le T-t \\ rE[A^2]\int_0^{t+\tau} p(\lambda-\tau)p(\lambda)d\lambda + r^2\mu_A^2\left(1-\frac{1}{N}\right)\int_0^{t+\tau} p(\lambda)d\lambda \int_0^t p(\lambda)d\lambda & -t \le \tau < 0 \end{cases} \tag{9.42}$$

For the case where $N$ is large,

$$R_{XX}(t+\tau,t) \approx \begin{cases} rE[A^2]\int_0^t p(\lambda+\tau)p(\lambda)d\lambda + \mu_X(t+\tau)\mu_X(t) & 0 \le \tau \le T-t \\ rE[A^2]\int_0^{t+\tau} p(\lambda-\tau)p(\lambda)d\lambda + \mu_X(t+\tau)\mu_X(t) & -t \le \tau < 0 \end{cases} \tag{9.43}$$

For the case where the amplitude random variables have zero mean, that is, $\mu_A = 0$, the following simplified results hold:

$$\mu_X(t) = 0 \quad \sigma_X^2(t) = r\sigma_A^2 \int_0^t p^2(\lambda)d\lambda \qquad 0 \le t \le T \tag{9.44}$$

$$R_{XX}(t+\tau,\, t) = \begin{cases} r\sigma_A^2 \int_0^t p(\lambda+\tau)p(\lambda)d\lambda & 0 \le t \le T,\, 0 \le \tau \le T-t \\ r\sigma_A^2 \int_0^{t+\tau} p(\lambda-\tau)p(\lambda)d\lambda & 0 \le t \le T,\, -t \le \tau \le 0 \end{cases} \tag{9.45}$$

**Proof**

These results follow from Theorem 9.8 and the results detailed in Appendix 9.C.

***9.6.1.1 Nature of Mean, Variance, and Autocorrelation*** Clearly, from the results for the mean, variance, and autocorrelation, as detailed in Theorem 9.9 for the case of a causal pulse function, the nature of these functions depends on the nature of the pulse function. For the case of a nonnegative pulse function with finite energy, the form of the mean and variance of a generalized shot noise random process is illustrated in Figure 9.4. The general form for the autocorrelation function is illustrated in Figure 9.5.

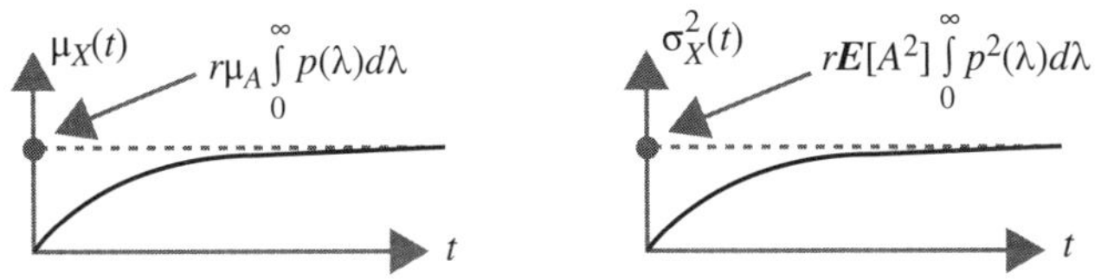


**FIGURE 9.4** Illustration of the form for the mean and variance of a generalized shot noise random process for the nonnegative, finite energy, causal pulse function case.

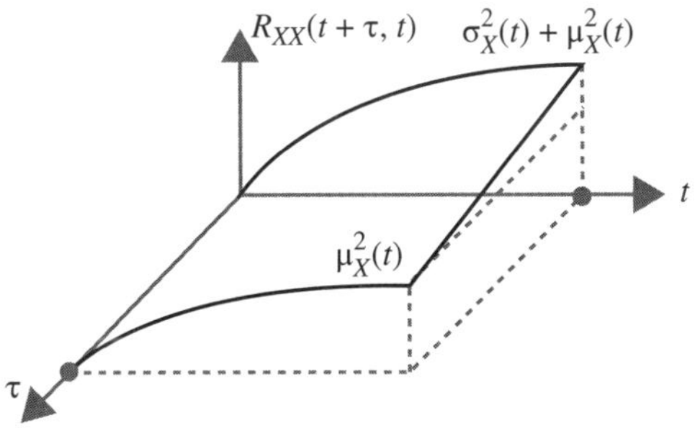


**FIGURE 9.5** Illustration of the form for the autocorrelation of a generalized shot noise random process for the nonnegative, finite energy, causal pulse function case.

### 9.6.2 Correlation Coefficient and Correlation Time

For a generalized shot noise random process and with the assumptions consistent with the results in Theorem 9.9, a simple expression for the correlation coefficient, based on the times $t$ and $t+\tau$, can be established.

**Theorem 9.10 Correlation Coefficient**
For the case of a causal pulse function, and when $N \gg 1$, the correlation coefficient, between the random variables defined by a generalized shot noise random process at the times $t$ and $t+\tau$, is

$$\rho_{XX}(t+\tau,t) = \frac{R_{XX}(t+\tau,t) - \mu_X(t+\tau)\mu_X(t)}{\sqrt{\sigma_X^2(t+\tau)\sigma_X^2(t)}} \approx \frac{\int_0^t p(\lambda+\tau)p(\lambda)d\lambda}{\sqrt{\int_0^{t+\tau} p^2(\lambda)d\lambda \int_0^t p^2(\lambda)d\lambda}} \tag{9.46}$$

$$0 \le t+\tau \le T,\ t,\ \tau > 0$$

For the steady-state case, where $t, T \to \infty$, it follows, for the case of a pulse function with finite energy, that

$$\rho_{XX}(t+\tau, t) \approx \int_0^{\infty} p(\lambda+\tau)p(\lambda)d\lambda \Big/ \int_0^{\infty} p^2(\lambda)d\lambda \qquad \tau<0 \tag{9.47}$$

**Proof**

This result follows directly from the definition of the correlation coefficient (Eq. 7.83) and Theorem 9.9.

***9.6.2.1 Effective Correlation Time*** The effective correlation time function $C_t : \boldsymbol{R} \to \boldsymbol{R}$ of a random process, at a time $t$, has been defined (Eq. 7.86) according to

$$C_t(t) = \tau \qquad \tau : \rho_{XX}(t+\tau, t) = \rho_o,\ \tau>0 \tag{9.48}$$

for a set value of $\rho_o$.

**Theorem 9.11 Effective Correlation Time**

The effective correlation time for a generalized shot noise random process, based on a causal pulse function, is

$$C_t(t) = \tau \qquad \tau : \frac{\int_0^{t} p(\lambda+\tau)p(\lambda)d\lambda}{\sqrt{\int_0^{t+\tau} p^2(\lambda)d\lambda \int_0^{t} p^2(\lambda)d\lambda}} = \rho_o, \tau>0 \tag{9.49}$$

For the steady-state case, consistent with $t, T \to \infty$, the effective correlation time is

$$C_t(t) = \tau \qquad \tau : \int_0^{\infty} p(\lambda+\tau)p(\lambda)d\lambda = \rho_o \int_0^{\infty} p^2(\lambda)d\lambda \qquad \tau>0 \tag{9.50}$$

**Proof**

These results follow directly from the definition of the effective correlation time and Theorem 9.10.

***9.6.2.2 Normalized Pulse Function Based on Effective Correlation Time*** It can be useful, when defining a normalized generalized shot noise random process, to choose a pulse function that yields a correlation time of unity in steady state. Results for the case of a pulse function defined by the impulse response of an $i$th order filter are stated in the following theorem:

**Theorem 9.12 Normalized Pulse Function for Effective Correlation Time**

Consider the case of a generalized shot noise random process based on a pulse function

$$p(t) = k_p t^i e^{-t/\tau_p} u(t) \qquad i \in \{0, 1, 2, \ldots\} \tag{9.51}$$

The time constant $\tau_p$, which is the solution of the following nonlinear equation,

$$\sum_{k=0}^{i} \frac{i!(2i-k)!}{(2i)!(i-k)!k!}\left(\frac{2\tau_C}{\tau_p}\right)^k = \rho_o e^{\tau_C/\tau_p} \tag{9.52}$$

yields a correlation time $C_t(t) = \tau_C$ in steady state assuming a correlation level of $\rho_o$.

**Proof**
The proof is detailed in Appendix 9.D.

***9.6.2.3 Example*** For example, for the case of $\rho_o = 0.1$, numerical analysis yields

$$\begin{aligned}
i=0: &\quad 1 = 0.1e^{\tau_C \tau_p} \quad \Rightarrow \quad \tau_p/\tau_C = 0.434294 \\
i=1: &\quad 1 + \frac{t_C}{\tau_p} = 0.1e^{\tau_C/\tau_p} \quad \Rightarrow \quad \tau_p/\tau_C = 0.257088 \\
i=2: &\quad 1 + \frac{\tau_C}{\tau_p} + \frac{\tau_C^2}{3\tau_p^2} = 0.1e^{\tau_C/\tau_p} \quad \Rightarrow \quad \tau_p/\tau_C = 0.20202 \\
i=3: &\quad 1 + \frac{\tau_C}{\tau_p} + \frac{2\tau_C^2}{5\tau_p^2} + \frac{\tau_C^3}{15\tau_p^3} = 0.1e^{\tau_C/\tau_p} \quad \Rightarrow \quad \tau_p/\tau_C = 0.172081
\end{aligned} \tag{9.53}$$

### 9.6.3 Normalization of Random Process Power

In steady state, the power of a random process is given by $\boldsymbol{E}\left[|X(\Omega,t)|^2\right] = R_{XX}(t,t)$, and in certain instances, it is useful if this is normalized. The issue is to determine the pulse amplitude such that

$$R_{XX}(t,t)\big|_{t\to\infty} = 1 \tag{9.54}$$

**Theorem 9.13 Normalization of Power**
Consider a generalized shot noise random process defined by

$$X(\Omega,t) = \sum_{k=1}^{\infty} A_k p(t-\Gamma_k) \tag{9.55}$$

where $A_k$, $k \in \{1,2,\ldots\}$ are identical independent random variables, $\Gamma_k$, $k \in \{1,2,\ldots\}$ are identical independent random variables, and the pulse function is causal and with finite energy. The following associated generalized shot noise random process

$$X_N(\Omega,t) = \sum_{k=1}^{\infty} A_k k_N p(t-\Gamma_k) \tag{9.56}$$

has unity power in steady state consistent with $R_{X_N X_N}(t,t)|_{t\to\infty} = 1$ when

$$k_N = \frac{1}{\sqrt{r(\sigma_A^2 + \mu_A^2)\int_0^\infty p^2(\lambda)d\lambda + r^2\mu_A^2\left[\int_0^\infty p(\lambda)d\lambda\right]^2}} \tag{9.57}$$

Here, $\mu_A$ and $\sigma_A^2$, respectively, are the mean and variance of the identical random variables defining the pulse amplitudes.

**Proof**

This result follows from Equation 9.42 with $N, T \to \infty$:

$$R_{XX}(t,t)|_{t\to\infty} = r\boldsymbol{E}\left[A^2\right]\int_0^\infty p^2(\lambda)d\lambda + r^2\mu_A^2\left[\int_0^\infty p(\lambda)d\lambda\right]^2 \tag{9.58}$$

### *9.6.3.1 Example: Gaussian Variation of Amplitudes around a Mean Level*

Consider a generalized shot noise random process, consistent with Equation 9.56, where the probability density function of $A_k$, $k \in \{1, 2, \ldots\}$ is defined according to

$$f_A(a) = \frac{1}{\sqrt{2\pi}\sigma_A}\exp\left[\frac{-(a-\mu_A)^2}{2\sigma_A^2}\right] \qquad \mu_A = 1,\ \sigma_A = 1/4 \tag{9.59}$$

and the pulse function is assumed to have the form

$$p(t) = t^2 e^{-t/\tau_p}u(t) \tag{9.60}$$

The pulse rate is assumed to be $r = 10$. The goal is to find the pulse time constant $\tau_p$ such that the correlation time in steady state is 1 second and to determine the constant $k_N$ such that the random process has unity power in steady state.

The correlation time, as defined by

$$C_t(t) = \tau \qquad t : \rho_{XX}(t+\tau, t) = \rho_o \tag{9.61}$$

results, for the case of $\rho_o = 0.1$ and the specification of $C_t(t)|_{t\to\infty} = 1$, in $\tau_p = 0.20202$ (see Theorem 9.12 and Eq. 9.53). Using Theorem 9.13, the pulse constant $k_N$ is $k_N = 5.7858$.

Using these parameter values, the pulse function and one waveform from the random process are shown in Figure 9.6. The autocorrelation function is graphed in Figure 9.7. The mean and variance are plotted in Figure 9.8. The correlation coefficient between the times $t+\tau$ and $t$ is shown in Figure 9.9. The effective correlation time function is shown in Figure 9.10.

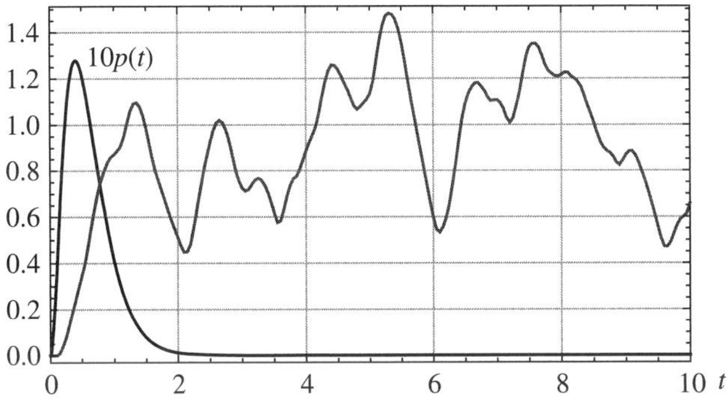


**FIGURE 9.6** Graph of pulse function scaled by 10 and one waveform from the defined generalized shot noise random process.

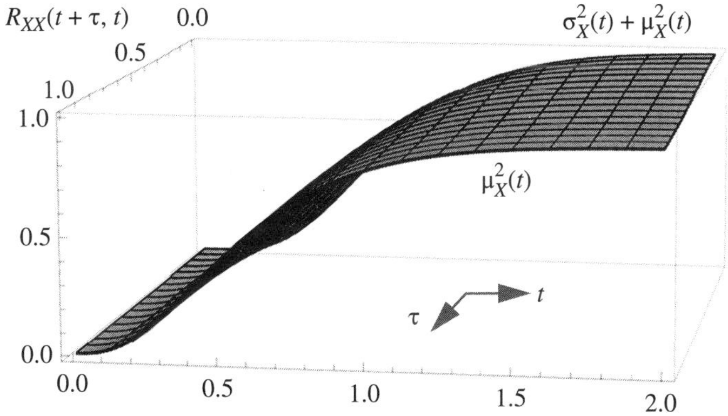


**FIGURE 9.7** Graph of autocorrelation function of the defined generalized shot noise random process.

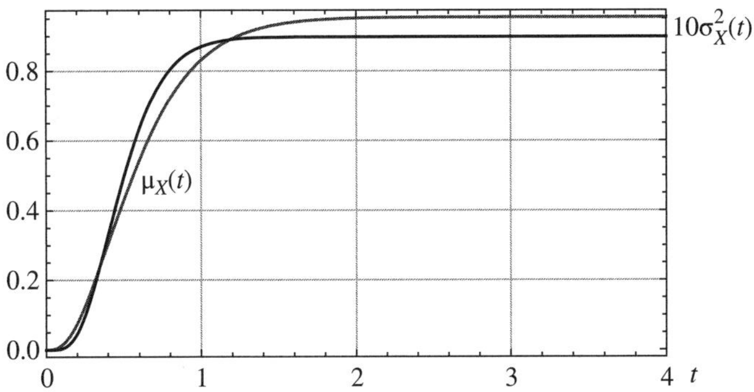


**FIGURE 9.8** Mean and variance (scaled by 10) of the defined generalized shot noise random process.

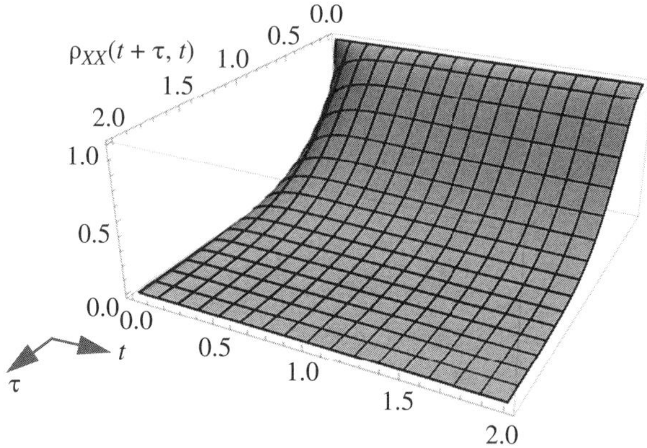


**FIGURE 9.9** Graph of the correlation coefficient of the defined generalized shot noise random process.

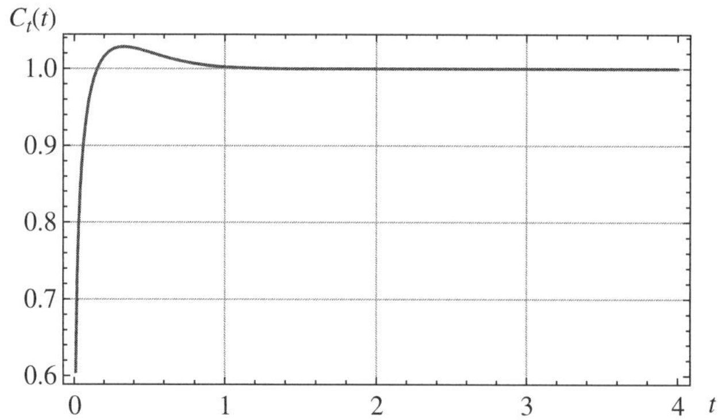


**FIGURE 9.10** Graph of the effective correlation time for the defined generalized shot noise random process.

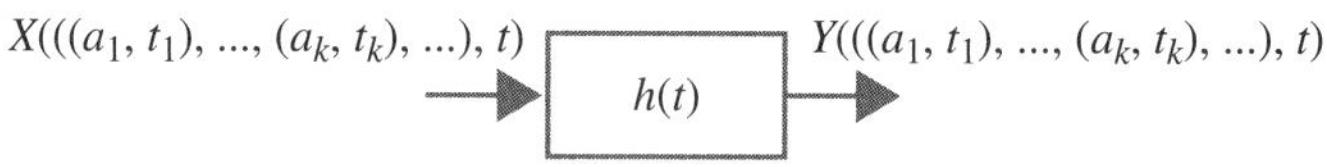


**FIGURE 9.11** Linear filtering by a filter with impulse response *h*.

### 9.6.4 Linear Filtering

Consider the case of a generalized shot noise random process $X: S \to S_X$, as defined by Equation 9.34, which is subject to linear filtering by a causal system, as illustrated in Figure 9.11, with an impulse response $h$. Associated with each input signal

$$X(((a_1, t_1),\ \dots,\ (a_k, t_k),\ \dots), t) = \sum_{k=1}^{\infty} a_k p(t - t_k) \qquad t \geq 0 \tag{9.62}$$

is an output signal

$$Y(((a_1, t_1), \dots (a_k, t_k), \dots), t) = \int_0^t X(((a_1, t_1), \dots, (a_k, t_k), \dots), \lambda) h(t-\lambda) d\lambda \quad (9.63)$$

and the collection of output signals defines a random process $Y : S \to S_Y$. This random process is also a generalized shot noise random process.

**Theorem 9.14 Linear Filtering of a Generalized Shot Noise Random Process** Linear filtering of a generalized shot noise random process, based on a causal pulse function $p$, by a causal linear filter with an impulse response $h$, leads to a generalized shot noise random process

$$Y(\Omega, t) = \sum_{k=1}^{\infty} A_k p_Y(t - \Gamma_k) \qquad t \geq 0 \quad (9.64)$$

with a pulse function

$$p_Y(t) = \int_0^t p(\lambda) h(t-\lambda) d\lambda \quad (9.65)$$

**Proof**
Substitution for the input signal into Equation 9.63 yields

$$Y(((a_1, t_1), \dots, (a_k, t_k), \dots), t) = \int_0^t \sum_{k=1}^{N_t} a_k p(\lambda - t_k) h(t-\lambda) d\lambda \quad (9.66)$$

where $N_t$ is such that $t_{N_t} < t$ and $t_{N_t+1} \geq t$. Interchanging the order of summation and integration yields

$$\begin{aligned} Y(((a_1, t_1), \dots, (a_k, t_k), \dots), t) &= \sum_{k=1}^{N_t} a_k \int_{t_k}^{t} p(\lambda - t_k) h(t-\lambda) d\lambda \\ &= \sum_{k=1}^{N_t} a_k \int_0^{t-t_k} p(\zeta) h(t - t_k - \zeta) d\zeta = \sum_{k=1}^{N_t} a_k p_Y(t - t_k) \end{aligned} \quad (9.67)$$

where

$$p_Y(t) = \int_0^t p(\lambda) h(t-\lambda) d\lambda \quad (9.68)$$

which is the required result.

***9.6.4.1 Notes*** The input and output pulse rates of the generalized shot noise processes are identical. As the input and output random processes have the same form, the results specified by Theorems 9.8–9.13 can be directly used.

***9.6.4.2 Relationship between Input and Output Variances*** With linear filtering of a generalized shot noise random process, it is of interest to know the relationship between the input and output variances.

**Theorem 9.15 Relationship between Input and Output Variances**
For the case of a causal pulse function, a causal linear filter as illustrated in Figure 9.11, and the input random process having a zero mean, the relationship between the input and output variances of the generalized shot noise processes is

$$\frac{\sigma_Y^2(t)}{\sigma_X^2(t)} = \frac{\int_0^t p_Y^2(\lambda)d\lambda}{\int_0^t p^2(\lambda)d\lambda} \qquad t>0 \tag{9.69}$$

For the special case where the input pulses become impulsive and have unity energy, the following holds:

$$\frac{\sigma_Y^2(t)}{\sigma_X^2(t)} = \int_0^t p_Y^2(\lambda)d\lambda = \int_0^t h^2(\lambda)d\lambda \qquad t>0 \tag{9.70}$$

**Proof**
The stated result arises from Theorem 9.14 and the variance relationship stated in Theorem 9.9.

## 9.7 SIGNALLING RANDOM PROCESS-FIXED PULSE CASE

Consider a signalling random process based on a fixed signalling pulse and defined according to

$$X(\Omega,t) = \sum_{k=0}^{N-1} A_k p(t-k\mathrm{D}) \qquad t \in [0,T], T = N\mathrm{D} \tag{9.71}$$

where $\Omega = (A_0, \ldots, A_{N-1})$ and the random variables $A_0, \ldots, A_{N-1}$ defining the amplitudes are all independent of one another. The mean, mean squared value, and variance of $A_k$ are denoted, respectively, $\mu_k, \boldsymbol{E}[A_k^2]$, and $\sigma_k^2$.

**Theorem 9.16 Mean, Variance, and Autocorrelation Function**
For the signalling random process defined by Equation 9.71, the mean, variance, and autocorrelation function, respectively, are

$$\mu_X(t) = \sum_{k=0}^{N-1} \mu_k p(t-k\mathrm{D}) \qquad \sigma_X^2(t) = \sum_{k=0}^{N-1} \sigma_k^2 p^2(t-k\mathrm{D}) \qquad t \in [0, N\mathrm{D}] \tag{9.72}$$

$$R_{XX}(t+\tau,t)=\sum_{i=0}^{N-1}E\left[A_i^2\right]p(t+\tau-i\mathrm{D})p(t-i\mathrm{D})$$
$$+\sum_{i=0}^{N-1}\sum_{\substack{j=0\\ j\neq i}}^{N-1}\mu_i\mu_j p(t+\tau-i\mathrm{D})p(t-j\mathrm{D}) \qquad t+\tau,t\in[0,N\mathrm{D}] \tag{9.73}$$

**Proof**

The proof is detailed in Appendix 9.E.

### 9.7.1 Finite Duration Pulse Case

For the case, as illustrated in Figure 9.12, where the pulse function is nonzero outside of an interval of duration $D_p$ seconds, Theorem 9.16 can be restated in a form, which facilitates evaluation, as specified in the following theorem.

**Theorem 9.17 Finite Duration Pulse Case**

For the case where the pulse function has a finite nonzero duration of $D_p$ seconds on the interval $\left[t_o,\ t_o+D_p\right]$, except on a set of zero measure, the autocorrelation function of the signalling random process defined by Equation 9.71 can be written in the following form which facilitates evaluation as the double summation is only over the nonzero terms:

$$R_{XX}(t+\tau,\ t)=u[N-1-i_{\min_1}]\cdot u[i_{\max_1}]\cdot u[i_{\max_1}-i_{\min_1}]\cdot$$
$$\sum_{i=\max\{i_{\min_1},0\}}^{\min\{i_{\max_1},\ N-1\}} E\left[A_i^2\right]p(t+\tau-i\mathrm{D})p(t-i\mathrm{D})$$
$$+u[N-1-i_{\min_2}]\cdot u[i_{\max_2}]\cdot u[N-1-j_{\min}]\cdot u[j_{\max}]\cdot u[i_{\max_2}-i_{\min_2}]\cdot \tag{9.74}$$
$$u[j_{\max}-j_{\min}]\cdot\sum_{i=\max\{i_{\min_2},0\}}^{\min\{i_{\max_2},N-1\}}\ \sum_{j=\max\{j_{\min},0\},\ j\neq i}^{\min\{j_{\max},N-1\}}\mu_i\mu_j p(t+\tau-i\mathrm{D})p(t-j\mathrm{D})$$
$$t+\tau,\ t\in[0,\ N\mathrm{D}]$$

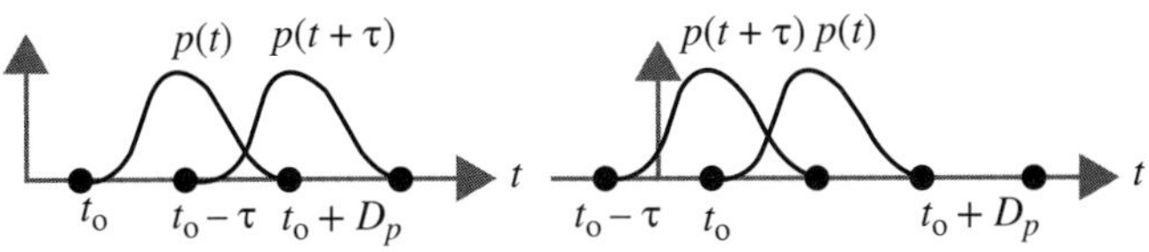


**FIGURE 9.12** Graphs of the pulse function for the finite duration case. Left: case for $\tau<0$. Right: case for $\tau>0$.

where

$$i_{\min_1} = \left\lfloor \max\left\{\frac{t-t_o-D_p}{D}, \frac{t+\tau-t_o-D_p}{D}\right\}\right\rfloor + 1 \qquad i_{\max_1} = \left\lfloor \min\left\{\frac{t-t_o}{D}, \frac{t+\tau-t_o}{D}\right\}\right\rfloor \tag{9.75}$$

$$\begin{aligned} i_{\min_2} &= \left\lfloor \frac{t+\tau-t_o-D_p}{D}\right\rfloor + 1 \qquad i_{\max_2} = \left\lfloor \frac{t+\tau-t_o}{D}\right\rfloor \\ j_{\min} &= \left\lfloor \frac{t-t_o-D_p}{D}\right\rfloor + 1 \qquad j_{\max} = \left\lfloor \frac{t-t_o}{D}\right\rfloor \end{aligned} \tag{9.76}$$

**Proof**

The proof is detailed in Appendix 9.F.

### 9.7.2 Examples

***9.7.2.1 Amplitude-Modulated Pulse Train*** Consider a random process $X$, which defines amplitude-modulated pulse trains according to

$$X(\Omega, t) = \sum_{k=0}^{N-1} A_k p(t-kD) \qquad t \in [0, T], \ T = ND \tag{9.77}$$

where $\Omega = (A_0, \ldots, A_{N-1})$ and the amplitudes of the pulses exhibit exponential decay consistent with

$$A_k = a_o B_k e^{-kD/\tau_A} \qquad k \in \{0, \ldots, N-1\} \tag{9.78}$$

Here, $a_0$ and $\tau_A$ are constants, and $(B_0, \ldots, B_{N-1})$ are identical and independent random variables with a Rayleigh distribution and a density function $f_B$:

$$f_B(b) = \frac{b}{\sigma_b^2}\cdot\exp\left[\frac{-b^2}{2\sigma_b^2}\right] \qquad b > 0 \tag{9.79}$$

The mean, variance, and mean squared value for $B_k$, respectively, are

$$\mu_{B_k} = \frac{\sqrt{\pi}\sigma_b}{\sqrt{2}} \qquad \sigma_{B_k}^2 = \left(2-\frac{\pi}{2}\right)\sigma_b^2 \qquad \boldsymbol{E}\left[B_k^2\right] = 2\sigma_b^2 \tag{9.80}$$

One waveform from the random process is shown in Figure 9.13 for the case of Gaussian pulses

$$p(t) = \frac{1}{\sqrt{2\pi}\sigma_t}\cdot\exp\left[\frac{-t^2}{2\sigma_t^2}\right] \tag{9.81}$$

and the values $D = 1$, $\sigma_t = D/4$, $a_o = 1$, $\sigma_b = 1/2$, and $\tau_A = 10$.

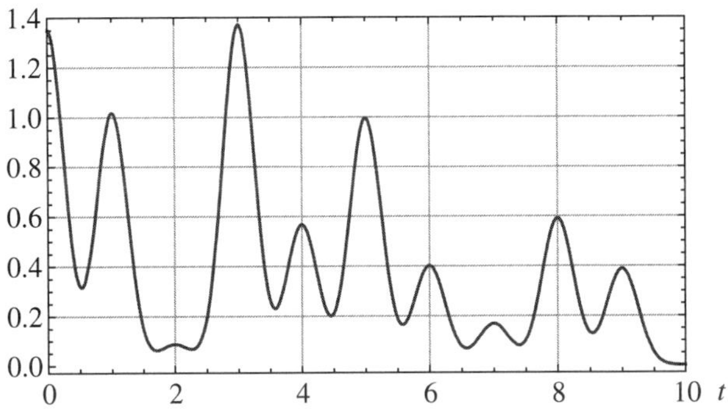


**FIGURE 9.13** Graph of one signal from the defined signalling random process.

It then follows, from the results specified in Theorem 9.16, that the mean, variance, and autocorrelation of this random process, respectively, are

$$\mu_X(t) = \frac{\sqrt{\pi}\sigma_b a_o}{\sqrt{2}} \sum_{k=0}^{N-1} e^{-kD/\tau_A} p(t-kD) \qquad t \in [0, ND] \tag{9.82}$$

$$\sigma_X^2(t) = \left(2-\frac{\pi}{2}\right)\sigma_b^2 a_o^2 \sum_{k=0}^{N-1} e^{-2kD/\tau_A} p^2(t-kD) \qquad t \in [0, ND] \tag{9.83}$$

$$\begin{aligned} R_{XX}(t+\tau,t) &= 2\sigma_b^2 a_o^2 \cdot \sum_{i=0}^{N-1} e^{-2iD/\tau_A} p(t+\tau-iD)p(t-iD) \\ &+ \frac{\pi\sigma_b^2 a_o^2}{2} \cdot \sum_{i=0}^{N-1} \sum_{\substack{j=0 \\ j\neq i}}^{N-1} e^{-(i+j)D/\tau_A} p(t+\tau-iD)p(t-jD) \qquad t+\tau, t \in [0, ND] \end{aligned} \tag{9.84}$$

To establish these results, the following have been used:

$$\begin{aligned} \mu_k &= \boldsymbol{E}[A_k] = a_o e^{-kD/\tau_A} \boldsymbol{E}[B_k] = \frac{\sqrt{\pi}\sigma_b a_o}{\sqrt{2}} \cdot e^{-kD/\tau_A} \\ \boldsymbol{E}\left[A_k^2\right] &= a_o^2 e^{-2kD/\tau_A} \boldsymbol{E}\left[B_k^2\right] = 2\sigma_o^2 \sigma_b^2 e^{-2kD/\tau_A} \end{aligned} \tag{9.85}$$

The mean and variance are shown in Figure 9.14. The autocorrelation function is shown in Figure 9.15.

***9.7.2.2 Amplitude Shift-Keyed Modulation*** Consider a random process $X$, which defines amplitude shift-keyed signals according to

$$X(\Omega, t) = \sum_{k=0}^{N-1} A_k p(t-kD) \qquad t \in [0, T],\ T = ND \tag{9.86}$$

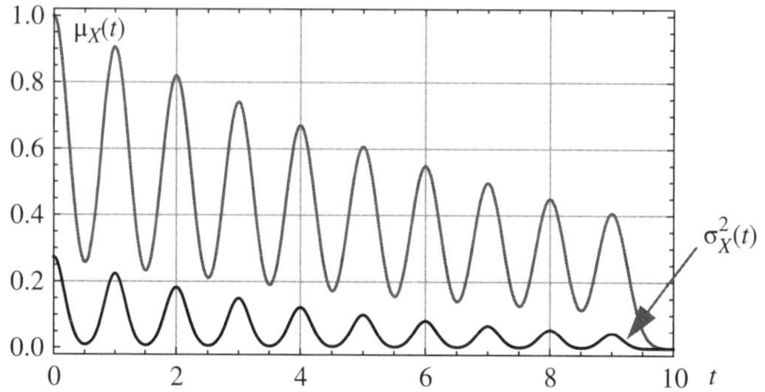


**FIGURE 9.14** Mean and variance of the defined signalling random process.

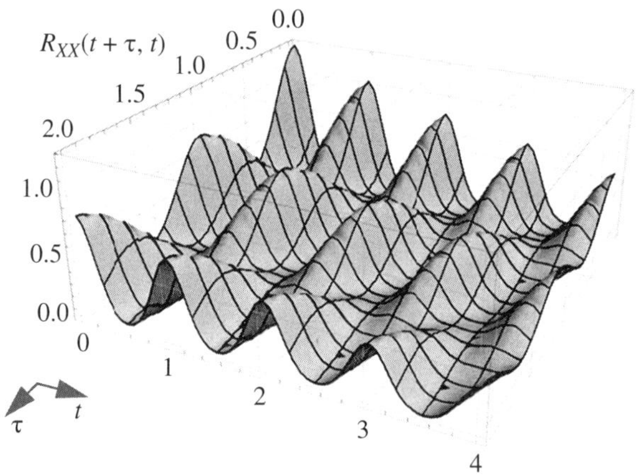


**FIGURE 9.15** Autocorrelation function of the defined signalling random process.

where $\Omega = (A_0, \ldots, A_{N-1})$ is a vector of independent and identical random variables with a sample space of $\{0, a_o\}$ and the pulse function is sinusoidal with a duration $D$ and a frequency $f_C$:

$$p(t) = \sin(2\pi f_C t)[u(t) - u(t-D)] \qquad f_C = nr,\ n \in \mathbf{Z}^+,\ r = 1/D \tag{9.87}$$

For the case of $P[A_k = 0] = P[A_k = a_o] = 0.5$, the mean, variance and mean squared value for $A_k$, respectively, are

$$\mu_{A_k} = \frac{a_o}{2} \qquad \sigma_{A_k}^2 = \frac{a_o^2}{4} \qquad \boldsymbol{E}\left[A_k^2\right] = \frac{a_o^2}{2} \tag{9.88}$$

One waveform from the random process is shown in Figure 9.16 for the case of $D = 1$, $a_o = 1$, and $f_C = 2$.

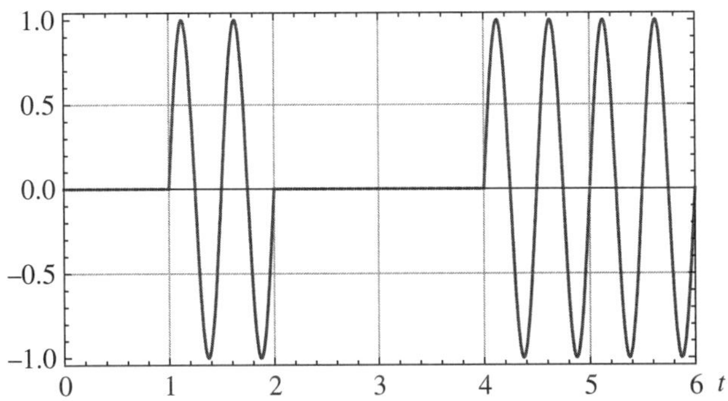


**FIGURE 9.16** Graph of one signal from the defined amplitude shift-keyed signalling random process for the case where the amplitudes are 0, 1, 0, 0, 1, 1.

It then follows, from the results specified in Theorem 9.16, that the mean, variance, and autocorrelation of the defined amplitude shift-keyed random process, respectively, are

$$\mu_X(t) = \frac{a_o}{2}\sum_{k=0}^{N-1} p(t-k\mathrm{D}) \qquad \sigma_X^2(t) = \frac{a_o^2}{4}\sum_{k=0}^{N-1} p^2(t-k\mathrm{D}) \qquad t \in [0,\ N\mathrm{D}] \tag{9.89}$$

$$\begin{aligned} R_{XX}(t+\tau,t) &= \frac{a_o^2}{2}\cdot\sum_{i=0}^{N-1} p(t+\tau-i\mathrm{D})p(t-i\mathrm{D}) \\ &+ \frac{a_o^2}{4}\cdot\sum_{i=0}^{N-1}\sum_{\substack{j=0 \\ j\neq i}}^{N-1} p(t+\tau-i\mathrm{D})p(t-j\mathrm{D}) \qquad t+\tau,\ t\in[0,\ N\mathrm{D}] \end{aligned} \tag{9.90}$$

As the pulse function is time limited to [0, D], the autocorrelation can be written, consistent with Theorem 9.17, for $t+\tau,\ t \in [0, N\mathrm{D}]$, according to

$$\begin{aligned} R_{XX}(t+\tau,t) &= u[N-1-i_{\min_1}]\cdot u[i_{\max_1}]\cdot u[i_{\max_1}-i_{\min_1}] \\ &\cdot\frac{a_o^2}{2}\cdot\sum_{i=\max\{i_{\min_1},0\}}^{\min\{i_{\max_1},N-1\}} p(t+\tau-i\mathrm{D})p(t-i\mathrm{D}) \\ &+u[N-1-i_{\min_2}]\cdot u[i_{\max_2}]\cdot u[N-1-j_{\min}]\cdot u[j_{\max}]\cdot u[i_{\max_2}-i_{\min_2}]\cdot u[j_{\max}-j_{\min}] \\ &\cdot\frac{a_o^2}{4}\cdot\sum_{i=\max\{i_{\min_2},0\}}^{\min\{i_{\max_2},N-1\}}\ \sum_{j=\max\{j_{\min},0\},\ j\neq 1}^{\min\{j_{\max},N-1\}} p(t+\tau-i\mathrm{D})p(t-j\mathrm{D}) \end{aligned} \tag{9.91}$$

where

$$i_{\min_1} = \left\lfloor \max\left\{\frac{t-\mathrm{D}}{\mathrm{D}}, \frac{t+\tau-\mathrm{D}}{\mathrm{D}}\right\} \right\rfloor + 1 \qquad i_{\max_1} = \left\lfloor \min\left\{\frac{t}{\mathrm{D}}, \frac{t+\tau}{\mathrm{D}}\right\} \right\rfloor \tag{9.92}$$

$$i_{\min_2} \left\lfloor \frac{t+\tau-\mathrm{D}}{\mathrm{D}} \right\rfloor + 1,\ i_{\max_2} = \left\lfloor \frac{t+\tau}{\mathrm{D}} \right\rfloor,\ j_{\min} \left\lfloor \frac{t-\mathrm{D}}{\mathrm{D}} \right\rfloor + 1,\ j_{\max} = \left\lfloor \frac{t}{\mathrm{D}} \right\rfloor \tag{9.93}$$

For the case of $\mathrm{D}=1$, $a_o=1$, and $f_C=2$, the mean and variance are shown in Figure 9.17. The autocorrelation function is shown in Figure 9.18.

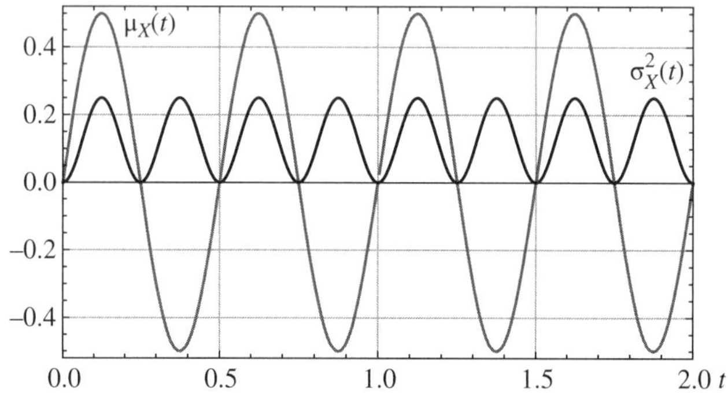


**FIGURE 9.17** Mean and variance of the defined amplitude shift-keyed signalling random process.

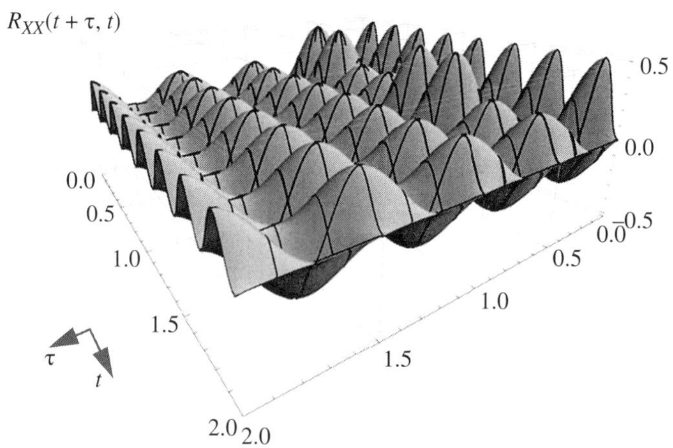


**FIGURE 9.18** Autocorrelation function of the defined amplitude shift-keyed signalling random process.

## 9.8 GENERALIZED SIGNALLING RANDOM PROCESS

Consider a generalized signalling random process $X: S \to S_X$, based on a signal set $S_\Phi$ and defined according to

$$X(\Omega, t) = \sum_{k=1}^{N} \phi(\Omega_k, t - \Gamma_k) \qquad \phi \in S_\Phi, t \in [0, T] \tag{9.94}$$

where $\Omega = ((\Omega_1, \Gamma_1), \ldots, (\Omega_N, \Gamma_N))$. Here, $\Omega_k$ is an index random variable, taking values from a set $S_{\Omega_k}$, and defines which waveform from the signal set $S_\Phi$ is associated with the time defined by $\Gamma_k$.

The usual case is where the index random variables $\Omega_1, \ldots, \Omega_N$ have an identical sample space $S_\Omega$ (which implies the same signal set for all waveforms) and $\Gamma_1, \ldots, \Gamma_N$ are a sequence of random variables consistent with the outcomes of a pseudo-Poisson experiment on $[0, T]$, with a rate $\lambda$, and $N = \lambda T$ assuming $T$ is consistent with $\lambda T$ being an integer.

The assumption is made that the index random variables define an uncountable sample space consistent with the signalling set comprising of an uncountable number of signals. Consistent with this assumption, the general case is where the vector of pairs of random variables are governed by the joint probability density function:

$$\begin{aligned} &P[\Omega_i \in [\omega_i, \omega_i + d\omega_i], \Gamma_i \in [\gamma_i, \gamma_i + d\gamma_i], i \in \{1, \ldots, N\}] \\ &\quad \approx f_{\Omega_1 \ldots \Omega_N \Gamma_1 \ldots \Gamma_N}(\omega_1, \ldots, \omega_N, \gamma_1, \ldots, \gamma_N) d\omega_1 \ldots d\omega_N d\gamma_1 \ldots d\gamma_N \end{aligned} \tag{9.95}$$

The ensemble of possible waveforms defined by $X$ is

$$S_X = \left\{ X(((\omega_1, \gamma_1), \ldots, (\omega_N, \gamma_N)), t) = \sum_{k=1}^{N} \phi(\omega_k, t - \gamma_k) : \quad \begin{matrix} \omega_k \in S_{\Omega_k} \\ \gamma_k \in S_{\Gamma_k} \\ \phi \in S_\Phi \end{matrix} \right\} \tag{9.96}$$

where $S_{\Omega_k}$ and $S_{\Gamma_k}$, respectively, are the sample spaces of $\Omega_k$ and $\Gamma_k$.

### 9.8.1 Basic Distinctions

***9.8.1.1 Degenerate Cases*** There are two degenerate cases. The first is the case where the times are fixed consistent with $S_{\Gamma_i} = \{(i-1)\mathrm{D}\}$ and a signalling random process is defined:

$$X((\Omega_1, \ldots, \Omega_N), t) = \sum_{k=1}^{N} \phi(\Omega_k, t - (k-1)\mathrm{D}) \qquad t \in [0, T] \tag{9.97}$$

with outcomes being defined by the joint probability density function

$$\begin{aligned} &P[\Omega_1 \in [\omega_1, \omega_1 + d\omega_1], \ldots, \Omega_N \in [\omega_N, \omega_N + d\omega_N]] \\ &\quad \approx f_{\Omega_1 \ldots \Omega_N}(\omega_1, \ldots, \omega_N) d\omega_1 \ldots d\omega_N \end{aligned} \tag{9.98}$$

The second is the case where the index random variable consists of a single outcome, that is, $S_{\Omega_k} = \{1\}$, which, in turn, is consistent with a signalling set comprising of a single waveform, that is, $S_{\Phi_k} = \{\phi : \boldsymbol{R}^+ \to \boldsymbol{R}\}$. For this case, the random process is

$$X((\Gamma_1, \ldots, \Gamma_N), t) = \sum_{k=1}^{N} \phi(t - \Gamma_k) \qquad t \in [0, T] \tag{9.99}$$

For the case where the random variables $\Gamma_1, \ldots, \Gamma_N$ are identical and independent, and with density functions $f_{\Gamma_i}(\gamma) = \frac{1}{T}$, a shot noise random process results.

***9.8.1.2 Separable and Nonseparable Signalling Set*** Consider the two signalling sets:

$$S_\Phi = \{\phi(\omega, t) = (1 + \omega)\sin(2\pi f_C t) : \ \omega \in (-1, 1)\} \tag{9.100}$$

$$S_\Phi = \{\phi(\omega, t) = A\sin(2\pi f_C t + \omega) : \ \omega \in (-\pi, \pi)\} \tag{9.101}$$

Consistent with the following definition, the former is consistent with a separable signalling random process, while the latter is consistent with a nonseparable signalling random process.

**Definition: Separable Signalling Random Process** A generalized signalling random process is separable if each signal $\phi$ in the signalling set $S_\Phi$ can be separated according to

$$\phi(\omega, t) = \phi_1(\omega)\phi_2(t) \qquad \phi \in S_\Phi \tag{9.102}$$

Thus, a generalized signalling random process is separable if the signal set consists of waveforms with a fixed time form but with amplitude scaling. For this case, the signalling set can be defined according to

$$S_\Phi = \{\phi(\omega, t) = \phi_1(\omega)\phi_2(t) : \ \omega \in S_{\Omega_1} \cup \ldots \cup S_{\Omega_N}\} \tag{9.103}$$

and the random process can be written as

$$X(\Omega, t) = \sum_{k=1}^{N} \phi_1(\Omega_k)\phi_2(t - \Gamma_k) \qquad t \in [0, T] \tag{9.104}$$

**Definition: Nonseparable Signalling Random Process** If each signal defined by a signalling random process cannot be separated in a manner consistent with Equation 9.102, then the random process is nonseparable.

The degenerate case, defined by Equation 9.99, is one example of a separable signalling random process.

***9.8.1.3 Distinct Cases Based on Dependence–Independence*** Consider the probability density function governing the outcomes of a signalling random process as defined by

$$\begin{aligned}&P[\Omega_i \in [\omega_i, \omega_i + d\omega_i], \Gamma_i \in [\gamma_i, \gamma_i + d\gamma_i], i \in \{1, \ldots, N\}] \\ &\approx f_{\Omega_1 \ldots \Omega_N \Gamma_1 \ldots \Gamma_N}(\omega_1, \ldots, \omega_N, \gamma_1, \ldots, \gamma_N) d\omega_1 \ldots d\omega_N d\gamma_1 \ldots d\gamma_N\end{aligned} \quad (9.105)$$

The nature of the dependence/independence between the random variables $\Omega_1, \ldots, \Omega_N$, $\Gamma_1, \ldots, \Gamma_N$ determines the nature of the waveforms defined by the signalling random process. The important cases are specified in Table 9.1.

Waveform dependence, consistent with $\Omega_1, \ldots, \Omega_N$ being correlated, arises, for example, in the communication context when coding is used.

Time dependence, consistent with $\Gamma_1, \ldots, \Gamma_N$ being correlated, can arise, for example, when the times are chosen from a fixed interval rather than consistent with a Poisson point process on the semi-infinite interval $[0, \infty]$. For example, the ordering of $N$ times chosen at random from the interval $[0, T]$ yields a joint probability density function

$$f_{\Gamma_1 \ldots \Gamma_N}(\gamma_1, \ldots, \gamma_N) = \frac{N!}{T^N} \qquad \gamma_1 < \gamma_2 < \ldots < \gamma_N \quad (9.106)$$

This result arises from order statistics (see Chapter 11).

Individual time-waveform dependence, consistent with $\Gamma_i$ being correlated with $\Omega_i$, arises, for example, when the waveform amplitude depends on the waveform time. This case can arise, for example, when the experiment underpinning the random process comprises a sequence of sequential subexperiments. One example is where the $i$th subexperiment is:

1. The $i$th time $\gamma_i$ is determined consistent with a specified probability density function $f_{\Gamma_i}$.

**TABLE 9.1 Demarcation of generalized signalling random processes according to the dependence/independence of the underlying random variables**

| Case | Level of dependency | Joint PDF |
|---|---|---|
| 1 | Full dependence | $f_{\Omega_1 \ldots \Omega_N \Gamma_1 \ldots \Gamma_N}(\omega_1, \ldots, \omega_N, \gamma_1, \ldots, \gamma_N)$ |
| 2 | Waveform dependence, time dependence, and time and waveform independence | $f_{\Omega_1 \ldots \Omega_N}(\omega_1, \ldots, \omega_N) f_{\Gamma_1 \ldots \Gamma_N}(\gamma_1, \ldots, \gamma_N)$ |
| 3 | Waveform dependence, time independence, and time and waveform independence | $f_{\Omega_1 \ldots \Omega_N}(\omega_1, \ldots, \omega_N) f_{\Gamma_1}(\gamma_1) \ldots f_{\Gamma_N}(\gamma_N)$ |
| 4 | Waveform independence, time dependence, and time and waveform independence | $f_{\Omega_1}(\omega_1) \ldots f_{\Omega_N}(\omega_N) f_{\Gamma_1 \ldots \Gamma_N}(\gamma_1, \ldots, \gamma_N)$ |
| 5 | Dependence of $i$th waveform on $i$th time only | $f_{\Omega_1 \Gamma_1}(\omega_1, \gamma_1) \ldots f_{\Omega_N \Gamma_N}(\omega_N, \gamma_N)$ |
| 6 | Full independence | $f_{\Omega_1}(\omega_1) \ldots f_{\Omega_N}(\omega_N) f_{\Gamma_1}(\gamma_1) \ldots f_{\Gamma_N}(\gamma_N)$ |

2. The waveform amplitude $\omega_i$ is then determined consistent with a conditional probability density function $f_{\Omega_i/\Gamma_i=\gamma_i}(\omega_i)$.
3. The joint probability density function $f_{\Omega_i\Gamma_i}$ is then given by

$$f_{\Omega_i\Gamma_i}(\omega_i,\gamma_i)=f_{\Omega_i/\Gamma_i=\gamma_i}(\omega_i)f_{\Gamma_i}(\gamma_i) \tag{9.107}$$

One example is where the time $\gamma_i$ is determined consistent with an exponential probability density function, that is,

$$f_{\Gamma_i}(\gamma_i)=\frac{e^{-\gamma_i/\tau}}{\tau} \qquad \gamma_i>0 \tag{9.108}$$

and the conditional probability density function governing $\omega_i$ is Gaussian according to

$$f_{\Omega_i/\Gamma_i=\gamma_i}(\omega_i)=\frac{e^{-\omega_i^2/2\sigma^2(\gamma_i)}}{\sqrt{2\pi}\sigma(\gamma_i)} \tag{9.109}$$

The dependency arises from the variance $\sigma^2(\gamma_i)$. The joint probability density function then is

$$f_{\Omega_i\Gamma_i}(\omega_i,\gamma_i)=\frac{e^{-\omega_i^2/2\sigma^2(\gamma_i)}}{\sqrt{2\pi}\sigma(\gamma_i)}\cdot\frac{e^{-\gamma_i/\tau}}{\tau} \tag{9.110}$$

### 9.8.2 Mean of Generalized Signalling Random Process

**Theorem 9.18 Mean of Generalized Signalling Random Process**
The mean of the generalized signalling random process $X$ defined by Equations 9.94, 9.95, and 9.96 is

$$\mu_X(t)=\sum_{k=1}^{N}\int_{S_{\Omega_k}}\int_{S_{\Gamma_k}}\phi(\omega_k,t-\gamma_k)f_{\Omega_k\Gamma_k}(\omega_k,\gamma_k)d\omega_k d\gamma_k \tag{9.111}$$

For the case of identical joint probability density functions, that is,

$$f_{\Omega_j\Gamma_j}(\omega,\gamma)=f_{\Omega_k\Gamma_k}(\omega,\gamma)=f_{\Omega\Gamma}(\omega,\gamma) \qquad j\neq k, j,k\in\{1,\ldots,N\} \tag{9.112}$$

the mean simplifies to

$$\mu_X(t)=N\int_{S_\Omega}\int_{S_\Gamma}\phi(\omega,t-\gamma)f_{\Omega\Gamma}(\omega,\gamma)d\omega d\gamma \tag{9.113}$$

**Proof**

By definition of the mean,

$$\mu_X(t) = \boldsymbol{E}\left[\sum_{k=1}^{N}\phi(\Omega_k,\ t-\Gamma_k)\right]$$

$$= \int\limits_{S_{\Omega_N}}\cdots\int\limits_{S_{\Omega_1}}\int\limits_{S_{\Gamma_N}}\cdots\int\limits_{S_{\Gamma_1}}\sum_{k=1}^{N}\phi(\omega_k,\ t-\gamma_k)f_{\Omega_1\ldots\Omega_N\Gamma_1\ldots\Gamma_N}(\omega_1,\ \ldots,\ \omega_N,\gamma_1,\ \ldots,\ \gamma_N)$$
$$d\gamma_1\ldots d\gamma_N d\omega_1\ldots d\omega_N$$

$$= \sum_{k=1}^{N}\int\limits_{S_{\Omega_k}}\int\limits_{S_{\Gamma_k}}\phi(\omega_k,t-\gamma_k)f_{\Omega_k\Gamma_k}(\omega_k,\gamma_k)d\omega_k d\gamma_k = \sum_{k=1}^{N}\boldsymbol{E}[\phi(\Omega_k,\ t-\Gamma_k)] \qquad (9.114)$$

***9.8.2.1 Mean for Different Cases of Dependency*** Expressions for the mean are tabulated in Table 9.2 for various cases of dependency. Results for the separable case are detailed in Table 9.3.

### 9.8.3 Autocorrelation of Generalized Signalling RP

**Theorem 9.19 Autocorrelation of Generalized Signalling RP**

The autocorrelation function of the generalized signalling random process $X$ defined by Equations 9.94, 9.95, and 9.96 is

$$R_{XX}(t+\tau,t) = \sum_{i=1}^{N}\int\limits_{S_{\Omega_i}}\int\limits_{S_{\Gamma_i}}\phi(\omega_i,t+\tau-\gamma_i)\phi(\omega_i,t-\gamma_i)f_{\Omega_i\Gamma_i}(\omega_i,\gamma_i)d\gamma_i d\omega_i$$
$$+\sum_{i=1}^{N}\sum_{\substack{k=1\\k\neq i}}^{N}\int\limits_{S_{\Omega_k}}\int\limits_{S_{\Omega_i}}\int\limits_{S_{\Gamma_k}}\int\limits_{S_{\Gamma_i}}\phi(\omega_i,t+\tau-\gamma_i)\phi(\omega_k,t-\gamma_k)f_{\Omega_i\Omega_k\Gamma_i\Gamma_k}(\omega_i,\omega_k,\gamma_i,\gamma_k).$$
$$d\gamma_i d\gamma_k d\omega_i d\omega_k \qquad t,t+\tau>0 \qquad (9.115)$$

**TABLE 9.2 Mean of a generalized signalling random process for the nonseparable case**

| Dependence | $\mu_X(t)$ for $t>0$ |
|---|---|
| $\Omega_k, \Gamma_k$ dependent | $\sum_{k=1}^{N}\int\limits_{S_{\Omega_k}}\int\limits_{S_{\Gamma_k}}\phi(\omega_k,t-\gamma_k)f_{\Omega_k\Gamma_k}(\omega_k,\gamma_k)d\gamma_k d\omega_k$ |
| $\Omega_k$ is independent of $\Gamma_k$ | $\sum_{k=1}^{N}\int\limits_{S_{\Omega_k}}\int\limits_{S_{\Gamma_k}}\phi(\omega_k,t-\gamma_k)f_{\Omega_k}(\omega_k)f_{\Gamma_k}(\gamma_k)d\gamma_k d\omega_k$ |
| $\Omega_k$ is independent of $\Gamma_k$, $f_{\Omega_i}(\omega)=f_{\Omega_k}(\omega)=f_{\Omega}(\omega)$, and $f_{\Gamma_k}(\gamma_k)=\frac{1}{T}$ | $r\int\limits_{S_\Omega}\int\limits_{0}^{T}\phi(\omega,t-\gamma)f_{\Omega}(\omega)d\gamma d\omega \qquad r=\frac{N}{T}$ |

**TABLE 9.3 Mean of a generalized signalling random process for the separable case where $\phi(\omega,t) = \phi_1(\omega)\phi_2(t)$**

| Dependence | $\mu_X(t)$ for $t>0$ |
|---|---|
| $\Omega_k, \Gamma_k$ dependent | $\sum_{k=1}^{N} \int_{S_{\Omega_k}} \int_{S_{\Gamma_k}} \phi_1(\omega_k)\phi_2(t-\gamma_k) f_{\Omega_k\Gamma_k}(\omega_k,\gamma_k)d\gamma_k d\omega_k$ |
| $\Omega_k$ is independent of $\Gamma_k$ | $\sum_{k=1}^{N} \int_{S_{\Omega_k}} \phi_1(\omega_k) f_{\Omega_k}(\omega_k)d\omega_k \int_{S_{\Gamma_k}} \phi_2(t-\gamma_k) f_{\Gamma_k}(\gamma_k)d\gamma_k$ |
| $\Omega_k$ is independent of $\Gamma_k$, $f_{\Omega_i}(\omega) = f_{\Omega_k}(\omega) = f_\Omega(\omega)$, and $f_{\Gamma_k}(\gamma_k) = \frac{1}{T}$ | $r\int_{S_\Omega} \phi_1(\omega) f_\Omega(\omega)d\omega \int_0^T \phi_2(t-\gamma)d\gamma \qquad r = \frac{N}{T}$ |

For the case where all signals in the signalling set $S_\Phi$ are causal, the autocorrelation function is

$$R_{XX}(t+\tau,t) = \sum_{i=1}^{N} \int_{S_{\Omega_i}} \int_0^{\min\{t,t+\tau\}} \phi(\omega_i, t+\tau-\gamma_i)\phi(\omega_i, t-\gamma_i) f_{\Omega_i\Gamma_i}(\omega_i,\gamma_i)d\gamma_i d\omega_i$$
$$+ \sum_{i=1}^{N} \sum_{\substack{k=1 \\ k\neq i}}^{N} \int_{S_{\Omega_k}} \int_{S_{\Omega_i}} \int_0^{t} \int_0^{t+\tau} \phi(\omega_i, t+\tau-\gamma_i)\phi(\omega_k, t-\gamma_k) f_{\Omega_i\Omega_k\Gamma_i\Gamma_k}(\omega_i,\omega_k,\gamma_i,\gamma_k).$$
$$d\gamma_i d\gamma_k d\omega_i d\omega_k \qquad t, t+\tau > 0 \tag{9.116}$$

**Proof**

The proof is detailed in Appendix 9.G.

***9.8.3.1 Autocorrelation: Causal Case*** The autocorrelation for the different cases, as defined by the dependence/independence between the random variables, is specified in Table 9.4. For the separable case, the autocorrelation function is specified in Table 9.5. In Table 9.5, the following definitions have been used:

$$\boldsymbol{E}[\phi_1(\Omega_i)] = \int_{S_{\Omega_i}} \phi_1(\omega_i) f_{\Omega_i}(\omega_i)d\omega_i \quad \boldsymbol{E}\left[\phi_1^2(\Omega_i)\right] = \int_{S_{\Omega_i}} \phi_1^2(\omega_i) f_{\Omega_i}(\omega_i)d\omega_i$$
$$R_{\phi_1\phi_1}(\Omega_i,\Omega_k) = \boldsymbol{E}[\phi_1(\Omega_i)\phi_1(\Omega_k)] = \int_{S_{\Omega_k}} \int_{S_{\Omega_i}} \phi_1(\omega_i)\phi_1(\omega_k) f_{\Omega_i\Omega_k}(\omega_i,\omega_k)d\omega_i d\omega_k \tag{9.117}$$

***9.8.3.2 Mean Squared Value and Variance*** The mean squared value $\boldsymbol{E}[X^2(\Omega, t)]$ can readily be obtained from $R_{XX}(t, t)$. The variance can then be obtained from

$$\sigma_X^2(t) = \boldsymbol{E}\left[X^2(\Omega,t)\right] - \mu_X^2(t) \tag{9.118}$$

**TABLE 9.4 Autocorrelation function for the nonseparable case. The assumptions are that all signals in the signalling set are causal and $t, t+\tau > 0$**

$R_{XX}(t+\tau, t) = \boldsymbol{E}[(\Omega, t+\tau)X(\Omega, t)]$

1. Full dependence:

$$\sum_{i=1}^{N}\int_{S_{\Omega_i}}\int_{0}^{\min\{t,t+\tau\}}\phi(\omega_i, t+\tau-\gamma_i)\phi(\omega_i, t-\gamma_i)f_{\Omega_i\Gamma_i}(\omega_i,\gamma_i)d\gamma_i d\omega_i +$$

$$\sum_{i=1}^{N}\sum_{\substack{k=1\\k\neq 1}}^{N}\int_{S_{\Omega_i}}\int_{S_{\Omega_k}}\int_{0}^{t+\tau}\int_{0}^{t}\phi(\omega_i, t+\tau-\gamma_i)\phi(\omega_k, t-\gamma_k)f_{\Omega_i\Omega_k\Gamma_i\Gamma_k}(\omega_i,\omega_k,\gamma_i,\gamma_k)d\gamma_k d\gamma_i d\omega_k d\omega_i$$

2. Waveform dependence, time dependence, and time and waveform independence:

$$\sum_{i=1}^{N}\int_{S_{\Omega_i}}\int_{0}^{\min\{t.t+\tau\}}\phi(\omega_i, t+\tau-\gamma_i)\phi(\omega_i, t-\gamma_i)f_{\Omega_i}(\omega_i)f_{\Gamma_i}(\gamma_i)d\gamma_i d\omega_i +$$

$$\sum_{i=1}^{N}\sum_{\substack{k=1\\k\neq i}}^{N}\int_{S_{\Omega_i}}\int_{S_{\Omega_k}}\int_{0}^{t+\tau}\int_{0}^{t}\phi(\omega_i, t+\tau-\gamma_i)\phi(\omega_k, t-\gamma_k)f_{\Omega_i\Omega_k}(\omega_i,\omega_k)f_{\Gamma_i\Gamma_k}(\gamma_i,\gamma_k)d\gamma_k d\gamma_i d\omega_k d\omega_i$$

3. Waveform dependence, time independence, and time and waveform independence:

$$\sum_{i=1}^{N}\int_{S_{\Omega_i}}\int_{0}^{\min\{t,t+\tau\}}\phi(\omega_i, t+\tau-\gamma_i)\phi(\omega_i, t-\gamma_i)f_{\Omega_i}(\omega_i)f_{\Gamma_i}(\gamma_i)d\gamma_i d\omega_i +$$

$$\sum_{i=1}^{N}\sum_{\substack{k=1\\k\neq i}}^{N}\int_{S_{\Omega_t}}\int_{S_{\Omega_k}}\int_{0}^{t+\tau}\int_{0}^{t}\phi(\omega_i, t+\tau-\gamma_i)\phi(\omega_k, t-\gamma_k)f_{\Omega_i\Omega_k}(\omega_i,\omega_k)f_{\Gamma_i}(\gamma_i)f_{\Gamma_k}(\gamma_k)d\gamma_k d\gamma_i d\omega_k d\omega_i$$

4. Waveform independence, time dependence, and time and waveform independence:

$$\sum_{i=1}^{N}\int_{S_{\Omega_i}}\int_{0}^{\min\{t,t+\tau\}}\phi(\omega_i, t+\tau-\gamma_i)\phi(\omega_i, t-\gamma_i)f_{\Omega_i}(\omega_i)f_{\Gamma_i}(\gamma_i)d\gamma_i d\omega_i +$$

$$\sum_{i=1}^{N}\sum_{\substack{k=1\\k\neq i}}^{N}\int_{S_{\Omega_t}}\int_{S_{\Omega_k}}\int_{0}^{t+\tau}\int_{0}^{t}\phi(\omega_i, t+\tau-\gamma_i)\phi(\omega_k, t-\gamma_k)f_{\Omega_i}(\omega_i)f_{\Omega_k}(\omega_k)f_{\Gamma_i\Gamma_k}(\gamma_i,\gamma_k)d\gamma_k d\gamma_i d\omega_k d\omega_i$$

5. Dependence of *ith* time and *ith* waveform:

$$\sum_{i=1}^{N}\int_{S_{\Omega_i}}\int_{0}^{\min\{t,t+\tau\}}\phi(\omega_i, t+\tau-\gamma_i)\phi(\omega_i, t-\gamma_i)f_{\Omega_i\Gamma_i}(\omega_i,\gamma_i)d\gamma_i d\omega_i +$$

$$\sum_{i=1}^{N}\int_{S_{\Omega_i}}\int_{0}^{t+\tau}\phi(\omega_i, t+\tau-\gamma_i)f_{\Omega_i\Gamma_i}(\omega_i,\gamma_i)d\gamma_i d\omega_i.$$

$$\sum_{\substack{k=1\\k\neq i}}^{N}\int_{S_{\Omega_k}}\int_{0}^{t}\phi(\omega_k, t-\gamma_k)f_{\Omega_k\Gamma_k}(\omega_k,\gamma_k)d\gamma_k d\omega_k$$

*(continued)*

**TABLE 9.4** *(continued)*

$R_{XX}(t+\tau, t) = \boldsymbol{E}[(\Omega, t+\tau)X(\Omega, t)]$

6. Full independence:

$$\sum_{i=1}^{N}\int_{S_{\Omega_i}}\int_{0}^{\min\{t,t+\tau\}} \phi(\omega_i, t+\tau-\gamma_i)\phi(\omega_i, t-\gamma_i)f_{\Omega_i}(\omega_i)f_{\Gamma_i}(\gamma_i)d\gamma_i d\omega_i +$$

$$\sum_{i=1}^{N}\int_{S_{\Omega_i}}\int_{0}^{t+\tau} \phi(\omega_i, t+\tau-\gamma_i)f_{\Omega_i}(\omega_i)f_{\Gamma_i}(\gamma_i)d\gamma_i d\omega_i.$$

$$\sum_{\substack{k=1\\k\neq i}}^{N}\int_{S_{\Omega_k}}\int_{0}^{t} \phi(\omega_k, t-\gamma_k)f_{\Omega_k}(\omega_k)f_{\Gamma_k}(\gamma_k)d\gamma_k d\omega_k$$

**TABLE 9.5 Autocorrelation function for the causal and separable case where $\boldsymbol{\phi(\omega,t) = \phi_1(\omega)\phi_2(t)}$**

$R_{XX}(t+\tau, t) = \boldsymbol{E}[X(\Omega, t+\tau)X(\Omega, t)]$

1. Full dependence:

$$\sum_{i=1}^{N}\int_{S_{\Omega_i}}\int_{0}^{\min\{t,t+\tau\}} \phi_1^2(\omega_i)\phi_2(t+\tau-\gamma_i)\phi_2(t-\gamma_i)f_{\Omega_i\Gamma_i}(\omega_i,\gamma_i)d\gamma_i d\omega_i +$$

$$\sum_{i=1}^{N}\sum_{\substack{k=1\\k\neq i}}^{N}\int_{S_{\Omega_i}}\int_{S_{\Omega_k}}\int_{0}^{t+\tau}\int_{0}^{t} \phi_1(\omega_i)\phi_1(\omega_k)\phi_2(t+\tau-\gamma_i)\phi_2(t-\gamma_k)f_{\Omega_i\Omega_k\Gamma_i\Gamma_k}(\omega_i,\omega_k,\gamma_i,\gamma_k)d\gamma_k d\gamma_i d\omega_k d\omega_i$$

2. Waveform dependence, time dependence, and time and waveform independence:

$$\sum_{i=1}^{N}\boldsymbol{E}\left[\phi_1^2(\Omega_i)\right]\int_{0}^{\min\{t,t+\tau\}} \phi_2(t+\tau-\gamma_i)\phi_2(t-\gamma_i)f_{\Gamma_i}(\gamma_i)d\gamma_i +$$

$$\sum_{i=1}^{N}\sum_{\substack{k=1\\k\neq i}}^{N}R_{\phi_1\phi_1}(\Omega_i, \Omega_k)\int_{0}^{t+\tau}\int_{0}^{t} \phi_2(t+\tau-\gamma_i)\phi_2(t-\gamma_k)f_{\Gamma_i\Gamma_k}(\gamma_i,\gamma_k)d\gamma_k d\gamma_i$$

3. Waveform dependence, time independence, and time and waveform independence:

$$\sum_{i=1}^{N}\boldsymbol{E}\left[\phi_1^2(\Omega_i)\right]\int_{0}^{\min\{t,t+\tau\}} \phi_2(t+\tau-\gamma_i)\phi_2(t-\gamma_i)f_{\Gamma_i}(\gamma_i)d\gamma_i +$$

$$\sum_{i=1}^{N}\sum_{\substack{k=1\\k\neq i}}^{N}R_{\phi_1\phi_1}(\Omega_i, \Omega_k)\int_{0}^{t+\tau} \phi_2(t+\tau-\gamma_i)f_{\Gamma_i}(\gamma_i)d\gamma_i\int_{0}^{t}\phi_2(t-\gamma_k)f_{\Gamma_k}(\gamma_k)d\gamma_k$$

**TABLE 9.5** *(continued)*

| $R_{XX}(t+\tau,t)=\boldsymbol{E}[X(\Omega,t+\tau)X(\Omega,t)]$ |
|---|
| 4. Waveform independence, time dependence, and time and waveform independence: |
| $\sum_{i=1}^{N}\boldsymbol{E}[\phi_1^2(\Omega_i)]\int_0^{\min\{t,t+\tau\}}\phi_2(t+\tau-\gamma_i)\phi_2(t-\gamma_i)f_{\Gamma_i}(\gamma_i)d\gamma_i+$ |
| $\sum_{i=1}^{N}\sum_{\substack{k=1\\k\neq i}}^{N}\boldsymbol{E}[\phi_1(\Omega_i)]\boldsymbol{E}[\phi_1(\Omega_k)]\int_0^{t+\tau}\int_0^{t}\phi_2(t+\tau-\gamma_i)\phi_2(t-\gamma_k)f_{\Gamma_i\Gamma_k}(\gamma_i,\gamma_k)d\gamma_k d\gamma_i$ |
| 5. Dependence of *ith* time and *ith* waveform: |
| $\sum_{i=1}^{N}\int_{S_{\Omega_i}}\int_0^{\min\{t,t+\tau\}}\phi_1^2(\omega_i)\phi_2(t+\tau-\gamma_i)\phi_2(t-\gamma_i)f_{\Omega_i\Gamma_i}(\omega_i,\gamma_i)d\gamma_i d\omega_i+$ |
| $\sum_{i=1}^{N}\sum_{\substack{k=1\\k\neq i}}^{N}\int_{S_{\Omega_i}}\int_0^{t+\tau}\phi_1(\omega_i)\phi_2(t+\tau-\gamma_i)f_{\Omega_i\Gamma_i}(\omega_i,\gamma_i)d\gamma_i d\omega_i\int_{S_{\Omega_k}}\int_0^{t}\phi_1(\omega_k)\phi_2(t-\gamma_k)f_{\Omega_k\Gamma_k}(\omega_k,\gamma_k)d\gamma_k d\omega_k$ |
| 6. Full Independence: |
| $\sum_{i=1}^{N}\boldsymbol{E}[\phi_1^2(\Omega_i)]\int_0^{\min\{t,t+\tau\}}\phi_2(t+\tau-\gamma_i)\phi_2(t-\gamma_i)f_{\Gamma_i}(\gamma_i)d\gamma_i+$ |
| $\sum_{i=1}^{N}\sum_{\substack{k=1\\k\neq i}}^{N}\boldsymbol{E}[\phi_1(\Omega_i)]\boldsymbol{E}[\phi_1(\Omega_k)]\int_0^{t+\tau}\phi_2(t+\tau-\gamma_i)f_{\Gamma_i}(\gamma_i)d\gamma_i\int_0^{t}\phi_2(t-\gamma_k)f_{\Gamma_k}(\gamma_k)d\gamma_k$ |

### 9.8.4 Example

Consider the separable case with time and waveform dependence (case 5 in Table 9.5), that is,

$$X(\Omega,t)=\sum_{k=1}^{N}\phi_1(\Omega_k)\phi_2(t-\Gamma_k) \tag{9.119}$$

where $\Omega=((\Omega_1,\Gamma_1),\ldots,(\Omega_N,\Gamma_N))$, $\Gamma_1,\ldots,\Gamma_N$ are a sequence of random variables consistent with a Poisson point process with a rate $\lambda$, the amplitude modulations of the pulse function are defined by $\phi_1(\omega_k)=\dfrac{a_o}{1+\lambda\omega_k}$, and the pulse function is defined according to

$$\phi_2(t)=p(t)=\frac{t^2e^{-t/\tau_A}}{2\tau_A^3}\cdot u(t) \tag{9.120}$$

Substitution yields

$$X(\Omega,t)=\sum_{k=1}^{N}\frac{a_o}{1+\lambda\Omega_k}\cdot\frac{t^2e^{-(t-\Gamma_k)/\tau_A}}{2\tau_A^2}\cdot u(t-\Gamma_k) \tag{9.121}$$

The dependency between $\Omega_i$ and $\Gamma_i$ is specified by the conditional probability density function $f_{\Omega_i/\Gamma_i=\gamma_i}$, and the joint probability density function $f_{\Omega_i\Gamma_i}$ is given by $f_{\Omega_i\Gamma_i}(\Omega_i,\gamma_i)=f_{\Omega_i/\Gamma_i=\gamma_i}(\omega_i)f_{\Gamma_i}(\gamma_i)$. The probability density function of the $i$th ordered time of a Poisson point process (see Chapter 11) is

$$f_{\Gamma_i}(\gamma_i)=\frac{\lambda(\lambda\gamma_i)^{i-1}e^{-\lambda\gamma_i}}{(i-1)!} \qquad i\ge 1,\gamma_i>0 \tag{9.122}$$

which is a Gamma distribution with parameters ($\lambda$, $i$). The amplitude random variable $\Omega_i$ is assumed to vary inversely with the pulse time consistent with the conditional probability density function:

$$f_{\Omega_i/\Gamma_i=\gamma_i}(\omega_i)=\begin{cases}0 & \omega_i\le 0,\ \omega_i>\gamma_i\\ \dfrac{1}{\gamma_i} & 0<\omega_i\le\gamma_i\end{cases} \tag{9.123}$$

Thus,

$$f_{\Omega_i\Gamma_i}(\omega_i,\gamma_i)=f_{\Omega_i/\Gamma_i=\gamma_i}(\omega_i)f_{\Gamma_i}(\gamma_i)=\begin{cases}0 & \omega_i\le 0,\ \omega_i>\gamma_i\\ \dfrac{\lambda(\lambda\gamma_i)^{i-1}e^{-\lambda\gamma_i}}{(i-1)!}\cdot\dfrac{1}{\gamma_i} & 0<\omega_i\le\gamma_i\end{cases} \tag{9.124}$$

One waveform from the random process, assuming parameters $a_o=1$, $\lambda=2$, $\tau_A=1/10\lambda$, is shown in Figure 9.19.

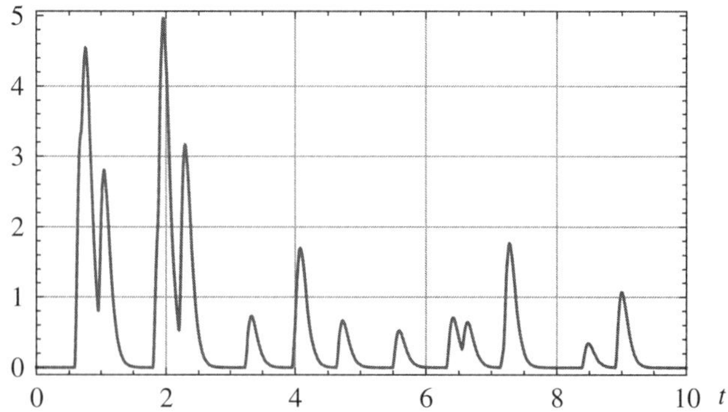


**FIGURE 9.19** Graph of one signal from the defined generalized signalling random process.

The mean of the random process is

$$\mu_X(t) = \boldsymbol{E}\left[\sum_{i=1}^{N} \frac{a_o}{1+\lambda\Omega_i} \cdot \frac{(t-\Gamma_i)^2 e^{-(t-\Gamma_i)/\tau_A}}{2\tau_A^3} \cdot u(t-\Gamma_i)\right] \qquad t \in [0, ND]$$

$$= \sum_{i=1}^{N} \int_0^t \int_0^{\gamma_i} \frac{a_o}{1+\lambda\omega_i} \cdot \frac{(t-\gamma_i)^2 e^{-(t-\gamma_i)/\tau_A}}{2\tau_A^3} \cdot f_{\Omega_i\Gamma_i}(\omega_i, \gamma_i) d\omega_i d\gamma_i \tag{9.125}$$

$$= \sum_{i=1}^{N} \int_0^t \int_0^{\gamma_i} \frac{a_o}{1+\lambda\omega_i} \cdot \frac{(t-\gamma_i)^2 e^{-(t-\gamma_i)/\tau_A}}{2\tau_A^3} \cdot \frac{\lambda(\lambda\gamma_i)^{i-1} e^{-\lambda\gamma_i}}{(i-1)!} \cdot \frac{1}{\gamma_i} \cdot d\omega_i d\gamma_i$$

From Table 9.5 (case 5), the autocorrelation is

$$R_{XX}(t+\tau, t) = \sum_{i=1}^{N} \int_{S_{\Omega_i}} \int_0^{\min\{t,t+\tau\}} \phi_1^2(\omega_i)\phi_2(t+\tau-\gamma_i)\phi_2(t-\gamma_i) f_{\Omega_i\Gamma_i}(\omega_i, \gamma_i) d\gamma_i d\omega_i$$
$$+ \sum_{i=1}^{N} \sum_{\substack{k=1 \\ k\neq i}}^{N} \int_{S_{\Omega_i}} \int_0^{t+\tau} \phi_1(\omega_i)\phi_2(t+\tau-\gamma_i) f_{\Omega_i\Gamma_i}(\omega_i, \gamma_i) d\gamma_i d\omega_i \cdot$$
$$\int_{S_{\Omega_k}} \int_0^{t} \phi_1(\omega_k)\phi_2(t-\gamma_k) f_{\Omega_k\Gamma_k}(\omega_k, \gamma_k) d\gamma_k d\omega_k \tag{9.126}$$

Substitution for the various functions, and integral limits, yields

$$R_{XX}(t+\tau, t) = \sum_{i=1}^{N} \int_0^{\min\{t,t+\tau\}} \int_0^{\gamma_i}$$
$$\frac{a_o^2}{(1+\lambda\omega_i)^2} \cdot \frac{(t+\tau-\gamma_i)^2 e^{-(t+\tau-\gamma_i)/\tau_A}}{2\tau_A^3} \cdot \frac{(t-\gamma_i)^2 e^{-(t-\gamma_i)/\tau_A}}{2\tau_A^3} \cdot f_{\Omega_i\Gamma_i}(\omega_i, \gamma_i) d\omega_i d\gamma_i$$
$$+ \sum_{i=1}^{N} \sum_{\substack{k=1 \\ k\neq i}}^{N} \int_0^{t+\tau} \int_0^{\gamma_i} \frac{a_o}{1+\lambda\omega_i} \cdot \frac{(t+\tau-\gamma_i)^2 e^{-(t+\tau-\gamma_i)/\tau_A}}{2\tau_A^3} \cdot f_{\Omega_i\Gamma_i}(\omega_i, \gamma_i) d\omega_i d\gamma_i \cdot$$
$$\int_0^{t} \int_0^{\gamma_k} \frac{a_o}{1+\lambda\omega_k} \cdot \frac{(t-\gamma_k)^2 e^{-(t-\gamma_k)/\tau_A}}{2\tau_A^3} \cdot f_{\Omega_k\Gamma_k}(\omega_k, \gamma_k) d\omega_k d\gamma_k \tag{9.127}$$

The variance can then be determined from

$$\sigma_X^2(t) = R_{XX}(t,t) - \mu_X^2(t) \tag{9.128}$$

The numerical evaluation of the autocorrelation function for each value of $t$ and $\tau$ requires the numerical evaluation of $N^2-N$ double integrals for the second term. To reduce the computation time, interchanging the order of summation and integration can be used according to

$$\begin{aligned} R_{XX}(t+\tau,t) = &\int_0^{t_M}\int_0^{t_M}\sum_{i=1}^{N} u(\gamma_i-\omega_i)\cdot\frac{a_o^2}{(1+\lambda\omega_i)^2}\cdot\frac{(t+\tau-\gamma_i)^2(t-\gamma_i)^2 e^{-2(t-\gamma_i)/\tau_A} e^{-\tau/\tau_A}}{4\tau_A^6} \\ &\cdot f_{\Omega_i\Gamma_i}(\omega_i,\gamma_i)d\omega_i d\gamma_i \\ &+\int_0^{t}\int_0^{t}\int_0^{t_2}\int_0^{t_2}\sum_{i=1}^{N}\sum_{k=1,k\neq i}^{N}\left[u(\gamma_i-\omega_i)\cdot\frac{a_o}{1+\lambda\omega_i}\cdot\frac{(t+\tau-\gamma_i)^2 e^{-(t+\tau-\gamma_i)/\tau_A}}{2\tau_A^3}\right] \\ &\cdot\left[u(\gamma_k-\omega_k)\cdot\frac{a_o}{1+\lambda\omega_k}\cdot\frac{(t-\gamma_k)^2 e^{-(t-\gamma_k)/\tau_A}}{2\tau_A^3}\right]\cdot f_{\Omega_i\Gamma_i}(\omega_i,\gamma_i)d\omega_i d\gamma_i \\ &\cdot f_{\Omega_k\Gamma_k}(\omega_k,\gamma_k)d\omega_k d\gamma_k \end{aligned} \tag{9.129}$$

which results in one double integral and one fourth-order integral for each value of $t$ and $\tau$. Here, $t_M = \min\{t, t+\tau\}$, $t_2 = t+\tau$ and the use of the unit step function allows the specification of the integrals over a rectangular region.

The mean and variance are shown in Figure 9.20. The autocorrelation function is shown in Figure 9.21. The parameters $a_o = 1, \lambda = 2$, and $\tau_A = 1/10\lambda$ have been assumed.

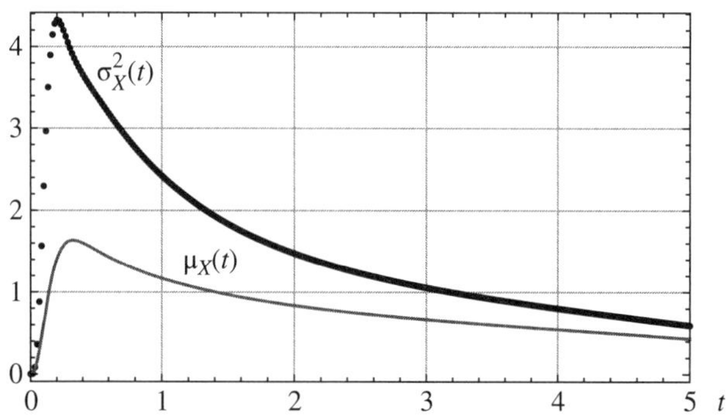


**FIGURE 9.20** Mean and variance of the defined generalized signalling random process.

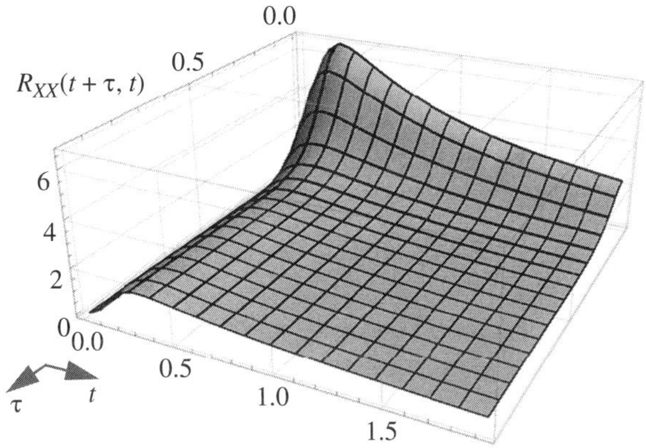


**FIGURE 9.21** Autocorrelation function of the defined generalized signalling random process based on a sample resolution for $t$ and $\tau$, respectively, of 1/16 and1/30.

## 9.9 AUTOCORRELATION: JITTERED RANDOM PROCESSES

Consider a jittered random process $X$ defined according to

$$X(\Omega, t) = A\sum_{k=1}^{N}(-1)^{k+1}u(t-kD-J_k) \tag{9.130}$$

where $D$ is the signal period and $\Omega = (J_1, \ldots, J_N)$ is a vector of random variables whose outcomes are consistent with a joint probability density function $f_{J_1, \ldots, J_N}$. One outcome of the random process is illustrated in Figure 9.22.

For the independent case where

$$f_{J_1, \ldots, J_N}(j_1, \ldots, j_N) = f_{J_1}(j_1)\ldots f_{J_N}(j_N) \tag{9.131}$$

the mean, variance, and autocorrelation function of this random process are stated in the following theorem.

**Theorem 9.20 Mean, Variance, and Autocorrelation Function**
The mean, variance, and autocorrelation of the jittered random process, for the independent case, respectively, are

$$\mu_X(t) = A\sum_{k=1}^{N}(-1)^{k+1}F_{J_k}(t-k\mathrm{D}) \qquad 0 \le t < N\mathrm{D} \tag{9.132}$$

$$\sigma_A^2(t) = A^2\sum_{k=1}^{N}\left[F_{J_k}(t-k\mathrm{D}) - F_{J_k}^2(t-k\mathrm{D})\right] \qquad 0 \le t < N\mathrm{D} \tag{9.133}$$

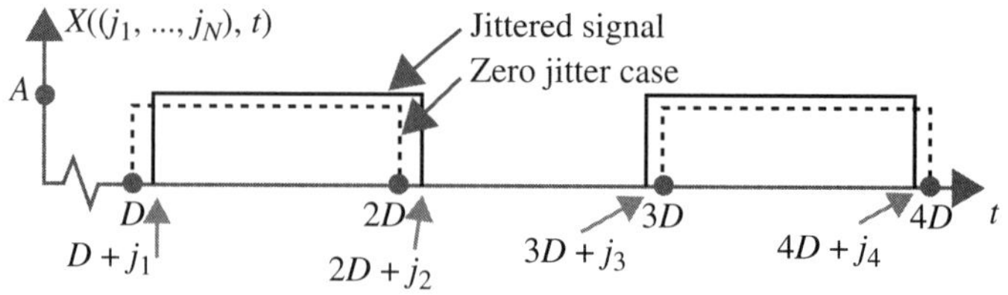


**FIGURE 9.22** Illustration of one signal from the defined jitter random process.

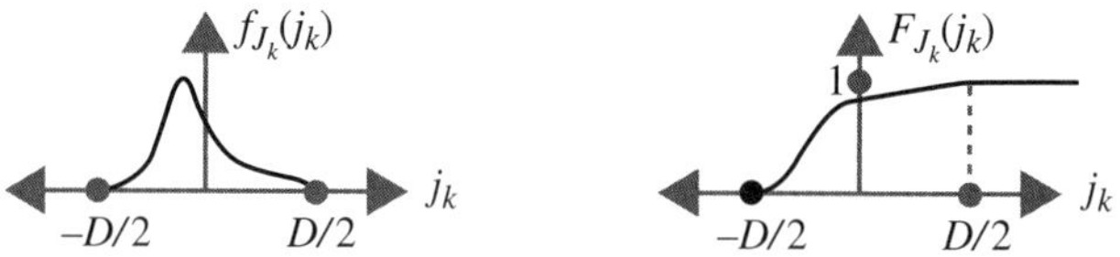


**FIGURE 9.23** Illustration of the probability density function $f_{J_k}$ and the associated cumulative distribution function.

$$
\begin{aligned}
R_{XX}(t+\tau,t) &= A^2\sum_{i=1}^{N}\min\{F_{J_k}(t-i\mathrm{D}),F_{J_i}(t+\tau-i\mathrm{D})\}\\
&\quad + A^2\sum_{i=1}^{N}\sum_{\substack{k=1\\k\neq i}}^{N}(-1)^{i+k}F_{J_i}(t+\tau-i\mathrm{D})F_{J_k}(t-k\mathrm{D}) \qquad 0\le t,t+\tau<N\mathrm{D}
\end{aligned}
\tag{9.134}
$$

Here, $F_{J_k}$ is the cumulative distribution associated with the $k$th jittered random variable $J_k$. For the usual case where the jitter probability density functions are nonzero only on the interval $[-\mathrm{D}/2, \mathrm{D}/2]$, as illustrated in Figure 9.23, it follows that

$$
\mu_X(t) = A\sum_{k=1}^{\lfloor t/\mathrm{D}+1/2\rfloor}(-1)^{k+1}F_{J_k}(t-k\mathrm{D}) \qquad t>\frac{\mathrm{D}}{2} \tag{9.135}
$$

$$
\sigma_X^2(t) = A^2\left[F_{J_k}(t-k\mathrm{D})-F_{J_k}^2(t-k\mathrm{D})\right] \qquad k=\left\lfloor\frac{t}{\mathrm{D}}+\frac{1}{2}\right\rfloor,\ t>\mathrm{D}/2 \tag{9.136}
$$

**Proof**

The proof is detailed in Appendix 9.H.

For the case of identical jitter random variables, the mean, variance, and autocorrelation functions have the forms illustrated in Figures 9.24, 9.25, and 9.26.

### 9.9.1 Example

Consider the jitter random process, as defined by Equation 9.130, with the parameters $\mathrm{D}=1, A=2$, and with identical and independent jitter random variables with a Gaussian probability density function:

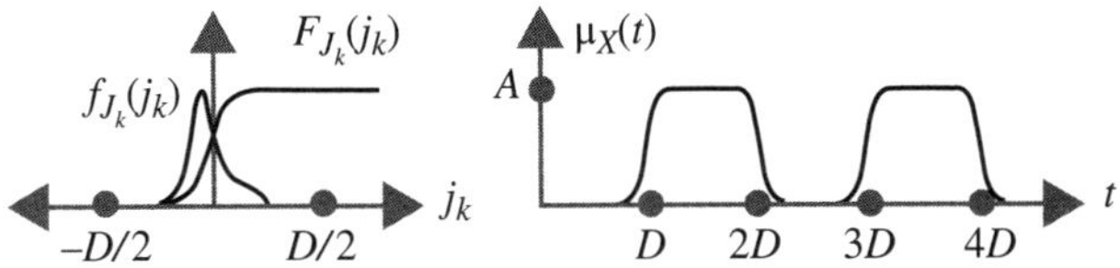


**FIGURE 9.24** Illustration of the mean of a jittered random process.

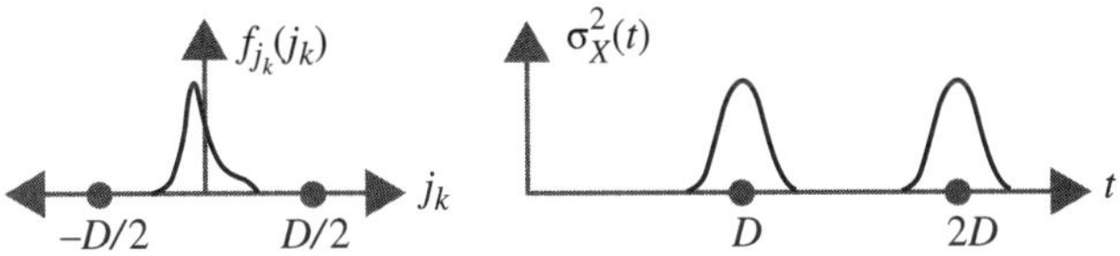


**FIGURE 9.25** Illustration of the variance of a jittered random process.

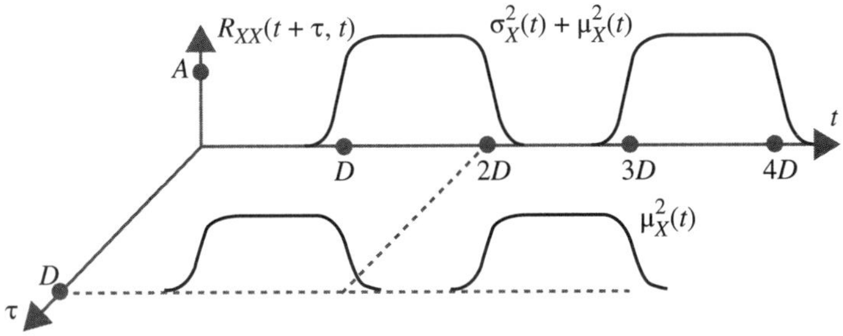


**FIGURE 9.26** Illustration of the autocorrelation of the jittered random process. The autocorrelation function, approximately, is periodic with $\tau$ with a period of 2D.

$$f_{J_k}(j) = \frac{e^{-(j-\mu_J)^2/2\sigma_J^2}}{\sqrt{2\pi}\sigma_J} \qquad k \in \{1, \ldots, N\} \tag{9.137}$$

$$\begin{aligned} F_{J_k}(j) &= \int_{-\infty}^{j} \frac{e^{-(\lambda-\mu_J)^2/2\sigma_J^2}}{\sqrt{2\pi}\sigma_J} d\lambda = 0.5 + \mathrm{erf}_1\left[\frac{j-\mu_J}{\sigma_J}\right] \\ \mathrm{erf}_1(x) &= \frac{1}{\sqrt{2\pi}} \int_0^x e^{-\lambda^2/2} d\lambda \qquad \mathrm{erf}_1(\infty) = 0.5 \end{aligned} \tag{9.138}$$

where $\mu_J = 0$ and $\sigma_J = D/10$. The mean, variance, and $R_{XX}(t, t)$ are shown in Figure 9.27. The autocorrelation function is shown in Figure 9.28.

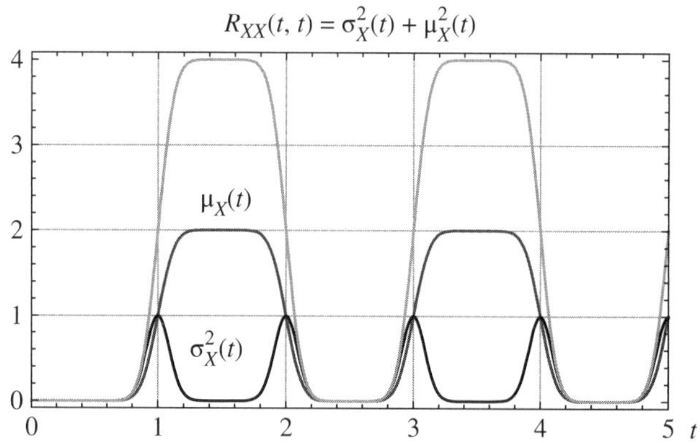


**FIGURE 9.27** Graph of the mean, variance, and $R_{XX}(t, t)$ for the defined jitter random process.

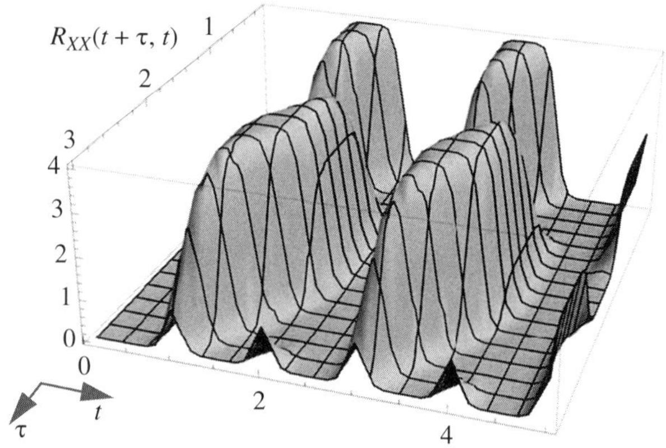


**FIGURE 9.28** Graph of the autocorrelation function for the defined jitter random process.

## 9.10 RANDOM WALK

The autocorrelation of a random walk, at the instants of change, is specified in the following theorem:

**Theorem 9.21 Autocorrelation and Correlation Coefficient of a Random Walk** For a random walk defined by

$$X(\Omega, n) = \sum_{i=1}^{n} X_i \qquad X_i \in \{\pm A_o\}, n \in \mathbf{Z}^+ \tag{9.139}$$

where $\Omega = (X_1, \ldots, X_n), X_i$ is independent of $X_j$, $i \neq j$ and $P[X_i = \pm A_o] = 0.5$, the autocorrelation function and correlation coefficient are given by

$$R_{XX}(n+m,n) = n\sigma^2 \quad m,n \in Z^+$$
$$\rho_{XX}(n+m,n) = \frac{1}{\sqrt{1+m/n}} \tag{9.140}$$

where $\sigma^2 = A_0^2$ is the variance of $X_i$. The autocorrelation is independent of $m$ and varies linearly with $n$.

**Proof**
First

$$E(X(\Omega,n)) = 0 \qquad \text{var}(X(\Omega,n)) = n\sigma^2 \tag{9.141}$$

It then follows that

$$\begin{aligned} R_{XX}(n+m,n) &= E[X(\Omega,n+m)X(\Omega,n)] = E\left[\sum_{i=1}^{n+m}\sum_{j=1}^{n} X_i X_j\right] \\ &= E\left[\sum_{j=1}^{n}\sum_{j=1}^{n} X_i X_j + \sum_{i=n+1}^{m}\sum_{j=1}^{n} X_i X_j\right] = n\sigma^2 \end{aligned} \tag{9.142}$$

The last result follows from the independence of $X_i$ and $X_j$ for $i \neq j$.
The correlation coefficient is given by

$$\begin{aligned} \rho_{XX}(n+m,n) &= \frac{R_{XX}(n+m,n) - E[X(\Omega,n+m)]E[X(\Omega,n)]}{\sqrt{\text{var}(X(\Omega,n+m))}\sqrt{\text{var}(X(\Omega,n))}} \\ &= \frac{n\sigma^2}{\sigma^2\sqrt{n+m}\sqrt{n}} = \frac{1}{\sqrt{1+m/n}} \end{aligned} \tag{9.143}$$

## 9.11 PROBLEMS

**9.1** Consider a random process defined according to

$$X(\Omega,t) = \sum_{k=0}^{\lfloor t/D \rfloor} A_k p(t-kD) \tag{9.144}$$

where $p: \boldsymbol{R} \to \boldsymbol{R}$ is a causal pulse function, $\Omega = (A_0, A_1, \ldots)$, and $A_i$ is independent of $A_k$ for $i \neq k$.

a. Determine analytical expressions for the mean, variance, and autocorrelation of this random process. For the autocorrelation function, assume $\tau > 0$.
   For the following parts, assume $p(t) = u(t) - u(t - \text{D}/2)$, $\text{D} = 1$ and identical zero mean Gaussian random variables with a variance $\sigma_A^2 = 1$.
b. Generate several signals from the random process.
c. Graph the mean, variance, and autocorrelation.
d. Specify the duration of dependence for $t = k\text{D} + 3\text{D}/8, k \in \{0, 1, \ldots\}$.

**9.2** A random process $X$, which models the transient behavior of a class of oscillators, is defined according to

$$X(\Omega,t) = A\sin(2\pi f_C t + g_1(t)\Phi) \qquad t \geq 0 \tag{9.145}$$

where $\Omega = (A, \Phi)$.

a. Determine general expressions for the mean, variance, and autocorrelation function of the random process for the case of $A$ and $\Phi$ being correlated and for the independent case.
For the following parts, assume $f_\Phi(\phi) = 1/2\pi$, $\phi \in [0, 2\pi)$, $A$ has a Gaussian distribution with $\mu_A = A_o$ and $\sigma_A = A_o/10$, $A_o = 1$, $g_1(t) = \tanh(t/\tau_1)$, $f_C = 1$, and $\tau_1 = 1$.
b. Graph a few signals from the random process.
c. Use numerical integration to graph the mean, variance, and autocorrelation function of the random process.

**9.3** Consider the random process defined according to

$$X(\Omega,t) = \mathrm{sgn}[X_1(\Omega_1,t) - X_2(\Omega_2,t)] \tag{9.146}$$

where $\Omega = \Omega_1 \times \Omega_2$ and $X_1(\Omega_1, t)$ and $X_2(\Omega_2, t)$ are causal random telegraph signals based on independent Poisson experiments that generate points, respectively, with rates $\lambda_1$ and $\lambda_2$.

a. Determine expressions for the mean, mean squared value, variance, and autocorrelation function of $X$.
For the case of $\lambda_1 = 1 + \Delta$, $\lambda_2 = 1 - \Delta$ and $\Delta = 0.2$:
b. Graph a few signals from the random process.
c. Graph the mean, variance, and autocorrelation.

**9.4** Consider a generalized shot noise random process defined according to

$$X(\Omega,t) = \sum_{k=1}^{\infty} A_k p(t - \Gamma_k) \qquad t \geq 0 \tag{9.147}$$

where $\Omega = ((A_1, \Gamma_1), \ldots)$, $\Gamma_1, \Gamma_2, \ldots$ are a sequence of random variables consistent with the outcomes of a Poisson experiment with a rate $\lambda$ and $(A_1, A_2, \ldots)$ is a vector of random variables that define the amplitudes. Assume $p$ is a Gaussian pulse:

$$p(t) = \frac{1}{\sqrt{2\pi}\sigma} \cdot \exp\left(-\frac{t^2}{2\sigma^2}\right) \qquad \sigma^2 = \frac{t_{\mathrm{fwhm}}^2}{8\ln(2)} \tag{9.148}$$

Assume $t_{\mathrm{fwhm}} = 1/2$, $\lambda = 1$ and identical amplitude random variables with a Rayleigh probability density function: $f_A(a) = \frac{a}{\sigma_R^2} \cdot \exp\left[\frac{-a^2}{2\sigma_R^2}\right] u(a)$ with $\sigma_R = \sqrt{2}/\sqrt{\pi}$.

a. Graph several signals from the random process.
b. Use established expressions to define explicit expressions for the mean, variance, and autocorrelation functions for the interval $[0, \infty]$. Graph these expressions.
c. Graph the correlation coefficient variation with $t$ and $\tau$.
d. Graph the effective correlation time for $\rho_o = 0.1$.
e. Determine a normalization constant for the random process such that the average power in steady state is unity.

**9.5** Consider a random process that defines pulse signals that occur during one phase of a cycle of a system. The random process is defined according to

$$X(\Omega, t) = \sum_{k=1}^{N} \phi(\Omega_k, t - \Gamma_k) \qquad \phi \in S_\Phi \tag{9.149}$$

where $\Omega = ((\Omega_k, \Gamma_k), \ldots, (\Omega_N, \Gamma_N))$, $\Gamma_1, \ldots, \Gamma_N$ are a sequence of identical and independent random variables with a probability density function

$$f_\Gamma(\gamma) = \frac{\pi}{2} \cdot \sin(\pi\gamma) \qquad \gamma \in [0, 1] \tag{9.150}$$

The random variables $\Omega_1, \ldots, \Omega_N$ are assumed to be identical and independent with a probability density function

$$f_\Omega(\omega) = \frac{\pi}{4\Delta} \cdot \cos\left[\pi\left(\frac{\omega - w_o}{2\Delta}\right)\right]$$
$$\omega \in [\omega_o - \Delta, \omega_o + \Delta], \Delta < \omega_o \tag{9.151}$$

The signalling set is defined according to

$$S_\Phi = \left\{ \begin{array}{l} \sin(2\pi\omega_k t)^2 \sin(0.5\pi\omega_k t) \left[u(t) - u\left(t - \frac{2}{\omega_k}\right)\right] : \\ \qquad\qquad \omega_k \in [\omega_o - \Delta, \ \omega_o + \Delta] \end{array} \right\} \tag{9.152}$$

Assume that $w_o = 10$, $\Delta = w_o/4$ and $N = 4$.

a. Graph a few signals from the random process.
b. Using established expressions, specify integral expressions for the mean and autocorrelation function of the random process. Use numerical integration to graph the mean, variance, and autocorrelation of the random process.

**9.6** The model for a random process $X: S \to S_X$, which defines jittered pulse trains, is

$$X(\Omega, t) = \sum_{k=1}^{N} A_k p(t-(k-1)\mathrm{D}-J_k) \qquad t \in [0, N\mathrm{D}] \tag{9.153}$$

where $\Omega = ((A_1, J_1), \ldots, (A_N, J_N))$.

a. Obtain general expressions for the mean and autocorrelation function of $X$ for the case of independence of all random variables.
Consider the case of

$$p(t) = \frac{1}{\sqrt{2\pi}\sigma_P} \cdot \exp\left[\frac{-(t-\mathrm{D}/2)^2}{2\sigma_P^2}\right] \qquad \sigma_P^2 = \frac{t_{\mathrm{fwhm}}^2}{8\ln(2)} \tag{9.154}$$

$$f_A(a) = \frac{1}{\sqrt{2\pi}\sigma_A} \cdot \exp\left[\frac{-(a-\mu_A)^2}{2\sigma_A^2}\right] \qquad f_J(j) = \frac{1}{\sqrt{2\pi}\sigma_J} \cdot \exp\left[\frac{-j^2}{2\sigma_J^2}\right] \tag{9.155}$$

and with the following parameters: $N = 10$, $\mathrm{D} = 1$, $\mu_A = 1$, $\sigma_A = \mu_A/10$, $\sigma_J = \mathrm{D}/10$, $t_{\mathrm{fwhm}} = \mathrm{D}/4$.

b. Establish analytical expressions for the mean and autocorrelation function in terms of the parameters $\mu_A$, $\sigma_A$, $\sigma_P$, and $\sigma_J$. The result

$$\begin{aligned} &\int_{-\infty}^{\infty} \frac{1}{\sqrt{2\pi}\sigma_1} \cdot \exp\left[\frac{-(x-k_1)^2}{2\sigma_1^2}\right] \cdot \frac{1}{\sqrt{2\pi}\sigma_2} \cdot \exp\left[\frac{-x^2}{2\sigma_2^2}\right] \cdot dx \\ &= \frac{1}{\sqrt{2\pi}\sqrt{\sigma_1^2+\sigma_2^2}} \cdot \exp\left[\frac{-k_1^2}{2(\sigma_1^2+\sigma_2^2)}\right] \end{aligned} \tag{9.156}$$

is useful.

c. Graph signals from the random process.
d. Graph the mean, variance, and autocorrelation function of the random process.

## APPENDIX 9.A PROOF OF THEOREM 9.6

By definition

$$\mu_X(t) = \boldsymbol{E}[X(\Omega,t)] = \int_{-\infty}^{\infty}\int_{-\infty}^{\infty} a\sin(2\pi f_C t+\phi) f_{A\Phi}(a,\phi)\,da\,d\phi \qquad (9.157)$$

and for the independent case

$$\begin{aligned}\mu_X(t) &= \int_{-\infty}^{\infty} a f_A(a)\,da \int_{-\infty}^{\infty}\sin(2\pi f_C t+\phi) f_{\Phi}(\phi)\,d\phi \\ &= \mu_A \int_{-\infty}^{\infty}\sin(2\pi f_C t+\phi) f_{\Phi}(\phi)\,d\phi\end{aligned} \qquad (9.158)$$

For the variance:

$$\sigma_X^2(t) = \int_{-\infty}^{\infty}\int_{-\infty}^{\infty} [a\sin(2\pi f_C t+\phi)]^2 f_{A\Phi}(a,\phi)\,da\,d\phi - \mu_X^2(t) \qquad (9.159)$$

As $\sin(2\pi f_C t+\phi)^2 = 0.5[1-\cos(4\pi f_C t+2\phi)]$, it follows that

$$\sigma_X^2(t) = \frac{\boldsymbol{E}[A^2]}{2} - \frac{1}{2}\int_{-\infty}^{\infty}\int_{-\infty}^{\infty} a^2\cos(4\pi f_C t+2\phi) f_{A\Phi}(a,\phi)\,da\,d\phi - \mu_X^2(t) \qquad (9.160)$$

and for the independent case

$$\begin{aligned}\sigma_X^2(t) &= \frac{\boldsymbol{E}[A^2]}{2} - \frac{1}{2}\int_{-\infty}^{\infty} a^2 f_A(a)\,da \int_{-\infty}^{\infty}\cos(4\pi f_C t+2\phi) f_{\Phi}(\phi)\,d\phi - \mu_X^2(t) \\ &= \frac{\boldsymbol{E}[A^2]}{2}\left[1 - \int_{-\infty}^{\infty}\cos(4\pi f_C t+2\phi) f_{\Phi}(\phi)\,d\phi\right] - \mu_X^2(t)\end{aligned} \qquad (9.161)$$

By definition, the autocorrelation function is

$$R_{XX}(t+\tau,t) = \int_{-\infty}^{\infty}\int_{-\infty}^{\infty} a^2 \sin(2\pi f_C(t+\tau)+\phi)\sin(2\pi f_C t+\phi) f_{A\Phi}(a,\phi)\,da\,d\phi \qquad (9.162)$$

As $\sin(A)\sin(B) = 0.5[\cos(A-B) - \cos(A+B)]$, it follows that

$$R_{XX}(t+\tau,t) = \frac{\boldsymbol{E}[A^2]}{2}\cos(2\pi f_C\tau) - \frac{1}{2}\int_{-\infty}^{\infty}\int_{-\infty}^{\infty}\frac{a^2}{2}[\cos(2\pi f_C(2t+\tau)+2\phi)]f_{A\Phi}(a,\phi)da\,d\phi \tag{9.163}$$

For the independent case,

$$R_{XX}(t+\tau,t) = \frac{\boldsymbol{E}[A^2]}{2}\left[\cos(2\pi f_C\tau) - \int_{-\infty}^{\infty}\cos(2\pi f_C(2t+\tau)+2\phi)f_\Phi(\phi)d\phi\right] \tag{9.164}$$

## APPENDIX 9.B PROOF OF THEOREM 9.7

By definition of the expectation operator:

$$\begin{aligned}\mu_X(t) &= \boldsymbol{E}[X(\Omega,t)] = 1\cdot P[X(\Omega,t)=1] + (-1)\cdot P[X(\Omega,t)=-1] \\ &= P[\text{even \# points in } (0,t)] - P[\text{odd \# points in } (0,t)]\end{aligned} \tag{9.165}$$

For a Poisson point process,

$$P[k \text{ points in } (0,t)] = \frac{(\lambda t)^k e^{-\lambda t}}{k!} \tag{9.166}$$

and hence,

$$\begin{aligned}\mu_X(t) &= \left[e^{-\lambda t} + \frac{(\lambda t)^2 e^{-\lambda t}}{2!} + \cdots \frac{(\lambda t)^{2i} e^{-\lambda t}}{(2i)!} + \cdots\right] \\ &\quad - \left[\lambda t e^{-\lambda t} + \frac{(\lambda t)^3 e^{-\lambda t}}{3!} + \cdots \frac{(\lambda t)^{2i+1} e^{-\lambda t}}{(2i+1)!} + \cdots\right] \\ &= e^{-\lambda t}\left[1 - \lambda t + \frac{(\lambda t)^2}{2!} - \frac{(\lambda t)^3}{3!} + \cdots\right] = e^{-2\lambda t}\end{aligned} \tag{9.167}$$

The autocorrelation function is

$$\begin{aligned}R_{XX}(t+\tau,t) &= \boldsymbol{E}[X(\Omega,t+\tau)X^*(\Omega,t)] \\ &= (-1)(-1)\cdot P[X(\Omega,t+\tau)=-1, X^*(\Omega,t)=-1] \\ &\quad + (-1)(1)\cdot P[X(\Omega,t+\tau)=-1, X^*(\Omega,t)=1] \\ &\quad + (1)(-1)\cdot P[X(\Omega,t+\tau)=1, X^*(\Omega,t)=-1] \\ &\quad + (1)(1)\cdot P[X(\Omega,t+\tau)=1, X^*(\Omega,t)=1]\end{aligned} \tag{9.168}$$

Consider the case of $\tau > 0$ and the probability

$$\begin{aligned} P[X(\Omega,t+\tau) = -1, X^*(\Omega,t) = -1] \\ = P\left[X(\Omega,t+\tau) = -1|_{X^*(\Omega,t)=-1}\right] \cdot P[X^*(\Omega,t) = -1] \\ = P[\text{even \#points in } (t,t+\tau)] \cdot P[\text{odd \#points in } (0,t)] \end{aligned} \tag{9.169}$$

For a Poisson point process,

$$P[\text{even \#points in } (t,t+\tau)] = P[\text{even \#points in } (0,\tau)] \tag{9.170}$$

and thus,

$$\begin{aligned} P[X(\Omega,t+\tau) = -1, X^*(\Omega,t) = -1] = \\ P[\text{even \#points in } (0,\tau)] = P[\text{even \#points in } (0,t)] \end{aligned} \tag{9.171}$$

Similar considerations lead to the result

$$\begin{aligned} R_{XX}(t+\tau,t) = P[\text{even \#points in } (0,\tau)]. \\ (P[\text{odd \#points in } (0,t)] + P[\text{even \#points in } (0,t)]) - \\ P[\text{odd \#points in } (0,\tau)]. \\ (P[\text{odd \#points in } (0,\tau)] + P[\text{even \#points in } (0,t)]) \end{aligned} \tag{9.172}$$

As $P[\text{odd \#points in } (0,t)] + P[\text{even \#points in } (0,t)] = 1$, it follows that

$$\begin{aligned} R_{XX}(t+\tau,t) = P[\text{even \#points in } (0,\tau)] - P[\text{odd \#points in } (0,\tau)] \\ = e^{-\lambda\tau}\left[1 - \lambda\tau + \frac{(\lambda\tau)^2}{2!} - \frac{(\lambda\tau)}{3!} + \cdots\right] = e^{-2\lambda\tau} \end{aligned} \tag{9.173}$$

For the case of $\tau < 0$, consider the probability

$$\begin{aligned} P[X(\Omega,\, t+\tau) = -1,\, X^*(\Omega,t) = -1] = \\ P\left[X^*(\Omega,\, t) = -1|_{X^*(\Omega,t+\tau)=-1}\right] \cdot P[X^*(\Omega,\, t+\tau) = -1] \\ = P[\text{even \#points in } (t+\tau,t)] \cdot P[\text{odd \#points in } (0,\, t+\tau)] \end{aligned} \tag{9.174}$$

and thus,

$$\begin{aligned} P[X(\Omega,\, t+\tau) = -1, X^*(\Omega,\, t) = -1] = \\ P[\text{even \#points in } (0,\, |\tau|)] \cdot P[\text{odd \#points in } (0,\mathrm{t}+\tau)] \end{aligned} \tag{9.175}$$

The proof proceeds as for the case of $\tau > 0$ with the restriction of $t + \tau > 0$.

The mean squared value then follows according to

$$\boldsymbol{E}\left[X^2(\Omega,t)\right] = \lim_{\tau\to 0} R_{XX}(t+\tau,t) = 1 \tag{9.176}$$

which is the expected result. From the definition $\sigma_X^2(t) = \boldsymbol{E}[X^2(\Omega,t)] - \mu_X^2(t)$, the variance is

$$\sigma_X^2(t) = 1 - e^{-4\lambda t} \tag{9.177}$$

## APPENDIX 9.C PROOF OF THEOREMS 9.8 AND 9.9

With the assumption of independence, the joint probability density function of the random variables $A_1, A_2, \ldots,\ \Gamma_1, \Gamma_2, \ldots$ is

$$f_{A_1,\ldots A_N,\Gamma_1,\ldots,\Gamma_N}(a_1,\ldots,a_N,t_1,\ldots t_N) = \frac{1}{T^N} f_A(a_1)\ldots f_A(a_N) \tag{9.178}$$

It then follows that

$$R_{XX}(t+\tau,t) = \boldsymbol{E}[X(\Omega,t+\tau)X^*(\Omega,t)] =$$

$$\int_0^T \cdots \int_0^T \int_{-\infty}^{\infty} \cdots \int_{-\infty}^{\infty} X(((a_1,t_1),\ldots,(a_N,t_N)),t+\tau)X(((a_1,t_1),\ldots,(a_N,t_N)),t)\cdot$$

$$f_{A_1,\ldots,A_N,\Gamma_1,\ldots,\Gamma_N}(a_1,\ldots,a_N,t_1,\ldots,t_N)da_1\ldots da_N dt_1\ldots dt_N \tag{9.179}$$

Substitution for $X$ and $f_{A_1,\ldots,A_N,\Gamma_1,\ldots,\Gamma_N}$ yields

$$R_{XX}(t+\tau,t) = \int_0^T \cdots \int_0^T \int_{-\infty}^{\infty} \cdots \int_{-\infty}^{\infty} \sum_{i=1}^{N}\sum_{j=1}^{N} a_i a_j p(t+\tau-t_i)p(t-t_j)\cdot$$
$$\frac{1}{T^N}\cdot f_A(a_1)\ldots f_A(a_N)da_1\ldots da_N dt_1\ldots dt_N \tag{9.180}$$

Interchanging the order of summation and integration yields

$$R_{XX}(t+\tau,t) = \frac{1}{T^N}\sum_{i=1}^{N}\sum_{j=1}^{N}\int_0^T \cdots \int_0^T \int_{-\infty}^{\infty} \cdots \int_{-\infty}^{\infty} a_i a_j p(t+\tau-t_i)p(t-t_j)\cdot$$
$$f_A(a_1)\ldots f_A(a_N)da_1\ldots da_N dt_1\ldots dt_N \tag{9.181}$$

As

$$\int_0^T \frac{1}{T} dt_k = 1 \qquad \int_{-\infty}^{\infty} f_A(a_k)da_k = 1 \qquad k \neq i,j \tag{9.182}$$

it follows, after splitting the double summation up into the diagonal terms and the nondiagonal terms, that

$$\begin{aligned} R_{XX}(t+\tau,t) &= \frac{1}{T}\sum_{i=1}^{N}\int_0^T\int_{-\infty}^{\infty} a_i^2 p(t+\tau-t_i)p(t-t_i)f_A(a_i)da_i dt_i \\ &+ \frac{1}{T^2}\sum_{i=1}^{N}\sum_{\substack{j=1\\j\neq i}}^{N}\int_0^T\int_0^T\int_{-\infty}^{\infty}\int_{-\infty}^{\infty} a_i a_j p(t+\tau-t_i)p(t-\tau_j)f_A(a_i)f_A(a_j)da_i da_j dt_i dt_j \end{aligned} \tag{9.183}$$

Thus,

$$\begin{aligned} R_{XX}(t+\tau,t) &= \sum_{i=1}^{N}\frac{1}{T}\int_0^T p(t+\tau-t_i)p(t-t_i)dt_i\left[\int_{-\infty}^{\infty} a_i^2 f_A(a_i)da_i\right] \\ &+ \sum_{i=1}^{N}\sum_{\substack{j=1\\j\neq i}}^{N}\frac{1}{T^2}\int_0^T\int_0^T p(t+\tau-t_i)p(t-t_j)dt_i dt_j\left[\int_{-\infty}^{\infty}\int_{-\infty}^{\infty} a_i a_j \cdot f_A(a_i)f_A(a_j)da_i da_j\right] \end{aligned} \tag{9.184}$$

By definition,

$$\int_{-\infty}^{\infty} a_i^2 f_A(a_i)da_i = \boldsymbol{E}\left[A^2\right] \qquad \int_{-\infty}^{\infty}\int_{-\infty}^{\infty} a_i a_j f_A(a_i)f_A(a_j)da_i da_j = \mu_A^2 \tag{9.185}$$

and it then follows that

$$\begin{aligned} R_{XX}(t+\tau,t) &= \boldsymbol{E}[A^2]\sum_{i=1}^{N}\frac{1}{T}\int_0^T p(t+\tau-t_i)p(t-t_i)dt_i \\ &+ \mu_A^2\sum_{i=1}^{N}\sum_{\substack{j=1\\j\neq i}}^{N}\frac{1}{T^2}\int_0^T\int_0^T p(t+\tau-t_i)p(t-t_j)dt_i dt_j \end{aligned} \tag{9.186}$$

The variables of integration $t_i$ and $t_j$ have arisen from outcomes of identical random variables, and it follows that the integrals in Equation 9.186 are identical for all values of the indexes as specified by the summations. Thus,

$$R_{XX}(t+\tau,t) = rE[A^2]\int_0^T p(t+\tau-t_i)p(t-t_i)dt_i + r^2\mu_A^2\left(1-\frac{1}{N}\right)\int_0^T\int_0^T p(t+\tau-t_i)p\left(t-t_j\right)dt_i dt_j \quad (9.187)$$

where $r=N/T$ is the mean waveform rate.

### 9.C.1 Causal Pulse Case

For the case where $p$ is causal, it follows, from the illustrations in Figure 9.29, that

$$\int_0^T\int_0^T p(t+\tau-t_i)p\left(t-t_j\right)dt_i dt_j = \int_0^{t+\tau} p(t+\tau-t_i)dt_i \int_0^t p\left(t-t_j\right)dt_j \qquad 0 \le t, t+\tau \le T \quad (9.188)$$

$$\int_0^T p(t+\tau-t_i)p(t-t_i)dt_i = \begin{cases} \int_0^t p(t+\tau-t_i)p(t-t_i)dt_i & \tau>0, t>0 \\ & 0 \le t+\tau \le T \\ \int_0^{t+\tau} p(t+\tau-t_i)p(t-t_i)dt_i & \tau<0, t>0 \\ & 0 \le t+\tau \le T \end{cases} \quad (9.189)$$

The changes of variables, $\lambda = t-t_i$ and $\lambda = t+\tau-t_i$, leads to

$$R_{XX}(t+\tau,t) = \begin{cases} rE[A^2]\int_0^t p(\lambda+\tau)p(\lambda)d\lambda + r^2\mu_A^2\left(1-\frac{1}{N}\right)\int_0^{t+\tau} p(\lambda)d\lambda \int_0^t p(\lambda)d\lambda & 0 \le \tau \le T-t \\ rE[A^2]\int_0^{t+\tau} p(\lambda-\tau)p(\lambda)d\lambda + r^2\mu_A^2\left(1-\frac{1}{N}\right)\int_0^{t+\tau} p(\lambda)d\lambda \int_0^t p(\lambda)d\lambda & -t \le \tau < 0 \end{cases} \quad (9.190)$$

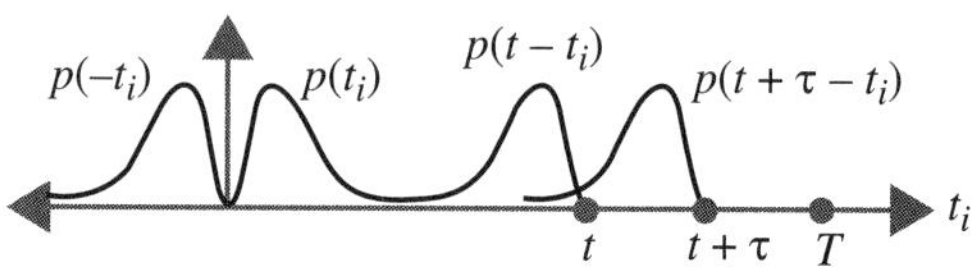


**FIGURE 9.29** Illustration of pulse functions for the causal case.

### 9.C.2 Mean and Variance

The mean, by definition, is

$$\mu_X(t) = E[X(\Omega,t)] = \int_0^T \cdots \int_0^T \int_{-\infty}^{\infty} \cdots \int_{-\infty}^{\infty} X(((a_1,t_1),\ldots,(a_N,t_N)),t)$$
$$f_{A_1,\ldots,A_N,\Gamma_1,\ldots,\Gamma_N}(a_1,\ldots,a_N,t_1,\ldots,t_N)da_1\ldots da_N dt_1\ldots dt_N \quad (9.191)$$

Substitution for $X$ and $f_{A_1,\ldots,A_N,\Gamma_1,\ldots,\Gamma_N}$, from Equation 9.178, yields

$$\mu_X(t) = \int_0^T \ldots \int_0^T \int_{-\infty}^{\infty} \ldots \int_{-\infty}^{\infty} \sum_{i=1}^{N} a_i p(t-t_i) \cdot$$
$$\frac{1}{T^N} \cdot f_A(a_1)\ldots f_A(a_N)da_1\ldots da_N dt_1\ldots dt_N \quad (9.192)$$

Interchanging the order of summation and integration yields

$$\mu_X(t) = \frac{1}{T^N}\sum_{i=1}^{N}\int_0^T \ldots \int_0^T \int_{-\infty}^{\infty} \ldots \int_{-\infty}^{\infty} a_i p(t-t_i) \cdot f_A(a_1)\ldots f_A(a_N)da_1\ldots da_N dt_1\ldots dt_N \quad (9.193)$$

Using the results detailed in Equation 9.182, it follows that

$$\mu_X(t) = \frac{1}{T}\sum_{i=1}^{N}\int_0^T \int_{-\infty}^{\infty} a_i p(t-t_i) \cdot f_A(a_i)da_i dt_i$$
$$= \sum_{i=1}^{N}\frac{1}{T}\int_0^T p(t-t_i)dt_i \left[\int_{-\infty}^{\infty} a_i f_A(a_i)da_i\right]$$
$$= \sum_{i=1}^{N}\frac{\mu_A}{T}\int_0^T p(t-t_i)dt_i = r\mu_A \int_0^T p(t-t_i)dt_i \quad (9.194)$$

where $\mu_A$ is the mean of the random variables $A_1, \ldots, A_N$. For the causal pulse case,

$$\mu_X(t) = r\mu_A \int_0^t p(t-t_i)dt_i = r\mu_A \int_0^t p(\lambda)d\lambda \quad (9.195)$$

where the last result arises from a change of variable $\lambda = t - t_i$.

Using the definition $\sigma_X^2(t) = \boldsymbol{E}\left[X^2(\Omega,t) - \mu_X^2(t)\right]$, it follows, from Equations 9.187 and 9.194, that

$$\begin{aligned}\sigma_X^2(t) &= r\boldsymbol{E}[A^2]\int_0^T p^2(t-t_i)dt_i + r^2\mu_A^2\left(1-\frac{1}{N}\right)\left[\int_0^T p(t-t_i)dt_i\right]^2 - r^2\mu_A^2\left[\int_0^T p(t-t_i)dt_i\right]^2 \\ &= r\boldsymbol{E}[A^2]\int_0^T p^2(t-t_i)dt_i - \frac{r^2\mu_A^2}{N}\left[\int_0^T p(t-t_i)dt_i\right]^2 \qquad 0 \le t \le T\end{aligned} \tag{9.196}$$

For the causal pulse case,

$$\begin{aligned}\sigma_X^2(t) &= r\boldsymbol{E}[A^2]\int_0^t p^2(t-t_i)dt_i - \frac{r^2\mu_A^2}{N}\left[\int_0^t p(t-t_i)dt_i\right]^2 \qquad 0 \le t \le T \\ &= r\boldsymbol{E}[A^2]\int_0^t p^2(\lambda)d\lambda - \frac{r^2\mu_A^2}{N}\left[\int_0^t p(\lambda)d\lambda\right]^2 \qquad 0 \le t \le T\end{aligned} \tag{9.197}$$

## APPENDIX 9.D PROOF OF THEOREM 9.12

For the case of $p(t) = k_p t^i e^{-t/\tau_p}$ and for the steady-state case, it follows, from Equation 9.50, that $\tau_p$ is defined by the equation

$$\int_0^\infty k_p(\lambda_1 + \tau_C)^i e^{-(\lambda_1+\tau_C)/\tau_p} k_p \lambda_1^i e^{-\lambda_1/\tau_p} d\lambda_1 = \rho_o \int_0^\infty k_p^2 \lambda_1^{2i} e^{-2\lambda_1/\tau_p} d\lambda_1 \tag{9.198}$$

A change of variable $\lambda = 2\lambda_1/\tau_p$, that is, $\lambda_1 = \tau_p\lambda/2$, yields

$$\begin{aligned} & e^{-\tau_C/\tau_p}\int_0^\infty \left(\frac{\tau_p\lambda}{2} + t_C\right)^i \left(\frac{\tau_p\lambda}{2}\right)^i e^{-\lambda}\left(\frac{\tau_p d\lambda}{2}\right) = \rho_o \int_0^\infty \left(\frac{\tau_p\lambda}{2}\right)^{2i} e^{-\lambda}\left(\frac{\tau_p d\lambda}{2}\right) \\ & \Rightarrow e^{-\tau_C/\tau_p}\int_0^\infty \left(\lambda + \frac{2\tau_C}{\tau_p}\right)^i \lambda^i e^{-\lambda} d\lambda = \rho_o \int_0^\infty \lambda^{2i} e^{-\lambda} d\lambda\end{aligned} \tag{9.199}$$

Using the expansion

$$(x+y)^n = x^n + nx^{n-1}y + \frac{n(n-1)}{2!}x^{n-2}y^2 + \cdots + y^n \tag{9.200}$$

it follows that

$$\rho_o e^{\tau_C/\tau_p}\Gamma(2i+1)=\int_0^\infty \lambda^i\left[\lambda^i+i\left(\frac{2\tau_C}{\tau_p}\right)\lambda^{i-1}+\frac{i(i-1)}{2!}\left(\frac{2\tau_C}{\tau_p}\right)^2\lambda^{i-2}\right.$$
$$\left.+\frac{i(i-1)(i-2)}{3!}\left(\frac{2\tau_C}{\tau_p}\right)^3\lambda^{i-3}+\cdots+\left(\frac{2\tau_C}{\tau_p}\right)^i\right]e^{-\lambda}d\lambda \tag{9.201}$$

where the definition for the Gamma function

$$\Gamma(n)=\int_0^\infty t^{n-1}e^{-t}dt=(n-1)! \qquad n\geq 1, n\in \mathbf{Z}^+ \tag{9.202}$$

has been used. Using this definition repeatedly yields

$$\rho_o e^{\tau_C/\tau_p}\Gamma(2i+1)=\Gamma(2i+1)+i\left(\frac{2\tau_C}{\tau_p}\right)\Gamma(2i)+\frac{i(i-1)}{2!}\left(\frac{2\tau_C}{\tau_p}\right)^2\Gamma(2i-1)$$
$$+\frac{i(i-1)(i-2)}{3!}\left(\frac{2\tau_C}{\tau_p}\right)^3\Gamma(2i-2)+\cdots+\left(\frac{2\tau_C}{\tau_p}\right)^i\Gamma(i+1) \tag{9.203}$$

that is,

$$\sum_{k=0}^{i}\frac{i!}{(i-k)!k!}\left(\frac{2\tau_C}{\tau_p}\right)^k\Gamma(2i-k+1)=\rho_o e^{\tau_C/\tau_p}\Gamma(2i+1) \tag{9.204}$$

As $\Gamma(n)=(n-1)!$ the required result follows:

$$\sum_{k=0}^{i}\frac{i!(2i-k)!}{(i-k)!k!}\left(\frac{2\tau_C}{\tau_p}\right)^k=\rho_o e^{\tau_C/\tau_p}(2i)!$$
$$\Rightarrow\sum_{k=0}^{i}\frac{i!(2i-k)!}{(2i)!(i-k)!k!}\left(\frac{2\tau_C}{\tau_p}\right)^k=\rho_o e^{\tau_C/\tau_p} \tag{9.205}$$

## APPENDIX 9.E PROOF OF THEOREM 9.16

Continuous random variables are assumed. With the assumption of independence, the joint probability density function of the random variables $A_0, A_1, \ldots$ is

$$f_{A_0,\ldots,A_{N-1}}(a_0,\ldots,a_{N-1})=f_{A_0}(a_0)\ldots f_{A_{N-1}}(a_{N-1}) \tag{9.206}$$

It then follows that the autocorrelation function is

$$R_{XX}(t+\tau,t)=\boldsymbol{E}[X(\Omega,t+\tau)X^*(\Omega,t)]=\int_{-\infty}^{\infty}\cdots\int_{-\infty}^{\infty}X((a_0,\ldots,a_{N-1}),t+\tau)\cdot X((a_0,\ldots,a_{N-1}),t)\cdot f_{A_0,\ldots,A_{N-1}}(a_0,\ldots,a_{N-1})da_0\ldots da_{N-1} \tag{9.207}$$

Substitution for $X$ and $f_{A_0,\ldots,A_{N-1}}$ yields

$$R_{XX}(t+\tau,t)=\int_{-\infty}^{\infty}\cdots\int_{-\infty}^{\infty}\sum_{i=0}^{N-1}\sum_{j=0}^{N-1}a_ia_jp(t+\tau-iD)p(t-jD)\cdot f_{A_0}(a_0)\ldots f_{A_{N-1}}(a_{N-1})da_0\ldots da_{N-1} \tag{9.208}$$

Interchanging the order of summation and integration, and splitting the double summation into the diagonal and nondiagonal terms, yields

$$\begin{aligned}R_{XX}(t+\tau,t)&\\=\sum_{i=0}^{N-1}\int_{-\infty}^{\infty}\cdots\int_{-\infty}^{\infty}&a_i^2p(t+\tau-iD)p(t-iD)\cdot f_{A_0}(a_0)\ldots f_{A_{N-1}}(a_{N-1})da_0\ldots da_{N-1}\\+\sum_{i=0}^{N-1}\sum_{\substack{j=0\\j\neq i}}^{N-1}\int_{-\infty}^{\infty}\cdots\int_{-\infty}^{\infty}&a_ia_jp(t+\tau-iD)p(t-jD)\cdot f_{A_0}(a_0)\ldots f_{A_{N-1}}(a_{N-1})da_0\ldots da_{N-1}\end{aligned} \tag{9.209}$$

With the result $\int_{-\infty}^{\infty}f_{A_k}(a_k)da_k=1$, it follows that

$$\begin{aligned}R_{XX}(t+\tau,t)=&\sum_{i=0}^{N-1}p(t+\tau-i\mathrm{D})p(t-i\mathrm{D})\int_{-\infty}^{\infty}a_i^2f_{A_i}(a_i)da_i\\&+\sum_{i=0}^{N-1}\sum_{\substack{j=0\\j\neq i}}^{N-1}p(t+\tau-i\mathrm{D})p(t-j\mathrm{D})\int_{-\infty}^{\infty}\int_{-\infty}^{\infty}a_ia_j\cdot f_{A_i}(a_i)f_{A_j}(a_j)da_ida_j\end{aligned} \tag{9.210}$$

and then

$$\begin{aligned}R_{XX}(t+\tau,t)&\\=\sum_{i=0}^{N-1}\boldsymbol{E}\left[A_i^2\right]p(t+\tau-i\mathrm{D})p(t-i\mathrm{D})&+\sum_{i=0}^{N-1}\sum_{\substack{j=0\\j\neq i}}^{N-1}\mu_i\mu_jp(t+\tau-i\mathrm{D})p(t-j\mathrm{D})\end{aligned} \tag{9.211}$$

where $\boldsymbol{E}\left[A_i^2\right]$ is the mean squared value of $A_i$ and $\mu_i$ is the mean of $A_i$.

The mean, by definition, is

$$\mu_X(t) - \boldsymbol{E}[X(\Omega,t)] = \int_{-\infty}^{\infty} \cdots \int_{-\infty}^{\infty} X((a_0, \ldots, a_{N-1}), t). \\ f_{A_0, \ldots, A_{N-1}}(a_0, \ldots, a_{N-1}) da_0 \ldots da_{N-1} \tag{9.212}$$

Substitution for $X$ and $f_{A_0, \ldots, A_{N-1}}$ yields

$$\mu_X(t) = \int_{-\infty}^{\infty} \cdots \int_{-\infty}^{\infty} \sum_{i=0}^{N-1} a_i p(t - i\mathrm{D}) \cdot f_{A_0}(a_0) \ldots f_{A_{N-1}}(a_{N-1}) da_0 \ldots da_{N-1} = \sum_{i=0}^{N-1} \mu_i p(t - i\mathrm{D}) \tag{9.213}$$

The variance is readily determined from the relationship $\sigma_X^2(t) = \boldsymbol{E}[X^2(\Omega,t)] - \mu_X^2(t)$. As $\boldsymbol{E}[X^2(\Omega,t)] = R_{XX}(t,t)$, it follows that

$$\begin{aligned} \boldsymbol{E}[X^2(\Omega,t)] &= \sum_{i=0}^{N-1} \boldsymbol{E}\left[A_i^2\right] p^2(t - i\mathrm{D}) + \sum_{i=0}^{N-1} \sum_{j=0, j\neq i}^{N-1} \mu_i \mu_j p(t - i\mathrm{D}) p(t - j\mathrm{D}) \\ &= \sum_{i=0}^{N-1} \boldsymbol{E}\left[A_i^2\right] p^2(t - i\mathrm{D}) + \mu_X^2(t) - \sum_{i=0}^{N-1} \mu_i^2 p^2(t - i\mathrm{D}) \end{aligned} \tag{9.214}$$

whereupon

$$\sigma_X^2(t) = \sum_{i=0}^{N-1} \left(\boldsymbol{E}\left[A_i^2\right] - \mu_i^2\right) p^2(t - i\mathrm{D}) = \sum_{i=0}^{N-1} \sigma_i^2 p^2(t - i\mathrm{D}) \tag{9.215}$$

## APPENDIX 9.F PROOF OF THEOREM 9.17

The results below depend on the result: if the pulse function $p$ is only nonzero on the interval $\left[t_o, t_o + \mathrm{D}_p\right]$, then $p(t + t_1) p(t + t_2)$ is only nonzero when $t + t_1$, $t + t_2 \in \left[t_o, t_o + \mathrm{D}_p\right)$.

First, consider the function $p(t + \tau - i\mathrm{D}) p(t - i\mathrm{D})$ in the first summation in the autocorrelation function defined in Theorem 9.16. This function is only nonzero when

$$t + \tau - i\mathrm{D}, t - i\mathrm{D} \in \left[t_o, t_o + \mathrm{D}_p\right) \tag{9.216}$$

which implies

$$\begin{aligned} & t_o \le t+\tau-i\mathrm{D} < t_o+\mathrm{D}_p && \text{and} && t_o \le t-i\mathrm{D} < t_o+\mathrm{D}_p \\ \Rightarrow\ & t_o-t-\tau \le i\mathrm{D} < t_o+\mathrm{D}_p-t-\tau && \text{and} && t_o-t \le -i\mathrm{D} < t_o+\mathrm{D}_p-t \\ \Rightarrow\ & \frac{-t_o+t+\tau}{\mathrm{D}} \ge i > \frac{-t_o-\mathrm{D}_p+t+\tau}{\mathrm{D}} && \text{and} && \frac{-t_o+t}{\mathrm{D}} \ge i > \frac{-t_o-\mathrm{D}_p+t}{\mathrm{D}} \end{aligned} \tag{9.217}$$

Thus,

$$\max\left\{\frac{t-t_o-\mathrm{D}_p}{\mathrm{D}}, \frac{t+\tau-t_o-\mathrm{D}_p}{\mathrm{D}}\right\} < i \le \min\left\{\frac{t-t_o}{\mathrm{D}}, \frac{t+\tau-t_o}{\mathrm{D}}\right\} \tag{9.218}$$

and

$$\begin{aligned} i_{\min_1} = \left\lfloor \max\left\{\frac{t-t_o-\mathrm{D}_p}{\mathrm{D}}, \frac{t-\tau-t_o-\mathrm{D}_p}{\mathrm{D}}\right\}\right\rfloor + 1 \le i \\ i \le \left\lfloor \min\left\{\frac{t-t_o}{\mathrm{D}}, \frac{t-\tau-t_o}{\mathrm{D}}\right\}\right\rfloor = i_{\max_1} \end{aligned} \tag{9.219}$$

The assumption here is

$$\left\lfloor \max\left\{\frac{t-t_o-\mathrm{D}_p}{\mathrm{D}}, \frac{t-\tau-t_o-\mathrm{D}_p}{\mathrm{D}}\right\}\right\rfloor + 1 \le \left\lfloor \min\left\{\frac{t-t_o}{\mathrm{D}}, \frac{t+\tau-t_o}{\mathrm{D}}\right\}\right\rfloor \tag{9.220}$$

and this constraint can be ensured by use of $u[i_{\max_1} - i_{\min_1}]$.

With these definitions for $i_{\min_1}$ and $i_{\max_1}$, the first summation in the autocorrelation function can be written as

$$\sum_{i=0}^{N-1} \boldsymbol{E}[A_i^2] p(t+\tau-i\mathrm{D})p(t-i\mathrm{D}) = u[i_{\max_1}] \cdot u[N-1-i_{\min_1}] \cdot u[i_{\max_1} - i_{\min_1}].$$

$$\sum_{i=\max\{i_{\min_1},0\}}^{\min\{i_{\max_1},N-1\}} \boldsymbol{E}[A_i^2] p(t+\tau-i\mathrm{D})p(t-i\mathrm{D})$$

where $u[i_{\max_1}] \cdot u[N-1-i_{\min_1}]$ ensures the summation is zero when $i_{\max_1} < 0$ or $i_{\min_1} > N-1$.

Second, consider the function $p(t+\tau-i\mathrm{D})p(t-j\mathrm{D})$ in the double summation in the autocorrelation function. This function is only nonzero when

$$p(t+\tau-i\mathrm{D})p(t-j\mathrm{D}) \in [t_o, t_o+\mathrm{D}_p) \tag{9.221}$$

which implies, consistent with Equation 9.217,

$$\frac{-t_0+t+\tau}{\mathrm{D}} \geq i > \frac{-t_o-\mathrm{D}_p+t+\tau}{\mathrm{D}} \quad \text{and} \quad \frac{-t_o+t}{\mathrm{D}} \geq j > \frac{-t_o-\mathrm{D}_p+t}{\mathrm{D}} \tag{9.222}$$

Thus,

$$\begin{aligned} i_{\min_2} &= \left\lfloor \frac{t+\tau-t_o-\mathrm{D}_p}{\mathrm{D}} \right\rfloor + 1 \leq i \leq \left\lfloor \frac{t+\tau-t_o}{\mathrm{D}} \right\rfloor = i_{\max_2} \\ j_{\min} &= \left\lfloor \frac{t-t_o-\mathrm{D}_p}{\mathrm{D}} \right\rfloor + 1 \leq j \leq \left\lfloor \frac{t-t_o}{\mathrm{D}} \right\rfloor = j_{\max} \end{aligned} \tag{9.223}$$

provided

$$\left\lfloor \frac{t+\tau-t_o-\mathrm{D}_p}{\mathrm{D}} \right\rfloor + 1 \leq \left\lfloor \frac{t+\tau-t_o}{\mathrm{D}} \right\rfloor \qquad \left\lfloor \frac{t-t_o-\mathrm{D}_p}{\mathrm{D}} \right\rfloor + 1 \leq \left\lfloor \frac{t-t_o}{\mathrm{D}} \right\rfloor \tag{9.224}$$

These constraints may not be satisfied, as illustrated in Figure 9.30, when $D_p < D$. Use of

$$u\left[i_{\max_2} - i_{\min_2}\right] \cdot u\left[j_{\max} - j_{\min}\right] \tag{9.225}$$

ensures results consistent with the constraints.

Finally, the double summation is zero when $i<0$ or $i>N-1$ or $j<0$ or $j>N-1$ and can be written as

$$\begin{aligned} &\sum_{i=0}^{N-1} \sum_{j=0, j\neq i}^{N-1} \mu_i \mu_j p(t+\tau-i\mathrm{D}) p(t-j\mathrm{D}) = \\ &\quad u\left[i_{\max_2}\right] \cdot u\left[N-1-i_{\min_2}\right] \cdot u\left[j_{\max}\right] \cdot u\left[N-1-j_{\min}\right] \cdot u\left[i_{\max_2} - i_{\min_2}\right] \cdot \\ &\quad u\left[j_{\max} - j_{\min}\right] \cdot \sum_{i=\max\left\{i_{\min_2}, 0\right\}}^{\min\left\{i_{\max_2}, N-1\right\}} \sum_{j=\max\{j_{\min}, 0\}, j\neq i}^{\min\{j_{\max}, N-1\}} \mu_i \mu_j p(t+\tau-i\mathrm{D}) p(t-j\mathrm{D}) \end{aligned} \tag{9.226}$$

which completes the proof.

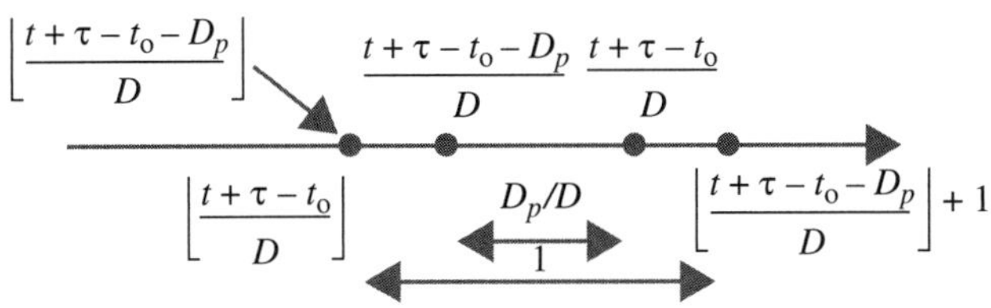

**FIGURE 9.30** Illustration of possible values when $\mathrm{D}_p < \mathrm{D}$.

## APPENDIX 9.G PROOF OF THEOREM 9.19

One outcome of the generalized signalling random process is

$$X(((\omega_1,\gamma_1),\ldots,(\omega_N,\gamma_N)),t)=\sum_{k=1}^{N}\phi(\omega_k,t-\gamma_k): \quad \begin{matrix}\omega_k\in S_{\Omega_k}\\ \gamma_k\in S_{\Gamma_k}\\ \phi\in S_\Phi\end{matrix} \tag{9.227}$$

and it then follows, by definition, that

$$\begin{aligned}R_{XX}(t+\tau,t)&=\boldsymbol{E}[X(\Omega,t+\tau)X^*(\Omega,t)]=\int_{S_{\Omega_N}}\cdots\int_{S_{\Omega_1}}\int_{S_{\Gamma_N}}\cdots\int_{S_{\Gamma_1}}\\ &X(((\omega_1,\gamma_1),\ldots,(\omega_N,\gamma_N)),t+\tau)\cdot X(((\omega_1,\gamma_1),\ldots,(\omega_N,\gamma_N)),t).\\ &f_{\Omega,\ldots\Omega_N,\Gamma_1,\ldots,\Gamma_N}(\omega_1,\ldots,\omega_N,\gamma_1,\ldots,\gamma_N)d\gamma_1\ldots d\gamma_N d\omega_1\ldots d\omega_N\end{aligned} \tag{9.228}$$

Substitution for $X$ yields

$$\begin{aligned}R_{XX}(t+\tau,t)=&\int_{S_{\Omega_N}}\cdots\int_{S_{\Omega_1}}\int_{S_{\Gamma_N}}\cdots\int_{S_{\Gamma_1}}\sum_{i=1}^{N}\sum_{k=1}^{N}\phi(\omega_i,t+\tau-\gamma_i)\phi(\omega_k,t-\gamma_k)\cdot\\ &f_{\Omega_1,\ldots,\Omega_N,\Gamma_1,\ldots,\Gamma_N}(\omega_1,\ldots,\omega_N,\gamma_1,\ldots,\gamma_N)d\gamma_1\ldots d\gamma_N d\omega_1\ldots d\omega_N\end{aligned} \tag{9.229}$$

Interchanging the order of integration and summation yields

$$\begin{aligned}R_{XX}(t+\tau,t)=&\sum_{i=1}^{N}\sum_{k=1}^{N}\int_{S_{\Omega_N}}\cdots\int_{S_{\Omega_1}}\int_{S_{\Gamma_N}}\cdots\int_{S_{\Gamma_1}}\phi(\omega_i,t+\tau-\gamma_i)\phi(\omega_k,t-\gamma_k).\\ &f_{\Omega_1,\ldots,\Omega_N,\Gamma_1,\ldots\Gamma_N}(\omega_1,\ldots,\omega_N,\gamma_1,\ldots,\gamma_N)d\gamma_1\ldots d\gamma_N d\omega_1\ldots d\omega_N\end{aligned} \tag{9.230}$$

Splitting the double summation into the diagonal and nondiagonal terms yields

$$\begin{aligned}R_{XX}(t+\tau,t)=&\sum_{i=1}^{N}\int_{S_{\Omega_N}}\cdots\int_{S_{\Omega_1}}\int_{S_{\Gamma_N}}\cdots\int_{S_{\Gamma_1}}\phi(\omega_i,t+\tau-\gamma_i)\phi(\omega_i,t-\gamma_i).\\ &f_{\Omega_1,\ldots\Omega_N,\Gamma_1,\ldots,\Gamma_N}(\omega_1,\ldots,\omega_N,\gamma_1,\ldots,\gamma_N)d\gamma_1\ldots d\gamma_N d\omega_1\ldots d\omega_N\\ &+\sum_{i=1}^{N}\sum_{\substack{k=1\\k\neq i}}^{N}\int_{S_{\Omega_N}}\cdots\int_{S_{\Omega_1}}\int_{S_{\Gamma_N}}\cdots\int_{S_{\Gamma_1}}\phi(\omega_i,t+\tau-\gamma_i)\phi(\omega_k,t-\gamma_k).\\ &f_{\Omega_1,\ldots,\Omega_N,\Gamma_1,\ldots,\Gamma_N}(\omega_1,\ldots,\omega_N,\gamma_1,\ldots,\gamma_N)d\gamma_1\ldots d\gamma_N d\omega_1\ldots d\omega_N\end{aligned} \tag{9.231}$$

Using the conditional result

$$f_{\Omega_1,\ldots,\Omega_N,\Gamma_1,\ldots,\Gamma_N}(\omega_1,\ldots,\omega_N,\gamma_1,\ldots,\gamma_N)= f_{\Omega_{-i}\Gamma_{-i}/\Omega_i\Gamma_i}(\omega_1,\ldots,\omega_{i-1},\omega_{i+1},\ldots,\omega_N,\gamma_1,\ldots,\gamma_{i-1},\gamma_{i+1},\ldots,\gamma_N)f_{\Omega_i\Gamma_i}(\omega_i,\gamma_i) \tag{9.232}$$

where

$$\Omega_{-i}\Gamma_{-i}=\Omega_1\ldots\Omega_{i-1}\Omega_{i+1}\ldots\Omega_N,\Gamma_1\ldots\Gamma_{i-1}\Gamma_{i+1}\ldots\Gamma_N \tag{9.233}$$

it follows that

$$\int_{S_{\Omega_N}}\cdots\int_{S_{\Omega_{i+1}}}\int_{S_{\Omega_{i-1}}}\cdots\int_{S_{\Omega_1}}\int_{S_{\Gamma_N}}\cdots\int_{S_{\Gamma_{i+1}}}\int_{S_{\Gamma_{i-1}}}\cdots\int_{S_{\Gamma_1}} f_{\Omega_{-i}\Gamma_{-i}/\Omega_i\Gamma_i}(\omega_1,\ldots,\omega_{i-1},\omega_{i+1},\ldots,\omega_N,\gamma_1,\ldots,\gamma_{i-1},\gamma_{i+1},\ldots,\gamma_N). \\ d\gamma_1\ldots d\gamma_{i-1}d\gamma_{i+1}\ldots d\gamma_N d\omega_1\ldots d\omega_{i-1}d\omega_{i+1}\ldots d\omega_N=1 \tag{9.234}$$

A similar result holds for the case of conditioning on $\Omega_i\Omega_k\Gamma_i\Gamma_k$. Using these results, it follows that

$$R_{XX}(t+\tau,t)=\sum_{i=1}^{N}\int_{S_{\Omega_i}}\int_{S_{\Gamma_i}}\phi(\omega_i,t+\tau-\gamma_i)\phi(\omega_i,\tau-\gamma_i)f_{\Omega_i\Gamma_i}(\omega_i,\gamma_i)d\gamma_i d\omega_i \\ +\sum_{i=1}^{N}\sum_{\substack{k=1\\k\neq i}}^{N}\int_{S_{\Omega_k}}\int_{S_{\Omega_i}}\int_{S_{\Gamma_i}}\int_{S_{\Gamma_k}}\phi(\omega_i,t+\tau-\gamma_i)\phi(\omega_k,\tau-\gamma_k)f_{\Omega_i\Omega_k\Gamma_i\Gamma_k}(\omega_i,\omega_k,\gamma_i,\gamma_k)d\gamma_i d\gamma_k d\omega_i d\omega_k \tag{9.235}$$

For the case where all the functions $\phi\in S_\Phi$ are causal, the graphs in Figure 9.31 allow the limits on the integrals, with respect to time, to be specified and the autocorrelation function can then be written as

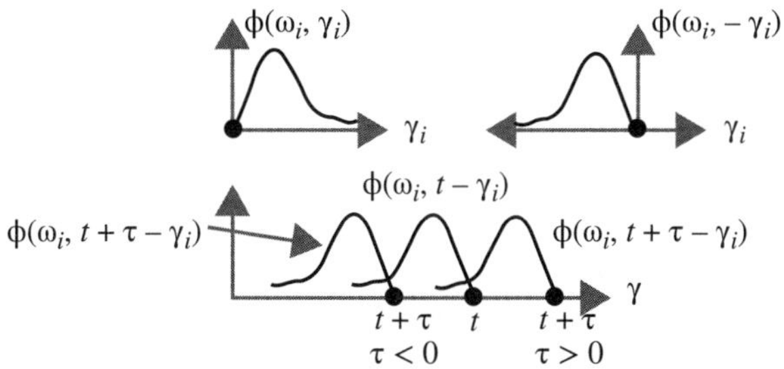


**FIGURE 9.31** Causal signal forms to illustrate the limits on the integrals of the autocorrelation function.

$$R_{XX}(t+\tau,t)=\sum_{i=1}^{N}\int_{S_{\Omega_i}}\int_{0}^{\min\{t,t+\tau\}}\phi(\omega_i,t+\tau-\gamma_i)\phi(\omega_i,t-\gamma_i)f_{\Omega_i\Gamma_i}(\omega_i,\gamma_i)d\gamma_i d\omega_i$$
$$+\sum_{i=1}^{N}\sum_{\substack{k=1\\k\neq i}}^{N}\int_{S_{\Omega_k}}\int_{S_{\Omega_i}}\int_{0}^{t}\int_{0}^{t+\tau}\phi(\omega_i,t+\tau-\gamma_i)\phi(\omega_k,t-\gamma_k)f_{\Omega_i\Omega_k\Gamma_i\Gamma_k}(\omega_i,\omega_k,\gamma_i,\gamma_k)d\gamma_i d\gamma_k d\omega_i d\omega_k \tag{9.236}$$

## APPENDIX 9.H PROOF OF THEOREM 9.20

### 9.H.1 Mean

By definition, $\mu_X(t)=\boldsymbol{E}[X(\Omega,t)]$ and for the independent case,

$$\begin{aligned}\mu_X(t)&=\int_{-\infty}^{\infty}\cdots\int_{-\infty}^{\infty}X((j_1,\ldots,j_N),t)\cdot f_{J_1,\ldots,J_N}(j_1,\ldots,j_N)dj_1\ldots dj_N\\&=A\sum_{k=1}^{N}(-1)^{k+1}\int_{-\infty}^{\infty}\cdots\int_{-\infty}^{\infty}u(t-k\mathrm{D}-j_k)\cdot f_{J_1}(j_1)\ldots f_{J_N}(j_N)dj_1\ldots dj_N\end{aligned} \tag{9.237}$$

where the last result arises from interchanging the order of summation and integration. For a fixed value of $k$, the $N$ integrals include $N-1$ integrals of the form

$$\int_{-\infty}^{\infty}f_{J_i}(j)dj=1 \qquad j\in\{1,\ \ldots,\ k-1,k+1,N\} \tag{9.238}$$

and it then follows that

$$\mu_X(t)=A\sum_{k=1}^{N}(-1)^{k+1}\int_{-\infty}^{\infty}u(t-k\mathrm{D}-j_k)\cdot f_{J_k}(j_k)dj_k \tag{9.239}$$

By considering the graphs shown in Figure 9.32, the limits of the integral can be specified, and it follows that

$$\mu_X(t)=A\sum_{k=1}^{N}(-1)^{k+1}\int_{-\infty}^{t-kD}f_{J_k}(j_k)dj_k=A\sum_{k=1}^{N}(-1)^{k+1}F_{J_k}(t-k\mathrm{D}) \tag{9.240}$$

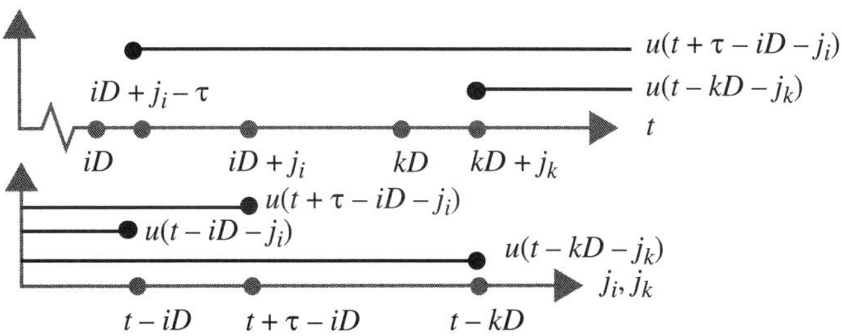


**FIGURE 9.32** Illustration of the functions $u(t+\tau-i\mathrm{D}-j_i)$ and $u(t-k\mathrm{D}-j_k)$ as a function of $t$ and as a function of, respectively, $j_i$ and $j_k$. The illustrations are for $\tau>0$.

where

$$F_{J_k}(j)=P[J_k\le j]=\int_{-\infty}^{j} f_{J_k}(j_k)dj_k \quad 1-F_{J_k}(j)=P[J_k>j]=\int_{j}^{\infty} f_{J_k}(j_k)dj_k \tag{9.241}$$

For the case where the probability density function is zero outside the interval $(-\mathrm{D}/2,\mathrm{D}/2)$, it follows that

$$F_{J_k}(j)=\begin{cases}0 & j\le -\mathrm{D}/2\\ F_{J_k}(j) & -\mathrm{D}/2<j<\mathrm{D}/2\\ 1 & j\ge \mathrm{D}/2\end{cases} \tag{9.242}$$

which implies

$$F_{J_k}(t-k\mathrm{D})=\begin{cases}0 & t\le k\mathrm{D}-\mathrm{D}/2\\ F_{J_k}(t-k\mathrm{D}) & k\mathrm{D}-\mathrm{D}/2<t<k\mathrm{D}+\mathrm{D}/2\\ 1 & t\ge k\mathrm{D}+\mathrm{D}/2\end{cases} \tag{9.243}$$

By considering the graphs illustrated in Figure 9.33, it follows, for $t>\mathrm{D}/2$, that

$$\mu_X(t)=A\sum_{k=1}^{\lfloor t/\mathrm{D}+1/2\rfloor}(-1)^{k+1}F_{J_k}(t-k\mathrm{D}) \tag{9.244}$$

### 9.H.2 Autocorrelation Function

By definition, the autocorrelation function is

$$R_{XX}(t+\tau,t)=\boldsymbol{E}[X(\Omega,t+\tau)X^*(\Omega,t)]=$$
$$\int_{-\infty}^{\infty}\cdots\int_{-\infty}^{\infty} X((j_1,\ldots,j_N),t+\tau)X((j_1,\ldots,j_N),t)\cdot f_{J_1,\ldots,J_N}(j_1,\ldots,j_N)dj_1\ldots dj_N \tag{9.245}$$

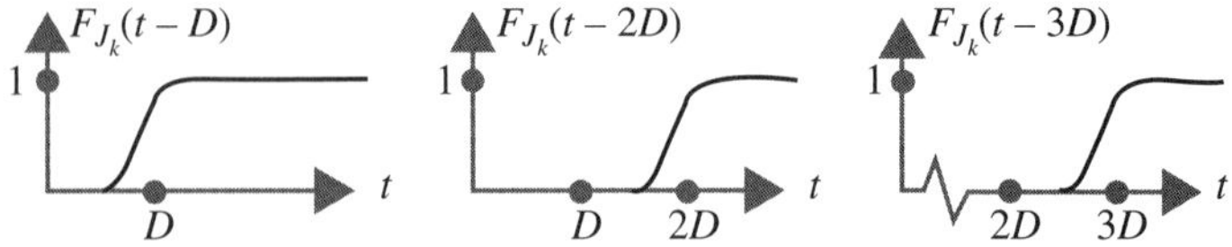


**FIGURE 9.33** Illustration of the cumulative density function $F_{J_k}$ for delays of $D$, $2D$, and $3D$.

Substitution for $X$ and interchanging the order of integration and summation yields

$$R_{XX}(t+\tau,t)=A^2\sum_{i=1}^{N}\sum_{k=1}^{N}(-1)^{i+1}(-1)^{k+1}\int_{-\infty}^{\infty}\cdots\int_{-\infty}^{\infty} u(t+\tau-i\mathrm{D}-j_i)u(t-k\mathrm{D}-j_k)\cdot f_{J_1}(j_1)\ldots f_{J_N}(j_N)dj_1\ldots dj_N \tag{9.246}$$

Splitting the double summation into the diagonal and nondiagonal terms yields

$$\begin{aligned}R_{XX}(t+\tau,t)=&A^2\sum_{i=1}^{N}\int_{-\infty}^{\infty}\cdots\int_{-\infty}^{\infty}u(t+\tau-i\mathrm{D}-j_i)u(t-i\mathrm{D}-j_i)\cdot f_{J_1}(j_1)\ldots f_{J_N}(j_N)dj_1\ldots dj_N\\&+A^2\sum_{i=1}^{N}\sum_{k=1,k\neq i}^{N}(-1)^{i+k}\int_{-\infty}^{\infty}\cdots\int_{-\infty}^{\infty}u(t+\tau-i\mathrm{D}-j_i)u(t-k\mathrm{D}-j_k)\cdot\\&\qquad f_{J_1}(j_1)\ldots f_{J_N}(j_N)dj_1\ldots dj_N\end{aligned} \tag{9.247}$$

Using Equation 9.238, it follows that

$$\begin{aligned}R_{XX}(t+\tau,t)=&A^2\sum_{i=1}^{N}\int_{-\infty}^{\infty}u(t+\tau-i\mathrm{D}-j_i)u(t-i\mathrm{D}-j_i)\cdot f_{J_i}(j_i)dj_i+\\&A^2\sum_{i=1}^{N}\sum_{\substack{k=1\\k\neq i}}^{N}(-1)^{i+k}\int_{-\infty}^{\infty}\int_{-\infty}^{\infty}u(t+\tau-i\mathrm{D}-j_i)u(t-k\mathrm{D}-j_k)\cdot f_{J_i}(j_i)f_{J_k}(j_k)dj_idj_k\end{aligned} \tag{9.248}$$

For the first integral, consider the graphs shown in Figure 9.32, whereupon it follows that

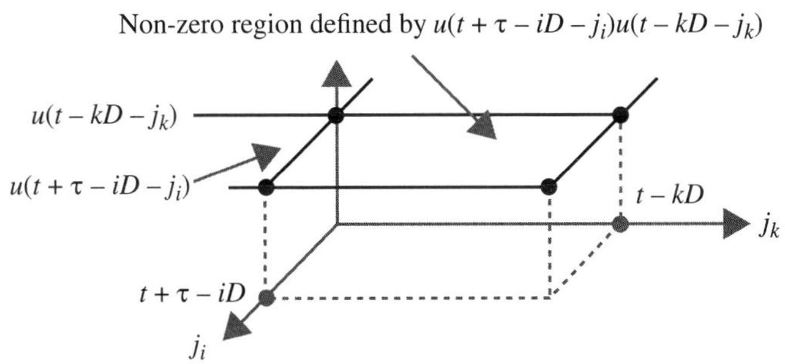


**FIGURE 9.34** Illustration of nonzero region defined by $u(t+\tau-i\text{D}-j_i)$ and $u(t-k\text{D}-j_k)$.

$$\int_{-\infty}^{\infty} u(t+\tau-i\text{D}-j_i)u(t-i\text{D}-j_i)\cdot f_{J_i}(j_i)dj_i = \begin{cases} \int_{-\infty}^{t+\tau-i\text{D}} f_{J_i}(j_i)dj_i & \tau<0 \\ \int_{-\infty}^{t-i\text{D}} f_{J_i}(j_i)dj_i & \tau\geq 0 \end{cases} \tag{9.249}$$

It then follows, as $f_{J_i}\geq 0$ and $F_{J_i}$ is monotonically increasing, that

$$\int_{-\infty}^{\min\{t-i\text{D},t+\tau-i\text{D}\}} f_{J_i}(j_i)dj_i = \min\{F_{J_i}(t-i\text{D}),F_{J_i}(t+\tau-i\text{D})\} \tag{9.250}$$

To evaluate the double integral, consider the graph in Figure 9.34, whereupon it follows that

$$\begin{aligned} &\int_{-\infty}^{\infty}\int_{-\infty}^{\infty} u(t+\tau-i\text{D}-j_i)u(t-k\text{D}-j_k)\cdot f_{J_i}(j_i)f_{J_k}(j_k)dj_idj_k \\ &= \int_{-\infty}^{t+\tau-i\text{D}} f_{J_i}(j_i)dj_i \int_{-\infty}^{t-k\text{D}} f_{J_k}(j_k)dj_k = F_{J_i}(t+\tau-i\text{D})F_{J_k}(t-k\text{D}) \end{aligned} \tag{9.251}$$

With these results, it then follows that

$$\begin{aligned} R_{XX}(t+\tau,t) &= A^2\sum_{i=1}^{N}\min\{F_{J_i}(t-i\text{D}),F_{J_i}(t+\tau-i\text{D})\} \\ &\quad + A^2\sum_{i=1}^{N}\sum_{\substack{k=1\\k\neq i}}^{N}(-1)^{i+k}F_{J_i}(t+\tau-i\text{D})F_{J_k}(t-k\text{D}) \end{aligned} \tag{9.252}$$

which is the required result.

### 9.H.3 Variance

The expected squared value of the random process is

$$\boldsymbol{E}[X^2(\Omega,t)] = R_{XX}(t,t) = A^2\sum_{i=1}^{N} F_{J_i}(t-i\mathrm{D}) + A^2\sum_{i=1}^{N}\sum_{\substack{k=1\\k\neq i}}^{N}(-1)^{i+k}F_{J_i}(t-i\mathrm{D})F_{J_k}(t-k\mathrm{D}) \tag{9.253}$$

and the variance can then be established according to

$$\begin{aligned}\sigma_X^2(t) &= \boldsymbol{E}[X^2(\Omega,t)] - \mu_X^2(t)\\ &= A^2\sum_{i=1}^{N}F_{\Omega_i}(t-i\mathrm{D}) + A^2\sum_{i=1}^{N}\sum_{\substack{k=1\\k\neq i}}^{N}(-1)^{i+k}F_{\Omega_i}(t-i\mathrm{D})F_{\Omega_k}(t-k\mathrm{D})\\ &\quad - A^2\sum_{i=1}^{N}\sum_{k=1}^{N}(-1)^{i+k}F_{\Omega_i}(t-i\mathrm{D})F_{\Omega_k}(t-k\mathrm{D})\\ &= A^2\sum_{i=1}^{N}F_{J_i}(t-i\mathrm{D}) - A^2\sum_{i=1}^{N}F_{J_i}^2(t-i\mathrm{D})\end{aligned} \tag{9.254}$$

For the case where the probability density function is zero outside the interval $(-\mathrm{D}/2, \mathrm{D}/2)$, it can be inferred from Figure 9.33 that

$$\sigma_X^2(t) = A^2\left[F_{J_i}(t-i\mathrm{D}) - F_{J_i}^2(t-iD)\right] \quad i = \left\lfloor \frac{t}{\mathrm{D}} + \frac{1}{2}\right\rfloor, \quad t > \mathrm{D}/2 \tag{9.255}$$

which is the last required result.

# 10

# POWER SPECTRAL DENSITY THEORY

## 10.1 INTRODUCTION

The most widely used approach in the physical sciences for characterizing a random process is via a power spectral density, that is, the decomposition of the signals defined by a random process into their constituent sinusoidal components. The power spectral density for a random process has been introduced in Chapter 7. This chapter extends the discussion of a power spectral density with the definition of the cross power spectral density and the consideration of the power spectral density of a sum of random processes. The power spectral density of a periodic pulse train, signalling random processes, quantization error signals, shot noise random processes, $1/f$ noise, and a correlated jittered random process are established. Further theory and results for the power spectral density, consistent with the approach taken, can be found in Howard (2002).

## 10.2 POWER SPECTRAL DENSITY THEORY

Consider a random process $X: S \rightarrow S_X$ defined on $[0, T]$, with a signal sample space

$$S_X = \{X(\omega, t): \quad \omega \in S\} \tag{10.1}$$

that is based on the sample space $S$ of experimental outcomes underpinning the random process. The sample space $S$ defines a random variable $\Omega$ with a probability

*A Signal Theoretic Introduction to Random Processes*, First Edition. Roy M. Howard.

mass function $p_\Omega(\omega) = P[\Omega = \omega]$ for the countable case and a probability density function $P[\Omega \in [\omega, \omega + d\omega]] \approx f_\Omega(\omega)d\omega$ for the uncountable case.

In terms of notation, for a random process $X$, a subscript of $X$ is added to the power spectral density, that is, $G(T,f)$ is written as $G_X(T,f)$.

The following subsections detail important results related to the power spectral density of a random process.

### 10.2.1 Bounded Signal Energy Implies Bounded PSD

**Theorem 10.1 Bounded Signal Energy Implies Bounded PSD**
If the signal energy, and average signal power, is bounded on $[0, T]$ for all signals defined by the random process, that is,

$$\sup\left\{\int_0^T |X(\omega, t)|^2 dt : \quad \omega \in S\right\} < \infty \tag{10.2}$$

then the magnitude of the Fourier transforms of individual signals, and individual signal power spectral densities, remains bounded, that is,

$$\sup\left\{ |X(\omega, T, f)| = \left|\int_0^T X(\omega, t)e^{-j2\pi ft} dt\right| : \quad \omega \in S, f \in \boldsymbol{R}\right\} < \infty \tag{10.3}$$

$$\sup\left\{ G_X(\omega, T, f) = \frac{|X(\omega, T, f)|^2}{T} : \quad \omega \in S, f \in \boldsymbol{R}\right\} < \infty \tag{10.4}$$

With individual power spectral densities being bounded, the power spectral density

$$G_X(T,f) = \boldsymbol{E}\left[\frac{|X(\Omega, T, f)|^2}{T}\right] \tag{10.5}$$

is bounded for $f \in \boldsymbol{R}$.

**Proof**
The proof is detailed in Appendix 10.A.

### 10.2.2 Cross Power Spectral Density

The following definition is fundamental.

**Definition: Cross Power Spectral Density** The cross power spectral density of two random processes $X_1(\Omega_1, t)$ and $X_2(\Omega_2, t)$, on the interval $[0, T]$, is defined according to

$$G_{X_1X_2}(T,f) = \begin{cases} \frac{1}{T}\sum_{i=1}^{\infty}\sum_{j=1}^{\infty} p_{\Omega_1\Omega_2}(\omega_{1i},\omega_{2j}) X_1(\omega_{1i},T,f) X_2^*(\omega_{2j},T,f) & \text{countable case} \\ \frac{1}{T}\int_{-\infty}^{\infty}\int_{-\infty}^{\infty} X_1(\omega_1,T,f) X_2^*(\omega_2,T,f) f_{\Omega_1\Omega_2}(\omega_1,\omega_2) d\omega_1\omega_2 & \text{uncountable case} \end{cases} \tag{10.6}$$

where the Fourier transforms, over $[0, T]$, of the $i$th signals defined by the random processes, are

$$X_1(\omega_{1i},T,f) = \int_0^T X_1(\omega_{1i},t)e^{-j2\pi ft}dt \qquad X_2(\omega_{2i},T,f) = \int_0^T X_2(\omega_{2i},t)e^{-j2\pi ft}dt \tag{10.7}$$

and the probability of occurrence of $(\omega_{1i}, \omega_{2i})$ is governed by $p_{\Omega_1\Omega_2}$ for the countable case and $f_{\Omega_1\Omega_2}$ for the uncountable case.

#### *10.2.2.1 Cross random processes: Independent Case*

**Theorem 10.2 Cross Power Spectral Density: Independent Case**
If two random processes $X_1$ and $X_2$ are independent, then their cross power spectral density is

$$G_{X_1X_2}(T,f) = \frac{\mu_{X_1}(T,f)\mu_{X_2}^*(T,f)}{T} \tag{10.8}$$

where $\mu_{X_1}(T,f)$ and $\mu_{X_2}(T,f)$, respectively, are the Fourier transforms of the mean signals $\mu_{X_1}(t)$ and $\mu_{X_2}(t)$ evaluated on the interval $[0, T]$, that is,

$$\begin{aligned} \mu_{X_1}(T,f) &= \int_0^T e^{-j2\pi ft}\left[\int_{-\infty}^{\infty} X_1(\omega_1,t) f_{\Omega_1}(\omega_1)d\omega_1\right]dt \\ \mu_{X_2}(T,f) &= \int_0^T e^{-j2\pi ft}\left[\int_{-\infty}^{\infty} X_2(\omega_2,t) f_{\Omega_2}(\omega_2)d\omega_2\right]dt \end{aligned} \tag{10.9}$$

The uncountable case is assumed here along with the assumption of bounded and integrable signals. For the case of a constant mean on the interval $[0, T]$, with respective means of $\mu_{X_1}$ and $\mu_{X_2}$, the cross power spectral density $G_{X_1X_2}(T,f)$ is

$$G_{X_1X_2}(T,f) = \mu_{X_1}\mu_{X_2}^* \cdot T\,\text{sinc}(fT)^2 \tag{10.10}$$

$$G_{X_1X_2}(\infty,f) = \mu_{X_1}\mu_{X_2}^*\delta(f) \tag{10.11}$$

If the mean of at least one of the random processes is zero, then

$$G_{X_1X_2}(T,f)=0 \tag{10.12}$$

**Proof**
The proof is detailed in Appendix 10.B.

### 10.2.3 Power Spectral Density of a Sum of Random Processes

Consider $N$ random processes $X_1(\Omega_1, t), \ldots, X_N(\Omega_N, t)$ defined on an experiment that defines a vector of random variables $(\Omega_1, \ldots, \Omega_N)$ with

$$P[\Omega_1=\omega_1,\ldots,\Omega_N=\omega_N]=p_{\Omega_1\ldots\Omega_N}(\omega_1,\ldots,\omega_N) \tag{10.13}$$

for the countable case and

$$\begin{aligned}P[\Omega_1\in[\omega_1,\omega_1+d\omega_1],\ldots,\Omega_N\in[\omega_N,\omega_N+d\omega_N]]\approx\\ f_{\Omega_1\ldots\Omega_N}(\omega_1,\ldots,\omega_N)d\omega_1\ldots d\omega_N\end{aligned} \tag{10.14}$$

for the uncountable case. The following theorems state the power spectral density for the weighted sum of such random processes defined according to

$$Z(\Omega,t)=\lim_{N\to\infty}\sum_{i=1}^{N}w_iX_i(\Omega_i,t) \tag{10.15}$$

**Theorem 10.3 Power Spectral Density of a Sum of Random Processes**

For the case of $\sum_{i=1}^{\infty}|w_i|$ finite and where each of the random processes $X_1, \ldots, X_N$ has bounded energy on a fixed interval $[0, T]$ consistent with

$$\sup\left\{\int_0^T|X_i(\omega,t)|^2dt:\ \ \omega\in S_i\right\}<\infty \qquad X_i:S_i\to S_{X_i} \tag{10.16}$$

the power spectral density of the random processes $Z$ is

$$\begin{aligned}G_Z(T,f)&=\sum_{i=1}^{\infty}|w_i|^2G_{X_i}(T,f)+\sum_{i=1}^{\infty}\sum_{\substack{j=1\\ j\neq i}}^{\infty}w_iw_j{}^*G_{X_iX_j}(T,f)\\ &=\sum_{i=1}^{\infty}|w_i|^2G_{X_i}(T,f)+\sum_{i=1}^{\infty}\sum_{\substack{j=1\\ j\neq i}}^{\infty}2\mathrm{Re}\left[w_iw_j{}^*G_{X_iX_j}(T,f)\right]\end{aligned} \tag{10.17}$$

Here, $G_{X_i}$ is the power spectral density of $X_i$ and $G_{X_iX_j}$ is the cross power spectral density between $X_i$ and $X_j$.

**Proof**

The proof is detailed in Appendix 10.C.

#### *10.2.3.1 PSD of a Sum of Independent Random Processes*

**Theorem 10.4 PSD: Sum of Independent Random Processes**

The power spectral density of the random process $Z(\Omega, t) = \sum_{i=1}^{\infty} w_i X_i(\Omega, t)$, for the case where $X_1, \ldots$ are independent and satisfy the conditions as specified in Theorem 10.3, is

$$G_Z(T,f) = \sum_{i=1}^{\infty} |w_i|^2 G_{X_i}(T,f) + \frac{1}{T}\sum_{i=1}^{\infty}\sum_{\substack{j=1\\ j\neq i}}^{\infty} w_i w_j^* \mu_{X_i}(T,f)\mu_{X_j}^*(T,f) \tag{10.18}$$

where $\mu_{X_i}(T,f)$, defined by Equation 10.9, is the Fourier transform of the mean of the $i$th random process evaluated on $[0, T]$. For the constant mean case,

$$G_Z(T,f) = \sum_{i=1}^{\infty} |w_i|^2 G_{X_i}(T,f) + \sum_{i=1}^{\infty}\sum_{\substack{j=1\\ j\neq 1}}^{\infty} w_i w_j^* \mu_{X_i}\mu_{X_j}^* \cdot T\,\text{sinc}(fT)^2 \tag{10.19}$$

$$G_Z(\infty,f) = \sum_{i=1}^{\infty} |w_i|^2 G_{X_i}(\infty,f) + \delta(f)\sum_{i=1}^{\infty}\sum_{\substack{j=1\\ j\neq 1}}^{\infty} w_i w_j^* \mu_{X_i}\mu_{X_j}^* \tag{10.20}$$

For the constant and zero-mean case, the power spectral density of the weighted sum of random processes equals the weighted sum of the individual power spectral density processes, that is,

$$G_Z(T,f) = \sum_{i=1}^{\infty} |w_i|^2 G_{X_i}(T,f) \tag{10.21}$$

**Proof**

The proof follows from Theorem 10.2 and Theorem 10.3.

## 10.3 POWER SPECTRAL DENSITY OF A PERIODIC PULSE TRAIN

Consider a periodic pulse train with period $D$, as illustrated in Figure 10.1, and defined according to

$$x(t) = \sum_{k=0}^{(N-1)D} Ap(t-kD) \qquad t \in [0, ND] \tag{10.22}$$

The power spectral density of such a pulse train is specified in the following theorem.

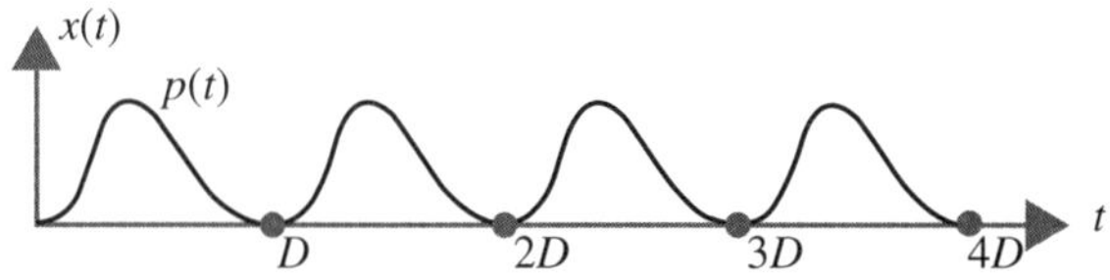


**FIGURE 10.1** Illustration of a periodic pulse train with period D.

**Theorem 10.5 Power Spectral Density of a Periodic Pulse Train**
The power spectral densities of a pulse train, as defined by Equation 10.22, on the finite and infinite intervals, respectively, are

$$G_X(ND,f) = rA^2|P(f)|^2\left[\frac{1}{N}\cdot\frac{\sin(\pi Nf/r)^2}{\sin(\pi f/r)^2}\right] \tag{10.23}$$

$$G_X(\infty,f) = r^2A^2|P(f)|^2\sum_{n=-\infty}^{\infty}\delta(f-nr) \tag{10.24}$$

where $r = 1/D$ is the pulse rate and $P(f)$ is the Fourier transform of $p(t)$ evaluated on the infinite interval. Here, the assumption is that the interval $[0, ND]$ does not significantly affect the Fourier transform of the pulse functions.

**Proof**
The proof is detailed in Appendix 10.D.

### 10.3.1 Example

Consider two pulse trains: the first with a pulse function defined by

$$p(i,t) = \frac{t^{i-1}}{\tau^i(i-1)!}\cdot e^{-t/\tau}u(t) \Leftrightarrow P(i,t) = \frac{1}{(1+j2\pi f\tau)^i} \tag{10.25}$$

for $i \in \mathbf{Z}^+$ and the second based on a Gaussian pulse:

$$p(t) = \frac{e^{-t^2/2\sigma_t^2}}{\sqrt{2\pi}\sigma_t} \Leftrightarrow P(f) = e^{-2\pi^2f^2\sigma_t^2} \tag{10.26}$$

Time domain signals from these two pulse trains are shown in Figure 10.2 for the case of $D = 1$, $i = 2$, $\tau = 0.1D$, and $\sigma_t = 0.1D$. The power spectral density of the two pulse trains is shown in Figure 10.3 assuming the infinite interval.

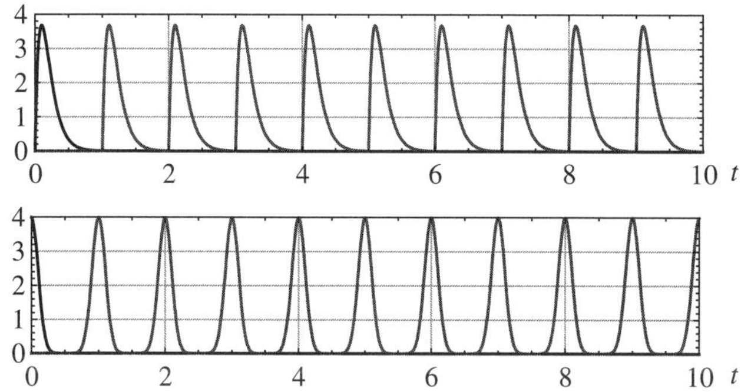


**FIGURE 10.2** Pulse train. Top: exponential decaying pulse. Bottom: Gaussian pulse.

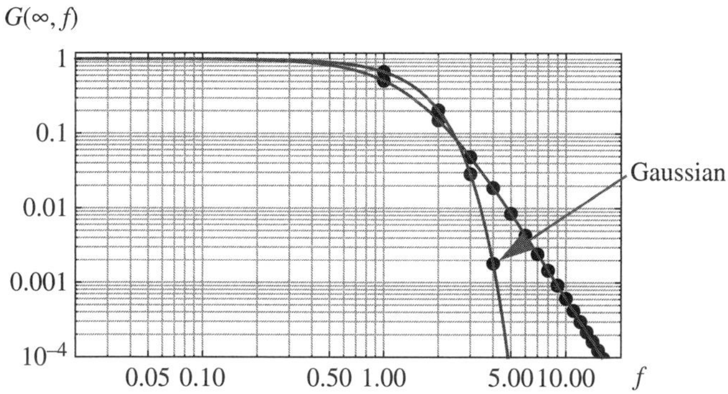


**FIGURE 10.3** Power spectral density of the two pulse trains. The dots represent the power in the impulsive components. For both pulse trains, the power in the impulse at $f = 0$ is unity. The continuous curve is the magnitude squared of the Fourier transform of the pulse function. The powers in the impulses at $f = 0, r, 2r, 3r, 4r, 5r$, as represented by the dots, are, respectively, 1, 0.514, 0.150, $4.82 \times 10^{-2}$, $1.87 \times 10^{-2}$ and $8.46 \times 10^{-3}$ for the exponential pulse case and 1, 0.674, 0.206, $2.86 \times 10^{-2}$, $1.81 \times 10^{-3}$, and $5.17 \times 10^{-5}$ for the Gaussian pulse case.

## 10.4 PSD OF A SIGNALLING RANDOM PROCESS

Signalling random processes underpin communication of information. With a signalling random process, the information is encoded in signals from a signal set, and one important aspect of choosing a signal set for conveyance of information is the resultant spectral efficiency (bits per second of information per Hertz of bandwidth). The power spectral density allows the determination of such efficiency.

### 10.4.1 Signalling Set

Consider a signalling set $S_\Phi$ of integrable functions:

$$S_\Phi = \begin{cases} \{\phi(\omega): \boldsymbol{R} \to \boldsymbol{R}, & \omega \in \{1,\ldots,N_\Phi\}\} \quad \text{finite case} \\ \{\phi(\omega): \boldsymbol{R} \to \boldsymbol{R}, & \omega \in \boldsymbol{Z}^+\} \quad \text{countable case} \\ \{\phi(\omega): \boldsymbol{R} \to \boldsymbol{R}, & \omega \in \boldsymbol{R}\} \quad \text{uncountable case} \end{cases} \tag{10.27}$$

The notation $S_\Phi = \{\phi(\omega, t): \ \omega \in \boldsymbol{R}\}$ for the uncountable case, for example, is convenient.

By taking the Fourier transform of each signal in the signal set, the associated set

$$S_{\Phi(f)} = \left\{ \Phi(\omega, f): \ \Phi(\omega, f) = \int_{-\infty}^{\infty} \phi(\omega, t) e^{-j2\pi ft} dt, \ \omega \in \boldsymbol{R} \right\} \tag{10.28}$$

is defined for the uncountable case, and similar sets can be defined for the countable and finite cases.

### 10.4.2 Signalling Random Process

Consider an experiment that defines the sample space of vectors comprising of $N$ elements:

$$S = \begin{cases} \{(\omega_1,\ldots,\omega_N): & \omega_i \in \{1,\ldots,N_\Phi\}\} \quad \text{finite case} \\ \{(\omega_1,\ldots,\omega_N): & \omega_i \in \{1,2,\ldots\}\} \quad \text{countable case} \\ \{(\omega_1,\ldots,\omega_N): & \omega_i \in \boldsymbol{R}\} \quad \text{uncountable case} \end{cases} \tag{10.29}$$

and where the outcome $(\omega_1, \ldots, \omega_N)$ is governed by the probability mass function $p_\Omega(\omega_1, \ldots, \omega_N)$ for the countable case and the probability density function $f_\Omega(\omega_1, \ldots, \omega_N)$ for the uncountable case. A signalling random process $X: S \to S_X$, based on the signalling set $S_\Phi$ (Eq. 10.27), and for the interval $[0, T]$, $T = ND$, is defined according to

$$X(\Omega, t) = \sum_{k=1}^{N} \phi(\Omega_k, t-(k-1)D) \qquad t \in [0, ND] \tag{10.30}$$

where $\Omega = (\Omega_1,\ldots,\Omega_N)$ is a vector of random variables, with outcomes $(\omega_1,\ldots,\omega_N) \in S$, which govern the waveforms from the signalling set that are associated with each signalling interval. The signalling rate is $r = 1/D$, and the signal sample space is

$$S_X = \left\{ X((\omega_1,\ldots,\omega_N), t) = \sum_{k=1}^{N} \phi(\omega_k, t-(k-1)D): \ (\omega_1,\ldots,\omega_N) \in S \right\} \tag{10.31}$$

### 10.4.3 Signalling Random Process as a Summation of Random Processes

A signalling random process can be considered to be the sum of $N$ component random processes:

$$X(\Omega, t) = \sum_{k=1}^{N} X_k(\Omega_k, t) \qquad X_k(\Omega_k, t) = \phi(\Omega_k, t-(k-1)D), t \in [0, ND] \tag{10.32}$$

The $k$th component random process is defined by the set of waveforms associated with the $k$th signalling interval as governed by the random variable $\Omega_k$, which has a sample space denoted $S_k$. The Fourier transform of the $i$th waveform from such a random process is

$$X_k(\omega_i, t) = \phi(\omega_i, t-(k-1)D) \rightarrow X_k(\omega_i, f) = \Phi(\omega_i, f)e^{-j2\pi f(k-1)D} \tag{10.33}$$

where $\Phi(\omega_i, f)$ is the Fourier transform of $\phi(\omega_i, t)$. The assumption here is that $T = ND$ is sufficiently large such that the Fourier transform evaluated on the interval $[0, T]$ approximates that from the infinite interval.

The $k$th component random process has a power spectral density on the interval $[0, T]$ given by

$$G_k(f) = \begin{cases} \dfrac{1}{T}\displaystyle\sum_{\omega_i \in S_k} p_{\Omega_k}(\omega_i)|\Phi(\omega_i, f)|^2 & \text{countable case} \\ \dfrac{1}{T}\displaystyle\int_{-\infty}^{\infty} |\Phi(\omega, f)|^2 f_{\Omega_k}(\omega)d\omega & \text{uncountable case} \end{cases} \tag{10.34}$$

where $p_{\Omega_k}$ and $f_{\Omega_k}$ are, respectively, the probability mass function and the probability density function for $\Omega_k$. The cross power spectral density between the $i$th and $k$th component random processes on the interval $[0, T]$ is

$$G_{ik}(f) = \begin{cases} \dfrac{1}{T}\displaystyle\sum_{\omega_i \in S_i}\sum_{\omega_k \in S_k} p_{\Omega_i\Omega_k}(\omega_i, \omega_k)\Phi(\omega_i, f)\Phi^*(\omega_k, f)e^{-j2\pi f(i-k)D} & \text{c.c.} \\ \dfrac{1}{T}\displaystyle\int_{-\infty}^{\infty}\int_{-\infty}^{\infty} \Phi(\omega_i, f)\Phi^*(\omega_k, f)e^{-j2\pi f(i-k)D} f_{\Omega_i\Omega_k}(\omega_i, \omega_k)d\omega_i d\omega_k & \text{u.c.} \end{cases} \tag{10.35}$$

where $p_{\Omega_i\Omega_k}$ and $f_{\Omega_i\Omega_k}$ are, respectively, the joint probability mass function and the joint probability density function for $\Omega_i$ and $\Omega_k$.

***10.4.3.1 Power Spectral Density of Signals in Signalling Set*** It is useful to define the mean Fourier transform, $\mu_{\Phi,i}$, and power spectral density, $G_{\Phi,i}$, of the $i$th component random process and the cross power spectral density, $G_{\Phi,ik}$, between the $i$th and $k$th component random processes, $i \neq k$, assuming there is no delay between the signals in these random process and assuming an interval consistent with the signalling interval, namely, $D$. Effectively, these are definitions based directly on the signalling set and assuming the interval $[0, D]$. The definitions are

$$\mu_{\Phi,i}(f) = \begin{cases} \displaystyle\sum_{\omega_i \in S_i} p_{\Omega_i}(\omega_i)\Phi(\omega_i, f) & \text{countable case} \\ \displaystyle\int_{-\infty}^{\infty} \Phi(\omega, f) f_{\Omega_i}(\omega)d\omega & \text{uncountable case} \end{cases} \tag{10.36}$$

$$G_{\Phi,i}(f)=\begin{cases}\dfrac{1}{D}\sum\limits_{\omega_i\in S_i}p_{\Omega_i}(\omega_i)|\Phi(\omega_i,f)|^2 & \text{countable case}\\ \dfrac{1}{D}\int\limits_{-\infty}^{\infty}|\Phi(\omega,f)|^2 f_{\Omega_i}(\omega)d\omega & \text{uncountable case}\end{cases} \quad (10.37)$$

$$G_{\Phi,ik}(f)=\begin{cases}\dfrac{1}{D}\sum\limits_{\omega_i\in S_i}\sum\limits_{\omega_k\in S_k}p_{\Omega_i\Omega_k}(\omega_i,\omega_k)\Phi(\omega_i,f)\Phi^*(\omega_k,f) & \text{c.c.}\\ \dfrac{1}{D}\int\limits_{-\infty}^{\infty}\int\limits_{-\infty}^{\infty}\Phi(\omega_i,f)\Phi^*(\omega_k,f)f_{\Omega_i\Omega_k}(\omega_i,\omega_k)d\omega_i d\omega_k & \text{u.c.}\end{cases} \quad (10.38)$$

For the independent case,

$$G_{\Phi,ik}(f)=r\mu_{\Phi,i}(f)\mu_{\Phi,k}{}^*(f) \quad (10.39)$$

### 10.4.4 Power Spectral Density of a Signalling Random Process

The power spectral density of a signalling random process, based on the previous definitions, is specified in the following theorem under the assumption that the energy, from outside of the interval $[0, ND]$, of the constituent waveforms defining a signal from the random process is zero or negligible.

**Theorem 10.6 Power Spectral Density of a Signalling Random Process** The power spectral density of a signalling random process $X$ defined on the interval $[0, T]$ with $T=ND$, $r=1/D$, is given by

$$G_X(T,f)=\frac{1}{N}\sum_{i=1}^{N}G_{\Phi,i}(f)+\frac{1}{N}\sum_{i=1}^{N}\sum_{\substack{k=1\\k\neq i}}^{N}e^{-j2\pi f(i-k)D}G_{\Phi,ik}(f) \quad (10.40)$$

For the case where the signalling waveforms in different signalling intervals are independent, consistent with

$$\begin{aligned}p_{\Omega_1\ldots\Omega_N}(\omega_1,\ldots,\omega_N)&=p_{\Omega_1}(\omega_1)\cdot\ldots\cdot p_{\Omega_N}(\omega_N) && \text{countable case}\\ f_{\Omega_1\ldots\Omega_N}(\omega_1,\ldots,\omega_N)&=f_{\Omega_1}(\omega_1)\cdot\ldots\cdot f_{\Omega_N}(\omega_N) && \text{uncountable case}\end{aligned} \quad (10.41)$$

it is the case that

$$G_X(T,f)=\frac{1}{N}\sum_{i=1}^{N}G_{\Phi,i}(f)+\frac{1}{T}\sum_{i=1}^{N}\sum_{\substack{k=1\\k\neq i}}^{N}\mu_{\Phi,i}(f)\mu_{\Phi,k}{}^*(f)e^{-j2\pi f(i-k)D} \quad (10.42)$$

For independent case and when the waveforms in the signalling set have a zero mean, consistent with $\mu_{\Phi,i}(f)=0$ for $i=1,\ldots,N$, it follows that

$$G_X(T,f)=\frac{1}{N}\sum_{i=1}^{N}G_{\Phi,i}(f) \tag{10.43}$$

For the independent and identical case consistent with

$$\begin{aligned}&p_{\Omega_i}(\omega_r)=p_{\Omega_k}(\omega_r) \quad f_{\Omega_i}(\omega)=f_{\Omega_k}(\omega) \quad i,k\in\{1,\ldots,N\}\\ &G_{\Phi,i}(f)=G_{\Phi,k}(f)=G_\Phi(f) \quad \mu_{\Phi,i}(f)=\mu_{\Phi,k}(f)=\mu_\Phi(f) \quad i,k\in\{1,\ldots,N\}\end{aligned} \tag{10.44}$$

it follows that

$$G_X(T,f)=G_\Phi(f)-r|\mu_\Phi(f)|^2+r|\mu_\Phi(f)|^2\cdot\frac{1}{N}\cdot\frac{\sin(\pi Nf/r)^2}{\sin(\pi f/r)^2} \tag{10.45}$$

$$G_X(\infty,f)=G_\Phi(f)-r|\mu_\Phi(f)|^2+r^2|\mu_\Phi(f)|^2\sum_{n=-\infty}^{\infty}\delta(f-nr) \tag{10.46}$$

For the independent and identical case and when the mean of the waveforms in the signalling set is zero, that is, $\mu_\Phi(f)=0$, the following simple result holds:

$$G_X(T,f)=G_\Phi(f) \tag{10.47}$$

**Proof**
The proof is detailed in Appendix 10.E.

***10.4.4.1 Results for Subcases*** Several important subcases can be defined: first, the case where the dependency between signalling waveforms depends on the difference between the location of the signalling intervals and not on their absolute location, that is, $p_{\Omega_i\Omega_k}(\omega_i,\omega_k)=p_{\Omega_{i+q}\Omega_{k+q}}(\omega_i,\omega_k)$, $q\in\{1,2,\ldots\}$, and, second, the case of finite dependency where the $i$th signalling waveform is independent of the $j$th signalling waveform for $|i-j|>m$.

**Theorem 10.7 Dependency on Difference between Signalling Intervals**
For the case where the dependency between signalling waveforms depends on the difference between the location of the signalling intervals and not on their absolute location, the power spectral density of a signalling random process is

$$G_X(ND,f)=\frac{1}{N}\sum_{i=1}^{N}G_{\Phi,i}(f)+2\sum_{i=1}^{N-1}\left[1-\frac{i}{N}\right]\mathrm{Re}\left[e^{j2\pi if/r}G_{\Phi,1,1+i}(f)\right] \tag{10.48}$$

For the case where the $i$th signalling waveform is independent of the $j$th signalling waveform for $|i-j|>m$, the power spectral density is

$$G_X(ND,f) = \frac{1}{N}\sum_{i=1}^{N} G_{\Phi,i}(f) + 2r\sum_{i=1}^{N-1}\left[1-\frac{i}{N}\right]\mathrm{Re}\left[e^{j2\pi if/r}\mu_{\Phi,1}(f)\mu_{\Phi,1+i}^{*}(f)\right]$$
$$+2\sum_{i=1}^{m}\left[1-\frac{i}{N}\right]\mathrm{Re}\left[e^{j2\pi if/r}\left[G_{\Phi,1,1+i}(f)-r\mu_{\Phi,1}(f)\mu_{\Phi,1+i}^{*}(f)\right]\right] \quad (10.49)$$

For the case of statistically identical signalling waveforms on each interval, consistent with $G_{\Phi,i}(f) = G_{\Phi,k}(f) = G_{\Phi}(f)$ and $\mu_{\Phi,i}(f) = \mu_{\Phi,k}(f) = \mu_{\Phi}(f)$, it follows that

$$G_X(ND,f) = G_{\Phi}(f) - r|\mu_{\Phi}(f)|^2 + r|\mu_{\Phi}(f)|^2 \cdot \frac{1}{N} \cdot \frac{\sin(\pi Nf/r)^2}{\sin(\pi f/r)^2}$$
$$+2\sum_{i=1}^{m}\left[1-\frac{i}{N}\right]\mathrm{Re}\left[e^{j2\pi if/r}\left[G_{\Phi,1,1+i}(f)-r|\mu_{\Phi}(f)|^2\right]\right] \quad (10.50)$$

$$G_X(\infty,f) = G_{\Phi}(f) - r|\mu_{\Phi}(f)|^2 + r^2|\mu_{\Phi}(f)|^2 \sum_{n=-\infty}^{\infty}\delta(f-nr)$$
$$+2\sum_{i=1}^{m}\mathrm{Re}\left[e^{j2\pi if/r}\left[G_{\Phi,1,1+i}(f)-r|\mu_{\Phi}(f)|^2\right]\right] \quad (10.51)$$

**Proof**
The proof is detailed in Appendix 10.F.

***10.4.4.2 Notes*** The results stated in this theorem for the independent, countable, and infinite interval case are consistent with those of van den Elzen (1970) and Howard (2002, ch. 5).

### 10.4.5 Information Encoded in Pulse Amplitudes

In communication signalling, it is common to encode information in the amplitudes of the signalling waveforms that have a set form. For this case, the signalling set is

$$S_{\Phi} = \begin{cases} \{\phi(\omega_i,t) = A(\omega_i)p(t): \quad \omega_i \in \{1,\ldots,N_{\Phi}\}\} & \text{finite case} \\ \{\phi(\omega_i,t) = A(\omega_i)p(t): \quad \omega_i \in \boldsymbol{Z}^{+}\} & \text{countable case} \\ \{\phi(\omega_i,t) = A(\omega_i)p(t): \quad \omega_i \in \boldsymbol{R}\} & \text{uncountable case} \end{cases} \quad (10.52)$$

where $p:\boldsymbol{R}\to\boldsymbol{R}$ is a suitable signalling waveform that is assumed to be integrable with a Fourier transform on the infinite interval denoted $P$.

The associated signalling random process $X:S\to S_X$ on the interval $[0,ND]$ is defined according to

$$X(\Omega,t) = \sum_{k=1}^{N} A(\Omega_k)p(t-(k-1)D) \quad (10.53)$$

where $\Omega = (\Omega_1, \ldots, \Omega_N)$ is a vector of index random variables, with outcomes $(\omega_1, \ldots, \omega_N) \in S$, that govern the waveforms from the signalling set that are associated with each of the signalling intervals. The index random variable $\Omega_i$ defines a sample space $S_i$ with outcomes according to a probability mass function $p_{\Omega_i}$ for the countable case and a probability density function $f_{\Omega_i}$ for the uncountable case.

The mean and mean squared amplitudes for the $i$th signalling interval are

$$\mu_{A,i} = \begin{cases} \sum\limits_{\omega_i \in S_i} p_{\Omega_i}(\omega_i) A(\omega_i) & \text{countable case} \\ \int\limits_{-\infty}^{\infty} A(\omega_i) f_{\Omega_i}(\omega_i) d\omega_i & \text{uncountable case} \end{cases} \tag{10.54}$$

$$\overline{|A_i|^2} = \begin{cases} \sum\limits_{\omega_i \in S_i} p_{\Omega_i}(\omega_i) |A(\omega_i)|^2 & \text{countable case} \\ \int\limits_{-\infty}^{\infty} |A(\omega_i)|^2 f_{\Omega_i}(\omega_i) d\omega_i & \text{uncountable case} \end{cases} \tag{10.55}$$

The correlation between the amplitudes in the $i$th and $k$th intervals is

$$\boldsymbol{E}[A_i A_k^*] = \begin{cases} \sum\limits_{\omega_i \in S_i} \sum\limits_{\omega_k \in S_k} p_{\Omega_i \Omega_k}(\omega_i, \omega_k) A(\omega_i) A^*(\omega_k) & \text{c.c.} \\ \int\limits_{-\infty}^{\infty} \int\limits_{-\infty}^{\infty} A(\omega_i) A^*(\omega_k) f_{\Omega_i \Omega_k}(\omega_i, \omega_k) d\omega_i d\omega_k & \text{u.c.} \end{cases} \tag{10.56}$$

***10.4.5.1 Power Spectral Density*** The power spectral density of a signalling random process with information encoded in the signalling waveform amplitude is specified in the following two theorems.

**Theorem 10.8 PSD: Information Encoded in Pulse Amplitudes**
The power spectral density of a signalling random process, with information encoded in the pulse amplitudes, is given by

$$G_X(T,f) = \frac{r|P(f)|^2}{N} \sum_{i=1}^{N} \overline{|A_i|^2} + \frac{r|P(f)|^2}{N} \sum_{i=1}^{N} \sum_{k=1, k \neq i}^{N} e^{-j2\pi f(i-k)D} \boldsymbol{E}[A_i A_k^*] \tag{10.57}$$

where $T = ND$ and $r = 1/D$. For the case where the signalling waveforms in different signalling intervals are independent, the power spectral density is

$$G_X(T,f) = \frac{r|P(f)|^2}{N} \sum_{i=1}^{N} \overline{|A_i|^2} + \frac{r|P(f)|^2}{N} \sum_{i=1}^{N} \sum_{k=1, k \neq i}^{N} \mu_{A,i} \mu_{A,k}^* e^{-j2\pi f(i-k)D} \tag{10.58}$$

For the independent and zero-mean case, that is, $\mu_{A,i}=0$ for $i=1,\ldots,N$, it follows that

$$G_X(T,f)=\frac{r|P(f)|^2}{N}\sum_{i=1}^{N}\sigma_{A_i}^2 \tag{10.59}$$

where $\sigma_{A_i}^2$ is the variance of the amplitude associated with the $i$th signalling interval.

For the independent and identical case consistent with $\overline{|A_i|^2}=\overline{|A_k|^2}=\overline{|A|^2}$ and $\mu_{A,i}=\mu_{A,k}=\mu_A$ for $i,k\in\{1,\ldots,N\}$, it follows that

$$G_X(T,f)=r|P(f)|^2\left[\overline{|A|^2}-|\mu_A|^2+|\mu_A|^2\cdot\frac{1}{N}\cdot\frac{\sin(\pi Nf/r)^2}{\sin(\pi f/r)^2}\right] \tag{10.60}$$

$$G_X(\infty,f)=r|P(f)|^2\left[\overline{|A|^2}-|\mu_A|^2+r|\mu_A|^2\sum_{n=-\infty}^{\infty}\delta(f-nr)\right] \tag{10.61}$$

For the independent and identical case and when the mean is zero, that is, $\mu_A=0$, the following simple result holds:

$$G_X(T,f)=r\sigma_A^2|P(f)|^2 \tag{10.62}$$

**Proof**

The proof is detailed in Appendix 10.G.

**Theorem 10.9 Dependency on Difference between Signalling Intervals**

For the case where the dependency between signalling waveforms depends on the difference between the location of the signalling intervals and not on their absolute location, the power spectral density is

$$G_X(ND,f)=r|P(f)|^2\left[\frac{1}{N}\sum_{i=1}^{N}\overline{|A_i|^2}+2\sum_{i=1}^{N-1}\left[1-\frac{i}{N}\right]\mathrm{Re}\left[e^{j2\pi if/r}\boldsymbol{E}[A_1A_{1+i}^*]\right]\right] \tag{10.63}$$

For the case where the $i$th signalling waveform is independent of the $j$th signalling waveform for $|i-j|>m$, the power spectral density is

$$\begin{aligned}G_X(ND,f)= r|P(f)|^2&\left[\frac{1}{N}\sum_{i=1}^{N}|A_i|^2+2\sum_{i=1}^{N-1}\left[1-\frac{i}{N}\right]\mathrm{Re}\left[e^{j2\pi if/r}\mu_{A,1}\mu_{A,1+i}^*\right]\right.\\&\left.+2\sum_{i=1}^{m}\left[1-\frac{i}{N}\right]\mathrm{Re}\left[e^{j2\pi iDf}\left[\boldsymbol{E}[A_1A_{1+i}^*]-\mu_{A,1}\mu_{A,1+i}^*\right]\right]\right]\end{aligned} \tag{10.64}$$

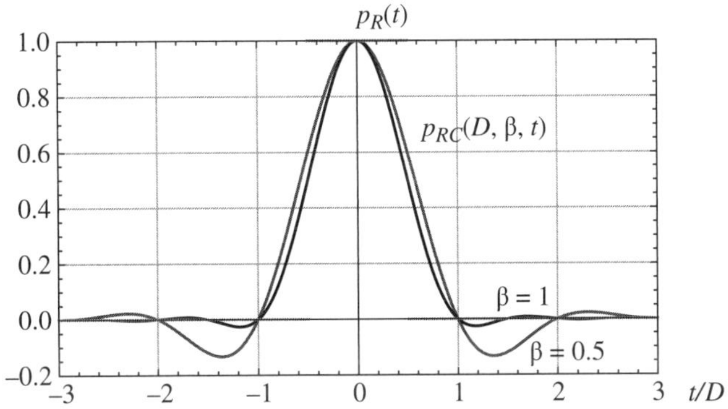


**FIGURE 10.4** Graph of a rectangular function and a raised cosine pulse.

For the case of statistically identical signalling waveforms on each interval, consistent with $\overline{|A_i|^2} = \overline{|A_k|^2} = \overline{|A|^2}$ and $\mu_{A,i} = \mu_{A,k} = \mu_A$ for $i, k \in \{1, \ldots, N\}$, it follows that

$$G_X(ND, f) = r|P(f)|^2 \left[ \overline{|A|^2} - |\mu_A|^2 + |\mu_A|^2 \cdot \frac{1}{N} \cdot \frac{\sin(\pi N f/r)^2}{\sin(\pi f/r)^2} + 2\sum_{i=1}^{m} \left[1 - \frac{i}{N}\right] \mathrm{Re}\left[e^{j2\pi i f/r}\left[\boldsymbol{E}[A_1 A_{1+i}^*] - |\mu_A|^2\right]\right]\right] \tag{10.65}$$

$$G_X(ND, f) = r|P(f)|^2 \left[ \overline{|A|^2} - |\mu_A|^2 + r|\mu_A|^2 \cdot \sum_{n=-\infty}^{\infty} \delta(f - nr) + 2\sum_{i=1}^{m} \mathrm{Re}\left[e^{j2\pi i f/r}\left[\boldsymbol{E}[A_1 A_{1+i}^*] - |\mu_A|^2\right]\right]\right] \tag{10.66}$$

**Proof**

These results follow directly from Theorem 10.7 and the results noted in Appendix 10.G of $\mu_{\Phi,i}(f) = \mu_{A,i} P(f)$, $G_{\Phi,i}(f) = r|P(f)|^2 \overline{|A_i|^2}$, and $G_{\Phi,ik}(f) = r|P(f)|^2 \boldsymbol{E}[A_i A_k^*]$.

***10.4.5.2 Pulse Shape and Spectral Efficiency*** It is clear from Theorems 10.8 and 10.9 that the power spectral density associated with communication of information via encoding of information in a pulse amplitude depends on the nature of the pulse used. In Figure 10.4, a rectangular pulse

$$p_R(t) = \begin{cases} 1 & -D/2 \le t < D/2 \\ 0 & \text{elsewhere} \end{cases} \Leftrightarrow P_R(f) = D\,\mathrm{sinc}(fD) \tag{10.67}$$

and a raised cosine pulse defined according to

$$p_{RC}(D,\beta,t) = \mathrm{sinc}\left(\frac{t}{D}\right) \cdot \frac{\cos(\pi\beta t/D)}{1-(2\beta t/D)^2} \qquad 0<\beta\leq 1 \tag{10.68}$$

are shown. Their respective Fourier transforms are shown in Figure 10.5. The raised cosine pulse is the inverse Fourier transform of a raised cosine spectrum defined according to (Carlson, 1986, p. 406; Proakis and Salehi, 2008, p. 607):

$$P_{RC}(r,\beta,f) = \begin{cases} \frac{1}{r} & |f| < \frac{r}{2}(1-\beta) \\ \frac{1}{r}\cos\left[\frac{\pi}{2}\cdot\frac{|f|-r(1-\beta)/2}{\beta r}\right]^2 & \frac{r}{2}(1-\beta) \leq |f| < \frac{r}{2}(1+\beta) \\ 0 & |f| \geq \frac{r}{2}(1+\beta) \end{cases} \tag{10.69}$$

Here, the parameter $\beta$ is the roll-off factor with the constraint $0<\beta\leq 1$, and $r = 1/D$ is the signalling rate.

Figure 10.5 clearly shows the spectral inefficiency with communication of binary information by a rectangular pulse and, in comparison, the potential spectral efficiency of signalling using a raised cosine pulse.

### 10.4.6 Examples

The following two subsections use the results established previously for a signalling random process to establish the power spectral density associated, first, with nonreturn to zero (NRZ) signalling and, second, with coded mark inversion (CMI) signalling. These signalling protocols are illustrated in Figure 10.6 for the case of signalling with rectangular pulses.

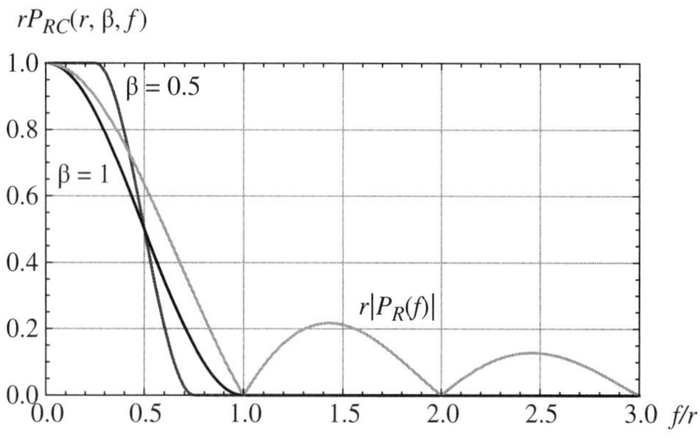


**FIGURE 10.5** Fourier transform of a rectangular pulse and a raised cosine pulse.

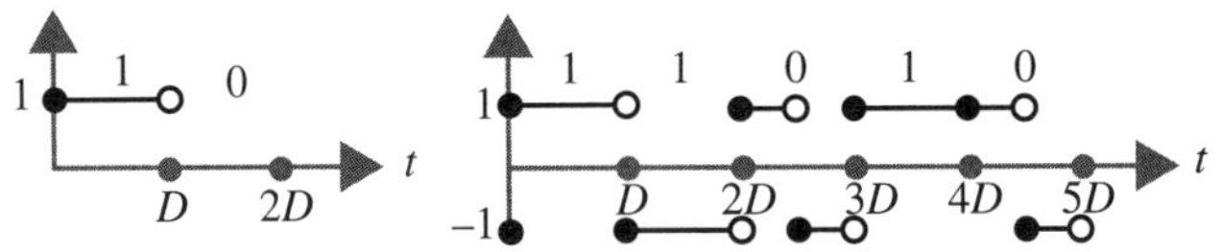


**FIGURE 10.6** Illustration of NRZ (left) and CMI (right) signalling for the case of rectangular pulses. With CMI signalling, a logic one is alternatively represented by positive and negative amplitudes.

***10.4.6.1 NRZ Signalling*** Consider the case of a signalling random process $X: S \to S_X$ defined, for the interval $[0, ND]$, according to

$$X(\Omega, t) = \sum_{k=1}^{N} A(\Omega_k) p(t-(k-1)D) \tag{10.70}$$

where $\Omega = (\Omega_1, \ldots, \Omega_N)$ is a vector of identical and independent random variables with $A(\Omega_k) = \Omega_k A_o$ for $k \in \{1, \ldots, N\}$ and where the sample space associated with $\Omega_k$ is $\{0, 1\}$. Equally probable amplitudes are assumed consistent with $P[\Omega_k = 0] = P[\Omega_k = 1] = 1/2$. The signalling set is $S_\Phi = \{0, A_o p(t)\}$ and $\overline{|A|^2} = A_o^2/2$, $\mu_A = A_o/2$. From Theorem 10.8, it follows that the power spectral density of $X$ is

$$G_X(T, f) = \frac{rA_o^2|P(f)|^2}{4}\left[1 + \frac{1}{N} \cdot \frac{\sin(\pi Nf/r)^2}{\sin(\pi f/r)^2}\right] \tag{10.71}$$

$$G_X(\infty, f) = \frac{rA_o^2|P(f)|^2}{4}\left[1 + r\sum_{n=-\infty}^{\infty} \delta(f - nr)\right] \tag{10.72}$$

The power spectral density is shown in Figure 10.7 for the cases of

$$p(t) = p_R(t - D/2) \Leftrightarrow P(f) = D\operatorname{sinc}(fD)e^{-j\pi fD} \tag{10.73}$$

$$p(t) = p_{RC}(D, 1, t - D/2) \Leftrightarrow P(f) = P_{RC}(r, 1, f)e^{-j\pi f/r} \tag{10.74}$$

$$p(t) = \left\{\begin{matrix} \sin(2\pi f_C t) & 0 \le t < D \\ 0 & \text{elsewhere} \end{matrix}\right\} \Leftrightarrow$$
$$P(f) = \frac{jDe^{-j\pi fD}}{2}\left[-e^{j\pi f_C D}\operatorname{sinc}((f - f_C)D) + e^{-j\pi f_C D}\operatorname{sinc}((f + f_C)D)\right] \tag{10.75}$$

Signalling with the first pulse waveform is termed *nonreturn to zero* signalling and is illustrated in Figure 10.6. The second case is signalling with the spectrally efficient raised cosine pulse defined in Equations 10.68 and 10.69. The last case is amplitude shift keyed modulation (on–off keying), and it is assumed that $f_C = kr$ for $k \in \{1, 2, \ldots\}$ consistent with an integer multiple of sinusoidal periods in each signalling interval.

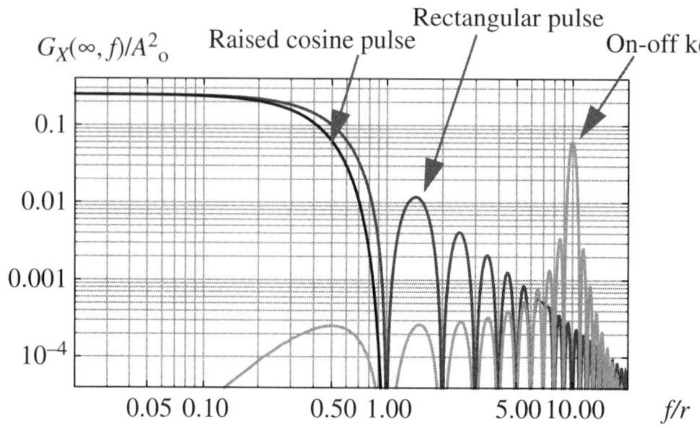


**FIGURE 10.7** Power spectral densities as defined by Equations 10.76, 10.77 ($\beta = 1$) and 10.78 ($f_C = 10r$). There are impulses of magnitude $A_o^2/4$ at $f = 0$ for the baseband power spectral densities and of magnitude $A_o^2/16$ at $f = \pm f_C$ for the on–off keying case.

For this case, $e^{\pm j\pi f_C D} = -1^k$. The respective power spectral densities for these three cases, and the infinite interval, are

$$G_X(\infty,f) = \frac{A_o^2 \text{sinc}^2(f/r)}{4r} + \frac{A_o^2 \delta(f)}{4} \tag{10.76}$$

$$G_X(\infty,f) = \frac{rA_o^2 |P_{RC}(r,\beta,f)|^2}{4} + \frac{A_o^2}{4}\delta(f) \tag{10.77}$$

$$G_X(\infty,f) = \frac{A_o^2}{16r}\cdot\left[\text{sinc}^2\left[\frac{f-f_C}{r}\right] + \text{sinc}^2\left[\frac{f+f_C}{r}\right]\right.$$
$$\left. -2\cos\left[\frac{2\pi f_C}{r}\right]\text{sinc}\left[\frac{f-f_C}{r}\right]\text{sinc}\left[\frac{f+f_C}{r}\right]\right] + \frac{A_o^2}{16}[\delta(f-f_C)+\delta(f-f_C)] \tag{10.78}$$

For the first and third cases, the sinc function is zero, except, respectively, at $f = 0, \pm f_C$ resulting in all the impulses, apart from those at $f = 0, \pm f_C$ being zero. For the second case, $P_{RC}(r,\beta,f) = 0$ for $|f| > r$, which results in the one nonzero impulse at $f = 0$. These power spectral densities are shown in Figure 10.7.

***10.4.6.2 Example: CMI Signalling*** With CMI signalling, binary data is encoded, consistent with the illustration in Figure 10.6, with three waveforms, and the signalling set comprises of three signals:

$$S_\Phi = \{\phi(-1,t) = -A_o p_1(t), \phi(0,t) = A_o p_0(t), \phi(1,t) = A_o p_1(t)\} \tag{10.79}$$

where $p_0(t)$ and $p_1(t)$ are the normalized signalling waveforms, respectively, for a logic zero and a logic one. The encoding is such that alternative logic ones are a pulse and its negative. This encoding leads to correlation between adjacent signalling

**TABLE 10.1 Possible outcomes, and probabilities, for two signalling intervals**

| Data | Signalling waveforms | Probabilities: $p_{\Omega_1\Omega_2}(\omega_1,\omega_2)$ | Probabilities: $p_{\Omega_1\Omega_i}(\omega_1,\omega_i), i\geq 3$ |
|---|---|---|---|
| 00 | $\phi(0,t), \phi(0,t)$ | $p_{\Omega_1\Omega_2}(0,0)=0.25$ | $p_{\Omega_1\Omega_i}(0,0)=0.25$ |
| 01 | $\phi(0,t),\phi(-1,t)$ | $p_{\Omega_1\Omega_2}(0,-1)=0.125$ | $p_{\Omega_1\Omega_i}(0,-1)=0.125$ |
| | or $\phi(0,t), \phi(1,t)$ | $p_{\Omega_1\Omega_2}(0,1)=0.125$ | $p_{\Omega_1\Omega_i}(0,1)=0.125$ |
| 10 | $\phi(-1,t),\phi(0,t)$ | $p_{\Omega_1\Omega_2}(-1,0)=0.125$ | $p_{\Omega_1\Omega_i}(-1,0)=0.125$ |
| | or $\phi(1,t), \phi(0,t)$ | $p_{\Omega_1\Omega_2}(1,0)=0.125$ | $p_{\Omega_1\Omega_i}(1,0)=0.125$ |
| 11 | $\phi(-1,t),\phi(1,t)$ | $p_{\Omega_1\Omega_2}(-1,1)=0.125$ | $p_{\Omega_1\Omega_i}(-1,1)=0.0625$ |
| | or $\phi(1,t), \phi(-1,t)$ | $p_{\Omega_1\Omega_2}(1,-1)=0.125$ | $p_{\Omega_1\Omega_i}(1,-1)=0.0625$ |
| | or $\phi(-1,t), \phi(-1,t)$ | $p_{\Omega_1\Omega_2}(-1,-1)=0$ | $p_{\Omega_1\Omega_i}(-1,-1)=0.0625$ |
| | or $\phi(1,t), \phi(1,t)$ | $p_{\Omega_1\Omega_2}(1,1)=0$ | $p_{\Omega_1\Omega_i}(1,1)=0.0625$ |

intervals. The probability mass function governing the signalling waveforms is detailed in Table 10.1 assuming equally probable binary data consistent with $p_{\Omega_k}(-1)=p_{\Omega_k}(1)=0.25$, $p_{\Omega_k}(0)=0.5$.

From Theorem 10.7 (Eq. 10.50), the power spectral density is given by

$$G_X(ND,f)= G_\Phi(f)-r|\mu_\Phi(f)|^2+r|\mu_\Phi(f)|^2\cdot\frac{1}{N}\cdot\frac{\sin(\pi Nf/r)^2}{\sin(\pi f/r)^2}$$
$$+2\sum_{i=1}^{m}\left[1-\frac{i}{N}\right]\text{Re}\left[e^{j2\pi if/r}\left[G_{\Phi,1,1+i}(f)-r|\mu_\Phi(f)|^2\right]\right] \tag{10.80}$$

where, as $\phi(-1,t)=-\phi(1,t)$,

$$\mu_\Phi(f)=\sum_{\omega_i=-1}^{1}p_{\Omega_i}(\omega_i)\Phi(\omega_i,f)=\frac{\Phi(0,f)}{2} \tag{10.81}$$

$$G_\Phi(f)=\frac{1}{D}\sum_{\omega_i=-1}^{1}p_{\Omega_i}(\omega_i)|\Phi(\omega_i,f)|^2=\frac{r}{2}\left[|\Phi(0,f)|^2+|\Phi(1,f)|^2\right] \tag{10.82}$$

For the case of $i=1$,

$$G_{\Phi,1,1+i}(f)|_{i=1}=\frac{1}{D}\sum_{\omega_1=-1}^{1}\sum_{\omega_i=-1}^{1}p_{\Omega_1\Omega_{1+i}}(\omega_1,\omega_{1+i})\Phi(\omega_1,f)\Phi^*(\omega_{1+i},f)$$
$$=\frac{r}{8}[0+\Phi(-1,f)\Phi^*(0,f)+\Phi(-1,f)\Phi^*(1,f)+\Phi(0,f)\Phi^*(-1,f)$$
$$+2|\Phi(0,f)|^2+\Phi(0,f)\Phi^*(1,f)+\Phi(1,f)\Phi^*(-1,f)$$
$$+\Phi(1,f)\Phi^*(0,f)+0]\;=\frac{r}{4}\left[|\Phi(0,f)|^2-|\Phi(1,f)|^2\right] \tag{10.83}$$

For $i \geq 2$, it is the case that $G_{\Phi,1,1+i}(f) = r|\mu_\Phi(f)|^2$, and this is consistent with the upper limit of the summation in Equation 10.80 being $m = 1$. Hence,

$$G_X(ND,f) = \frac{r}{4}\left[|\Phi(0,f)|^2 + 2|\Phi(1,f)|^2\right] + \frac{r|\Phi(0,f)|^2}{4} \cdot \frac{1}{N} \cdot \frac{\sin(\pi Nf/r)^2}{\sin(\pi f/r)^2} - \frac{r}{2}\left[1 - \frac{1}{N}\right] \mathrm{Re}\left[e^{j2\pi f/r}|\Phi(1,f)|^2\right] \quad (10.84)$$

With raised cosine signalling consistent with

$$\phi(0,t) = A_o\left[p_{RC}\left(\frac{D}{2},\beta,t-D/4\right) - p_{RC}\left(\frac{D}{2},\beta,t-3D/4\right)\right] \Leftrightarrow \Phi(0,f) = A_o\left[P_{RC}(2r,\beta,f)e^{-j\pi f/2r} - P_{RC}(2r,\beta,f)e^{-j3\pi f/2r}\right] \quad (10.85)$$

$$\phi(1,t) = A_o p_{RC}(D,\beta,t-D/2) \Leftrightarrow \Phi(1,f) = A_o P_{RC}(r,\beta,f)e^{-j\pi f/r} \quad (10.86)$$

$$\phi(-1,t) = -\phi(1,t) \Leftrightarrow \Phi(-1,f) = -A_o P_{RC}(r,\beta,f)e^{-j\pi f/r} \quad (10.87)$$

it follows that the power spectral density is

$$G_X(ND,f) = \frac{rA_o^2}{2}\left[P_{RC}^2(2r,\beta,f) - P_{RC}^2(2r,\beta,f)\cos(\pi f/r) + P_{RC}^2(r,\beta,f)\right] + \frac{rA_o^2}{2}\left[P_{RC}^2(2r,\beta,f) - P_{RC}^2(2r,\beta,f)\cos(\pi f/r)\right] \cdot \frac{1}{N} \cdot \frac{\sin(\pi fN/r)^2}{\sin(\pi f/r)^2} - \frac{rA_o^2}{2}\left[1 - \frac{1}{N}\right] P_{RC}^2(r,\beta,f)\cos(2\pi f/r) \quad (10.88)$$

One signal from a CMI signalling random process with such signalling waveforms is shown in Figure 10.8 for the case of $\beta = 1$. The power spectral density is plotted in Figure 10.9 for the case of $N \to \infty$ ($N = 10000$). The power in the impulsive component at $f = r$, from Equation 10.51, is given by

$$r^2|\mu_\Phi(r)|^2 = r^2|\Phi(0,r)|^2/4 = r^2 A_o^2 P_{RC}^2(2r,\beta,r) = A_o^2/16 \quad (10.89)$$

for the cases of $\beta = 1/2$ and $\beta = 1$.

***10.4.6.3 Issues for Communication Signalling*** Two important issues for communication signalling are, first, spectral efficiency and, second, transitions in the time domain signal to ensure synchronization and signal recovery at the communication receiver. Clearly, from the baseband power spectral densities shown in Figures 10.7 and 10.8, the use of raised cosine pulses greatly facilitates spectral efficiency. As is evident from Figures 10.6, 10.7, and10.8, signalling using the CMI protocol results in

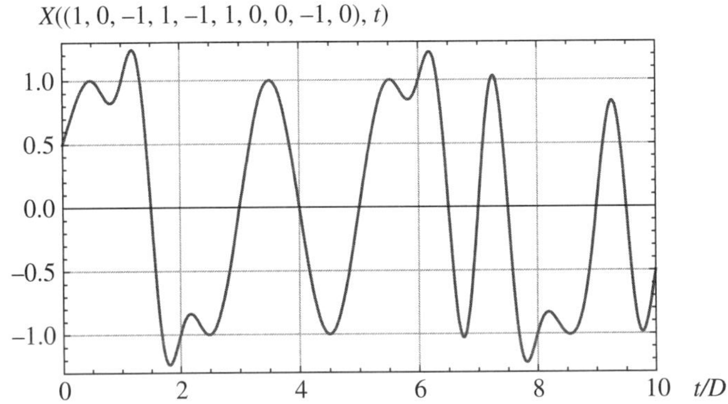


**FIGURE 10.8** One signal from a signalling random process with CMI encoding of equally probable binary data and with raised cosine signalling pulses with unit amplitude.

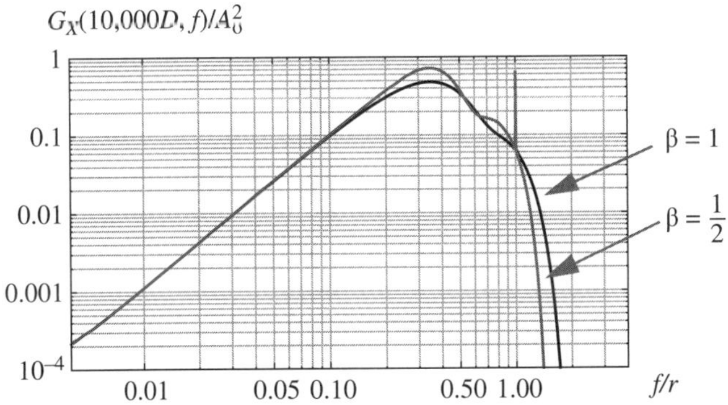


**FIGURE 10.9** Power spectral density of a signalling random process using pulses associated with a raised cosine spectrum and CMI coding. The power in the impulsive component at $f = r$ is $A_o^2/16$ for the case of $\beta = 1/2$ and $\beta = 1$.

guaranteed zero crossings, and this facilitates synchronization at the receiver. Finally, the encoding with CMI signalling ensures that the power spectral density is zero at zero frequency that ensures that such a signal can be passed by a communication system whose transfer function has zero response at DC.

## 10.5 DIGITAL TO ANALOGUE CONVERSION

With the rise of digital information processing technology, signal generation for physical applications is usually implemented by generating accurate signal sample values and outputting these via a $M$-bit digital-to-analogue converter (DAC) at a constant rate of $r = 1/D$ samples per second. An ideal $M$-bit DAC has $2^M$ equally spaced levels

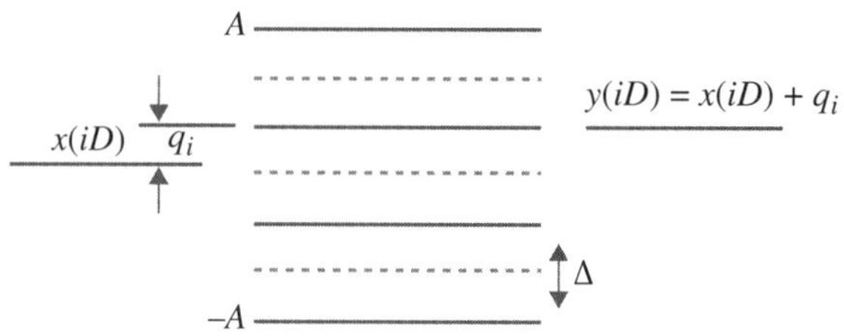


**FIGURE 10.10** Illustration of quantization error with a 2-bit DAC (4 levels).

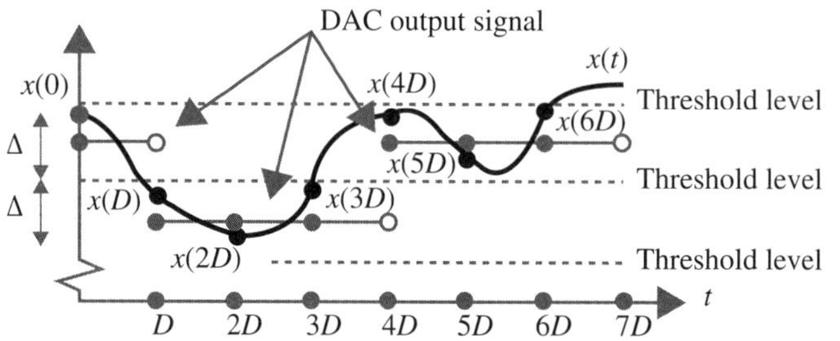


**FIGURE 10.11** Illustration of signal, signal samples, and ideal DAC output signal.

between, and including, $\pm A$, and the output resolution is $\Delta = 2A/\left(2^M - 1\right)$. The finite output resolution results in signal quantization, and the generated signal can be modelled as the desired signal plus a time-varying error signal.

For the *ith* interval $[iD, (i+1)D]$, the signal sample, ideally, is $x(iD)$, but the quantization inherent in the DAC converter results in an output level of $y(iD) = x(iD) + q_i$, as illustrated in Figure 10.10, where $q_i$ is the quantization error. The ideal signal, the signal samples, and the ideal DAC output signal are illustrated in Figure 10.11.

For a specific signal $x: \boldsymbol{R} \to \boldsymbol{R}$, the DAC output signal, on the interval $[0, ND]$, is

$$\begin{aligned} y(t) &= \sum_{i=1}^{N} [x((i-1)D) + q_i] p(t-(i-1)D) \\ &= \sum_{i=1}^{N} x((i-1)D) p(t-(i-1)D) + \sum_{i=1}^{N} q_i p(t-(i-1)D) \\ &= x_S(t) + q(t) \end{aligned} \tag{10.90}$$

Here, $x_S(t)$ is a step approximation to the signal $x(t)$, $q(t)$ is the quantization error signal, and the pulse function $p(t)$ is defined according to

$$p(t) = \begin{cases} 1 & 0 \le t < D \\ 0 & \text{elsewhere} \end{cases} \Leftrightarrow P(f) = D\text{sinc}(fD)e^{-j\pi fD} \tag{10.91}$$

For a set signal, and a given DAC operating at set conditions, the DAC output signal, ideally, is deterministic. However, for a set signal, and an ensemble of DACs, the variations in the DACs will lead to an ensemble of different quantization signals and, hence, output signals. For a given ensemble of DACs and the experiment of choosing a DAC at random and then specifying the quantization errors in the interval $[0, ND]$, the experimental sample space is

$$S = \{(q_1,\ldots,q_N): \quad q_i \in [-\Delta/2, \Delta/2]\} \tag{10.92}$$

where $q_i$ is the outcome of a random variable $Q_i$ with a probability density function

$$f_Q(q) = \begin{cases} 0 & \text{elsewhere} \\ 1/\Delta & -\Delta/2 \le q < \Delta/2 \end{cases} \tag{10.93}$$

The random process $Y: S \to S_Y$ defined by the possible DAC output signals is

$$Y(\Omega, t) = x_S(t) + \sum_{i=1}^{N} Q_i p(t-(i-1)D) \tag{10.94}$$

where $\Omega = (Q_1,\ldots,Q_N)$ is a vector of random variables, with outcomes $(q_1,\ldots,q_N) \in S$ and with a joint probability density function $f_{Q_1\ldots Q_N}(q_1,\ldots,q_N)$. The signal sample space is

$$S_Y = \left\{ Y((q_1,\ldots,q_N),t) = x_S(t) + \sum_{i=1}^{N} q_i p(t-(i-1)D): \quad (q_1,\ldots,q_N) \in S \right\} \tag{10.95}$$

The usual situation is where the change in the signal from one sample to the next is such that the quantization errors are uncorrelated, and independent of the signal, such that $f_{Q_1\ldots Q_N}(q_1,\ldots,q_N) = f_{Q_1}(q_1)\cdot\ldots\cdot f_{Q_N}(q_N)$. For this case, the random process is the sum of a signal and an independent zero-mean random process, $Q$, as defined by the quantization signals. For such a case and consistent with Theorem 10.4, the power spectral densities add

$$G_Y(T,f) = G_{X_S}(T,f) + G_Q(T,f) \tag{10.96}$$

The quantization random process $Q: S \to S_Q$ with the signal sample space

$$S_Q = \left\{ Q((q_1,\ldots,q_N),t) = \sum_{i=1}^{N} q_i p(t-(i-1)D): \quad (q_1,\ldots,q_N) \in S \right\} \tag{10.97}$$

is that of a signalling random process with information encoded in the pulse amplitudes. Accordingly, its power spectral density is given by Theorem 10.8 and specifically Equation 10.62, that is,

$$G_Q(ND,f) = \frac{r\Delta^2|P(f)|^2}{12} = \frac{\Delta^2 \text{sinc}^2(f/r)}{12r} = \frac{(2A)^2 \text{sinc}^2(f/r)}{12(2^M-1)^2 r} \tag{10.98}$$

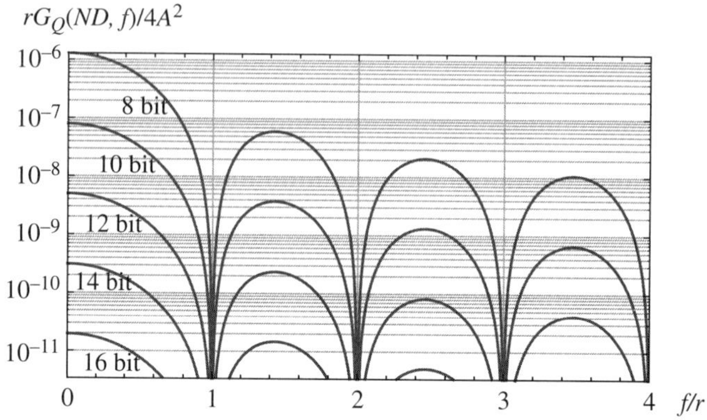


**FIGURE 10.12** Normalized power spectral density of the quantization noise random process arising from a DAC due to quantization. Upper to lower curves, respectively, are for 8-, 10-, 12-, 14-, and 16-bit DACs.

Here, the result that the variance of a uniform random variable on the interval $[-\Delta/2, \Delta/2)$ is $\Delta^2/12$ has been used.

This power spectral density, normalized by the DAC range of $2A$ and the output rate $r$, is shown in Figure 10.12 for DACs with 8, 10, 12, 14, and 16 bits.

### 10.5.1 Notes

Consider a fixed DAC range. It is clear from Equation 10.98 that the power spectral density due to quantization is inversely proportional to the number of levels and inversely proportional to the sample rate $r$.

Of interest is the signal-to-noise ratio achievable with a DAC. To this end, consider a DAC range of $[-A, A]$ that is optimally used such that the signal power is of the order of $A^2/2$. This power level is consistent with the power in a sinusoid with an amplitude $A$. The DAC quantization noise power, as given by the integral of the DAC quantization power spectral density, is

$$\frac{(2A)^2}{12\left(2^M-1\right)^2} \tag{10.99}$$

which arises as the integral of $\text{sinc}^2(f/r)$ on $(-\infty, \infty)$ equals $r$. The signal-to-noise ratio achievable then is

$$\text{snr}(M) = \frac{A^2}{2} \cdot \frac{12\left(2^M-1\right)^2}{(2A)^2} = 1.5\left(2^M-1\right)^2 \approx 1.5\left(2^{2M}\right) \tag{10.100}$$

$$\text{snr}_{dB}(M) = 10\log[\text{snr}(M)] \approx 1.76 + 6.05M \quad dB \tag{10.101}$$

For 8-, 10-, 12-, 14-, and 16-bit DACs, the respective achievable signal-to-noise ratios are 50, 62, 74, 86, and 98 dB.

## 10.6 PSD OF SHOT NOISE RANDOM PROCESSES

Consider a generalized shot noise random process defined according to

$$X(\Omega, t) = \sum_{k=1}^{\infty} A_k p(t - \Gamma_k) \tag{10.102}$$

where $\Omega = ((A_1, \Gamma_1), \ldots, (A_k, \Gamma_k), \ldots)$ with $\Gamma_1, \Gamma_2, \ldots$ being a sequence of unordered random variables consistent with the outcomes of a Poisson experiment with a rate of $\lambda$ per second and $(A_1, A_2, \ldots)$ is a vector of random variables defining the pulse amplitudes. For the case of $A_1 = A_2 = \cdots = 1$, a shot noise random process arises.

### 10.6.1 Power Spectral Density

For the interval $[0, T]$, where $T$ is sufficiently large such that the number of pulses arising can be approximated by $N = \lambda T$, $\Gamma_i$ is approximately independent of $\Gamma_k$, and the energy of the pulses outside the interval $[0, T]$ is negligible, the following power spectral density results can be proved.

**Theorem 10.10 Power Spectral Density: Shot Noise Random Process**
The power spectral density of a shot noise process is

$$G_X(T, f) \approx \lambda |P(f)|^2 + \lambda^2 |P(f)|^2 \left(1 - \frac{1}{\lambda \mathrm{T}}\right) T \operatorname{sinc}(fT)^2 \tag{10.103}$$

$$G_{X_\infty}(f) = \lambda |P(f)|^2 + \lambda^2 |P(0)|^2 \delta(f) \tag{10.104}$$

Here, $P(f)$ is the Fourier transform of $p(t)$.

**Proof**
The proof is detailed in Appendix 10.H.

**Theorem 10.11 Power Spectral Density: Generalized Shot Noise Process**
For the case of independent and identical amplitude random variables, the power spectral density of a generalized shot noise process is

$$G_X(T, f) = \lambda |P(f)|^2 \left[\sigma_A^2 + \mu_A^2 + \lambda \mu_A^2 \left(1 - \frac{1}{N}\right) T \operatorname{sinc}(fT)^2\right] \tag{10.105}$$

$$G_{X_\infty}(f) = \lambda \left[\sigma_A^2 + \mu_A^2\right] |P(f)|^2 + \lambda^2 \mu_A^2 |P(0)|^2 \delta(f) \tag{10.106}$$

where $\mu_A$ and $\sigma_A^2$ are, respectively, the mean and variance of $A_k$, $k \in \{1, \ldots, N\}$.

**Proof**
The proof follows in a similar manner to the proof of Theorem 10.10.

***10.6.1.1 Example*** Consider two shot noise processes: the first based on the pulse function

$$p(i,t)=\frac{t^{i-1}}{\tau^i(i-1)!}\cdot e^{-t/\tau}\cdot u(t)\Leftrightarrow P(i,f)=\frac{1}{(1+j2\pi f\tau)^i} \tag{10.107}$$

for $i\in\mathbf{Z}^+$ and the second based on the Gaussian pulse

$$p(t)=\frac{e^{-t^2/2\sigma_t^2}}{\sqrt{2\pi}\sigma_t}\Leftrightarrow P(f)=e^{-2\pi^2 f^2\sigma_t^2} \tag{10.108}$$

For the case of $\lambda=1$, $i=2$, $\tau=0.1$ and $\sigma_t=0.1$, the time domain signals of the two random processes are shown in Figure 10.13. The power spectral density for the interval [0, 10] is shown in Figure 10.14. The impulsive term in the power spectral density—the second term in Equation 10.103—is clearly evident in the power spectral density in the range $f<0.1$.

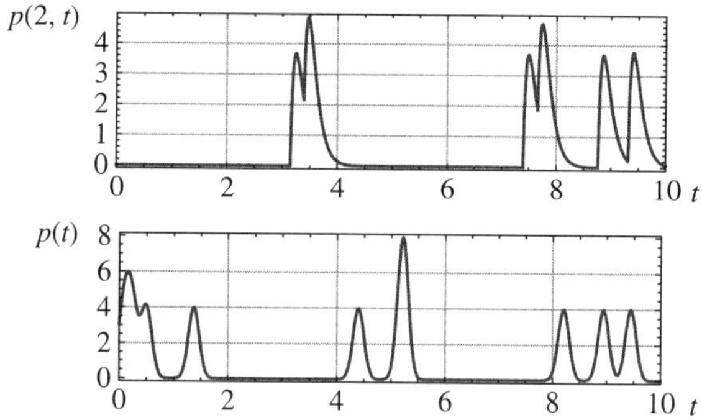


**FIGURE 10.13** Signals from a shot noise process. Upper: exponential decaying pulse. Lower: Gaussian pulse.

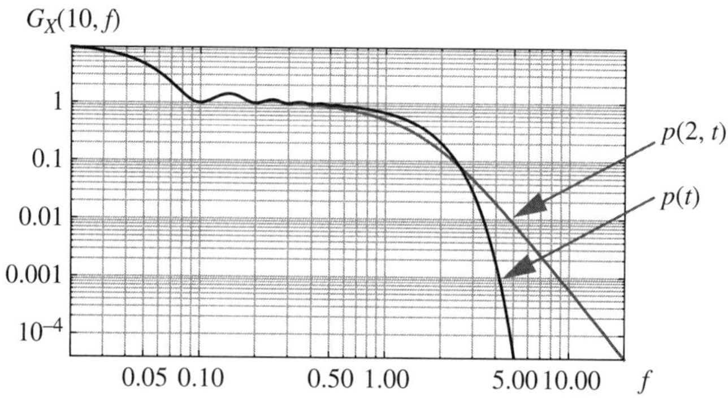


**FIGURE 10.14** Power spectral density of a shot noise process, with a pulse rate of $\lambda=1$, for the interval [0, 10].

### 10.6.2 Shot Noise due to Electrons Crossing a Barrier

Consider a classical description of electrons moving through an entity due to some mechanism and, at random times, crossing a boundary $x = x_o$ as illustrated in Figure 10.15.

With a classical description, the electrons behave as particles with finite dimensions, and an electron will take $dt$ seconds to cross the boundary. The charge, $q_i$, passing a boundary due to the $i$th electron is illustrated in Figure 10.16, along with the corresponding current flow, $i_i$ through the boundary. The relationship $i(t) = dq/dt$ implies that $\int_0^\infty i_i(t)dt = -q$.

It is convenient to define a normalized pulse function, $h$, as illustrated in Figure 10.16, according to

$$h(t) = \frac{-i_i(t+t_i)}{q} \qquad H(0) = \int_0^\infty h(t)dt = 1 \tag{10.109}$$

where $H$ is the Fourier transform of $h$. Assuming all electrons behave in a similar manner, the current generated by electrons passing the boundary can be written as

$$i(t) = -q\sum_{i=1}^{\infty} h(t-t_i) \tag{10.110}$$

and this signal represents one possible signal from a shot noise process.

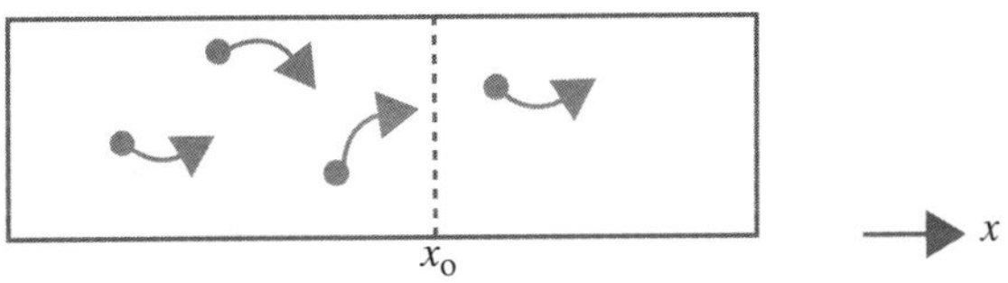


**FIGURE 10.15** Illustration of electron movement.

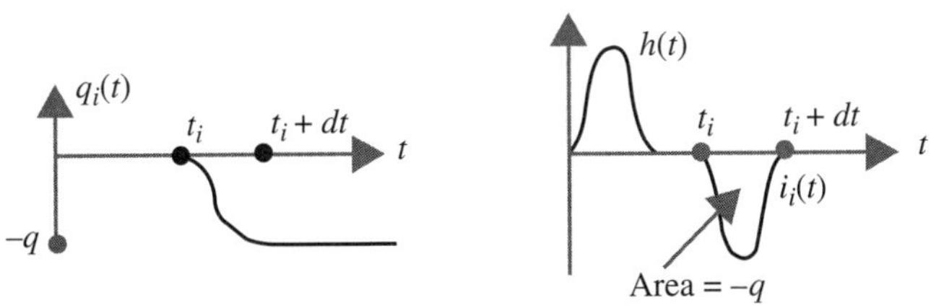


**FIGURE 10.16** Illustration of a charge $-q$ crossing a boundary in $dt$ seconds and the resulting contribution to current flow through the boundary. For the case of an electron $q = 1.6 \times 10^{-9}C$.

If, on average, there are $\lambda$ electrons per second passing the boundary, it then follows from Theorem 10.10, and the result $H(0) = 1$, that the power spectral density of the random process associated with shot noise current flow is

$$G_\infty(f) = q^2\lambda|H(f)|^2 + q^2\lambda^2\delta(f) \tag{10.111}$$

As $\bar{I} = q\lambda$ is the magnitude of the mean current flowing through the boundary, it follows that the power spectral density can be written as

$$G_\infty(f) = q\bar{I}|H(f)|^2 + \bar{I}^2\delta(f) \tag{10.112}$$

If the transition time for electrons is short relative to the measurement response time, then, for frequencies less than the bandwidth of the measuring system, $|H(f)|^2 \approx H^2(0) = 1$, and the most commonly used result

$$G_\infty(f) \approx q\bar{I} + \bar{I}^2\delta(f) \tag{10.113}$$

is obtained. Ignoring the impulse at zero frequency, Schottky's formula $G_\infty(f) = q\bar{I}$ results. Note that for frequencies less than the reciprocal of the transition time, the spectral density is flat, that is, white.

### 10.6.3 Noise Modelling of Devices

Shot noise arises in electronic devices and, in particular, with charge flow through a *PN* junction. An example of this is the base and collector current flow in a bipolar junction transistor (BJT). Accordingly, the base and collector currents exhibit shot noise with respective power spectral densities $qI_B$ and $qI_C$ (ignoring the impulsive power spectral density components at DC) where $I_B$ and $I_C$ are, respectively, the mean base and collector currents. When the effect of noise of an electronic device has to be established, an equivalent noise model is required, and the noise equivalent model for the bipolar junction transistor when operating in a small signal manner, and in the forward active region, is detailed in Figure 10.17. A detailed discussion of

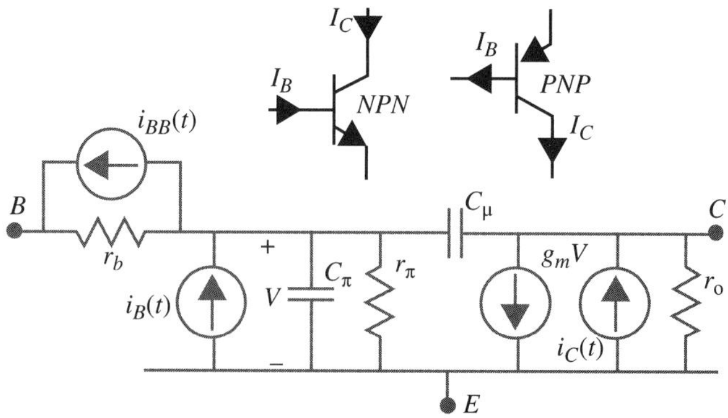


**FIGURE 10.17** Small signal equivalent noise model for a NPN or PNP BJT operating in the forward active region. The current sources $i_B(t)$ and $i_C(t)$ account for shot noise, respectively, in the base and collector currents and have respective power spectral densities of $qI_B$ and $qI_C$. The current source $i_{BB}(t)$ accounts for thermal noise arising in the base spreading resistance $r_b$.

characterizing noise in electronic devices, in a manner consistent with the discussion in this book, can be found in Howard (2002, ch. 9).

## 10.7 WHITE NOISE

The following theorem states the power spectral density of a white noise random process defined according to

$$X(\Omega, t) = \sum_{k=1}^{N} \frac{\sqrt{2}A_o}{\sqrt{N}} \sin(2\pi F_k t + \Phi_k) \qquad t > 0 \tag{10.114}$$

where $\Omega = ((\Phi_1, F_1), \ldots, (\Phi_N, F_N))$, $\Phi_1, \ldots, \Phi_N$ are identical and independent random variables with a uniform distribution on $[0, 2\pi)$ and $F_1, \ldots, F_N$ are identical and independent random variables with a uniform distribution on $[0, f_{max}]$.

**Theorem 10.12 Power Spectral Density**
The power spectral density of the white noise random process defined by Equation 10.114 is

$$\begin{aligned} G_X(T,f) = {} & \frac{A_o^2}{2\pi f_{max}} \cdot \left[ \frac{\sin(\pi(f - f_{max})T)^2}{\pi(f - f_{max})T} - \frac{\sin(\pi(f + f_{max})T)^2}{\pi(f + f_{max})T} \right] \\ & + \frac{A_o^2}{2\pi f_{max}} \cdot [\mathrm{Si}[2\pi(f + f_{max})T] - \mathrm{Si}[2\pi(f - f_{max})T]] \end{aligned} \tag{10.115}$$

$$G_X(\infty, f) = \begin{cases} \dfrac{A_o^2}{2f_{max}} & -f_{max} < f < f_{max} \\ \dfrac{A_o^2}{4f_{max}} & f = \pm f_{max} \\ 0 & \text{elsewhere} \end{cases} \tag{10.116}$$

With the definition of $\eta = A_o^2/f_{max}$, the power spectral density, for the infinite interval, can be written as

$$G_X(\infty, f) = \begin{cases} \eta/2 & -f_{max} < f < f_{max} \\ \eta/4 & f = \pm f_{max} \\ 0 & \text{elsewhere} \end{cases} \tag{10.117}$$

**Proof**
See problem 10.10.

### 10.7.1 Results

The power spectral density is shown in Figure 10.18 for the cases of $f_{max} = 1$, $A_o = 1$ and for $T = 1, 2, 10, 100$.

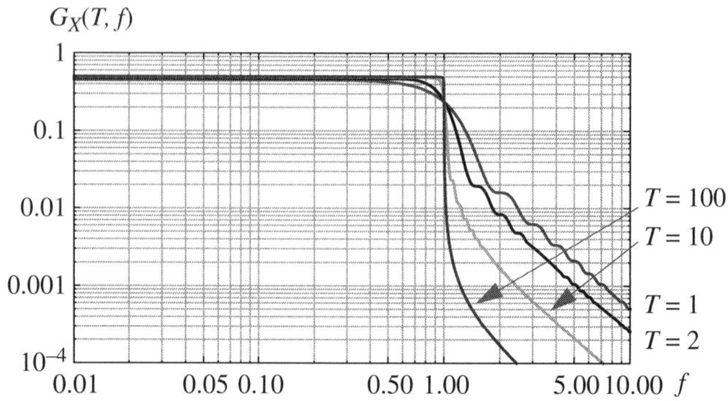


**FIGURE 10.18** Graph of the power spectral density of a white noise random process for the case of $f_{max} = 1$ and $A_o = 1$.

## 10.8 1/*f* NOISE

There are many models consistent with 1/*f* noise. One useful and simple model, which is detailed in the following, is that of a summation of equal power signalling random processes with rates that form a geometric series with a ratio of 2 and with pulse functions that have a measure of duration that is inversely proportional to the pulse rate (Howard and Raffel, 2003).

### 10.8.1 Synthesis of a 1/*f* PSD

A 1/*f* power spectral density form can be synthesized over a finite frequency range from a summation of distinct power spectral densities. Consider the case of a random process $X$ that is a summation of $N+1$ independent random processes $X_0,\ldots, X_N$:

$$X(\Omega, t) = \sum_{i=0}^{N} X_i(\Omega_i, t) \tag{10.118}$$

where $\Omega = (\Omega_0,\ldots,\Omega_N)$. To illustrate the synthesis process and the associated model for 1/*f* noise, consider the case where $X_0, \ldots, X_N$ are independent zero-mean signalling random processes based on pulse functions $p_0, \ldots, p_N$ such that their power spectral densities, from Equation 10.62, are given by

$$G_i(T,f) = r_i\sigma_i^2 |P_i(f)|^2 \quad i \in \{0,\ldots,N\} \tag{10.119}$$

Here, for the $i$th random process, $r_i$ is the pulse rate, $\sigma_i^2$ is the variance of the random variable defining the amplitudes in each signalling period, and $P_i(f)$ is the Fourier transform of the pulse $p_i(t)$. For the zero-mean and independent random process case, the power spectral density of a sum, consistent with Theorem 10.4, equals the summation of the individual power spectral densities, that is,

$$G_X(T,f) = \sum_{i=0}^{N} G_i(T,f) \tag{10.120}$$

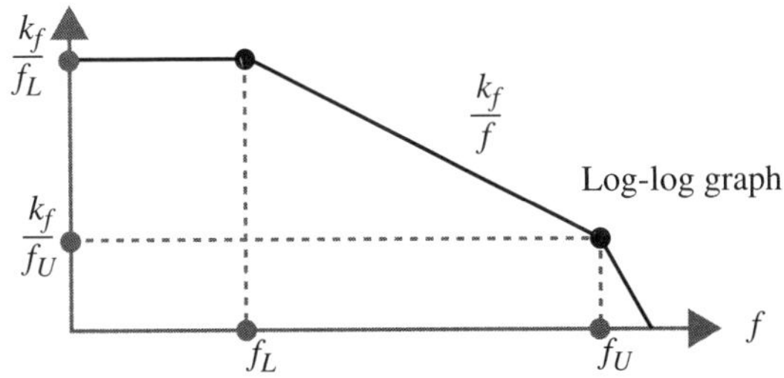


**FIGURE 10.19** Power spectral density form consistent with 1/*f* noise over the range $[f_L, f_U]$.

The goal of synthesis is to find $N+1$ practical random processes such that $G_X(T,f)$ approximates $k_f/f$ over a set frequency range $[f_L, f_U]$ as illustrated in Figure 10.19. Mathematically, the synthesis goal is

$$\sum_{i=0}^{N} G_i(T,f) \approx \frac{k_f}{f} \qquad f_L < f < f_U \quad \text{or} \quad \int_{f_L}^{f_U} \left| \sum_{i=0}^{N} G_i(T,f) - \frac{k_f}{f} \right| df << \int_{f_L}^{f_U} \frac{k_f}{f} df \tag{10.121}$$

The average power in a random process that has the power spectral density form illustrated in Figure 10.19, and over the range $-f_U < f < f_U$, is

$$\bar{P} = 2\int_0^{f_L} \frac{k_f}{f_L} df + 2\int_{f_L}^{f_U} \frac{k_f}{f} df = 2k_f \left[1 + \ln\left(\frac{f_U}{f_L}\right)\right] \tag{10.122}$$

For the case of equal power in all the random processes, the pulse rate, amplitude variance, and pulse form should satisfy the constraint

$$r_i \sigma_i^2 \int_{-\infty}^{\infty} |P_i(f)|^2 df = \frac{2k_f}{N+1}\left[1 + \ln\left(\frac{f_U}{f_L}\right)\right] \tag{10.123}$$

***10.8.1.1 Example 1*** Consider the case of zero-mean signalling random processes based on rectangular pulses where $p_i(t)$ equals unity for $0 \le t < D_i$ and zero elsewhere, that is,

$$p_i(t) = u(t) - u(t - D_i) \leftrightarrow P_i(f) = D_i \text{sinc}(D_i f) e^{-j\pi f D_i} \tag{10.124}$$

Consistent with Equation 10.119, the power spectral density of the *i*th random process is

$$G_i(T,f) = r_i \sigma_i^2 D_i^2 \text{sinc}^2(D_i f) \tag{10.125}$$

The assumption here is that $T >> D_i$. The power associated with this random process is $\bar{P}_i = r_i \sigma_i^2 D_i$ as $\int_{-\infty}^{\infty} \text{sinc}^2(kx)dx = \frac{1}{k}$. For the case of equal power in all the random

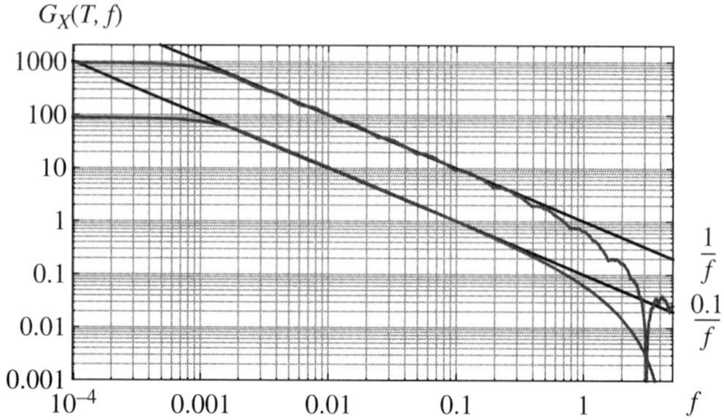


**FIGURE 10.20** Power spectral density from the summation of 11 independent signalling random processes with signalling rates of 1/1024, 1/512, …, 1 times the base rate of $r_o$. The synthesis is over the frequency range of $f_L$ = 1/1024 to $f_U$ = 1 Hz. The upper graph is the case of a rectangular pulse; the lower graph is for the case of a Gaussian pulse.

processes and where the pulse rate $r_i$ is the reciprocal of the pulse duration, that is, $r_i = 1/D_i$, the amplitude variance is constrained according to Equation 10.123, that is,

$$\sigma_i^2 = \frac{2k_f}{N+1}\left[1 + \ln\left(\frac{f_U}{f_L}\right)\right] = \sigma_o^2 \tag{10.126}$$

and is a constant for all the constituent random processes. Synthesis is based on a fundamental period of $D_o$ and a fundamental rate $r_o = 1/D_o$. When the signalling rates, and pulse durations, form a geometric series with a ratio of two, consistent with

$$D_i = 2^i D_o \qquad r_i = \frac{r_o}{2^i} \qquad r_i D_i = r_o D_o \tag{10.127}$$

then synthesis of the 1/*f* form can be achieve as illustrated in Figure 10.20. Here, the parameters $f_L = 1/1024$, $f_U = 1$, $N = 10$, $k_f = 1$, $D_o = 0.32/f_U$, $r_o = 3.125$ and $\sigma_o^2 = 1.442$ have been used. The value of $D_o = 0.32/f_U$ centers the synthesized form between $f_L$ and $f_U$.

***10.8.1.2 Example 2*** Consider a signalling random process with a period of $D_i$ seconds, a rate $r_i = 1/D_i$, and based upon a Gaussian pulse with a full width at half maximum of $w_i$ seconds, that is,

$$p_i(t) = e^{-\ln(2)(t-0.5D_i)^2/w_i^2} \leftrightarrow P_i(f) = \frac{\sqrt{\pi}w_i}{\sqrt{\ln(2)}} \cdot e^{-\pi^2 w_i^2 f^2/\ln(2)} \cdot e^{-j\pi f D_i} \tag{10.128}$$

The power spectral density of such a random process is

$$G_i(T,f) \approx \frac{\pi r_i \sigma_i^2 w_i^2}{\ln(2)} \cdot e^{-2\pi^2 w_i^2 f^2/\ln(2)} \tag{10.129}$$

assuming $T >> w_i$. As $\int_{-\infty}^{\infty} e^{-kx^2}dx = \frac{\sqrt{\pi}}{\sqrt{k}}$, the average power associated with this random process is $\bar{P}_i = \frac{\pi r_i \sigma_i^2 w_i^2}{\ln(2)} \cdot \frac{\sqrt{\pi}\sqrt{\ln(2)}}{\sqrt{2\pi^2 w_i^2}} = \frac{\sqrt{\pi} r_i \sigma_i^2 w_i}{\sqrt{2\ln(2)}}$. Synthesis of the 1/*f* form can be achieved when all random processes have equal power and the rates and pulse durations are consistent with a geometric series with a ratio of two:

$$w_i = 2^i w_o \qquad r_i = \frac{r_o}{2^i} \qquad r_i w_i = r_o w_o \tag{10.130}$$

For this case, the variance of the random variable defining the amplitudes, from Equation 10.123, is

$$\sigma_i^2 = \frac{\sqrt{2\ln(2)}}{\sqrt{\pi} r_o w_o} \cdot \frac{2k_f}{N+1}\left[1 + \ln\left(\frac{f_U}{f_L}\right)\right] = \sigma_o^2 \tag{10.131}$$

The power spectral density is illustrated in Figure 10.20 for the case of $f_L = 1/1024$, $f_U = 1, N = 10$, $k_f = 0.1$, $w_o = 0.1/f_U$, $r_o = 1$, and $\sigma_o^2 = 0.96$.

## 10.9 PSD OF A JITTERED BINARY RANDOM PROCESS

Consider a jittered binary random process defined according to

$$X(\Omega, t) = A\sum_{k=1}^{N}(-1)^{k+1}u(t - kD - J_k) \tag{10.132}$$

where $D$ is the signal period and $\Omega = (J_1, \ldots, J_N)$ is a vector of random variables defining the jitter and whose outcomes are according to the joint probability density function $f_{J_1,\ldots,J_N}$. One signal defined by the random process is illustrated in Figure 10.21.

This jittered random process can be written in several alternative ways: first, as a summation of pulses according to

$$X(\Omega, t) = A\sum_{k=1}^{N/2} u[t - (2k-1)D - J_{2k-1}] - u[t - 2kD - J_{2k}] \tag{10.133}$$

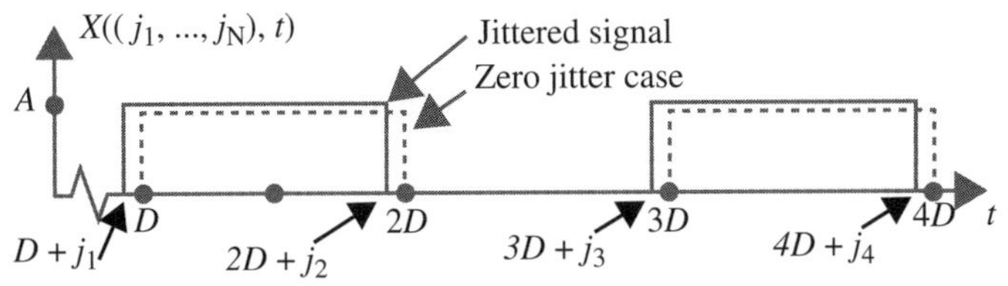


**FIGURE 10.21** Illustration of one signal from a jittered binary random process.

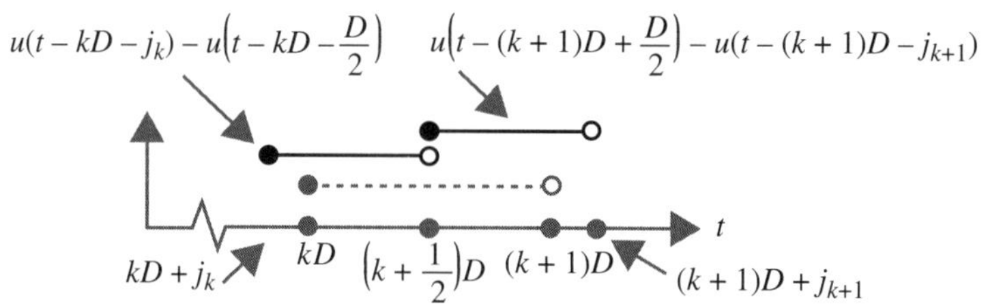


**FIGURE 10.22** Illustration of the unit step functions that define the two half pulses associated with the $k$th ($k$ being odd) and $(k+1)$th jittered times for a signal from the jittered binary random process.

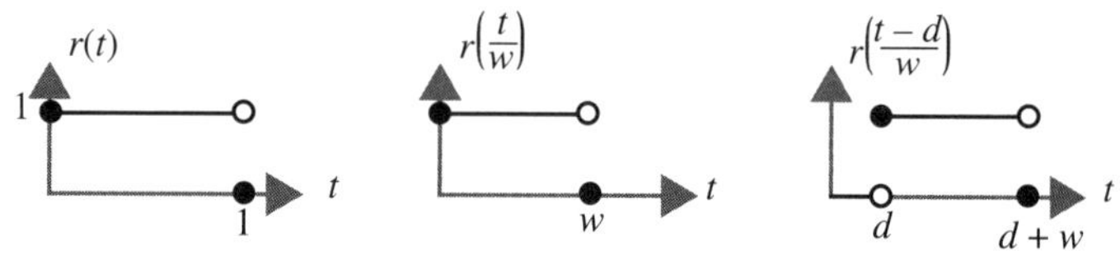


**FIGURE 10.23** Pulse functions based on $r(t) = u(t) - u(t-1)$.

and, second, consistent with Figure 10.22, as a summation of half pulses based on splitting the pulse at its midpoint of $(k+0.5)D$ for $k$ odd. The model is

$$X(\Omega,t)=A\sum_{k=1}^{N}(-1)^{k+1}\left[u(t-kD-J_k)-u\left(t-kD+(-1)^k\cdot\frac{D}{2}\right)\right] \tag{10.134}$$

Third, consistent with the pulse functions defined in Figure 10.23 and with

$$\begin{aligned} &d=(2k-1)D+J_{2k-1} \qquad w=D+J_{2k}-J_{2k-1} \\ &r(t)=u(t)-u(t-1) \end{aligned} \tag{10.135}$$

the model is

$$\begin{aligned} X(\Omega,t)&=A\sum_{k=1}^{N/2}\left[u\left(\frac{t-[(2k-1)D+J_{2k-1}]}{D+J_{2k}-J_{2k-1}}\right)-u\left(\frac{t-[(2k-1)D+J_{2k-1}]}{D+J_{2k}-J_{2k-1}}-1\right)\right] \\ &=A\sum_{k=1}^{N/2}\left[u\left(\frac{t-(2k-1)D-J_{2k-1}}{D+J_{2k}-J_{2k-1}}\right)-u\left(\frac{t-2kD-J_{2k}}{D+J_{2k}-J_{2k-1}}\right)\right] \end{aligned} \tag{10.136}$$

### 10.9.1 Power Spectral Density: Independent Case

The first form above

$$X(\Omega,t)=A\sum_{k=1}^{N/2}u[t-(2k-1)D-J_{2k-1}]-u[t-2kD-J_{2k}] \tag{10.137}$$

is consistent with a signalling random process with a period of $2D$. Using this form, the power spectral density of a jittered binary random process can be established.

**Theorem 10.13 Power Spectral Density of Jitter: Independent Case**
The power spectral density of the jittered random process, for the interval $[0, ND]$ and for the case of identical and independent jitter random variables with a jitter probability density function $f_J$, is

$$\begin{aligned} G_X(ND,f) &= \frac{rA^2}{2}\boldsymbol{E}\left[|Z(\Omega,f)|^2\right] - \frac{rA^2}{2}|\boldsymbol{E}[Z(\Omega,f)]|^2 \\ &\quad + rA^2\left[\frac{1}{N}\cdot\frac{\sin(\pi Nf/r)^2}{\sin(2\pi f/r)^2}\right]|\boldsymbol{E}[Z(\Omega,f)]|^2 \end{aligned} \tag{10.138}$$

$$\begin{aligned} G_X(\infty,f) &= \frac{rA^2}{2}\boldsymbol{E}\left[|Z(\Omega,f)|^2\right] - \frac{rA^2}{2}|\boldsymbol{E}[Z(\Omega,f)]|^2 \\ &\quad + \frac{r^2A^2}{4}\cdot|\boldsymbol{E}[Z(\Omega,f)]|^2\sum_{n=-\infty}^{\infty}\delta\left(f-\frac{nr}{2}\right) \end{aligned} \tag{10.139}$$

where $r = 1/D$ and

$$\begin{aligned} Z(\Omega,f) &= (D+J_v-J_{v-1})\cdot\text{sinc}[f(D+J_v-J_{v-1})]\cdot e^{-j\pi f(J_v+J_{v-1})} \\ &\qquad v\in\{2,4,\ldots,N/2\} \end{aligned} \tag{10.140}$$

$$\begin{aligned} \boldsymbol{E}[Z(\Omega,f)] = \int_{-\infty}^{\infty}\int_{-\infty}^{\infty} & (D+j_v-j_{v-1})\cdot\text{sinc}[f(D+j_v-j_{v-1})]\cdot e^{-j\pi f(j_v+j_{v-1})} \\ & f_J(j_v)f_J(j_{v-1})dj_v dj_{v-1} \end{aligned} \tag{10.141}$$

$$\begin{aligned} &\boldsymbol{E}\left[|Z(\Omega,f)|^2\right] = \\ &\int_{-\infty}^{\infty}\int_{-\infty}^{\infty}(D+j_v-j_{v-1})^2\cdot\text{sinc}[f(D+j_v-j_{v-1})]^2\cdot f_J(j_v)f_J(j_{v-1})dj_v dj_{v-1} \end{aligned} \tag{10.142}$$

**Proof**
The proof is detailed in Appendix 10.I.

***10.9.1.1 Examples*** For the identical, independent, zero-mean Gaussian jitter case consistent with

$$f_J(j) = \frac{1}{\sqrt{2\pi}\sigma_J}\cdot\exp\left(\frac{-j^2}{2\sigma_J^2}\right) \tag{10.143}$$

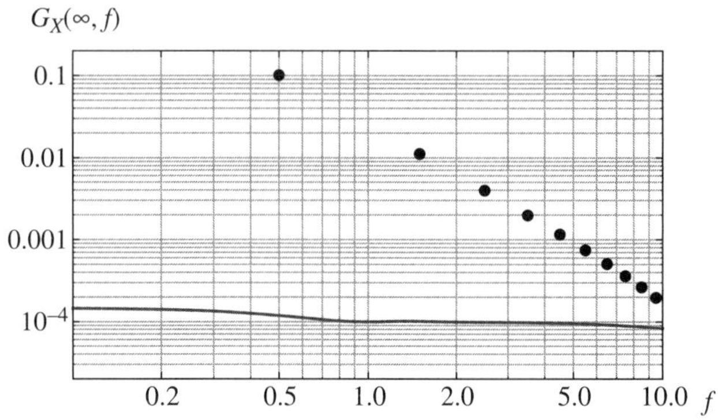


**FIGURE 10.24** Power spectral density of a jittered random process for the case of $A = 1$, $D = 1$, $r = 1/D$, and Gaussian jitter with zero mean and variance $\sigma_J = 10^{-2}D$. The powers in the impulsive components at $f = 0$, $r/2$, $3r/2$, $5r/2$, $7r/2$, respectively, are $0.25, 0.101$, $1.11 \times 10^{-2}, 3.95 \times 10^{-3}$, and $1.97 \times 10^{-3}$.

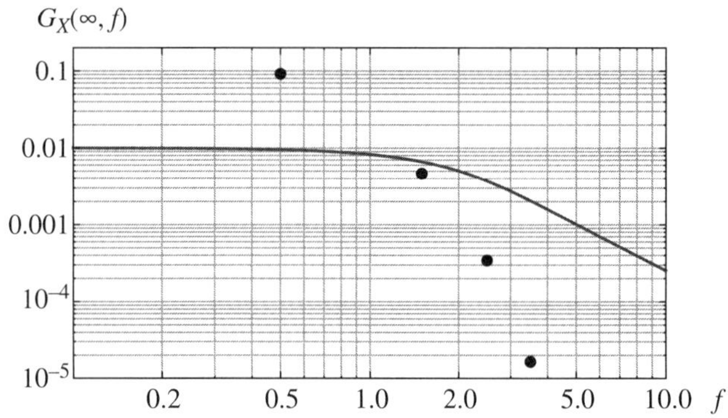


**FIGURE 10.25** Power spectral density of a jittered random process for the case of $A = 1$, $D = 1$, $r = 1/D$, and Gaussian jitter with zero mean and variance $\sigma_J = 10^{-1}D$. The powers in the impulsive components at $f = 0$, $r/2$, $3r/2$, $5r/2$, $7r/2$, respectively, are $0.25, 0.0918$, $4.63 \times 10^{-3}, 3.44 \times 10^{-4}$, and $1.64 \times 10^{-5}$.

the power spectral density, as specified by Equation 10.139, is shown in Figures 10.24 and 10.25 for, respectively, low and high jitter levels. For the low jitter case, the impulsive components are consistent with that of a periodic pulse train. With high jitter levels, the power in the impulsive components decreases, and there is a corresponding increase in the continuous part of the power spectral density.

### 10.9.2 Power Spectral Density: Correlated Case

**Theorem 10.14 Power Spectral Density of Jitter: Correlated Case**
The power spectral density of the jittered random process, for the interval $[0, ND]$ and for the case of correlated jitter random variables, is

$$G_X(ND,f) = \sum_{k=1}^{N/2} \frac{A^2}{ND} \int_{-\infty}^{\infty} \int_{-\infty}^{\infty} (D + j_{2k} - j_{2k-1})^2 \cdot$$

$$\text{sinc}\left[f(D + j_{2k} - j_{2k-1})\right]^2 f_{J_{2k-1}J_{2k}}(j_{2k-1}, j_{2k}) dj_{2k} dj_{2k-1} +$$

$$\frac{A^2}{ND} \sum_{i=1}^{N/2} \sum_{\substack{k=1 \\ k \neq i}}^{N/2} e^{-j\pi 4f(i-k)D} \int_{-\infty}^{\infty} \int_{-\infty}^{\infty} \int_{-\infty}^{\infty} \int_{-\infty}^{\infty} (D + j_{2i} - j_{2i-1}) \cdot (D + j_{2k} - j_{2k-1}) \tag{10.144}$$

$$\text{sinc}\left[f(D + j_{2i} - j_{2i-1})\right] \cdot \text{sinc}\left[f(D + j_{2k} - j_{2k-1})\right] \cdot e^{-j\pi f(j_{2i} + j_{2i-1})} \cdot$$

$$e^{j\pi f(j_{2k} + j_{2k-1})} f_{J_{2i-1}J_{2i}J_{2k-1}J_{2k}}(j_{2i-1}, j_{2i}, j_{2k-1}, j_{2k}) dj_{2i-1} dj_{2i} dj_{2k-1} dj_{2k}$$

**Proof**

The proof is detailed in Appendix 10.I.

## 10.10 PSD OF A JITTERED PULSE TRAIN

Consider a signal, as illustrated in Figure 10.26, from a random process that defines pulse trains with both amplitude and timing jitter.

The model for a random process $X : S \to S_X$ that defines jittered pulse trains is

$$X(\Omega, t) = \sum_{k=1}^{N} A_k p(t - (k-1)D - J_k) \quad t \in [0, ND] \tag{10.145}$$

where $\Omega = ((A_1, J_1), \ldots, (A_N, J_N))$ and with an underlying experimental sample space of

$$S = \{((a_1, j_1), \ldots, (a_N, j_N)) : \quad a_k \in S_{A_k}, j_k \in S_{J_k}\} \tag{10.146}$$

### 10.10.1 Power Spectral Density

The power spectral of such a random process is specified in the following theorem.

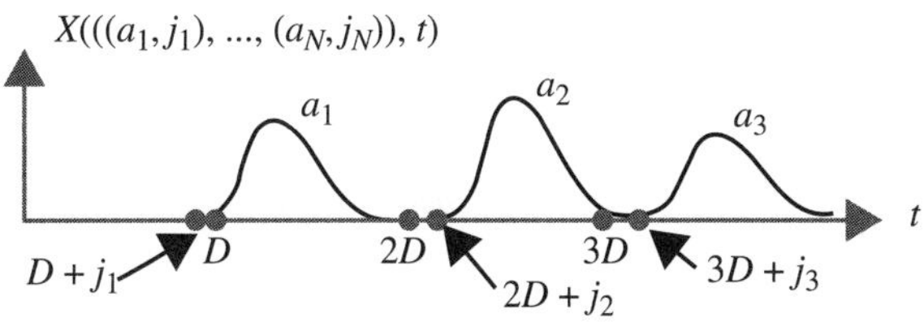


**FIGURE 10.26** Illustration of one signal from a pulse train with both amplitude and timing jitter.

**Theorem 10.15 PSD of a Jittered Pulse Train**
For the case where the amplitude variations are independent of the jitter variations, the power spectral density of a jittered pulse train, as specified by Equation 10.145, is

$$G_X(ND,f) = \frac{r|P(f)|^2}{N}\left[\sum_{i=1}^{N} \boldsymbol{E}\left[|A_i|^2\right] + \sum_{i=1}^{N}\sum_{k=1,\,k\neq i}^{N} e^{-j2\pi f(i-k)D}\cdot \boldsymbol{E}[A_i A_k^*]\cdot \boldsymbol{E}\left[e^{-j2\pi f(J_i-J_k)}\right]\right] \tag{10.147}$$

Here, $P(f)$ is the Fourier transform of $p(t)$ assuming the interval $[0, ND]$ does not significantly affect the Fourier transform and $r = 1/D$ is the pulse rate.

For the case where the correlation between amplitudes and jitter times depends only on the spacing between the pulses, and not on their absolute location, consistent with the assumptions

$$\begin{aligned} f_{A_iA_k}(a_i,a_k) &= f_{A_{i+r}A_{k+r}}(a_i,a_k) \\ f_{J_iJ_k}(j_i,j_k) &= f_{J_{i+r}J_{k+r}}(j_i,j_k) \qquad i,k,i+r,k+r\in \boldsymbol{Z}^+ \end{aligned} \tag{10.148}$$

and where the individual random variables are identically distributed, that is, $f_{A_i}(a_i) = f_{A_{i+r}}(a_i)$ and $f_{J_i}(j_i) = f_{J_{i+r}}(j_i)$, the power spectral density is

$$\begin{aligned} G_X(ND,f) = r|P(f)|^2\overline{|A|^2} + 2r|P(f)|^2\sum_{i=1}^{N-1}\left[1-\frac{i}{N}\right]\cdot \\ \mathrm{Re}\left[e^{j2\pi iDf}\boldsymbol{E}[A_1A_{1+i}^*]\cdot \boldsymbol{E}\left[e^{-j2\pi f(J_1-J_{1+i})}\right]\right] \end{aligned} \tag{10.149}$$

where $\overline{|A|^2} = \boldsymbol{E}\left[|A_i|^2\right]$ is the mean squared amplitude. For the finite and infinite interval $[0,\infty]$, this power spectral density can be written as

$$\begin{aligned} G_X(ND,f) = r|P(f)|^2\Bigg[&\overline{|A|^2} - |\mu_A|^2|F_J(f)|^2 + |\mu_A|^2|F_J(f)|^2\cdot\frac{1}{N}\cdot\frac{\sin(\pi Nf/r)^2}{\sin(\pi f/r)^2} \\ &+2\sum_{i=1}^{M-1}\left[1-\frac{i}{N}\right]\cdot\mathrm{Re}\left[e^{j2\pi iDf}\cdot\left[\boldsymbol{E}[A_1A_{1+i}^*]-|\mu_A|^2\right]\cdot\left[\boldsymbol{E}\left[e^{-j2\pi f(J_1-J_{1+i})}\right]-|F_J(f)|^2\right]\right] \\ &+2|\mu_A|^2\sum_{i=1}^{M-1}\left[1-\frac{i}{N}\right]\mathrm{Re}\left[e^{j2\pi iDf}\left[\boldsymbol{E}\left[e^{-j2\pi f(J_1-J_{1+i})}\right]-|F_J(f)|^2\right]\right] \\ &+2|F_J(f)|^2\sum_{i=1}^{M-1}\left[1-\frac{i}{N}\right]\mathrm{Re}\left[e^{j2\pi iDf}\left(\boldsymbol{E}[A_1A_{1+i}^*]-|\mu_A|^2\right)\right]\Bigg] \end{aligned} \tag{10.150}$$

$$G_X(\infty,f) = r|P(f)|^2 \left[ \overline{|A|^2} - |\mu_A|^2|F_J(f)|^2 + r|\mu_A|^2|F_J(f)|^2 \cdot \sum_{n=-\infty}^{\infty} \delta(f-nr) \right.$$

$$+2\sum_{i=1}^{M-1} \mathrm{Re}\left[e^{j2\pi iDf} \cdot \left[\boldsymbol{E}[A_1A_{1+i}^*] - |\mu_A|^2\right] \cdot \left[\boldsymbol{E}\left[e^{-j2\pi f(J_1-J_{1+i})}\right] - |F_J(f)|^2\right]\right]$$

$$+2|\mu_A|^2\sum_{i=1}^{M-1} \mathrm{Re}\left[e^{j2\pi iDf}\left[\boldsymbol{E}\left[e^{-j2\pi f(J_1-J_{1+i})}\right] - |F_J(f)|^2\right]\right]$$

$$\left. +2|F_J(f)|^2\sum_{i=1}^{M-1} \mathrm{Re}\left[e^{j2\pi iDf}\left(\boldsymbol{E}[A_1A_{1+i}^*] - |\mu_A|^2\right)\right]\right] \quad (10.151)$$

Here, $\mu_A = \boldsymbol{E}[A_i]$, $F_J(f) = \boldsymbol{E}\left[e^{-j2\pi f J_i}\right] = \phi_J[-f]$ is the Fourier transform of $f_{J_i}$ with $\phi_J$ being the characteristic function of $J_i$, and the correlation is assumed to extend over $M$ pulses, $M < N$, consistent with $A_1$ and $A_{M+1}$, as well as $J_1$ and $J_{M+1}$, being independent.

**Proof**

The proof is detailed in Appendix 10.J.

***10.10.1.1 Notes*** The first three terms in Equations 10.150 and 10.151 are associated with the independent random variable case, and the other terms are, respectively, the contribution due to both the amplitude and jitter random variables being correlated, the jitter random variables being correlated, and the amplitude random variables being correlated.

### *10.10.1.2 Independent Case*

**Theorem 10.16 PSD of a Jittered Pulse Train: Independent Case**

For the case where all the random variables are independent of one another, the power spectral density on the interval $[0, ND]$ is

$$G_X(ND,f) =$$

$$\frac{r|P(f)|^2}{N}\left[\sum_{i=1}^{N}\overline{|A_i|^2} + \sum_{i=1}^{N}\sum_{k=1,\,k\neq i}^{N} e^{-j2\pi f(i-k)D}\mu_{A_i}\mu_{A_k}^* F_{J_i}(f)F_{J_k}^*(f)\right] \quad (10.152)$$

where $\overline{|A_i|^2} = \boldsymbol{E}\left[|A_i|^2\right]$, $\mu_{A_i} = \boldsymbol{E}[A_i]$ and $F_{J_i}(f) = \boldsymbol{E}\left[e^{-j2\pi f J_i}\right]$.

For the identical and independent random variable case,

$$G_X(ND,f) = r|P(f)|^2 \left[ \overline{|A|^2} - |\mu_A|^2 |F_J(f)|^2 + |\mu_A|^2 |F_J(f)|^2 \cdot \frac{1}{N} \cdot \frac{\sin(\pi N f/r)^2}{\sin(\pi f/r)^2} \right] \tag{10.153}$$

$$G_X(\infty,f) = r|P(f)|^2 \left[ \overline{|A|^2} - |\mu_A|^2 |F_J(f)|^2 + r|\mu_A|^2 |F_J(f)|^2 \sum_{n=-\infty}^{\infty} \delta(f-nr) \right] \tag{10.154}$$

**Proof**

These results follow directly from Theorem 10.15.

### 10.10.2 Example

Consider the case of a jittered pulse train based on a Gaussian pulse function

$$p(t) = \frac{e^{-t^2/2\sigma_t^2}}{\sqrt{2\pi}\sigma_t} \Leftrightarrow P(f) = e^{-2\pi^2 f^2 \sigma_t^2} \tag{10.155}$$

and identical Gaussian probability density functions for the jitter and amplitude random variables:

$$\begin{aligned} f_A(a) &= \frac{e^{-(a-\mu_A)^2/2\sigma_A^2}}{\sqrt{2\pi}\sigma_A} \\ f_J(j) &= \frac{e^{-(j-\mu_J)^2/2\sigma_J^2}}{\sqrt{2\pi}\sigma_J} \Leftrightarrow F_J(f) = e^{-j2\pi f \mu_J - 2\pi^2 f^2 \sigma_J^2} \end{aligned} \tag{10.156}$$

Results are given for the case of $D = 1/r = 1$, $\sigma_t = 0.1D$, $\mu_A = A_o = 1$, $\sigma_A = 0.1A_0$, $\mu_J = 0$ and $\sigma_J = 0.1D$. One signal from the random process is shown in Figure 10.27.

***10.10.2.1 Identical and Independent RV Case*** The power spectral density, for the independent and identical case, is

$$G_X(ND,f) =$$

$$re^{-4\pi^2 f^2 \sigma_t^2} \left[ \sigma_A^2 + \mu_A^2 - \mu_A^2 e^{-4\pi^2 f^2 \sigma_J^2} + \mu_A^2 e^{-4\pi^2 f^2 \sigma_J^2} \cdot \frac{1}{N} \cdot \frac{\sin(\pi N f/r)^2}{\sin(\pi f/r)^2} \right] \tag{10.157}$$

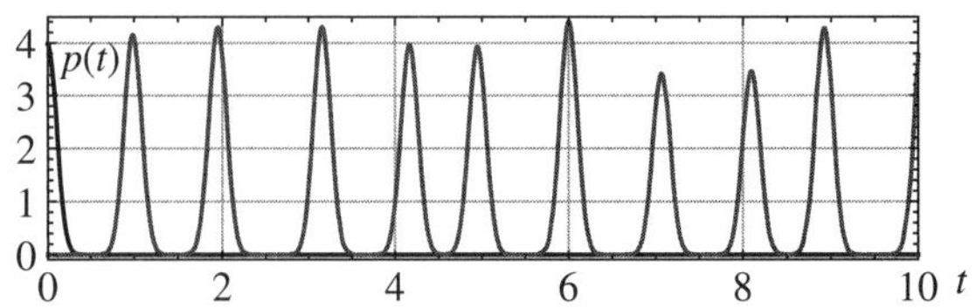


**FIGURE 10.27** One waveform from the defined jittered pulse train.

$$G_X(\infty, f) = \\ re^{-4\pi^2 f^2 \sigma_t^2}\left[\sigma_A^2 + \mu_A^2 - \mu_A^2 e^{-4\pi^2 f^2 \sigma_J^2} + r\mu_A^2 e^{-4\pi^2 f^2 \sigma_J^2} \sum_{n=-\infty}^{\infty} \delta(f - nr)\right] \tag{10.158}$$

The power spectral density for the case of amplitude variations only ($\sigma_J = 0$), and jitter variations only ($\sigma_A = 0$), is shown in Figure 10.28. The power spectral density for the case of both amplitude and jitter variations is shown in Figure 10.29.

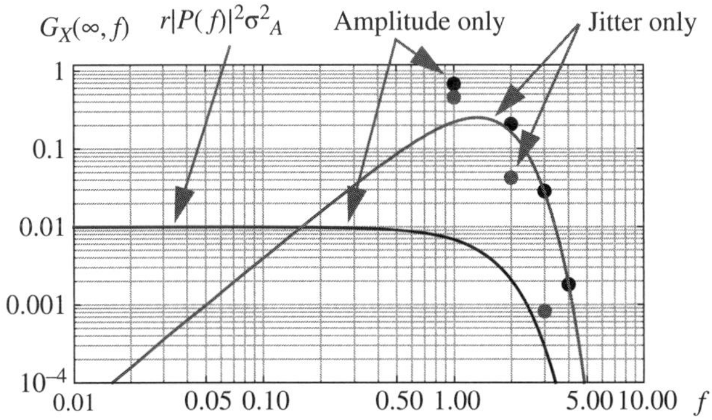


**FIGURE 10.28** Power spectral density for the two cases of amplitude-only variations and jitter-only variations. The powers in the impulses at $f = 0$, $r$, $2r$, $3r$, $4r$, $5r$, as represented by the dots, are, respectively, 1, 0.674, 0.206, $2.86 \times 10^{-2}$, $1.81 \times 10^{-3}$, and $5.17 \times 10^{-5}$ for the amplitude case and 1, 0.454, $4.25 \times 10^{-2}$, $8.20 \times 10^{-4}$, $3.26 \times 10^{-6}$, and $2.68 \times 10^{-9}$ for the jitter case.

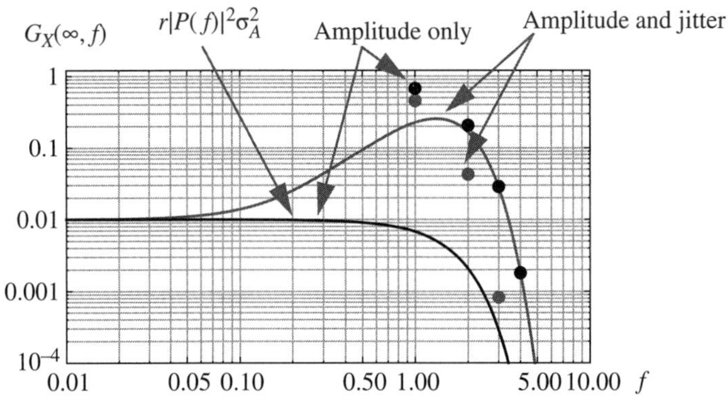


**FIGURE 10.29** Power spectral density for the case of both amplitude and jitter variations. The powers in the impulses at $f = 0$, $r$, $2r$, $3r$, $4r$, $5r$, as represented by the dots, respectively, are 1, 0.454, $4.25 \times 10^{-2}$, $8.20 \times 10^{-4}$, $3.26 \times 10^{-6}$, and $2.68 \times 10^{-9}$.

***10.10.2.2 Correlated Case*** Consider the case where the amplitude and jitter random variables are governed by a multinormal distribution. Denote a vector $X=(X_1,X_2,\ldots,X_N)$ that represents either the amplitude or jitter random variables. The joint probability density function of the random variables is defined by

$$f_X(x)=\frac{1}{(2\pi)^{M/2}|\boldsymbol{\Sigma}|^{1/2}}\cdot\exp\left[\frac{-(x-\mu)^T\boldsymbol{\Sigma}^{-1}(x-\mu)}{2}\right] \qquad (10.159)$$
$$x=(x_1,x_2,\ldots,x_N)$$

where $\boldsymbol{\mu}=\{\mu_1,\ldots,\mu_N\},\mu_i=\boldsymbol{E}[X_i]$ and the correlation matrix $\Sigma$ is

$$\boldsymbol{\Sigma}=\left[C_{ij}\right]=\left[\rho_{ij}\sigma_i\sigma_j\right] \quad i,j\in\{1,\ldots,N\} \qquad (10.160)$$

Here, $C_{ij}=\boldsymbol{E}\left[(X_i-\mu_i)\;(X_j-\mu_j)\right]$, $\sigma_i^2=\mathrm{var}(X_i)$, and $\rho_{ij}$ is the correlation coefficient of $X_i$ and $X_j$. The marginal density functions are normal and are given by Equation 10.156. The joint probability density function is the bivariate normal:

$$f_{X_iX_j}(x_i,x_j)=\frac{1}{2\pi\sigma_i\sigma_j\sqrt{1-\rho_{ij}^2}}\cdot\exp\left[\frac{-1}{1-\rho_{ij}^2}\cdot\left[\frac{(x_i-\mu_i)^2}{2\sigma_i^2}-\frac{\rho_{ij}(x_i-\mu_i)\,(x_j-\mu_j)}{\sigma_i\sigma_j}+\frac{(x_j-\mu_j)^2}{2\sigma_j^2}\right]\right] \qquad (10.161)$$

It then follows that $\boldsymbol{E}\left[X_iX_j^*\right]=\mu_i\mu_j+\rho_{ij}\sigma_i\sigma_j$ and for the case of $\mu_i=\mu_k=0$:

$$\boldsymbol{E}\left[e^{-j2\pi f(X_i-X_k)}\right]=\exp\left[-2\pi^2f^2\left(\sigma_i^2+\sigma_k^2-2\rho_{ik}\sigma_i\sigma_k\right)\right] \qquad (10.162)$$

The amplitude random variables are assumed to be identical consistent with $\mu_i=\mu_j=\mu_A,\sigma_i=\sigma_j=\sigma_A$, and exponential correlation is assumed according to $\rho_{ij}=\exp(-|i-j|)$. The correlation matrix then is

$$\boldsymbol{\Sigma}_A=\left[\rho_{ij}\sigma_A^2\right]=\begin{bmatrix}\sigma_A^2 & e^{-1}\sigma_A^2 & \ldots & e^{-N+1}\sigma_A^2\\ \ldots & & & \ldots\\ e^{-N+1}\sigma_A^2 & \ldots & e^{-1}\sigma_A^2 & \sigma_A^2\end{bmatrix} \qquad (10.163)$$

Analogous definitions, and results, are assumed for the jittered random variables $J_1,\ldots,J_N$.

The power spectral density, as given by Equation 10.149, becomes

$$G_X(ND,f)=r|P(f)|^2\overline{|A|^2}+2r|P(f)|^2\sum_{i=1}^{N-1}\left[1-\frac{i}{N}\right]\cdot\left(\mu_A^2+\rho_{1,1+i}^A\sigma_A^2\right)\exp\left[-4\pi^2f^2\sigma_J^2\left(1-\rho_{1,1+i}^J\right)\right]\cos\left[2\pi iDf\right] \qquad (10.164)$$

where $\overline{|A|^2} = \mu_A^2 + \sigma_A^2$ and superscripts have been used to distinguish between the amplitude and jitter correlation coefficients. For the infinite interval, the power spectral density, as given by Equation 10.151, is

$$\begin{aligned} G_X(\infty, f) = r|P(f)|^2 \Bigg[ & \overline{|A|^2} - \mu_A^2 e^{-4\pi^2 f^2 \sigma_J^2} + r\mu_A^2 e^{-4\pi^2 f^2 \sigma_J^2} \sum_{n=-\infty}^{\infty} \delta(f - nr) \\ & + 2\sigma_A^2 \sum_{i=1}^{M-1} \cos(2\pi i D f) \rho_{1,1+i}^A e^{-4\pi^2 f^2 \sigma_J^2} \cdot \left[ \exp\left(4\pi^2 f^2 \sigma_J^2 \rho_{1,1+i}^J\right) - 1 \right] \\ & + 2\mu_A^2 \sum_{i=1}^{M-1} \cos(2\pi i D f) e^{-4\pi^2 f^2 \sigma_J^2} \left[ \exp\left(4\pi^2 f^2 \sigma_J^2 \rho_{1,1+i}^J\right) - 1 \right] \\ & + 2e^{-4\pi^2 f^2 \sigma_J^2} \sigma_A^2 \sum_{i=1}^{M-1} \cos(2\pi i D f) \rho_{1,1+i}^A \Bigg] \end{aligned} \qquad (10.165)$$

Consider the previously assumed parameter values and with the additional definitions $\rho_{1,1+i}^A = e^{-i}$, $\rho_{1,1+i}^J = e^{-i}$ and $M = 10$. One signal from the correlated and jittered random process is shown in Figure 10.30, and the power spectral density is shown in Figure 10.31.

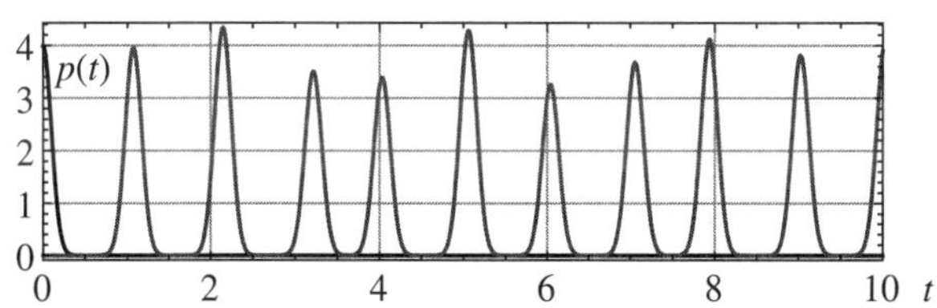


**FIGURE 10.30** One waveform from the jittered pulse train for the correlated case.

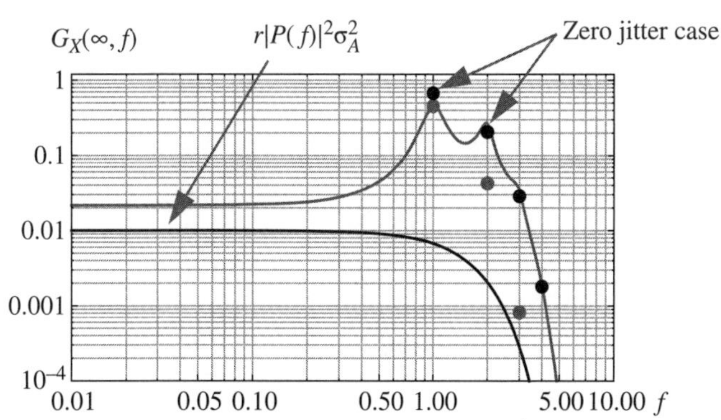


**FIGURE 10.31** Power spectral density for the defined correlated and jittered pulse train random process. The powers in the impulses at $f = 0$, $r$, $2r$, $3r$, $4r$, $5r$, as represented by the dots, are, respectively, 1, 0.454, $4.25 \times 10^{-2}$, $8.20 \times 10^{-4}$, $3.26 \times 10^{-6}$, and $2.68 \times 10^{-9}$.

Evidence for the validity of the power spectral density, as given by Equation 10.165, is shown in Figure 10.32 where simulation results using the FFT are displayed after averaging 100 jittered signals on the interval $[0, 200D]$.

***10.10.2.3 Components of Power Spectral Density*** The five nonimpulsive components of the power spectral density, as specified by Equation 10.165, are shown in Figure 10.33. The first two components are associated with the independent random variable case. For the parameters chosen, the dominant term (fourth term) is due to the correlated jitter that leads to spectral broadening around the harmonic components. The correlated amplitudes (term 5) lead to an increase in the spectrum level at low frequencies. The third term, due to both the amplitude and jitter random variables being correlated, is small. Results for the case of a pulse train based on decaying exponential pulses can be found in Howard (2012b).

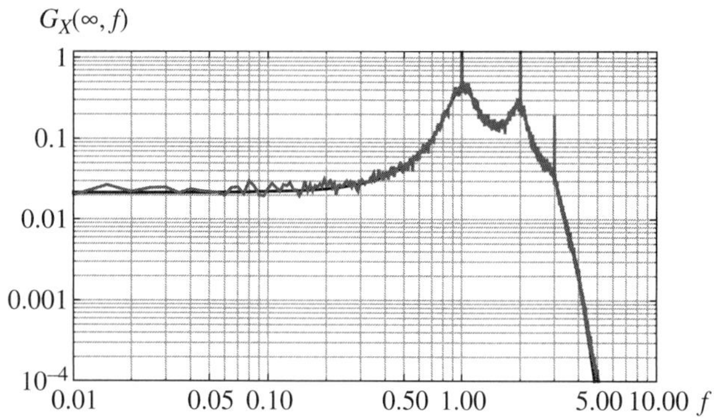


**FIGURE 10.32** Simulation and theoretical results for the power spectral density of a correlated and jittered pulse train.

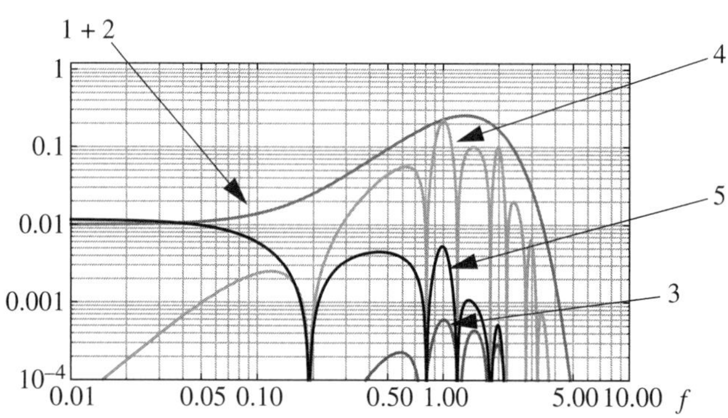


**FIGURE 10.33** Magnitude of nonimpulsive components of power spectral density.

## 10.11 PROBLEMS

**10.1** Consider a random process defined by a signal plus additive noise that has a nonzero mean. The model is

$$X(\Omega,t)=s(t)+N(\Omega,t) \qquad (10.166)$$

a. Determine a general expression for the power spectral density of $X$.

b. For the case of a sinusoidal signal with an additive random DC offset:

$$\begin{aligned} &s(t)=A\sin(2\pi f_C t) \\ &N(\Omega,t)=\Omega \qquad f_\Omega(\omega)=\frac{\omega}{\sigma_N^2}\cdot\exp\left[\frac{-\omega^2}{2\sigma_N^2}\right] \end{aligned} \qquad (10.167)$$

determine the power spectral density for the interval $[0,T]$ and the interval $[0,\infty]$.

c. Graph the power spectral density for the case of $f_C=1, T=4.5, A=1$ and $\sigma_N=A/4$.

**10.2** Consider a periodic pulse train defined according to

$$x(t)=\sum_{k=0}^{N-1}Ap(t-kD) \quad t\in[0,ND] \qquad (10.168)$$

where

$$p(t)=\sin(2\pi f_D t)^{\mathrm{v}}[u(t)-u(t-D)] \quad f_D=1/2D \qquad (10.169)$$

a. Show that the Fourier transform of $p$ is

$$P(f)=\frac{De^{-j\pi fD}}{2^{\mathrm{v}}}\sum_{k=0}^{\mathrm{v}}\frac{\mathrm{v}!}{(\mathrm{v}-k)!k!}\cdot\mathrm{sinc}[fD-(\mathrm{v}/2-k)] \qquad (10.170)$$

and the average power in the pulse train is $\bar{P}=\frac{A^2}{2^{2\mathrm{v}}}\cdot\frac{(2\mathrm{v})!}{(\mathrm{v}!)^2}$.

To prove these results, use the exponential form for the sine function and the expansion

$$(1+x)^{\mathrm{v}}=1+\mathrm{v}x+\frac{\mathrm{v}(\mathrm{v}-1)x^2}{2!}+\cdots+\frac{\mathrm{v}!x^k}{(\mathrm{v}-k)!k!}+\cdots+x^{\mathrm{v}} \qquad (10.171)$$

b. Determine the power spectral density of the periodic pulse train for the general case.

c. For the case of $\mathrm{v}=8$, $N\to\infty$, $A=1$ and $D=1$, graph the pulse train and its power spectral density. Check that the power in the impulsive components of the power spectral density equals the average power.

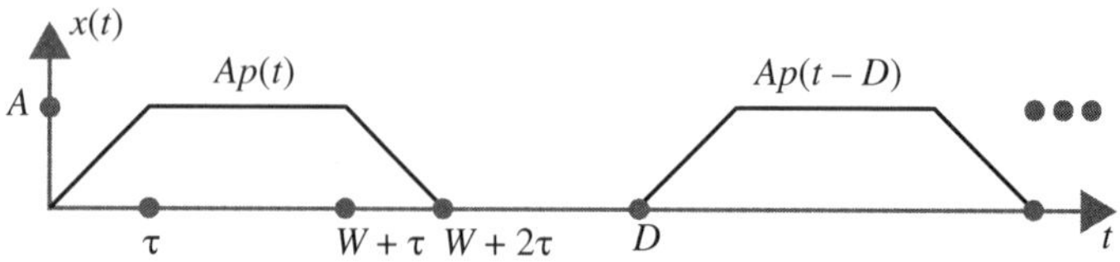


**FIGURE 10.34** First-order model for a periodic digital signal.

**10.3** Consider a first-order model for a periodic digital signal as illustrated in Figure 10.34.

a. Determine an expression for the power spectral density of this signal using the result that the Fourier transform of the normalized pulse, $p$, defined on the interval $[0, D]$ is

$$P(f) = (W+\tau)e^{-j2\pi(W/2+\tau)f} \cdot \text{sinc}(\tau f) \cdot \text{sinc}[(W+\tau)f] \qquad (10.172)$$

b. Consider the normalized case of $D=1$, $W+\tau=1/2$ and $A=1$. Graph the digital signals, and their power spectral densities, for the three cases of $W=9/20$, $\tau=1/20$, $W=3/8$, $\tau=1/8$, and $W=\tau=1/4$.

c. Check that the power in the significant impulsive components of the power spectral density closely approximates the average power.

**10.4** Determine the power spectral density of a random process $X$ that is based on a digital pulse, as defined in the previous exercise, and defined according to

$$X(\Omega, t) = \sum_{k=1}^{N} A_k p(t-(k-1)D) \quad t \in [0, ND] \qquad (10.173)$$

where $\Omega = (A_1, \ldots, A_N)$, $A_1, \ldots, A_N$ are a sequence of identical and independent random variables with a sample space $\{-A_o, 0, A_o\}$ and with $P[A_k = -A_o] = P[A_k = 0] = P[A_k = A_o] = 1/3$.

a. Determine an expression for the power spectral density of $X$.

b. Graph several signals from the random process, and the power spectral density, for the case of $D=1$, $A_o=1$ and the three cases of $W=9/20$, $\tau=1/20$, $W=3/8$, $\tau=1/8$, and $W=\tau=1/4$.

**10.5** Determine the power spectral density of a signalling random process defined according to

$$X(\Omega, t) = \sum_{k=1}^{N} A_k \phi(t-(k-1)D) \quad t \in [0, ND]$$
$$\phi(t) = u(t) - u(t-D/2) \qquad (10.174)$$

where $\Omega=(A_1,\ldots,A_N)$, $A_1,\ldots,A_N$ are a sequence of identical and independent random variables with a sample space $\{0,A\}$ and with $P[A_k=0]=P[A_k=A]=1/2$.

a. Determine an analytical expression for the power spectral density of $X$.
b. Graph the power spectral density for the case of $N=64$, $D=1$, and $A=1$.
c. Graph the power spectral density for the infinite interval case and with $D=1$ and $A=1$.

**10.6** Consider the case of bipolar signalling consistent with a random process defined according to

$$X(\Omega,t)=\sum_{k=1}^{N}\phi(\Omega_k,t-(k-1)D) \qquad t\in[0,ND] \tag{10.175}$$

where $\Omega=(\Omega_1,\ldots,\Omega_N)$, the sample space for $\Omega_k$, $k\in\{1,\ldots,N\}$, is $S_k=\{-1,0,1\}$ and the signalling set is

$$S_\Phi=\{\phi(\omega_i,t)=\omega_i A p_{RC}(D,\beta,t-D/2): \quad \omega_i\in\{-1,0,1\}\} \tag{10.176}$$

Here, $p_{RC}(D,\beta,t)$ is the raised cosine pulse with roll-off factor $\beta$. Bipolar signalling is when binary data is encoded with no waveform corresponding to a logic 0 and with logic 1 being encoded alternatively with $p_{RC}$ and $-p_{RC}$. Assume binary data with equal likelihood of logic 0 and with logic 1.

a. For the case of $\beta=1$, $A=1$, and $D=1$, graph a few signals from the random process.
b. Determine an analytic expression for the power spectral density of the random process for the finite and infinite intervals.
c. For the case of $\beta=1$, $A=1$, and $D=1$, graph the power spectral density for the infinite interval case.

**10.7** A bounded random walk with upper and lower bounds of $\pm B$ can be defined according to

$$X(\Omega,t)=\sum_{k=1}^{N}\Omega_k p(t-(k-1)D) \qquad t\in[0,T] \tag{10.177}$$

where $p(t)=u(t)-u(t-BD)$, $\Omega=(\Omega_1,\ldots,\Omega_N)$, $\Omega_1,\ldots,\Omega_N$ are independent and identical random variables with a sample space of $\{-1,1\}$, $P[\Omega_k=1]=p$, $P[\Omega_k=-1]=1-p$, and $T=(N+B)D$.

a. Graph a few signals from the random process assuming $B=10$, $D=1$, and $p=0.6$.
b. Determine a general expression for the power spectral density of the random process assuming a general pulse function $p$.

c. Graph the power spectral density, assuming the infinite interval, for the case of $B=10$, $D=1$, $p(t)=u(t)-u(t-BD)$, and $p=0.6$.

**10.8** Consider a generalized shot noise random process

$$X(\Omega,t)=\sum_{k=1}^{N}A_k p(t-\Gamma_k) \qquad t\in[0,T] \tag{10.178}$$

where $\Omega=((A_1,\Gamma_1),\ldots,(A_k,\Gamma_k),\ldots)$, $\Gamma_1,\Gamma_2,\ldots$ is a sequence of independent random variables with a uniform distribution on $[0,T]$ and $(A_1,A_2,\ldots)$ is a vector of identical and independent random variables that define the pulse amplitudes. Show that

$$\begin{aligned} G_X(T,f)&=\lambda|P(f)|^2\left[\sigma_A^2+\mu_A^2+\lambda\mu_A^2\cdot\left(1-\frac{1}{N}\right)\cdot T\cdot \operatorname{sinc}(fT)^2\right]\\ G_{X_\infty}(f)&=\lambda\left[\sigma_A^2+\mu_A^2\right]\ |P(f)|^2+\lambda^2\mu_A^2|P(0)|^2\delta(f)\end{aligned} \tag{10.179}$$

where $\lambda=N/T$ is the mean pulse rate, $\sigma_A^2$ is the variance of $A_k$, and $\mu_A$ is the mean of $A_k$.

**10.9** Consider a generalized shot noise random process

$$X(\Omega,t)=\sum_{k=1}^{\infty}A_k p(t-\Gamma_k) \qquad t>0 \tag{10.180}$$

arising from the output of a system with an impulse response

$$p(t)=\frac{t^2}{2\tau^3}\cdot\exp(-t/\tau)\cdot u(t) \tag{10.181}$$

Here, $\Omega=((A_1,\Gamma_1),\ldots,(A_k,\Gamma_k),\ldots)$, $\Gamma_1,\Gamma_2,\ldots$ is a sequence of independent random variables as defined by the outcomes of a Poisson experiment, and $(A_1,A_2,\ldots)$ is a vector of identical and independent random variables where each random variable has an exponential distribution:

$$f_{A_k}(a)=\lambda_A e^{-\lambda_A a} \qquad a>0 \tag{10.182}$$

Assume that $\lambda=1$, $\tau=1/2\pi$ and $\lambda_A=1$.

a. Graph a few signals from the random process.
b. Graph the power spectral density for the case of $T=1000$.

**10.10** Consider a white noise random process

$$X(\Omega,t)=\sum_{k=1}^{N}\frac{\sqrt{2}A_o}{\sqrt{N}}\sin(2\pi F_k t+\Phi_k) \qquad t>0 \tag{10.183}$$

where $\Omega = ((\Phi_1, F_1), \ldots, (\Phi_N, F_N)), \Phi_1, \ldots, \Phi_N$ are identical and independent random variables with a uniform distribution on $[0, 2\pi]$ and $F_1, \ldots, F_N$ are identical and independent random variables with a uniform distribution on $[0, f_{max}]$.

a. Show that

$$G_X(T,f) = \frac{A_o^2}{2f_{max}} \cdot T \int_0^{f_{max}} \left[ \text{sinc}[(f - f_i)T]^2 + \text{sinc}[(f + f_i)T]^2 \right] \cdot df_i \tag{10.184}$$

and, hence, that:

$$\begin{aligned} G_X(T,f) = {} & \frac{A_o^2}{2\pi f_{max}} \cdot \left[ \frac{\sin(\pi(f - f_{max})T)^2}{\pi(f - f_{max})T} - \frac{\sin(\pi(f + f_{max})T)^2}{\pi(f + f_{max})T} \right] \\ & + \frac{A_o^2}{2\pi f_{max}} \cdot [\text{Si}[2\pi(f + f_{max})T] - \text{Si}[2\pi(f - f_{max})T]] \end{aligned} \tag{10.185}$$

$$G_X(\infty, f) = \begin{cases} \dfrac{A_o^2}{2f_{max}} & -f_{max} < f < f_{max} \\ \dfrac{A_o^2}{4f_{max}} & f = \pm f_{max} \\ 0 & \text{elsewhere} \end{cases} \tag{10.186}$$

The following integral result (equation 14.374, Spiegel, M. R., 'Mathematical Handbook of Formulas and Tables', McGraw Hill, 1968) is useful:

$$\int \frac{1}{x^2} \cdot \cos(ax)\, dx = -\frac{\cos(ax)}{x} - a \int \frac{1}{x} \cdot \frac{\sin(ax)}{x} dx \tag{10.187}$$

b. For the case of $f_{max} = 1$, $N = 10$, $A_o = 1$, graph several signals defined by the random process.

c. Graph the power spectral density for the case of $f_{max} = 1$, $N = 10$, $A_o = 1$, and $T = 1, 2, 10, 100$.

**10.11** Consider a pulse train that is subject to amplitude and timing jitter and is defined by the random process

$$X(\Omega, t) = \sum_{k=1}^{N} A_k p(t - kD - J_k) \qquad t \in [0, ND] \tag{10.188}$$

where $\Omega = ((A_1, J_1), \ldots, (A_N, J_N))$, $A_1, \ldots, A_N$ are independent and identical random variables, as are $J_1, \ldots, J_N$, and the pulse function is defined according to

$$p(t) = \sin(2\pi f_D t)^{v} [u(t) - u(t - D)] \qquad f_D = 1/2D \tag{10.189}$$

The probability density function for $A_k$ is

$$f_A(a) = \frac{a}{\sigma_A^2} \cdot \exp\left[\frac{-a^2}{2\sigma_A^2}\right] \qquad \sigma_A = \sqrt{2}a_o/\sqrt{\pi} \tag{10.190}$$

The jitter arises from two independent sources, and

$$J_k = \begin{cases} J_A \text{ probability } p_1 & f_{J_A}(j) = \dfrac{1}{\sqrt{2\pi}\sigma_1} \cdot \exp\left[\dfrac{-j^2}{2\sigma_1^2}\right] \\ J_B \text{ probability } p_2 & f_{J_B}(j) = \dfrac{1}{\sqrt{2\pi}\sigma_2} \cdot \exp\left[\dfrac{-j^2}{2\sigma_2^2}\right] \end{cases} \tag{10.191}$$

Assume $D = 1$, $a_o = 1, v = 4$, $\sigma_1 = D/10$, $p_1 = 0.8$, $\sigma_2 = D/4$, and $p_2 = 0.2$.

a. Graph a few signals from the random process.
b. Assuming a constant pulse amplitude of $a_o$, determine and graph the power spectral density for the zero jitter case.
c. Determine and graph the power spectral density for the case of amplitude-only jitter, timing-only jitter, and both timing and amplitude jitter.

## APPENDIX 10.A PROOF OF THEOREM 10.1

First, the result from Theorem 2.13: if $x \in \boldsymbol{L}^2[0, T]$ then $x \in L[0, T]$, implies that finite energy is consistent with absolute integrability, that is,

$$\int_0^T |X(\omega, t)|^2 dt < \infty \Rightarrow \int_0^T |X(\omega, t)| dt < \infty \qquad \omega \in S \tag{10.192}$$

Thus, the assumption of boundedness of the signal energy according to Equation 10.2 implies

$$\sup\left\{\int_0^T |X(\omega, t)| dt : \quad \omega \in S\right\} < \infty \tag{10.193}$$

Second, absolute integrability implies the magnitude of the Fourier transform of an individual signal is bounded for all $f \in \boldsymbol{R}$:

$$|X(\omega, T, f)| = \left|\int_0^T X(\omega, t) e^{-j2\pi ft} dt\right| \leq \int_0^T |X(\omega, t)| dt < \infty \qquad f \in \boldsymbol{R}, \omega \in S \tag{10.194}$$

It then follows, for an individual signal, that $|X(\omega, T, f)|^2$ and, hence, $G(\omega, T, f)$, is finite for all $f \in \boldsymbol{R}$, $\omega \in S$. Thus, Equation 10.193 implies

$$\sup\left\{ |X(\omega,T,f)| : \begin{matrix} f \in \boldsymbol{R} \\ \omega \in S \end{matrix} \right\} < \infty \qquad \sup\left\{ G_X(\omega,T,f) : \begin{matrix} f \in \boldsymbol{R} \\ \omega \in S \end{matrix} \right\} < \infty \tag{10.195}$$

Third, the power spectral density of the random process, for the countable case, is

$$G_X(T,f) = \sum_{i=1}^{\infty} p_\Omega(\omega_i) G_X(\omega_i, T, f) \tag{10.196}$$

Convergence is guaranteed for all $f \in \boldsymbol{R}$ if $\sup\{G_X(\omega_i, T, f) : \ \omega_i \in S, f \in \boldsymbol{R}\} < \infty$, and this holds as stated in Equation 10.195. An analogous result holds for the uncountable case.

## APPENDIX 10.B PROOF OF THEOREM 10.2

Consider the uncountable case. Independence of the two random processes implies that $f_{\Omega_1\Omega_2}(\omega_1, \omega_2) = f_{\Omega_1}(\omega_1) f_{\Omega_2}(\omega_2)$, and it then follows that the cross power spectral density, as defined by Equation 10.6, is

$$G_{X_1X_2}(T,f) = \frac{1}{T} \int_{-\infty}^{\infty} f_{\Omega_1}(\omega_1) X_1(\omega_1, T, f) d\omega_1 \int_{-\infty}^{\infty} X_2^*(\omega_2, T, f) f_{\Omega_2}(\omega_2) d\omega_2 \tag{10.197}$$

Substituting for the Fourier transform of the individual signals results in

$$\begin{aligned} G_{X_1X_2}(T,f) = \frac{1}{T} \int_{-\infty}^{\infty} f_{\Omega_1}(\omega_1) \left[ \int_0^T X_1(\omega_1, t) e^{-j2\pi f t} dt \right] d\omega_1 \cdot \\ \int_{-\infty}^{\infty} \left[ \int_0^T X_2^*(\omega_2, t) e^{j2\pi f t} dt \right] f_{\Omega_2}(\omega_2)\, d\omega_2 \end{aligned} \tag{10.198}$$

Interchanging the order of integration results in

$$\begin{aligned} G_{X_1X_2}(T,f) = \frac{1}{T} \int_0^T e^{-j2\pi f t} \left[ \int_{-\infty}^{\infty} X_1(\omega_1, t) f_{\Omega_1}(\omega_1)\, d\omega_1 \right] dt \cdot \\ \int_0^T e^{j2\pi f t} \left[ \int_{-\infty}^{\infty} X_2^*(\omega_2, t) f_{\Omega_2}(\omega_2)\, d\omega_2 \right] dt \end{aligned} \tag{10.199}$$

Thus,

$$G_{X_1X_2}(T,f) = \frac{1}{T}\int_0^T e^{-j2\pi ft}\mu_{X_1}(t)dt \cdot \int_0^T e^{j2\pi ft}\mu_{X_2}{}^*(t)dt = \frac{\mu_{X_1}(T,f)\mu_{X_2}{}^*(T,f)}{T} \qquad (10.200)$$

For the constant mean case,

$$G_{X_1X_2}(T,f) = \frac{\mu_{X_1}\mu_{X_2}{}^*}{T} \cdot \int_0^T e^{-j2\pi ft}dt \cdot \int_0^T e^{j2\pi ft}dt = \frac{\mu_{X_1}\mu_{X_2}{}^*}{T}\left|\frac{e^{-j2\pi fT}-1}{-j2\pi f}\right|^2 \qquad (10.201)$$

Using the result

$$\frac{e^{-j2\pi fT}-1}{-j2\pi f} = e^{-j\pi fT} \cdot \frac{e^{-j\pi fT}-e^{j\pi fT}}{-j2\pi f} = Te^{-j\pi fT} \cdot \text{sinc}(f\text{T}) \qquad (10.202)$$

it follows that

$$G_{X_1X_2}(T,f) = \mu_{X_1}\mu_{X_2}{}^* \cdot T\text{sinc}(fT)^2 \qquad (10.203)$$

It is the case that $\lim_{T\to\infty} T\text{sinc}(fT)^2 = \delta(f)$ (Theorem 2.33) and the result for the infinite interval follows.

## APPENDIX 10.C PROOF OF THEOREM 10.3

Define $Z_N(\Omega, t) = \sum_{i=1}^{N} w_i X_i(\Omega_i, t)$ and it then follows that $G_Z(T,f) = \lim_{N\to\infty} G_{Z_N}(T,f)$. Assume the experiment underpinning the $N$ random processes $X_1(\Omega_1, t), \ldots, X_N(\Omega_2, t)$ defines a vector of random variables $(\Omega_1, \ldots, \Omega_N)$ and the sample space

$$S = \{(\omega_1,\ldots,\omega_N): \ \omega_i \in S_i\} \qquad (10.204)$$

with $S_i$ being the sample space of the random variable $\Omega_i$. The proof is given for the countable case where

$$P[\Omega_1 = \omega_1,\ldots,\Omega_N = \omega_N] = p_{\Omega_1\ldots\Omega_N}(\omega_1,\ldots,\omega_N) \qquad (10.205)$$

The proof for the continuous case follows in an analogous manner.

By definition, the power spectral density of $Z_N$ is

$$G_{Z_N}(T,f) = \frac{1}{T}\sum_{S_1}\cdots\sum_{S_N} p_{\Omega_1\ldots\Omega_N}(\omega_1,\ldots,\omega_N)|Z_N((\omega_1,\ldots,\omega_N),T,f)|^2 \qquad (10.206)$$

where

$$Z_N((\omega_1,\ldots,\omega_N),T,f)=\int_0^T Z_N((\omega_1,\ldots,\omega_N),t)e^{-j2\pi ft}dt \qquad (10.207)$$

$$Z_N((\omega_1,\ldots,\omega_N),t)=\sum_{i=1}^{N} w_i X_i(\omega_i,t) \qquad (10.208)$$

Substitution of Equation 10.208 into Equation 10.207 yields

$$Z_N((\omega_1,\ldots,\omega_N),T,f)=\int_0^T\left[\sum_{i=1}^{N} w_i X_i(\omega_i,t)\right]e^{-j2\pi ft}dt=\sum_{i=1}^{N} w_i X_i(\omega_i,T,f) \qquad (10.209)$$

after interchanging the order of summation and integral. Consistent with Theorem 10.1 and the assumption of finite average signal power, $Z_N$ is finite for all values of $f$. Thus,

$$|Z_N((\omega_1,\ldots,\omega_N),T,f)|^2=\sum_{i=1}^{N}\sum_{j=1}^{N} w_i w_j^* X_i(\omega_i,T,f)X_j^*(\omega_j,T,f) \qquad (10.210)$$

and the power spectral density is

$$G_{Z_N}(T,f)= \frac{1}{T}\sum_{S_1}\cdots\sum_{S_N}\sum_{i=1}^{N}\sum_{j=1}^{N} p_{\Omega_1\ldots\Omega_N}(\omega_1,\ldots,\omega_N)w_i w_j^* X_i(\omega_i,T,f)X_j^*(\omega_j,T,f) \qquad (10.211)$$

The summations are absolutely convergent, and hence, according to the Fubini–Tonelli theorem, the order of summation can be interchanged to yield

$$G_{Z_N}(T,f)=\sum_{i=1}^{N}\sum_{S_1}\cdots\sum_{S_N} p_{\Omega_1\ldots\Omega_N}(\omega_1,\ldots,\omega_N)\frac{|w_i|^2|X_i(\omega_i,T,f)|^2}{T}$$
$$+\sum_{i=1}^{N}\sum_{\substack{j=1\\ j\neq i}}^{N}\sum_{S_1}\cdots\sum_{S_N} p_{\Omega_1\ldots\Omega_N}(\omega_1,\ldots,\omega_N)\frac{w_i w_j^*}{T}X_i(\omega_i,T,f)X_j^*(\omega_j,T,f) \qquad (10.212)$$

Using the results from conditional probability:

$$p_{\Omega_1\ldots\Omega_N}(\omega_1,\ldots,\omega_N)=p_{\Omega_i}(\omega_i)\cdot p_{\Omega_1\ldots\Omega_{i-1}\,\Omega_{i+1}\ldots\Omega_N/\Omega_i=\omega_i}(\omega_1,\ldots,\omega_{i-1},\omega_{i+1},\ldots,\omega_N) \qquad (10.213)$$

$$\sum_{S_1}\cdots\sum_{S_{i-1}}\sum_{S_{i+1}}\cdots\sum_{S_N} p_{\Omega_1\ldots\Omega_{i-1}\ \Omega_{i+1}\ldots\Omega_N/\Omega_i=\omega_i}(\omega_1,\ldots,\omega_{i-1},\omega_{i+1},\ldots,\omega_N)=1 \tag{10.214}$$

it follows that the first summation in Equation 10.212 can be written as

$$\begin{aligned}&\sum_{i=1}^{N}|w_i|^2\sum_{S_i}\sum_{S_1}\cdots\sum_{S_{i-1}}\sum_{S_{i+1}}\cdots\sum_{S_N} p_{\Omega_i}(\omega_i)\cdot\frac{|X_i(\omega_i,T,f)|^2}{T}\cdot\\ &p_{\Omega_1\ldots\Omega_{i-1}\ \Omega_{i+1}\ldots\Omega_N/\Omega_i=\omega_i}(\omega_1,\ldots,\omega_{i-1},\omega_{i+1},\ldots,\omega_N)=\\ &\sum_{i=1}^{N}|w_i|^2\sum_{S_i}p_{\Omega_i}(\omega_i)\frac{|X_i(\omega_i,T,f)|^2}{T}=\sum_{i=1}^{N}|w_i|^2G_{X_i}(T,f)\end{aligned} \tag{10.215}$$

Similarly, using the result:

$$\begin{aligned}&p_{\Omega_1\ldots\Omega_N}(\omega_1,\ldots,\omega_N)=p_{\Omega_i\Omega_j}(\omega_i,\omega_j)\cdot\\ &p_{\Omega_{1N/ij}}(\omega_1,\ldots,\omega_{i-1},\omega_{i+1},\ldots,\omega_{j-1},\omega_{j+1},\ldots,\omega_N)\end{aligned} \tag{10.216}$$

where the following notation has been used:

$$\Omega_{1N/ij}=\Omega_1\ldots\Omega_{i-1}\Omega_{i+1}\ldots\Omega_{j-1}\Omega_{j+1}\ldots\Omega_N/\Omega_i=\omega_i,\Omega_j=\omega_j \tag{10.217}$$

it follows that the summations with respect to $\omega_1,\ldots,\omega_N$ in the second term of Equation 10.212 can be written as

$$\begin{aligned}&\sum_{S_i}\sum_{S_j}\sum_{S_1}\cdots\sum_{S_{i-1}}\sum_{S_{i+1}}\cdots\sum_{S_{j-1}}\sum_{S_{j+1}}\cdots\sum_{S_N}p_{\Omega_i\Omega_j}(\omega_i,\omega_j)\cdot\frac{X_i(\omega_i,T,f)X_j^*(\omega_j,T,f)}{T}\cdot\\ &p_{\Omega_{1N/ij}}(\omega_1,\ldots,\omega_{i-1},\omega_{i+1},\ldots,\omega_{j-1},\omega_{j+1},\ldots,\omega_N)]=G_{X_iX_j}(T,f)\end{aligned} \tag{10.218}$$

Substitution of these two results into Equation 10.212 yields the required result, namely,

$$G_{Z_N}(T,f)=\sum_{i=1}^{N}|w_i|^2G_{X_i}(T,f)+\sum_{i=1}^{N}\sum_{\substack{j=1\\ j\neq i}}^{N}w_iw_j^*G_{X_iX_j}(T,f) \tag{10.219}$$

## APPENDIX 10.D PROOF OF THEOREM 10.5

Consider the Fourier transform of the pulse train on the interval $[0, ND]$:

$$\begin{aligned}X(ND,f)&=\int_0^{ND}\left[\sum_{k=0}^{N-1}Ap(t-kD)\right]e^{-j2\pi ft}dt\\ &=\sum_{k=0}^{N-1}A\int_0^{ND}p(t-kD)e^{-j2\pi ft}dt\end{aligned} \tag{10.220}$$

Assuming the restriction of the interval $[0, ND]$ does not significantly affect the Fourier transform of the pulse functions, it follows that

$$X(ND,f) \approx \sum_{k=0}^{N-1} AP(f)e^{-j2\pi fkD} = AP(f)\sum_{k=1}^{N} e^{-j2\pi f(k-1)D} \tag{10.221}$$

It then follows that

$$\begin{aligned} |X(ND,f)|^2 &\approx A^2|P(f)|^2 \sum_{i=1}^{N}\sum_{k=1}^{N} e^{-j2\pi f(i-1)D} e^{j2\pi f(k-1)D} \\ &= A^2|P(f)|^2 N + A^2|P(f)|^2 \sum_{i=1}^{N}\sum_{k=1,\, k\neq i}^{N} e^{-j2\pi f(i-k)D} \end{aligned} \tag{10.222}$$

Using a result from Theorem 2.33, it follows that

$$\begin{aligned} \sum_{i=1}^{N}\sum_{k=1,\, k\neq i}^{N} e^{-j2\pi f(i-k)D} &= \left|\sum_{i=1}^{N} e^{-j2\pi fiD}\right|^2 - N \\ &= \frac{\sin(\pi Nf/r)^2}{\sin(\pi f/r)^2} - N \end{aligned} \tag{10.223}$$

where $r = 1/D$. The power spectral density then is

$$\begin{aligned} G(ND,f) = \frac{|X(ND,f)|^2}{ND} &= \frac{A^2|P(f)|^2 N}{ND} + \frac{A^2|P(f)|^2}{ND}\left[\frac{\sin(\pi Nf/r)^2}{\sin(\pi f/r)^2} - N\right] \\ &= rA^2|P(f)|^2\left[\frac{1}{N}\cdot\frac{\sin(\pi Nf/r)^2}{\sin(\pi f/r)^2}\right] \end{aligned} \tag{10.224}$$

As $\lim_{N\to\infty} \frac{1}{N}\cdot\frac{\sin(\pi Nf/r)^2}{\sin(\pi f/r)^2} = r\sum_{n=-\infty}^{\infty}\delta(f-nr)$ (Theorem 2.33), it follows that

$$G(\infty,f) = r^2A^2|P(f)|^2 \sum_{n=-\infty}^{\infty}\delta(f-nr) \tag{10.225}$$

## APPENDIX 10.E PROOF OF THEOREM 10.6

Consider the uncountable case. First, the Fourier transform of one signal from the random process is

$$X(\omega,t) = \sum_{k=1}^{N}\phi(\omega_k, t-(k-1)D) \to X(\omega,f) = \sum_{k=1}^{N}\Phi(\omega_k,f)e^{-j2\pi f(k-1)D} \tag{10.226}$$

where $\Phi(\omega_i, f)$ is the Fourier transform of $\phi(\omega_i, t)$ evaluated over $[0, \infty]$. The power spectral density (Eq. 10.5) then is

$$G_X(T,f) = \frac{1}{T}\int_{-\infty}^{\infty}\cdots\int_{-\infty}^{\infty}\left[\sum_{i=1}^{N}\sum_{k=1}^{N}\Phi(\omega_i,f)e^{-j2\pi f(i-1)D}\Phi^*(\omega_k,f)e^{j2\pi f(k-1)D}\right]\cdot f_{\Omega_1\ldots\Omega_N}(\omega_1,\ldots,\omega_N)d\omega_1\ldots d\omega_N \tag{10.227}$$

The assumption here is that use of the interval $[0, ND]$, rather than the infinite interval $[0, \infty]$, has negligible effect on the Fourier transform of the signals from the signal set. This is generally acceptable for practical signalling waveforms and the case of $N$ large. Interchanging the order of summation and integrations yields

$$G_X(T,f) = \frac{1}{T}\sum_{i=1}^{N}\sum_{k=1}^{N}e^{-j2\pi f(i-1)D}\int_{-\infty}^{\infty}\cdots\int_{-\infty}^{\infty}\Phi(\omega_i,f)\Phi^*(\omega_k,f)\cdot f_{\Omega_1\ldots\Omega_N}(\omega_1,\ldots,\omega_N)d\omega_1\ldots d\omega_N \tag{10.228}$$

Using the result from conditional probability:

$$f_{\Omega_1\ldots\Omega_N}(\omega_1,\ldots,\omega_N) = f_{\Omega_i\Omega_k}(\omega_i,\omega_k)\cdot f_{\Omega_{1N/ik}}(\omega_1,\ldots,\omega_{i-1},\omega_{i+1},\ldots,\omega_{k-1},\omega_{k+1},\ldots,\omega_N) \tag{10.229}$$

where the following notation has been used:

$$\Omega_{1N/ik} = \Omega_1\ldots\Omega_{i-1}\Omega_{i+1}\ldots\Omega_{k-1}\Omega_{k+1}\ldots\Omega_N/\Omega_i = \omega_i, \Omega_k = \omega_k \tag{10.230}$$

it follows that

$$G_X(T,f) = \frac{1}{N}\sum_{i=1}^{N}\sum_{k=1}^{N}e^{-j2\pi f(i-k)D}\left[\frac{1}{D}\int_{-\infty}^{\infty}\int_{-\infty}^{\infty}\Phi(\omega_i,f)\Phi^*(\omega_k,f)\cdot f_{\Omega_i\Omega_k}(\omega_i,\omega_k)\right]\cdot \int_{-\infty}^{\infty}\cdots\int_{-\infty}^{\infty}f_{\Omega_{1N/ik}}(\omega_1,\ldots,\omega_{i-1},\omega_{i+1},\ldots,\omega_{k-1},\omega_{k+1},\ldots,\omega_N)\cdot d\omega_1,\ldots,d\omega_{i-1},d\omega_{i+1},\ldots,d\omega_{k-1},d\omega_{k+1},\ldots,d\omega_N \tag{10.231}$$

The last $N-2$ integrals are of a density function with the result of unity. The inner two integrals, by definition, are the cross power spectral density of the random processes defined by the $i$th and $k$th signalling waveforms. Hence, using the definition specified in Equation 10.38, the power spectral density can be written as

$$G_X(T,f) = \frac{1}{N}\sum_{i=1}^{N}\sum_{k=1}^{N}e^{-j2\pi f(i-k)D}G_{\Phi,ik}(f) \tag{10.232}$$

Splitting the double summation into the diagonal and nondiagonal terms yields the required result:

$$G_X(T,f) = \frac{1}{N}\sum_{i=1}^{N} G_{\Phi,i}(f) + \frac{1}{N}\sum_{i=1}^{N}\sum_{\substack{k=1\\k\neq i}}^{N} e^{-j2\pi f(i-k)D} G_{\Phi,ik}(f) \tag{10.233}$$

### 10.E.1 Independent Case

The power spectral density result for the independent case, as given by Equation 10.42, follows from Equation 10.39.

For the independent and identical case, it follows from Equation 10.42 that

$$G_X(T,f) = G_\Phi(f) + \frac{|\mu_\Phi(f)|^2}{T}\sum_{i=1}^{N}\sum_{k=1,\,k\neq i}^{N} e^{-j2\pi f(i-k)D} \tag{10.234}$$

It then follows—see Equation 10.223—that

$$G_X(T,f) = G_\Phi(f) + r|\mu_\Phi(f)|^2\left[\frac{1}{N}\cdot\frac{\sin(\pi fN/r)^2}{\sin(\pi f/r)^2} - 1\right] \tag{10.235}$$

As $\lim_{N\to\infty}\frac{1}{N}\cdot\frac{\sin(\pi Nf/r)^2}{\sin(\pi f/r)^2} = r\sum_{n=-\infty}^{\infty}\delta(f-nr)$ (Theorem 2.33), it follows that

$$G_X(\infty,f) = G_\Phi(f) - r|\mu_\Phi(f)|^2 + r^2|\mu_\Phi(f)|^2\sum_{n=-\infty}^{\infty}\delta(f-nr) \tag{10.236}$$

which is the second last required result.

## APPENDIX 10.F PROOF OF THEOREM 10.7

Consider the uncountable case and the case where correlations only depend on the difference between the location of the signalling intervals, and not on their absolute location, consistent with $f_{\Omega_i\Omega_k}(\omega_i,\omega_k) = f_{\Omega_{i+q}\Omega_{k+q}}(\omega_i,\omega_k)$. The general form for the power spectral density, as given by Equation 10.40 in Theorem 10.6, applies. First, consider the double summation, denoted $S_2$, in this result:

$$\begin{aligned} S_2 &= \frac{1}{N}\sum_{i=1}^{N}\sum_{\substack{k=1\\k\neq i}}^{N} e^{-j2\pi(i-k)Df} G_{\Phi,ik}(f) \\ &= \frac{1}{T}\sum_{i=1}^{N}\sum_{\substack{k=1\\k\neq i}}^{N} e^{-j2\pi(i-k)Df}\int_{-\infty}^{\infty}\int_{-\infty}^{\infty}\Phi(\omega_i,f)\Phi^*(\omega_k,f)f_{\Omega_i\Omega_k}(\omega_i,\omega_k)\,d\omega_i d\omega_k \end{aligned} \tag{10.237}$$

Expanding the double summation out yields

$$\begin{aligned} S_2 &= \frac{e^{j2\pi Df}}{T}\int_{-\infty}^{\infty}\int_{-\infty}^{\infty}\Phi(\omega_1,f)\Phi^*(\omega_2,f)f_{\Omega_1\Omega_2}(\omega_1,\omega_2)d\omega_1 d\omega_2 \\ &+\frac{e^{-j2\pi Df}}{T}\int_{-\infty}^{\infty}\int_{-\infty}^{\infty}\Phi(\omega_2,f)\Phi^*(\omega_1,f)f_{\Omega_2\Omega_1}(\omega_2,\omega_1)d\omega_2 d\omega_1 \\ &+\ldots\frac{e^{j2\pi(N-1)Df}}{T}\int_{-\infty}^{\infty}\int_{-\infty}^{\infty}\Phi(\omega_1,f)\Phi^*(\omega_N,f)f_{\Omega_1\Omega_N}(\omega_1,\omega_N)d\omega_1 d\omega_N \\ &+\frac{e^{-j2\pi(N-1)Df}}{T}\int_{-\infty}^{\infty}\int_{-\infty}^{\infty}\Phi(\omega_N,f)\Phi^*(\omega_1,f)f_{\Omega_N\Omega_1}(\omega_N,\omega_1)d\omega_N d\omega_1+\ldots \end{aligned} \tag{10.238}$$

To further simplify this equation, first, note that there are $N-1$ terms where $k=i+1$, $k=i-1$, $N-2$ terms where $k=i+2$, $k=i-2\ \ldots$ and 1 term where $k=i+(N-1)$, $k=i-(N-1)$. Second, note the assumption $f_{\Omega_i\Omega_k}(\omega_i,\omega_k)=f_{\Omega_{i+q}\Omega_{k+q}}(\omega_i,\omega_k)$. Third, note that

$$\begin{aligned} &f_{\Omega_i\Omega_k}(\omega_i,\omega_k)d\omega_i d\omega_k \approx P[\Omega_i\in[\omega_i,\omega_i+d\omega_i],\Omega_k\in[\omega_k,\omega_k+d\omega_k]] \\ &=P[\Omega_k\in[\omega_k,\omega_k+d\omega_k],\Omega_i\in[\omega_i,\omega_i+d\omega_i]]\approx f_{\Omega_k\Omega_i}(\omega_k,\omega_i)d\omega_k d\omega_i \end{aligned} \tag{10.239}$$

It then follows that

$$\begin{aligned} S_2 &= \frac{2N}{T}\left[1-\frac{1}{N}\right]\mathrm{Re}\left[e^{j2\pi Df}\int_{-\infty}^{\infty}\int_{-\infty}^{\infty}\Phi(\omega_1,f)\Phi^*(\omega_2,f)f_{\Omega_1\Omega_2}(\omega_1,\omega_2)d\omega_1 d\omega_2\right] \\ &+\frac{2N}{T}\left[1-\frac{2}{N}\right]\mathrm{Re}\left[e^{j2\pi 2Df}\int_{-\infty}^{\infty}\int_{-\infty}^{\infty}\Phi(\omega_1,f)\Phi^*(\omega_3,f)f_{\Omega_1\Omega_3}(\omega_1,\omega_3)d\omega_1 d\omega_3\right]+\cdots \\ &+\frac{2N}{T}\left[1-\frac{N-1}{N}\right]\mathrm{Re}\left[e^{j2\pi(N-1)Df}\int_{-\infty}^{\infty}\int_{-\infty}^{\infty}\Phi(\omega_1,f)\Phi^*(\omega_N,f)f_{\Omega_1\Omega_N}(\omega_1,\omega_N)d\omega_1 d\omega_N\right] \end{aligned} \tag{10.240}$$

which can be rewritten as

$$S_2 = 2r\sum_{i=1}^{N-1}\left[1-\frac{i}{N}\right]\cdot$$

$$\mathrm{Re}\left[e^{j2\pi iDf}\int_{-\infty}^{\infty}\int_{-\infty}^{\infty}\Phi(\omega_1,f)\Phi^*(\omega_{1+i},f)f_{\Omega_1\Omega_{1+i}}(\omega_1,\omega_{1+i})d\omega_1 d\omega_{1+i}\right] \quad (10.241)$$

$$= 2\sum_{i=1}^{N-1}\left[1-\frac{i}{N}\right]\mathrm{Re}\left[e^{j2\pi iDf}G_{\Phi,1,1+i}(f)\right]$$

Hence, the power spectral density, as given by Equation 10.40, becomes

$$G_X(ND,f) = \frac{1}{N}\sum_{i=1}^{N}G_{\Phi,i}(f) + 2\sum_{i=1}^{N-1}\left[1-\frac{i}{N}\right]\mathrm{Re}\left[e^{j2\pi iDf}G_{\Phi,1,1+i}(f)\right] \quad (10.242)$$

which is the first required result.

For the independent case, the double summation, as given by Equation 10.241, can be simplified using Equation 10.39, $G_{\Phi,ik}(f) = r\mu_{\Phi,i}(f)\mu_{\Phi,k}^*(f)$, according to

$$S_2 = r\sum_{i=1}^{N-1}\left[1-\frac{i}{N}\right]\mathrm{Re}\left[e^{j2\pi iDf}\mu_{\Phi,1}(f)\mu_{\Phi,1+i}^*(f)\right] \quad (10.243)$$

It then follows, for the case where the $i$th signalling waveform is independent of the $j$th signalling waveform for $|i-j|>m$, that

$$G_X(ND,f) = \frac{1}{N}\sum_{i=1}^{N}G_{\Phi,i}(f) + 2\sum_{i=1}^{m}\left[1-\frac{i}{N}\right]\mathrm{Re}\left[e^{j2\pi iDf}G_{\Phi,1,1+i}(f)\right]$$
$$+ 2r\sum_{i=m+1}^{N-1}\left[1-\frac{i}{N}\right]\mathrm{Re}\left[e^{j2\pi iDf}\mu_{\Phi,1}(f)\mu_{\Phi,1+i}^*(f)\right] \quad (10.244)$$

which can be written as

$$G_X(ND,f) = \frac{1}{N}\sum_{i=1}^{N}G_{\Phi,i}(f) + 2r\sum_{i=1}^{N-1}\left[1-\frac{i}{N}\right]\mathrm{Re}\left[e^{j2\pi iDf}\mu_{\Phi,1}(f)\mu_{\Phi,1+i}^*(f)\right]$$
$$+ 2\sum_{i=1}^{m}\left[1-\frac{i}{N}\right]\mathrm{Re}\left[e^{j2\pi iDf}\left[G_{\Phi,1,1+i}(f) - r\mu_{\Phi,1}(f)\mu_{\Phi,1+i}^*(f)\right]\right]$$
$$(10.245)$$

For the identical case where $G_{\Phi,i}(f) = G_{\Phi,k}(f) = G_{\Phi}(f)$ and $\mu_{\Phi,i}(f) = \mu_{\Phi,k}(f) = \mu_{\Phi}(f)$, it follows that

$$\begin{aligned} G_X(ND,f) = {} & G_{\Phi}(f) - r|\mu_{\Phi}(f)|^2 + r|\mu_{\Phi}(f)|^2 \cdot \frac{1}{N} \cdot \frac{\sin(\pi Nf/r)^2}{\sin(\pi f/r)^2} \\ & + 2\sum_{i=1}^{m}\left[1 - \frac{i}{N}\right] \mathrm{Re}\left[e^{j2\pi iDf}\left[G_{\Phi,1,1+i}(f) - r|\mu_{\Phi}(f)|^2\right]\right] \end{aligned} \tag{10.246}$$

The second and third terms in this equation arise from the result:

$$\begin{aligned} 2\sum_{i=1}^{N-1}\left[1 - \frac{i}{N}\right]\mathrm{Re}\left[e^{j2\pi iDf}\right] &= \frac{1}{N} \cdot \sum_{i=1}^{N} \sum_{k=1,\ k\neq i}^{N} e^{-j2\pi(i-k)Df} \\ &= \frac{1}{N} \cdot \frac{\sin(\pi Nf/r)^2}{\sin(\pi f/r)^2} - 1 \end{aligned} \tag{10.247}$$

The first result in this equation follows from expanding out the double summation and collecting terms which are the complex conjugate of one another. The second result in this equation is consistent with Equation 10.223.

As $\lim_{N\to\infty} \frac{1}{N} \cdot \frac{\sin(\pi NfD)^2}{\sin(\pi fD)^2} = r \sum_{n=-\infty}^{\infty} \delta(f - nr)$ (Theorem 2.33), it follows that

$$\begin{aligned} G_X(\infty,f) = {} & G_{\Phi}(f) - r|\mu_{\Phi}(f)|^2 + r^2|\mu_{\Phi}(f)|^2 \sum_{n=-\infty}^{\infty} \delta(f - nr) \\ & + 2\sum_{i=1}^{m} \mathrm{Re}\left[e^{j2\pi iDf}\left[G_{\Phi,1,1+i}(f) - r|\mu_{\Phi}(f)|^2\right]\right] \end{aligned} \tag{10.248}$$

which is the required result for the infinite interval.

## APPENDIX 10.G PROOF OF THEOREM 10.8

The power spectral density for the general case is given by Theorem 10.6. The required results follow by evaluating $\mu_{\Phi,i}(f)$, $G_{\Phi,i}(f)$ and $G_{\Phi,ik}(f)$ as given by Equations 10.36–10.38 for the case

$$\phi(\omega_i, t) = A(\omega_i)p(t) \leftrightarrow \Phi(\omega_i, f) = A(\omega_i)P(f) \tag{10.249}$$

The following results hold: first, from Equation 10.36,

$$\mu_{\Phi,i}(f) = \left\{ \begin{array}{ll} P(f)\displaystyle\sum_{\omega_i \in S_i} p_{\Omega_i}(\omega_i)A(\omega_i) & \text{countable} \\ P(f)\displaystyle\int_{-\infty}^{\infty} A(\omega_i) f_{\Omega_i}(\omega_i)\,d\omega_i & \text{uncountable} \end{array} \right\} = \mu_{A,i}P(f) \tag{10.250}$$

Similarly, Equations 10.37 and 10.38 become:

$$G_{\Phi,i}(f)=\left\{\begin{array}{ll}\dfrac{|P(f)|^2}{D}\displaystyle\sum_{\omega_i\in S_i}p_{\Omega_i}(\omega_i)|A(\omega_i)|^2 & \text{c.c.}\\ \dfrac{|P(f)|^2}{D}\displaystyle\int_{-\infty}^{\infty}|A(\omega_i)|^2 f_{\Omega_i}(\omega_i)d\omega_i & \text{u.c.}\end{array}\right\}=r|P(f)|^2\overline{|A_i|^2} \tag{10.251}$$

$$G_{\Phi,ik}(f)=\left\{\begin{array}{ll}\dfrac{|P(f)|^2}{D}\displaystyle\sum_{\omega_i\in S_i}\sum_{\omega_k\in S_k}p_{\Omega_i\Omega_k}(\omega_i,\omega_k)A(\omega_i)A^*(\omega_k) & \text{c.c.}\\ \dfrac{|P(f)|^2}{D}\displaystyle\int_{-\infty}^{\infty}\int_{-\infty}^{\infty}A(\omega_i)A^*(\omega_k)f_{\Omega_i\Omega_k}(\omega_i,\omega_k)d\omega_i d\omega_k & \text{u.c.}\end{array}\right. \tag{10.252}$$

$$=r|P(f)|^2\boldsymbol{E}[A_iA_k^*]$$

Substitution of these results into Equation 10.40 yields the first required result. Subsequent results follow by using simplifications that arise from the identical and independent cases and the result

$$G_{\Phi,ik}(f)=r|P(f)|^2\mu_{A,i}\mu_{A,k}^* \tag{10.253}$$

which holds for the independent case.

## APPENDIX 10.H PROOF OF THEOREM 10.10

By definition, the power spectral density of $X(\Omega,t)=\sum_{k=1}^{N}p(t-\Gamma_k)$ on the interval $[0, T]$ is

$$G_X(T,f)=\frac{1}{T}\int_0^T\cdots\int_0^T|X((\gamma_1,\ldots,\gamma_N),T,f)|^2 f_{\Gamma_1\ldots\Gamma_N}(\gamma_1,\ldots,\gamma_N)d\gamma_1\ldots d\gamma_N \tag{10.254}$$

where

$$f_{\Gamma_1\ldots\Gamma_N}(\gamma_1,\ldots,\gamma_N)=f_{\Gamma_1}(\gamma_1)\cdot\ldots\cdot f_{\Gamma_N}(\gamma_N)=\frac{1}{T^N} \tag{10.255}$$

$$X((\gamma_1,\ldots,\gamma_N),T,f)=\int_{-\infty}^{\infty}\left[\sum_{k=1}^{N}p(t-\gamma_k)\right]e^{-j2\pi ft}\approx P(f)\sum_{k=1}^{N}e^{-j2\pi f\gamma_k} \tag{10.256}$$

Here, $P(f)$ is the Fourier transform of $p(t)$ assuming the energy of the pulses outside the interval $[0, T]$ is negligible. It then follows that

$$G_X(T,f) \approx \frac{|P(f)|^2}{T^{N+1}} \int_0^T \cdots \int_0^T \sum_{i=1}^{N} \sum_{k=1}^{N} e^{-j2\pi f(\gamma_i - \gamma_k)} d\gamma_1 \ldots d\gamma_N \tag{10.257}$$

Interchanging the order of summation and splitting the double summation into its diagonal, and non-diagonal, terms yield

$$\begin{aligned} G_X(T,f) \approx {} & \frac{|P(f)|^2}{T^{N+1}} \cdot \sum_{i=1}^{N} \int_0^T \cdots \int_0^T d\gamma_1 \ldots d\gamma_N \\ & + \frac{|P(f)|^2}{T^{N+1}} \cdot \sum_{i=1}^{N} \sum_{\substack{k=1 \\ k \neq i}}^{N} \int_0^T \cdots \int_0^T e^{-j2\pi f(\gamma_i - \gamma_k)} d\gamma_1 \ldots d\gamma_N \end{aligned} \tag{10.258}$$

Using the definition of $N = \lambda T$, it follows that

$$G_X(T,f) \approx \lambda |P(f)|^2 + \frac{|P(f)|^2}{T^3} \cdot \sum_{i=1}^{N} \sum_{\substack{k=1 \\ k \neq i}}^{N} \int_0^T \int_0^T e^{-j2\pi f(\gamma_i - \gamma_k)} d\gamma_i d\gamma_k \tag{10.259}$$

Evaluating the integrals yields

$$G_X(T,f) \approx \lambda |P(f)|^2 + \frac{|P(f)|^2}{T^3} \cdot \sum_{i=1}^{N} \sum_{\substack{k=1 \\ k \neq i}}^{N} \left| \frac{e^{-j2\pi fT} - 1}{-j2\pi f} \right|^2 \tag{10.260}$$

Using the result

$$\begin{aligned} \frac{e^{-j2\pi fT} - 1}{-j2\pi f} & = e^{-j\pi fT} \cdot \left[ \frac{e^{-j\pi fT} - e^{j\pi fT}}{-j2\pi f} \right] = e^{-j\pi fT} \cdot \frac{\sin(\pi fT)}{\pi f} \\ & = T e^{-j\pi fT} \cdot \operatorname{sinc}(fT) \end{aligned} \tag{10.261}$$

it follows that

$$\begin{aligned} G_X(T,f) & \approx \lambda |P(f)|^2 + \frac{|P(f)|^2}{T} \cdot (N^2 - N) \cdot \operatorname{sinc}(fT)^2 \\ & = \lambda |P(f)|^2 + \lambda^2 |P(f)|^2 \cdot \left(1 - \frac{1}{N}\right) \cdot T \operatorname{sinc}(fT)^2 \end{aligned} \tag{10.262}$$

which is the required result.

For the infinite interval, the result $\lim_{T\to\infty} T\,\mathrm{sinc}(fT)^2 = \delta(f)$ (Theorem 2.33) yields the required result, namely,

$$G_{X_\infty}(f) = \lim_{T\to\infty} G_X(T,f) = \lambda|P(f)|^2 + \lambda^2|P(0)|^2\delta(f) \tag{10.263}$$

## APPENDIX 10.I PROOF OF THEOREM 10.13

Consider one signal defined by the random process:

$$X((j_1,\ldots,j_N),t) = A\sum_{k=1}^{N/2} u[t-(2k-1)D-j_{2k-1}] - u[t-2kD-j_{2k}] \tag{10.264}$$

As

$$u[t-\alpha]-u[t-\beta] \leftrightarrow (\beta-\alpha)\mathrm{sinc}[f(\beta-\alpha)]e^{-j\pi f(\beta+\alpha)} \quad \beta>\alpha \tag{10.265}$$

it follows that the Fourier transform of this signal is

$$X((j_1,\ldots,j_N),ND,f) = A\sum_{k=1}^{N/2}(D+j_{2k}-j_{2k-1})\,\mathrm{sinc}[f(D+j_{2k}-j_{2k-1})]e^{-j\pi f(j_{2k}-j_{2k-1})}e^{-j\pi f(4k-1)D} \tag{10.266}$$

and

$$\begin{aligned}|X((j_1,\ldots,j_N),ND,f|^2 &= A^2\sum_{k=1}^{N/2}(D+j_{2k}-j_{2k-1})^2\mathrm{sinc}[f(D+j_{2k}-j_{2k-1})]^2 \\ &+ A^2\sum_{i=1}^{N/2}\sum_{k=1,k\neq i}^{N/2} e^{-j4\pi f(i-k)D}\cdot(D+j_{2i}-j_{2i-1})\cdot(D+j_{2k}-j_{2k-1}) \\ &\mathrm{sinc}[f(D+j_{2i}-j_{2i-1})]\mathrm{sinc}[f(D+j_{2k}-j_{2k-1})]\cdot e^{-j\pi f(j_{2i}+j_{2i-1})}e^{j\pi f(j_{2k}+j_{2k-1})}\end{aligned} \tag{10.267}$$

It then follows that the power spectral density of the random process is

$$\begin{aligned}G_X(ND,f) = \boldsymbol{E}\left[\frac{|X(\Omega,T,f)|^2}{ND}\right] &= \int_{-\infty}^{\infty}\cdots\int_{-\infty}^{\infty}\left[\frac{A^2}{ND}\sum_{k=1}^{N/2}(D+j_{2k}-j_{2k-1})^2\cdot\right. \\ &\mathrm{sinc}[f(D+j_{2k}-j_{2k-1})]^2 + \frac{A^2}{ND}\sum_{i=1}^{N/2}\sum_{\substack{k=1\\k\neq i}}^{N/2} e^{-j4\pi f(i-k)D}\cdot(D+j_{2i}-j_{2i-1})\cdot \\ &(D+j_{2k}-j_{2k-1})\cdot\mathrm{sinc}[f(D+j_{2i}-j_{2i-1})]\cdot\mathrm{sinc}[f(D+j_{2k}-j_{2k-1})]\cdot \\ &\left. e^{-j\pi f(j_{2i}+j_{2i-1})}e^{j\pi f(j_{2k}+j_{2k-1})}\right]f_{J_1\ldots J_N}(j_1,\ldots,j_N)dj_1\ldots dj_N\end{aligned} \tag{10.268}$$

Interchanging the order of integration and summation yields, for the independent case where $f_{J_1 \ldots J_N}(j_1, \ldots, j_N) = f_{J_1}(j_1) \ldots f_{J_N}(j_N)$:

$$G_X(ND, f) = \sum_{k=1}^{N/2} \frac{A^2}{ND} \int_{-\infty}^{\infty} \int_{-\infty}^{\infty} (D + j_{2k} - j_{2k-1})^2 \cdot$$

$$\text{sinc}[f(D + j_{2k} - j_{2k-1})]^2 f_{J_{2k}}(j_{2k}) f_{J_{2k-1}}(j_{2k-1}) dj_{2k} dj_{2k-1} +$$

$$\frac{A^2}{ND} \sum_{i=1}^{N/2} \sum_{\substack{k=1 \\ k \neq i}}^{N/2} e^{-j\pi 4f(i-k)D} \int_{-\infty}^{\infty} \int_{-\infty}^{\infty} \int_{-\infty}^{\infty} \int_{-\infty}^{\infty} (D + j_{2i} - j_{2i-1}) \cdot (D + j_{2k} - j_{2k-1}) \cdot \tag{10.269}$$

$$\text{sinc}[f(D + j_{2i} - j_{2i-1})] \cdot \text{sinc}[f(D + j_{2k} - j_{2k-1})] \cdot e^{-j\pi f(j_{2i} + j_{2i-1})} \cdot$$

$$e^{j\pi f(j_{2k} + j_{2k-1})} f_{J_{2i}}(j_{2i}) f_{J_{2i-1}}(j_{2i-1}) f_{J_{2k}}(j_{2k}) f_{J_{2k-1}}(j_{2k-1}) dj_{2i} dj_{2i-1} dj_{2k} dj_{2k-1}$$

For the identical and independent random variable case with a jitter probability density function $f_J$, all the double and quadruple integrals are identical resulting in

$$G_X(ND, f) =$$

$$\frac{rA^2}{ND} \int_{-\infty}^{\infty} \int_{-\infty}^{\infty} (D + j_v - j_{v-1})^2 \cdot \text{sinc}[f(D + j_v - j_{v-1})]^2 \cdot f_J(j_v) f_J(j_{v-1}) dj_v dj_{v-1} +$$

$$\left[ \int_{-\infty}^{\infty} \int_{-\infty}^{\infty} \int_{-\infty}^{\infty} \int_{-\infty}^{\infty} (D + j_v - j_{v-1}) \cdot (D + j_w - j_{w-1}) \cdot \text{sinc}[f(D + j_v - j_{v-1})] \cdot \right.$$

$$\text{sinc}[f(D + j_w - j_{w-1})] \cdot e^{-j\pi f(j_v + j_{v-1})} e^{j\pi f(j_w + j_{w-1})} \cdot f_J(j_v) f_J(j_{v-1}) \cdot$$

$$\left. f_J(j_w) f_J(j_{w-1}) dj_v dj_{v-1} dj_w dj_{w-1} \right] \cdot \frac{A^2}{ND} \sum_{i=1}^{N/2} \sum_{k=1, k \neq i}^{N/2} e^{-j4\pi f(i-k)D} \tag{10.270}$$

where $r = 1/D$. Further simplification relies on establishing a simplified expression for the double summation, which can be written as

$$\begin{aligned} \sum_{i=1}^{N/2} \sum_{k=1, k \neq i}^{N/2} e^{-j4\pi f(i-k)D} &= \sum_{i=1}^{N/2} \sum_{k=1}^{N/2} e^{-j4\pi f(i-k)D} - \frac{N}{2} \\ &= \left| \sum_{i=1}^{N/2} e^{-j4\pi f i D} \right|^2 - \frac{N}{2} \end{aligned} \tag{10.271}$$

Using a result from Theorem 2.33:

$$\left|\sum_{i=1}^{M} e^{-j2\pi if/f_o}\right|^2 = \frac{\sin(\pi Mf/f_o)^2}{\sin(\pi f/f_o)^2} \tag{10.272}$$

it follows, with $M \to N/2$ and $f_o \to 1/2D = r/2$, that

$$\sum_{i=1}^{N/2} \sum_{k=1,k\neq i}^{N/2} e^{-j\pi 4f(i-k)D} = \frac{\sin(\pi Nf/r)^2}{\sin(2\pi f/r)^2} - \frac{N}{2} \tag{10.273}$$

Hence,

$$\begin{aligned} &G_X(ND,f) = \\ &\frac{rA^2}{2}\int_{-\infty}^{\infty}\int_{-\infty}^{\infty}(D+j_v-j_{v-1})^2 \cdot \mathrm{sinc}[f(D+j_v-j_{v-1})]^2 \cdot f_J(j_v)f_J(j_{v-1})dj_v dj_{v-1} + \\ &\frac{rA^2}{2}\left[\frac{\sin(\pi Nf/r)^2}{\sin(2\pi f/r)^2} - \frac{N}{2}\right] \cdot \left|\int_{-\infty}^{\infty}\int_{-\infty}^{\infty}(D+j_v-j_{v-1}) \cdot \right. \\ &\left. \mathrm{sinc}[f(D+j_v-j_{v-1})] \cdot e^{-j\pi f(j_v+j_{v-1})} \cdot f_J(j_v)f_J(j_{v-1})dj_v dj_{v-1}\right|^2 \end{aligned} \tag{10.274}$$

With the definition of a random process $Z$ based on the jitter random variables $J_v$ and $J_{v-1}$ and defined according to

$$Z(\Omega,f) = (D+J_v-J_{v-1}) \cdot \mathrm{sinc}[f(D+J_v-J_{v-1})] \cdot e^{-j\pi f(J_v+J_{v-1})} \tag{10.275}$$

it follows that

$$\begin{aligned} G_X(ND,f) = {} & \frac{rA^2}{2}\boldsymbol{E}\left[|Z(\Omega,f)|^2\right] - \frac{rA^2}{2}|\boldsymbol{E}[Z(\Omega,f)]|^2 \\ & + rA^2\left[\frac{1}{N} \cdot \frac{\sin(\pi Nf/r)^2}{\sin(2\pi f/r)^2}\right]|\boldsymbol{E}[Z(\Omega,f)]|^2 \end{aligned} \tag{10.276}$$

The result for the infinite interval follows from the result (Theorem 2.33)

$$\lim_{N\to\infty}\frac{1}{N} \cdot \frac{\sin(\pi Nf/r)^2}{\sin(2\pi f/r)^2} = \lim_{N_o\to\infty}\frac{1}{2N_o} \cdot \frac{\sin(\pi N_o f/r_o)^2}{\sin(\pi f/r_o)^2} = \frac{r_o}{2}\sum_{n=-\infty}^{\infty}\delta(f-nr_o) \tag{10.277}$$

where $r_o = r/2$ and $N_o = N/2$. The power spectral density for the infinite interval is

$$G_X(\infty,f) = \frac{rA^2}{2}\boldsymbol{E}\left[|Z(\Omega,f)|^2\right] - \frac{rA^2}{2}|\boldsymbol{E}[Z(\Omega,f)]|^2 + \frac{r^2A^2}{4}\cdot|\boldsymbol{E}[Z(\Omega,f)]|^2\sum_{n=-\infty}^{\infty}\delta\left(f-\frac{nr}{2}\right) \quad (10.278)$$

### 10.I.1 Correlated Case

From Equation 10.268 and after interchanging the order of summation and integration, the power spectral density is

$$\begin{aligned}G_X(ND,f) = &\sum_{k=1}^{N/2}\frac{A^2}{ND}\int_{-\infty}^{\infty}\cdots\int_{-\infty}^{\infty}(D+j_{2k}-j_{2k-1})^2\cdot \mathrm{sinc}[f(D+j_{2k}-j_{2k-1})]^2\cdot\\ &f_{J_1\ldots J_N}(j_1,\ldots,j_N)dj_1\ldots dj_N + \\ &\frac{A^2}{ND}\sum_{i=1}^{N/2}\sum_{\substack{k=1\\k\neq i}}^{N/2}e^{-j4\pi f(i-k)D}\cdot\int_{-\infty}^{\infty}\cdots\int_{-\infty}^{\infty}[(D+j_{2i}-j_{2i-1})\cdot\\ &(D+j_{2k}-j_{2k-1})\cdot\mathrm{sinc}[f(D+j_{2i}-j_{2i-1})]\cdot\mathrm{sinc}[f(D+j_{2k}-j_{2k-1})]\cdot\\ &e^{-j\pi f(j_{2i}+j_{2i-1})}e^{j\pi f(j_{2k}+j_{2k-1})}]f_{J_1\ldots J_N}(j_1,\ldots,j_N)dj_1\ldots dj_N\end{aligned} \quad (10.279)$$

By using the results

$$f_{J_1\ldots J_N}(j_1,\ldots,j_N) = f_{J_{2k-1}J_{2k}}(j_{2k-1},j_{2k})\cdot f_{1N/2k-1,2k}(j_1,\ldots,j_{2k-2},j_{2k+1},\ldots,j_N) \quad (10.280)$$

$$\begin{aligned}f_{J_1\ldots J_N}(j_1,\ldots,j_N) = &f_{J_{2i-1}J_{2i}J_{2k-1}J_{2k}}(j_{2i-1},j_{2i},j_{2k-1},j_{2k})\cdot\\ &f_{1N/2i-1,2i,2k-1,2k}(j_1,\ldots,j_{2i-2},j_{2i+1},\ldots,j_{2k-2},j_{2k+1},\ldots,j_N)\end{aligned} \quad (10.281)$$

where the following notation has been used:

$$\begin{aligned}1N/2k-1,2k &= J_1\ldots J_{2k-2}J_{2k+1}\ldots J_N/J_{2k-1}=j_{2k-1},J_{2k}=j_{2k}\\ 1N/2i-1,2i,2k-1,2k &= J_1\ldots J_{2i-2}J_{2i+1}\ldots J_{2k-2}J_{2k+1}\ldots J_N/\\ &\quad J_{2i-1}=j_{2i-1},J_{2i}=j_{2i},J_{2k-1}=j_{2k-1},J_{2k}=j_{2k}\end{aligned} \quad (10.282)$$

the power spectral density, after integrating the conditional probability density functions, becomes

$$G_X(ND,f)=\sum_{k=1}^{N/2}\frac{A^2}{ND}\int_{-\infty}^{\infty}\int_{-\infty}^{\infty}(D+j_{2k}-j_{2k-1})^2\cdot$$

$$\text{sinc}[f(D+j_{2k}-j_{2k-1})]^2 f_{J_{2k-1}J_{2k}}(j_{2k-1},j_{2k})dj_{2k-1}dj_{2k}+$$

$$\frac{A^2}{ND}\sum_{i=1}^{N/2}\sum_{\substack{k=1\\k\neq i}}^{N/2}e^{-j\pi 4f(i-k)D}\int_{-\infty}^{\infty}\int_{-\infty}^{\infty}\int_{-\infty}^{\infty}\int_{-\infty}^{\infty}(D+j_{2i}-j_{2i-1})\cdot(D+j_{2k}-j_{2k-1})\cdot \tag{10.283}$$

$$\text{sinc}[f(D+j_{2i}-j_{2i-1})]\cdot\text{sinc}[f(D+j_{2k}-j_{2k-1})]\cdot e^{-j\pi f(j_{2i}+j_{2i-1})}\cdot$$

$$e^{j\pi f(j_{2k}+j_{2k-1})}f_{J_{2i-1}J_{2i}J_{2k-1}J_{2k}}(j_{2i-1},j_{2i},j_{2k-1},j_{2k})dj_{2i-1}dj_{2i}dj_{2k}dj_{2k-1}$$

which is the required result.

## APPENDIX 10.J PROOF OF THEOREM 10.15

The proof follows that given in Howard (2012b) and relies on the following result.

**Theorem 10.17 Summation of Matrix of Hermitian Terms**
For the case where $F_{ik}=F_{ki}^*$ and $F_{ik}=F_{i+r,k+r}$, $r\in\mathbf{Z}$, it follows that

$$S_2=\sum_{i=1}^{N}\sum_{k=1,k\neq i}^{N}F_{ik}=2N\sum_{i=1}^{N-1}\left[1-\frac{i}{N}\right]\text{Re}[F_{1,1+i}] \tag{10.284}$$

**Proof**
By writing out the summation according to

$$\begin{aligned}S_2=&[F_{12}+F_{23}+\cdots+F_{N-1,N}]+[F_{21}+F_{32}+\cdots+F_{N,N-1}]\\&+[F_{13}+F_{24}+\cdots+F_{N-2,N}]+[F_{31}+F_{42}+\cdots+F_{N,N-2}]+\cdots\\&+F_{1,N}+F_{N,1}\end{aligned} \tag{10.285}$$

it is clear that there are $N-1$ terms where $k=i+1$, $k=i-1$; $N-2$ terms where $k=i+2$, $k=i-2$ ...; and 1 term where $k=i+(N-1)$, $k=i-(N-1)$. Hence,

$$S_2=\sum_{i=1}^{N-1}[N-i]F_{1,1+i}+\sum_{i=1}^{N-1}[N-i]F_{1+i,1} \tag{10.286}$$

and as $F_{1+i,1} = F_{1,1+i}^*$, it follows that

$$S_2 = 2N\sum_{i=1}^{N-1}\left[1-\frac{i}{N}\right]\text{Re}[F_{1,1+i}] \tag{10.287}$$

### 10.J.1 Proof of First Result

First, with the underlying experimental sample space being

$$S = \{((a_1, j_1), \ldots, (a_N, j_N)): \; a_i \in S_{A_i}, j_i \in S_{J_i}\} \tag{10.288}$$

and with the amplitude variations being independent of the jitter variations, the outcomes of the experiment are governed by the probability density function

$$\begin{aligned} P[A_1 \in (a_1 + da_1), \ldots, A_N \in (a_N + da_N), J_1 \in (j_1 + dj_1), \ldots, J_N \in (j_N + dj_N)] \approx \\ f_{A_1 \ldots A_N}(a_1, \ldots, a_N) f_{J_1 \ldots J_N}(j_1, \ldots, j_N) da_1 \ldots da_N dj_1 \ldots dj_N \end{aligned} \tag{10.289}$$

Second, the Fourier transform of a single outcome

$$X(((a_1, j_1), \ldots, (a_N, j_N)), t) = \sum_{k=1}^{N} a_k p(t-(k-1)D-j_k) \tag{10.290}$$

is

$$X(((a_1, j_1), \ldots, (a_N, j_N)), T, f) = P(f)\sum_{k=1}^{N} a_k e^{-j2\pi f[(k-1)D+j_k)} \tag{10.291}$$

where $P(f)$ is the Fourier transform of $p(t)$ assuming the interval $[0, T]$, $T = ND$, does not significantly affect the Fourier transform. Thus,

$$\begin{aligned} &|X(((a_1, j_1), \ldots, (a_N, j_N)), T, f)|^2 = \\ &\qquad |P(f)|^2 \sum_{i=1}^{N}\sum_{k=1}^{N} a_i a_k^* e^{-j2\pi f[(i-1)D+j_i]} e^{j2\pi f[(k-1)D+j_k]} \end{aligned} \tag{10.292}$$

Splitting the double summation into the diagonal and off-diagonal terms yields

$$\begin{aligned} |X(((a_1, j_1), \ldots, (a_N, j_N)), T, f|^2 &= |P(f)|^2 \sum_{i=1}^{N} |a_i|^2 \\ &\quad + |P(f)|^2 \sum_{i=1}^{N}\sum_{k=1, k\neq i}^{N} a_i a_k^* e^{-j2\pi f[(i-k)D+j_i-j_k]} \end{aligned} \tag{10.293}$$

Using Equation 10.289 and the definition of a power spectral density, the power spectral density of $X$ is

$$G_X(T,f) = \frac{1}{T}\int_{-\infty}^{\infty}\cdots\int_{-\infty}^{\infty}\int_{-\infty}^{\infty}\cdots\int_{-\infty}^{\infty}|X(((a_1,j_1),\ldots,(a_N,j_N)),T,f)|^2 f_{A_1\ldots A_N}(a_1,\ldots,a_N)f_{J_1\ldots J_N}(j_1,\ldots,j_N)da_1\ldots da_N dj_1\ldots dj_N \quad (10.294)$$

Substituting for the magnitude squared of the Fourier transform yields

$$G_X(T,f) = \frac{|P(f)|^2}{T}\int_{-\infty}^{\infty}\cdots\int_{-\infty}^{\infty}\sum_{i=1}^{N}|a_i|^2 f_{A_1\ldots A_N}(a_1,\ldots,a_N)da_1\ldots da_N + \frac{|P(f)|^2}{T}\int_{-\infty}^{\infty}\cdots\int_{-\infty}^{\infty}\int_{-\infty}^{\infty}\cdots\int_{-\infty}^{\infty}\sum_{i=1}^{N}\sum_{k=1,k\neq i}^{N} a_i a_k^{*} e^{-j2\pi f[(i-k)D+j_i-j_k]} f_{A_1\ldots A_N}(a_1,\ldots,a_N)f_{J_1\ldots J_N}(j_1,\ldots,j_N)da_1\ldots da_N dj_1\ldots dj_N \quad (10.295)$$

Interchanging the order of summation and integration leads to

$$G_X(T,f) = \frac{|P(f)|^2}{T}\sum_{i=1}^{N}\int_{-\infty}^{\infty}\cdots\int_{-\infty}^{\infty}|a_i|^2 f_{A_1\ldots A_N}(a_1,\ldots,a_N)da_1\ldots da_N + \frac{|P(f)|^2}{T}\sum_{i=1}^{N}\sum_{k=1,k\neq i}^{N} e^{-j2\pi f(i-k)D}\int_{-\infty}^{\infty}\cdots\int_{-\infty}^{\infty}\int_{-\infty}^{\infty}\cdots\int_{-\infty}^{\infty} a_i a_k^{*} e^{-j2\pi f(j_i-j_k)} f_{A_1\ldots A_N}(a_1,\ldots,a_N)f_{J_1\ldots J_N}(j_1,\ldots,j_N)da_1\ldots da_N dj_1\ldots dj_N \quad (10.296)$$

Simplification requires the use of the following conditional probability theory results:

$$f_{A_1\ldots A_N}(a_1,\ldots,a_N) = f_{A_i}(a_i)\cdot f_{A_1\ldots A_{i-1}A_{i+1}\ldots A_N|A_i=a_i}(a_1,\ldots,a_{i-1},a_{i+1},\ldots,a_N) \quad (10.297)$$

$$f_{A_1\ldots A_N}(a_1,\ldots,a_N) = f_{A_iA_k}(a_i,a_k)\cdot f_{A_{1N/ik}}(a_1,\ldots,a_{i-1},a_{i+1},\ldots,a_{k-1},a_{k+1},\ldots,a_N) \quad (10.298)$$

$$f_{J_1\ldots J_N}(j_1,\ldots,j_N) = f_{J_iJ_k}(j_i,j_k)\cdot f_{J_{1N/ik}}(j_1,\ldots,j_{i-1},j_{i+1},\ldots,j_{k-1},j_{k+1},\ldots,j_N) \quad (10.299)$$

where the following notation has been used:

$$\begin{aligned} A_{1N/ik} &= A_1\ldots A_{i-1}A_{i+1}\ldots A_{k-1}A_{k+1}\ldots A_N/A_i=a_i,A_k=a_k \\ J_{1N/ik} &= J_1\ldots J_{i-1}J_{i+1}\ldots J_{k-1}J_{k+1}\ldots J_N/J_i=j_i,J_k=j_k \end{aligned} \quad (10.300)$$

Use of these results yields

$$G_X(T,f)=\frac{|P(f)|^2}{T}\sum_{i=1}^{N}\int_{-\infty}^{\infty}|a_i|^2 f_{A_i}(a_i)da_i+$$
$$\frac{|P(f)|^2}{T}\sum_{i=1}^{N}\sum_{k=1,k\neq i}^{N}e^{-j2\pi f(i-k)D}\int_{-\infty}^{\infty}\int_{-\infty}^{\infty}a_i a_k^* f_{A_iA_k}(a_i,a_k)da_i da_k\cdot \tag{10.301}$$
$$\int_{-\infty}^{\infty}\int_{-\infty}^{\infty}e^{-j2\pi f(j_i-j_k)}f_{J_iJ_k}(j_i,j_k)dj_i dj_k$$

Using expectation notation, this result can be written as

$$G_X(T,f)=\frac{|P(f)|^2}{T}\left[\sum_{i=1}^{N}\boldsymbol{E}\left[|A_i|^2\right]\right.$$
$$\left.+\sum_{i=1}^{N}\sum_{k=1,k\neq i}^{N}e^{-j2\pi f(i-k)D}\cdot\boldsymbol{E}[A_iA_k^*]\cdot\boldsymbol{E}\left[e^{-j2\pi f(J_i-J_k)}\right]\right] \tag{10.302}$$

which is the first required result.

### 10.J.2 Proof of Second Result

The second result arises by using Theorem 10.17 on the double summation in Equation 10.302. Before this result can be used, it is necessary to confirm that $F_{ik}=F_{ki}^*$ and $F_{ik}=F_{i+r,k+r}$ where

$$F_{ik}=e^{-j2\pi f(i-k)D}\cdot\boldsymbol{E}[A_iA_k^*]\cdot\boldsymbol{E}\left[e^{-j2\pi f(J_i-J_k)}\right] \tag{10.303}$$

As $\boldsymbol{E}[A_iA_k^*]=\boldsymbol{E}[A_k^*A_i]$, the order of conjugation and expectation can be interchanged to yield

$$F_{ki}=e^{-j2\pi f(k-i)D}\cdot\boldsymbol{E}[A_kA_i^*]\cdot\boldsymbol{E}\left[e^{-j2\pi f(J_k-J_i)}\right]\Rightarrow$$
$$F_{ki}^*=e^{-j2\pi f(i-k)D}\cdot\boldsymbol{E}[A_k^*A_i]\cdot\boldsymbol{E}\left[e^{-j2\pi f(J_i-J_k)}\right]=F_{ik} \tag{10.304}$$

To confirm $F_{ik}=F_{i+r,k+r}$ consider,

$$F_{i+r,k+r}=e^{-j2\pi f(i-k)D}\cdot\boldsymbol{E}[A_{i+r}A_{k+r}^*]\cdot\boldsymbol{E}\left[e^{-j2\pi f(J_{i+r}-J_{k+r})}\right]=F_{ik} \tag{10.305}$$

where the required result follows as the correlation is assumed to depend only on the spacing between the pulses, not on their absolute location:

$$\begin{aligned} f_{A_iA_k}(a_i,a_k) &= f_{A_{i+r}A_{k+r}}(a_i,a_k) \\ f_{j_ij_k}(j_i,j_k) &= f_{J_{i+r}J_{k+r}}(j_i,j_k) \quad i,k,i+r,k+r \in \mathbf{Z}^+ \end{aligned} \tag{10.306}$$

Use of Theorem 10.17 in Equation 10.302 yields the required result:

$$\begin{aligned} G_X(T,f) = r|P(f)|^2\overline{|A|^2} + 2r|P(f)|^2 \sum_{i=1}^{N-1}\left[1-\frac{i}{N}\right]\cdot \\ \mathrm{Re}\left[e^{j2\pi iDf}\boldsymbol{E}[A_1A_{1+i}^*]\cdot\boldsymbol{E}\left[e^{-j2\pi f(J_1-J_{1+i})}\right]\right] \end{aligned} \tag{10.307}$$

### 10.J.3 Final Form

The final form arises by adding and subtracting terms consistent with the independent case in the power spectral density as given by Equation 10.307. Consider the summation in Equation 10.307, which is denoted $S_2$:

$$S_2 = \sum_{i=1}^{N-1}\left[1-\frac{i}{N}\right]\cdot\mathrm{Re}\left[e^{j2\pi iDf}\cdot\boldsymbol{E}[A_1A_{1+i}^*]\cdot\boldsymbol{E}\left[e^{-j2\pi f(J_1-J_{1+i})}\right]\right] \tag{10.308}$$

By adding and subtracting terms associated with the independent case, $S_2$ can be written as

$$\begin{aligned} S_2 = \sum_{i=1}^{N-1}\left[1-\frac{i}{N}\right]\cdot\mathrm{Re}\Big[e^{j2\pi iDf}\cdot\left(\boldsymbol{E}[A_1A_{1+i}^*]-|\mu_A|^2+|\mu_A|^2\right)\cdot \\ \left(\boldsymbol{E}\left[e^{-j2\pi f(J_1-J_{1+i})}\right]-|F_J(f)|^2+|F_J(f)|^2\right)\Big] \end{aligned} \tag{10.309}$$

and hence as

$$\begin{aligned} S_2 = I_1 + I_2 + I_3 + \sum_{i=1}^{N-1}\left[1-\frac{i}{N}\right]\cdot\mathrm{Re}\Big[e^{j2\pi iDf}\cdot\left(\boldsymbol{E}[A_1A_{1+i}^*]-|\mu_A|^2\right)\cdot \\ \left(\boldsymbol{E}\left[e^{-j2\pi f(J_1-J_{1+i})}\right]-|F_J(f)|^2\right)\Big] \end{aligned} \tag{10.310}$$

where

$$I_1 = |F_J(f)|^2\sum_{i=1}^{N-1}\left[1-\frac{i}{N}\right]\cdot\mathrm{Re}\left[e^{j2\pi iDf}\cdot\left[\boldsymbol{E}[A_1A_{1+k}^*]-|\mu_A|^2\right]\right] \tag{10.311}$$

$$I_2 = |\mu_A|^2 \sum_{i=1}^{N-1} \left[1 - \frac{i}{N}\right] \cdot \mathrm{Re}\left[e^{j2\pi iDf} \cdot \left[\boldsymbol{E}\left[e^{-j2\pi f(J_1 - J_{1+i})}\right] - |F_J(f)|^2\right]\right] \tag{10.312}$$

$$I_3 = |\mu_A|^2 \; |F_J(f)|^2 \sum_{i=1}^{N-1} \left[1 - \frac{i}{N}\right] \cdot \mathrm{Re}\left[e^{j2\pi iDf}\right] \tag{10.313}$$

By using Theorem 10.17, $I_3$ can be written as

$$\begin{aligned} I_3 &= \frac{|\mu_A|^2 |F_J(f)|^2}{2N} \sum_{i=1}^{N} \sum_{k=1,\ k \neq i}^{N} e^{-j2\pi f(i-k)D} \\ &= \frac{|\mu_A|^2 |F_J(f)|^2}{2} \cdot \left[\frac{1}{N} \cdot \frac{\sin(\pi N f/r)^2}{\sin(\pi f/r)^2} - 1\right] \end{aligned} \tag{10.314}$$

where the last result arises from Theorem 2.33. Substitution of $I_1$, $I_2$, and $I_3$ into Equation 10.310 yields

$$\begin{aligned} G_X(ND,f) = r|P(f)|^2 \Bigg[ & \overline{|A|^2} - |\mu_A|^2 |F_J(f)|^2 + |\mu_A|^2 |F_J(f)|^2 \cdot \frac{1}{N} \cdot \frac{\sin(\pi N f/r)^2}{\sin(\pi f/r)^2} \\ & + 2\sum_{i=1}^{N-1} \left[1 - \frac{i}{N}\right] \cdot \mathrm{Re}\left[e^{j2\pi iDf} \cdot \left[\boldsymbol{E}[A_1 A_{1+i}^*] - |\mu_A|^2\right] \cdot \left[\boldsymbol{E}\left[e^{-j2\pi f(J_1 - J_{1+i})}\right] - |F_J(f)|^2\right]\right] \\ & + 2|\mu_A|^2 \sum_{i=1}^{N-1} \left[1 - \frac{i}{N}\right] \mathrm{Re}\left[e^{j2\pi iDf} \cdot \left[\boldsymbol{E}\left[e^{-j2\pi f(J_1 - J_{1+i})}\right] - |F_J(f)|^2\right]\right] \\ & + 2|F_J(f)|^2 \sum_{i=1}^{N-1} \left[1 - \frac{i}{N}\right] \mathrm{Re}\left[e^{j2\pi iDf} \left(\boldsymbol{E}[A_1 A_{1+i}^*] - |\mu_A|^2\right)\right] \Bigg] \end{aligned} \tag{10.315}$$

Using the result $\lim_{N \to \infty} \frac{1}{N} \cdot \frac{\sin(\pi N f/r)^2}{\sin(\pi f/r)^2} = r \sum_{n=-\infty}^{\infty} \delta(f - nr)$ (Theorem 2.33), the power spectral density for the infinite interval then follows:

$$\begin{aligned} G_X(\infty,f) = r|P(f)|^2 \Bigg[ & \overline{|A|^2} - |\mu_A|^2 |F_J(f)|^2 + r|\mu_A|^2 |F_J(f)|^2 \cdot \sum_{n=-\infty}^{\infty} \delta(f - nr) \\ & + 2\sum_{i=1}^{\infty} \mathrm{Re}\left[e^{j2\pi iDf} \cdot \left[\boldsymbol{E}[A_1 A_{1+i}^*] - |\mu_A|^2\right] \cdot \left[\boldsymbol{E}\left[e^{-j2\pi f(J_1 - J_{1+i})}\right] - |F_J(f)|^2\right]\right] \\ & + 2|\mu_A|^2 \sum_{i=1}^{\infty} \mathrm{Re}\left[e^{j2\pi iDf} \left[\boldsymbol{E}\left[e^{-j2\pi f(J_1 - J_{1+i})}\right] - |F_J(f)|^2\right]\right] \\ & + 2|F_J(f)|^2 \sum_{i=1}^{\infty} \mathrm{Re}\left[e^{j2\pi iDf} \left(\boldsymbol{E}[A_1 A_{1+i}^*] - |\mu_A|^2\right)\right] \Bigg] \end{aligned} \tag{10.316}$$

The results stated in the theorem are for the finite correlation case where the correlation extends over $M$ pulses consistent with $A_1$ and $A_{M+1}$, as well as $J_1$ and $J_{M+1}$, being independent.

# 11

# ORDER STATISTICS

## 11.1 INTRODUCTION

Order statistics leads to the specification of the statistics of a specific outcome based on a set of experimental outcomes that have been ordered, for example, from the lowest to the highest level. Ordering of experimental outcomes is of interest in many areas, and one prominent case is when the maximum outcome is required. Of interest to the development of material in this book is the ordering of the outcomes of a Poisson experiment to obtain an ordered set of times that underpin Poisson processes, shot noise random process, etc.

This chapter provides an introduction to ordered random variable theory including general results for the cumulative distribution function, probability density function, and joint probability density function of such random variables. The independent random variable case is assumed. As the difference between ordered outcomes is of importance to Poisson processes, the probability density function and joint probability density function of the difference random variables arising from ordering are considered. Specific results are established for both the finite and infinite intervals assuming that the underlying random variables have a uniform distribution. This assumption is consistent with the outcomes underpinning Poisson point and associated random processes. Useful references for order statistics include Arnold et al. (2008) and David and Nagaraja (2003).

*A Signal Theoretic Introduction to Random Processes*, First Edition. Roy M. Howard.

### 11.1.1 Background Results

The following subsections detail necessary mathematical results for subsequent analysis:

*11.1.1.1 Binomial Probability Mass Function* The probability of $k$ successes in $N$ independent trials of an experiment, where the probability of success in a given trial is $p$, is stated in Theorem 4.24:

$$b(k,N,p) = \frac{N!}{(N-k)!k!} \cdot p^k (1-p)^{N-k} \tag{11.1}$$

*11.1.1.2 Probability of Outcomes in Set Categories*

**Theorem 11.1 Probability of Outcomes in Set Categories**
Consider an experiment that yields, for each trial, a vector of $N$ outcomes. Each outcome is assumed to be from an independent subexperiment and can be categorized as being in one of $k$ categories. The probability of an outcome being in the $r$th category is denoted $p_r$. The probability of $N_1$ of the outcomes being in category one, …, and $N_k$ of the outcomes being in category $k$, with

$$N_k = N - \sum_{r=1}^{k-1} N_r \tag{11.2}$$

is

$$P[N_r \text{ in category } r : r \in \{1,\ldots,k\}] = \frac{N!}{(N_1)!(N_2)!\ldots(N_k)!} \cdot \prod_{r=1}^{k} p_r^{N_r} \tag{11.3}$$

**Proof**
The proof is detailed in Appendix 11.A.

*11.1.1.3 Binomial Formula and Combination Formula*

**Theorem 11.2 Binomial Formula**
The well-known result, called the binomial formula, is

$$(x+y)^N = x^N + \binom{N}{1} x^{N-1} y + \cdots + \binom{N}{k} x^{N-k} y^k + \cdots + \binom{N}{N} y^N \tag{11.4}$$

where $\binom{N}{k} = \frac{N!}{(N-k)!k!}$ is the combination formula for the number of ways $k$ distinct entities can be chosen from $N$. It then follows that

$$(1+y)^N = 1 + \binom{N}{1} y + \cdots + \binom{N}{k} y^k + \cdots + y^N \tag{11.5}$$

**Theorem 11.3 Normalized Combination Formula**

The combination formula $\binom{N}{k}$, when normalized by $N^k$, can be written in the form

$$\frac{N!}{(N-k)!k!} \cdot \frac{1}{N^k} = \begin{cases} 1 & k=0,1 \\ \frac{1}{k!} \cdot \left(1-\frac{1}{N}\right) \cdots \left(1-\frac{k-1}{N}\right) = \frac{1}{k!} \prod_{j=0}^{k-1} \left(1-\frac{j}{N}\right) & k \geq 1 \\ \frac{1}{k!} \cdot \left(1-\frac{1}{N}\right) \cdots \left(1-\frac{k-1}{N}\right) = \frac{1}{k!} \prod_{j=1}^{k-1} \left(1-\frac{j}{N}\right) & k \geq 2 \end{cases} \tag{11.6}$$

**Proof**

The stated result arises from expanding out the factorial terms.

### *11.1.1.4 Approximation for an Affine Function to a Power*

**Theorem 11.4 Approximation for an Affine Function to a Power**

The following limit holds:

$$\lim_{\substack{N,T\to\infty \\ N/T\to\lambda}} \left(1-\frac{t}{T}\right)^{N-k} = e^{-\lambda t} \quad k \in \mathbf{Z}^+,\ t \in (0,\infty) \tag{11.7}$$

and the approximation

$$\left(1-\frac{t}{T}\right)^{N-k} \approx e^{-\lambda t} \quad t \in (0,T), k \in \{0,1,\ldots,N-1\},\ \lambda = \frac{N}{T} \tag{11.8}$$

is valid with a relative error

$$\text{r.e.} = \frac{kt}{T} - N\left(\frac{t}{T}\right)^2 \left(1-\frac{k}{N}\right)\left[\frac{1}{2} + \frac{1}{3}\left(\frac{t}{T}\right) + \frac{1}{4}\left(\frac{t}{T}\right)^2 + \cdots\right] \tag{11.9}$$

The relative error is low when $k << T/t$ and $\sqrt{N} << T/t$. For this case, and when $k << N$, $t << T$, the relative error can be approximated according to

$$\text{r.e.} \approx \frac{kt}{T} - \frac{Nt^2}{2T^2} \tag{11.10}$$

The relative error is plotted in Figure 11.1 for the case of $t/T = 0.01$.

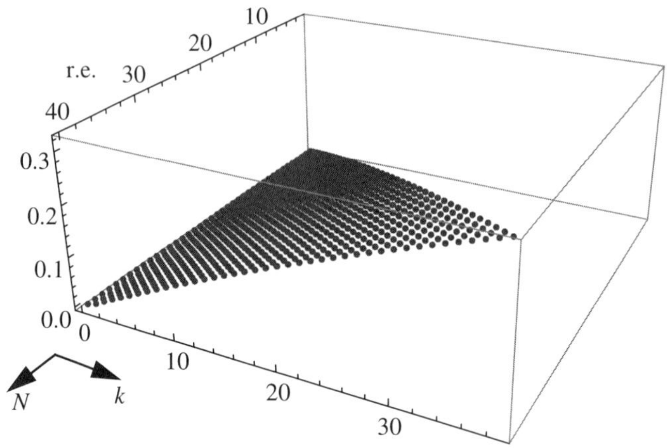


**FIGURE 11.1** Graph of relative error in the approximation stated in Equation 11.8 for the case of $t/T = 0.01$.

**Proof**
The proof is detailed in Appendix 11.B.

#### *11.1.1.5 Joint PDF of Sum of Two Random Variables*

**Theorem 11.5 Joint PDF of $Z_1$ and $Z_2$ where $Z_1 = X_1$ and $Z_2 = X_1 + X_2$**
With the transformation

$$\begin{aligned} Z_1 &= g_1(X_1, X_2) = X_1 \\ Z_2 &= g_2(X_1, X_2) = X_1 + X_2 \end{aligned} \tag{11.11}$$

the joint probability density function of $Z_1$ and $Z_2$ is

$$f_{Z_1Z_2}(z_1, z_2) = f_{X_1X_2}(z_1, z_2 - z_1) \tag{11.12}$$

where $f_{X_1X_2}$ is the joint probability density function of $X_1$ and $X_2$.

**Proof**
The proof is detailed in Appendix 11.C.

#### *11.1.1.6 PDF of a Conditioned Sum*

**Theorem 11.6 Conditioned Sum**
The conditional probability density function of $Z = X + Y/Y \in [y, y + dy]$, $f_{X+Y/Y\in[y,y+dy]}(z)$, which for notational convenience is written as $f_{X+Y/Y=y}(z)$, is given by

$$f_{X+Y/Y=y}(z) = \frac{f_{XY}(z-y, y)}{f_Y(y)} \tag{11.13}$$

The assumption here is that $f_Y(y) \neq 0$.

**Proof**

The proof is detailed in Appendix 11.D.

## 11.2 ORDERED RANDOM VARIABLE THEORY

Consider an experiment comprising of $N$ independent subexperiments and yielding, for each trial, a vector of $N$ numbers. Depending on the random phenomena being modelled, the probability density/mass function of the maximum, minimum, difference between the maximum and minimum, median and $i$th ordered outcome, etc. may be of interest. The probability density/mass function of the maximum is of importance, for example, in reliability theory. With robust statistics, the probability density/mass function of the median is of interest. With Poisson point processes, and related random processes, the probability density function of the $i$th ordered outcome is of importance.

### 11.2.1 Events

***11.2.1.1 Random Occurrence of Events*** Consider the case, commencing at $t=0$ and as illustrated in Figure 11.2, of the random arrival of events with a set rate $\lambda$. The outcomes illustrated in Figure 11.2 can be modelled by considering the selection, at random, of a point in the interval [0, $T$], where $T = N/\lambda$, to yield a time $\tau_1$. Repeating this $N$ times yields the set of times $\tau_1, \ldots, \tau_N$. Reordering these times yields a set of increasing times consistent with the random arrival of events. Order statistics yields statistical information about the time of the $i$th outcome.

A general requirement is that the duration of the interval [0, $T$] should be large relative to the time interval of interest to ensure the statistical independence of the times. The model where $T$ is not large, relative to the time interval of interest, does

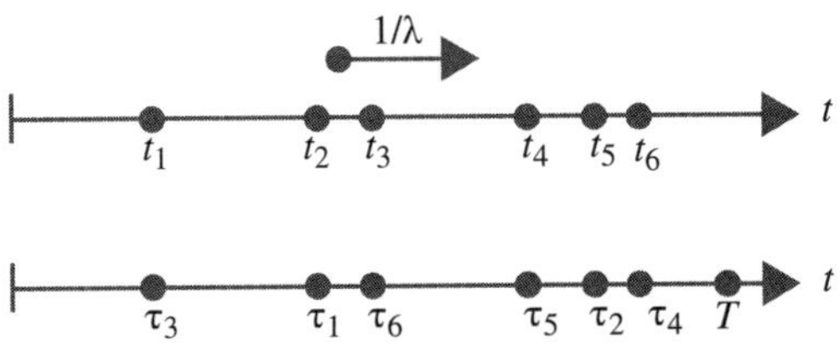


**FIGURE 11.2** Top: illustration of the random arrival times of events occurring at a uniform rate $\lambda$. Bottom: selection of times, at random, from the interval [0, $T$]. The reordering of these times yields the sequence of times as illustrated in the top graph.

not yield statistical independence of times as the number of times on $[0, T]$ equals $N$ and is not random.

***11.2.1.2 Fixed Number of Events: Case of Finite Resources*** Consider the case of the occurrence of events at a fixed rate $\lambda$ but where the physical constraint is that the number of events that can occur is limited to $N_{max}$. This case can be modelled by the random arrival of $N$ events on the interval $[0,T]$, $N \to \infty$, $T \to \infty$, $\lambda = N/T$ fixed and by taking the times of occurrence of the first $N_{max}$ events.

***11.2.1.3 Fixed Number of Events on a Fixed Interval*** The following situations are consistent with $N$ events occurring, at random, in a set interval $[0, T]$: First, the placement of $N$ entities, at random, in $M$ bins where the $i$th bin is associated with an outcome in the interval $\left[\frac{(i-1)T}{M}, \frac{iT}{M}\right)$. Second, the selection, at random, of $N$ items from a range of $M$ where the $i$th item is associated with an outcome in the interval $\left[\frac{(i-1)T}{M}, \frac{iT}{M}\right)$. Third, the occurrence of $N$ roundoff errors in the interval $[-\varepsilon/2, \varepsilon/2]$ where the $i$th error is defined according to

$$i\text{th error} = \frac{-\varepsilon}{2} + \frac{(i\text{th outcome})\varepsilon}{T} \tag{11.14}$$

The more general case is where $N$ events occur in a set time interval but with a nonuniform rate. One example is the arrival times of $N$ students for a scheduled lecture.

***11.2.1.4 Minimum and Maximum Random Variable*** Consider the case of $N$ events occurring in the interval $[0, T]$, for the case of both $N$ and $T$ fixed. The minimum and the maximum outcomes are, respectively, the first and the $N$th ordered outcomes. For the case where the joint probability density function of the first and $N$th ordered outcomes is known, the probability density function of the random variable defined by the difference between the minimum and maximum outcomes can readily be established.

### 11.2.2 Notation

Consistent with Figure 11.2, the times of the first, second, etc. events are consistent with the outcomes of ordered random variables, and the first goal is to establish their cumulative distribution and probability density functions. For the $i$th and $j$th ordered random variables, and for the uncountable case, the following notation is used:

$$\begin{array}{ll} F_{(i)}(x) & \text{CDF of ith ordered random variable} \\ f_{(i)}(x) & \text{PDF of ith ordered random variable} \\ f_{(i)(j)}(x,y) & \text{Joint PDF of ith and jth ordered random variables} \end{array} \tag{11.15}$$

The goal is to establish the above-noted density and joint density functions for random events, at a uniform rate of $\lambda$ per second, on the interval $[0, T]$. Simplified results, with important practical application, can be obtained for the case of $T \to \infty$.

Consider an experiment based on $N$ independent subexperiments where each trial yields a vector of outcomes $(\gamma_1, \ldots, \gamma_N)$. The experiment is assumed to define a vector of random variables $X_A = (X_1,\ldots,X_N)$ where

$$\begin{array}{l} X_i : S_i \to S_{X_i} \\ S_i = \{\gamma_i : \gamma_i \in I_i\}, I_i \subseteq \boldsymbol{R} \qquad S_{X_i} = \{X_i(\gamma_i) : \ \gamma_i \in S_i\} \end{array} \tag{11.16}$$

The cumulative distribution function and the probability density function for the $i$th random variable, respectively, are denoted $F_{X_i}$ and $f_{X_i}$.

If the outcomes of each of the random variables for each given trial of the experiment are ordered from lowest to highest, then a vector of ordered random variables $X = (X_{(1)},\ldots,X_{(N)})$ is defined. The procedure, and associated notation, is illustrated in Figure 11.3.

The sample space of the experiment is

$$S = \{\omega : \ \omega = (\gamma_1, \gamma_2, \ldots, \gamma_N), \ \gamma_1 \in S_1, \ldots, \gamma_N \in S_N\} \tag{11.17}$$

and on this sample space, a random vector $X : S \to S_X$ is defined according to

$$X(\omega) = \left(X_{(1)}(\omega), X_{(2)}(\omega), \ldots, X_{(N)}(\omega)\right) \tag{11.18}$$

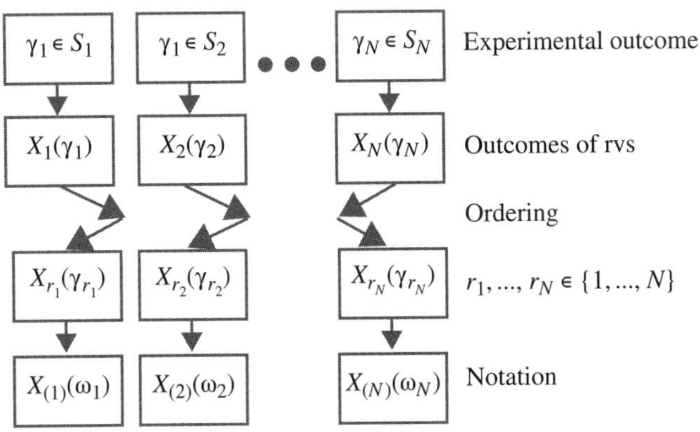


**FIGURE 11.3** Illustration of notation and procedure associated with defining an ordered vector of outcomes from an experiment that defines $N$ random variables.

where

$$\begin{aligned} X_1(\omega) &= \min\{X_1(\gamma_1), X_2(\gamma_2), \ldots, X_N(\gamma_N)\} \\ X_2(\omega) &= \min\left\{\{X_1(\gamma_1), X_2(\gamma_2), \ldots, X_N(\gamma_N)\} \cap \{X_{(1)}(\omega)\}^C\right\} \\ X_3(\omega) &= \min\left\{\{X_1(\gamma_1), X_2(\gamma_2), \ldots, X_N(\gamma_N)\} \cap \{X_{(1)}(\omega), X_{(2)}(\omega)\}^C\right\} \\ &\vdots \\ X_N(\omega) &= \max\{X_1(\gamma_1), X_2(\gamma_2), \ldots, X_N(\gamma_N)\} \end{aligned} \tag{11.19}$$

Here, the complementation operator is operating in the context of the set $\{X_1(\gamma_1), X_2(\gamma_2), \ldots, X_N(\gamma_N)\}$. The sample space of the random vector is

$$S_X = \left\{ \begin{matrix} (X_{(1)}(\omega), X_{(2)}(\omega), \ldots, X_{(N)}(\omega)) : \ \omega = (\gamma_1, \gamma_2, \ldots, \gamma_N) \in S \\ X_{(1)}(\omega) \le X_{(2)}(\omega) \le \ldots \le X_{(N)}(\omega) \end{matrix} \right\} \tag{11.20}$$

### 11.2.3 Cumulative Distribution Function

The first measure of importance for a set of ordered random variables is the cumulative distribution function.

The normal notation to use for the cumulative distribution function of the $i$th ordered random variable $X_{(i)}$ is $F_{X_{(i)}}(x)$. For notational simplicity, and as noted in Equation 11.15, the notation $F_{(i)}(x)$ is used. Similar notation is used for the probability density function.

**Definition: Cumulative Distribution Function** By definition,

$$F_{(i)}(x) = P\left[X_{(i)} \le x\right] = 1 - P\left[X_{(i)} > x\right] \tag{11.21}$$

**Theorem 11.7 Cumulative Distribution Function**
The cumulative distribution function of the $i$th ordered random variable, $1 \le i \le N$, has the following equivalent forms:

$$\begin{aligned} F_{(i)}(x) &= P[i, i+1, \ldots, N \text{ outcomes of } X_1 \ldots X_N \text{ are less than, or equal, to } x] \\ &= \sum_{k=1}^{N} P[k \text{ out of } N \text{ r.v.} \le x] = \sum_{r=0}^{N-i} P[N-r \text{ out of } N \text{ r.v.} \le x] \end{aligned} \tag{11.22}$$

$$\begin{aligned} F_{(i)}(x) &= P[0, 1, \ldots, N-i \text{ outcomes of } X_1 \ldots X_N \text{ are greater than } x] \\ &= \sum_{k=0}^{N-i} P[k \text{ out of } N \text{ r.v.} > x] \end{aligned} \tag{11.23}$$

$$\begin{aligned} F_{(i)}(x) &= 1 - P[0, 1, \ldots, i-1 \text{ outcomes of } X_1 \ldots X_N \text{ are less than, or equal, to } x] \\ &= 1 - \sum_{k=0}^{i-1} P[k \text{ out of } N \text{ r.v.} \le x] \end{aligned} \tag{11.24}$$

$$F_{(i)}(x) = 1 - P[N-i+1,\ldots,N \text{ outcomes of } X_1 \ldots X_N \text{ are greater than } x]$$
$$= 1 - \sum_{k=N-i+1}^{N} P[k \text{ out of } N \text{ r.v.} > x] = 1 - \sum_{k=0}^{i-1} P[N-k \text{ out of } N \text{ r.v.} > x] \quad (11.25)$$

**Proof**

Consider the illustration shown in Figure 11.4, and note that the events of $i$, $i+1, \ldots$ outcomes of the random variables $X_1 \ldots X_N$ being greater or less than a set number are disjoint events.

The event of the $i$th ordered outcome being less than or equal to $x$ is consistent with a minimum of $i$ random variables having values that are less than or equal to $x$ and the events:

$$\begin{array}{l} i \text{ outcomes} \leq x \text{ or } \ldots N \text{ outcomes} \leq x \\ N-i \text{ outcomes} > x \text{ or } \ldots 0 \text{ outcomes} > x \end{array} \quad (11.26)$$

These events are consistent with the first and second results.

The event of the $i$th ordered outcome being greater than $x$ is consistent with a maximum of $i-1$ random variables having values that are less than or equal to $x$ and the events:

$$\begin{array}{l} 0 \text{ outcomes} \leq x \text{ or } \ldots i-1 \text{ outcomes} \leq x \\ N \text{ outcomes} > x \text{ or } \ldots N-i+1 \text{ outcomes} > x \end{array} \quad (11.27)$$

These events are consistent with the third and fourth results.

***11.2.3.1 CDF: Independent Case*** For the case where $X_1 \ldots X_N$ are independent random variables, and $X_i$ has a cumulative distribution function $F_{X_i}$, the following result holds for the first and last ordered random variables:

**Theorem 11.8 CDF for Minimum and Maximum Random Variables**

For $N$ independent random variables—whether discrete or continuous—it is the case that

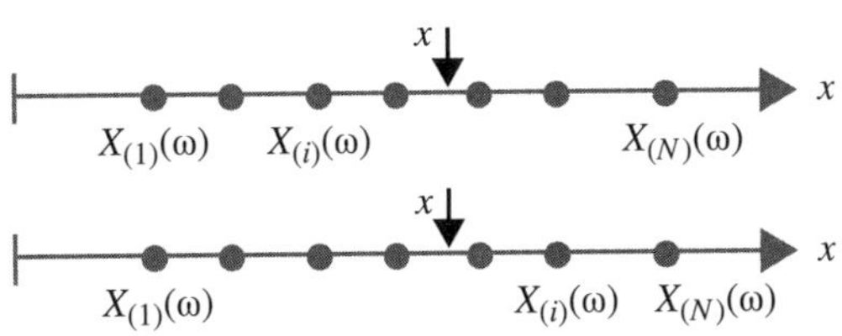


**FIGURE 11.4** Top: illustration of the random variable outcomes consistent with the $i$th ordered random variable being less than or equal to a set number $x$. Bottom: illustration for the $i$th ordered random variable being greater than a set number $x$.

$$F_{(1)}(x) = P\left[X_{(1)} \le x\right] = 1 - \prod_{i=1}^{N}(1 - F_{X_i}(x)) \tag{11.28}$$

$$F_{(N)}(x) = P\left[X_{(N)} \le x\right] = 1 - \prod_{i=1}^{N} F_{X_i}(x) \tag{11.29}$$

**Proof**

The proof is detailed in Appendix 11.E.

For the case where $X_1 \ldots X_N$ are identical and independent random variables, where each has a set cumulative probability density function $F_X$, the event of $i, i+1, \ldots, N$ outcomes of the random variables $X_1 \ldots X_N$ being less than, or equal to, $x$ is equivalent to at least $i$ successes in $N$ trials where the probability of success is $F_X(x)$. This leads to the following result:

**Theorem 11.9 CDF for Order Random Variables: Independent Case**

For the case of $N$ independent and identical random variables where $F_{X_i}(x) = F_X(x)$, $i \in \{1, \ldots, N\}$, the following results are equivalent:

$$F_{(i)}(x) = \sum_{k=i}^{N} \frac{N!}{(N-k)!k!} F_X^k(x)[1 - F_X(x)]^{N-k} \quad 1 \le i \le N \tag{11.30}$$

$$F_{(i)}(x) = \sum_{k=0}^{N-i} \frac{N!}{(N-k)!k!} [1 - F_X(x)]^k F_X^{N-k}(x) \quad 1 \le i \le N \tag{11.31}$$

$$F_{(i)}(x) = 1 - \sum_{k=0}^{i-1} \frac{N!}{(N-k)!k!} F_X^k(x)[1 - F_X(x)]^{N-k} \quad 1 \le i \le N \tag{11.32}$$

$$F_{(i)}(x) = 1 - \sum_{k=N-i+1}^{N} \frac{N!}{(N-k)!k!} [1 - F_X(x)]^k F_X^{N-k}(x) \quad 1 \le i \le N \tag{11.33}$$

$$F_{(i)}(x) = F_X^i(x) \sum_{k=0}^{N-i} \frac{(i-1+k)!}{(i-1)!k!} \cdot [1 - F_X(x)]^k \quad 1 \le i \le N \tag{11.34}$$

The following algorithm can be used for calculating $F_{(i)}(x)$:

$$\begin{aligned} &F_{(1)}(x) = 1 - [1 - F_X(x)]^N \\ &F_{(i)}(x) = F_{(i-1)}(x) - \frac{N!}{(N-i+1)!(i-1)!} F_X^{i-1}(x)[1 - F_X(x)]^{N-i+1} \quad 2 \le i \le N \end{aligned} \tag{11.35}$$

**Proof**

The proof is detailed in Appendix 11.E.

### 11.2.4 Probability Density Function

For the case of continuous random variables, the probability density function can be obtained, by differentiation, from the results given in Theorem 11.9 to yield, for example,

$$f_{(i)}(x) = f_X(x)\sum_{k=i}^{N}\frac{N!}{(N-k)!k!}F_X^{k-1}(x)[1-F_X(x)]^{N-k-1}[k-NF_X(x)] \qquad 1 \le i \le N \tag{11.36}$$

This result arises from the cumulative distribution function given in Equation 11.30. However, direct application of first principles leads directly to a simpler form:

**Theorem 11.10 PDF of *i*th Order RV: Identical and Independent Case**
For the identical, independent, and continuous random variable case, the probability density functions of the first, *N*th, and *i*th ordered random variables, respectively, are:

$$\begin{aligned} f_{(1)}(x) &= N[1-F_X(x)]^{N-1}f_X(x) \\ f_{(N)}(x) &= NF_X^{N-1}(x)f_X(x) \end{aligned} \tag{11.37}$$

$$f_{(i)}(x) = \frac{N!}{(N-i)!(i-1)!}F_X^{i-1}(x)[1-F_X(x)]^{N-i}f_X(x) \qquad 1 \le i \le N \tag{11.38}$$

**Proof**
The proof is detailed in Appendix 11.F.

***11.2.4.1 Example*** Consider the case of $N = 10$ identical and independent random variables that have a uniform distribution on the interval [0, 1], that is,

$$F_X(x) = \begin{cases} 0 & x<0 \\ x & 0 \le x < 1 \\ 1 & x \ge 1 \end{cases} \qquad f_X(x) = \begin{cases} 0 & x<0, x\ge 1 \\ 1 & 0 \le x < 1 \end{cases} \tag{11.39}$$

These functions are shown in Figure 11.5. The cumulative distribution functions of the ordered random variables are shown in Figure 11.6, while the corresponding probability density functions are shown in Figure 11.7.

### 11.2.5 Joint Probability Density Function

The marginal density functions of the ordered random variables can also be determined from the joint probability density function, which is defined in the following theorem:

**Theorem 11.11 Joint PDF of *N* Ordered Independent Random Variables**
For the case of *N* identical and independent random variables, the joint probability density function of $X_{(1)}, \ldots, X_{(N)}$ is

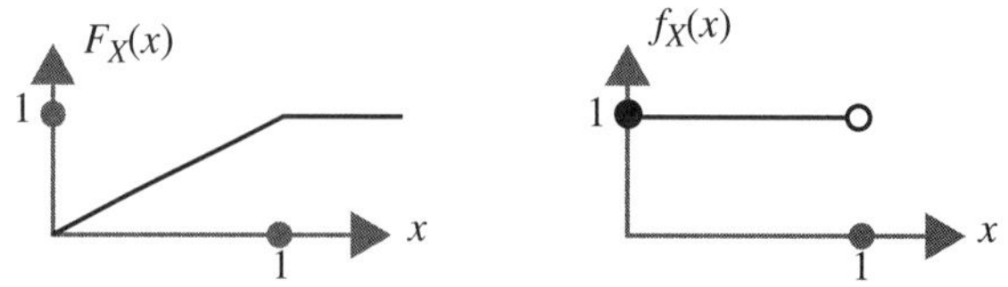


**FIGURE 11.5** Graph of $F_X(x)$ and $f_X(x)$ for the case of random variables with a uniform distribution on the interval [0, 1].

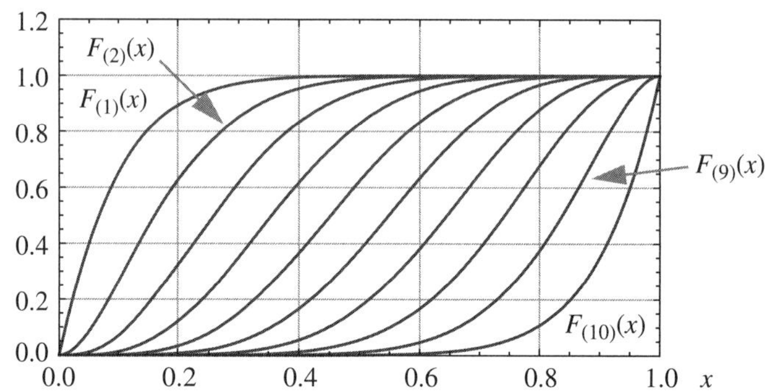


**FIGURE 11.6** Graph of $F_{(i)}(x)$, $i \in \{1, \ldots, 10\}$, for the case of 10 independent and identical random variables with a uniform distribution on the interval [0, 1].

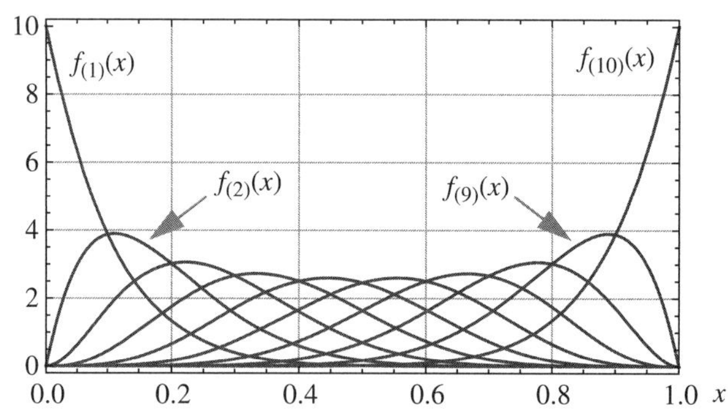


**FIGURE 11.7** Graph of $f_{(i)}(x)$, $i \in \{1, \ldots, 10\}$, for the case of 10 independent and identical random variables with a uniform distribution on the interval [0, 1].

$$f_{(1)\ldots(N)}(x_1,\ldots,x_N) = \begin{cases} N!\prod_{i=1}^{N} f_X(x_i) & x_1 < x_2 < \ldots < x_N \\ 0 & \text{otherwise} \end{cases} \tag{11.40}$$

**Proof**
By definition,

$$f_{(1)\ldots(N)}(x_1,\ldots,x_N)dx_1\ldots dx_N = P\big[X_{(1)} \in [x_1, x_1+dx_1], \ldots, X_{(N)} \in [x_N, x_N+dx_N]\big] \tag{11.41}$$

as $dx_1 \to 0, \ldots, dx_N \to 0$. Consider the event comprising of outcomes such that

$$X_1(\omega) \in [x_1, x_1+dx_1], \ldots, X_N(\omega) \in [x_N, x_N+dx_N] \tag{11.42}$$

where $x_1 < x_2 < \ldots < x_N$. The probability of such an event is

$$P[X_1(\omega) \in [x_1, x_1+dx_1], \ldots, X_N(\omega) \in [x_N, x_N+dx_N]] \approx \prod_{i=1}^{N} f_X(x_i)dx_i \tag{11.43}$$

There are $N! - 1$ equally probable other events leading to outcomes in the same distinct intervals. Hence,

$$P\big[X_{(1)} \in [x_1, x_1+dx_1], \ldots, X_{(N)} \in [x_N, x_N+dx_N]\big] \approx N! \prod_{i=1}^{N} f(x_i)dx_i \tag{11.44}$$

as required.

***11.2.5.1 Marginal PDF from Joint PDF*** As detailed in Appendix 11.G, the $i$th marginal density function, as given by Equation 11.38, follows from the joint probability density function, that is,

$$\begin{aligned} f_{(i)}(x_i) &= \int_{-\infty}^{\infty} \cdots \int_{-\infty}^{\infty} f_{X_{(1)}\ldots X_{(N)}}(x_1,\ldots,x_N)dx_1\ldots dx_{i-1}dx_{i+1}\ldots dx_N \\ &= \frac{N!}{(N-i)!\,(i-1)!} F_X^{i-1}(x_i)[1-F_X(x_i)]^{N-i} f_X(x_i) \qquad 1 \le i \le N \end{aligned} \tag{11.45}$$

### 11.2.6 Joint PDF of Two Ordered Random Variables

To determine measures of the difference between two ordered random variables, their joint probability density function is required.

**Theorem 11.12 Joint PDF of *i*th and *j*th Ordered RVs**
For $N$ identical and independent random variables, where each has a continuous probability density function $f_X$, the joint probability density function for the $i$th and $j$th ordered random variables can be written in two forms:

$$f_{(i)(j)}(x_i,x_j) = \frac{N!}{(i-1)!(j-i-1)!(N-j)!} \cdot F_X^{i-1}(x_i)\left[F_X(x_j)-F_X(x_i)\right]^{j-i-1} \cdot \left[1-F_X(x_j)\right]^{N-j} f_X(x_i) f_X(x_j)$$

$$= \frac{N^j}{(i-1)!\,(j-i-1)!} \cdot \prod_{k=1}^{j-1}\left(1-\frac{k}{N}\right) \cdot F_X^{i-1}(x_i)\left[F_X(x_j)-F_X(x_i)\right]^{j-i-1} \cdot \left[1-F_X(x_j)\right]^{N-j} f_X(x_i) f_X(x_j) \qquad x_i<x_j,\ 1\le i<j\le N \tag{11.46}$$

The following specific results then follow:

$$f_{(1)(2)}(x_1,x_2) = N(N-1)[1-F_X(x_2)]^{N-2} f_X(x_1) f_X(x_2) \qquad x_1<x_2 \tag{11.47}$$

$$f_{(1)(j)}(x_1,x_j) = \frac{N!}{(j-2)!\,(N-j)!}\left[F_X(x_j)-F_X(x_1)\right]^{j-2}\left[1-F_X(x_j)\right]^{N-j} \cdot f_X(x_1) f_X(x_j) \qquad x_1<x_j,\ 2\le j\le N \tag{11.48}$$

$$f_{(1)(N)}(x_1,x_N) = N(N-1)[F_X(x_N)-F_X(x_1)]^{N-2} f_X(x_1) f_X(x_N) \qquad x_1<x_N \tag{11.49}$$

$$f_{(i)(i+1)}(x_i,x_{i+1}) = \frac{N!}{(i-1)!(N-i-1)!} F_X^{i-1}(x_i)[1-F_X(x_{i+1})]^{N-i-1} \cdot f_X(x_i) f_X(x_{i+1}) \qquad x_i<x_{i+1},\ 1\le i\le N-1 \tag{11.50}$$

$$f_{(i-1)(i)}(x_{i-1},x_i) = \frac{N!}{(i-2)!(N-i)!} F_X^{i-2}(x_{i-1})[1-F_X(x_i)]^{N-i} \cdot f_X(x_{i-1}) f_X(x_i) \qquad x_{i-1}<x_i,\ 2\le i\le N \tag{11.51}$$

**Proof**

The proof is detailed in Appendix 11.H.

***11.2.6.1 Example*** The joint probability density function for the second and third ordered random variables, that is, $f_{(2)(3)}(x_2, x_3)$, is shown in Figure 11.8 for the case of 10 identical and independent random variables with a uniform distribution on the interval [0, 1].

### 11.2.7 Joint PDF of $(i-1)$ith, $i$th, $(j-1)$th, and $j$th Ordered RV

To establish the joint probability density function of differences between ordered random variables, higher level joint probability density functions are required:

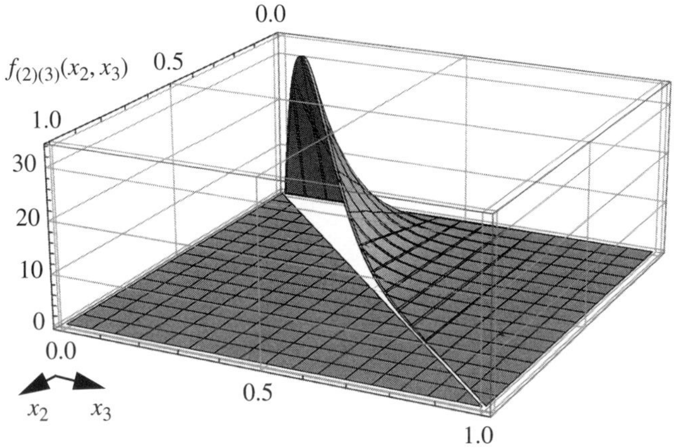


**FIGURE 11.8** Graph of $f_{(2)(3)}(x_2, x_3)$ for the case of 10 independent and identical random variables with a uniform distribution on the interval [0, 1].

**Theorem 11.13 Joint PDF of $(i-1)$ith, $i$th, $(j-1)$th, and $j$th Ordered RVs** For $N$ identical and independent random variables, where each has a continuous probability density function $f_X$, it is the case that

$$\begin{gathered} f_{(i-1)(i)(j-1)(j)}(x_{i-1},x_i,x_{j-1},x_j) = \frac{N!}{(i-2)!(j-i-2)!(N-j)!} \cdot F_X(x_{i-1})^{i-2} \cdot \\ \left[F_X(x_{j-1}) - F_X(x_i)\right]^{j-i-2}\left[1-F_X(x_j)\right]^{N-j} f_X(x_{i-1}) f_X(x_i) f_X(x_{j-1}) f_X(x_j) \\ x_{i-1} < x_i < x_{j-1} < x_j, 2 \le i < j-1 \le N-1 \end{gathered} \tag{11.52}$$

An alternative form is

$$\begin{gathered} f_{(i-1)(i)(j-1)(j)}(x_{i-1},x_i,x_{j-1},x_j) = \frac{N^j}{(i-2)!(j-i-2)!} \cdot \prod_{k=1}^{j-1}\left(1-\frac{k}{N}\right) \cdot F_X(x_{i-1})^{i-2} \cdot \\ \left[F_X(x_{j-1}) - F_X(x_i)\right]^{j-i-2} \cdot \left[1-F_X(x_j)\right]^{N-j} f_X(x_{i-1}) f_X(x_i) f_X(x_{j-1}) f_X(x_j) \\ x_{i-1} < x_i < x_{j-1} < x_j, 2 \le i < j-1 \le N-1 \end{gathered} \tag{11.53}$$

For the case of $j = i+1$ and for $x_{i-1} < x_i < x_{i+1}, 2 \le i \le N-1$,

$$\begin{gathered} f_{(i-1)(i)(i+1)}(x_{i-1},x_i,x_{i+1}) = \frac{N!}{(i-2)!(N-i-1)!} \cdot F_X(x_{i-1})^{i-2} \cdot \\ \left[1-F_X(x_{i+1})\right]^{N-i-1} \cdot f_X(x_{i-1}) f_X(x_i) f_X(x_{i+1}) \end{gathered} \tag{11.54}$$

and an alternative form is

$$f_{(i-1)(i)(i+1)}(x_{i-1},x_i,x_{i+1}) = \frac{N^{i+1}}{(i-2)!} \cdot \prod_{k=1}^{i}\left(1-\frac{k}{N}\right) \cdot F_X(x_{i-1})^{i-2} \cdot [1-F_X(x_{i+1})]^{N-i-1} \cdot f_X(x_{i-1}) f_X(x_i) f_X(x_{i+1}) \tag{11.55}$$

For the case of $i=1, 3 \le j \le N$ and for $x_1 < x_{j-1} < x_j$,

$$f_{(1)(j-1)(j)}(x_1,x_{j-1},x_j) = \frac{N!}{(j-3)!(N-j)!} \cdot f_X(x_1) f_X(x_{j-1}) f_X(x_j) \cdot [F_X(x_{j-1}) - F_X(x_1)]^{j-3} [1-F_X(x_j)]^{N-j} \tag{11.56}$$

or

$$f_{(1)(j-1)(j)}(x_1,x_{j-1},x_j) = \frac{N^j}{(j-3)!} \cdot \prod_{k=1}^{j-1}\left(1-\frac{k}{N}\right) \cdot f_X(x_1) f_X(x_{j-1}) f_X(x_j) \cdot [F_X(x_{j-1}) - F_X(x_1)]^{j-3} [1-F_X(x_j)]^{N-j} \tag{11.57}$$

**Proof**
The proof is detailed in Appendix 11.I.

### 11.2.8 Difference Random Variables

In many applications, the difference between the outcomes of two ordered random variables, as illustrated in Figure 11.9, is useful.

**Definition: *i*th Ordered Difference Random Variable** The *i*th ordered difference random variable, $\Delta X_{(i)}$, or *i*th difference random variable for short, is defined according to

$$\Delta X_{(i)} = X_{(i)} - X_{(i-1)} \quad 1 \le i \le N \tag{11.58}$$

where $X_{(0)} = x_o$ with $x_o$ being the minimum possible outcome.

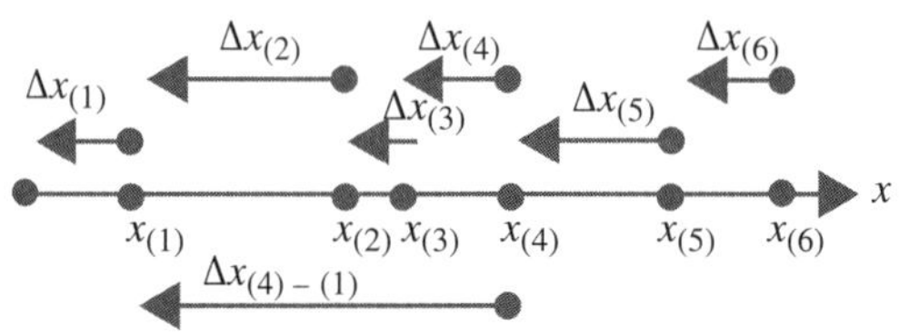


**FIGURE 11.9** Illustration of the outcomes of difference random variables defined on ordered random variables.

**Definition: *j*th to *i*th Difference Random Variables** The random variable $\Delta X_{(j)-(i)} = X_{(j)} - X_{(i)}$ defined according to

$$\Delta X_{(j)-(i)} = X_{(j)} - X_{(i)} \qquad 1 \le i < j \le N \tag{11.59}$$

is the difference between the *j*th and *i*th ordered random variables or the *j*th−*i*th difference random variable for short. Note that

$$X_{(j)} = X_{(i)} + \Delta X_{(j)-(i)} \tag{11.60}$$

***11.2.8.1 Notation*** The probability density function of $\Delta X_{(i)}$ is normally written as $f_{\Delta X_{(i)}}(\Delta x)$. For notational convenience, the notation $f_{\Delta_{(i)}}(\Delta x)$ is used. Similarly, the probability density function of $\Delta X_{(j)-(i)}$ is normally written as $f_{\Delta X_{(j)-(i)}}(\Delta x)$ and, for notational convenience, $f_{\Delta_{(j)-(i)}}(\Delta x)$ is used.

### *11.2.8.2 Probability Density Functions*

**Theorem 11.14 PDF of Difference Random Variables**
The probability density function of the *i*th difference random variable $\Delta X_{(i)} = X_{(i)} - X_{(i-1)}$ is

$$f_{\Delta_{(1)}}(\Delta x) = f_{(1)}(\Delta x)$$

$$f_{\Delta_{(i)}}(\Delta x) = \begin{cases} \displaystyle\int_{-\infty}^{\infty} f_{(i-1)(i)}(u, u + \Delta x)\, du & 0 < \Delta x, 2 \le i \le N \\ 0 & \Delta x \le 0 \end{cases} \tag{11.61}$$

**Proof**
This result is a subcase of the more general result given in Theorem 11.15.

**Theorem 11.15 PDF of *j*th to *i*th Difference Random Variables**
For the case of *N* ordered random variables, the probability density function of the *j*th−*i*th difference random variable $\Delta X_{(j)-(i)} = X_{(j)} - X_{(i)}$ is

$$f_{\Delta_{(j)-(i)}}(\Delta x) = \begin{cases} \displaystyle\int_{-\infty}^{\infty} f_{(i)(j)}(u, u + \Delta x)\, du & 0 < \Delta x, 1 \le i < j \le N \\ 0 & \Delta x \le 0 \end{cases} \tag{11.62}$$

For the case of *N* ordered random variables defined on the interval $[\alpha, \beta]$, the *j*th−*i*th difference random variable $\Delta X_{(j)-(i)} = X_{(j)} - X_{(i)}$ is given by

$$f_{\Delta_{(j)-(i)}}(\Delta x)=\begin{cases}\displaystyle\int_{\alpha}^{\beta-\Delta x} f_{(i)(j)}(u,u+\Delta x)\,du & 0<\Delta x<\beta-\alpha, 1\le i<j\le N\\ 0 & \Delta x\le 0, \Delta x\ge \beta-\alpha\end{cases}\tag{11.63}$$

and for the case of the difference between $X_{(i)}$ and $X_{(i-1)}$,

$$f_{\Delta_{(i)-(i-1)}}(\Delta x)=f_{\Delta_{(i)}}(\Delta x)=\begin{cases}\displaystyle\int_{\alpha}^{\beta-\Delta x} f_{(i-1)(i)}(u,u+\Delta x)\,du\\ 0<\Delta x<\beta-\alpha, 2\le i\le N\\ 0 \quad \Delta x\le 0, \Delta x\ge \beta-\alpha\end{cases}\tag{11.64}$$

For the case of $N$ ordered random variables defined on the interval $[0, T]$, the probability density function of the $i$th difference random variable is

$$f_{\Delta_{(i)}}(\Delta x)=\begin{cases}\displaystyle\int_{0}^{T-\Delta x} f_{(i-1)(i)}(u,u+\Delta x)\,du & 0<\Delta x<T, 2\le i\le N\\ 0 & \Delta x\le 0, \Delta x\ge T\end{cases}\tag{11.65}$$

**Proof**

The proof is detailed in Appendix 11.J.

The difference probability density functions $f_{\Delta_{(i)}}(\Delta x), i\in\{1,\ldots,10\}$, are shown in Figure 11.10 for the case of 10 uniformly distributed random variables on the interval

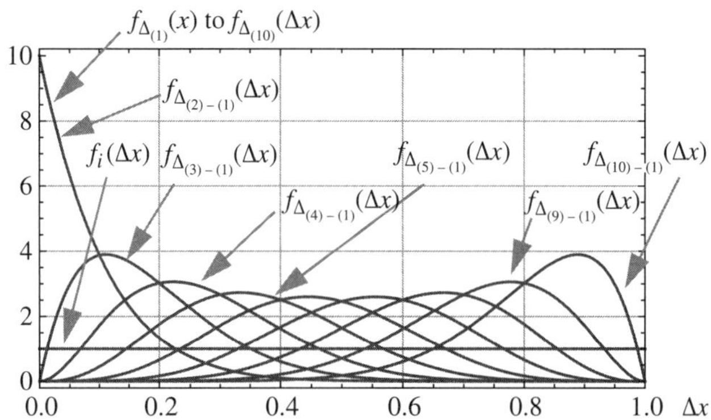


**FIGURE 11.10** Graph of $f_{\Delta_{(i)}}(\Delta x), i\in\{1,\ldots,10\}$, and graph of $f_{\Delta_{(j)-(1)}}(\Delta x), j\in\{2,\ldots,10\}$, for the case of 10 identical and independent random variables with a uniform distribution on the interval [0, 1].

[0, 1]. Note that the difference probability density functions $f_{\Delta_{(1)}}(x)$ to $f_{\Delta_{(10)}}(\Delta x)$ are identical as expected. Also shown in Figure 11.10, for the same case, are the graphs of $f_{\Delta_{(j)-(1)}}(\Delta x), j \in \{2,\ldots,10\}$.

***11.2.8.3 PDF of Difference Between Maximum to Minimum RVs*** Define the random variable $X_{\text{MaxMin}} = X_{(N)} - X_{(1)}$ as the difference between the maximum and minimum ordered random variables. It then follows from Theorem 11.15 that

$$f_{\text{MaxMin}}(\Delta x) = f_{\Delta_{(N)-(1)}}(\Delta x) = \begin{cases} \int_{-\infty}^{\infty} f_{(1)(N)}(u, u+\Delta x)\, du & \Delta x > 0 \\ 0 & \Delta x \le 0 \end{cases} \tag{11.66}$$

and for the case of random variables restricted to the interval [α, β]:

$$f_{\text{MaxMin}}(\Delta x) = f_{\Delta_{(N)-(1)}}(\Delta x) = \begin{cases} \int_{\alpha}^{\beta-\Delta x} f_{(1)(N)}(u, u+\Delta x)\, du & 0 < \Delta x < \beta - \alpha \\ 0 & \Delta x \le 0, \Delta x \ge \beta - \alpha \end{cases} \tag{11.67}$$

***11.2.8.4 Joint PDF of Difference Random Variables*** The joint probability density function between the $i$th and $j$th difference random variables

$$\Delta X_{(i)} = X_{(i)} - X_{(i-1)} \qquad \Delta X_j = X_{(j)} - X_{(j-1)} \tag{11.68}$$

is specified in the following theorem:

**Theorem 11.16 Joint PDF of Difference Random Variables**
For the case of $N$ ordered random variables defined on the interval [α, β], and with a zeroth ordered random variable defined by $X_{(0)} = \alpha$, the following results hold for the joint probability density function of the $i$th and $j$th difference random variables:

First, for the case of $i = 1$,

$$f_{\Delta_{(1)}\Delta_{(2)}}(\Delta x_1, \Delta x_2) = f_{(1)(2)}(\alpha + \Delta x_1, \alpha + \Delta x_1 + \Delta x_2) \tag{11.69}$$

$$f_{\Delta_{(1)}\Delta_{(j)}}(\Delta x_1, \Delta x_j) = \int_0^{\beta-\alpha-\Delta x_j-\Delta x_1} f_{(1)(j-1)(j)}(\alpha + \Delta x_1, \alpha + \Delta x_1 + u, \alpha + \Delta x_1 + \Delta x_j + u)\, du \qquad 3 \le j \le N \tag{11.70}$$

Second, for the case $i \ge 2$ and $j = i + 1$,

$$f_{\Delta_{(i)}\Delta_{(i+1)}}(\Delta x_i, \Delta x_{i+1}) = \int_0^{\beta-\alpha-\Delta x_i-\Delta x_{i+1}} f_{(i-1)\,(i)\,(i+1)}(\alpha+u, \alpha+\Delta x_i+u, \alpha+\Delta x_i+\Delta x_{i+1}+u)\,du$$

$$0<\Delta x_i<\beta-\alpha-\Delta x_{i+1}, 0<\Delta x_{i+1}<\beta-\alpha-\Delta x_i, \Delta x_i+\Delta x_{i+1}<\beta-\alpha, 2\le i\le N-1 \tag{11.71}$$

or alternatively,

$$f_{\Delta_{(i)}\Delta_{(i+1)}}(\Delta x_i, \Delta x_{i+1}) = \int_{\alpha+\Delta x_i}^{\beta-\Delta x_{i+1}} f_{(i-1)\,(i)\,(i+1)}(u-\Delta x_i, u, u+\Delta x_{i+1})\,du$$

$$0<\Delta x_i<\beta-\alpha-\Delta x_{i+1}, 0<\Delta x_{i+1}<\beta-\alpha-\Delta x_i, \Delta x_i+\Delta x_{i+1}<\beta-\alpha, 2\le i\le N-1 \tag{11.72}$$

Third, for the case of $i\ge 2$ and $j\ge i+2$,

$$f_{\Delta_{(i)}\Delta_{(j)}}(\Delta x_i, \Delta x_j) = \int_{\alpha+\Delta x_i}^{\beta-\Delta x_j}\left[\int_{u+\Delta x_j}^{\beta} f_{(i-1)\,(i)\,(j-1)\,(j)}(u-\Delta x_i, u, v-\Delta x_j, v)\,dv\right]du \tag{11.73}$$

$$0<\Delta x_i<\beta-\alpha,\ 0<\Delta x_j<\beta-\alpha, \Delta x_i+\Delta x_j<\beta-\alpha, 2\le i<N-2, i+2\le j\le N$$

or alternatively,

$$f_{\Delta_{(i)}\Delta_{(j)}}(\Delta x_i, \Delta x_j) = \int_{\alpha}^{\beta-\Delta x_i-\Delta x_j}\left[\int_{u}^{\beta-\Delta x_j} f_{(i-1)\,(i)\,(j-1)\,(j)}(u, u+\Delta x_i, v, v+\Delta x_j)\,dv\right]du \tag{11.74}$$

$$0<\Delta x_i<\beta-\alpha, 0<\Delta x_j<\beta-\alpha,\ \Delta x_i+\Delta x_j<\beta-\alpha,\ 2\le i\le N-2,\ i+2\le j\le N$$

**Proof**
The proof is detailed in Appendix 11.K.

***11.2.8.5 Example*** The joint probability density function for the first and tenth difference random variables, $f_{\Delta_{(1)}\Delta_{(10)}}(\Delta x_1, \Delta x_{10})$, is shown in Figure 11.11 for the case of 10 identical and independent random variables with a uniform distribution on [0, 1].

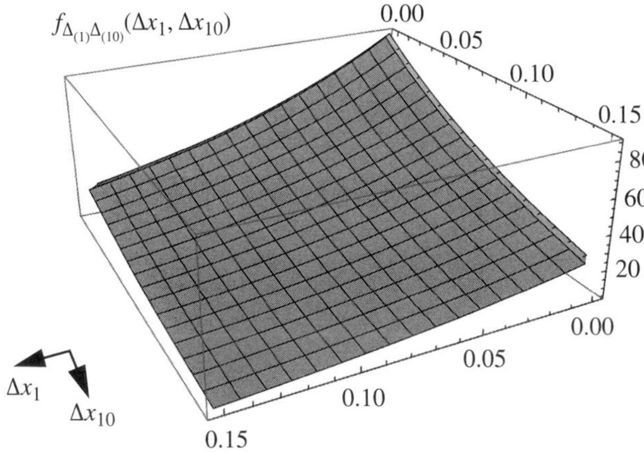


**FIGURE 11.11** Graph of $f_{\Delta_{(1)}\Delta_{(10)}}(\Delta x_1, \Delta x_{10})$ for the case of 10 independent and identical random variables with a uniform distribution on [0, 1].

#### *11.2.8.6 Joint PDF of Difference Random Variables: Infinite Interval Case*

The following theorem holds for the case of random outcomes on the interval $(0, \infty)$:

**Theorem 11.17 Joint PDF of Interarrival Times: Infinite Interval Case**
For the case of $N$ ordered random variables defined on the interval $[0, \infty]$, the joint probability density function of the $i$th and $j$th difference random variables is, first, for the case of $i = 1$,

$$f_{\Delta_{(1)}\Delta_{(2)}}(\Delta x_1, \Delta x_2) = f_{(1)(2)}(\Delta x_1, \Delta x_1 + \Delta x_2) \quad (11.75)$$

$$f_{\Delta_{(1)}\Delta_{(j)}}(\Delta x_1, \Delta x_j) = \int_0^\infty f_{(1)(j-1)(j)}(\Delta x_1, \Delta x_1 + u,\ \Delta x_1 + \Delta x_j + u)\, du \quad 3 \le j \le N \quad (11.76)$$

Second, for the case $i \ge 2$ and $j = i + 1$,

$$f_{\Delta_{(i)}\Delta_{(i+1)}}(\Delta x_i, \Delta x_{i+1}) = \int_0^\infty f_{(i-1)(i)(i+1)}(u, u + \Delta x_i, u + \Delta x_i + \Delta x_{i+1})\, du$$
$$0 < \Delta x_i,\ 0 < \Delta x_{i+1}, 2 \le i \le N - 1 \quad (11.77)$$

or alternatively,

$$f_{\Delta_{(i)}\Delta_{(i+1)}}(\Delta x_i, \Delta x_{i+1}) = \int_{\Delta x_i}^\infty f_{(i-1)(i)(i+1)}(u - \Delta x_i, u, u + \Delta x_{i+1})\, du$$
$$0 < \Delta x_i,\ 0 < \Delta x_{i+1}, 2 \le i \le N - 1 \quad (11.78)$$

Third, for the case of $i \geq 2$ and $j \geq i+2$,

$$f_{\Delta_{(i)}\Delta_{(j)}}(\Delta x_i, \Delta x_j) = \int_{\Delta x_i}^{\infty}\left[\int_{u+\Delta x_j}^{\infty} f_{(i-1)(i)(j-1)(j)}(u-\Delta x_i, u, v-\Delta x_j, v)\,dv\right]du \tag{11.79}$$

$$0<\Delta x_i, 0<\Delta x_j, 2 \leq i < N-2, i+2 \leq j \leq N$$

or alternatively,

$$f_{\Delta_{(i)}\Delta_{(j)}}(\Delta x_i, \Delta x_j) = \int_{0}^{\infty}\left[\int_{u}^{\infty} f_{(i-1)(i)(j-1)(j)}(u, u+\Delta x_i, v, v+\Delta x_j)\,dv\right]du \tag{11.80}$$

$$0<\Delta x_i, 0<\Delta x_j, 2 \leq i < N-2, i+2 \leq j \leq N$$

For the case of the infinite interval $(-\infty, \infty)$, the lower limit in the outer integral is $-\infty$.

**Proof**
The stated results follow from Theorem 11.16 for the case of $T \to \infty$.

## 11.3 IDENTICAL RVs WITH UNIFORM DISTRIBUTION

Many applications of ordered random variables arise for the case where the random variables have a uniform distribution on the interval [0, $T$] and the interval relates to time. Accordingly, the outcomes of the random variables will be assumed to have a dependence on time, and the variable $t$ will be used. Further, the notation of arrival time and interarrival time will be used. For the case of $N$ events on the interval [0, $T$], the average event rate is defined as:

**Definition: Average Event Rate** For the case of $N$ events on the interval [0, $T$], the average event rate, or rate for short, is defined as

$$\lambda = \frac{N}{T} \tag{11.81}$$

### 11.3.1 Results for the Finite Interval

For the finite interval [0, $T$], and for the case of $N$ identical and independent random variables with a uniform distribution on this interval, that is,

$$F_X(t) = \begin{cases} 1 & T \leq t \\ t/T & 0 \leq t < T \\ 0 & t < 0 \end{cases} \qquad f_X(t) = \begin{cases} 1/T & 0 \leq t < T \\ 0 & \text{elsewhere} \end{cases} \tag{11.82}$$

the following results hold:

**Theorem 11.18 Cumulative Distribution Function**
For the case of $N$ identical and independent random variables with a uniform distribution on the interval [0, $T$], the cumulative distribution functions of the first and last ordered random variables, respectively, are

$$F_{(1)}(t)=\begin{cases}0 & t<0\\ 1-\left(1-\frac{t}{T}\right)^N=1-\left(1-\frac{\lambda t}{N}\right)^N & 0\le t\le T\\ 1 & t\ge T\end{cases} \tag{11.83}$$

$$F_{(N)}(t)=\begin{cases}0 & t<0\\ \left(\frac{t}{T}\right)^N=\frac{(\lambda t)^N}{N^N} & 0\le t\le T\\ 1 & t\ge T\end{cases} \tag{11.84}$$

For the $i$th ordered random variable and for $0\le t\le T$:

$$F_{(i)}(t)=\sum_{k=i}^{N}\frac{N!}{(N-k)!k!}\left(\frac{t}{T}\right)^k\left[1-\frac{t}{T}\right]^{N-k} \qquad 1\le i\le N \tag{11.85}$$

$$F_{(i)}(t)=\sum_{k=0}^{N-i}\frac{N!}{(N-k)!k!}\left[1-\frac{t}{T}\right]^k\left(\frac{t}{T}\right)^{N-k} \qquad 1\le i\le N \tag{11.86}$$

$$F_{(i)}(t)=1-\sum_{k=0}^{i-1}\frac{N!}{(N-k)!k!}\left(\frac{t}{T}\right)^k\left[1-\frac{t}{T}\right]^{N-k} \qquad 1\le i\le N \tag{11.87}$$

$$F_{(i)}(t)=1-\sum_{k=N-i+1}^{N}\frac{N!}{(N-k)!k!}\left[1-\frac{t}{T}\right]^k\left(\frac{t}{T}\right)^{N-k} \qquad 1\le i\le N \tag{11.88}$$

$$F_{(i)}(t)=\left(\frac{t}{T}\right)^i\sum_{k=0}^{N-i}\frac{(i-1+k)!}{(i-1)!k!}\cdot\left[1-\frac{t}{T}\right]^k \qquad 1\le i\le N \tag{11.89}$$

The following algorithm can be used for calculating $F_{(i)}(t)$ for $0\le t\le T$:

$$\begin{aligned}&F_{(1)}(t)=1-\left[1-\frac{t}{T}\right]^N\\ &F_{(i)}(t)=F_{(i-1)}(t)-\frac{N!}{(N-i-1)!(i-1)!}\left(\frac{t}{T}\right)^{i-1}\left[1-\frac{t}{T}\right]^{N-i-1} \qquad 2\le i\le N-1\end{aligned} \tag{11.90}$$

Alternative forms arise by replacing $t/T$ by $\lambda t/N$.

**Proof**
These results follow directly from Theorems 11.8 and 11.9.

**Theorem 11.19 Alternative Forms for CDF**
For the case of $N$ identical and independent random variables with a uniform distribution on the interval $[0, T]$, the following alternative forms for the cumulative distribution function of the ordered random variables hold:

$$F_{(1)}(t) = \sum_{k=1}^{N} \frac{(-1)^{k-1}\lambda^k t^k}{k!} \prod_{j=0}^{k-1}\left(1-\frac{j}{N}\right)$$

$$= \lambda t - \frac{1}{2}\left(1-\frac{1}{N}\right)\lambda^2 t^2 + \frac{1}{3!}\left(1-\frac{1}{N}\right)\left(1-\frac{2}{N}\right)\lambda^3 t^3 - \cdots \quad (11.91)$$

$$\frac{(-1)^{k-1}\lambda^k t^k}{k!} \prod_{j=1}^{k-1}\left(1-\frac{j}{N}\right) + \cdots + (-1)^{N-1}\left(\frac{t}{T}\right)^N \qquad 0 \le t \le T$$

For the general case and for $0 \le t \le T$, $1 \le i \le N$:

$$F_{(i)}(t) = 1 - \sum_{k=0}^{i-1} \frac{\lambda^k t^k}{k!}\left[1-\frac{t}{T}\right]^{N-k}\left[\prod_{j=0}^{\max\{0,k-1\}}\left(1-\frac{j}{N}\right)\right] \quad (11.92)$$

$$F_{(i)}(t) = \sum_{k=i}^{N} \frac{\lambda^k t^k}{k!}\left[1-\frac{t}{T}\right]^{N-k}\left[\prod_{j=0}^{k-1}\left(1-\frac{j}{N}\right)\right] \quad (11.93)$$

This last equation, for the case of $i = 2$, yields

$$F_{(2)}(t) = \frac{(\lambda t)^2}{2}\left(1-\frac{1}{N}\right)\left(1-\frac{t}{T}\right)^{N-2} + \cdots$$

$$+ \frac{(\lambda t)^{N-1}}{(N-1)!}\left(1-\frac{1}{N}\right)\cdots\left(1-\frac{N-2}{N}\right)\left(1-\frac{t}{T}\right) + \frac{(\lambda t)^N}{N!}\left(1-\frac{1}{N}\right)\cdots\left(1-\frac{N-1}{N}\right) \quad (11.94)$$

and similar expansions result for $3 \le i \le N$.

**Proof**
The proof is detailed in Appendix 11.L.

### *11.3.1.1 Probability Density Function*

**Theorem 11.20 Probability Density Function**
For the finite interval $[0, T]$ and for the case of $N$ identical and independent random variables with a uniform distribution on this interval, the probability density function has the following forms:

$$f_{(1)}(t) = \lambda\left(1-\frac{t}{T}\right)^{N-1} = \lambda\left(1-\frac{\lambda t}{N}\right)^{N-1} \qquad 0 \le t < T$$
$$f_{(N)}(t) = \lambda\left(\frac{t}{T}\right)^{N-1} = \lambda\left(\frac{\lambda t}{N}\right)^{N-1} \qquad 0 \le t < T \tag{11.95}$$

$$f_{(i)}(t) = \frac{\lambda(\lambda t)^{i-1}}{(i-1)!} \cdot \left(1-\frac{t}{T}\right)^{N-i} \cdot \prod_{j=0}^{i-1}\left(1-\frac{j}{N}\right) \qquad 1 \le i \le N, 0 \le t < T \tag{11.96}$$

Alternative forms arise by replacing $t/T$ by $\lambda t/N$.

**Proof**

The proof is detailed in Appendix 11.M.

***11.3.1.2 Approximations*** For $N$ large, the factorial terms in the cumulative distribution function and probability density function become very large, and approximations for this case are useful.

**Theorem 11.21 CDF and PDF Approximation: $N >> i$**

For the case of $N >> i$, the following approximations are valid:

$$F_{(i)}(t) \approx \frac{1}{(i-1)!}[\Gamma(i) - \Gamma(i, \lambda t)] \qquad 0 \le t < T, 1 \le i << N \tag{11.97}$$

$$F_{(i)}(t) \approx \frac{\lambda(\lambda t)^{i-1}}{(i-1)!} e^{-\lambda t} \qquad 0 \le t < T, 1 \le i << N \tag{11.98}$$

where $\lambda = N/T$ and

$$\Gamma(x) = \int_0^\infty u^{x-1} \exp(-u) du \qquad \Gamma(x,a) = \int_a^\infty u^{x-1} \exp(-u) du \tag{11.99}$$

are, respectively, the complete and incomplete Gamma functions. The relative error in the probability density function approximation is

$$\text{re}_{(i)}(t) = 1 - \frac{e^{-\lambda t}}{\left(1-\frac{t}{T}\right)^{N-i} \cdot \prod_{j=0}^{i-1}\left(1-\frac{j}{N}\right)} \tag{11.100}$$

**Proof**

The assumption of $i << N$ in Equation 11.96 yields

$$f_{(i)}(t) \approx \frac{\lambda(\lambda t)^{i-1}}{(i-1)!} \cdot \left(1-\frac{t}{T}\right)^{N-i} = \frac{\lambda(\lambda t)^{i-1}}{(i-1)!} \cdot e^{-\lambda t} \tag{11.101}$$

where the last result follows from Theorem 11.4. The relative error follows from this equation and Equation 11.96. Using the approximation defined by Equation 11.101, it then follows that

$$\begin{aligned} F_{(i)}(t) &\approx \frac{\lambda}{(i-1)!}\int_0^t (\lambda x)^{i-1} e^{-\lambda x}dx = \frac{1}{(i-1)!}\int_0^{\lambda t} u^{i-1}e^{-u}du \\ &= \frac{1}{(i-1)!}\left[\int_0^{\infty} u^{i-1}e^{-u}du - \int_{\lambda t}^{\infty} u^{i-1}e^{-u}du\right] = \frac{1}{(i-1)!}[\Gamma(i)-\Gamma(i,\lambda t)] \end{aligned} \quad (11.102)$$

where a change of variable $u=\lambda x$ has been used.

***11.3.1.3 Example*** The relative error in the probability density function approximation is plotted in Figure 11.12 for the case of $n=10$, $\lambda=10$, and $i=1,2,3$. The true probability density functions are shown in Figure 11.7.

### *11.3.1.4 Expected Values*

**Theorem 11.22 Expected Values**

For the case of identical and independent random variables with a uniform distribution on the interval [0, *T*], the *k*th moment of the ordered random variable $X_{(i)}$ is

$$m_{(i)}(k) = \frac{\lambda^i}{(i-1)!}\cdot\prod_{j=0}^{i-1}\left(1-\frac{j}{N}\right)\cdot\int_0^T t^{k+i-1}\left(1-\frac{t}{T}\right)^{N-i}dt \qquad \begin{matrix} 1\le i\le N \\ k\in\mathbf{Z}^+ \end{matrix} \quad (11.103)$$

**Proof**

This result follows directly from Equation 11.96.

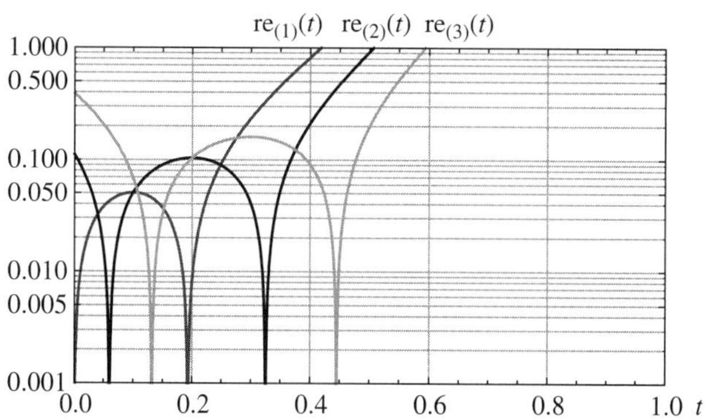


**FIGURE 11.12** Graph of the relative error in the approximation to the probability density function, as given in Equation 11.98, for the case of $i\in\{1, 2, 3\}$ and 10 independent random variables ($\lambda=10$) with a uniform distribution on the interval [0, 1].

### 11.3.2 Joint Probability Density Functions

**Theorem 11.23 Joint Probability Density Function**
For the case of identical and independent random variables with a uniform distribution on the interval [0, T], the joint probability density function of the ordered random variables is

$$f_{X_{(1)}\ldots X_{(N)}}(t_1,\ldots,t_N)=\begin{cases}0 & \text{otherwise}\\ \dfrac{N!}{T^N}=\lambda^N\left[\prod_{k=1}^{N-1}\left(1-\dfrac{k}{N}\right)\right] & 0<t_1<t_2<\cdots<t_N<T\end{cases} \tag{11.104}$$

**Proof**
From Theorem 11.11, it follows that

$$f_{(1)\ldots(N)}(t_1,\ldots,t_N)=N!\prod_{i=1}^{N}f_X(t_i)=\frac{N!}{T^N} \qquad 0<t_1<t_2<\cdots<t_N<T \tag{11.105}$$

Expanding out $N!$ yields

$$N!=N(N-1)\ldots2\cdot1=N^N\left(1-\frac{1}{N}\right)\left(1-\frac{2}{N}\right)\ldots\left(1-\frac{N-2}{N}\right)\left(1-\frac{N-1}{N}\right) \tag{11.106}$$

and substitution yields the required result.

**Theorem 11.24 Joint PDF of Two Ordered Random Variables**
For the case of identical and independent random variables with a uniform distribution on the interval [0, T], the joint probability density function of two ordered random variables is

$$\begin{aligned}f_{(i)(j)}(t_i,t_j) &=\frac{\lambda^j}{(i-1)!(j-i-1)!}\cdot\prod_{k=1}^{j-1}\left(1-\frac{k}{N}\right)\cdot t_i^{i-1}t_j^{j-i-1}\left[1-\frac{t_i}{t_j}\right]^{j-i-1}\\ &\left[1-\frac{t_j}{T}\right]^{N-j} \qquad 0\le t_i<t_j<T,1\le i<j\le N\end{aligned} \tag{11.107}$$

It then follows that

$$f_{(1)(2)}(t_1,t_2)=\lambda^2\left(1-\frac{1}{N}\right)\left[1-\frac{t_2}{T}\right]^{N-2} \qquad t_1<t_2<T \tag{11.108}$$

$$f_{(1)(j)}(t_1,t_j)=\frac{\lambda^j}{(j-2)!}\cdot\prod_{k=1}^{j-1}\left(1-\frac{k}{N}\right)\cdot t_j^{j-2}\left[1-\frac{t_1}{t_j}\right]^{j-2}\left[1-\frac{t_j}{T}\right]^{N-j} \qquad t_1<t_j<T,2\le j\le N \tag{11.109}$$

$$f_{(1)(N)}(t_1,t_N) = \lambda^2\left(1-\frac{1}{N}\right)\left[\frac{t_N}{T}-\frac{t_1}{T}\right]^{N-2}$$
$$= \frac{\lambda^N}{(N-2)!}\cdot\prod_{k=1}^{N-1}\left(1-\frac{k}{N}\right)\cdot t_N^{N-2}\left[1-\frac{t_1}{t_N}\right]^{N-2} \qquad 0 \le t_1 < t_N < T \tag{11.110}$$

$$f_{(i)(i+1)}(t_i,t_{i+1}) = \frac{\lambda^{i+1}}{(i-1)!}\cdot\prod_{k=1}^{i}\left(1-\frac{k}{N}\right)\cdot t_i^{i-1}\left[1-\frac{t_{i+1}}{T}\right]^{N-i-1}$$
$$0 \le t_i < t_{i+1} < T, 1 \le i < N-1 \tag{11.111}$$

$$f_{(i-1)(i)}(t_{i-1},t_i) = \frac{\lambda^{i}}{(i-2)!}\cdot\prod_{k=1}^{i-1}\left(1-\frac{k}{N}\right)\cdot t_{i-1}^{i-2}\left[1-\frac{t_i}{T}\right]^{N-i}$$
$$0 \le t_{i-1} < t_i < T, 2 \le i < N \tag{11.112}$$

By comparing Equation 11.107 with Equation 11.96, it is clear that

$$f_{(i)(j)}(t_i,t_j) \ne f_{(i)}(t_i) f_{(j)}(t_j) \tag{11.113}$$

and, hence, the $i$th and $j$th ordered random variables are not independent.

**Proof**

The proof is detailed in Appendix 11.N.

**Theorem 11.25 Joint PDF of (*i* – 1)*i*th, *i*th, (*j* – 1)th, and *j*th Ordered RVs**

For the case of identical and independent random variables with a uniform distribution on the interval [0, *T*], the joint probability density function of the $(i-1)$th, $i$th, $(j-1)$th, and $j$th ordered random variables is

$$f_{(i-1)(i)(j-1)(j)}(t_{i-1},t_i,t_{j-1},t_j) = \frac{\lambda^j}{(i-2)!(j-i-2)!}\cdot\prod_{k=1}^{j-1}\left(1-\frac{k}{N}\right)\cdot t_{i-1}^{i-2}\cdot$$
$$t_{j-1}^{j-i-2}\left[1-\frac{t_i}{t_{j-1}}\right]^{j-i-2}\cdot\left[1-\frac{t_j}{T}\right]^{N-j} \quad \begin{cases} t_{i-1} < t_i < t_{j-1} < t_j \\ 2 \le i < j \le N \end{cases} \tag{11.114}$$

For the case of $j = i+1$,

$$f_{(i-1)(i)(i+1)}(t_{i-1},t_i,t_{i+1}) = \frac{\lambda^{i+1}}{(i-2)!}\cdot\prod_{k=1}^{i}\left(1-\frac{k}{N}\right)\cdot t_{i-1}^{i-2}\cdot\left[1-\frac{t_{i+1}}{T}\right]^{N-i-1}$$
$$t_{i-1} < t_i < t_{i+1}, 2 \le i \le N-1 \tag{11.115}$$

**Proof**

The proof is detailed in Appendix 11.O.

### 11.3.3 Interarrival Times

Analogous to the general definitions for the *i*th ordered difference random variable and the *j*th−*i*th difference random variable, the following definitions can be made for the uniform random variable case:

**Definition: *i*th Interarrival Time and *j*th−*i*th Interarrival Time**

Define the random variable $\Delta X_{(i)} = X_{(i)} - X_{(i-1)}$ as the *i*th interarrival time where $X_{(0)} = 0$. Define the random variable $\Delta X_{(j)-(i)} = X_{(j)} - X_{(i)}$ as the *j*th−*i*th interarrival time assuming $j > i$.

**Theorem 11.26 PDF of Interarrival Time**

For the case of identical and independent random variables with a uniform distribution on the interval $[0, T]$, the probability density function of the *i*th interarrival time $\Delta X_{(i)} = X_{(i)} - X_{(i-1)}$ is given by

$$f_{\Delta_{(i)}}(\Delta t) = \lambda\left(1 - \frac{\Delta t}{T}\right)^{N-1} \qquad 0 < \Delta t < T, 1 \leq i \leq N \tag{11.116}$$

The probability density function for the *i*th interarrival time random variable is graphed in Figure 11.13 for various values of *N*.

**Proof**

The proof is detailed in Appendix 11.P.

***11.3.3.1 Notes*** First, all the interarrival random variables have identical probability density functions.

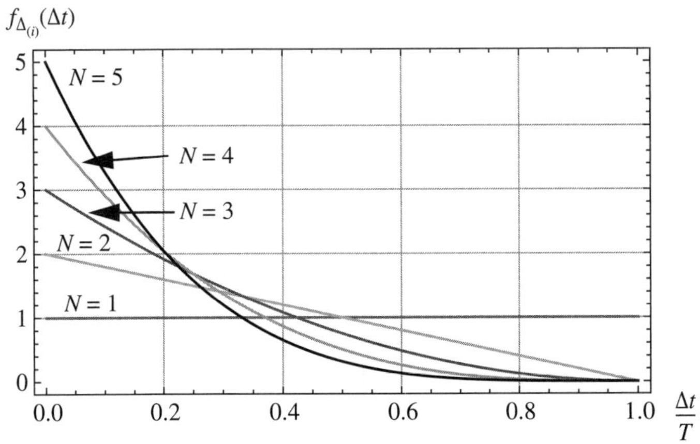


**FIGURE 11.13** Probability density function of the interarrival times as *N* varies.

Second, as $X_{(i)}$ and $X_{(i+k)}$, $i, i+k \in \{1,\ldots,N\}$, are not independent, and $\sum_{k=1}^{i} \Delta X_{(k)} = X_{(i)}$, it is the case that $\Delta X_{(i)}$ is not independent of $\Delta X_{(i-1)}$, for example, $\Delta X_{(1)} + \Delta X_{(2)} = X_{(2)}, \Delta X_{(i-1)} + \Delta X_{(i)} = X_{(i)} - X_{(i-2)}$.

### 11.3.4 PDF of *j*th to *i*th Interarrival Time

**Theorem 11.27 PDF of *j*th−*i*th Interarrival Time**
For the case of identical and independent random variables with a uniform distribution on the interval [0, *T*], the probability density function of the difference between the arrival times of the *j*th and *i*th ordered random variables, $\Delta X_{(j)-(i)} = X_{(j)} - X_{(i)}$, is given by

$$f_{\Delta_{(j)-(i)}}(\Delta t) = \frac{\lambda^j \Delta t^{j-i-1}}{(i-1)!(j-i-1)!} \cdot \prod_{k=1}^{j-1}\left(1-\frac{k}{N}\right) \cdot \int_0^{T-\Delta t} t_i^{i-1}\left[1-\frac{t_i+\Delta t}{T}\right]^{N-j} dt_i \qquad (11.117)$$

$$1 \le i < j \le N$$

and is insensitive to an integer shift in the ordered outcomes being considered, that is,

$$f_{\Delta_{(j+k)-(i+k)}}(\Delta t) = f_{\Delta_{(j)-(i)}}(\Delta t) \qquad \begin{matrix} 1 \le i < j \le N, 1 \le i+k < j+k \le N \\ 0 < \Delta t < T \end{matrix} \qquad (11.118)$$

Thus, the difference between the ordered outcomes depends only of the difference in position between the ordered outcomes considered and not on their absolute position.

**Proof**
The proof is detailed in Appendix 11.Q.

***11.3.4.1 Implications*** From Theorem 11.27, it follows that

$$f_{\Delta_{(j)-(i)}}(\Delta t) = f_{\Delta_{(j-i+1)-(1)}}(\Delta t) \qquad 1 \le i < j \le N, 0 < \Delta t < T \qquad (11.119)$$

and, hence, all required information for the probability density function is contained in $f_{\Delta_{(j)-(1)}}(\Delta t)$, which is explicitly defined in the following theorem:

**Theorem 11.28 Difference PDF and Order PDF**
The probability density function for the difference between the arrival times of the *j*th and 1st ordered random variable is equal to the probability density function of the $(j-1)$th ordered random variable, that is,

$$f_{\Delta_{(j)-(1)}}(\Delta t)=f_{(j-1)}(\Delta t)=\frac{\lambda^{j-1}\Delta t^{j-2}}{(j-2)!}\cdot\prod_{k=1}^{j-2}\left(1-\frac{k}{N}\right)\cdot\left[1-\frac{\Delta t}{T}\right]^{N-j+1}$$
$$2\le j\le N, 0<\Delta t<T \tag{11.120}$$

The probability density function for the difference between the arrival times of the $j$th and $i$th ordered random variable is equal to the probability density function of the $(j-i)$th ordered random variable, that is,

$$f_{\Delta_{(j)-(i)}}(\Delta t)=f_{(j-i)}(\Delta t)=\frac{\lambda^{j-i}\Delta t^{j-i-1}}{(j-i-1)!}\cdot\left[\prod_{k=1}^{j-i-1}\left(1-\frac{k}{N}\right)\right]\cdot$$
$$\left[1-\frac{\Delta t}{T}\right]^{N-j+i} \qquad 1\le i<j\le N, 0<\Delta t<T \tag{11.121}$$

**Proof**

The proof is detailed in Appendix 11.R.

***11.3.4.2 Difference between First and Last Ordered Random Variables*** Define the random variable $X_{\text{MaxMin}}=X_{(N)}-X_{(1)}$ as the difference between the first and last ordered random variables. It then follows from Equation 11.120 that

$$f_{\text{MaxMin}}(\Delta t)=f_{\Delta_{(N)-(1)}}(\Delta t)$$
$$=\begin{cases}\dfrac{\lambda^{N-1}\Delta t^{N-2}}{(N-2)!}\cdot\left[\prod_{k=1}^{N-2}\left(1-\dfrac{k}{N}\right)\right]\cdot\left[1-\dfrac{\Delta t}{T}\right] & 0<\Delta t<T\\ 0 & \Delta t\le 0, \Delta t\ge T\end{cases} \tag{11.122}$$

### 11.3.5 Joint PDF for Interarrival Times

The final function of interest is the joint probability density function of the two difference random variables $\Delta X_{(i)}$ and $\Delta X_{(j)}$. The region where these two random variables are nonzero, consistent with the restriction of $\Delta X_{(i)}+\Delta X_{(j)}<T$, is illustrated in Figure 11.14.

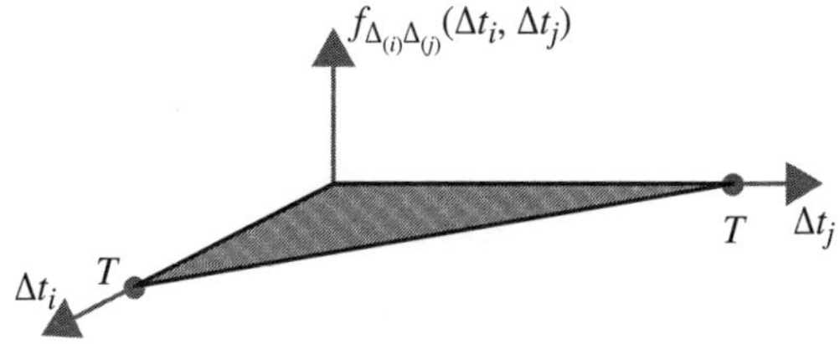


**FIGURE 11.14** Region where the joint probability density function of the interarrival random variables $\Delta X_{(i)}$ and $\Delta X_{(j)}$ is potentially nonzero.

**Theorem 11.29 Joint PDF of Interarrival Times**
For the case of identical and independent random variables with a uniform distribution on the interval [0, $T$], the following results hold for the joint probability density function of two interarrival time random variables:

$$f_{\Delta_{(1)}\Delta_{(2)}}(\Delta t_1, \Delta t_2) = \lambda^2\left(1-\frac{1}{N}\right)\left[1-\frac{\Delta t_1 + \Delta t_2}{T}\right]^{N-2} \qquad (11.123)$$
$$0 < \Delta t_1 + \Delta t_2 < T$$

In general,

$$f_{\Delta_{(i)}\Delta_{(i+1)}}(\Delta t_i, \Delta t_{i+1}) = \lambda^2 \cdot \left(1-\frac{1}{N}\right)\cdot\left[1-\frac{\Delta t_i + \Delta t_{i+1}}{T}\right]^{N-2} \qquad (11.124)$$
$$0 < \Delta t_i + \Delta t_{i+1} < T, 1 \le i \le N-1$$

For the general case,

$$f_{\Delta_{(i)}\Delta_{(j)}}(\Delta t_i, \Delta t_j) = \frac{\lambda^j}{(i-2)!(j-i-2)!}\cdot\prod_{k=1}^{j-1}\left(1-\frac{1}{N}\right)\cdot$$
$$\int_{\Delta t_i}^{T-\Delta t_j}\left(\int_{u+\Delta t_j}^{T}(u-\Delta t_i)^{i-2}\cdot\left[v-\Delta t_j-u\right]^{j-i-2}\cdot\left[1-\frac{v}{T}\right]^{N-j}dv\right)du \qquad (11.125)$$
$$0 < \Delta t_i < T, 0 < \Delta t_j < T, \Delta t_i + \Delta t_j < T, 1 \le i < N-2, i+2 \le j \le N$$

**Proof**
The proof is detailed in Appendix 11.S.

Note that the integrand in Equation 11.125, when expanded, yields a polynomial that can be integrated, and this leads, in general, to a polynomial expression for the joint probability density function.

## 11.4 UNIFORM DISTRIBUTION AND INFINITE INTERVAL

For the case of identical and independent random variables with a uniform distribution on the interval [0, $T$], the above results can be simplified for the case of $N, T \to \infty$ and with $N/T \to \lambda$. For this case, the ordered random variables correspond to the ordered random variables arising from a Poisson experiment with a rate $\lambda$.

### 11.4.1 PDF and Joint PDF of Ordered Times

#### *11.4.1.1 Individual PDFs: Infinite Interval*

**Theorem 11.30 CDF and PDF for *i*th Ordered Time**
For the case of $N, T \to \infty$, and with $N/T \to \lambda$, the cumulative distribution function and the probability density function for the $i$th ordered time, respectively, are

$$F_{(i)}(t) = 1 - e^{-\lambda t} \sum_{k=0}^{i-1} \frac{(\lambda t)^k}{k!} \qquad i \geq 1, t > 0 \tag{11.126}$$

$$f_{(i)}(t) = \frac{\lambda^i t^{i-1} e^{-\lambda t}}{(i-1)!} \qquad i \geq 1, t > 0 \tag{11.127}$$

**Proof**

The result for the cumulative distribution function arises from the expression for $F_{(i)}$ specified in Theorems 11.19 and 11.4 and by using the result

$$\lim_{N \to \infty} \prod_{j=0}^{i-1} \left(1 - \frac{j}{N}\right) = 1 \qquad i \in \mathbf{Z}^+ \tag{11.128}$$

The result for the probability density function arises from the expression for $f_{(i)}$ specified in Theorems 11.20 and 11.4 and the result stated in Equation 11.128.

The probability density functions for the first to tenth ordered times are shown in Figure 11.15 for the case of $\lambda = 10$.

### *11.4.1.2 Joint PDFs: Infinite Interval*

**Theorem 11.31 PDF for *i*th and *j*th Ordered Time**

For the case of $N, T \to \infty$, and with $N/T \to \lambda$, the probability density function for the $i$th and the $j$th ordered time is

$$f_{(i)(j)}(t_i, t_j) = \frac{\lambda^j}{(i-1)!(j-i-1)!} \cdot t_i^{i-1} t_j^{j-i-1} \left[1 - \frac{t_i}{t_j}\right]^{j-i-1} e^{-\lambda t_j} \qquad 0 < t_i < t_j, 1 \leq i < j \tag{11.129}$$

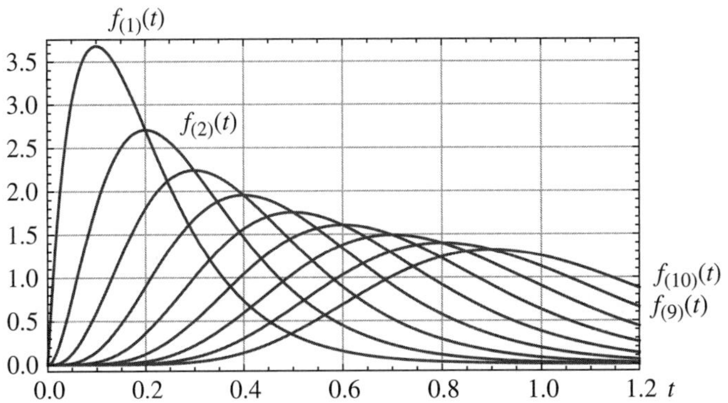


**FIGURE 11.15** Graph of $f_{(i)}(t)$, $i \in \{1, \ldots, 10\}$, for the infinite interval case with a rate of $\lambda = 10$.

and it then follows that

$$f_{(i)(j)}\left(t_i,t_j\right)=f_{(i)}\left(t_i\right)f_{(j-i)}\left(t_j-t_i\right)$$
$$=\left[\frac{\lambda^i t_i^{i-1}e^{-\lambda t_i}}{(i-1)!}\right]\cdot\left[\frac{\lambda^{j-i}\left(t_j-t_i\right)^{j-i-1}e^{-\lambda\left(t_j-t_i\right)}}{(j-i-1)!}\right] \qquad t_j>t_i, 1\le i<j \tag{11.130}$$

It also follows that

$$f_{(i)(j)}\left(t_i,t_i+\Delta t\right)=f_{(i)\Delta_{(j)-(i)}}\left(t_i,\Delta t\right) \qquad \Delta t>0 \tag{11.131}$$

where $f_{(i)\Delta_{(j)-(i)}}$ is the joint probability density function between the $i$th ordered random variable and the random variable defined by the difference between the $j$th and $i$th ordered random variables.

With the definition $\Delta t=t_j-t_i$, the following alternative forms are valid:

$$f_{(i)(j)}(t,t+\Delta t)=f_{(i)}(t)f_{(j-i)}(\Delta t)=f_{(i)\Delta_{(j)-(i)}}(t,\Delta t)=f_{(i)}(t)f_{\Delta_{(j)-(i)}}(\Delta t)$$
$$0<t, 0<\Delta t, 1\le i<j \tag{11.132}$$

The proof is detailed in Appendix 11.T.

***11.4.1.3 Notes*** The joint probability density function $f_{(i)(j)}(t,t+\Delta t)$ is shown in Figure 11.16 for the case of $\lambda=10$ and $i=1$, $j=10$.

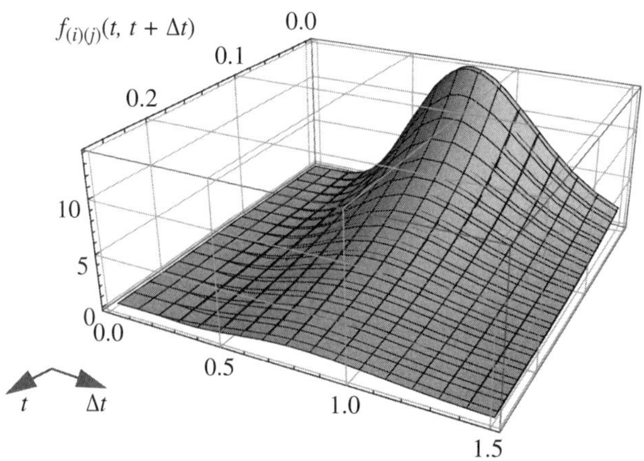


**FIGURE 11.16** Graph of $f_{(i)(j)}(t, t+\Delta t)$, $i=1$, $j=10$, for the infinite interval case with a rate of $\lambda=10$.

As an example of the joint probability density function specified by Equation 11.129, consider the specific case of $i=1$:

$$f_{(1)(j)}(t_1,t_j)=\frac{\lambda^j}{(j-2)!}\cdot\left[t_j-t_1\right]^{j-2}e^{-\lambda t_j}\quad t_j>t_1, j\geq 2$$
$$=\lambda e^{-\lambda t_1}\cdot\frac{\lambda^{j-1}\left(t_j-t_1\right)^{j-2}e^{-\lambda(t_j-t_1)}}{(j-2)!}=f_{(1)}(t_1)f_{(j-1)}(t_j-t_1) \tag{11.133}$$

The results stated in Theorem 11.31 (Eq. 11.132) imply, for the infinite interval, that $\Delta X_{(j)-(i)}=X_{(j)}-X_{(i)}$ is independent of $X_{(i)}$ as the joint probability density function between these random variables is

$$f_{(i)\Delta_{(j)-(i)}}(t,\Delta t)=f_{(i)}(t)f_{\Delta_{(j)-(i)}}(\Delta t)\qquad 0<t, 0<\Delta t, 1\leq i<j \tag{11.134}$$

Thus, the change from the $i$th ordered random variable and the change between the $i$th and $j$th random variables are independent when the infinite interval is assumed. It then follows that

$$f_{(j)/X_{(i)}=t_i}(t_j)=\frac{f_{(i)(j)}(t_i,t_j)}{f_{(i)}(t_i)}=f_{(j-i)}(t_j-t_i)=f_{\Delta_{(j)-(i)}}(t_j-t_i) \tag{11.135}$$

### *11.4.1.4 Joint PDF of (i – 1)ith, ith, (j – 1)th, and jth Ordered RVs*

**Theorem 11.32 Joint PDF of (*i* – 1)*i*th, *i*th, (*j* – 1)th, and *j*th Ordered RVs**
For the case of $N,T\to\infty$, and with $N/T\to\lambda$, $t_{i-1}<t_i<t_{j-1}<t_j$, and $2\leq i<j-1$, it follows that

$$f_{(i-1)(i)(j-1)(j)}(t_{i-1},t_i,t_{j-1},t_j)=$$
$$\frac{\lambda^j e^{-\lambda t_j}}{(i-2)!(j-i-2)!}\cdot t_{i-1}^{i-2}\cdot t_{j-1}^{j-i-2}\left[1-\frac{t_i}{t_{j-1}}\right]^{j-i-2} \tag{11.136}$$

For the case of $j=i+1$,

$$f_{(i-1)(i)(i+1)}(t_{i-1},t_i,t_{i+1})=\frac{\lambda^{i+1}e^{-\lambda t_{i+1}}}{(i-2)!}\cdot t_{i-1}^{i-2}$$
$$t_{i-1}<t_i<t_{i+1}, 2\leq i \tag{11.137}$$

**Proof**
These results follow from the equations for $f_{(i-1)(i)(j-1)(j)}$ and $f_{(i-1)(i)(i+1)}$ specified in Theorem 11.25, by using Theorem 11.4, and the result:

$$\lim_{N\to\infty}\prod_{k=0}^{j-1}\left(1-\frac{j}{N}\right)=1\qquad j\in\mathbf{Z}^+ \tag{11.138}$$

### 11.4.2 PDF and Joint PDF of Interarrival Times

The following important result holds for the infinite interval:

**Theorem 11.33 PDF and Joint PDF of Interarrival Times**
For the case of $N, T \to \infty$, and with $N/T \to \lambda$, the probability density function of the $i$th interarrival time, as defined by $\Delta X_{(i)} = X_{(i)} - X_{(i-1)}$, has an exponential distribution according to

$$f_{\Delta_{(i)}}(\Delta t) = \begin{cases} \lambda e^{-\lambda \Delta t} & i \geq 1, \Delta t > 0 \\ 0 & \Delta t < 0 \end{cases} \tag{11.139}$$

Further, the $i$th and $j$th interarrival times are independent, that is,

$$f_{\Delta_{(i)}\Delta_{(j)}}\left(\Delta t_i, \Delta t_j\right) = f_{\Delta_{(i)}}(\Delta t_i) f_{\Delta_{(j)}}\left(\Delta t_j\right) \qquad \Delta t_i, \Delta t_j > 0, 1 \leq i < j \tag{11.140}$$

**Proof**
Consider the result stated in Theorem 11.26 of $f_{\Delta_{(i)}}(\Delta t) = \lambda(1 - \Delta t/T)^{N-1}$. Using the approximation stated in Theorem 11.4 yields the required result. Proof of the independence of the $i$th and $j$th interarrival times follows from Theorem 11.36.

**Theorem 11.34 PDF of $j$th $-$ $i$th Interarrival Time**
For the case of $N, T \to \infty$, and with $N/T \to \lambda$, the probability density function of $\Delta X_{(j)-(i)} = X_{(j)} - X_{(i)}$ is insensitive to an integer shift in the ordered outcomes being considered, that is, it depends only of the difference in position between the ordered outcomes considered and not on their absolute position. Further, the probability density function of the interarrival time between the $i$th and $j$th ordered random variables is equal to the probability density function of the $(j-i)$th ordered pair.

$$f_{\Delta_{(j)-(i)}}(\Delta t) = f_{(j-i)}(\Delta t) = \frac{\lambda^{j-i} \Delta t^{j-i-1} e^{-\lambda \Delta t}}{(j-i-1)!} \qquad 1 \leq i < j, 0 < \Delta t \tag{11.141}$$

**Proof**
The proof follows from the result in Theorem 11.28 of

$$f_{\Delta_{(j)-(i)}}(\Delta t) = f_{(j-i)}(\Delta t) \qquad 1 \leq i < j \leq N, 0 < \Delta t < T \tag{11.142}$$

the expression for $f_{\Delta_{(j)-(i)}}(\Delta t)$ given in this theorem, the approximation stated in Theorem 11.4, and the result stated in Equation 11.138.

#### *11.4.2.1 Joint PDF of Interarrival Times*

**Theorem 11.35 Independence of Ordered Time and Interarrival Time**
For the case of $N,T \to \infty$, and with $N/T \to \lambda$, the $i$th arrival time $X_{(i)}$ and the subsequent interarrival time $\Delta X_{(i+1)}$ are independent random variables, that is,

$$f_{(i)\Delta_{(i+1)}}(t,\Delta t) = f_{(i)}(t) f_{\Delta_{(i+1)}}(\Delta t) = \frac{\lambda^i t^{i-1} e^{-\lambda t}}{(i-1)!} \cdot \lambda e^{-\lambda \Delta t} \tag{11.143}$$

**Proof**
From Theorem 11.31

$$f_{(i)\Delta_{(j)-(i)}}(t,\Delta t) = f_{(i)}(t) f_{\Delta_{(j)-(i)}}(\Delta t) \tag{11.144}$$

which implies

$$f_{(i)\Delta_{(i+1)}}(t,\Delta t) = f_{(i)}(t) f_{\Delta_{(i+1)}}(\Delta t) \tag{11.145}$$

Using the result for $f_{(i)}(t)$ from Equation 11.127 and the result for $f_{\Delta_{(j)-(i)}}(\Delta t)$ from Equation 11.141 for the case of $j = i + 1$, the required result follows, namely,

$$f_{(i)\Delta_{(i+1)}}(t,\Delta t) = \frac{\lambda^i t^{i-1} e^{-\lambda t}}{(i-1)!} \cdot \lambda e^{-\lambda \Delta t} \tag{11.146}$$

The joint probability density function $f_{(i)\Delta_{(i+1)}}(t,\Delta t)$ is shown in Figure 11.17 for the case of $\lambda = 10$ and $i = 10$.

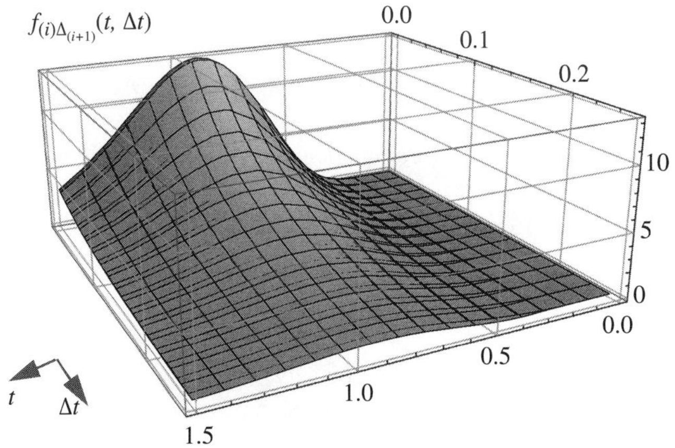


**FIGURE 11.17** Graph of $f_{(i)\Delta_{(i+1)}}(t,\Delta t)$, $i = 10$, for the infinite interval case with a rate of $\lambda = 10$.

**Theorem 11.36 Independence of *i*th and Subsequent Interarrival Times**
For the case of $N,T \to \infty$, and with $N/T \to \lambda$, the *i*th interarrival time $\Delta X_{(i)}$ and the subsequent interarrival time $\Delta X_{(i+1)}$ are independent, that is,

$$f_{\Delta_{(i)}\Delta_{(i+1)}}(\Delta t_i, \Delta t_{i+1}) = f_{\Delta_{(i)}}(\Delta t_i) f_{\Delta_{(i+1)}}(\Delta t_{i+1}) = \lambda e^{-\lambda \Delta t_i} \cdot \lambda e^{-\lambda \Delta t_{i+1}} \qquad 1 \le i, \Delta t_i > 0, \Delta t_{i+1} > 0 \tag{11.147}$$

Further, the *i*th and the *j*th interarrival times are independent and

$$f_{\Delta_{(i)}\Delta_{(j)}}(\Delta t_i, \Delta t_j) = f_{\Delta_{(i)}}(\Delta t_i) f_{\Delta_{(j)}}(\Delta t_j) = \lambda e^{-\lambda \Delta t_i} \cdot \lambda e^{-\lambda \Delta t_j} \qquad 1 \le i < j, \Delta t_i > 0, \Delta t_j > 0 \tag{11.148}$$

**Proof**
The proof is detailed in Appendix 11.U.

## 11.5 PROBLEMS

For Exercises 1–4, consider an experiment comprising of $N$ independent subexperiments. Each subexperiment yields a number from the interval $[\alpha, \beta]$ that is chosen consistent with a probability density function

$$f_X(x) = \frac{\pi}{2(\beta-\alpha)} \cdot \sin\left[\frac{\pi}{(\beta-\alpha)}(x-\alpha)\right] \quad x \in [\alpha, \beta] \tag{11.149}$$

Assume $N = 10$, $\alpha = 1$, and $\beta = 3$.

**11.1**

a. Determine the cumulative distribution function, $F_X$, associated with $f_X$. Graph $f_X$ and $F_X$.
b. Determine and graph the cumulative distribution function for the *i*th ordered outcome, $i \in \{1, \ldots, N\}$.
c. Determine and graph the probability density function for the *i*th ordered outcome, $i \in \{1, \ldots, N\}$.

**11.2**

a. Specify the joint probability density function between the *i*th and *j*th ordered random variables. Graph results for $(i=1, j=2)$, $(i=2, j=3)$ and $(i=1, j=10)$.
b. Determine an explicit expression for $f_{(i-1)(i)}(x_{i-1}, x_i)$ for arbitrary values of $N$, $\alpha$, and $\beta$.

**11.3**

a. Specify and graph the difference probability density function $f_{\Delta_{(i)}}(\Delta x)$ for $i \in \{2,\ldots,10\}$.
b. Specify and graph the probability density function of the $j$th – 1st difference random variable, $\Delta X_{(j)-(1)} = X_{(j)} - X_{(1)}$, for $j \in \{2,\ldots,10\}$.

**11.4** Specify the joint probability density function of the $i$th and $j$th ordered difference random variables. Graph this joint probability density function for the case of $(i=1, j=2)$, $(i=1, j=3)$, $(i=1, j=10)$, $(i=5, j=6)$ and $(i=5, j=10)$.

**11.5** Consider the case of $N$ identical and independent random variables with uniform distributions on the interval $[0, T]$ and one outcome from each random variable. Graph the probability density function, $f_{\Delta_{(j)-(i)}}(\Delta t)$, for the difference between the $j$th and $i$th ordered values for the cases of $\lambda = N/T = 1$ and $T=N=4$, $T=N=8$ and $T=N=16$. Use $j=4$ and $i=1$. Compare the graphs with the case of $\lambda = 1$ and $N, T \to \infty$.

## APPENDIX 11.A PROOF OF THEOREM 11.1

Consider the ordered outcome $(x_1, \ldots, x_N)$ of a trial of an experiment based on $N$ independent subexperiments. Assume $x_i$ is in a specified category $r_j \in \{1,\ldots,k\}$ and, in total, there are $N_1$ subexperimental outcomes in category one, …, $N_k$ in category $k$. With an outcome, $x_i$, of a subexperiment having a probability of $p_r$ of being in a category $r$, it follows for a single ordered outcome $(x_1, \ldots, x_N)$ that

$$P\begin{bmatrix} x_i \text{ in category } r_j, N_r \text{ in category } r \\ i \in \{1,\ldots,N\}, r_j \in \{1,\ldots,k\} \end{bmatrix} = \prod_{r=1}^{k} p_r^{N_r} \qquad N_k = N - \sum_{r=1}^{k-1} N_r \tag{11.150}$$

There are $\dfrac{N!}{(N_1)!(N_2)!\ldots(N_k)!} - 1$ other orderings of $(x_1, \ldots, x_N)$, corresponding to distinct experimental outcomes, which lead to $N_r$ subexperimental outcomes being in category $r$ for $r \in \{1,\ldots,k\}$. Each of these outcomes has a probability of occurrence as specified by Equation 11.150. Thus, it is the case that

$$P[N_r \text{ in } S_r : r \in \{1,\ldots,k\}] = \frac{N!}{(N_1)!(N_2)!\ldots(N_k)!} \cdot \prod_{r=1}^{k} p_r^{N_r} \tag{11.151}$$

## APPENDIX 11.B PROOF OF THEOREM 11.4

First, consider

$$Y = (1 - t/T)^{N-k} \Rightarrow \ln(Y) = (N-k)\ln(1 - t/T) \tag{11.152}$$

and assume $t \in (0,T)$. Second, consider a Taylor series expansion for the natural logarithm:

$$\ln(1+x) = x - \frac{x^2}{2} + \frac{x^3}{3} - \frac{x^4}{3} + \cdots \quad |x| < 1 \tag{11.153}$$

Using this expansion, it then follows that

$$\ln(Y) = -(N-k)\left[\frac{t}{T} + \frac{1}{2}\left(\frac{t}{T}\right)^2 + \frac{1}{3}\left(\frac{t}{T}\right)^3 + \frac{1}{4}\left(\frac{t}{T}\right)^4 + \cdots\right] \qquad \left|\frac{t}{T}\right| < 1 \tag{11.154}$$

Hence,

$$Y = \exp\left[-\lambda t\left(1 - \frac{k}{N}\right)\left[1 + \frac{1}{2}\left(\frac{t}{T}\right) + \frac{1}{3}\left(\frac{t}{T}\right)^2 + \frac{1}{4}\left(\frac{t}{T}\right)^3 + \cdots\right]\right] \tag{11.155}$$

where $\lambda = N/T$. For any fixed $k \in \mathbf{Z}^+$, and fixed $t \in (0, \infty)$, it is the case that

$$\lim_{N,T\to\infty}\left(1 - \frac{k}{N}\right)\left[1 + \frac{1}{2}\left(\frac{t}{T}\right) + \frac{1}{3}\left(\frac{t}{T}\right)^2 + \frac{1}{4}\left(\frac{t}{T}\right)^3 + \cdots\right] = 1 \tag{11.156}$$

Finally, interchanging the order of the limit and function operator (Theorem 2.2) yields the required result:

$$\lim_{N,T\to\infty} Y = \exp\left[\lim_{N,T\to\infty} -\lambda t\left(1 - \frac{k}{N}\right)\left[1 + \frac{1}{2}\left(\frac{t}{T}\right) + \frac{1}{3}\left(\frac{t}{T}\right)^2 + \frac{1}{4}\left(\frac{t}{T}\right)^3 + \cdots\right]\right] = e^{-\lambda t} \tag{11.157}$$

### 11.B.1 Relative Error

First, consider the general result:

$$\ln(B) = A + \alpha \;\Rightarrow\; B = e^A e^\alpha \approx e^A(1+\alpha) \quad |\alpha| << 1 \tag{11.158}$$

The relative error in the approximation $B \approx e^A$, consistent with the approximation $\ln(B) \approx A$, is

$$\text{r.e.} = 1 - \frac{e^A}{e^A e^\alpha} \approx 1 - \frac{1}{1+\alpha} \approx 1 - (1-\alpha) = \alpha \tag{11.159}$$

Second, consider Equation 11.154, which can be written as

$$\ln\left[\left(1-\frac{t}{T}\right)^{N-k}\right] = -\lambda t\left(1-\frac{k}{N}\right)\left[1+\frac{1}{2}\left(\frac{t}{T}\right)+\frac{1}{3}\left(\frac{t}{T}\right)^2+\frac{1}{4}\left(\frac{t}{T}\right)^3+\cdots\right]$$
$$= -\lambda t+\frac{\lambda tk}{N}-\lambda t\left(1-\frac{k}{N}\right)\left[\frac{1}{2}\left(\frac{t}{T}\right)+\frac{1}{3}\left(\frac{t}{T}\right)^2+\frac{1}{4}\left(\frac{t}{T}\right)^3+\cdots\right] \tag{11.160}$$

With $B=(1-t/T)^{N-k}$, $A=-\lambda t$, it then follows, from Equations 11.158–11.160, that the relative error in the approximation $(1-t/T)^{N-k}\approx\exp(-\lambda t)$ is

$$\text{r.e.} = 1-\frac{e^{-\lambda t}}{(1-t/T)^{N-k}} \approx \frac{\lambda tk}{N}-\lambda t\left(1-\frac{k}{N}\right)\left[\frac{1}{2}\left(\frac{t}{T}\right)+\frac{1}{3}\left(\frac{t}{T}\right)^2+\frac{1}{4}\left(\frac{t}{T}\right)^3+\cdots\right]$$
$$= \frac{kt}{T}-N\left(\frac{t}{T}\right)^2\left(1-\frac{k}{N}\right)\left[\frac{1}{2}+\frac{1}{3}\left(\frac{t}{T}\right)+\frac{1}{4}\left(\frac{t}{T}\right)^2+\cdots\right] \tag{11.161}$$

## APPENDIX 11.C PROOF OF THEOREM 11.5

Consider the joint probability:

$$f_{Z_1Z_2}(z_1,z_2)dz_1dz_2 \approx P[Z_1\in(z_1,z_1+dz_1),Z_2\in(z_2,z_2+dz_2)]$$
$$= P[X_1\in(z_1,z_1+dz_1),X_1+X_2\in(z_2,z_2+dz_2)] \tag{11.162}$$
$$= P[X_1\in(z_1,z_1+dz_1),X_2\in(z_2-X_1,z_2+dz_2-X_1)]$$

The area of interest is illustrated in Figure 11.18. From this figure, it follows that

$$P[X_1\in(z_1,z_1+dz_1),X_2\in(z_2-X_1,z_2+dz_2-X_1)]\approx f_{X_1X_2}(z_1,z_2-z_1)dz_1dz_2 \tag{11.163}$$

and the required result of $f_{Z_1Z_2}(z_1,z_2)=f_{X_1X_2}(z_1,z_2-z_1)$ then follows.

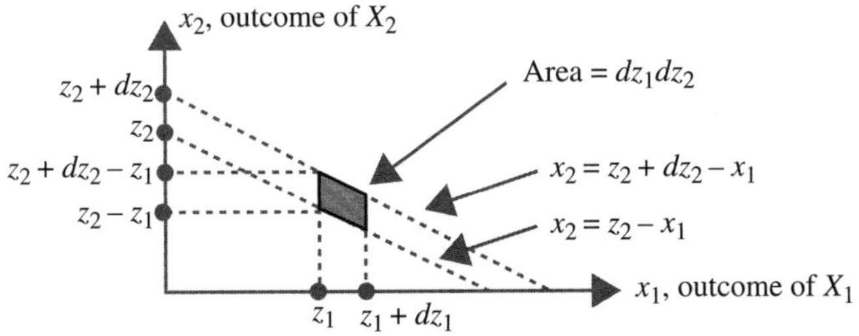


**FIGURE 11.18** The region defined by the constraints $x_1\in(z_1,z_1+dz_1)$ and $x_2\in(z_2-x_1,z_2+dz_2-x_1)$.

## APPENDIX 11.D PROOF OF THEOREM 11.6

By definition,

$$f_{X+Y/Y\in[y,y+dy]}(z)dz \approx P\left[X+Y\in(z,z+dz)|_{Y\in[y,y+dy]}\right] \\ = \frac{P[X+Y\in(z,z+dz),Y\in[y,y+dy]]}{P[Y\in[y,y+dy]]} \tag{11.164}$$

The joint probability can be written as

$$P[X+Y\in(z,z+dz),Y\in[y,y+dy]] = \\ P[Y\in[y,y+dy],X\in(z-Y,z+dz-Y)] \tag{11.165}$$

Consistent with the area illustrated in Figure 11.19, it then follows that

$$f_{X+Y/Y\in[y,y+dy]}(z)dz \approx \frac{f_{XY}(z-y,y)dzdy}{f_Y(y)dy} = \frac{f_{XY}(z-y,y)dz}{f_Y(y)} \tag{11.166}$$

as required.

## APPENDIX 11.E PROOF OF THEOREMS 11.8 AND 11.9

First, the assumption of independent random variables yields

$$P\left[X_{(N)}\le x\right] = P[X_1\le x,\ldots,X_N\le x] = P[X_1\le x]\ldots P[X_N\le x] \\ = \prod_{i=1}^{N} F_{X_i}(x) \tag{11.167}$$

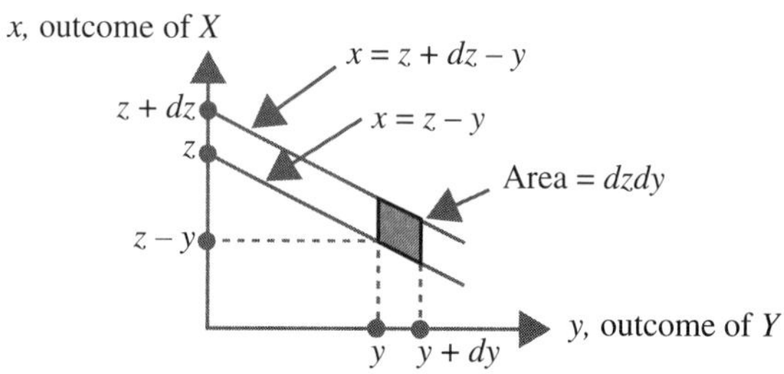


**FIGURE 11.19** Illustration of the area in the $x$–$y$ plane consistent with the joint event $Y\in[y,y+dy],X\in(z-Y,z+dz-Y)$.

Similarly,

$$\begin{aligned} P\left[X_{(1)} \le x\right] &= 1 - P\left[X_{(1)} > x\right] = 1 - P[X_1 > x, \ldots, X_N > x] \\ &= 1 - P[X_1 > x] \ldots P[X_N > x] = 1 - (1 - P[X_1 \le x]) \ldots (1 - P[X_N \le x]) \\ &= 1 - \prod_{i=1}^{N} (1 - F_{X_i}(x)) \end{aligned} \tag{11.168}$$

For the identical and independent random variable case, the event of $k, k+1, \ldots, N$ outcomes of the random variables $X_1, \ldots, X_N$ being less than, or equal to, $x$ can be viewed as the outcome of at least $k$ successes in $N$ trials where the probability of success is $F_X(x)$. It then follows from Equation 11.1 that

$$\begin{aligned} &P[k \text{ outcomes of the r.v. } X_1 \ldots X_N \text{ are less than, or equal, to } x] \\ &= \frac{N!}{(N-k)!k!} F_X^k(x) [1 - F_X(x)]^{N-k} \end{aligned} \tag{11.169}$$

and the results as stated in Equations 11.30–11.33 follow, respectively, from Equations 11.22 to 11.25.

The result specified in Equation 11.34 follows by noting the equivalence of the events:

$$\begin{aligned} &\{i, i+1, \ldots, N \text{ r.v.} \ge x\} = \{\text{first } i \text{ r.v} \le x\} \cup \\ &\{i-1 \text{ r.v. out of first } i \text{ r.v.} \le x, (i+1)\text{th r.v.} \le x\} \cup \\ &\{i-1 \text{ r.v. out of first } i+1 \text{ r.v.} \le x, (i+2)\text{th r.v.} \le x\} \cup \ldots \cup \\ &\{i-1 \text{ r.v. out of first } N-1 \text{ r.v.} \le x, N\text{th r.v.} \le x\} \end{aligned} \tag{11.170}$$

Hence, with a probability of success of $F_X(x)$,

$$\begin{aligned} F_{(i)}(x) &= \sum_{k=0}^{N-i} P[i-1 \text{ successes in } i-1+k \text{ trials}] P[\text{success at } (i+k)\text{th trial}] \\ &= F_X(x) \sum_{k=0}^{N-i} \frac{(i-1+k)!}{k!(i-1)!} \cdot F_X^{i-1}(x) [1 - F_X(x)]^k \end{aligned} \tag{11.171}$$

Finally, the algorithm for $F_{(i)}(x)$ arises by noting that Equation 11.32

$$F_{(i)}(x) = 1 - \sum_{k=0}^{i-1} \frac{N!}{(N-k)!k!} \cdot F_X^k(x) [1 - F_X(x)]^{N-k} \qquad 1 \le i \le N \tag{11.172}$$

implies

$$\begin{aligned} F_{(i+1)}(x) &= 1-\sum_{k=0}^{i}\frac{N!}{(N-k)!k!}F_X^k(x)[1-F_X(x)]^{N-k} \quad 0\le i\le N-1 \\ &= 1-\sum_{k=0}^{i-1}\frac{N!}{(N-k)!k!}F_X^k(x)[1-F_X(x)]^{N-k} - \\ &\quad \frac{N!}{(N-i)!i!}F_X^i(x)[1-F_X(x)]^{N-i} \\ &= F_{(i)}(x)-\frac{N!}{(N-i)!i!}F_X^i(x)[1-F_X(x)]^{N-i} \end{aligned} \tag{11.173}$$

Thus, $F_{(i)}(x)=F_{(i-1)}(x)-\frac{N!}{(N-i+1)!(i-1)!}F_X^{i-1}(x)[1-F_X(x)]^{N-i+1}$ for $2\le i\le N$, with $F_{(1)}(x)=1-[1-F_X(x)]^N$.

## APPENDIX 11.F PROOF OF THEOREM 11.10

By definition and for $dx$ sufficiently small,

$$\begin{aligned} f_{(i)}(x)dx &\approx P[i\text{th ordered outcome on } [x,\ x+dx]] \\ &= P[i-1 \text{ outcomes on} (-\infty,x), 1 \text{ outcome on } [x,\ x+dx], \\ &\qquad N-i \text{ outcomes on } (x+dx,\ \infty)] \end{aligned} \tag{11.174}$$

First, consider the case where each of the outcomes in the vector $(x_1, \ldots, x_N)$ is allocated into three distinct categories and the probability of an individual outcome being in these categories is denoted, respectively, $p_1$, $p_2$, and $p_3$. Second, consider the event where specific outcomes are allocated to specific categories (the ordered case), and overall there are $i-1$ outcomes in category one, 1 outcome in category two, and $N-i$ outcomes in category three. The probability of such an event is

$$p_1^{i-1}p_2p_3^{N-i} \tag{11.175}$$

Third, there are $\frac{N!}{(i-1)!(1)!(N-i)!}-1$ other events with $i-1$ outcomes in category one, 1 outcome in category two, and $N-i$ outcomes in category three. Fourth, by defining the categories, respectively, as the intervals $(-\infty,x), [x,x+dx]$ and $(x+dx,\infty)$, it follows that $p_1=F_X(x)$, $p_2=f_X(x)dx$, and $p_3=1-F_X(x+dx)$. The required result follows, namely,

$$f_{(i)}(x)=\frac{N!}{(i-1)!(N-i)!}\cdot F_X^{i-1}(x)f_X(x)[1-F_X(x)]^{N-i} \tag{11.176}$$

## APPENDIX 11.G PROOF: MARGINAL PDF FROM JOINT PDF

By definition,

$$f_{(i)}(x_i) = \int_{-\infty}^{\infty} \cdots \int_{-\infty}^{\infty} f_{X_{(1)}\ldots X_{(N)}}(x_1,\ldots,x_N)dx_1\ldots dx_{i-1}dx_{i+1}\ldots dx_N \tag{11.177}$$

and substitution of the joint probability density function as given by Equation 11.40 yields

$$\begin{aligned} f_{(i)}(x_i) &= N!f_X(x_i)\cdot \\ &\int_{R_{N-1}} f_X(x_1)\ldots f_X(x_{i-1})f_X(x_{i+1})\ldots f_X(x_N)dx_1\ldots dx_{i-1}dx_{i+1}\ldots dx_N \end{aligned} \tag{11.178}$$

where $R_{N-1}$ is the region in $N-1$ -dimensional space specified by $x_1 < x_2 < \ldots < x_{i-1} < x_i < x_{i+1} \ldots < x_N$. For $x_i$ fixed, these constraints imply two subsets of constraints: $x_1 < x_2 < \ldots < x_{i-1} < x_i$ and $x_i < x_{i+1} \ldots < x_N$. The marginal probability density function then is

$$\begin{aligned} f_{(i)}(x_i) &= N!f_X(x_i)\cdot \\ &\int_{-\infty}^{x_i}\left(\int_{-\infty}^{x_{i-1}}\left(\int_{-\infty}^{x_{i-2}}\cdots\left(\int_{-\infty}^{x_2} f_X(x_1)dx_1\right)f_X(x_{i-3})dx_{i-3}\right)f_X(x_{i-2})dx_{i-2}\right)f_X(x_{i-1})dx_{i-1}\cdot \\ &\int_{x_i}^{\infty}\left(\int_{x_{i+1}}^{\infty}\left(\int_{x_{i+2}}^{\infty}\cdots\left(\int_{x_{N-1}}^{\infty} f_X(x_N)dx_N\right)f_X(x_{i+3})dx_{i+3}\right)f_X(x_{i+2})dx_{i+2}\right)f_X(x_{i+1})dx_{i+1} \\ &= N!f(x_i)\cdot I_1\cdot I_2 \end{aligned} \tag{11.179}$$

Direct evaluation of the sequence of integrals defining $I_1$ yields $\frac{F_X^{i-1}(x_i)}{(i-1)!}$ as can be seen from the evaluation of the first few inner integrals:

$$\begin{aligned} \int_{-\infty}^{x_3}\left(\int_{-\infty}^{x_2} f_X(x_1)dx_1\right)f_X(x_2)dx_2 &= \int_{-\infty}^{x_3} F_X(x_2)f_X(x_2)dx_2 = \left.\frac{F_X^2(x_2)}{2}\right|_{-\infty}^{x_3} \\ &= \frac{F_X^2(x_3)}{2} \end{aligned} \tag{11.180}$$

Similarly, $I_2 = \frac{[1-F_X(x_i)]^{N-i}}{(N-i)!}$, which can be seen from the evaluation of the first few inner integrals:

$$\int_{x_{N-2}}^{\infty}\left(\int_{x_{N-1}}^{\infty} f_X(x_N)dx_N\right)f_X(x_{N-1})dx_{N-1} = \int_{x_{N-2}}^{\infty}[1-F_X(x_{N-1})]f_X(x_{N-1})dx_{N-1}$$
$$= \frac{-[1-F_X(x_{N-1})]^2}{2}\Bigg|_{x_{N-2}}^{\infty} = \frac{[1-F_X(x_{N-2})]^2}{2} \quad (11.181)$$

Substitution of these results yields the required result, namely,

$$f_{(i)}(x_i) = \frac{N!}{(N-i)!(i-1)!}\cdot f_X(x_i)F_X^{i-1}(x_i)[1-F_X(x_i)]^{N-i} \quad (11.182)$$

## APPENDIX 11.H PROOF OF THEOREM 11.12

By definition and as $dx_i \to 0, dx_j \to 0$,

$$f_{(i)(j)}(x_i,x_j)dx_i dx_j = P\left[X_{(i)} \in [x_i,x_i+dx_i], X_{(j)} \in [x_j,x_j+dx_j]\right] \quad (11.183)$$

For the case of $j>i$, this probability is associated with the event defined by

$$\begin{aligned}&i-1 \text{ outcomes on } (-\infty,x_i), 1 \text{ outcome on } [x_i,x_i+dx_i],\\ &j-i-1 \text{ outcomes on } (x_i+dx_i,x_j), 1 \text{ outcome on } [x_j,x_j+dx_j],\\ &N-j \text{ outcome on } (x_j+dx_j, \infty)\end{aligned} \quad (11.184)$$

The probability of an individual outcome of $(x_1, \ldots, x_N)$ being in the intervals $(-\infty,x_i)$, $[x_i,x+dx_i]$, $(x+dx_i,x_j)$, $[x_j,x+dx_j]$, and $(x_j+dx_j,\infty)$, respectively, is $p_1=F_X(x_i)$, $p_2=f_X(x_i)dx_i$, $p_3=F_X(x_j)-F_X(x_i+dx_i)$, $p_4=f_X(x_j)dx_j$, and $p_5=1-F_X(x_j+dx_j)$. The probability of $i-1$ individual outcomes in $(-\infty,x_i)$, 1 in $[x_i,x+dx_i]$, $j-i-1$ in $(x_i+dx_i,x_j)$, 1 in $[x_j,x_j+dx_j]$, and $N-j$ in $(x_j+dx_j,\infty)$, where the order of allocation into the five categories is specified, occurs with a probability

$$p_1^{i-1}\,p_2\,p_3^{j-i-1}\,p_4\,p_5^{N-j} \quad (11.185)$$

There are $\frac{N!}{(i-1)!(1)!(j-i-1)!(1)!(N-j)!}-1$ other vectors of outcomes that lead to the same number of individual outcomes in each of the defined intervals. The required result then follows, namely,

$$f_{(i)(j)}\left(x_i,x_j\right)dx_idx_j=\frac{N!}{(i-1)!(j-i-1)!(N-j)!}\cdot \tag{11.186}$$
$$F_X^{i-1}(x_i)f_X(x_i)dx_i\left[F_X\left(x_j\right)-F_X(x_i)\right]^{j-i-1}f_X\left(x_j\right)dx_j\left[1-F_X\left(x_j\right)\right]^{N-j}$$

As

$$\begin{aligned}\frac{N!}{(N-j)!}&=N(N-1)\dots(N-j+1)=N^j\left(1-\frac{1}{N}\right)\cdots\left(1-\frac{j-1}{N}\right)\\&=N^j\prod_{k=0}^{j-1}\left(1-\frac{k}{N}\right)\quad\left[=N^j\prod_{k=1}^{j-1}\left(1-\frac{k}{N}\right)\ j\geq 2\right]\end{aligned} \tag{11.187}$$

it follows that

$$f_{(i)(j)}\left(x_i,x_j\right)=\frac{N^j}{(i-1)!(j-i-1)!}\cdot\prod_{k=1}^{j-1}\left(1-\frac{k}{N}\right)\cdot \tag{11.188}$$
$$F_X^{i-1}(x_i)\left[F_X\left(x_j\right)-F_X(x_i)\right]^{j-i-1}\left[l-F_X\left(x_j\right)\right]^{N-j}f_X(x_i)f_X\left(x_j\right)$$

## APPENDIX 11.I PROOF OF THEOREM 11.13

By definition and as $dx_{i-1}\to 0, dx_i\to 0, dx_{j-1}\to 0, dx_j\to 0$,

$$\begin{aligned}&f_{(i-1)(i)(j-1)(j)}\left(x_{i-1},x_i,x_{j-1},x_j\right)dx_{i-1}dx_idx_{j-1}dx_j=\\&P\left[X_{(i-1)}\in[x_{i-1},x_{i-1}+dx_{i-1}],X_{(i)}\in[x_i,x_i+dx_i],\right.\\&\left.X_{(j-1)}\in\left[x_{j-1},x_{j-1}+dx_{j-1}\right],X_{(j)}\in\left[x_j,x_j+dx_j\right]\right]\end{aligned} \tag{11.189}$$

This probability is associated with the number of individual outcomes in the nine intervals as detailed in Table 11.1 assuming $x_j>x_{j-1}>x_i>x_{i-1}$.

**TABLE 11.1 Interval, outcomes, and probabilities**

| Interval | # outcomes | Probability for individual outcome |
|---|---|---|
| $(-\infty, x_{i-1})$ | $i-2$ | $p_1=F_X(x_{i-1})$ |
| $[x_{i-1},x_{i-1}+dx_{i-1})$ | 1 | $p_2=f_X(x_{i-1})dx_{i-1}$ |
| $[x_{i-1}+dx_{i-1},x_i)$ | 0 | $p_3=F_X(x_i)-F_X(x_{i-1}+dx_{i-1})$ |
| $[x_i,x_i+dx_i)$ | 1 | $p_4=f_X(x_i)dx_i$ |
| $\left[x_i+dx_i,x_{j-1}\right)$ | $j-i-2$ | $p_5=F_X\left(x_{j-1}\right)-F_X(x_i+dx_i)$ |
| $\left[x_{j-1},x_{j-1}+dx_{j-1}\right)$ | 1 | $p_6=f_X\left(x_{j-1}\right)dx_{j-1}$ |
| $\left[x_{j-1}+dx_{j-1},x_j\right)$ | 0 | $p_7=F_X\left(x_j\right)-F_X\left(x_{j-1}+dx_{j-1}\right)$ |
| $\left[x_j,x_j+dx_j\right)$ | 1 | $p_8=f_X\left(x_j\right)dx_j$ |
| $\left[x_j+dx_j,\infty\right)$ | $N-j$ | $p_9=1-F_X\left(x_j+dx_j\right)$ |

The probability of the specified number of entities in the set categories as detailed in Table 11.1, where the order of allocation is set, occurs with a probability

$$p_1^{i-2}\, p_2\, p_3^0\, p_4\, p_5^{j-i-2}\, p_6\, p_7^0\, p_8\, p_9^{N-j} \tag{11.190}$$

There are

$$\frac{N!}{(i-2)!(1)!(0)!(1)!(j-i-2)!(1)!(0)!(1)!(N-j)!}-1 \tag{11.191}$$

other vectors of outcomes that lead to the same number of outcomes in each of the categories. The required result then follows, namely,

$$\begin{aligned} f_{(i-1)(i)(j-1)(j)}\left(x_{i-1},x_i,x_{j-1},x_j\right) &= \frac{N!}{(i-2)!(j-i-2)!(N-j)!}\cdot F_X\left(x_{i-1}\right)^{i-2}\cdot \\ & f_X(x_{i-1})f_X(x_i)\left[F_X\left(x_{j-1}\right)-F_X\left(x_i\right)\right]^{j-i-2} f_X\left(x_{j-1}\right)f_X\left(x_j\right)\left[1-F_X\left(x_j\right)\right]^{N-j} \end{aligned} \tag{11.192}$$

The alternative form

$$\begin{aligned} f_{(i-1)(i)(j-1)(j)}\left(x_{i-1},x_i,x_{j-1},x_j\right) &= \frac{N}{(i-2)!(j-i-2)!}\cdot\prod_{k=1}^{j-1}\left(1-\frac{k}{N}\right)\cdot \\ & F_X(x_{i-1})^{i-2} f_X(x_{i-1})f_X(x_i)\left[F_X\left(x_{j-1}\right)-F_X\left(x_i\right)\right]^{j-i-2} f_X\left(x_{j-1}\right)f_X\left(x_j\right)\cdot \\ & \left[1-F_X\left(x_j\right)\right]^{N-j} \end{aligned} \tag{11.193}$$

follows from Equation 11.187.

### 11.I.1 Proof for Case of 1, $j-1$, and $j$

Consider $dx_1$, $dx_{j-1}$, and $dx_j$ sufficiently small and a vector of outcomes such that $X_{(1)} \in [x_1, x_1+dx_1]$, $X_{(j-1)} \in [x_{j-1}, x_{j-1}+dx_{j-1}]$, and $X_{(j)} \in [x_j, x_j+dx_j]$, assuming $x_j > x_{j-1} > x_1$ and $j \geq 3$. Such a vector of outcomes is consistent with individual outcomes as follows:

$$\begin{aligned} &0 \text{ outcomes on } (-\infty, x_1), 1 \text{ outcome on } [x_1, x_1+dx_1), \\ &j-3 \text{ outcomes on } [x_1+dx_1, x_{j-1}), 1 \text{ outcome on } [x_{j-1}, x_{j-1}+dx_{j-1}), \\ &0 \text{ outcomes on } [x_{j-1}+dx_{j-1}, x_j), 1 \text{ outcome on } [x_j, x_j+dx_j), \\ &N-j \text{ outcomes on } [x_j+dx_j, \infty) \end{aligned} \tag{11.194}$$

Using the probabilities tabulated in Table 11.1 and noting that there are

$$\frac{N!}{(0)!(1)!(j-3)!(1)!(0)!(1)!(N-j)!}-1 \tag{11.195}$$

other vector outcomes leading to the same partitioning of the individual outcomes, the required result follows, namely,

$$f_{(1)(j-1)(j)}\left(x_1,x_{j-1},x_j\right)dx_1dx_{j-1}dx_j = \frac{N!}{(j-3)!(N-j)!}\cdot f_X(x_1)dx_1\cdot$$
$$\left[F_X\left(x_{j-1}\right)-F_X(x_1+dx_1)\right]^{j-3} f_X\left(x_{j-1}\right)dx_{j-1}f_X\left(x_j\right)dx_j\left[1-F_X\left(x_j+dx_j\right)\right]^{N-j} \tag{11.196}$$

The alternative form

$$f_{(1)(j-1)(j)}\left(x_1,x_{j-1},x_j\right) = \frac{N^j}{(j-3)!}\cdot\prod_{k=1}^{j-1}\left(1-\frac{k}{N}\right)\cdot f_X(x_1)\cdot$$
$$\left[F_X\left(x_{j-1}\right)-F_X(x_1)\right]^{j-3} f_X\left(x_{j-1}\right)f_X\left(x_j\right)\left[1-F_X\left(x_j\right)\right]^{N-j} \tag{11.197}$$

follows from Equation 11.187.

### 11.I.2 Proof for Case of $i-1$, $i$, and $i+1$

Consider $dx_{i-1}$, $dx_i$, and $dx_{i+1}$ sufficiently small and a vector of outcomes such that $X_{(i-1)}\in\left[x_{i-1},x_{i-1}+dx_{i-1}\right]$, $X_{(i)}\in\left[x_i,x_i+dx_i\right]$, and $X_{(i+1)}\in\left[x_{i+1},x_{i+1}+dx_{i+1}\right]$, assuming $x_{i+1}>x_i>x_{i-1}$. Such a vector of outcomes is consistent with individual outcomes as follows:

$$\begin{aligned}
&i-2 \text{ outcomes on } \left(-\infty,x_{i-1}\right), 1 \text{ outcome on } \left[x_{i-1},x_{i-1}+dx_{i-1}\right),\\
&0 \text{ outcomes on } \left[x_{i-1}+dx_{i-1},x_i\right), 1 \text{ outcome on } \left[x_i,x_i+dx_i\right),\\
&0 \text{ outcomes on } \left[x_i+dx_i,x_{i+1}\right), 1 \text{ outcome on } \left[x_{i+1},x_{i+1}+dx_i\right),\\
&N-i-1 \text{ outcomes on } \left[x_{i+1}+dx_{i+1},\infty\right)
\end{aligned} \tag{11.198}$$

Using the probabilities tabulated in Table 11.1 and noting that there are

$$\frac{N!}{(i-2)!(1)!(0)!(1)!(0)!(1)!(N-i-1)!}-1 \tag{11.199}$$

other vector outcomes leading to the same partitioning of the individual outcomes, the required result follows, namely,

$$f_{(i-1)(i)(i+1)}\left(x_{i-1},x_i,x_{i+1}\right)dx_{i-1}dx_idx_{i+1} = \frac{N!}{(i-2)!(N-i-1)!}\cdot F_X\left(x_{i-1}\right)^{i-2}\cdot$$
$$f_X\left(x_{i-1}\right)dx_{i-1}f_X\left(x_i\right)dx_if_X\left(x_{i+1}\right)dx_{i+1}\left[1-F_X\left(x_{i+1}+dx_{i+1}\right)\right]^{N-i-1} \tag{11.200}$$

The alternative form

$$f_{(i-1)(i)(i+1)}(x_{i-1},x_i,x_{i+1}) = \frac{N^{i+1}}{(i-2)!}\cdot\prod_{k=1}^{i}\left(1-\frac{k}{N}\right)\cdot F_X(x_{i-1})^{i-2}\cdot \\ f_X(x_{i-1})f_X(x_i)f_X(x_{i+1})[1-F_X(x_{i+1})]^{N-i-1} \tag{11.201}$$

follows from Equation 11.187.

## APPENDIX 11.J PROOF OF THEOREM 11.15

Consider the case of random variables defined on the interval [α, β]. By definition, $\Delta X_{(j)-(i)} = X_{(j)} - X_{(i)}$, and this implies

$$f_{\Delta_{(j)-(i)}}(\Delta x)dx = \\ \begin{cases} P[\Delta X_{(j)-(i)} \in [\Delta x, \Delta x + dx]] = P[X_{(j)} - X_{(i)} \in [\Delta x, \Delta x + dx]] & \Delta x > 0 \\ 0 & \Delta x \leq 0 \end{cases} \tag{11.202}$$

Consider a pair of outcomes $(x_i, x_j)$ from the random variables $X_{(i)}$, $X_{(j)}$ and the set

$$\{(x_i,x_j):\ x_j - x_i \in [\Delta x, \Delta x + dx], \alpha \leq x_i < x_j \leq \beta\} \\ = \{(x_i,x_j): x_j \in [x_i + \Delta x, x_i + \Delta x + dx], x_i \in [\alpha, \beta - \Delta x], \alpha \leq x_i < x_j \leq \beta\} \tag{11.203}$$

The region defined by this set is illustrated in Figure 11.20, and it then follows that

$$f_{\Delta_{(j)-(i)}}(\Delta x)dx = \int_{\alpha}^{\beta-\Delta x}\left[\int_{x_i+\Delta x}^{x_i+\Delta x+dx} f_{(i)(j)}(x_i,x_j)dx_j\right]dx_i \\ \approx dx\int_{\alpha}^{\beta-\Delta x} f_{(i)(j)}(x_i, x_i+\Delta x)dx_i \tag{11.204}$$

and the required result follows assuming $dx \to 0$. The result for the infinite interval case follows in a similar manner.

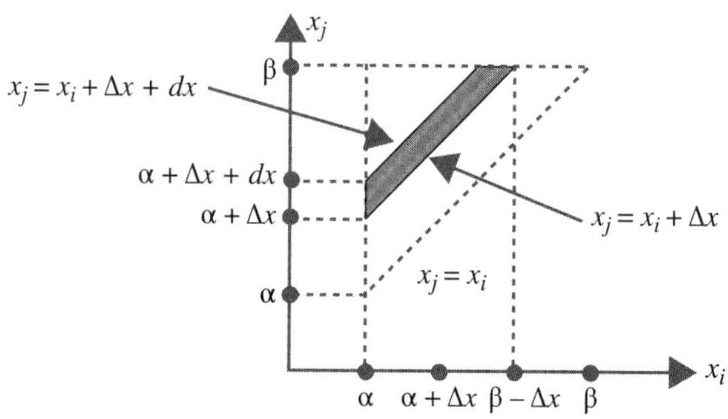


**FIGURE 11.20** Illustration of the set defined by $x_j - x_i \in [\Delta x, \Delta x + dx]$.

## APPENDIX 11.K PROOF OF THEOREM 11.16

### 11.K.1 Part 1: Case of $i=1$ and $j=2$

For the case of $i=1$ and $j=2$, the ordered difference random variables are $\Delta X_{(1)}=X_{(1)}-X_{(0)}=X_{(1)}-\alpha$ and $\Delta X_{(2)}=X_{(2)}-X_{(1)}$. Their joint probability density function is specified by the joint probability

$$f_{\Delta_{(1)}\Delta_{(2)}}(\Delta x_1,\Delta x_2)dx_1dx_2 \approx P\big[\Delta X_{(1)}\in(\Delta x_1,\Delta x_1+dx_1),\Delta X_{(2)}\in(\Delta x_2,\Delta x_2+dx_2)\big] \quad (11.205)$$

for $dx_1$ and $dx_2$ sufficiently small. To establish an expression for this joint probability, consider the event of

$$\Delta X_{(1)}\in(\Delta x_1,\Delta x_1+dx_1) \quad \text{and} \quad \Delta X_{(2)}\in(\Delta x_2,\Delta x_2+dx_2) \quad (11.206)$$

which implies the outcomes of the ordered random variables $X_{(1)}$ and $X_{(2)}$ as illustrated in Figure 11.21. It then follows that

$$P\big[\Delta X_{(1)}\in(\Delta x_1,\Delta x_1+dx_1),\Delta X_{(2)}\in(\Delta x_2,\Delta x_2+dx_2)\big] = P\big[X_{(1)}\in(\alpha+\Delta x_1,\alpha+\Delta x_1+dx_1),X_{(2)}\in\big(X_{(1)}+\Delta x_2,X_{(1)}+\Delta x_2+dx_2\big)\big] \quad (11.207)$$

which is consistent with outcomes of $X_{(1)}$ and $X_{(2)}$ being in the area defined in Figure 11.22. Thus,

$$\begin{aligned} f_{\Delta_{(1)}\Delta_{(2)}}(\Delta x_1,\Delta x_2)dx_1dx_2 &\approx P\big[X_{(1)}\in(\alpha+\Delta x_1,\alpha+\Delta x_1+dx_1),X_{(2)}\in\big(X_{(1)}+\Delta x_2,X_{(1)}+\Delta x_2+dx_2\big)\big] \\ &= f_{(1)(2)}(\alpha+\Delta x_1,\alpha+\Delta x_1+\Delta x_2)dx_1dx_2 \end{aligned} \quad (11.208)$$

and in the limit as $dx_1 \to 0, dx_2 \to 0$, the required result follows.

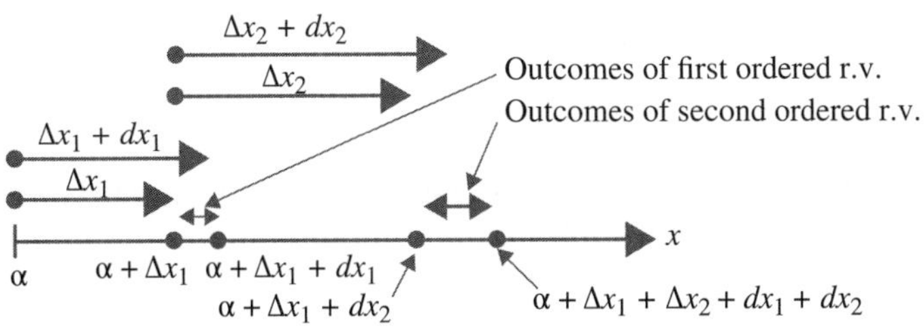


**FIGURE 11.21** Illustration of the outcomes of $X_{(1)}$ and $X_{(2)}$ consistent with the constraints on the difference random variables as given in Equation 11.206.

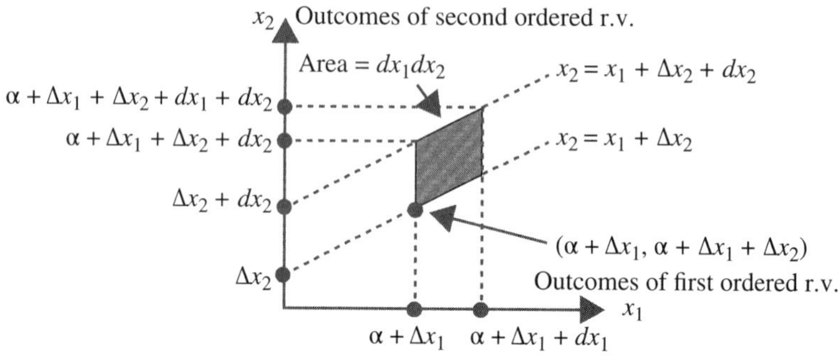


**FIGURE 11.22** Illustration of the outcomes of the first two ordered random variables given the constraints on the outcomes of the first two difference random variables of $(\Delta x_1, \Delta x_1 + dx_1)$ and $(\Delta x_2, \Delta x_2 + dx_2)$.

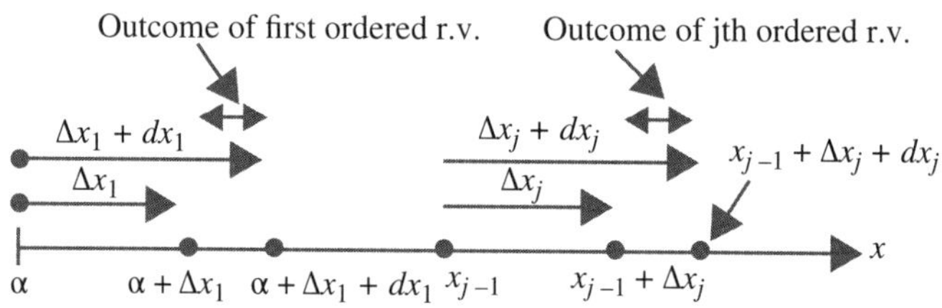


**FIGURE 11.23** Illustration of the outcomes of the 1*st* and *j*th ordered random variables assuming a fixed outcome for the $(j-1)$th ordered random variable of $x_{j-1}$ and constraints on the first and *j*th difference random variables of $(\Delta x_1, \Delta x_1 + dx_1)$ and $(\Delta x_j, \Delta x_j + dx_j)$.

## 11.K.2 Part 2: Case of $i = 1$ and $j \geq 3$

For the case of $i = 1$ and $j \geq 3$, the ordered difference random variables are $\Delta X_{(1)} = X_{(1)} - X_{(0)} = X_{(1)} - \alpha$ and $\Delta X_{(j)} = X_{(j)} - X_{(j-1)}$. Their joint probability density function is specified by the joint probability

$$\begin{aligned} &f_{\Delta_{(1)}\Delta_{(j)}}(\Delta x_1, \Delta x_j)dx_1 dx_j \approx \\ &\quad P\left[\Delta X_{(1)} \in (\Delta x_1, \Delta x_1 + dx_1), \Delta X_{(j)} \in (\Delta x_j, \Delta x_j + dx_j)\right] \end{aligned} \tag{11.209}$$

To establish this joint probability, consider the event of

$$\Delta X_{(1)} \in (\Delta x_1, \Delta x_1 + dx_1) \text{ and } \Delta X_{(j)} \in (\Delta x_j, \Delta x_j + dx_j) \tag{11.210}$$

which implies the outcomes of the ordered random variables $X_{(1)}$ and $X_{(j)}$ as illustrated in Figure 11.23. It then follows that

$$\begin{aligned} &P\left[\Delta X_{(1)} \in (\Delta x_1, \Delta x_1 + dx_1), \Delta X_{(j)} \in (\Delta x_j, \Delta x_j + dx_j)\right] = \\ &\quad P\left[X_{(1)} \in (\Delta x_1, \Delta x_1 + dx_1), X_{(j)} \in (X_{(j-1)} + \Delta x_j, X_{(j-1)} + \Delta x_j + dx_j)\right] \end{aligned} \tag{11.211}$$

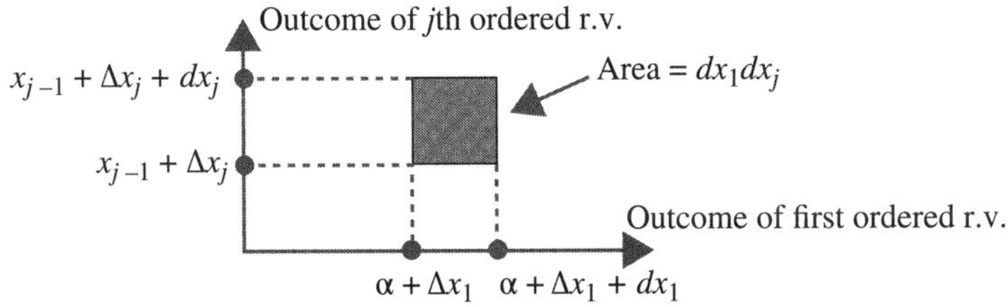


**FIGURE 11.24** Illustration of the outcomes of the 1st and $j$th ordered random variables assuming a fixed outcome for the $(j-1)$th ordered random variable and constraints on the first and $j$th difference random variables of $(\Delta x_1, \Delta x_1 + dx_1)$ and $(\Delta x_j, \Delta x_j + dx_j)$.

Here, the restrictions of $X_{(j-1)} > \alpha + \Delta x_1 + dx_1$ and $X_{(j-1)} + \Delta x_j + dx_j < \beta$ apply. The area defined by the constraints specified in Equation 11.211 is illustrated in Figure 11.24, and it then follows that

$$f_{\Delta_{(1)}\Delta_{(j)}}(\Delta x_1, \Delta x_j)dx_1dx_j \approx \\ P\left[X_{(1)} \in (\alpha+\Delta x_1, \alpha+\Delta x_1+dx_1), X_{(j)} \in \left(X_{(j-1)}+\Delta x_j, X_{(j-1)}+\Delta x_j+dx_j\right)\right] \tag{11.212}$$

Consider partitioning the interval of possible nonzero values for $X_{(j-1)}$, that is, the interval $[\alpha+\Delta x_1+dx_1, \beta-\Delta x_j-dx_j]$, into $M$ equal regions with length $d\delta$. Using the Theorem of Total Probability, it follows, with $x_o = \alpha+\Delta x_1+dx_1$, that

$$\begin{aligned} f_{\Delta_{(1)}\Delta_{(j)}}(\Delta x_1, \Delta x_j)dx_1dx_j &\approx \\ &\sum_{i=1}^{M} P\left[\begin{matrix} X_{(1)} \in (\alpha+\Delta x_1, \alpha+\Delta x_1+dx_1), \\ X_{(j)} \in \left(X_{(j-1)}+\Delta x_j, X_{(j-1)}+\Delta x_j+dx_j\right) \end{matrix} \middle| _{X_{(j-1)}\in(x_o+(i-1)d\delta, x_o+id\delta)}\right]. \\ &P\left[X_{(j-1)} \in (x_o+(i-1)d\delta, x_o+id\delta)\right] \\ &\approx \sum_{i=1}^{M} P\left[X_{(1)} \in (\alpha+\Delta x_1, \alpha+\Delta x_1+dx_1), X_{(j-1)} \in (x_o+(i-1)d\delta, x_o+id\delta),\right. \\ &\left. X_{(j)} \in \left(x_o+(i-1)d\delta+\Delta x_j, x_o+id\delta+\Delta x_j+dx_j\right)\right] \end{aligned} \tag{11.213}$$

The last result has assumed that $d\delta << dx_j$. It has also been assumed that $dx_1$ and $dx_j$ are arbitrarily small. In the limit of $d\delta \to 0$, it follows that

$$f_{\Delta_{(1)}\Delta_{(j)}}(\Delta x_1, \Delta x_j)dx_1dx_j \approx \\ \int_0^{\beta-\alpha-\Delta x_j-\Delta x_1} f_{(1)(j-1)(j)}\left(\alpha+\Delta x_1, \alpha+\Delta x_1+u, \alpha+\Delta x_1+\Delta x_j+u\right)dx_1dx_jdu \tag{11.214}$$

and the required result follows.

### 11.K.3 Part 3: Case of $j = i + 1$

For the case of $j = i + 1$, the joint probability density function of $\Delta X_{(i)} = X_{(i)} - X_{(i-1)}$ and $\Delta X_{(i+1)} = X_{(i+1)} - X_{(i)}$ is specified by the joint probability

$$\begin{aligned} &f_{\Delta_{(i)}\Delta_{(i+1)}}(\Delta x_i, \Delta x_{i+1})dx_i dx_{i+1} \\ &\approx P\left[\Delta X_{(i)} \in (\Delta x_i, \Delta x_i + dx_i), \Delta X_{(i+1)} \in (\Delta x_{i+1}, \Delta x_{i+1} + dx_{i+1})\right] \end{aligned} \tag{11.215}$$

The event of

$$\Delta X_{(i)} \in (\Delta x_i, \Delta x_i + dx_i) \quad \text{and} \quad \Delta X_{(i+1)} \in (\Delta x_{i+1}, \Delta x_{i+1} + dx_{i+1}) \tag{11.216}$$

implies the *i*th and $(i+1)$th ordered random variables, as illustrated in Figure 11.25, are in the intervals

$$\begin{aligned} &X_{(i)} \in (x_{i-1} + \Delta x_i, x_{i-1} + \Delta x_i + dx_i) \\ &X_{(i+1)} \in (x_{i-1} + \Delta x_i + \Delta x_{i+1}, x_{i-1} + \Delta x_i + \Delta x_{i+1} + dx_i + dx_{i+1}) \end{aligned} \tag{11.217}$$

for the case of $X_{(i-1)} = x_{i-1}$. The area defined by the outcomes of $X_{(i)}$ and $X_{(i+1)}$ is illustrated in Figure 11.26, and, consistent with this figure, it follows that

$$\begin{aligned} &f_{\Delta_{(i)}\Delta_{(i+1)}}(\Delta x_i, \Delta x_{i+1})dx_i dx_{i+1} \approx \\ &P\begin{bmatrix} X_{(i)} \in \left(X_{(i-1)} + \Delta x_i, X_{(i-1)} + \Delta x_i + dx_i\right) \\ X_{(i+1)} \in \left(X_{(i)} + \Delta x_{i+1}, X_{(i)} + \Delta x_{i+1} + dx_{i+1}\right) \end{bmatrix} \end{aligned} \tag{11.218}$$

The following constraints apply: $x_{i-1} > \alpha$ and $x_{i-1} + \Delta x_i + \Delta x_{i+1} < \beta$, that is, $\alpha < x_{i-1} < \beta - \Delta x_i - \Delta x_{i+1}$. Consider the partition of the interval of possible nonzero values for $X_{(i-1)}$, that is, the interval $[\alpha, \beta - \Delta x_i - \Delta x_{i+1}]$, into $M$ equal regions with length $d\delta$. The Theorem of Total Probability then yields

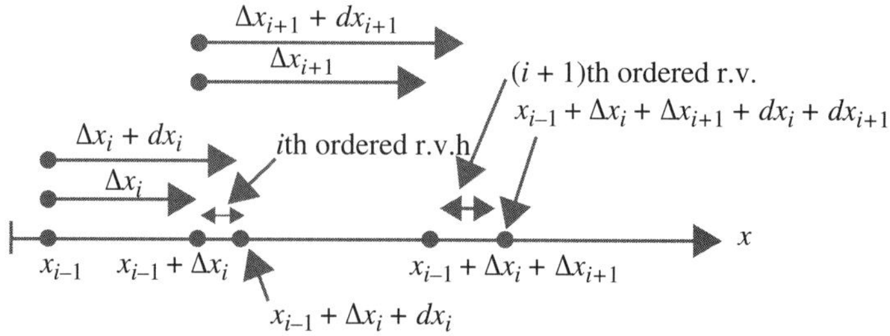


**FIGURE 11.25** Illustration of the outcomes of the *i*th and $(i+1)$th ordered random variables assuming a fixed outcome for the $(i-1)$th ordered random variable of $x_{i-1}$ and outcomes for the *i*th and $(i+1)$th difference random variables of $(\Delta x_i, \Delta x_i + dx_i)$ and $(\Delta x_{i+1}, \Delta x_{i+1} + dx_{i+1})$.

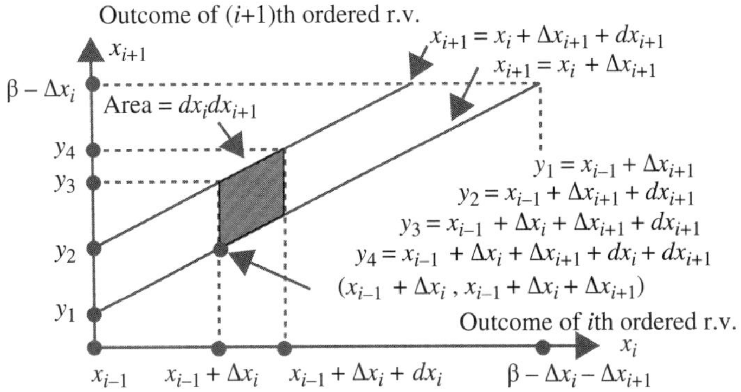


**FIGURE 11.26** Illustration of the outcomes of the $i$th and $(i + 1)$th ordered random variables assuming a fixed outcome for the $(i - 1)$th ordered random variable and constraints on the $i$th and $(i + 1)$th difference random variables of $(\Delta x_i, \Delta x_i + dx_i)$ and $(\Delta x_{i+1}, \Delta x_{i+1} + dx_{i+1})$.

$$
\begin{aligned}
&f_{\Delta_{(i)}\Delta_{(i+1)}}(\Delta x_i, \Delta x_{i+1})dx_i dx_{i+1} \approx \\
&\sum_{k=1}^{M} P\Big[X_{(i)} \in \big(X_{(i-1)} + \Delta x_i, X_{(i-1)} + \Delta x_i + dx_i\big), \\
&X_{(i+1)} \in \big(X_{(i)} + \Delta x_{i+1}, X_{(i)} + \Delta x_{i+1} + dx_{i+1}\big)\big|_{X_{(i-1)} \in (\alpha + (k-1)d\delta, \alpha + kd\delta)}\Big] \cdot \\
&P\big[X_{(i-1)} \in (\alpha + (k-1)d\delta, \alpha + kd\delta)\big]
\end{aligned}
\tag{11.219}
$$

Assuming $d\delta$ is chosen such that $d\delta << dx_i$, it follows that

$$
f_{\Delta_{(i)}\Delta_{(i+1)}}(\Delta x_i, \Delta x_{i+1})dx_i dx_{i+1} \approx
\sum_{k=0}^{M} P\begin{bmatrix} X_{(i-1)} \in (\alpha + (k-1)d\delta, \alpha + kd\delta), \\ X_{(i)} \in (\alpha + kd\delta + \Delta x_i, \alpha + kd\delta + \Delta x_i + dx_i), \\ X_{(i+1)} \in \big(X_{(i)} + \Delta x_{i+1}, X_{(i)} + \Delta x_{i+1} + dx_{i+1}\big) \end{bmatrix}
\tag{11.220}
$$

For the case of $dx_i << \Delta x_i$, it follows that

$$
f_{\Delta_{(i)}\Delta_{(i+1)}}(\Delta x_i, \Delta x_{i+1})dx_i dx_{i+1} \approx
\sum_{k=0}^{M} P\begin{bmatrix} X_{(i-1)} \in (\alpha + (k-1)d\delta, \alpha + kd\delta), \\ X_{(i)} \in (\alpha + kd\delta + \Delta x_i, \alpha + kd\delta + \Delta x_i + dx_i), \\ X_{(i+1)} \in (\alpha + kd\delta + \Delta x_i + \Delta x_{i+1}, \alpha + kd\delta + \Delta x_i + \Delta x_{i+1} + dx_{i+1}) \end{bmatrix}
\tag{11.221}
$$

Hence, as $d\delta \to 0, dx_i \to 0, dx_{i+1} \to 0$, it follows that

$$f_{\Delta_{(i)}\Delta_{(i+1)}}(\Delta x_i, \Delta x_{i+1})dx_i dx_{i+1} \approx dx_i dx_{i+1} \cdot \int_0^{\beta-\alpha-\Delta x_i-\Delta x_{i+1}} f_{(i-1)(i)(i+1)}(\alpha+u, \alpha+u+\Delta x_i, \alpha+u+\Delta x_i+\Delta x_{i+1})du \quad (11.222)$$

which is the required result.

### 11.K.4 Part 4: Case of $j = i+2, j > i+2$

For the case of $j \geq i+2$, the joint probability density function of $\Delta X_{(i)} = X_{(i)} - X_{(i-1)}$ and $\Delta X_{(j)} = X_{(j)} - X_{(j-1)}$ is specified by the joint probability

$$f_{\Delta_{(i)}\Delta_{(j)}}(\Delta x_i, \Delta x_j)\ dx_i dx_j \approx P\left[\Delta X_{(i)} \in (\Delta x_i, \Delta x_i + dx_i), \Delta X_{(j)} \in \left(\Delta x_j, \Delta x_j + dx_j\right)\right] \quad (11.223)$$

for $dx_i$ and $dx_j$ sufficiently small. The event of

$$\Delta X_{(i)} \in (\Delta x_i, \Delta x_i + dx_i) \text{ and } \Delta X_{(j)} \in \left(\Delta x_j, \Delta x_j + dx_j\right) \quad (11.224)$$

implies the $(i-1)$th and $(j-1)$th ordered random variables, as illustrated in Figure 11.27, are in the intervals

$$\begin{aligned} X_{(i-1)} &\in (x_i - \Delta x_i - dx_i, x_i - \Delta x_i) \\ X_{(j-1)} &\in \left(x_j - \Delta x_j - dx_j, x_j - \Delta x_j\right) \end{aligned} \quad (11.225)$$

for the case of $X_{(i)} = x_i$ and $X_{(j)} = x_j$. The restrictions are $x_i - \Delta x_i - dx_i > \alpha$, $x_i < x_j - \Delta x_j - dx_j$ and $x_j < \beta$.

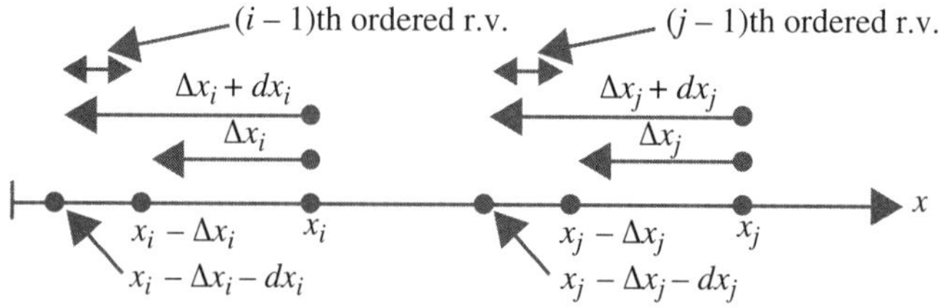


**FIGURE 11.27** Illustration of the possible outcomes of the $(i-1)$th and $(j-1)$th difference random variables assuming a fixed outcome for the $i$th and $j$th ordered random variable of $x_i$ and $x_j$ and constraints on the $i$th and $j$th difference random variables of $(\Delta x_i, \Delta x_i + dx_i)$ and $\left(\Delta x_j, \Delta x_j + dx_j\right)$.

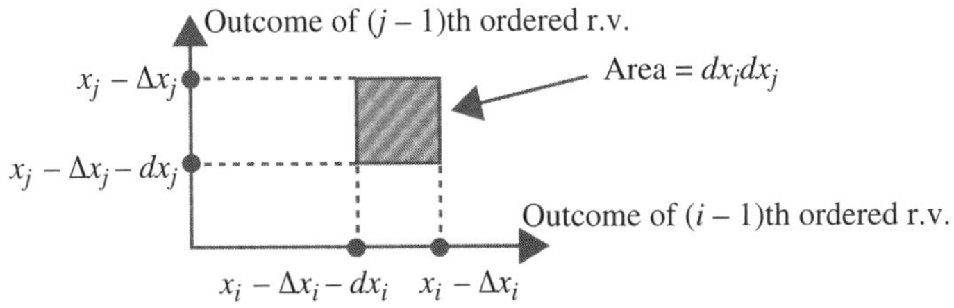


**FIGURE 11.28** Illustration of the outcomes of the $(i-1)$th and $(j-1)$th ordered random variables assuming a fixed outcome for the $i$th and $j$th ordered random variable and constraints on the $i$th and $j$th difference random variables of $(\Delta x_i, \Delta x_i + dx_i)$ and $(\Delta x_j, \Delta x_j + dx_j)$.

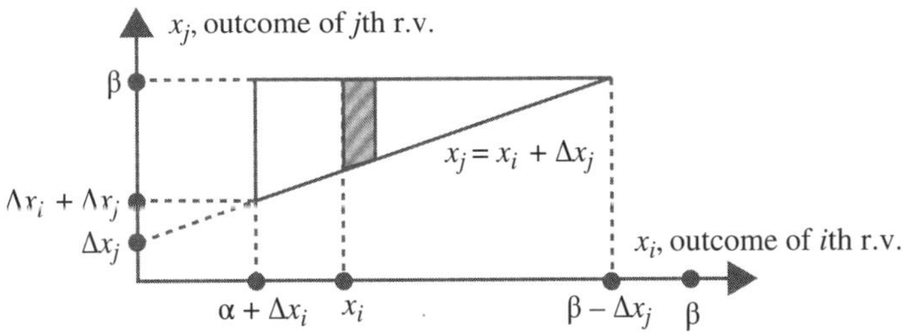


**FIGURE 11.29** The region defined by the constraints of $\alpha + \Delta x_i < x_i < \beta - \Delta x_j, x_j > x_i + \Delta x_j$ and $x_j < \beta$.

The area defined by the outcomes of $X_{(i-1)}$ and $X_{(j-1)}$ is illustrated in Figure 11.28, and, consistent with this figure, it follows that

$$f_{\Delta_{(i)}\Delta_{(j)}}(\Delta x_i, \Delta x_j)\ dx_i dx_j \approx P\begin{bmatrix} X_{(i-1)} \in (X_{(i)} - \Delta x_i - dx_i, X_{(i)} - \Delta x_i), \\ X_{(j-1)} \in (X_{(j)} - \Delta x_j - dx_j, X_{(j)} - \Delta x_j) \end{bmatrix} \tag{11.226}$$

The constraints of $\alpha + \Delta x_i < x_i < \beta - \Delta x_j, x_j > x_i + \Delta x_j$ and $x_j < \beta$ lead to the region illustrated in Figure 11.29 for the outcomes of the $i$th and $j$th ordered random variables. The Theorem of Total Probability then yields

$$\begin{aligned} f_{\Delta_{(i)}\Delta_{(j)}}(\Delta x_i, \Delta x_j)\ dx_i dx_j &\approx \\ \sum_{x_i}\sum_{x_j} P&\left[\begin{matrix} X_{(i-1)} \in (X_{(i)} - \Delta x_i - dx_i, X_{(i)} - \Delta x_i), \\ X_{(j-1)} \in (X_{(j)} - \Delta x_j - dx_j, X_{(j)} - \Delta x_j) \end{matrix} \middle| \begin{matrix} X_{(i)} \in (x_i, x_i + \delta x_i) \\ X_{(j)} \in (x_j, x_j + \delta x_j) \end{matrix}\right] \cdot \\ P&\left[X_{(i)} \in (x_i, x_i + \delta x_i), X_{(j)} \in (x_j, x_j + \delta x_j)\right] \\ \approx \sum_{x_i}\sum_{x_j} P&\begin{bmatrix} X_{(i-1)} \in (x_i - \Delta x_i - dx_i, x_i - \Delta x_i), X_{(i)} \in (x_i, x_i + \delta x_i), \\ X_{(j-1)} \in (x_j - \Delta x_j - dx_j, x_j - \Delta x_j), X_{(j)} \in (x_j, x_j + \delta x_j) \end{bmatrix} \end{aligned} \tag{11.227}$$

where the summation is over the values of $(x_i, x_j)$ consistent with the area defined in Figure 11.29 and the last result follows assuming $d\delta_i << dx_i$ and $d\delta_j << dx_j$. It then follows that

$$f_{\Delta_{(i)}\Delta_{(j)}}\left(\Delta x_i, \Delta x_j\right) dx_i dx_j \approx$$
$$dx_i dx_j \int_{\alpha+\Delta x_i}^{\beta-\Delta x_j} \left[ \int_{u+\Delta x_j}^{\beta} f_{(i-1)(i)(j-1)(j)}\left(u-\Delta x_i, u, v-\Delta x_j, v\right) dv \right] du \tag{11.228}$$

and the required result then follows. The alternative form follows from a double change of variable.

## APPENDIX 11.L PROOF OF THEOREM 11.18

The use of the binomial formula stated in Theorem 11.2 (Eq. 11.5) on the form $F_{X_{(1)}}(t) = 1 - \left(1 - \frac{\lambda t}{N}\right)^N$ yields

$$\begin{aligned} F_{X_{(1)}}(t) &= 1 - \left[1 - \binom{N}{1}\frac{\lambda t}{N} + \binom{N}{2}\left(\frac{\lambda t}{N}\right)^2 - \cdots + (-1)^k \binom{N}{k}\left(\frac{\lambda t}{N}\right)^k + \cdots + (-1)^N \left(\frac{\lambda t}{N}\right)^N\right] \\ &= \lambda t - \frac{N(N-1)}{2}\left(\frac{\lambda t}{N}\right)^2 + \cdots + \frac{(-1)^{k+1} N!}{(N-k)!k!}\left(\frac{\lambda t}{N}\right)^k + \cdots + (1)^{N+1}\left(\frac{\lambda t}{N}\right)^N \\ &= \lambda t - \frac{1}{2}\left(1 - \frac{1}{N}\right)\lambda^2 t^2 + \cdots + \frac{(-1)^{k-1}\lambda^k t^k}{k!}\prod_{j=1}^{k-1}\left(1 - \frac{j}{N}\right) + \cdots + (-1)^{N-1}\left(\frac{\lambda t}{N}\right)^N \end{aligned} \tag{11.229}$$

This last result follows from Theorem 11.3, and the first result then follows.

The second and third results, as given by Equations 11.92 and 11.93, follow, respectively, from Equations 11.87 and 11.85 by using Theorem 11.3.

## APPENDIX 11.M PROOF OF THEOREM 11.20

From Theorem 11.10, the results for $f_{(1)}(t)$ and $f_{(N)}(t)$ follow directly. For the general case, the general result stated in Theorem 11.10 (Eq. 11.38) yields

$$\begin{aligned} f_{(i)}(t) &= \frac{N!}{(N-i)!(i-1)!}\left(\frac{t}{T}\right)^{i-1}\left(1 - \frac{t}{T}\right)^{N-i}\frac{1}{T} \\ &= \frac{i}{T}\cdot\frac{N!}{(N-i)!(i)!}\cdot\frac{N^i}{N^i}\cdot\left(\frac{t}{T}\right)^{i-1}\left(1 - \frac{t}{T}\right)^{N-i} \end{aligned} \tag{11.230}$$

The use of Theorem 11.3 then yields

$$\begin{aligned} f_{(i)}(t) &= \frac{iN^i}{T^i}\cdot\left[\frac{1}{i!}\cdot\prod_{j=0}^{i-1}\left(1-\frac{j}{N}\right)\right]t^{i-1}\left(1-\frac{t}{T}\right)^{N-i} \\ &= \frac{\lambda(\lambda t)^{i-1}}{(i-1)!}\left[\prod_{j=0}^{i-1}\left(1-\frac{j}{N}\right)\right]\left(1-\frac{t}{T}\right)^{N-i} \end{aligned} \tag{11.231}$$

To check the validity of the probability density function, consider its integral, that is,

$$\int_0^T f_{(1)}(t)dt = \int_0^T \lambda\left(1-\frac{t}{T}\right)^{N-1}dt = \frac{-\lambda T}{N}\left(1-\frac{t}{T}\right)^N\Bigg|_0^T = -\left(1-\frac{t}{T}\right)^N\Bigg|_0^T = 1 \tag{11.232}$$

In general, integration by parts can be used to prove $\int_0^T f_{(i)}(t)dt = 1$.

## APPENDIX 11.N PROOF OF THEOREM 11.24

It follows from Theorem 11.12, and for the case of $1 \le i < j \le N$ and $0 \le t_i < t_j < T$, that

$$\begin{aligned} f_{(i)(j)}(t_i,t_j) &= \frac{N^j}{(i-1)!(j-i-1)!}\cdot\prod_{k=1}^{j-1}\left(1-\frac{k}{N}\right)\cdot\left(\frac{t_i}{T}\right)^{i-1}\left[\frac{t_j}{T}-\frac{t_i}{T}\right]^{j-i-1}\left[1-\frac{t_j}{T}\right]^{N-j}\frac{1}{T^2} \\ &= \frac{N^j}{(i-1)!(j-i-1)!}\cdot\prod_{k=1}^{j-1}\left(1-\frac{k}{N}\right)\cdot\frac{1}{T^j}\cdot t_i^{i-1}\left[t_j-t_i\right]^{j-i-1}\left[1-\frac{t_j}{T}\right]^{N-j} \\ &= \frac{\lambda^j}{(i-1)!(j-i-1)!}\cdot\prod_{k=1}^{j-1}\left(1-\frac{k}{N}\right)\cdot t_i^{i-1}t_j^{j-i-1}\left[1-\frac{t_i}{t_j}\right]^{j-i-1}\left[1-\frac{t_j}{T}\right]^{N-j} \end{aligned} \tag{11.233}$$

The first form in Equation 11.110 results from Equation 11.49. The second form follows from Equation 11.109. The results in Equations 11.111 and 11.112 follow from Equation 11.107.

## APPENDIX 11.O PROOF OF THEOREM 11.25

From Theorem 11.13, it follows, for $t_{i-1} < t_i < t_{j-1} < t_j, 2 \le i < j-1 \le N-1$, that

$$f_{(i-1)(i)(j-1)(j)}(t_{i-1},t_i,t_{j-1},t_j)=\frac{N^j}{(i-2)!(j-i-2)!}\cdot\prod_{k=1}^{j-1}\left(1-\frac{k}{N}\right)\cdot$$

$$\left[\frac{t_{i-1}}{T}\right]^{i-2}\cdot\left[\frac{t_{j-1}}{T}-\frac{t_i}{T}\right]^{j-i-2}\cdot\left[1-\frac{t_j}{T}\right]^{N-j}\cdot\frac{1}{T}\cdot\frac{1}{T}\cdot\frac{1}{T}\cdot\frac{1}{T}$$

$$=\frac{N^j}{(i-2)!(j-i-2)!}\cdot\prod_{k=1}^{j-1}\left(1-\frac{k}{N}\right)\cdot\frac{1}{T^j}\cdot t_{i-1}^{i-2}\left[t_{j-1}-t_i\right]^{j-i-2}\left[1-\frac{t_j}{T}\right]^{N-j} \tag{11.234}$$

$$=\frac{\lambda^j}{(i-2)!(j-i-2)!}\cdot\prod_{k=1}^{j-1}\left(1-\frac{k}{N}\right)\cdot t_{i-1}^{i-2}\cdot t_{j-1}^{j-i-2}\left[1-\frac{t_i}{t_{j-1}}\right]^{j-i-2}\cdot\left[1-\frac{t_j}{T}\right]^{N-j}$$

For the case of $j=i+1$, it follows from Theorem 11.13 (Eq. 11.55), and for the case of $t_{i-1}<t_i<t_{i+1}, 2\le i\le N-1$, that

$$f_{(i-1)(i)(i+1)}(t_{i-1},t_i,t_{i+1})=\frac{N^{i+1}}{(i-2)!}\cdot\prod_{k=1}^{i}\left(1-\frac{k}{N}\right)\cdot\left(\frac{t_{i-1}}{T}\right)^{i-2}\cdot$$

$$\left[1-\frac{t_{i+1}}{T}\right]^{N-i-1}\cdot\frac{1}{T}\cdot\frac{1}{T}\cdot\frac{1}{T}$$

$$=\frac{N^{i+1}}{(i-2)!}\cdot\prod_{k=1}^{i}\left(1-\frac{k}{N}\right)\cdot\frac{1}{T^{i+1}}\cdot t_{i-1}^{i-2}\cdot\left[1-\frac{t_{i+1}}{T}\right]^{N-i-1} \tag{11.235}$$

$$=\frac{\lambda^{i+1}}{(i-2)!}\cdot\prod_{k=1}^{i}\left(1-\frac{k}{N}\right)\cdot t_{i-1}^{i-2}\left[1-\frac{t_{i+1}}{T}\right]^{N-i-1}$$

which is the required result.

## APPENDIX 11.P PROOF OF THEOREM 11.26

The first interarrival time equals the first arrival time. Hence, from Theorem 11.20,

$$f_{\Delta_{(1)}}(\Delta t)=f_{(1)}(\Delta t)=\lambda\left(1-\frac{\Delta t}{T}\right)^{N-1} \tag{11.236}$$

For the case of $i\ge 2$, consider the result stated in Theorem 11.15:

$$f_{\Delta_{(i)}}(\Delta t)=\int_0^{T-\Delta t} f_{(i-1)(i)}(u,u+\Delta t)\,du \qquad 0<\Delta t<T, 2\le i\le N \tag{11.237}$$

Substitution of $f_{(i-1)(i)}(t_{i-1},t_i)$, as given by Equation 11.112, yields

$$f_{\Delta_{(i)}}(\Delta t)=\frac{\lambda^i}{(i-2)!}\cdot\prod_{k=1}^{i-1}\left(1-\frac{k}{N}\right)\cdot\int_0^{T-\Delta t}u^{i-2}\left[1-\frac{u+\Delta t}{T}\right]^{N-i}du \tag{11.238}$$

$$f_{\Delta_{(i+1)}}(\Delta t)=\frac{\lambda^{i+1}}{(i-1)!}\cdot\prod_{k=1}^{i}\left(1-\frac{k}{N}\right)\cdot\int_0^{T-\Delta t}w^{i-1}\left[1-\frac{w+\Delta t}{T}\right]^{N-i-1}dw \tag{11.239}$$

The goal is to find the relation between $f_{\Delta_{(i)}}(\Delta t)$ and $f_{\Delta_{(i+1)}}(\Delta t)$. To this end, consider integration by parts with $u=w^{i-1}$ and $dv=\left[1-\frac{\Delta t}{T}-\frac{w}{T}\right]^{N-i-1}dw$ on the integral defining $f_{\Delta_{(i+1)}}(\Delta t)$. The results $du=(i-1)w^{i-2}dw$ and $v=\frac{-T}{N-i}\left[1-\frac{\Delta t}{T}-\frac{w}{T}\right]^{N-i}$ yield, for the case of $i\geq 2$,

$$\int_0^{T-\Delta T}w^{i-1}\left[1-\frac{w+\Delta t}{T}\right]^{N-i-1}dw=uv\Big|_0^{T-\Delta t}-\int_0^{T-\Delta t}vdu$$

$$=\frac{-w^{i-1}T}{N-i}\left[1-\frac{\Delta t}{T}-\frac{w}{T}\right]^{N-i}\Bigg|_0^{T-\Delta T}+\int_0^{T-\Delta t}\frac{(i-1)Tw^{i-2}}{N-i}\left[1-\frac{\Delta t}{T}-\frac{w}{T}\right]^{N-i}dw \tag{11.240}$$

$$=\frac{(i-1)}{\lambda\left(1-\frac{i}{N}\right)}\int_0^{T-\Delta t}w^{i-2}\left[1-\frac{\Delta t}{T}-\frac{w}{T}\right]^{N-i}dw$$

Hence, for $0<\Delta t<T, 2\leq i\leq N-1$,

$$\begin{aligned}f_{\Delta_{(i+1)}}(\Delta t)&=\frac{\lambda^{i+1}}{(i-1)!}\cdot\prod_{k=1}^{i}\left(1-\frac{k}{N}\right)\cdot\frac{(i-1)}{\lambda\left(1-\frac{i}{N}\right)}\int_0^{T-\Delta t}w^{i-2}\left[1-\frac{w+\Delta t}{T}\right]^{N-i}dw\\&=\frac{\lambda^i}{(i-2)!}\cdot\prod_{k=1}^{i-1}\left(1-\frac{k}{N}\right)\cdot\int_0^{T-\Delta t}w^{i-2}\left[1-\frac{w+\Delta t}{T}\right]^{N-i}dw\end{aligned} \tag{11.241}$$

and, thus $f_{\Delta_{(i+1)}}(\Delta t)=f_{\Delta_{(i)}}(\Delta t)$. It remains to determine $f_{\Delta_{(2)}}(\Delta t)$. For the case of $i=2$, it follows, from Equation 11.238, that

$$f_{\Delta_{(2)}}(\Delta t) = \lambda^2 \cdot \left(1-\frac{1}{N}\right) \cdot \int_0^{T-\Delta t} \left[1-\frac{u+\Delta t}{T}\right]^{N-2} du$$

$$= \lambda^2 \cdot \left(1-\frac{1}{N}\right) \left[\frac{-T}{N-1}\left(1-\frac{u+\Delta t}{T}\right)^{N-1}\Bigg|_0^{T-\Delta t}\right] \tag{11.242}$$

$$= \lambda^2 \cdot \left(1-\frac{1}{N}\right) \left[\frac{T}{N-1}\left(1-\frac{\Delta t}{T}\right)^{N-1}\right] = \lambda\left(1-\frac{\Delta t}{T}\right)^{N-1}$$

as required.

## APPENDIX 11.Q PROOF OF THEOREM 11.25

Consider the result from Theorem 11.15, which is valid for $0 < \Delta x < T, 1 \le i < j \le N$:

$$f_{\Delta_{(j)-(i)}}(\Delta x) = \int_0^{T-\Delta x} f_{(i)(j)}(u, u+\Delta x)du \tag{11.243}$$

Substitution of $f_{(i)(j)}(t_i, t_j)$ from Theorem 11.24 yields

$$f_{\Delta_{(j)-(i)}}(\Delta t) = \frac{\lambda^j}{(i-1)!(j-i-1)!} \cdot \prod_{k=1}^{j-1}\left(1-\frac{k}{N}\right) \cdot$$

$$\int_0^{T-\Delta t} t_i^{i-1}(t_i+\Delta t)^{j-i-1}\left[1-\frac{t_i}{t_i+\Delta t}\right]^{j-i-1}\left[1-\frac{t_i+\Delta t}{T}\right]^{N-j} dt_i \tag{11.244}$$

$$= \frac{\lambda^j \Delta t^{j-i-1}}{(i-1)!(j-i-1)!} \cdot \prod_{k=1}^{j-1}\left(1-\frac{k}{N}\right) \cdot \int_0^{T-\Delta t} t_j^{i-1}\left[1-\frac{t_i+\Delta t}{T}\right]^{N-j} dt_i$$

which is the first required result. To prove the insensitivity to a change in the absolute location of the ordered random variables, consider

$$f_{\Delta_{(j+1)-(i+1)}}(\Delta t) =$$

$$\frac{\lambda^{j+1}\Delta t^{j-i-1}}{i!(j-i-1)!} \cdot \prod_{k=1}^{j}\left(1-\frac{k}{N}\right) \cdot \int_0^{T-\Delta t} t_i^i\left[1-\frac{t_i+\Delta t}{T}\right]^{N-j-1} dt_i \tag{11.245}$$

Consider integration by parts with

$$\begin{aligned} u &= t_i^i & &\Rightarrow\; du = i t_i^{i-1} dt_i \\ dv &= \left[1-\frac{t_i+\Delta t}{T}\right]^{N-j-1} dt_i & &\Rightarrow\; v = \frac{-T}{(N-j)}\left[1-\frac{t_i+\Delta t}{T}\right]^{N-j} \end{aligned} \tag{11.246}$$

which leads to

$$\begin{aligned} f_{\Delta_{(j+1)-(i+1)}}(\Delta t) &= \frac{\lambda^{j+1}\Delta t^{j-i-1}}{i!(j-i-1)!}\cdot\prod_{k=1}^{j}\left(1-\frac{k}{N}\right)\cdot \\ &= \left[\frac{-T t_i^i}{(N-j)}\left[1-\frac{t_i+\Delta t}{T}\right]^{N-j}\Bigg|_0^{T-\Delta T} - \int_0^{T-\Delta t} i t_i^{i-1}\cdot\frac{-T}{(N-j)}\left[1-\frac{t_i+\Delta t}{T}\right]^{N-j} dt_i\right] \\ &= \frac{\lambda^{j+1}\Delta t^{j-i-1}}{i!(j-i-1)}\cdot\prod_{k=1}^{j}\left(1-\frac{k}{N}\right)\cdot\left[\frac{iT}{N-j}\int_0^{T-\Delta t} t_i^{i-1}\left[1-\frac{t_i+\Delta t}{T}\right]^{N-j} dt_i\right] \\ &= \frac{\lambda^{j}\Delta t^{j-i-1}}{(i-1)!(j-i-1)!}\cdot\prod_{k=1}^{j-1}\left(1-\frac{k}{N}\right)\cdot\left[\int_0^{T-\Delta t} t_j^{i-1}\left[1-\frac{t_i+\Delta t}{T}\right]^{N-j} dt_i\right] \\ &= f_{\Delta_{(j)-(i)}}(\Delta t) \end{aligned} \tag{11.247}$$

It then follows that $f_{\Delta_{(j+2)-(i+2)}}(\Delta t) = f_{\Delta_{(j+1)-(i+1)}}(\Delta t) = f_{\Delta_{(j)-(i)}}(\Delta t)$ and, by iteration, $f_{\Delta_{(j+k)-(i+k)}}(\Delta t) = f_{\Delta_{(j)-(i)}}(\Delta t)$ as required.

## APPENDIX 11.R PROOF OF THEOREM 11.28

From Theorem 11.27, it follows that

$$f_{\Delta_{(j)-(1)}}(\Delta t)\frac{\lambda^{j}\Delta t^{j-2}}{(j-2)!}\cdot\prod_{k=1}^{j-1}\left(1-\frac{k}{N}\right)\cdot\int_0^{T-\Delta t}\left[1-\frac{t_1+\Delta t}{T}\right]^{N-j} dt_1 \tag{11.248}$$

The evaluation yields

$$\begin{aligned} f_{\Delta_{(j)-(1)}}(\Delta t) &= \frac{\lambda^{j}\Delta t^{j-2}}{(j-2)!}\cdot\prod_{k=1}^{j-1}\left(1-\frac{k}{N}\right)\cdot\frac{-T}{(N-j+1)}\left[1-\frac{t_1+\Delta t}{T}\right]^{N-j+1}\Bigg|_0^{T-\Delta t} \\ &= \frac{\lambda^{j-1}\Delta t^{j-2}}{(j-2)!}\cdot\prod_{k=1}^{j-2}\left(1-\frac{k}{N}\right)\cdot\left[1-\frac{\Delta t}{T}\right]^{N-j+1} \end{aligned} \tag{11.249}$$

The comparison of this result with the result from Theorem 11.20 of

$$f_{(i)}(t)=\frac{\lambda(\lambda t)^{i-1}}{(j-1)!}\left[\prod_{k=1}^{i-1}\left(1-\frac{k}{N}\right)\right]\left(1-\frac{t}{T}\right)^{N-i} \quad 1\leq i\leq N, 0\leq t<T \tag{11.250}$$

yields

$$f_{\Delta_{(j)-(i)}}(\Delta t)=f_{(j-1)}(\Delta t) \tag{11.251}$$

The general result then follows from Equation 11.119 according to

$$\begin{aligned} f_{\Delta_{(j)-(i)}}(\Delta t) &= f_{\Delta_{(j-i+1)-(1)}}(\Delta t)=f_{(j-i)}(\Delta t) \\ &= \frac{\lambda^{j-i}\Delta t^{j-i-1}}{(j-i-1)!}\cdot\left[\prod_{k=1}^{j-i-1}\left(1-\frac{k}{N}\right)\right]\cdot\left[1-\frac{\Delta t}{t}\right]^{N-j+1} \end{aligned} \tag{11.252}$$

## APPENDIX 11.S PROOF OF THEOREM 11.29

First, from Theorem 11.16,

$$f_{\Delta_{(1)}\Delta_{(2)}}(\Delta t_1,\Delta t_2)=f_{(1)(2)}(\Delta t_1,\Delta t_1+\Delta t_2) \tag{11.253}$$

and it then follows from Theorem 11.24 that

$$f_{\Delta_{(1)}\Delta_{(2)}}(\Delta t_1,\Delta t_2)=\lambda^2\left(1-\frac{1}{N}\right)\left[1-\frac{\Delta t_1+\Delta t_2}{T}\right]^{N-2} \quad 0<\Delta t_1+\Delta t_2<T \tag{11.254}$$

Second, consider the general result for the joint probability density function between the $i$th and $(i+1)$th interarrival times as given in Theorem 11.16 for the case of $0<\Delta t_i<T-\Delta t_{i+1}, 0<\Delta t_{i+1}<T-\Delta t_i$, $\Delta t_i+\Delta t_{i+1}<T, 2\leq i\leq N-1$:

$$f_{\Delta_{(i)}\Delta_{(i+1)}}(\Delta t_i,\Delta t_{i+1})= \int_0^{T-\Delta t_i-\Delta t_{i+1}} f_{(i-1)(i)(i+1)}(u,u+\Delta t_i,u+\Delta t_i+\Delta t_{i+1})du \tag{11.255}$$

Substitution of $f_{(i-1)(i)(i+1)}$ from Theorem 11.25 yields

$$f_{\Delta_{(i)}\Delta_{(i+1)}}(\Delta t_i,\Delta t_{i+1})= \frac{\lambda^{i+1}}{(i-2)!}\cdot\prod_{k=1}^{i}\left(1-\frac{k}{N}\right)\cdot\int_0^{T-\Delta t_i-\Delta t_{i+1}} u^{i-2}\cdot\left[1-\frac{u+\Delta t_i+\Delta t_{i+1}}{T}\right]^{N-i-1} du \tag{11.256}$$

which is valid for $t_{i-1}<t_i<t_{i+1}, 2\leq i\leq N-1$. To solve the integral, note that the integrand is a polynomial of order $N-3$, and, hence, an analytical solution can always

be obtained. A better approach is to write $f_{\Delta_{(i)}\Delta_{(i+1)}}(\Delta t_i, \Delta t_{i+1})$ in terms of $f_{\Delta_{(i-1)}\Delta_{(i)}}(\Delta t_i, \Delta t_{i+1})$, and to this end, consider

$$\begin{aligned} f_{\Delta_{(2)}\Delta_{(3)}}(\Delta t_2, \Delta t_3) &= \lambda^3 \cdot \prod_{k=1}^{2}\left(1-\frac{k}{N}\right) \cdot \int_0^{T-\Delta t_2-\Delta t_3} \left[1-\frac{u+\Delta t_2+\Delta t_3}{T}\right]^{N-3} du \\ &= \lambda^3 \cdot \prod_{k=1}^{2}\left(1-\frac{k}{N}\right) \cdot \frac{-T}{(N-2)}\left[1-\frac{u+\Delta t_2+\Delta t_3}{T}\right]^{N-2}\Bigg|_0^{T-\Delta t_2-\Delta t_3} \\ &= \lambda^2 \cdot \left(1-\frac{1}{N}\right) \cdot \left[1-\frac{\Delta t_2+\Delta t_3}{T}\right]^{N-2} = f_{\Delta_{(1)}\Delta_{(2)}}(\Delta t_2, \Delta t_3) \end{aligned} \tag{11.257}$$

For the general case, where $i \geq 3$, consider integration by parts on Equation 11.256 with

$$\begin{aligned} &w = u^{i-2} \;\Rightarrow\; dw = (i-2)u^{i-3}du \\ &dv = \left[1-\frac{u+\Delta t_i+\Delta t_{i+1}}{T}\right]^{N-i-1} du \Rightarrow v = \frac{-T}{N-i}\left[1-\frac{u+\Delta t_i+\Delta t_{i+1}}{T}\right]^{N-i} \end{aligned} \tag{11.258}$$

Hence,

$$\begin{aligned} &f_{\Delta_{(i)}\Delta_{(i+1)}}(\Delta t_i, \Delta t_{i+1}) = \\ &\frac{\lambda^{i+1}}{(i-2)!} \cdot \prod_{k=1}^{i}\left(1-\frac{k}{N}\right) \cdot \left[u^{i-2} \cdot \frac{-T}{N-i}\left[1-\frac{u+\Delta t_i+\Delta t_{i+1}}{T}\right]^{N-i}\Bigg|_0^{T-\Delta t_i-\Delta t_{i+1}} - \right. \\ &\left. \int_0^{T-\Delta t_i-\Delta t_{i+1}} (i-2)u^{i-3} \cdot \frac{-T}{N-i}\left[1-\frac{u+\Delta t_i+\Delta t_{i+1}}{T}\right]^{N-i} du \right] \end{aligned} \tag{11.259}$$

Simplification yields

$$\begin{aligned} &f_{\Delta_{(i)}\Delta_{(i+1)}}(\Delta t_i, \Delta t_{i+1}) \\ &= \frac{\lambda^i}{(i-3)!} \cdot \prod_{k=1}^{i-1}\left(1-\frac{k}{N}\right) \cdot \int_0^{T-\Delta t_i-\Delta t_{i+1}} u^{i-3} \cdot \left[1-\frac{u+\Delta t_i+\Delta t_{i+1}}{T}\right]^{N-i} du \\ &= f_{\Delta_{(i-1)}\Delta_{(i)}}(\Delta t_i, \Delta t_{i+1}) \end{aligned} \tag{11.260}$$

Hence, in general and for $1 \leq i \leq N-1$,

$$\begin{aligned} f_{\Delta_{(i)}\Delta_{(i+1)}}(\Delta t_i, \Delta t_{i+1}) &= f_{\Delta_{(1)}\Delta_{(2)}}(\Delta t_i, \Delta t_{i+1}) \\ &= \lambda^2 \cdot \left(1-\frac{1}{N}\right) \cdot \left[1-\frac{\Delta t_i+\Delta t_{i+1}}{T}\right]^{N-2} \end{aligned} \tag{11.261}$$

Third, consider the general result for the case of $j \geq i+2$, which, from Theorem 11.16, is

$$f_{\Delta_{(i)}\Delta_{(j)}}(\Delta t_i, \Delta t_j) = \int_{\Delta t_i}^{T-\Delta t_j} \left( \int_{u+\Delta t_j}^{T} f_{(i-1)(i)(j-1)(j)}(u-\Delta t_i, u, v-\Delta t_j, v)\, dv \right) du \tag{11.262}$$

$$0 < \Delta t_i < T, 0 < \Delta t_j < T, \Delta t_i + \Delta t_j < T, 2 \leq i < N-2, i+2 \leq j \leq N$$

Substituting $f_{(i-1)(i)(j-1)(j)}$ from Theorem 11.25 yields the required result:

$$f_{\Delta_{(i)}\Delta_{(j)}}(\Delta t_i, \Delta t_j) = \frac{\lambda^j}{(i-2)!(j-i-2)!} \cdot \prod_{k=1}^{j-1}\left(1-\frac{1}{N}\right) \cdot$$
$$\int_{\Delta t_i}^{T-\Delta t_j} \left( \int_{u+\Delta t_j}^{T} (u-\Delta t_i)^{i-2} \cdot (v-\Delta t_j)^{j-i-2} \left[1-\frac{u}{v-\Delta t_j}\right]^{j-i-2} \cdot \left[1-\frac{v}{T}\right]^{N-j} dv \right) du \tag{11.263}$$

## APPENDIX 11.T PROOF OF THEOREM 11.31

The first result follows from $f_{(i),(j)}$, as specified in Theorem 11.24, using Theorem 11.4 and the result $\lim_{N\to\infty} \prod_{k=1}^{j-1}\left(1-\frac{k}{N}\right) = 1$.

Second, the result $f_{(i)(j)}(t_i, t_j) = f_{(i)}(t_i) f_{(j-i)}(t_j - t_i)$ follows from the expression for $f_{(i)}$ given by Equation 11.127.

Third, as $X_{(j)} = X_{(i)} + \Delta X_{(j)-(i)}$, the result from Theorem 11.5:

$$\begin{aligned} Z_1 &= g_1(X_1, X_2) = X_1 \\ Z_2 &= g_2(X_1, X_2) = X_1 + X_2 \end{aligned} \Rightarrow f_{Z_1 Z_2}(z_1, z_2) = f_{X_1 X_2}(z_1, z_2 - z_1) \tag{11.264}$$

implies that

$$f_{(i)(j)}(t_i, t_j) = f_{(i)\Delta_{(j)-(i)}}(t_i, t_j - t_i) \tag{11.265}$$

With the definition $\Delta t = t_j - t_i$, it follows that

$$f_{(i)(j)}(t_i, t_i + \Delta t) = f_{(i)\Delta_{(j)-(i)}}(t_i, \Delta t) \tag{11.266}$$

Alternatively, this result can be shown by considering the joint probability density function between $\Delta X_{(j)-(i)}$ and $X_{(i)}$: Consider the probability

$$f_{(i)\Delta_{(j)-(i)}}(t,\Delta t)dt(ddt) \approx P\left[X_{(i)} \in [t,t+dt], \Delta X_{(j)-(i)} \in [\Delta t, \Delta t+ddt]\right]$$
$$= P\left[X_{(i)} \in [t,t+dt], X_{(j)} - X_{(i)} \in [\Delta t, \Delta t+ddt]\right]$$
$$= P\left[X_{(i)} \in [t,t+dt], X_{(j)} \in \left[X_{(i)}+\Delta t, X_{(i)}+\Delta t+ddt\right]\right] \quad (11.267)$$

and the area illustrated in Figure 11.30. It then follows that

$$f_{(i)\Delta_{(j)-(i)}}(t,\Delta t) = f_{(i)(j)}(t,t+\Delta t) \quad (11.268)$$

which is consistent with Equation 11.266.

The last result follows from Theorem 11.28, which states the result $f_{\Delta_{(j)-(i)}}(\Delta t) = f_{(j-i)}(\Delta t)$.

## APPENDIX 11.U PROOF OF THEOREM 11.36

Consider the result from Theorem 11.17, valid for $0<\Delta x_i, 0<\Delta x_j$, $2 \le i < N-2, i+2 \le j \le N$:

$$f_{\Delta_{(i)}\Delta_{(j)}}(\Delta x_i, \Delta x_j) = \int_{\Delta x_i}^{\infty}\left(\int_{u+\Delta x_j}^{\infty} f_{(i-1)(i)(j-1)(j)}\left(u-\Delta x_i, u, v-\Delta x_j, v\right)dv\right)du \quad (11.269)$$

Substitution of $f_{(i-1)(i)(j-1)(j)}$ from Theorem 11.32 yields

$$f_{\Delta_{(i)}\Delta_{(j)}}(\Delta t_i, \Delta t_j)$$
$$= \frac{\lambda^j}{(i-2)!(j-i-2)!}\int_{\Delta t_i}^{\infty}(u-\Delta t_i)^{i-2}\left(\int_{u+\Delta t_j}^{\infty} e^{-\lambda v}\cdot\left[v-\Delta t_j-u\right]^{j-i-2}dv\right)du \quad (11.270)$$

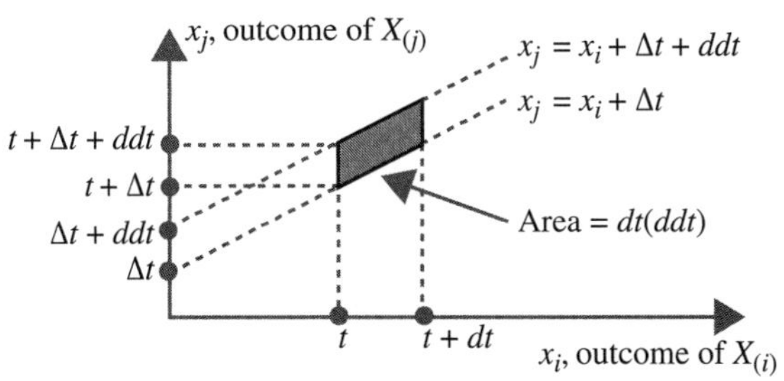


**FIGURE 11.30** The region defined by the constraints $x_i \in (t,t+dt)$ and $x_j \in (x_i+\Delta t, x_i+\Delta t+ddt)$.

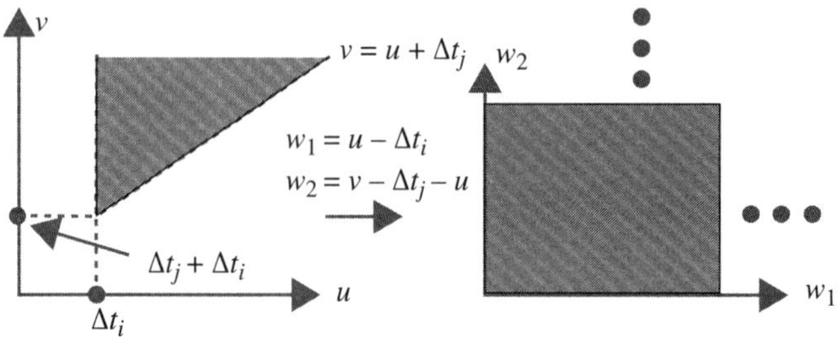


**FIGURE 11.31** The region of integration. Left: before transformation. Right: after transformation. Note: the line $v=u+\Delta t_j$ maps into the line $w_2=0$. The line $u=\Delta t_i$ maps into the line $w_1=0$.

The area of integration is over the region illustrated in Figure 11.31. Consider the integral

$$\int_{R_{uv}} f(u,v)dudv \qquad f(u,v)=(u-\Delta t_i)^{i-2}e^{-\lambda v}\left[v-\Delta t_j-u\right]^{j-i-2} \tag{11.271}$$

and the change of variable:

$$\begin{matrix} w_1=u-\Delta t_i \\ w_2=v-\Delta t_j-u \end{matrix} \Rightarrow \begin{matrix} u=w_1+\Delta t_i \\ v=w_2+\Delta t_j+u=w_2+\Delta t_j+w_1+\Delta t_i \end{matrix} \tag{11.272}$$

The region of integration changes as illustrated in Figure 11.31, and the joint probability density function becomes

$$\begin{aligned} f_{\Delta_{(i)}\Delta_{(j)}}\left(\Delta t_i,\Delta t_j\right) &= \frac{\lambda^j}{(i-2)!(j-i-2)!}\int_0^\infty\int_0^\infty w_1^{i-2}e^{-\lambda\left[w_1+\Delta t_i+w_2+\Delta t_j\right]}\cdot w_2^{j-i-2}dw_1dw_2 \\ &= \frac{\lambda^j e^{-\lambda\Delta t_i}e^{-\lambda\Delta t_j}}{(i-2)!(j-i-2)!}\cdot\left(\int_0^\infty w_1^{i-2}e^{-\lambda w_1}dw_1\right)\cdot\left(\int_0^\infty w_2^{j-i-2}e^{-\lambda w_2}dw_2\right) \end{aligned} \tag{11.273}$$

As $\int_0^\infty w^i e^{-\lambda w}dw=\frac{i!}{\lambda^{i+1}}$ it follows that

$$f_{\Delta_{(i)}\Delta_{(j)}}\left(\Delta t_i,\Delta t_j\right)=\frac{\lambda^j e^{-\lambda\Delta t_i}e^{-\lambda\Delta t_j}}{(i-2)!(j-i-2)!}\cdot\frac{(i-2)!}{\lambda^{i-1}}\cdot\frac{(j-i-2)!}{\lambda^{j-i-1}}=\lambda^2 e^{-\lambda\Delta t_i}e^{-\lambda\Delta t_j} \tag{11.274}$$

as required.

# 12

# POISSON POINT RANDOM PROCESSES

## 12.1 INTRODUCTION

Many phenomena are based, in part, on the random arrival of events, and when such events are independent of one another, and occur at a uniform rate, a Poisson point model is appropriate. One important example is that of shot noise. An important property of a Poisson point random process with a constant rate is that it is memoryless. Effectively, it is possible to start, stop, and then restart such a process without altering its statistics after allowing for the stopped time. This property is implicit in the modelling of birth–death random processes. One generalization of a Poisson point process is the Lévy random process. The theory of point random processes is vast and Dayley and Vere-Jones (2003) is a useful reference.

Using, in part, the results from order statistics established in Chapter 11, important results for Poisson point random processes are established for both the finite and infinite intervals. These results include the probability mass function for the number of points in a set interval, the joint probability mass function for the number of points in two disjoint intervals, the probability density function for the *i*th arrival time and *i*th interarrival time, and the joint probability density function for the *i*th and *j*th arrival times and the *i*th and *j*th interarrival times.

*A Signal Theoretic Introduction to Random Processes*, First Edition. Roy M. Howard.

### 12.1.1 Background Results

The following random variables and associated results are fundamental to the discussion of the Poisson point random process.

#### *12.1.1.1 Poisson Random Variable*

**Definition: Poisson Random Variable** A random variable $X$ has a Poisson distribution with a parameter $\lambda$, written as $X \sim \text{Poi}(\lambda)$, if

$$p_X[k] = P[X=k] = \frac{\lambda^k e^{-\lambda}}{k!} \qquad k \in \{0,1,2,\ldots\} \tag{12.1}$$

**Theorem 12.1 Mean, Variance, and Moments for a Poisson Random Variable** For a random variable $X$ with a Poisson distribution, it follows that

$$\begin{aligned} &\boldsymbol{E}[X] = \lambda && \text{var}(X) = \lambda \\ &\boldsymbol{E}[X^2] = \lambda^2 + \lambda && \boldsymbol{E}[X^i] = \lambda\left[\boldsymbol{E}[X^{i-1}] + \frac{d}{d\lambda}\boldsymbol{E}\left[X^{i-1}\right]\right] \quad i \geq 2 \end{aligned} \tag{12.2}$$

**Proof**

These are standard results readily proved by application of the expectation operator and use of moment generating functions. See, for example, Papoulis and Pillai (2002).

#### *12.1.1.2 Exponential Random Variable*

**Definition: Exponential Variable** A random variable $X$ is an exponential random variable with a parameter $\lambda$, if

$$f_X(x) = \begin{cases} \lambda e^{-\lambda x} & x \geq 0 \\ 0 & x < 0 \end{cases} \tag{12.3}$$

**Theorem 12.2 Mean, Variance, and Moments for an Exponential RV** For a exponential random variable $X$, it is the case that

$$\begin{aligned} &\boldsymbol{E}(X) = \frac{1}{\lambda} \quad \boldsymbol{E}\left[X^2\right] = \frac{2}{\lambda^2} \quad \text{var}(X) = \frac{1}{\lambda^2} \\ &\boldsymbol{E}[X^i] = \frac{i!}{\lambda^i} \quad i \geq 1 \end{aligned} \tag{12.4}$$

**Proof**

The proof for the $i$th moment arises from direct integration and is detailed in part of the analysis in Appendix 12.E.

#### *12.1.1.3 Gamma Random Variable*

**Definition: Gamma Variable** A random variable $X$ is said to have a gamma distribution, with parameters $\alpha$, $\beta > 0$, if its probability density function is of the form

$$f_{\Gamma(\alpha,\beta)}(x) = \begin{cases} 0 & x<0 \\ \dfrac{\alpha(\alpha x)^{\beta-1}e^{-\alpha x}}{\Gamma(\beta)} & x>0 \end{cases} \tag{12.5}$$

where $\Gamma$ is the gamma function. $\Gamma(\beta) = (\beta-1)!$ for the case of $\beta \in \boldsymbol{Z}^+$.

**Theorem 12.3 Mean, Variance, and Moments for a Gamma Random Variable** The first three moments and the variance of a random variable with a Gamma distribution, for the case where $\beta$ is a positive integer, are

$$\begin{gathered} \boldsymbol{E}[X] = \frac{\beta}{\alpha} \quad \boldsymbol{E}\left[X^2\right] = \frac{\beta(\beta+1)}{\alpha^2} \quad \boldsymbol{E}\left[X^3\right] = \frac{\beta(\beta+1)(\beta+2)}{\alpha^3} \\ \operatorname{var}(X) = \frac{\beta}{\alpha^2} \end{gathered} \tag{12.6}$$

**Proof**
These results arise from integration.

## 12.2 CHARACTERIZING POISSON RANDOM PROCESSES

Consider a Poisson experiment that defines a sample space

$$S = \{(\gamma_1,\ldots,\gamma_k,\ldots) : \gamma_k \in \boldsymbol{R}^+, \gamma_1 \le \ldots \le \gamma_k \le \ldots\} \tag{12.7}$$

and the case where this experiment is based on selecting $N$ points, at random, from the interval $[0,T]$, $N,T \to \infty$ consistent with a point rate, or rate for short, of

$$\lambda = \frac{N}{T} \tag{12.8}$$

A Poisson point process has been defined in Chapter 6, on such a sample space, according to

$$X(\Omega,t) = \sum_{k=1}^{\infty} \delta_K(t-\Gamma_k) \quad \text{or} \quad X(\Omega,t) = \sum_{k=1}^{\infty} \delta(t-\Gamma_k) \tag{12.9}$$

where $\Omega = (\Gamma_1, \Gamma_2, \ldots)$ is a vector of random variables defined by the outcomes of a Poisson experiment. One signal from a Poisson point process is illustrated in Figure 12.1 for the Kronecker delta case.

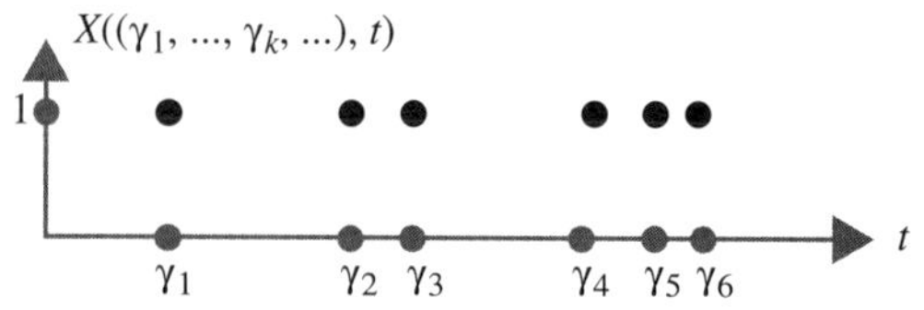


**FIGURE 12.1** One signal defined by a Poisson point process.

A Poisson random process defines a vector of random variables consistent with the arrival times and interarrival times associated with the points.

**Definition: Random Variable Defined by *k*th Arrival Time** The *k*th arrival time is defined by the random variable $X_k = \Gamma_k$.

**Definition: Interarrival Times** The *k*th interarrival time is defined by the random variable $\Delta X_k = X_k - X_{k-1}$ for $k \geq 2$.

### 12.2.1 Points in a Subset of $[0,T]$

Consider a pseudo-Poisson experiment that leads to $N$ points on the interval $[0,T]$.

**Definition: Random Variable: Number of Points in Subset of $[0,T]$** The random variable $N_A : S \to S_{N_A}$ defined by the number of points in the set $A \subseteq [0,T]$, and arising from a pseudo-Poisson experiment with $N$ points placed on $[0,T]$, is defined according to

$$N_A(\gamma_1,\ldots,\gamma_N) = \sum_{k=1}^{N} I_A(\gamma_k) \qquad I_A(\gamma_k) = \begin{cases} 1 & \gamma_k \in A \\ 0 & \gamma_k \notin A \end{cases}$$
$$S_{N_A} = \left\{ \sum_{k=1}^{N} I_A(\gamma_k) : \ (\gamma_1,\ldots,\gamma_N) \in S \right\} \tag{12.10}$$

For the case where the subset $A$ of $[0,T]$ is an interval $A = [\alpha, \beta)$, the random variable $N_A : S \to S_{N_A}$ can be defined according to

$$N_A(\gamma_1,\ldots,\gamma_N) = \sum_{k=1}^{N} [u(\gamma_k - \alpha) - u(\gamma_k - \beta)]$$
$$S_{N_A} = \left\{ \sum_{k=1}^{N} [u(\gamma_k - \alpha) - u(\gamma_k - \beta)] : \ (\gamma_1,\ldots,\gamma_N) \in S \right\} \tag{12.11}$$

### 12.2.2 Summation of Poisson Random Processes

The summation of a sum of Poisson point processes is also a Poisson point process.

**Theorem 12.4 Summation of Poisson Point Processes**
Consider $M$ independent pseudo-Poisson point processes $X_1(\Omega_1, t), \ldots, X_M(\Omega_M, t)$ with respective rates $\lambda_1, \ldots, \lambda_M$ and where

$$X_i(\Omega_i,t)=\sum_{k=1}^{N_i}\delta_K(t-\Gamma_{ik}) \qquad t\in[0,T],\lambda_i=\frac{N_i}{T} \tag{12.12}$$

Here, $\Omega_i=(\Gamma_{i1},\ldots,\Gamma_{iN_i})$ and $\Gamma_{i1},\ldots,\Gamma_{iN_i}$ are independent random variables with a uniform distribution on [0, $T$]. The summation of these random processes is a Poisson point process $X(\Omega, t)$ with a rate $\lambda=\lambda_1+\cdots+\lambda_M$, that is,

$$X(\Omega,t)=\sum_{i=1}^{M}X_i(\Omega_i,t)=\sum_{i=1}^{M}\sum_{k=1}^{N_i}\delta_K(t-\Gamma_{ik})=\sum_{k=1}^{N}\delta_K(t-\Gamma_k) \tag{12.13}$$

where $N=N_1+\cdots+N_M$, $\Omega=(\Gamma_{11},\ldots,\Gamma_{MN_M})=(\Gamma_1,\ldots,\Gamma_N)$, and $\Gamma_1, \ldots, \Gamma_N$ are independent random variables with a uniform distribution on [0,$T$].

**Proof**
One outcome of the random variable $X$ is based on $N_1$ points chosen, at random, from [0,$T$], …, $N_M$ points chosen, at random, from [0,$T$]. In total, one outcome is based on $N=N_1+\cdots+N_M$ points chosen at random from [0,$T$] at a rate $\lambda=N/T$.

## 12.3 PMF: NUMBER OF POINTS IN A SUBSET OF AN INTERVAL

The following results hold for the case of a pseudo-Poisson experiment with $N$ points independently placed on [0,$T$] consistent with a rate of $\lambda=N/T$.

### 12.3.1 Number of Points in a Subset of an Interval

#### *12.3.1.1 Number of Points in a Subset of an Interval*

**Theorem 12.5 PMF for Number of Points in a Set**
The probability mass function for the random variable, $N_A$ defined as the number of points in the set $A \subseteq [0,T]$, is

$$p_{N_A}(k)=b\left(k,N,\frac{m[A]}{T}\right)=\frac{N!}{(N-k)!k!}\cdot\left(\frac{m[A]}{T}\right)^k\cdot\left(1-\frac{m[A]}{T}\right)^{N-k} \qquad k\in\{0,1,\ldots,N\} \tag{12.14}$$

or, alternatively,

$$p_{N_A}(k)=\frac{\lambda^k m[A]^k}{k!}\cdot\prod_{j=0}^{\max\{k-1,0\}}\left(1-\frac{j}{N}\right)\cdot\left(1-\frac{m[A]}{T}\right)^{N-k} \qquad k\in\{0,1,\ldots,N\} \tag{12.15}$$

Here, $b$ is the binomial probability mass function, $m$ is the Lebesgue measure operator and, for the case of $A = [\alpha, \beta]$, it follows that $m[A] = \beta - \alpha$.

**Proof**

The proof is detailed in Appendix 12.A.

**Theorem 12.6 Probability of Zero or One Point in a Set**

For the case where $m[A] << 1/\lambda, A \subseteq [0,T]$ and $N \geq 1$, it is the case that

$$\begin{aligned} P[\text{one point in } A] &= \lambda m[A](1-\varepsilon_1) \\ P[\text{zero points in } A] &= 1-\lambda m[A]+\varepsilon_2 \end{aligned} \tag{12.16}$$

where $0 < \varepsilon_1 << 1, |\varepsilon_2| << \lambda m[A]$.

**Proof**

The proof is detailed in Appendix 12.B.

**Theorem 12.7 Poisson Approximation to PMF of Number of Points in a Set**

With the assumptions $N >> 1$, $m[A]/T << 1$, $\lambda m[A]^2/2T << 1$, the probability mass function for the random variable defined by the number of points in the set $A \subseteq [0, T]$ can be approximated according to

$$p_{N_A}(k) \approx \frac{\lambda^k m[A]^k e^{-\lambda m[A]}}{k!} \qquad k \in \{0,1,\ldots\}, k << \min\left\{\sqrt{2N}, T/m[A]\right\} \tag{12.17}$$

**Proof**

Poisson's theorem (Theorem 4.29) states: If $N >> 1$, $p << 1$ and $Np^2/2 << 1$, then

$$b(k,N,p) \approx \frac{(Np)^k e^{-Np}}{k!} \tag{12.18}$$

is valid for $0 \leq k << \min\left\{\sqrt{2N}, 1/p\right\}$. The required result follows with $p = m[A]/T$ and $\lambda = N/T$ whereupon $Np = \lambda m[A]$. The constraint of $Np^2/2 << 1$ is $Nm[A]^2/2T^2 << 1$ or $\lambda m[A]^2/2T << 1$.

### 12.3.2 Example

An example of the probability mass function of the number of points in an interval, along with the Poisson approximation, is detailed in Figure 12.2.

### 12.3.3 Mean and Variance

**Theorem 12.8 Mean, Variance, and Moments**

The mean and variance of the random variable $N_A$, defined as the number of points in the set $A \subseteq [0, T]$, is

$$\mu_{N_A} = \lambda m[A] \qquad \text{var}(N_A) = \sigma_{N_A}^2 = \lambda m[A] \cdot \left(1 - \frac{m[A]}{T}\right) \tag{12.19}$$

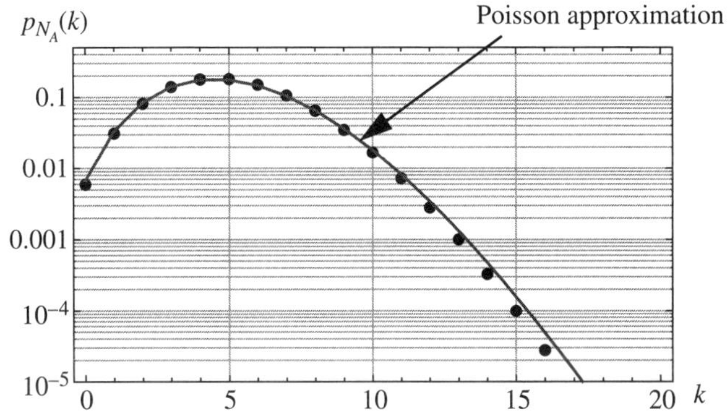


**FIGURE 12.2** Probability mass function for the number of points in an interval: $T=1$, $N=100$, $\lambda=100$, $m[A]=0.05$, $p=0.05$ and $\lambda m[A]^2/2T=0.125$. The upper limit of validity for the Poisson approximation, as given by $\min\{\sqrt{2N},\ T/m[A]\}$, is $\min\{14.14,\ 20\}=14$.

**Proof**

The follows from the mean and variance of a binomial random variable as given by Theorem 4.25 and by using the definitions of $p=m[A]/T$ and $\lambda=N/T$.

### 12.3.4 Number of Points in Two Disjoint Subsets of Interval

Consider the pair of random variables $N_A: S \to S_{N_A}$, $N_B: S \to S_{N_B}$ defined by the number of points in two disjoint subsets $A$ and $B$ of $[0,T]$ and based on the experimental outcomes $S$ of a pseudo-Poisson experiment:

$$
\begin{aligned}
&N_A(\gamma_1,\ldots,\gamma_N)=\sum_{k=1}^{N} I_A(\gamma_k) \qquad N_B(\gamma_1,\ldots,\gamma_N)=\sum_{k=1}^{N} I_B(\gamma_k) \\
&S_{N_A}=\left\{\sum_{k=1}^{N} I_A(\gamma_k):\quad (\gamma_1,\ldots,\gamma_N)\in S\right\} \\
&S_{N_B}=\left\{\sum_{k=1}^{N} I_B(\gamma_k):\quad (\gamma_1,\ldots,\gamma_N)\in S\right\}
\end{aligned}
\tag{12.20}
$$

Here, $I_A$ and $I_B$ are indicator functions for the sets $A$ and $B$. A single outcome of $N_A$ is consistent with the number of points in the set $A \subseteq [0,\ T]$ that arises from the corresponding experimental outcome of a pseudo-Poisson experiment. Similarly for a single outcome of $N_B$.

**Theorem 12.9 Probability of Points in Nonoverlapping Intervals**
The joint probability mass function for the random variables $N_A$ and $N_B$ is

$$
\begin{aligned}
p_{N_A N_B}(k_A,k_B)=&\frac{N!}{k_A!k_B!(N-k_A-k_B)!}\cdot\left(\frac{m[A]}{T}\right)^{k_A}\cdot\left(\frac{m[B]}{T}\right)^{k_B}\cdot \\
&\left(1-\frac{m[A]}{T}-\frac{m[B]}{T}\right)^{N-k_A-k_B} \qquad k_A,k_B\in\{0,1,\ \ldots,N\}, k_A+k_B\le N
\end{aligned}
\tag{12.21}
$$

or, alternatively,

$$p_{N_A N_B}(k_A, k_B) = \frac{\lambda^{k_A+k_B} \cdot m[A]^{k_A} \cdot m[B]^{k_B}}{k_A! k_B!} \cdot \left[ \prod_{j=0}^{\max\{0, k_A+k_B-1\}} \left(1 - \frac{j}{N}\right) \right] \cdot \left(1 - \frac{m[A]}{T} - \frac{m[B]}{T}\right)^{N-k_A-k_B} \quad k_A, k_B \in \{0, 1, \ldots, N\}, k_A + k_B \le N \tag{12.22}$$

**Proof**

The proof is detailed in Appendix 12.C.

### *12.3.4.1 Correlation between Number of Points in Disjoint Sets*

**Theorem 12.10 Correlation between Number of Points in Disjoint Sets**

The correlation between the random variables $N_A$ and $N_B$ is

$$\boldsymbol{E}[N_A N_B] = \sum_{k_A=1}^{N} \sum_{k_B=1}^{N} \frac{N!}{(k_A-1)!(k_B-1)!(N-k_A-k_B)!} \cdot \left(\frac{m[A]}{T}\right)^{k_A} \cdot \left(\frac{m[B]}{T}\right)^{k_B} \cdot \left(1 - \frac{m[A]}{T} - \frac{m[B]}{T}\right)^{N-k_A-k_B} \cdot u(N-k_A-k_B) \tag{12.23}$$

or, alternatively,

$$\boldsymbol{E}[N_A N_B] = \sum_{k_A=1}^{N} \sum_{k_B=1}^{N} \frac{\lambda^{k_A+k_B} \cdot m[A]^{k_A} \cdot m[B]^{k_B}}{(k_A-1)!(k_B-1)!} \cdot \left[ \prod_{j=0}^{k_A+k_B-1} \left(1 - \frac{j}{N}\right) \right] \cdot \left(1 - \frac{m[A]}{T} - \frac{m[B]}{T}\right)^{N-k_A-k_B} \cdot u(N-k_A-k_B) \tag{12.24}$$

**Proof**

This result follows by application of the expectation operator. The lower limit of one, rather than zero, on both the summations arises as one, or both, of $k_A$ and $k_B$ being zero results in zero being added to the double summation.

### *12.3.4.2 Correlation Coefficient between Number of Points in Disjoint Sets*

**Theorem 12.11 Correlation Coefficient: Number of Points in Disjoint Sets**

The correlation coefficient between the random variables $N_A$ and $N_B$ is

$$\rho_{N_A N_B} = \frac{\boldsymbol{E}[N_A N_B] - \mu_{N_A} \mu_{N_B}}{\sqrt{\operatorname{var}(N_A) \operatorname{var}(N_B)}} = \frac{\boldsymbol{E}[N_A N_B] - \lambda^2 m[A] m[B]}{\lambda \sqrt{m[A] m[B] \left(1 - \frac{m[A]}{T}\right)\left(1 - \frac{m[B]}{T}\right)}} \tag{12.25}$$

where $\boldsymbol{E}[N_A N_B]$ is specified by Theorem 12.10.

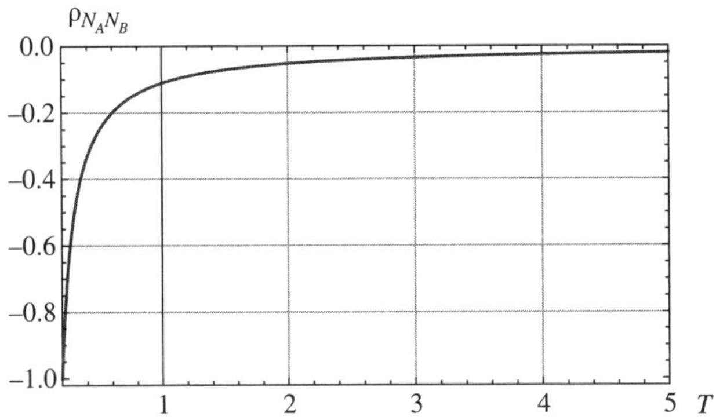


**FIGURE 12.3** Graph of the correlation coefficient of $N_A$ and $N_B$ for the case of $\lambda=10$, $m[A]=m[B]=0.1$ and for $T$ increasing between 0.2 and 5.

**Proof**
This result follows from the definition of the correlation coefficient and substitution of results from Theorem 12.8.

The correlation coefficient is plotted in Figure 12.3, for the case of $\lambda=10$, $m[A]=m[B]=0.1$ and for $T$ increasing between 0.2 and 5. The decreasing magnitude of the correlation coefficient, with increasing $T$, is consistent with the trend toward independence as the interval becomes increasingly large.

***12.3.4.3 Independence on the Number of Points in Disjoint Intervals*** As is evident from Figure 12.3, the random variables $N_A$ and $N_B$, in general, are dependent but become increasingly uncorrelated as $T$ increases and becomes much larger than $m[A]$, $m[B]$.

**Theorem 12.12 Independence of the Number of Points in Disjoint Intervals**
The random variables $N_A$ and $N_B$ become independent as $N\to\infty$, $T\to\infty$ and with $\lambda=N/T$ being fixed, that is,

$$p_{N_A N_B}(k_A,k_B)=p_{N_A}(k_A)p_{N_B}(k_B) \qquad N,T\to\infty,\ \lambda=\frac{N}{T}\text{ fixed} \tag{12.26}$$

**Proof**
The proof is detailed in Appendix 12.D.

### 12.3.5 Results for Infinite Interval

For the infinite interval, the following theorem states important results for the number of points in a set.

**Theorem 12.13 PMF for Number of Points in a Set: Infinite Interval**
For the case of $N\to\infty$, $T\to\infty$, and with $\lambda=N/T$ being fixed, the random variable defined by the number of points in a set $A$ is Poisson with parameter $\lambda m[A]$, that is,

$$p_{N_A}(k) = \frac{\lambda^k m[A]^k e^{-\lambda m[A]}}{k!} \quad k \in \{0, 1, \ldots\} \tag{12.27}$$

For two disjoint sets,

$$p_{N_A N_B}(k_A, k_B) = p_{N_A}(k_A) p_{N_B}(k_B) = \frac{\lambda^{k_A} m[A]^{k_A} e^{-\lambda m[A]}}{k_A!} \cdot \frac{\lambda^{k_B} m[B]^{k_B} e^{-\lambda m[B]}}{k_B!} \tag{12.28}$$

$$\rho_{N_A N_B} = 0 \tag{12.29}$$

The mean, variance, and moments are given by

$$\begin{aligned} &\mu_{N_A} = \mathrm{var}[N_A] = \lambda m[A] \quad \boldsymbol{E}\left[N_A^2\right] = \lambda^2 m[A]^2 + \lambda m[A] \\ &\boldsymbol{E}\left[N_A^i\right] = \lambda m[A]\left[\boldsymbol{E}\left[N_A^{i-1}\right] + \frac{\mathrm{d}}{\mathrm{d}\lambda m[A]}\boldsymbol{E}\left[N_A^{i-1}\right]\right] \quad i \geq 2 \end{aligned} \tag{12.30}$$

**Proof**
The results for the probability mass functions follow from Theorems 12.7 and 12.12. The results for the mean, variance, and moments follow from Theorem 12.1.

## 12.4 RESULTS FROM ORDER STATISTICS

Characterization of the $i$th arrival time and $i$th interarrival time of a Poisson point random process is facilitated by using results from order statistics theory, and the following subsections provide a summary of results from Chapter 11. The following notation is used:

$$\begin{aligned} &F_{(i)}(t) \ \text{ cumulative distribution function for } i\text{th arrival time} \\ &f_{(i)}(t) \ \text{ probability density function for } i\text{th arrival time} \\ &f_{(i)(j)}(t_i, t_j) \ \text{ joint PDF for } i\text{th and } j\text{th arrival time} \\ &f_{\Delta_{(i)}}(\Delta t) \ \text{ probability density function of } i\text{th inter-arrival time} \\ &f_{\Delta_{(i)}\Delta_{(j)}}(\Delta t_i, \Delta t_j) \ \text{ joint PDF for } i\text{th and } j\text{th inter-arrival time} \end{aligned} \tag{12.31}$$

### 12.4.1 Finite Interval

For a Poisson point process defined on the finite interval ($N$ points in $[0,T]$), the following theorem summarizes important statistical results.

**Theorem 12.14 Results for Finite Interval**
For a Poisson point process defined on $[0,T]$ with $\lambda = N/T$, the cumulative distribution function and the probability density function for the $i$th arrival time are given by

$$F_{(i)}(t) = 1 - \sum_{k=0}^{i-1} \frac{\lambda^k t^k}{k!} \left[1 - \frac{t}{T}\right]^{N-k} \left[\prod_{j=0}^{\max\{0,k-1\}} \left(1 - \frac{j}{N}\right)\right] \qquad \begin{matrix} 1 \le i \le N \\ 0 \le t < T \end{matrix} \tag{12.32}$$

$$f_{(i)}(t) = \frac{\lambda(\lambda t)^{i-1}}{(i-1)!} \cdot \left(1 - \frac{t}{T}\right)^{N-i} \cdot \prod_{j=0}^{i-1} \left(1 - \frac{j}{N}\right) \qquad 1 \le i \le N, 0 \le t < T \tag{12.33}$$

The $k$th moment of the $i$th arrival time is

$$m_{(i)}(k) = \frac{\lambda^i}{(i-1)!} \cdot \prod_{j=0}^{i-1} \left(1 - \frac{j}{N}\right) \cdot \int_0^T t^{k+i-1} \left(1 - \frac{t}{T}\right)^{N-i} dt \qquad \begin{matrix} 1 \le i \le N \\ k \in \mathbf{Z}^+ \end{matrix} \tag{12.34}$$

The joint probability density function of the $i$th and $j$th arrival times is

$$f_{(i)(j)}(t_i, t_j) = \frac{\lambda^j}{(i-1)!(j-i-1)!} \cdot \prod_{k=1}^{j-1} \left(1 - \frac{k}{N}\right) \cdot t_i^{i-1} t_j^{j-i-1} \cdot \left[1 - \frac{t_i}{t_j}\right]^{j-i-1} \cdot \left[1 - \frac{t_j}{T}\right]^{N-j} \qquad 0 \le t_i < t_j < T, 1 \le i < j \le N \tag{12.35}$$

All the interarrival times have an identical probability density function defined according to

$$f_{\Delta_{(i)}}(\Delta t) = \lambda \left(1 - \frac{\Delta t}{T}\right)^{N-1} \qquad 0 < \Delta t < T, 1 \le i \le N \tag{12.36}$$

The probability density function of the interarrival time between the $i$th and $j$th ordered times is equal to the probability density function of the $(j-i)$th ordered time:

$$f_{\Delta_{(j)-(i)}}(\Delta t) = f_{(j-i)}(\Delta t) = \frac{\lambda^{j-i} \Delta t^{j-i-1}}{(j-i-1)!} \cdot \left[\prod_{k=1}^{j-i-1} \left(1 - \frac{k}{N}\right)\right] \cdot \left[1 - \frac{\Delta t}{T}\right]^{N-j+i} \qquad 1 \le i < j \le N, 0 < \Delta t < T \tag{12.37}$$

The joint probability density function between the $i$th and $(i+1)$th interarrival times is

$$f_{\Delta_{(i)}\Delta_{(i+1)}}(\Delta t_i, \Delta t_{i+1}) = \lambda^2 \cdot \left(1 - \frac{1}{N}\right) \cdot \left[1 - \frac{\Delta t_i + \Delta t_{i+1}}{T}\right]^{N-2} \qquad (12.38)$$

$$1 < i < N-1, 0 < \Delta t_i + \Delta t_{i+1} < T$$

The joint probability density function for the $i$th and $j$th interarrival times is

$$f_{\Delta_{(i)}\Delta_{(j)}}(\Delta t_i, \Delta t_j) = \frac{\lambda^j}{(i-2)!(j-i-2)!} \cdot \prod_{k=1}^{j-1}\left(1 - \frac{1}{N}\right) \cdot$$

$$\int_{\Delta t_i}^{T-\Delta t_j}\left(\int_{u+\Delta t_j}^{T}(u - \Delta t_i)^{i-2} \cdot [v - \Delta t_j - u]^{j-i-2} \cdot \left[1 - \frac{v}{T}\right]^{N-j} dv\right) du \qquad (12.39)$$

$$0 < \Delta t_i < T, 0 < \Delta t_j < T, \Delta t_i + \Delta t_j < T, 1 \leq i < N-2, i+2 \leq j \leq N$$

### 12.4.2 Infinite Interval Case

For a Poisson point process defined on the infinite interval, the following theorem summarizes important statistical results.

**Theorem 12.15 Results for the Infinite Interval**
For the case of $N \to \infty$, $T \to \infty$ and with $\lambda = N/T$ being fixed, the cumulative distribution and probability density functions for the $i$th arrival time of a Poisson point process are

$$F_{(i)}(t) = 1 - e^{-\lambda t}\sum_{k=0}^{i-1}\frac{(\lambda t)^k}{k!} \qquad i \geq 1, t > 0 \qquad (12.40)$$

$$f_{(i)}(t) = \frac{\lambda(\lambda t)^{i-1} e^{-\lambda t}}{(i-1)!} \qquad i \geq 1, t > 0 \qquad (12.41)$$

The joint probability density function of the $i$th and $j$th arrival times is

$$f_{(i)(j)}(t_i, t_j) = f_{(i)}(t_i) f_{(j-i)}(t_j - t_i) = \frac{\lambda^i t_i^{i-1} e^{-\lambda t_i}}{(i-1)!} \cdot \frac{\lambda^{j-i}(t_j - t_i)^{j-i-1} e^{-\lambda(t_j - t_i)}}{(j-i-1)!} \qquad (12.42)$$

$$t_j > t_i, 1 \leq i < j$$

All the interarrival times are identical, independent, exponential random variables with parameter $\lambda$, that is,

$$f_{\Delta_{(i)}}(\Delta t) = \begin{cases} \lambda e^{-\lambda \Delta t} & i \geq 1, \Delta t > 0 \\ 0 & \Delta t < 0 \end{cases} \qquad (12.43)$$

The probability density function of the interarrival time between the $i$th and $j$th arrival times is equal to the probability density function of the $(j-i)$th arrival time:

$$f_{\Delta_{(j)-(i)}}(\Delta t) = f_{(j-i)}(\Delta t) = \frac{\lambda^{j-i}\Delta t^{j-i-1}e^{-\lambda\Delta t}}{(j-i-1)!} \qquad 1 \le i < j, 0 < \Delta t \tag{12.44}$$

The interarrival times are independent with

$$f_{\Delta_{(i)}\Delta_{(j)}}(\Delta t_i, \Delta t_j) = f_{\Delta_{(i)}}(\Delta t_i) f_{\Delta_{(j)}}(\Delta t_j) = \lambda e^{-\lambda\Delta t_i} \cdot \lambda e^{-\lambda\Delta t_j}$$
$$1 \le i < j, \quad \Delta t_i > 0, \quad \Delta t_j > 0 \tag{12.45}$$

Finally, the $i$th arrival time $X_{(i)}$ and the subsequent interarrival time $\Delta X_{(i+1)}$ are independent random variables, with

$$f_{(i)\Delta_{(i+1)}}(t, \Delta t) = f_{(i)}(t) f_{\Delta_{(i+1)}}(\Delta t) = \frac{\lambda^{i+1} t^{i-1} e^{-\lambda t} e^{-\lambda\Delta t}}{(i-1)!} \tag{12.46}$$

***12.4.2.1 Notes*** The probability density function of the $i$th arrival time is consistent with a gamma probability density function with parameters $\lambda$ and $i-1$, that is, $f_{(i)}(t) = f_{\Gamma(\lambda, i-1)}(t)$.

***12.4.2.2 Moments***

**Theorem 12.16 Moments for *i*th Arrival Time and Interarrival Time**
For the case of $N \to \infty$, $T \to \infty$ and with $\lambda = N/T$ being fixed, the $k$th moment of the $i$th arrival time is

$$\boldsymbol{E}\left[X_{(i)}^k\right] = \frac{(k+i-1)!}{\lambda^k (i-1)!} \tag{12.47}$$

and the mean, mean squared, and variance are

$$\boldsymbol{E}\left[X_{(i)}\right] = \frac{i}{\lambda} \quad \boldsymbol{E}\left[X_{(i)}^2\right] = \frac{(i+1)i}{\lambda^2} \quad \mathrm{var}\left(X_{(i)}\right) = \frac{i}{\lambda^2} \tag{12.48}$$

The $k$th moment of the $i$th interarrival time is

$$\boldsymbol{E}\left[\Delta X_{(i)}^k\right] = \boldsymbol{E}\left[X_{(1)}^k\right] = \frac{k!}{\lambda^k} \tag{12.49}$$

and the mean, mean squared, and variance are

$$\boldsymbol{E}\left[\Delta X_{(i)}\right] = \frac{1}{\lambda} \quad \boldsymbol{E}\left[\Delta X_{(i)}^2\right] = \frac{2}{\lambda^2} \quad \mathrm{var}\left(\Delta X_{(i)}\right) = \frac{1}{\lambda^2} \tag{12.50}$$

**Proof**

The proof is detailed in Appendix 12.E.

#### *12.4.2.3 Relationship between Number of Points and PDF of Arrival Time*

For the infinite interval, there is a simple relationship between the probability of $i$ points in the interval $[0, t]$ and the probability density function of the $i$th arrival time.

**Theorem 12.17 Relationship: Number of Points and PDF of Arrival Time**
The relationships between the probability of $i$ points in the interval $[0,t]$, denoted as $p_{N_{[0,t]}}(i)$, and probability density function of the $i$th arrival time are

$$f_{(i)}(t) = \frac{\lambda(\lambda t)^{i-1} e^{-\lambda t}}{(i-1)!} = \lambda p_{N_{[0,t]}}(i-1) = \lambda \cdot P[i-1 \text{ points on } [0,t]] \quad i \in \{1,2,\ldots\}$$

$$p_{N_{[0,t]}}(i) = \frac{\lambda^i t^i e^{-\lambda t}}{i!} = \frac{1}{\lambda} \cdot f_{(i+1)}(t) \quad i \in \{0,1,\ldots\} \tag{12.51}$$

**Proof**

The follows directly from Equations 12.41 and 12.27.

## 12.5 ALTERNATIVE CHARACTERIZATION FOR INFINITE INTERVAL

A Poisson point random process is often defined via the statement of properties as defined in the following. Grimmett and Stirzaker (1992, p. 228f) provides an appropriate overview.

**Definition: Poisson Random Process** Consider an experiment that defines points on the interval $[0, \infty]$ and whose outcomes defined the sample space $\Omega$. Define the random process $N(\Omega, t)$ as the number of points in the interval $[0, t]$. This random process is a Poisson random process with parameter $\lambda$ if:

- $N(\Omega,0) = 0$.
- If $t_1 < t_2$, then $N(\Omega,t_1) \le N(\Omega,t_2)$.
- For the case of $\Delta t << 1/\lambda$:

$$P[N(\Omega,t+\Delta t) = n+m|_{N(\Omega,t)=n}] = \begin{cases} 1-\lambda\Delta t + o(\Delta t) & m=0 \\ \lambda\Delta t + o(\Delta t) & m=1 \\ o(\Delta t) & m>1 \end{cases}$$

- $N(\Omega,t_2) - N(\Omega,t_1)$ is independent of $N(\Omega, t_1)$ when $t_2 > t_1$.

The last requirement states that the number of outcomes in the interval $(t_1, t_2]$ is independent of the number of outcomes in the interval $[0, t_1]$. This is a Markov property.

**Theorem 12.18 Probability Mass Function of $N(\Omega, t)$**
For a fixed time, the random process $N(\Omega, t)$ has a Poisson distribution with parameter $\lambda t$, that is,

$$P[N(\Omega,t)=k]=\frac{\lambda^k t^k}{k!}\cdot e^{-\lambda t} \quad k\in\{0,1,2,\ldots\} \tag{12.52}$$

The probability mass function for $k$ points in the interval $(t_1, t_2)$ is

$$P[N(\Omega,t_2)-N(\Omega,t_1)=k]=\frac{\lambda^k (t_2-t_1)^k}{k!}\cdot e^{-\lambda(t_2-t_1)} \quad t_2>t_1, k\in\{0,1,2,\ldots\} \tag{12.53}$$

**Proof**
The proof is detailed in Theorem 12.6.

**Theorem 12.19 Disjoint Intervals: Independence of Number of Outcomes**
The number of points in disjoint intervals, as defined by the experiment underpinning the random process $N(\Omega, t)$, are independent consistent with

$$\begin{gathered} P[N(\Omega,t_2)-N(\Omega,t_1)=k_1, N(\Omega,t_4)-N(\Omega,t_3)=k_2]= \\ \left[\frac{\lambda^{k_1}(t_2-t_1)^{k_1}}{k_1!}\cdot e^{-\lambda(t_2-t_1)}\right]\cdot\left[\frac{\lambda^{k_2}(t_4-t_3)^{k_2}}{k_2!}\cdot e^{-\lambda(t_4-t_3)}\right] \\ k_1,k_2\in\{0,1,2,\ldots\}, t_1<t_2<t_3<t_4 \end{gathered} \tag{12.54}$$

The joint probability mass function for the number of points in the intervals $(0, t_1)$ and $(0, t_2)$ is

$$P\begin{bmatrix} N(\Omega,t_1)=k_1, \\ N(\Omega,t_2)=k_2 \end{bmatrix}=\begin{cases} 0 & t_1>t_2, k_1>k_2 \\ \left[\frac{\lambda^{k_1}t_1^{k_1}}{k_1!}\cdot e^{-\lambda t_1}\right]\left[\frac{\lambda^{k_2-k_1}(t_2-t_1)^{k_2-k_1}}{(k_2-k_1)!}\cdot e^{-\lambda(t_2-t_1)}\right] \\ k_1,k_2\in\{0,\ 1,\ 2,\ldots\},\ k_2\geq k_1,\ t_2>t_1 \end{cases} \tag{12.55}$$

**Proof**
The proof is detailed in Appendix 12.G. The result specified in Equation 12.54 is consistent with that stated in Equation 12.28.

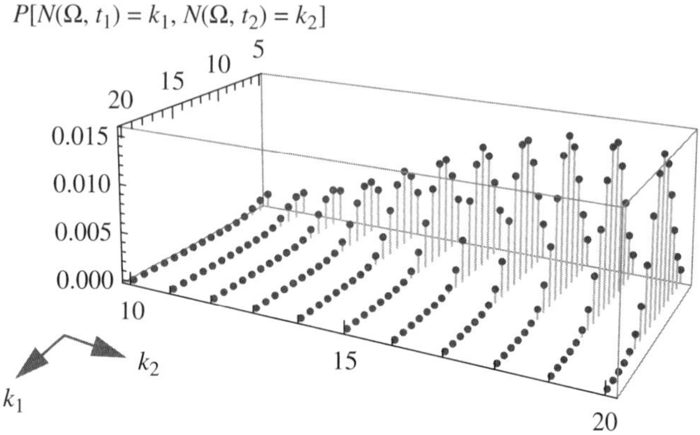


**FIGURE 12.4** Joint probability mass function for the case of $\lambda = 10,\ t_1 = 1,\ t_2 = 2$.

### 12.5.1 Example

The joint probability mass function for the number of points in the intervals $(0, t_1)$ and $(0, t_2)$ is illustrated in Figure 12.4 for the case of $\lambda = 10,\ t_1 = 1,\ t_2 = 2$.

## 12.6 MODELLING WITH UNORDERED OR ORDERED TIMES

For a random process based on independent points in a fixed interval $[0, T]$, the points can be considered to be either ordered or unordered. Naturally, the statistics of the random process should be independent of which approach is used. To illustrate this, consider a shot noise random process defined on the interval $[0, T]$ according to

$$X(\Omega, t) = \sum_{i=1}^{N} A_i p(t - \Gamma_i) \quad t \in [0, T] \tag{12.56}$$

where $\Omega = ((A_1, \Gamma_1), \ldots, (A_N, \Gamma_N))$, $A_1, \ldots, A_N$ are independent and identical random variables and $\Gamma_1, \ldots, \Gamma_N$ are random variables defining the times associated with a Poisson point random process with rate $\lambda = N/T$.

There are two forms for this random process. First is the case where $\Gamma_1, \ldots, \Gamma_N$ are unordered and independent and such that

$$f_{\Gamma_i}(t_i) = \frac{1}{T} \qquad f_{\Gamma_i\Gamma_j}(t_i, t_j) = \frac{1}{T^2} \qquad f_{\Gamma_1 \ldots \Gamma_N}(t_1, \ldots, t_N) = \left(\frac{1}{T}\right)^N \tag{12.57}$$

Second is the case where $\Gamma_1, \ldots, \Gamma_N$ are ordered and dependent. From order statistics (Eq. 12.33),

$$f_{\Gamma_i}(t_i) = \frac{\lambda(\lambda t_i)^{i-1}}{(i-1)!} \cdot \left(1 - \frac{t_i}{T}\right)^{N-i} \cdot \prod_{j=0}^{i-1}\left(1 - \frac{j}{N}\right) \tag{12.58}$$

The joint probability density function (Theorem 11.23) is

$$f_{\Gamma_1 \ldots \Gamma_N}(t_1, \ldots, t_N) = \frac{N!}{T^N} \quad t_1 < t_2 < \ldots < t_N \tag{12.59}$$

and from Equation 12.35, the joint probability density function of the $i$th and $j$th ordered times is

$$f_{\Gamma_i \Gamma_j}(t_i, t_j) = \frac{\lambda^j}{(i-1)!(j-i-1)!} \cdot \prod_{k=1}^{j-1}\left(1-\frac{k}{N}\right) \cdot t_i^{i-1} t_j^{j-i-1} \cdot \left[1-\frac{t_i}{t_j}\right]^{j-i-1} \cdot \left[1-\frac{t_j}{T}\right]^{N-j} \quad 0 \le t_i < t_j < T, 1 \le i < j \le N \tag{12.60}$$

### 12.6.1 Difference between Two Forms

First, both forms define the same time waveforms, that is, both define the same random process. The difference lies in whether the times are to be considered to be ordered or unordered. For the unordered and ordered cases, consider the autocorrelation function, which, from Table 9.5 (cases 6 and 4), respectively, are

$$R_{XX}(t+\tau, t) = \lambda \boldsymbol{E}[A^2] \int_0^{\min\{t, t+\tau\}} p(t+\tau-t_i) p(t-t_i) dt_i + \lambda^2 \left(1-\frac{1}{N}\right) \mu_A^2 \int_0^{t+\tau} p(t+\tau-t_i) dt_i \int_0^t p(t-t_k) dt_k \tag{12.61}$$

$$R_{XX}(t+\tau, t) = \boldsymbol{E}[A^2] \sum_{i=1}^{N} \int_0^{\min\{t, t+\tau\}} p(t+\tau-t_i) p(t-t_i) f_{\Gamma_i}(t_i) dt_i + \mu_A^2 \sum_{i=1}^{N} \sum_{\substack{j=1 \\ j \neq i}}^{N} \int_0^{t+\tau} \int_0^{t} p(t+\tau-t_i) p(t-t_j) f_{\Gamma_i \Gamma_j}(t_i, t_j) dt_j dt_i \tag{12.62}$$

where $\mu_A$ and $\boldsymbol{E}[A^2]$, respectively, are the mean and the mean squared value for the amplitude random variables and $\lambda$ is the pulse rate. For the last equation, consistent with ordered times, interchanging the order of summation and integration yields

$$R_{XX}(t+\tau, t) = \boldsymbol{E}[A^2] \int_0^{\min\{t, t+\tau\}} p(t+\tau-t_o) p(t-t_o) \left[\sum_{i=1}^{N} f_{\Gamma_i}(t_o)\right] dt_o + \mu_A^2 \int_0^{t} \int_0^{t+\tau} p(t+\tau-t_1) p(t-t_2) \left[\sum_{i=1}^{N} \sum_{\substack{j=1 \\ j \neq i}}^{N} f_{\Gamma_i \Gamma_j}(t_1, t_2)\right] dt_2 dt_1 \tag{12.63}$$

Equality between the unordered and ordered forms for the autocorrelation function is guaranteed if

$$\sum_{i=1}^{N} f_{\Gamma_i}(t_o) = \lambda \qquad \sum_{i=1}^{N}\sum_{\substack{j=1 \\ j\neq i}}^{N} f_{\Gamma_i\Gamma_j}(t_1,t_2) = \lambda^2\left(1-\frac{1}{N}\right) \tag{12.64}$$

That these results hold is shown in Appendix 12.H.

## 12.7 ZERO CROSSING TIMES OF RANDOM TELEGRAPH SIGNAL

Consider random telegraph signals of types I and II and defined according to

$$X_1(\Omega_1,t) = \sum_{j=0}^{\infty} a_j u(t-\Gamma_j) \qquad \begin{cases} a_0 = 1 \\ a_1, a_3, \ldots = -2 \\ a_2, a_4, \ldots = 2 \end{cases}$$
$$X_2(\Omega_2,t) = \sum_{j=0}^{\infty} A_j\left[u(t-\Gamma_j) - u(t-\Gamma_{j+1})\right], \tag{12.65}$$

where $\Omega_1 = (\Gamma_0, \Gamma_1, \ldots)$, $\Omega_2 = ((A_0,\Gamma_0),(A_1,\Gamma_1),\ldots)$, $\Gamma_1,\ldots$ is a sequence of random variables consistent with the outcomes of a Poisson experiment, $\Gamma_0 = 0$, $A_1,\ldots$ is a sequence of dependent random variables with a sample space of $\{-1,1\}$ and $A_0 = 1$. As a zero crossing coincides, potentially, with a Poisson point, the following result holds.

**Theorem 12.20 PDF of *i*th Zero Crossing**
For a random telegraph signal of type I, the probability density function associated with the *i*th zero crossing is

$$f_{(i)}(t) = \frac{\lambda(\lambda t)^{i-1}e^{-\lambda t}}{(i-1)!} \qquad i \in \{1,2,\ldots\} \tag{12.66}$$

The mean and variance of the *i*th zero crossing time are

$$\mu_i = \frac{i}{\lambda} \qquad \mathrm{var}_i = \frac{i}{\lambda^2} \tag{12.67}$$

For a random telegraph signal of type II, where the probability of a transition at a Poisson point is $p$, the probability density function associated with the *i*th zero crossing is

$$f_{(i)}(t) = \frac{p\lambda(p\lambda t)^{i-1}e^{-p\lambda t}}{(i-1)!} \qquad i \in \{1,2,\ldots\} \tag{12.68}$$

The mean and variance of the $i$th zero crossing time are

$$\mu_i = \frac{i}{p\lambda} \qquad \text{var}_i = \frac{i}{p^2\lambda^2} \tag{12.69}$$

**Proof**

The of the result for the random telegraph signal of type I follows directly from Equations 12.41 and 12.48. The proof of the results for the random telegraph signal of type II arises from the fact that selection of a transition with a probability $p$ creates an effective Poisson point process with a rate $\lambda p$.

## 12.8 POINT PROCESSES: THE GENERAL CASE

The experiment underpinning a point process may lead to the points having a non-uniform distribution and being correlated with one another, etc. Accordingly, the theory for point processes and associated random processes is vast, for example, Dayley and Vere–Jones (2003) and Ross (2000, p. 289f ).

## 12.9 PROBLEMS

**12.1** Consider an experiment comprising of $N$ independent subexperiments. Each subexperiment yields a number from the interval $[\alpha, \beta]$ that is chosen consistent with a probability density function

$$f_X(x) = \frac{\pi}{2(\beta-\alpha)} \cdot \sin\left[\frac{\pi}{\beta-\alpha}(x-\alpha)\right] \quad x \in [\alpha,\beta] \tag{12.70}$$

The number of points in a set $A \subseteq [\alpha,\beta]$ defines a random variable $N_A$ with a probability mass function $p_{N_A}$.

a. Specify the probability mass function $p_{N_A}$.
b. Specify an approximation to this probability mass function using Poisson's theorem.
c. For the case of $N=100$, $\alpha=0$, $\beta=1$ and $A=[0,0.05]$, graph $p_{N_A}$ and the Poisson approximation to $p_{N_A}$. Specify the true mean and variance of $N_A$ as well as the mean and variance implied by the Poisson approximation.
d. Repeat (c) for the case of $A=[0.45,0.5]$.

**12.2** Consider an experiment comprising of $N$ independent subexperiments. Each subexperiment yields a number from the interval $[\alpha,\beta]$ that is chosen consistent with a probability density function

$$f_X(x) = \frac{\pi}{2(\beta-\alpha)} \cdot \sin\left[\frac{\pi}{\beta-\alpha}(x-\alpha)\right] \quad x \in [\alpha,\beta] \tag{12.71}$$

The number of points in the disjoint sets $A \subseteq [\alpha, \beta]$ and $B \subseteq [\alpha, \beta]$ define the random variables $N_A$ and $N_B$ with respective probability mass functions $p_{N_A}$ and $p_{N_B}$ and a joint probability mass function $p_{N_A N_B}$.

a. Specify the joint probability mass function $p_{N_A N_B}$.
b. Graph $p_{N_A N_B}$ for the case of $N = 10$, $\alpha = 0$, $\beta = 1$, $A = [0.2, 0.4]$ and $B = [0.5, 0.7]$.
c. Specify an expression for the correlation coefficient between $N_A$ and $N_B$. For the parameters defined in (b), determine the correlation coefficient.

**12.3** Consider the case of two disjoint sets on the interval [0, $T$], which both have a measure of $m_o$, and consider the case of $T = k_T m_o \in \{3m_o, 4m_o, \ldots\}$. Points are placed on [0, $T$] at a uniform rate of $\lambda = k_\lambda \lambda_o$, where $\lambda_o = 1/m_o, k_\lambda \in \{1, 2, \ldots\}$, and consistent with $N = \lambda T = k_\lambda k_T$ being an integer. A measure of the dependence/independence of the number of points in the two sets is given by

$$\frac{p_{N_A}(k_A) \cdot p_{N_B}(k_B)}{p_{N_A N_B}(k_A, k_B)} \tag{12.72}$$

Graph this ratio for the case of $k_A = k_B = k_\lambda$ (the number of points in the sets equalling the mean number of points in the sets) as $T$ increases according to $T = k_T m_o \in \{3m_o, 4m_o, \ldots\}$.

**12.4** To obtain a set of ordered points for an interval [0, $T_o$], which are consistent with a Poisson point process on the infinite interval $[0, \infty]$ with a rate $\lambda$, one approach is to generate $kN_o = k\lambda T_o$ points, at random, from the interval $[0, kT_o], k \in \{1, 2, \ldots\}$, order the points and take only those that are on the interval [0, $T_o$]. Consider the case of $\lambda = 10$, $T_o = 1$ and the time of the fifth ordered point for the cases of $k = 1, 2, 5, 10$.

a. Graph the probability density function of the time of the fifth ordered point for the cases of $k = 1, 2, 5, 10$ and compare with the probability density function for the infinite interval case.
b. For $k = 1, 2, \ldots, 10$, graph the relative error in the mean and variance of the time of the fifth ordered point using as a basis the mean and variance for the infinite interval case.

**12.5** Consider a Poisson point random process defined on $[0, \infty)$. Determine an expression for the joint probability mass function for the number of points in the intervals $(t_o, t_o + \Delta_1)$ and $(t_o + \Delta_1, t_o + \Delta_1 + \Delta_2)$. Graph this probability mass function for the case of $\lambda = 1$, $t_o = 10$, $\Delta_1 = 1$, and $\Delta_2 = 1$.

**12.6** Consider two random processes defined according to

$$\begin{aligned} X(\Omega, t) &= X_1(\Omega_1, t) - X_2(\Omega_2, t) \\ X(\Omega, t) &= \operatorname{sgn}[X_1(\Omega_1, t) - X_2(\Omega_2, t)] \end{aligned} \tag{12.73}$$

where $\Omega = \Omega_1 \times \Omega_2$ and $X_1(\Omega_1, t)$ and $X_2(\Omega_2, t)$ are causal random telegraph signals based on independent Poisson experiments that generate points, respectively, at a rate of $\lambda_1$ and $\lambda_2$.

Are the points of change (zero crossings points for the first random process) of these random processes consistent with a Poisson point random process? Justify your answer by considering the probability mass function for the points of change of $X$.

## APPENDIX 12.A PROOF OF THEOREM 12.5

Consider a single trial of a pseudo-Poisson experiment for the interval $[0, T]$, where the outcome is the vector $(\gamma_1, \ldots, \gamma_N), \gamma_k \in [0, T], \gamma_1 \leq \gamma_2 \leq \ldots \leq \gamma_N$. This outcome is the result of the ordering of the outcomes of $N$ independent trials of a subexperiment of choosing a point, at random, from the interval $[0,T]$. It then follows that the probability $p$ of an individual point being in the set $A$ equals

$$p = \frac{m[A]}{T} \tag{12.74}$$

In $N$ independent trials, the probability of $k$ points being in the set $A$ is given by the probability of $k$ successes in $N$ trials where the probability of success in any one trial is $p$. Thus, from Theorem 4.24,

$$p_{N_A}(k) = \frac{N!}{(N-k)!k!} \cdot \left(\frac{m[A]}{T}\right)^k \cdot \left(1 - \frac{m[A]}{T}\right)^{N-k} \quad k \in \{0, 1, \ldots, N\} \tag{12.75}$$

The alternative form arises from Theorem 11.3 and the definition of $\lambda = N/T$.

## APPENDIX 12.B PROOF OF THEOREM 12.6

Using the approximation

$$(1+x)^p \approx 1 + px \quad |px| << 1 \tag{12.76}$$

it follows from Theorem 12.5, assuming $N \geq 2$, that

$$\begin{aligned} P[\text{one point in } A] &= \lambda m[A]\left(1 - \frac{m[A]}{T}\right)^{N-1} \approx \lambda m[A]\left[1 - \frac{(N-1)m[A]}{T}\right] \\ &= \lambda m[A]\left[1 - \lambda m[A]\left(1 - \frac{1}{N}\right)\right] \quad \lambda m[A]\left(1 - \frac{1}{N}\right) << 1 \end{aligned} \tag{12.77}$$

For $N \geq 2$, the assumption $\lambda m[A](1-1/N) << 1$ is satisfied by $\lambda m[A] << 1$. Thus, when $\lambda m[A] << 1$, it is the case that $P[\text{one point in } A] = \lambda m[A](1-\varepsilon_1)$ where $0 < \varepsilon_1 << 1$. For the case of $N = 1$, it is the case that $\varepsilon_1 = 0$.

Similarly, it follows that

$$P[\text{no points in } A] = \left(1 - \frac{m[A]}{T}\right)^N \approx 1 - \frac{Nm[A]}{T} = 1 - \lambda m[A] \tag{12.78}$$

provided $\lambda m[A] << 1$.

## APPENDIX 12.C PROOF OF THEOREM 12.9

Consistent with the discussion in Appendix 12.A, it is the case that the probability of an individual point being in the set $A$, in set $B$, or in neither set, respectively, equals

$$p_A = \frac{m[A]}{T} \qquad p_B = \frac{m[B]}{T} \qquad 1 - p_A - p_B = 1 - \frac{m[A]}{T} - \frac{m[B]}{T} \tag{12.79}$$

In $N$ independent trials, the outcome of $k_A$ points being in the set $A$, $k_B$ points being in the set $B$, and $N - k_A - k_B$ points being in neither set is given by the multinomial formula. Alternatively, consider $k_A$ points being in the set $A$, $k_B$ points being in the set $B$, and $N - k_A - k_B$ points being outside both sets where the order of allocation to the categories is set. The probability of such an outcome is

$$\left(\frac{m[A]}{T}\right)^{k_A} \left(\frac{m[B]}{T}\right)^{k_B} \left(1 - \frac{m[A]}{T} - \frac{m[B]}{T}\right)^{N-k_A-k_B} \tag{12.80}$$

There are $\dfrac{N!}{k_A!k_B!(N-k_A-k_B)!} - 1$ other outcomes that lead to $k_A$ points being in the set $A$, $k_B$ points being in the set $B$, and $N - k_A - k_B$ points being in neither set. The required result then follows as such outcomes are disjoint and occur with equal probability, that is,

$$p_{N_A N_B}(k_A, k_B) = \\ \frac{N!}{k_A!k_B!(N-k_A-k_B)!} \left(\frac{m[A]}{T}\right)^{k_A} \left(\frac{m[B]}{T}\right)^{k_B} \left(1 - \frac{m[A]}{T} - \frac{m[B]}{T}\right)^{N-k_A-k_B} \\ k_A, k_B \in \{0, 1, \ldots, N\}, k_A + k_B \leq N \tag{12.81}$$

The alternative form arises from Theorem 11.3 which implies

$$\frac{N!}{(N-k_A-k_B)!k_A!k_B!} \cdot \frac{1}{N^{k_A+k_B}} = \frac{1}{k_A!k_B!} \prod_{j=0}^{\max\{0, k_A+k_B-1\}} \left(1 - \frac{j}{N}\right) \tag{12.82}$$

for $k_A + k_B \geq 0$ whereupon

$$p_{N_A N_B}(k_A, k_B) = \frac{\lambda^{k_A + k_B}}{k_A! k_B!} \cdot \left[ \prod_{j=0}^{\max\{0, k_A + k_B - 1\}} \left(1 - \frac{j}{N}\right) \right] \cdot m[A]^{k_A} \cdot m[B]^{k_B} \cdot \left(1 - \frac{m[A]}{T} - \frac{m[B]}{T}\right)^{N - k_A - k_B} \quad k_A, k_B \in \{0, 1, \ldots, N\}, k_A + k_B \leq N \tag{12.83}$$

## APPENDIX 12.D PROOF OF THEOREM 12.12

Independence implies

$$p_{N_A N_B}(k_A, k_B) = p_{N_A}(k_A) p_{N_B}(k_B) \tag{12.84}$$

Using the expressions in Equations 12.15 and 12.22, it then follows that independence is consistent with

$$\prod_{j=0}^{\max\{0, k_A + k_B - 1\}} \left(1 - \frac{j}{N}\right) \cdot \left(1 - \frac{m[A]}{T} - \frac{m[B]}{T}\right)^{N - k_A - k_B} = \prod_{j=0}^{\max\{k_A - 1, 0\}} \left(1 - \frac{j}{N}\right) \cdot \prod_{j=0}^{\max\{k_B - 1, 0\}} \left(1 - \frac{j}{N}\right) \cdot \left(1 - \frac{m[A]}{T}\right)^{N - k_A} \cdot \left(1 - \frac{m[A]}{T}\right)^{N - k_B} \tag{12.85}$$

For the case of $N \to \infty$, $T \to \infty$ and with $k_A$, $k_B$, $m[A]$, and $m[B]$ fixed and finite, it follows that

$$\prod_{j=0}^{\max\{0, k_A + k_B - 1\}} \left(1 - \frac{j}{N}\right) \to 1 \qquad \prod_{j=0}^{\max\{k_A - 1, 0\}} \left(1 - \frac{j}{N}\right) \to 1 \tag{12.86}$$

Using this result and the result from Theorem 11.4, $(1 - t/T)^{N-k} \approx e^{-\lambda t}$, $\lambda = N/T$, it then follows that

$$\frac{\left(1 - \frac{m[A]}{T} - \frac{m[B]}{T}\right)^{N - k_A - k_B}}{\left(1 - \frac{m[A]}{T}\right)^{N - k_A} \cdot \left(1 - \frac{m[A]}{T}\right)^{N - k_B}} \approx \frac{e^{-\lambda[m[A] + m[B]]}}{e^{-\lambda m[A]} \cdot e^{-\lambda m[B]}} = 1 \tag{12.87}$$

as $N \to \infty$, $T \to \infty$. This is the result required for independence.

## APPENDIX 12.E PROOF OF THEOREM 12.16

By definition, the $k$th moment of the $i$th arrival time follows from Equation 12.41 according to

$$\boldsymbol{E}\left[X_{(i)}^k\right]=\int_0^\infty t^k\cdot\frac{\lambda(\lambda t)^{i-1}e^{-\lambda t}}{(i-1)!}dt=\frac{\lambda^i}{(i-1)!}\int_0^\infty t^{k+i-1}e^{-\lambda t}dt \tag{12.88}$$

A change of variable $\xi=\lambda t$ yields

$$\boldsymbol{E}\left[X_{(i)}^k\right]=\frac{\lambda^{i-1}}{\lambda^{k+i-1}(i-1)!}\int_0^\infty \xi^{k+i-1}e^{-\xi}d\xi=\frac{1}{\lambda^k(i-1)!}\Gamma(k+i) \tag{12.89}$$

where $\Gamma(n)$ is the Gamma function (Eq. 2.19). As $\Gamma(n)=(n-1)!$ for $n\in\boldsymbol{Z}^+$, it follows that

$$\boldsymbol{E}\left[X_{(i)}^k\right]=\frac{(k+i-1)!}{\lambda^k(i-1)!} \tag{12.90}$$

For the interarrival time, it follows from Equation 12.43 that

$$\boldsymbol{E}\left[\Delta X_{(i)}^k\right]=\int_0^\infty (\Delta t)^k\cdot\lambda e^{-\lambda\Delta t}d\Delta t \tag{12.91}$$

A change of variable $\xi=\lambda\Delta t$ yields

$$\boldsymbol{E}\left[X_{(i)}^k\right]=\frac{1}{\lambda^k}\int_0^\infty \xi^k e^{-\xi}d\xi=\frac{1}{\lambda^k}\Gamma(k+1)=\frac{k!}{\lambda^k} \tag{12.92}$$

## APPENDIX 12.F PROOF OF THEOREM 12.18

Consider the interval $[0,t+\Delta t]$, $\Delta t<<1/\lambda$ and use of the Theorem of Total Probability:

$$\begin{aligned}P[N(\Omega,t+\Delta t)=k]&=\sum_{i=0}^{k}P\left[N(\Omega,t+\Delta t)=k\middle|_{N(\Omega,t)=i}\right]P[N(\Omega,t)=i]\\&=P\left[N(\Omega,t+\Delta t)=k\middle|_{N(\Omega,t)=0}\right]P[N(\Omega,t)=0]\\&\quad+P\left[N(\Omega,t+\Delta t)=k\middle|_{N(\Omega,t)=1}\right]P[N(\Omega,t)=1]\\&\quad+\cdots+P\left[N(\Omega,t+\Delta t)=k\middle|_{N(\Omega,t)=k-1}\right]P[N(\Omega,t)=k-1]\\&\quad+P\left[N(\Omega,t+\Delta t)=k\middle|_{N(\Omega,t)=k}\right]P[N(\Omega,t)=k]\end{aligned} \tag{12.93}$$

As $P\left[N(\Omega,t+\Delta t)=k\big|_{N(\Omega,t)=0}\right]=P[k\,\text{points in}\,(t,t+\Delta t)]$, etc., it follows from the stated properties that

$$\begin{aligned}P[N(\Omega,t+\Delta t)=k]&=\mathrm{o}(\Delta t)P[N(\Omega,t)=0]+\mathrm{o}(\Delta t)P[N(\Omega,t)=1]+\cdots\\&+(\lambda\Delta t+\mathrm{o}(\Delta t))P[N(\Omega,t)=k-1]+P[1-\lambda\Delta t+\mathrm{o}(\Delta t)][N(\Omega,t)=k]\end{aligned}\tag{12.94}$$

As $\Delta t\to 0$, it follows that $\mathrm{o}(\Delta t)\to 0$ and

$$\begin{aligned}&P[N(\Omega,t+\Delta t)=0]\approx(1-\lambda\Delta t)P[N(\Omega,t)=0]\quad k=0\\&P[N(\Omega,t+\Delta t)=k]\approx(\lambda\Delta t)P[N(\Omega,t)=k-1]+(1-\lambda\Delta t)P[N(\Omega,t)=k]\quad k\geq 1\end{aligned}\tag{12.95}$$

Thus,

$$\begin{aligned}&\frac{P[N(\Omega,t+\Delta t)=0]-P[N(\Omega,t)=0]}{\Delta t}\approx-\lambda P[N(\Omega,t)=0]\quad k=0\\&\frac{P[N(\Omega,t+\Delta t)=k]-P[N(\Omega,t)=k]}{\Delta t}\approx\lambda P[N(\Omega,t)=k-1]\\&-\lambda P[N(\Omega,t)=k]\quad k\geq 1\end{aligned}\tag{12.96}$$

Hence, in the limit as $\Delta t\to 0$:

$$\begin{aligned}&\frac{d}{dt}P[N(\Omega,t)=0]=-\lambda P[N(\Omega,t)=0]\quad k=0\\&\frac{d}{dt}P[N(\Omega,t)=k]=\lambda P[N(\Omega,t)=k-1]-\lambda P[N(\Omega,t)=k]\quad k\geq 1\end{aligned}\tag{12.97}$$

which is a set of differential equations. The boundary conditions are $P[N(\Omega,0)=0]=1$ and $P[N(\Omega,0)=k]=0$ for $k\geq 1$.

This set of differential equations can be solved in two ways: first, by induction and, second, via Laplace transformation. Consider the solution by induction: the solution to the zeroth order, $k=0$, equation is

$$P[N(\Omega,t)=0]=e^{-\lambda t}\tag{12.98}$$

The differential equation for the case of $k=1$ then becomes

$$\frac{d}{dt}P[N(\Omega,t)=1]+\lambda P[N(\Omega,t)=1]=\lambda e^{-\lambda t}\qquad P[N(\Omega,0)=1]=0\tag{12.99}$$

The solution of this differential equation is $\lambda te^{-\lambda t}$. The differential equation for the case of $k=2$ then is

$$\frac{d}{dt}P[N(\Omega,t)=2]+\lambda P[N(\Omega,t)=2]=\lambda^2te^{-\lambda t}\qquad P[N(\Omega,0)=2]=0\tag{12.100}$$

For the general case, the solution is

$$P[N(\Omega,t)=k]=\frac{(\lambda t)^k e^{-\lambda t}}{k!} \tag{12.101}$$

To prove the second result, note, from the Theorem of Total Probability, that

$$\begin{aligned} &P[N(\Omega,t_2+\Delta)-N(\Omega,t_1)=k]= \\ &\sum_{i=0}^{k} P\left[N(\Omega,t_2+\Delta)-N(\Omega,t_1)=k|_{N(\Omega,t_2)-N(\Omega,t_1)=i}\right]\cdot \\ &\qquad\qquad P[N(\Omega,t_2)-N(\Omega,t_1)=i] \end{aligned} \tag{12.102}$$

Define $N(\Omega,t)=N(\Omega,t_1+t)-N(\Omega,t_1)$, where $t=t_2-t_1$, and it then follows, using the aforementioned argument for the first proved result, that

$$P[N(\Omega,t_2)-N(\Omega,t_1)=k]=\frac{\lambda^k(t_2-t_1)^k}{k!}\cdot e^{-\lambda(t_2-t_1)} \quad t_2>t_1, k\in\{0,1,2,\ldots\} \tag{12.103}$$

## APPENDIX 12.G PROOF OF THEOREM 12.19

Consider

$$\begin{aligned} &P[N(\Omega,t_2)-N(\Omega,t_1)=k_1, N(\Omega,t_4)-N(\Omega,t_3)=k_2]= \\ &P\left[N(\Omega,t_4)-N(\Omega,t_3)=k_2|_{N(\Omega,t_2)-N(\Omega,t_1)=k_1}\right]\cdot \\ &\qquad\qquad P[N(\Omega,t_2)-N(\Omega,t_1)=k_1] \\ &=P[N(\Omega,t_4)-N(\Omega,t_3)=k_2]\cdot P[N(\Omega,t_2)-N(\Omega,t_1)=k_1] \end{aligned} \tag{12.104}$$

The last result arises from the Markov assumption. The required result then follows from Theorem 12.18.

The second result arises by setting $t_1=0$ and $t_3=t_2$ in Equation 12.54 whereupon

$$\begin{aligned} &P[N(\Omega,t_2)-N(\Omega,t_1)=k_1, N(\Omega,t_4)-N(\Omega,t_3)=k_2]= \\ &P[N(\Omega,t_2)=k_1, N(\Omega,t_4)=k_1+k_2] \\ &=\left[\frac{\lambda^{k_1}t_2^{k_1}}{k_1!}\cdot e^{-\lambda t_2}\right]\left[\frac{\lambda^{k_2}(t_4-t_2)^{k_2}}{k_2!}\cdot e^{-\lambda(t_4-t_2)}\right] \\ &\qquad\qquad k_1,k_2\in\{0,1,2,\ \ldots\}, 0<t_2<t_4 \end{aligned} \tag{12.105}$$

The required result follows by changing the notation according to $t_2 \to t_1$, $t_4 \to t_2$, $k_1 + k_2 \to k_2$.

## APPENDIX 12.H EQUIVALENCE: ORDERED–UNORDERED TIMES

To confirm the required relationships stated in Equation 12.64, first, consider the summation

$$\sum_{i=1}^{N} f_{\Gamma_i}(t_o) = \sum_{i=1}^{N} \frac{\lambda(\lambda t_o)^{i-1}}{(i-1)!} \cdot \left(1 - \frac{t_o}{T}\right)^{N-i} \cdot \prod_{j=0}^{i-1} \left(1 - \frac{j}{N}\right) \tag{12.106}$$

For $N = 1$, the required result follows. For $N = 2$, with $T = N/\lambda = 2/\lambda$, it follows that

$$\begin{aligned} \sum_{i=1}^{N} \frac{\lambda(\lambda t_o)^{i-1}}{(i-1)!} \cdot \left(1 - \frac{t_o}{T}\right)^{N-i} \cdot \prod_{j=0}^{i-1} \left(1 - \frac{j}{N}\right) &= \lambda \left(1 - \frac{\lambda t_o}{2}\right) + \lambda^2 t_o \left(1 - \frac{1}{2}\right) \\ &= \lambda - \frac{\lambda^2 t_o}{2} + \frac{\lambda^2 t_o}{2} = \lambda \end{aligned} \tag{12.107}$$

For $N = 3$, with $T = N/\lambda = 3/\lambda$, it follows that

$$\begin{aligned} \sum_{i=1}^{N} \frac{\lambda(\lambda t_o)^{i-1}}{(i-1)!} \cdot \left(1 - \frac{t_o}{T}\right)^{N-i} \cdot \prod_{j=0}^{i-1} \left(1 - \frac{j}{N}\right) &= \lambda \left(1 - \frac{\lambda t_o}{3}\right)^2 \\ &+ \lambda^2 t_o \left(1 - \frac{\lambda t_o}{3}\right)\left(1 - \frac{1}{3}\right) + \frac{\lambda^3 t_o^2}{2}\left(1 - \frac{1}{3}\right)\left(1 - \frac{2}{3}\right) \\ &= \lambda \left[1 - \frac{2\lambda t_o}{3} + \frac{\lambda^2 t_o^2}{9} + \frac{2\lambda t_o}{3} - \frac{2\lambda^2 t_o^2}{9} + \frac{\lambda^2 t_o^2}{9}\right] = \lambda \end{aligned} \tag{12.108}$$

etc. Second, consider the summation

$$\begin{aligned} \sum_{i=1}^{N} \sum_{\substack{j=1 \\ j \neq i}}^{N} f_{\Gamma_i \Gamma_j}(t_1, t_2) &= \sum_{i=1}^{N} \sum_{\substack{j=1 \\ j \neq i}}^{N} \frac{\lambda^j}{(i-1)!(j-i-1)!} \cdot \prod_{k=1}^{j-1} \left(1 - \frac{k}{N}\right) \cdot t_1^{i-1} t_2^{j-i-1} \\ &\cdot \left[1 - \frac{t_1}{t_2}\right]^{j-i-1} \cdot \left[1 - \frac{t_2}{T}\right]^{N-j} \end{aligned} \tag{12.109}$$

and the required result clearly holds for $N = 1$. For $N = 2$ with $T - N/\lambda = 2/\lambda$, and for $N = 3$ with $T = N/\lambda = 3/\lambda$, etc., it can similarly be shown (preferably using analytical software, e.g., Mathematica) that

$$\sum_{i=1}^{N} \sum_{\substack{j=1 \\ j \neq i}}^{N} f_{\Gamma_i \Gamma_j}(t_1, t_2) = \lambda^2 \left(1 - \frac{1}{N}\right) \tag{12.110}$$

# 13

# BIRTH–DEATH RANDOM PROCESSES

## 13.1 INTRODUCTION

Many random phenomena can be characterized in a state manner and with changes of states defining the nature of the phenomena. For the case where the states are integers, the state represents the population of a set of entities, and the population changes in integer steps according to *births* and *deaths*, a birth–death random process is defined.

The usual manner for characterizing a birth–death random process is via a sequence of differential equations that define the change in state probability with time. Results for important cases—constant birth rate and zero death rate processes, state-dependent birth rate and zero death rate processes, zero birth rate and constant death rate processes, and constant birth and constant death rate processes—are discussed. A constant birth and constant death process underpins the simplest queueing system—the *M*/*M*/1 queueing system (a first-come–first-serve queue with waiting and service times).

## 13.2 DEFINING AND CHARACTERIZING BIRTH–DEATH PROCESSES

### 13.2.1 Notation and Definitions

Fundamental to birth–death random processes are the following definitions.

*A Signal Theoretic Introduction to Random Processes*, First Edition. Roy M. Howard.

**Definition: Population Being in State $E_k$** The population is said to be in state $E_k$ if the population consists of $k$ entities.

**Definition: Birth** A transition from state $E_k$ to state $E_{k+1}$ signifies a *birth.*

**Definition: Death** A transition from state $E_k$ to state $E_{k-1}$ signifies a *death.*

**Definition: State Probability** The following notation is used for the state probability:

$$P_k(t) = P[\text{population is } k \text{ (state } E_k) \text{ at time } t] \tag{13.1}$$

**Definition: Rate** The term *rate* implies a number per unit time.

**Definition: Birth Rate** The birth rate, denoted $\lambda_k$, is the rate at which births occur when the population is $k$.

**Definition: Death Rate** The death rate, denoted $\mu_k$, is the rate at which deaths occur when the population is $k$.

### 13.2.2 State Transition Representation

A birth–death random process is represented by a state transition diagram as illustrated in Figure 13.1. As is clear from this figure, the state $E_k$ can only be entered from state $E_{k-1}$ or from state $E_{k+1}$. Similarly, the state $E_k$ can only be left by entering state $E_{k-1}$ or state $E_{k+1}$.

### 13.2.3 Assumptions

The following assumptions underpin the analysis of birth–death random processes:

1. Birth and death rates are independent of time and only depend on the state, that is, they are only dependent on the population.
2. The death rate when the population is zero is zero, that is, $\mu_0 = 0$.

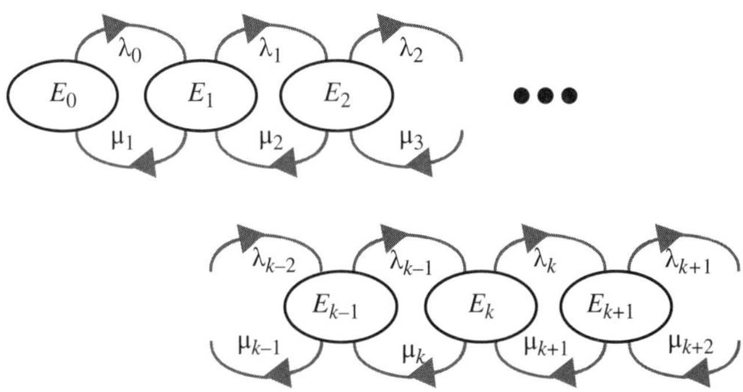


**FIGURE 13.1** State transition diagram consistent with a birth–death random process. For state $k$, the birth rate is $\lambda_k$ and the death rate is $\mu_k$.

3. The birth rate when the population is zero is potentially nonzero, that is, $\lambda_0 \geq 0$. A nonzero birth rate with zero population is consistent with spontaneous evolution.
4. Births and deaths occur at random times.
5. The probability of multiple births or deaths or a birth and a death in a small interval of time is negligible. Mathematically and for $\Delta t << 1/\lambda_k, 1/\mu_k$:

$$
\begin{aligned}
&P[\text{one birth in } (t, t+\Delta t)/\text{population is } k] = \lambda_k \Delta t + o(\Delta t) \\
&P[\text{one death in } (t, t+\Delta t)/\text{population is } k] = \mu_k \Delta t + o(\Delta t) \\
&P[\text{more than one death or one birth in } (t, t+\Delta t)/\text{population is } k] = o(\Delta t) \qquad (13.2) \\
&P[\text{no births in } (t, t+\Delta t)/\text{population is } k] = 1 - \lambda_k \Delta t + o(\Delta t) \\
&P[\text{no deaths in } (t, t+\Delta t)/\text{population is } k] = 1 - \mu_k \Delta t + o(\Delta t)
\end{aligned}
$$

### 13.2.4 Experiments for Birth–Death Random Processes

#### *13.2.4.1 Experiment: Birth–Death Random Process—Zero Initial Population*

One outcome of a birth–death random process is a sequence of states consistent with a sequence of changes in the population level, for example,

$$E_0, E_1, E_2, E_3, E_2, E_3, E_4, E_3, E_2, \ldots \qquad (13.3)$$

Such an outcome is consistent with the sequence of births, $B$, and deaths, $D$, according to

$$B, B, B, D, B, B, D, D, \ldots \qquad (13.4)$$

Such a sequence arises from the following algorithmic experiment where a state parameter $k$ represents the population with an initial population at $t = 0$ of $k_0 = 0$:

1. Set the state parameter $k$ to $k_0 = 0$. Set the time parameter $t_0$ to zero.
2. Define $t_1$ as the first time defined by a Poisson point random process commencing at $t = 0$ and with rate $\lambda_0$. Increment the state parameter, that is, $k_1 = k_0 + 1$.
3. Consider two Poisson random processes commencing at $t = t_1$: a birth process with rate $\lambda_1$ and a death process with rate $\mu_1$. Take the earliest time arising from these two random processes and denote this time as $t_2$. If the earliest time is associated with a birth, then increment the state parameter, that is, $k_2 = k_1 + 1$; if the earliest time is associated with a death, then decrement the state parameter, that is, $k_2 = k_1 - 1$.

...

*j*. Consider two Poisson random processes commencing at $t = t_j$ where the state parameter is $k_j$: a birth process with rate $\lambda_j$ and a death process with rate $\mu_j$. Take the earliest time arising from these two random processes and denote this time $t_{j+1}$. If the earliest time is associated with a birth, then increment the state parameter, that is, $k_{j+1} = k_j + 1$; if the earliest time is associated with a death, then decrement the state parameter, that is, $k_{j+1} = k_j - 1$.

...

If at any time the population drops to zero, that is, $\exists j$ such that $k_j = 0$, then the algorithm terminates and all future population values are zero.

The outcome of one trial of the experiment is the sequence of ordered pairs:

$$\{(t_0, k_0), (t_1, k_1), \ldots, (t_i, k_i), \ldots\} \tag{13.5}$$

and the sample space associated with the experiment is

$$S = \left\{ \begin{array}{l} ((t_0, k_0), (t_1, k_1), \ldots, (t_i, k_i) \ldots) : \\ (t_i, k_i) \text{ as defined by algorithm} \end{array} \right\} \tag{13.6}$$

The specification of the times for a given outcome depends on the memoryless property of a Poisson point random process. The birth and death rates vary according to the population and, hence, according to whether the last transition was a birth or a death. Such a characterization is an idealization as, physically, the birth and death rates depend on the prior population as well as on the current population.

***13.2.4.2 Experiment: Birth–Death RP—Nonzero Initial Population*** When the initial population at $t = 0$ is $k_0$, $k_0 > 0$, the algorithmic experiment underpinning a birth–death random process is modified in the first two steps according to the following:

1. Set the state parameter $k$ to $k_0 > 0$. Set the time parameter $t_0$ to zero.
2. Consider two Poisson point random processes commencing at $t = t_0$: a birth process with rate $\lambda_{k_0}$ and a death process with rate $\mu_{k_0}$. Take the earliest time arising from these two random processes and denote this time $t_1$. If the earliest time is associated with a birth, then increment the state parameter, that is, $k_1 = k_0 + 1$; if the earliest time is associated with a death, then decrement the state parameter, that is, $k_1 = k_0 - 1$.

   …

### 13.2.5 Definition: Birth–Death Random Process

Based on the experiment and sample space detailed previously, a birth–death random process can be defined according to

$$\begin{aligned} & X : S \to S_X \\ & S = \left\{ \begin{array}{l} ((t_0, k_0), (t_1, k_1), \ldots, (t_i, k_i), \ldots) : \\ (t_i, k_i) \text{ as defined by algorithm,} \end{array} \right\} \\ & X(((t_0, k_0), (t_1, k_1), \ldots, (t_i, k_i), \ldots), t) \\ & \quad = k_0 u(t - t_0) + \sum_{i=1}^{\infty} (k_i - k_{i-1}) u[t - t_i] \qquad k_0 = k_0, t_0 = 0 \end{aligned} \tag{13.7}$$

$$S_X = \left\{ \begin{array}{l} X(((t_0, k_0), (t_1, k_1), \ldots, (t_i, k_i), \ldots), t) = k_0 u(t - t_0) + \\ \sum_{i=1}^{\infty} (k_i - k_{i-1}) u[t - t_i] : ((t_0, k_0), (t_1, k_1), \ldots, (t_i, k_i), \ldots) \in S \end{array} \right\} \tag{13.8}$$

For notational convenience, a birth–death random process is written as

$$X(\Omega,t)=k_0u(t-t_0)+\sum_{j=1}^{\infty}(K_j-K_{j-1})u\left[t-\Gamma_j\right] \quad K_0=k_0 \tag{13.9}$$

where $\Omega=((\Gamma_1,K_1),\ldots),\Gamma_1,\Gamma_2,\ldots$ are a sequence of random variables consistent with the outcomes of the sequences of Poisson experiments underpinning a birth–death random process, and $K_1,K_2,\ldots$ are a sequence of random variables defining the population levels at the transition times.

One signal defined by a birth–death random process is illustrated in Figure 13.2.

### 13.2.6 Characterizing Birth–Death Random Processes

The characterization of birth–death random processes is not usually in terms of the time characteristics of the random process, such as the mean, variance, and autocorrelation, but in terms of the evolution of the probability of a state with time as illustrated in Figure 13.3. When such an evolution is known for all states, the probability mass function of the state level, at a specified time, can be specified.

In general, a birth–death random process is non-Markovian as the probability of transition from one state to the next depends on the prior state history and not only on the current state.

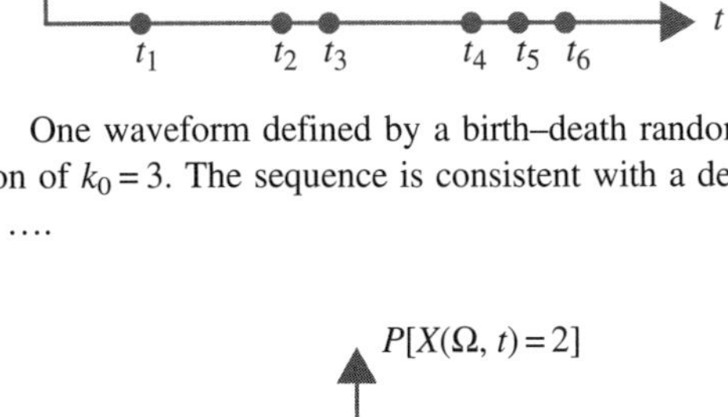


**FIGURE 13.2** One waveform defined by a birth–death random process for the case of an initial population of $k_0=3$. The sequence is consistent with a death, a birth, a death, a birth, a birth, a birth, ….

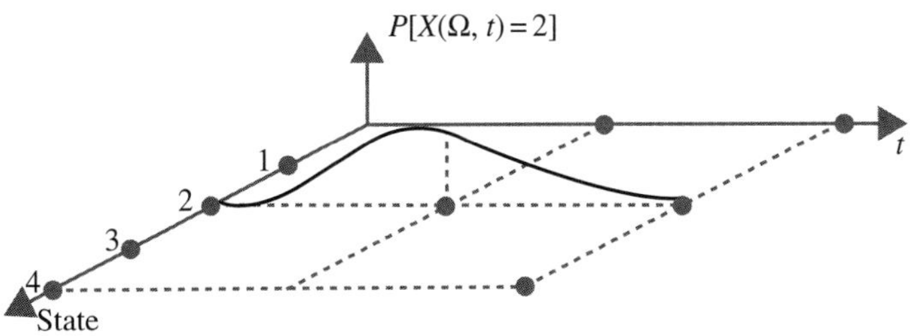


**FIGURE 13.3** Illustration of the evolution of probability of a state (state 2) with time for a birth–death random process.

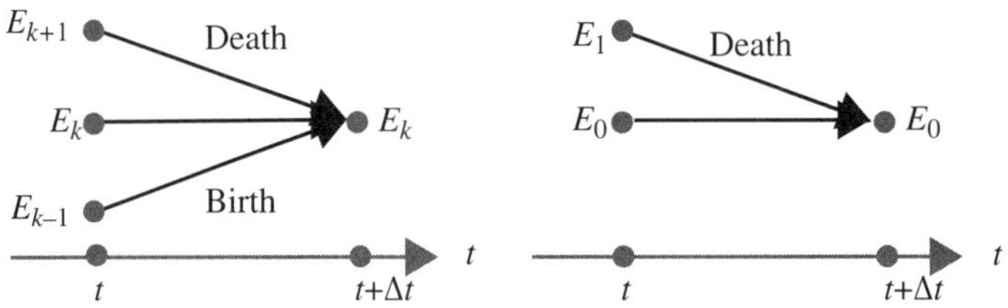


**FIGURE 13.4** Three possible transitions leading to the $k$th state. Left: case for $k>0$. Right: case for $k=0$.

### 13.2.7 Population Dynamics

Consider a birth–death random process that is in the state $E_k$ at time $t+\Delta t$. This outcome, as illustrated in Figure 13.4, is a result of three possible transitions assuming $\Delta t$ is sufficiently small such that multiple births or deaths occur with negligible probability.

It follows, for sufficiently small $\Delta t$, that being in state $E_k$ at time $t+\Delta t$ is consistent with one of the following:

1. At time $t$, the population is $k$ and there are no births or deaths in the interval $(t, t+\Delta t)$.
2. At time $t$, the population is $k-1$ and there is one birth in the interval $(t, t+\Delta t)$.
3. At time $t$, the population is $k+1$ and there is one death in the interval $(t, t+\Delta t)$.

These outcomes underpin the following theorem.

**Theorem 13.1 Differential Equation Defining State Probability**
The following sequence of differential equations defines the change in the state probability of a birth–death random process with time:

$$\frac{d}{dt}P[X(\Omega,t)=0] = -\lambda_0 P[X(\Omega,t)=0] + \mu_1 P[X(\Omega,t)=1] \quad k=0 \tag{13.10}$$

$$\begin{aligned}\frac{d}{dt}P[X(\Omega,t)=k] = -(\lambda_k+\mu_k)P[X(\Omega,t)=k] + \lambda_{k-1}P[X(\Omega,t)=k-1] \\ +\mu_{k+1}P[X(\Omega,t)=k+1] \quad k\geq 1\end{aligned} \tag{13.11}$$

**Proof**
The proof is detailed in Appendix 13.A.

***13.2.7.1 Notes*** The solution of this sequence of differential equations requires the initial conditions

$$P[X(\Omega,0)=k] \tag{13.12}$$

to be specified for all $k \in \{0,1, \ldots\}$. The solution must be subject to the constraint, for all fixed values of $t \geq 0$, of the sum of the state probabilities equalling unity, that is,

$$\sum_{k=0}^{\infty} P[X(\Omega,t) = k] = 1 \tag{13.13}$$

In general, solving the sequence of equations is analytically difficult. Solutions, however, can be obtained for simple cases.

It may not be possible to find analytical solutions to the state probability of a complex birth–death random process. However, it may be possible to find solutions to the steady-state probabilities, if these exist.

**Definition: Steady State—Statistical Equilibrium** If $\lim_{t\to\infty} P[X(\Omega,t) = k]$ exists, for all possible states $k$, then the system is in statistical equilibrium, or in steady state, as $t \to \infty$.

***13.2.7.2 Conservation of Probability*** An alternative approach to establishing the differential equation specifying the probability evolution of a set state is to consider the transitions, and associated probabilities, illustrated in Figure 13.5. By considering the possible outcomes at time $t$, it follows that the probability of being in state $E_{k+1}$ at time $t$, and a death occurring in the interval $(t, t+\Delta t)$, for $\Delta t$ sufficiently small, approximately is

$$P[X(\Omega,t) = k+1]\mu_{k+1}\Delta t \tag{13.14}$$

Similarly for other probabilities. It then follows that the probability of a change in the interval $(t, t+\Delta t)$ to the state $E_k$ is

$$\begin{aligned} P[\text{change to state } E_k \text{ in } (t,t+\Delta t)] \approx\ & P[X(\Omega,t) = k+1]\mu_{k+1}\Delta t \\ & + P[X(\Omega,t) = k-1]\lambda_{k-1}\Delta t \end{aligned} \tag{13.15}$$

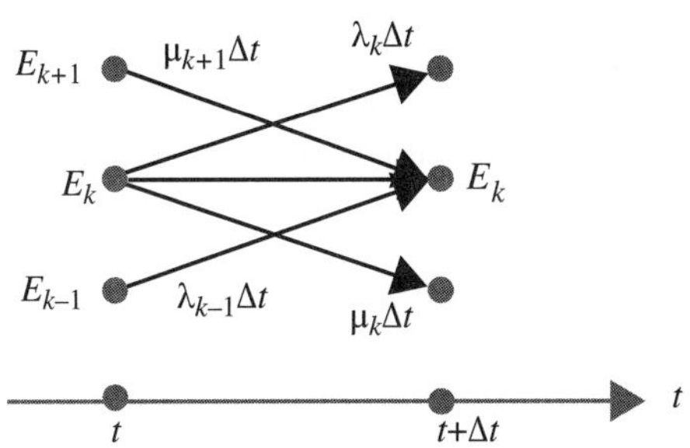


**FIGURE 13.5** Transitions, and associated probabilities, between states for the interval $(t, t + \Delta t)$.

and the probability of a change in the interval $(t, t+\Delta t)$ from the state $E_k$ is

$$P[\text{change from state } E_k \text{ in } (t, t+\Delta t)] \approx P[X(\Omega,t)=k]\mu_k \Delta t + P[X(\Omega,t)=k]\lambda_k \Delta t \tag{13.16}$$

These two results imply that the increase in probability of state $E_k$ in the interval $[t, t+\Delta t]$ is

$$\begin{aligned} &\text{change in probability of state } E_k \text{ in } (t, t+\Delta t) \approx \\ &P[X(\Omega,t)=k+1]\mu_{k+1}\Delta t + P[X(\Omega,t)=k-1]\lambda_{k-1}\Delta t \\ &-P[X(\Omega,t)=k]\mu_k \Delta t - P[X(\Omega,t)=k]\lambda_k \Delta t \end{aligned} \tag{13.17}$$

Therefore and consistent with Theorem 13.1,

$$\frac{P[X(\Omega,t+\Delta t)=k]-P[X(\Omega,t)=k]}{\Delta t} \approx P[X(\Omega,t)=k+1]\mu_{k+1} + P[X(\Omega,t)=k-1]\lambda_{k-1} - P[X(\Omega,t)=k]\mu_k - P[X(\Omega,t)=k]\lambda_k \tag{13.18}$$

## 13.3 CONSTANT BIRTH RATE, ZERO DEATH RATE PROCESS

Consider a constant birth rate, zero death rate, birth–death random process $X: S \to S_X$ with zero population at $t=0$. Such an random process is consistent with a Poisson counting process, that is, $X$ is an random process based on a Poisson experiment according to

$$\begin{aligned} &X: S \to S_X \\ &S = \{(t_1, \ldots, t_i, \ldots): t_i \in \boldsymbol{R}^+, t_1 \leq \ldots \leq t_i \leq \ldots\} \\ &X((t_1, \ldots, t_i, \ldots), t) = \sum_{i=1}^{\infty} u(t-t_i) \end{aligned} \tag{13.19}$$

At a time $t_o$, the level of one outcome of $X$ equals the number of points arising from the associated Poisson experiment that are in the interval $[0, t_o]$. One outcome of $X$ is illustrated in Figure 13.6.

The state probability, as a function of time, for a birth–death random process with a constant birth rate and zero death rate is specified in the following theorem for the case of zero population at $t=0$.

**Theorem 13.2 State Probability: Constant Birth, Zero Death Process**
For the case of $\lambda_k = \lambda, k \in \boldsymbol{Z}^+$, $\mu_k = 0, k \in \boldsymbol{Z}^+$, and $X(\Omega, 0) = 0$, it follows that

$$\begin{aligned} &\frac{d}{dt}P[X(\Omega,t)=0] = -\lambda P[X(\Omega,t)=0] \quad k=0 \\ &\frac{d}{dt}P[X(\Omega,t)=k] = -\lambda P[X(\Omega,t)=k] + \lambda P[X(\Omega,t)=k-1] \quad k \geq 1 \end{aligned} \tag{13.20}$$

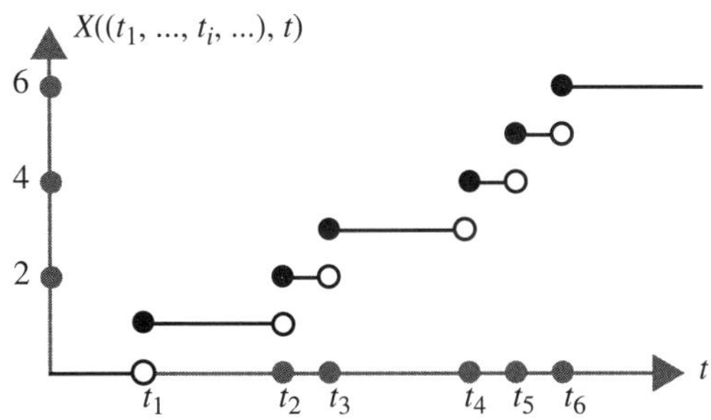


**FIGURE 13.6** One waveform defined by a constant birth rate, zero death rate, birth-death random process with zero population at $t = 0$.

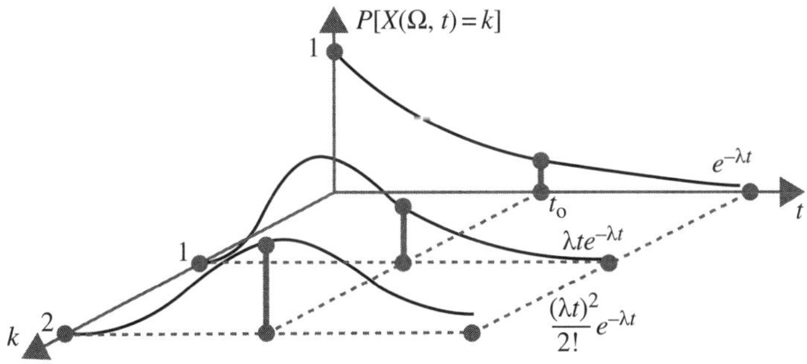


**FIGURE 13.7** Evolution of the probability of a state for a birth–death random process with a constant birth rate and a zero death rate. The population is assumed to be zero at $t = 0$.

The solution to this system of differential equations is

$$
\begin{aligned}
&P[X(\Omega,t)=0]=e^{-\lambda t} \qquad t\geq 0\\
&P[X(\Omega,t)=k]=\frac{(\lambda t)^k e^{-\lambda t}}{k!} \qquad t\geq 0, k\in \mathbf{Z}^+
\end{aligned}
\tag{13.21}
$$

**Proof**
The proof is detailed in Appendix 13.B.

### 13.3.1 Notes

The evolution of the state probabilities with time is illustrated in Figure 13.7.

***13.3.1.1 Link to Poisson Point and Poisson Counting Processes*** A birth–death random process with a constant birth rate λ, zero death rate, and zero initial condition, as illustrated in Figure 13.6, is consistent with a Poisson counting random process with constant rate $\lambda$. The state transition times are consistent with a Poisson point process with rate $\lambda$, and accordingly, it is the case that

$$\text{arrival of } k\text{th Poisson point} = \text{transition of state from } k-1 \text{ to } k \tag{13.22}$$

The following definition then is appropriate.

**Definition: Random Variables Defined by Time *k*th State Is Entered** Define the random variable at which the $k_o$th state is first entered as $X_{(k_o)}$. Denote the probability density function of this random variable as $f_{(k_o)}$.

**Theorem 13.3 PDF of Time *k*th State Is Entered**
The probability density function of the random variable $X_{(k_o)}$ is

$$f_{(k_o)}(t) = \frac{\lambda(\lambda t)^{k_o-1}e^{-\lambda t}}{(k_o-1)!} \qquad t>0,\ k_o \in \{1,2,\ldots\} \tag{13.23}$$

**Proof**
The $k_o$th state is entered at the time of the $k_o$th Poisson point. The probability density function of such a random variable is specified in Equation 12.41.

***13.3.1.2 Fixed Time: Poisson Random Variable*** For a fixed time $t_o$, the probability $P[X(\Omega,t_o)=k]$ is consistent with $k$ Poisson points and $k$ births in the interval $[0, t_o]$. As stated in Theorem 13.2, this probability is

$$P[X(\Omega,t)=k] = \frac{(\lambda t)^k e^{-\lambda t}}{k!} \qquad t>0,\ k\in \mathbf{Z}^+ \tag{13.24}$$

and thus, for a fixed time $t_o$, $P[X(\Omega,t_o)=k]$ is a Poisson random variable with parameter $\lambda t_o$.

The following theorem states the intuitive result: with an average birth rate of $\lambda$ per second, it is the case that the population at a time $t_o$ is expected to be $\lambda t_o$.

**Theorem 13.4 Mean and Variance of Population**
The expected, or mean, population at a time $t_o$ is

$$E[X(\Omega,t_o)] = \lambda t_o \tag{13.25}$$

The variance of the population at time $t_o$ is

$$\text{var}[X(\Omega,t_o)] = \lambda t_o \tag{13.26}$$

**Proof**
These results follow from the mean and variance of a Poisson random variable with parameter $\lambda t_o$.

It also follows, as required, that the sum of the state probabilities for a fixed time $t_o$ (probabilities as illustrated in Fig. 13.7) sums to one, that is,

$$\sum_{k=0}^{\infty} P[X(\Omega,t_o)=k]=1 \Rightarrow \sum_{k=0}^{\infty}\frac{(\lambda t_o)^k e^{-\lambda t_o}}{k!}=1 \tag{13.27}$$

The following is a direct proof of this result:

$$\sum_{k=0}^{\infty}\frac{(\lambda t_o)^k e^{-\lambda t_o}}{k!}=e^{-\lambda t_o}\sum_{k=0}^{\infty}\frac{(\lambda t_o)^k}{k!}=e^{-\lambda t_o}\left(e^{\lambda t_o}\right)=1 \tag{13.28}$$

***13.3.1.3 Fixed Population Level*** Consider the probability of the population being at a fixed level $k_o$ as specified in Theorem 13.2, that is,

$$P[X(\Omega,t)=k_o]=\frac{(\lambda t)^{k_o} e^{-\lambda t}}{k_o!} \qquad t>0,\ k_o \in \mathbf{Z}^+ \tag{13.29}$$

which is illustrated in Figure 13.8.

If the probability function specified by Equation 13.29 is normalized such that the area it defines is unity, then a probability density function is defined, and implicitly, a random variable is defined. The following theorem states the required normalization and the random variable defined by the resulting probability density function:

**Theorem 13.5 Relationship: Probability of State and State Transition Time** It is the case that

$$\int_0^{\infty} P[X(\Omega,t)=k_o]dt=\frac{1}{\lambda} \tag{13.30}$$

and the relationship between the probability of being in a set state $k_o$ at a time $t$ is directly proportional to the probability density function of the time for the transition from the $k_o$th to the $(k_o+1)$th state. The relationship is

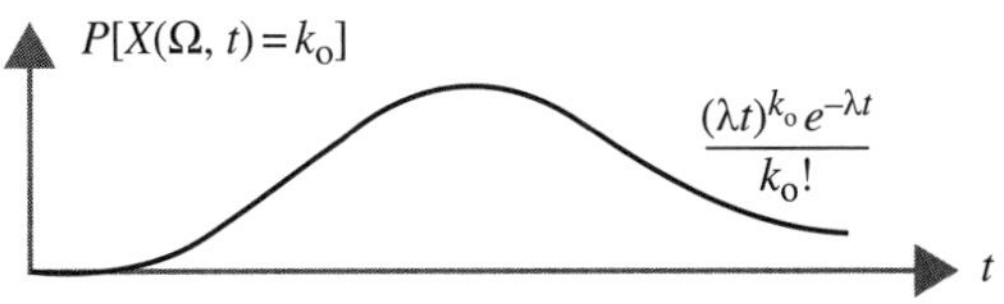


**FIGURE 13.8** Illustration of the variation with time of the probability of the $k_o$th state for a birth–death random process with a constant birth rate, a zero death rate, and zero population at $t=0$.

$$f_{(k_o+1)}(t) = \lambda P[X(\Omega,t)=k_o] = \frac{\lambda(\lambda t)^{k_o} e^{-\lambda t}}{k_o!} \qquad t>0,\, k_o \in \mathbf{Z}^+ \tag{13.31}$$

**Proof**

The proof of Equation 13.31 follows directly from Equations 13.23 and 13.24. A direct proof is detailed in Appendix 13.C.

***13.3.1.4 Mean Time for Increase in Population Level*** The mean time for the population to increase, consistent with moving from one state to another, is specified in the following theorem:

**Theorem 13.6 Mean Time for Increase in Population from a Set Level**

The mean time for the population to increase from the level of $k_o$ to $k_o+1$ is

$$\text{mean time for population to increase from } k_o \text{ to } k_o+1 = \frac{k_o+1}{\lambda} \tag{13.32}$$

The mean time for the transition from the $k$th state to the $(k_o+1)$th state, coinciding with the $(k_o+1)$th birth, is illustrated in Figure 13.9.

**Proof**

The population changes from $k_o$ to $k_o+1$ at the time of the $(k_o+1)$th Poisson point, which is specified by the random variable $X_{(k_o+1)}$. It then follows, from Theorem 13.3, that

$$\text{mean time for population to increasing from } k_o \text{ to } k_o+1 = \boldsymbol{E}\left[X_{(k_o+1)}\right]$$
$$= \int_0^\infty \frac{t\lambda(\lambda t)^{k_o} e^{-\lambda t}}{k_o!} dt = (k_o+1)\int_0^\infty \frac{(\lambda t)^{k_o+1} e^{-\lambda t}}{(k_o+1)!} dt = \frac{k_o+1}{\lambda} \tag{13.33}$$

This result follows from repetitious use of integration by parts or by using the definition and properties of the Gamma function (Eq. 2.19).

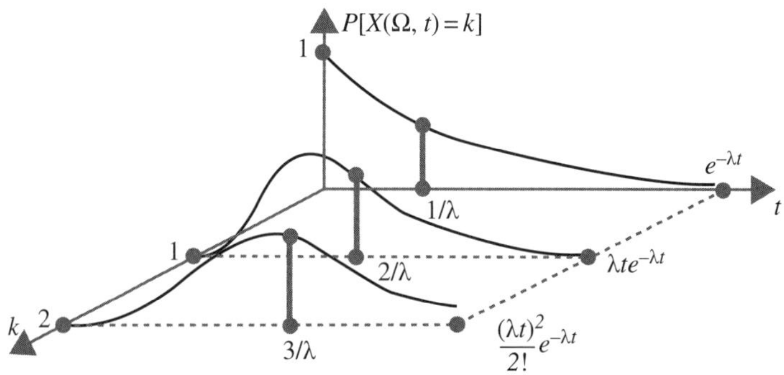


**FIGURE 13.9** Illustration of the mean time for the population to increase from $k_o$ to $k_o+1$ for a birth–death random process with a constant birth rate and a zero death rate. The population is assumed to be zero at $t=0$.

***13.3.1.5 Times between Births*** With a pure birth random process, the birth times are consistent with a Poisson point random process, and accordingly, the times between births are consistent with the interarrival times of a Poisson point random process.

**Theorem 13.7 Time between Births**

The times between births are independent and consistent with an exponential random variable, that is,

$$P[\text{time between births} < \tau] = 1 - e^{-\lambda\tau} \tag{13.34}$$

The probability density function of the time between births is

$$f_{\Delta}(\tau) = \lambda e^{-\lambda\tau} \quad \tau > 0 \tag{13.35}$$

and the mean time between births and the variance of this time, respectively, are

$$\text{mean time between births} = \frac{1}{\lambda} \tag{13.36}$$

$$\text{variance of times between births} = \frac{1}{\lambda^2} \tag{13.37}$$

**Proof**

This result follows from the results stated in Theorems 12.15 and 12.2.

***13.3.1.6 Example*** Consider the random arrival of electrons in a shot noise process, which commences at $t = 0$. The notation

$$P[n,t] = P[n \text{ eletrons have arrived by time } t] \tag{13.38}$$

is convenient. For an arrival rate of $\lambda$ and

$$P[\text{electron arrival in } \Delta t \text{ seconds}] \approx \lambda\Delta t \quad \Delta t << \frac{1}{\lambda} \tag{13.39}$$

the transitions with significant probability, for the interval $[t, t+\Delta t]$ and resulting in a population of $n$ at $t+\Delta t$, are illustrated in Figure 13.10. It then follows that

$$P[n, t+\Delta t] \approx (1-\lambda\Delta t)P[n,t] + \lambda\Delta t P[n-1,t] \tag{13.40}$$

which leads to the differential equation

$$\frac{d}{dt}P(n,t) = \lambda[P(n-1,t) - P(n,t)] \quad P(0,0) = 1 \tag{13.41}$$

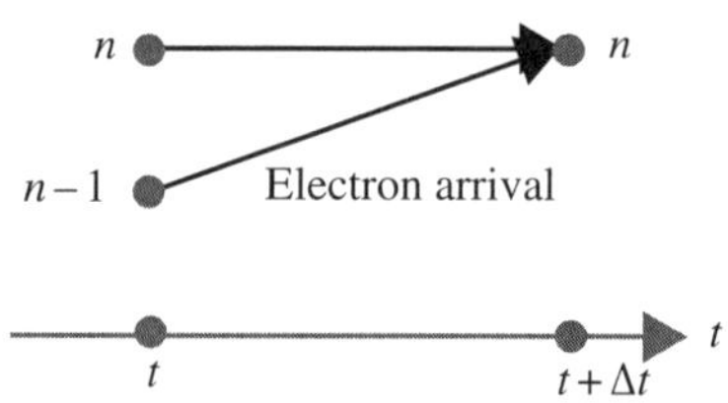


**FIGURE 13.10** Transitions for the interval $[t, t + \Delta t]$ leading to a population of $n$ at $t + \Delta t$. The assumption is $\Delta t \ll 1/\lambda$.

which is the same differential equation for a birth-only random process as specified in Theorem 13.2. It then follows from Theorem 13.2 that

$$P(n,t) = \frac{(\lambda t)^n e^{-\lambda t}}{n!} \qquad t \geq 0, n \in \mathbf{Z}^+ \tag{13.42}$$

## 13.4 STATE DEPENDENT BIRTH RATE - ZERO DEATH RATE

Consider a birth–death random process with a state-dependent birth rate and zero death rate. Such an random process is based on the following algorithmic experiment:

1. Set the state parameter $k$ to zero. Set the time parameter $t_0$ to zero.
2. Define $t_1$ as the first time of a Poisson random process commencing at $t = 0$ and with rate $\lambda_0$. Increment the state parameter, that is, $k = k + 1$.
3. Consider a Poisson random processes commencing at $t = t_1$ with rate $\lambda_1$. Define $t_2$ as the first time of this random process. Increment the state parameter, that is, $k = k + 1$.
4. Iterate.

The outcome of one trial of the experiment is the set of ordered times:

$$\{t_1, \ldots, t_i, \ldots\} \tag{13.43}$$

and at the $i$th time $t_i$, the state parameter $k$ changes from $i-1$ to $i$. The sample space associated with the experiment is

$$S = \{\omega : \omega = (t_1, \ldots, t_i, \ldots), t_i \text{ as defined by algorithm}\} \tag{13.44}$$

The birth–death random process $X : S \rightarrow S_X$ with a state-dependent birth rate and zero death rate is defined according to

$$\begin{aligned} &X : S \rightarrow S_X \\ &X((t_1, \ldots, t_i, \ldots), t) = \sum_{i=1}^{\infty} u(t - t_i) \\ &S_X = \{X((t_1, \ldots, t_i, \ldots), t) : \ (t_1, \ldots, t_i, \ldots) \in S\} \end{aligned} \tag{13.45}$$

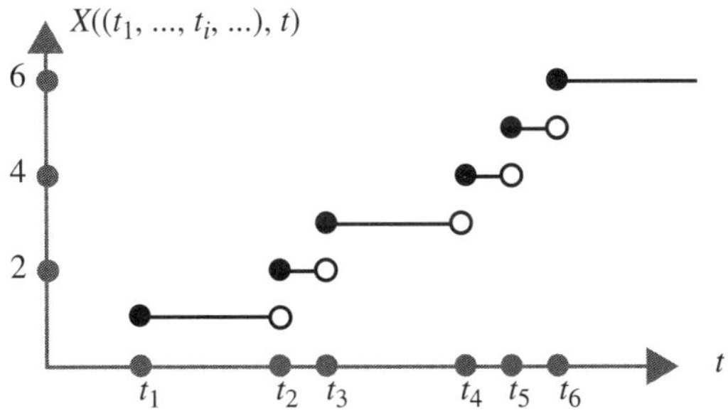


**FIGURE 13.11** One signal defined by a variable birth rate, zero death rate, birth–death random process.

At a time $t_o$, the level of one outcome of this random process equals the number of outcomes, as defined by the underlying experiment, that have occurred in the interval $[0, t_o]$. One signal defined by this process is illustrated in Figure 13.11.

The variation of the state probability with time for a birth–death random process with a variable birth rate and zero death rate is specified in the following theorem for the case of zero population at $t=0$.

**Theorem 13.8 State Probability: Variable Birth Rate, Zero Death Rate**
For the case of $\lambda_k > 0, k \in \mathbf{Z}^+$, $\mu_k = 0, k \in \mathbf{Z}^+$, and $X(\Omega,0)=0$, it follows that

$$\begin{aligned} &\frac{d}{dt}P[X(\Omega,t)=0] = -\lambda_0 P[X(\Omega,t)=0] \quad k=0 \\ &\frac{d}{dt}P[X(\Omega,t)=k] = -\lambda_k P[X(\Omega,t)=k] + \lambda_{k-1}P[X(\Omega,t)=k-1] \quad k \geq 1 \end{aligned} \tag{13.46}$$

The solution to this system of differential equations is

$$\begin{aligned} &P[X(\Omega,t)=0] = e^{-\lambda_0 t} \quad t>0 \\ &P[X(\Omega,t)=1] = \frac{\lambda_0\left(e^{-\lambda_0 t} - e^{-\lambda_1 t}\right)}{\lambda_1 - \lambda_0} \quad t>0, \lambda_1 \neq \lambda_0 \end{aligned} \tag{13.47}$$

$$P[X(\Omega,t)=2] = \lambda_0\lambda_1\left[\frac{e^{-\lambda_0 t}}{(\lambda_1-\lambda_0)\,(\lambda_2-\lambda_0)} - \frac{e^{-\lambda_1 t}}{(\lambda_1-\lambda_0)\,(\lambda_2-\lambda_1)} + \frac{e^{-\lambda_2 t}}{(\lambda_2-\lambda_0)\,(\lambda_2-\lambda_1)}\right]$$
$$t>0, \lambda_1 \neq \lambda_0, \lambda_2 \neq \lambda_1 \tag{13.48}$$

$$P[X(\Omega,t)=k] = e^{-\lambda_k t}\left[\lambda_{k-1}\int_0^t P[X(\Omega,T)=k-1]e^{\lambda_k T}\,dT + P[X(\Omega,0)=k]\right] \tag{13.49}$$
$$t>0, \lambda_i \neq \lambda_j, k \geq 1$$

**Proof**

The proof is detailed in Appendix 13.D.

### 13.4.1 Examples

The variation, with time, of the probability of the population being a set level is shown in Figures 13.12 and 13.13 for the case of the birth rate varying with the population according to $\lambda_0 = 1$, $\lambda_k = 0.1k$ and $\lambda_0 = 1$, $\lambda_k = 0.2k$. When the birth rate is proportional to the population, the population will grow *exponentially*. This can be seen in the times at which the probability of being at a set level is a maximum.

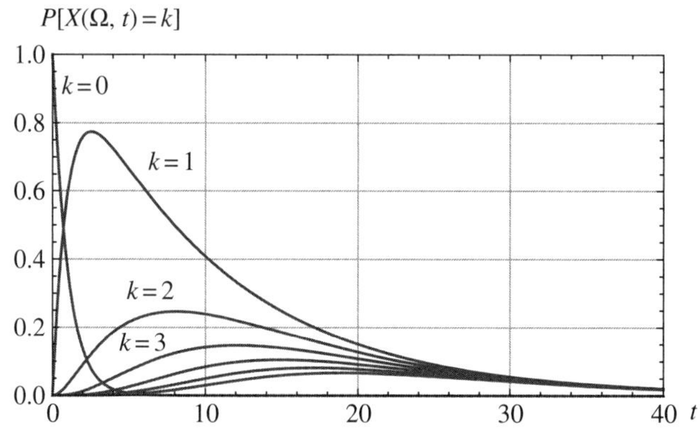


**FIGURE 13.12** Variation of the probability, with time, for a set population level $k$ for the case of $\lambda_0 = 1$, $\lambda_k = 0.1k$.

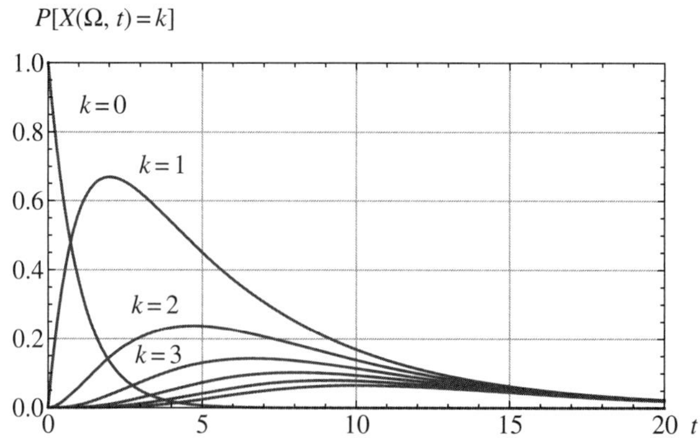


**FIGURE 13.13** Variation of the probability, with time, of a set population level $k$ for the case of $\lambda_0 = 1$, $\lambda_k = 0.2k$.

## 13.5 CONSTANT DEATH RATE, ZERO BIRTH RATE, PROCESS

Consider a birth–death random process with a constant death rate, zero birth rate, and an initial population of $N$ at $t=0$. Such an random process, based on the first $N$ outcomes of a Poisson experiment with a sample space $S$, is defined according to

$$X: S \to S_X$$
$$X((t_1,\dots,t_N),t) = N - \sum_{i=1}^{N} u(t-t_i) \tag{13.50}$$
$$S_X = \{X((t_1,\dots,t_N),t): \ (t_1,\dots,t_N) \in S\}$$

At a time $t_o$, the level of one outcome of this random process equals $N$ minus the number of outcomes of the underlying Poisson experiment that have occurred in the interval $[0, t_o]$. One signal from this process is illustrated in Figure 13.14.

The state probability as a function of time for a birth–death random process with a constant death rate and zero birth rate is specified in the following theorem for the case of a population of $N$ at $t=0$.

**Theorem 13.9 State Probability: Constant Death, Zero Birth Process**
For the case of $\mu_k = \mu, k \in \mathbf{Z}^+$, $\lambda_k = 0, k \in \mathbf{Z}^+$, and $X(\Omega,0) = N$, it follows that

$$\frac{d}{dt}P[X(\Omega,t)=N] = -\mu P[X(\Omega,t)=N] \quad k=N$$
$$\frac{d}{dt}P[X(\Omega,t)=k] = -\mu P[X(\Omega,t)=k] + \mu P[X(\Omega,t)=k+1] \quad 1 \le k < N \tag{13.51}$$
$$\frac{d}{dt}P[X(\Omega,t)=0] = \mu P[X(\Omega,t)=1] \quad k=0$$

The solution to this system of differential equations is

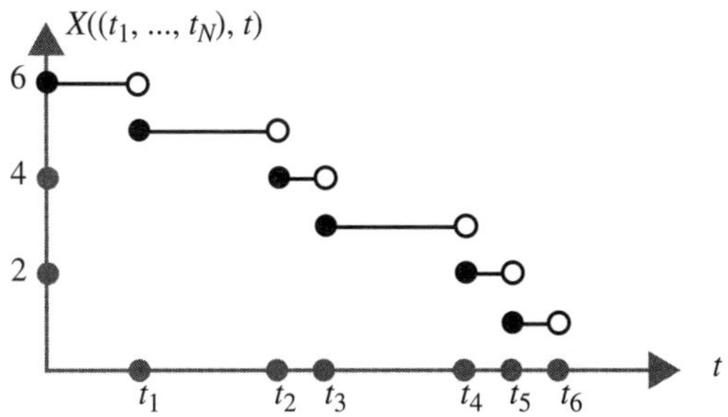


**FIGURE 13.14** One signal defined by a birth–death random process with a constant death rate, zero birth rate, and an initial population of $N=6$.

$$P[X(\Omega,t)=k]=\frac{(\mu t)^{N-k}e^{-\mu t}}{(N-k)!} \quad t>0, 1\le k\le N$$
$$P[X(\Omega,t)=0]=1-e^{-\mu t}\sum_{q=1}^{N}\frac{\mu^{q-1}t^{q-1}}{(q-1)!} \quad t>0, k=0 \tag{13.52}$$

**Proof**
The proof is detailed in Appendix 13.E.

### 13.5.1 Nature of State Probability Evolution

The evolution with time of the state probabilities is illustrated in Figure 13.15.

### 13.5.2 Example

The evolution with time of the state probabilities for a pure death random process with a death rate of unity is shown in Figure 13.16.

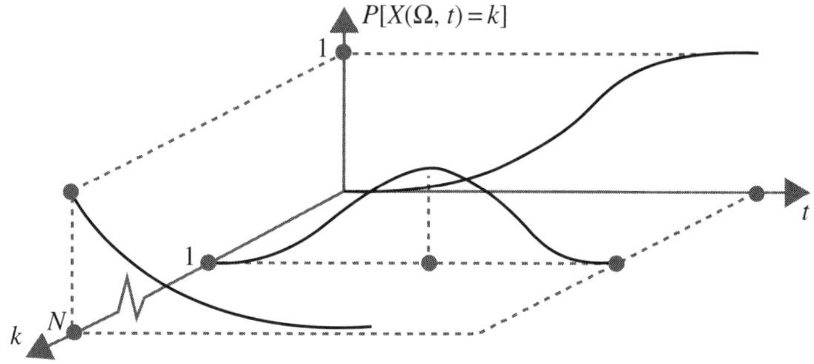


**FIGURE 13.15** Evolution with time of the state probability for a birth–death random process with a constant death rate, a zero birth rate, and an initial population of $N$.

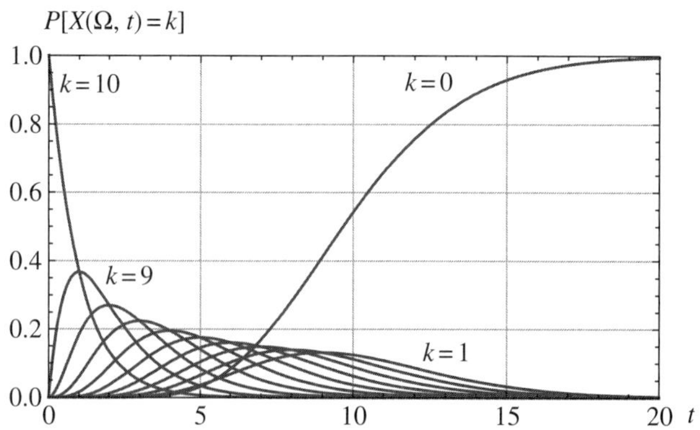


**FIGURE 13.16** Variation with time of the probability of a set population level $k$ for the case of $\mu = 1$ and an initial population of 10.

## 13.6 CONSTANT BIRTH AND CONSTANT DEATH RATE PROCESS

Consider a constant birth rate, constant death rate, birth–death random process $X:S\to S_X$ with an initial population of $N_0$ at $t=0$. Such an random process is based on the following algorithmic experiment:

1. Set the state parameter $k$ to $N_0$. Set the time parameter $t_0$ to zero.
2. Consider two Poisson random processes commencing at $t=t_i$: a birth process with rate $\lambda$ and a death process with rate $\mu$, except when $k=0$, whereupon $\mu=0$. Take the earliest time arising from these two random processes and denote this time $t_{i+1}$. If the earliest time is associated with a birth, then increment the state parameter, that is, $k=k+1$; if the earliest time is associated with a death, then decrement the state parameter, that is, $k=k-1$.
3. Iterate.

The outcome of one trial of the experiment is the set of ordered pairs:

$$\{(t_1,k_1),\ldots,(t_i,k_i),\ldots\} \tag{13.53}$$

and the sample space associated with the experiment is

$$S=\left\{\begin{array}{c}\omega:\omega=((t_1,k_1),\ldots,(t_i,k_i),\ldots)\\ t_i,k_i \text{ as defined by algorithm}\end{array}\right\} \tag{13.54}$$

The constant birth rate, constant death rate, birth–death random process is defined according to

$$\begin{aligned}&X:S\to S_X\\ &X(((t_1,k_1),\ldots,(t_i,k_i),\ldots),t)=N_0u(t)+\sum_{i=1}^{\infty}(k_i-k_{i-1})u[t-t_i]\\ &S_X=\left\{\begin{array}{c}X(((t_1,k_1),\ldots,(t_i,k_i),\ldots),t):\\ ((t_1,k_1),\ldots,(t_i,k_i),\ldots)\in S\end{array}\right\}\qquad k_0=N_0\end{aligned} \tag{13.55}$$

The evolution of the state probability with time is specified in the following theorem.

**Theorem 13.10 State Probability: Constant Birth and Death Rate Process** For the case of $\lambda_k=\lambda, k\in\{0,1,\ldots\}$, $\mu_k=\mu, k\in\{1,\ldots\}$, $\mu_0=0$, and $X(\Omega,0)=N_0$, it follows that

$$\begin{aligned}&\frac{d}{dt}P[X(\Omega,t)=0]=-\lambda P[X(\Omega,t)=0]+\mu P[X(\Omega,t)=1]\qquad k=0\\ &\frac{d}{dt}P[X(\Omega,t)=k]=-(\lambda+\mu)P[X(\Omega,t)=k]+\lambda P[X(\Omega,t)=k-1]\\ &\qquad\qquad +\mu P[X(\Omega,t)=k+1]\qquad k\geq 1\end{aligned} \tag{13.56}$$

The solution to this system of differential equations is

$$P[X(\Omega,t)=k]=e^{-(\lambda+\mu)t}\left[\rho^{(k-N_0)/2}I_{k-N_0}(at)+\rho^{(k-N_0-1)/2}I_{k+N_0+1}(at)\right.$$
$$\left.+(1-\rho)\rho^{k}\sum_{j=k+N_0+2}^{\infty}\rho^{-j/2}I_j(at)\right]\quad t>0,k\geq 0 \tag{13.57}$$

where $\rho=\lambda/\mu$, $a=2\mu\rho^{1/2}=2\sqrt{\mu\lambda}$, and $I_i$ is the modified Bessel function of the first kind and order $i$, that is,

$$I_i(x)=\sum_{m=0}^{\infty}\frac{(x/2)^{i+2m}}{(i+m)!m!}\qquad i\geq -1 \tag{13.58}$$

**Proof**
See Kleinrock (1975, pp. 74–78).

### 13.6.1 Example

The evolution of the state probabilities with time for a birth–death random process with an initial zero population and with nonzero birth and death rates is shown in Figure 13.17.

### 13.6.2 Application: M/M/1 Queueing System

Consider the *M*/*M*/1 (exponential interarrival time statistics, exponential service time statistics, single server) queueing system illustrated in Figure 13.18.

For such a system, the queue population is a birth–death random process with a constant birth rate and constant death (service) rate. The modelling of the *M*/*M*/1 queueing system, consistent with a birth–death random process, is as follows:

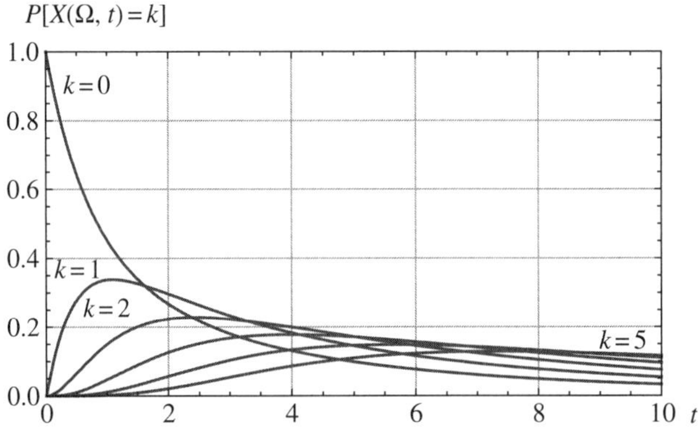


**FIGURE 13.17** Variation with time of the probability of a set population level $k$ for a birth–death random process with an initial zero population, a birth rate of $\lambda = 1$, and a death rate of $\mu = 0.5$.

**FIGURE 13.18** Illustration of the *M/M/*1 queueing system.

consider an initial queue population at $t=0$ of $N_0$. Consider arrivals at the queue in a manner consistent with a Poisson point random process with a rate $\lambda$. Consider entities in the queue being serviced, provided the queue population is at least one, in a manner consistent with a Poisson point random process with a rate $\mu$. The queue population defines a random process $X: S \to S_X$ as defined by Equation 13.55 and with the probability of the queue population varying with time as specified in Theorem 13.10.

The *M/M/*1 queueing system is one of the simplest possible queueing systems, and more complicated queueing systems are generally compared with the performance of such a system. Useful references for queueing theory include Allen (1978), Kleinrock (1975), and Gross and Harris (1998).

## 13.7 PROBLEMS

**13.1** Consider a birth–death random process with a constant birth rate $\lambda$, zero death rate, and a population at $t=0$ of $k_0$. Consider the case of $\lambda=1$ and $k_0=4$.

a. Graph several signals defined by the random process.
b. Specify a general expression for the variation of the state probabilities with time. Graph the evolution with time of the state probabilities.
c. Specify the mean time for the population to increase from 4 to 5 and from 5 to 6.

**13.2** Consider a birth–death random process with an initial population of $k_0=4$, zero death rate, and a birth rate that varies with population according to $\lambda_k = \exp[-(k-k_0)/4]$.

a. Graph several signals from this random process.
b. Graph the probability evolution, with time, of the population levels.

**13.3** Consider a birth–death random process with an initial population of $k_0=10$, zero birth rate, and a death rate of $\mu_k = 1/2$.

a. Graph several signals from this random process.
b. Graph the probability evolution, with time, of the population levels.

**13.4** Consider a birth–death random process with an initial population of $N_0=10$, a constant birth rate of $\lambda=1$, and a death rate of $\mu=0.9$.

a. Graph several signals from this random process.
b. Graph the probability evolution, with time, of the population levels.

**13.5** Consider a birth–death random process where the birth and death rates vary cyclically according to

$$\begin{aligned}\lambda_k &= 0.5\left[1+\sin\left(2\pi f_\lambda k\right)\right]\\ \mu_k &= 0.5\left[1-\sin\left(2\pi f_\mu k\right)\right]\end{aligned} \tag{13.59}$$

Graph signals from such a birth–death random process for the case of $f_\lambda = f_\mu = 0.1$ and an initial population of $k_0 = 10$ at $t = 0$. Specify, and justify, the two stable population levels.

## APPENDIX 13.A PROOF OF THEOREM 13.1

First, consider the case of $k \geq 1$. It follows from Figure 13.4 that the event of $X(\Omega, t+\Delta t) = k$ can be written as

$$\begin{aligned}\{X(\Omega,t+\Delta t)=k\} = \; &\{X(\Omega,t)=k, \text{no births and no deaths in } (t,t+\Delta t)\}\cup\\ &\{X(\Omega,t)=k+1, \text{a death in } (t,t+\Delta t)\}\cup\\ &\{X(\Omega,t)=k-1, \text{a birth in } (t,t+\Delta t)\}\cup\\ &\cdots\end{aligned} \tag{13.60}$$

As these are disjoint events, it follows that

$$\begin{aligned}&P[X(\Omega,t+\Delta t)=k]=\\ &\quad P[\text{no births, no deaths in } (t,t+\Delta t)/X(\Omega,t)=k]\cdot P[X(\Omega,t)=k]\\ &\quad + P[\text{a death in } (t,t+\Delta t)/X(\Omega,t)=k+1]\cdot P[X(\Omega,t)=k+1]\\ &\quad + P[\text{a birth in } (t,t+\Delta t)/X(\Omega,t)=k-1]\cdot P[X(\Omega,t)=k-1]+\cdots\end{aligned} \tag{13.61}$$

With the interval $\Delta t$ chosen such that the probability of multiple births, multiple deaths, or a birth and a death in $(t, t+\Delta t)$ is small, as well as the result for independent events $A$ and $B$ of $P[AB/C] = P[A/C]P[B/C]$, it follows from Equation 13.2 that

$$\begin{aligned}P[X(\Omega,t+\Delta t)=k] = P[X(\Omega,t)=k]\;&(1-\lambda_k\Delta t+o(\Delta t))\;(1-\mu_k\Delta t+o(\Delta t))\\ &+P[X(\Omega,t)=k+1]\;(\mu_{k+1}\Delta t+o(\Delta t))\\ &+P[X(\Omega,t)=k-1]\;(\lambda_{k-1}\Delta t+o(\Delta t))+\cdots\end{aligned} \tag{13.62}$$

Hence,

$$\begin{aligned}&P[X(\Omega,t+\Delta t)=k]\approx P[X(\Omega,t)=k]+\Delta t P[X(\Omega,t)=k]\;(-\lambda_k-\mu_k)\\ &\quad+\Delta t P[X(\Omega,t)=k+1]\mu_{k+1}+\Delta t P[X(\Omega,t)=k-1]\lambda_{k-1}\end{aligned} \tag{13.63}$$

and

$$\begin{aligned}&\frac{P[X(\Omega,t+\Delta t)=k]-P[X(\Omega,t)=k]}{\Delta t}\approx P[X(\Omega,t)=k]\;(-\lambda_k-\mu_k)\\ &\quad+P[X(\Omega,t)=k+1]\mu_{k+1}+P[X(\Omega,t)=k-1]\lambda_{k-1}\end{aligned} \tag{13.64}$$

As $\Delta t \to 0$, it follows that

$$\frac{d}{dt}P[X(\Omega,t)=k] = -(\lambda_k+\mu_k)P[X(\Omega,t)=k]+\mu_{k+1}P[X(\Omega,t)=k+1] \\ +\lambda_{k-1}P[X(\Omega,t)=k-1] \tag{13.65}$$

For the case of $k=0$, the death rate $\mu_0$ is zero and this equation becomes

$$\frac{d}{dt}P[X(\Omega,t)=0] = -\lambda_0 P[X(\Omega,t)=0]+\mu_1 P[X(\Omega,t)=1] \tag{13.66}$$

which completes the proof.

## APPENDIX 13.B PROOF OF THEOREM 13.2

First, from Theorem 13.1, the differential equations governing the change in probability of a given state with time, for the case of $\lambda_k=\lambda, k\in\mathbf{Z}^+$ and $\mu_k=0, k\in\mathbf{Z}^+$, become

$$\begin{aligned} &\frac{d}{dt}P[X(\Omega,t)=0] = -\lambda P[X(\Omega,t)=0] \qquad k=0 \\ &\frac{d}{dt}P[X(\Omega,t)=k] = -\lambda P[X(\Omega,t)=k]+\lambda P[X(\Omega,t)=k-1] \qquad k\ge 1 \end{aligned} \tag{13.67}$$

Second, consider the first differential equation with the initial condition $P[X(\Omega,0)=0]=1$, that is, zero population at $t=0$. The solution to this differential equation is

$$P[X(\Omega,t)=0]=e^{-\lambda t} \tag{13.68}$$

Next, consider the case of $k=1$. Using the result given by Equation 13.68, and the initial condition of $P[X(\Omega,0)=1]=0$ (population of zero at $t=0$), the differential equation for this case becomes

$$\frac{d}{dt}P[X(\Omega,t)=1] = -\lambda P[X(\Omega,t)=1]+\lambda e^{-\lambda t}, \ \ P[X(\Omega,0)=1]=0 \tag{13.69}$$

The solution to this differential equation is

$$P[X(\Omega,t)=1]=\lambda t e^{-\lambda t} \tag{13.70}$$

It then follows, for the case of $k=2$, that the differential equation is

$$\frac{d}{dt}P[X(\Omega,t)=2] = -\lambda P[X(\Omega,t)=2]+\lambda^2 t e^{-\lambda t}, \ \ P[X(\Omega,0)=2]=0 \tag{13.71}$$

The solution to this differential equation is

$$P[X(\Omega,t)=2]=\frac{(\lambda t)^2 e^{-\lambda t}}{2!} \tag{13.72}$$

The required result follows by induction.

## APPENDIX 13.C PROOF OF THEOREM 13.5

First, using the change of variable $q=\lambda t$, the integral of the probability can be written in the form

$$\int_0^\infty P[X(\Omega,t)=k_o]dt=\int_0^\infty \frac{(\lambda t)^{k_o} e^{-\lambda t}}{k_o!}dt=\frac{1}{\lambda k_o!}\int_0^\infty q^{k_o} e^{-q} dq=\frac{1}{\lambda} \tag{13.73}$$

The required result follows from repetitious use of integration by parts or by using the definition and properties of the Gamma function (Eq. 2.19).

Second, the change of state occurs at the times specified by the Poisson point process underpinning the birth process. The cumulative distribution function of $X_{(k_o)}$, which is the time the $k_o$th state is first entered, is

$$\begin{aligned} F_{(k_o)}(t) &= P\left[X_{(k_o)}\le t\right]=P[k_o\text{th outcome prior to, or at, time } t] \\ &= P[k_o, k_o+1, \ldots \text{outcomes in } [0,t]] \\ &= 1-P[0,1,\ldots,k_o-1, \text{ outcomes in } [0,t]] \end{aligned} \tag{13.74}$$

As the events of 0, 1, … outcomes in the interval [0, $t$] are disjoint, it follows that

$$\begin{aligned} F_{(k_o)}(t) &= 1-P[0 \text{ outcomes in } [0,t]]-\cdots-P[k_o-1 \text{ outcomes in } [0,t]] \\ &= 1-\sum_{i=0}^{k_o-1}\frac{(\lambda t)^i e^{-\lambda t}}{i!} \end{aligned} \tag{13.75}$$

Differentiating yields

$$\begin{aligned} f_{(k_o)}(t) &= -\frac{d}{dt}\left[e^{-\lambda t}+\lambda t e^{-\lambda t}+\frac{(\lambda t)^2 e^{-\lambda t}}{2}+\cdots+\frac{(\lambda t)^{k_o-2} e^{-\lambda t}}{(k_o-2)!}+\frac{(\lambda t)^{k_o-1} e^{-\lambda t}}{(k_o-1)!}\right] \\ &= -\left[-\lambda e^{-\lambda t}+\left(\lambda e^{-\lambda t}-\lambda^2 t e^{-\lambda t}\right)+\left(\lambda^2 t e^{-\lambda t}-\frac{\lambda^3 t^2 e^{-\lambda t}}{2}\right)+\cdots\right. \\ &\quad \left.+\left(\frac{\lambda^{k_o-2} t^{k_o-3} e^{-\lambda t}}{(k_o-3)!}-\frac{\lambda^{k_o-1} t^{k_o-2} e^{-\lambda t}}{(k_o-2)!}\right)+\left(\frac{\lambda^{k_o-1} t^{k_o-2} e^{-\lambda t}}{(k_o-2)!}-\frac{\lambda^{k_o} t^{k_o-1} e^{-\lambda t}}{(k_o-1)!}\right)\right] \\ &= \frac{\lambda^{k_o} t^{k_o-1} e^{-\lambda t}}{(k_o-1)!} \end{aligned} \tag{13.76}$$

as required.

## APPENDIX 13.D PROOF OF THEOREM 13.8

First, from Theorem 13.1, the differential equations governing the change in probability of a given state with time, for the case of $\mu_k = 0, k \in \mathbf{Z}^+$, become

$$\begin{aligned}\frac{d}{dt}P[X(\Omega,t)=0] &= -\lambda_0 P[X(\Omega,t)=0] \qquad k=0 \\ \frac{d}{dt}P[X(\Omega,t)=k] &= -\lambda_k P[X(\Omega,t)=k] + \lambda_{k-1}P[X(\Omega,t)=k-1] \qquad k\geq 1\end{aligned} \tag{13.77}$$

Second, consider the first differential equation with the initial condition $P[X(\Omega,0)=0]$, that is, zero population at $t=0$. The solution to this differential equation is

$$P[X(\Omega,t)=0] = e^{-\lambda_0 t} \tag{13.78}$$

Next, consider the case of $k=1$. Using the result given by Equation 13.78, and the initial condition of $P[X(\Omega,0)=1]=0$ (zero population at $t=0$), the differential equation for this case becomes

$$\frac{d}{dt}P[X(\Omega,t)=1] = -\lambda_1 P[X(\Omega,t)=1] + \lambda_0 e^{-\lambda_0 t}, \quad P[X(\Omega,0)=1]=0 \tag{13.79}$$

With $X_1(\Omega, s)$ defined as the Laplace transform of $P[X(\Omega,t)=1]$ and using the Laplace transform result

$$\frac{t^{n-1}e^{-at}}{(n-1)!}\cdot u(t) \Rightarrow \frac{1}{(s+a)^n} \tag{13.80}$$

it then follows, after Laplace transformation,

$$\begin{aligned}sX_1(\Omega,s) - P[X(\Omega,0)=1] &= -\lambda_1 X(\Omega,s) + \frac{\lambda_0}{s+\lambda_0} \\ \Rightarrow X_1(\Omega,s) &= \frac{1}{s+\lambda_1}\cdot\frac{\lambda_0}{s+\lambda_0} = \frac{\lambda_0}{\lambda_1-\lambda_0}\left[\frac{1}{s+\lambda_0} - \frac{1}{s+\lambda_1}\right] \\ \Rightarrow P[X(\Omega,t)=1] &= \frac{\lambda_0}{\lambda_1-\lambda_0}\left[e^{-\lambda_0 t} - e^{-\lambda_1 t}\right]\end{aligned} \tag{13.81}$$

Using Equation 13.81, the differential equation for the case of $k=2$ is

$$\begin{aligned}\frac{d}{dt}P[X(\Omega,t)=2] &= -\lambda_2 P[X(\Omega,t)=2] + \frac{\lambda_0\lambda_1}{\lambda_1-\lambda_0}\left[e^{-\lambda_0 t} - e^{-\lambda_1 t}\right] \\ P[X(\Omega,0)=2] &= 0\end{aligned} \tag{13.82}$$

Laplace transformation yields

$$sX_2(\Omega,s)-P[X(\Omega,0)=2]=-\lambda_2X(\Omega,s)+\frac{\lambda_0\lambda_1}{\lambda_1+\lambda_0}\left[\frac{1}{s+\lambda_0}-\frac{1}{s+\lambda_1}\right]\Rightarrow$$

$$X_2(\Omega,s)=\frac{1}{s+\lambda_2}\cdot\frac{\lambda_0\lambda_1}{\lambda_1-\lambda_0}\left[\frac{1}{s+\lambda_0}-\frac{1}{s+\lambda_1}\right]=\frac{\lambda_0\lambda_1}{(\lambda_1-\lambda_0)(\lambda_2-\lambda_0)(s+\lambda_0)} \quad (13.83)$$

$$-\frac{\lambda_0\lambda_1}{(\lambda_1-\lambda_0)(\lambda_2-\lambda_1)(s+\lambda_1)}+\frac{\lambda_0\lambda_1}{(\lambda_2-\lambda_0)(\lambda_2-\lambda_1)(s+\lambda_2)}$$

which implies

$$P[X(\Omega,t)=2]=$$
$$\lambda_0\lambda_1\left[\frac{e^{-\lambda_0 t}}{(\lambda_1-\lambda_0)(\lambda_2-\lambda_0)}-\frac{e^{-\lambda_1 t}}{(\lambda_1-\lambda_0)(\lambda_2-\lambda_1)}+\frac{e^{-\lambda_2 t}}{(\lambda_2-\lambda_0)(\lambda_2-\lambda_1)}\right] \quad (13.84)$$

For the general case, differentiation of the form

$$P[X(\Omega,t)=k]=e^{-\lambda_k t}\left[\lambda_{k-1}\int_0^t P[X(\Omega,T)=k-1]e^{\lambda_k T}dT+P[X(\Omega,0)=k]\right] \quad (13.85)$$

yields

$$\frac{d}{dt}P[X(\Omega,t)=k]=$$
$$-\lambda_k e^{-\lambda_k t}\left[\lambda_{k-1}\int_0^t P[X(\Omega,T)=k-1]e^{\lambda_k T}dT+P[X(\Omega,0)=k]\right] \quad (13.86)$$
$$+e^{-\lambda_k t}\lambda_{k-1}P[X(\Omega,t)=k-1]e^{\lambda_k t}$$

Using the form for the solution as given by Equation 13.85, the differential equation becomes

$$\frac{d}{dt}P[X(\Omega,t)=k]=-\lambda_k P[X(\Omega,t)=k]+\lambda_{k-1}P[X(\Omega,t)=k-1] \quad (13.87)$$

which proves the required result assuming $k\geq 1$.

## APPENDIX 13.E PROOF OF THEOREM 13.9

First, from Theorem 13.1, the differential equations governing the change in probability of a given state with time, for the case of $\lambda_k=0, k\in \mathbf{Z}^+$, $\mu_k=\mu, k\in\{1,\ldots,N\}$, and $\mu=0, k=0, k>N$, become

$$\frac{d}{dt}P[X(\Omega,t)=N]=-\mu P[X(\Omega,t)=N] \qquad k=N$$
$$\frac{d}{dt}P[X(\Omega,t)=k]=-\mu P[X(\Omega,t)=k]+\mu P[X(\Omega,t)=k+1] \qquad 1\le k<N \tag{13.88}$$
$$\frac{d}{dt}P[X(\Omega,t)=0]=\mu P[X(\Omega,t)=1] \qquad k=0$$

Second, consider the first differential equation with the initial condition $P[X(\Omega,0)=N]=1$, that is, an initial population of $N$ at $t=0$. The solution to this differential equation is

$$P[X(\Omega,t)=N]=e^{-\mu t} \qquad t\ge 0 \tag{13.89}$$

Next, consider the case of $k=N-1$. Using the result given by Equation 13.89, and the initial condition of $P[X(\Omega,0)=N-1]=0$ (population of $N$ at $t=0$), the differential equation for this case becomes

$$\begin{aligned}\frac{d}{dt}P[X(\Omega,t)=N-1]&=-\mu P[X(\Omega,t)=N-1]+\mu e^{-\mu t}\\ P[X(\Omega,0)=N-1]&=0\end{aligned} \tag{13.90}$$

With $X_{N-1}(\Omega,s)$ defined as the Laplace transform of $P[X(\Omega,t)=N-1]$, and using the result specified in Equation 13.80, it follows, after Laplace transformation, that

$$\begin{aligned}&sX_{N-1}(\Omega,s)-P[X(\Omega,0)=N-1]=-\mu X_{N-1}(\Omega,s)+\frac{\mu}{s+\mu}\\ &\Rightarrow\; X_{N-1}(\Omega,s)=\frac{\mu}{(s+\mu)^2} \;\Rightarrow\; P[X(\Omega,t)=N-1]=\mu te^{-\mu t}\end{aligned} \tag{13.91}$$

For the case of $k=N-2$, the differential equation is

$$\begin{aligned}\frac{d}{dt}P[X(\Omega,t)=N-2]&=-\mu P[X(\Omega,t)=N-2]+\mu^2te^{-\mu t}\\ P[X(\Omega,0)=N-2]&=0\end{aligned} \tag{13.92}$$

Laplace transformation yields

$$\begin{aligned}&sX_{N-2}(\Omega,s)-P[X(\Omega,0)=N-2]=-\mu X_{N-2}(\Omega,s)+\frac{\mu^2}{(s+\mu)^2}\\ &\Rightarrow\; X_{N-2}(\Omega,s)=\frac{\mu^2}{(s+\mu)^3} \;\Rightarrow\; P[X(\Omega,t)=N-2]=\frac{\mu^2t^2e^{-\mu t}}{2}\end{aligned} \tag{13.93}$$

Continuing in a similar manner yields

$$P[X(\Omega,t)=N-q]=\frac{\mu^q t^q e^{-\mu t}}{q!}\Rightarrow P[X(\Omega,t)=k]=\frac{\mu^{N-k}t^{N-k}e^{-\mu t}}{(N-k)!} \quad (13.94)$$

for $0 \le q \le N-1$ or $1 \le k \le N$. Finally, for the case of zero population state, the differential equation is

$$\begin{aligned} &\frac{d}{dt}P[X(\Omega,t)=0]=\mu P[X(\Omega,t)=1] \quad P[X(\Omega,0)=0]=0 \\ &\Rightarrow \frac{d}{dt}P[X(\Omega,t)=0]=\frac{\mu^N t^{N-1}e^{-\mu t}}{(N-1)!} \end{aligned} \quad (13.95)$$

Laplace transformation yields

$$\begin{aligned} &X_0(\Omega,s)=\frac{\mu^N}{s(s+\mu)^N}=\frac{1}{s}-\sum_{q=1}^{N}\frac{\mu^{q-1}}{(s+\mu)^q} \\ &\Rightarrow P[X(\Omega,t)=0]=\left[1-e^{-\mu t}\sum_{q=1}^{N}\frac{\mu^{q-1}t^{q-1}}{(q-1)!}\right]u(t) \end{aligned} \quad (13.96)$$

# 14

# THE FIRST PASSAGE TIME

## 14.1 INTRODUCTION

For signals defined by a random process $X(\Omega, t)$ on an interval $[0,T]$, the characteristics, illustrated in Figure 14.1, can be defined and, in an appropriate context, provide valuable information. These characteristics include the maximum and minimum level; the first passage time to, or first level crossing time of, a set level; the time spent above a set level; the number of level crossings of a set level; the signal path length; etc.

This chapter provides an introduction to first passage time theory. In general, useful ways of characterizing the first passage time for signals defined by a random process are analytically difficult with results obtainable for a few important cases including the random walk, one-dimensional Brownian motion, monotonically increasing signals, and the linear signal plus noise case under limiting, but useful, conditions. These cases are considered along with general approaches including a Rice series approach and an integral equation approach. Finally, the relationship for a random process, between the maximum level on an interval and the first passage time on the same interval, is defined.

## 14.2 FIRST PASSAGE TIME

When the system performance, in some manner, is affected by a system parameter reaching a set level for the first time, then a first passage (FP) time problem exists,

*A Signal Theoretic Introduction to Random Processes*, First Edition. Roy M. Howard.

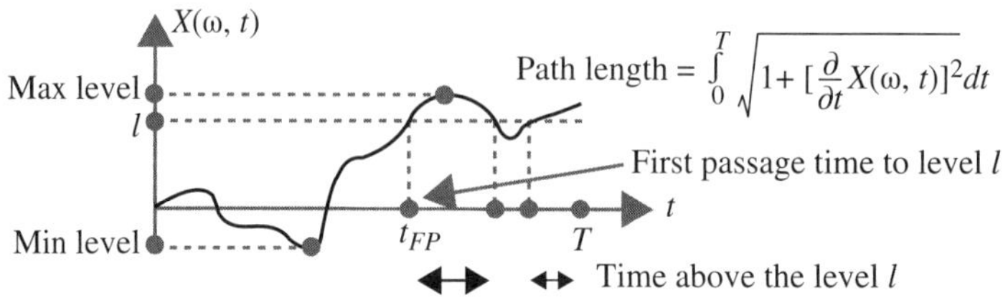


**FIGURE 14.1** Illustration of one signal, $X(\omega, t)$, from the random process $X(\Omega, t)$ and associated characteristics.

and the probability density function for the first passage time is of interest. Apart from a few simple cases, finding the solution of the first passage time to a set level $l$ is an unsolved problem.

The prototypical first passage time problem is that of a monotonically increasing signal $s: \boldsymbol{R} \rightarrow \boldsymbol{R}$, with an additive noise signal $n: \boldsymbol{R} \rightarrow \boldsymbol{R}$, leading to

$$x(t) = s(t) + n(t) \tag{14.1}$$

and the illustration of Figure 14.1 is appropriate.

**Definition: First Passage Time** The first passage time for a signal $x$ to the level $l$ on an interval $[0,T]$ is the time, $t_{FP}$, when the signal first reaches the level $l$, that is,

$$t_{FP} = \min\{t: \quad x(t) = l, t \in [0, T]\} \tag{14.2}$$

For the signal plus noise case with $x(t) = s(t) + n(t)$, the first passage time is

$$t_{FP} = \min\{t: \quad n(t) = l - s(t), t \in [0, T]\} \tag{14.3}$$

In general, there is no analytical solution for the first passage time.

### 14.2.1 Difficulty in Determining the First Passage Time

To determine the first passage time of a signal to a set level, on an interval $[0, T]$, requires systematic *sensing* of the value of the signal at all points (an uncountable number) in the interval. An integral of the signal, or the integral of a function of the signal, for example, which effectively senses a signal at all points, is not likely to lead to a unique determination of the first passage time as it is easy to construct two functions with the same area on an interval but with very different first passage times to a set level. When finite resolution sampling is used, it is, in general, possible to miss the first passage time of a signal.

### 14.2.2 Applications of First Passage Time Theory

The first passage time to a set level is of interest in many areas. The following subsections give an indication of the importance of first passage time theory.

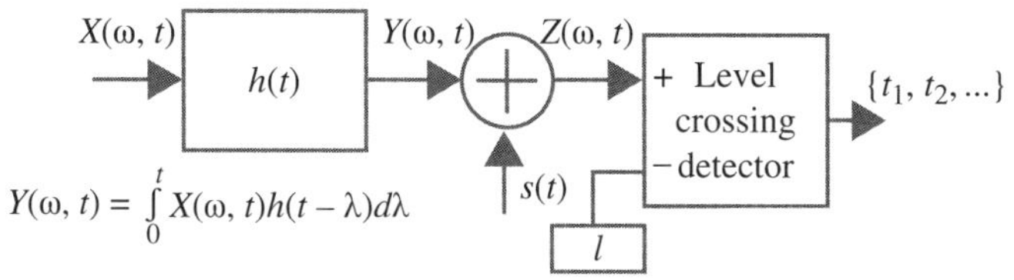


**FIGURE 14.2** Schematic diagram of prototypical jitter case. The signal $s(t)$ is assumed to be periodic and deterministic. The signal $X(\omega, t)$ is one outcome of a random process $X(\Omega, t)$ and similarly for the other signals. The level crossing detector output is the set of first passage times $\{t_1, t_2, \ldots\}$.

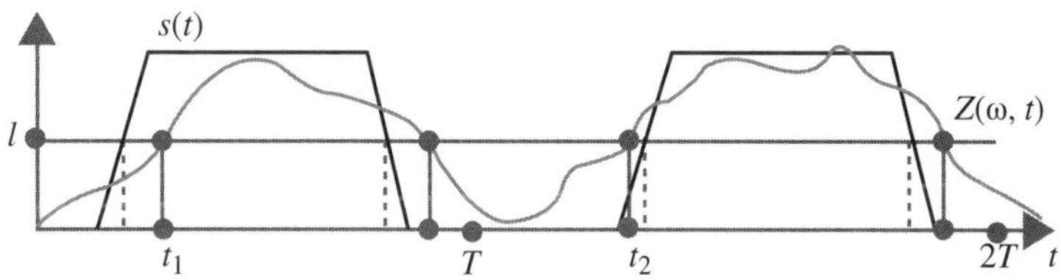


**FIGURE 14.3** Illustration of signals arising in a jitter situation consistent with Figure 14.2. The first passage times form the set $\{t_1, t_2, \ldots\}$.

***14.2.2.1 Jitter*** When noise is added to a deterministic signal and the resultant signal is input into a level crossing detector, a first passage time problem exists. When the deterministic signal is periodic, a repetitive first passage time problem exists, and the noise results in the first passage times to a set level being jittered relative to the zero noise case. The prototypical system is shown in Figure 14.2, and illustrative waveforms are shown in Figure 14.3 for the case of a periodic binary signal. For the case where the filtered noise has a correlation time greater than the characteristic repetition time period, the sequence of first passage times, referenced to a fixed point on the deterministic signal, are correlated.

***14.2.2.2 System Reliability: Risk Management*** The performance of most complex systems degrades with time due to the ageing of components/subsystems and random inputs that stress the system. For a complex system, various reliability (actually unreliability measures) measures can be proposed and monitored. When these measures exceed a predefined threshold, appropriate action is required. Appropriate actions include maintenance, rebuilding, replacement, etc.

One reliability measure is the failure time of a component where failure is defined according to some performance criteria or criterion. As illustrated in Figure 14.4, the time that a reliability measure first exceeds a threshold is a first passage time problem.

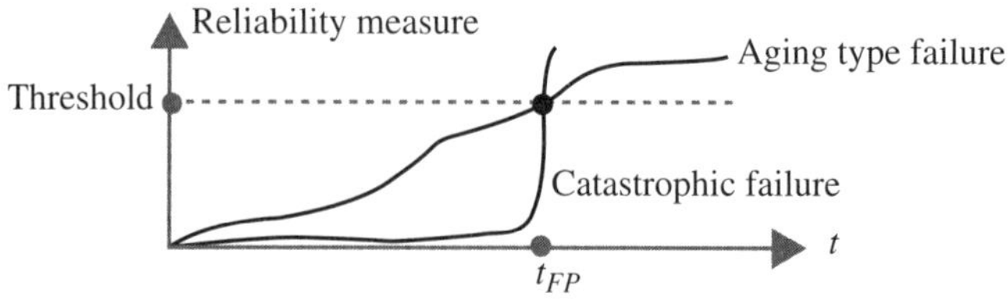


**FIGURE 14.4** Reliability measure as a function of time.

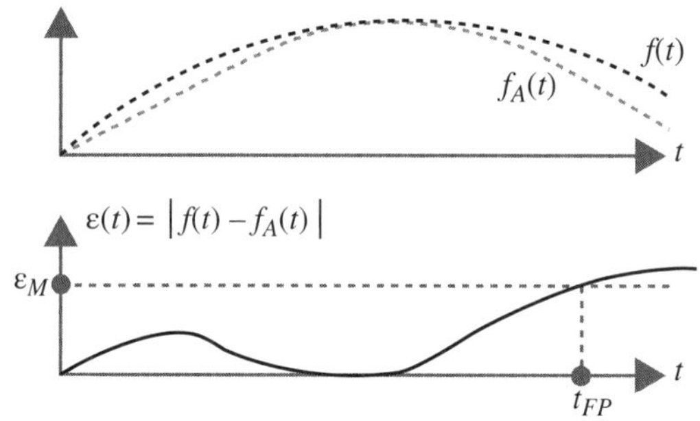


**FIGURE 14.5** Illustration of the error in an approximation to a function.

***14.2.2.3 Overload Recovery*** The time taken for a circuit to recover from overload, in response to a large input signal, as given by the output falling to a set level, is a first passage time problem. Similarly, the time it takes for a latch to recover from a metastable event, as given by the output rising or falling to a set level, is a first passage time problem.

***14.2.2.4 Determination of Update Interval for Signal Approximation*** Consider the case of adaptive signal approximation, as illustrated in Figure 14.5, based on estimating the time where the signal approximation $f_A$ differs from the true signal $f$ by a specified error level $\varepsilon_M$ The time to be estimated is the first passage time of $|f(t)-f_A(t)|$ to the level $\varepsilon_M$.

***14.2.2.5 Modelling Disease Progression*** Consider the case of a measure, $m$, of a disease in a patient after initial diagnosis, as given by, for example, the number of cancer cells, or the mass of a tumor, or the penetration of a disease in an organ, etc. Over a population of patients, the measure of disease progression defines a random process $M(\Omega, t)$ whose probability density function is illustrated in Figure 14.6.

It is useful to characterize the disease progression into stages as given by some level of the measure $m$. The disease can be characterized as being in state $i$ when the disease parameter is between level $i$ and level $i+1$. Of interest is the time, that is, the first passage time, when the transition from state $i$ to state $i+1$ occurs. The nature of the probability density function for the first passage time to a set disease level is illustrated in Figure 14.7.

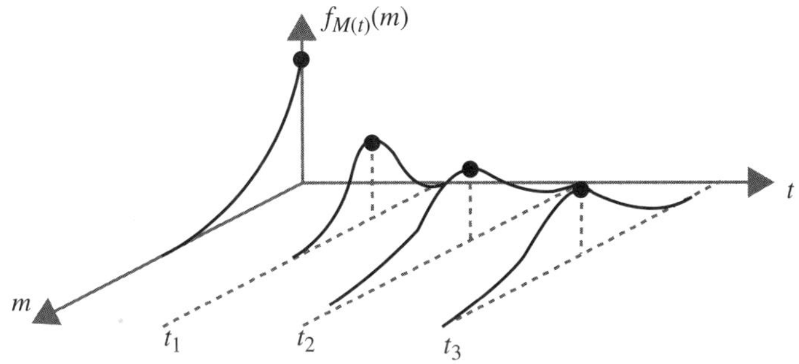


**FIGURE 14.6** Illustration of the probability density function of a measure of disease progression after diagnosis at $t = 0$.

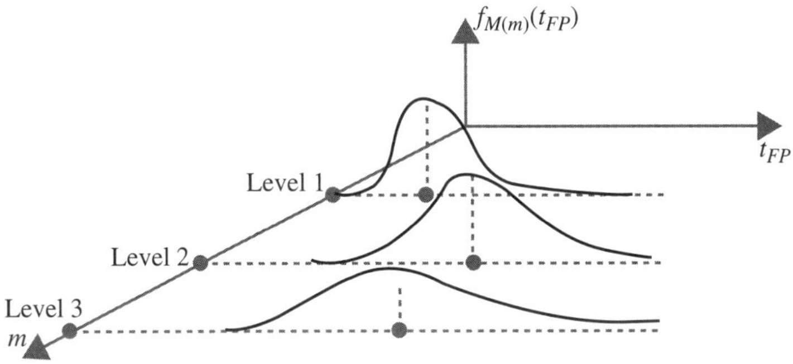


**FIGURE 14.7** Illustration of the probability density function for the first passage time of a disease measure to a specified level.

***14.2.2.6 Neuronal Firing*** The spikes in electrical potential emanating from neurons in complex networks can arise from the response to a large number ($10^4$ to $10^5$) of excitatory and inhibitory synaptic inputs (e.g. Ricciardi, 1977; Ricciardi et al., 1983; Shimokawa et al., 2000), and all these affect the neuronal membranes' potential. When this potential reaches a threshold, that is, the first passage time of the potential to the threshold, the neuron emits a spike. Prior to reaching the threshold, the membrane potential acts as a random walk under the influence of linear filtering consistent with the membrane time constant.

## 14.3 APPROACHES: ESTABLISHING THE FIRST PASSAGE TIME

The material in the following subsections is illustrative of different ways the first passage time can be defined and approached.

### 14.3.1 First Passage Time as the Lowest-Order Root

Consider a function $x:\mathbf{R}\rightarrow\mathbf{R}$ that is assumed to be continuous on an interval $[0, T]$. By definition, the first passage time, $t_{FP}$, to the level $l$ is the lowest real root of the function $x(t)-l$, that is,

$$t_{FP}=\min\{t: \quad x(t)-l=0, t\in[0,T]\} \tag{14.4}$$

### 14.3.2 Integral Specification for First Passage Time: Form

Consider a continuous signal $x:\mathbf{R}\rightarrow\mathbf{R}$. Define the derivative signals:

$$v_1(t)=u[x(t)-l] \qquad v_2(t)=\int_0^t v_1(q)dq=\int_0^t u[x(q)-l]dq \tag{14.5}$$

$$v_3(t)=u_M[v_2(t)]=u_M\left[\int_0^t u[x(q)-l]dq\right] \tag{14.6}$$

$$v_4(T)=\int_0^T v_3(t)dt=\int_0^T u_M\left[\int_0^t u[x(q)-l]dq\right]dt \tag{14.7}$$

where $u_M$ is a modified step function that is zero for the case of zero argument and is defined according to

$$u_M(t)=u(t)-\delta_K(t) \tag{14.8}$$

These functions are illustrated in Figure 14.8. The region of integration for the double integral that defines $v_4(T)$ is illustrated in Figure 14.9. The following can then be stated.

**Theorem 14.1 Integral Specification of First Passage Time**
The first passage time $t_{FP}$ of a continuous signal $x(t)$ to the level $l$ on $[0, T]$ is given by

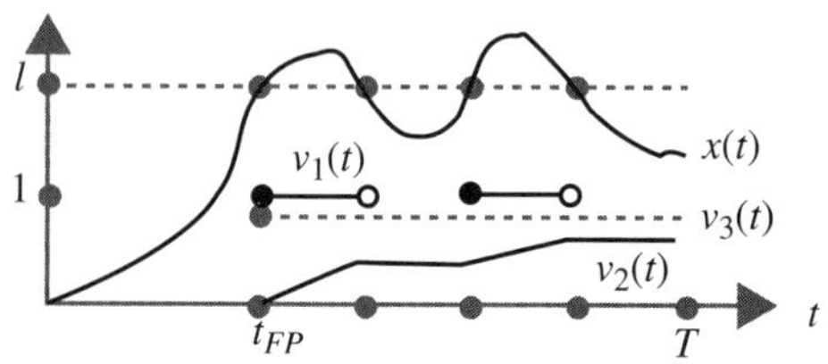


**FIGURE 14.8** Illustration of the functions $v_1$, $v_2$, and $v_3$.

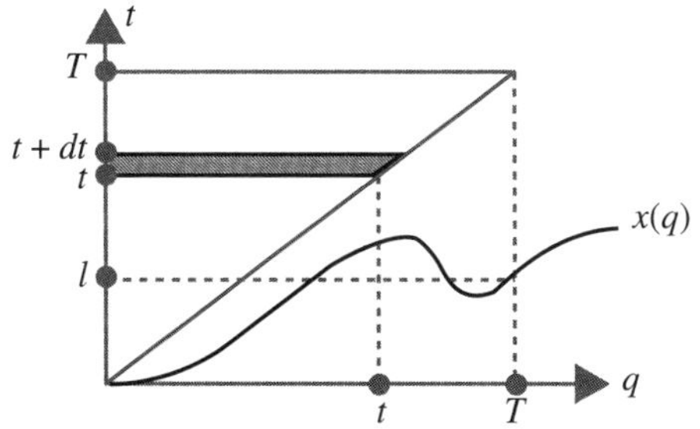


**FIGURE 14.9** Illustration of the region of integration in the double integral that defines $v_4(T)$.

$$t_{FP} = T - v_4(T) = T - \int_0^T u_M \left[ \int_0^t u[x(q-l)]dq \right] dt \tag{14.9}$$

**Proof**

The result follows from the construction of the component functions defined in Equations 14.5–14.7 and illustrated in Figure 14.8.

### 14.3.3 Integral Specification for First Passage Time: Form 2

The following theorem states an alternative integral expression for the first passage time of a signal on a finite interval.

**Theorem 14.2 First Passage Time: Sequence of Delayed Functions**

The first passage time of a continuous signal $x$ to the level $l$ on the interval $[0, T]$ is defined according to

$$t_{FP} = T - \lim_{\Delta \to 0} \int_0^T u_M \left[ \sum_{i=0}^{N-1} u[x(t - i\Delta) - l] \right] dt \tag{14.10}$$

where $T = N\Delta$. The assumption here is that the first passage time to the level $l$ is consistent with $x(t) - l \geq 0$ on a set of nonzero measure around the first passage time.

**Proof**

The proof follows from the construction of the summation of delayed functions as illustrated in Figure 14.10. With $\Delta \to 0$, the summation of delayed functions effectively covers the interval $[t_{FP}, T]$ with a level above, or equal to, unity.

***14.3.3.1 Note*** The integral forms of Theorems 14.1 and 14.2 indicate the potential difficulty in establishing, for a random process, the probability density function for the first passage time to a set level.

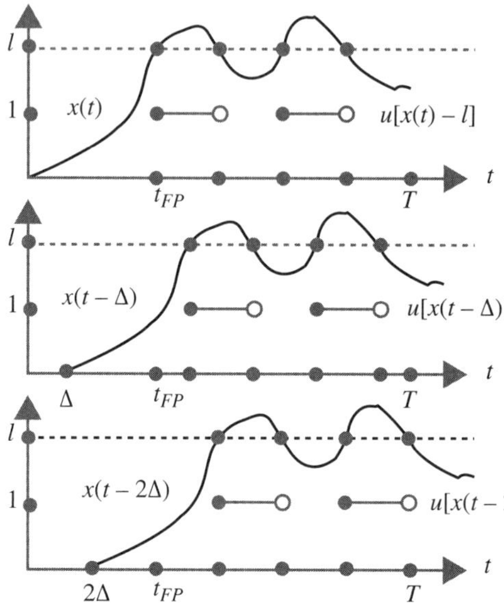


**FIGURE 14.10** Illustration of a function and its delayed versions.

### 14.3.4 Integral Equation for First Passage Time

Consider a random process $X$ comprising of differentiable signals and with an initial value of $x_o$ at time $t_o$. The conditional probability density function evolution with time for the random process is defined according to

$$f_{X(t)/(x_o,t_o)}(x)dx \approx P\left[X(\Omega,t) \in [x,x+dx] |_{X(\Omega,t_o)=x_o}\right] \tag{14.11}$$

for $dx$ sufficiently small. The conditional first passage time probability density function to a level $l$ is defined according to

$$f_{L(l)/(x_o,t_o)}(t)dt \approx P\left[\text{first passage time to } l \text{ in } [t,t+dt] |_{X(\Omega,t_o)=x_o}\right] \tag{14.12}$$

for $dt$ sufficiently small. The following theorem states a fundamental result relating these two probability density function, for example, Blake and Lindsey (1973, p. 297), Smith (2000, p. 431), and Ricciardi (1977, pp. 61f).

**Theorem 14.3 Integral Equation for First Passage Time PDF**
Consider a random process $X$ that defines differentiable signals with an upper bound on the rate of change and has the Markov property of

$$\begin{aligned} P&\left[X(\Omega,t_2) \in [x_2,x_2+dx_2] |_{X(\Omega,t_1)\in[x_1x_1+dx_1],X(\Omega,t_o)\in[x_0,x_o+dx_o]}\right] \\ &= P\left[X(\Omega,t_2) \in [x_2,x_2+dx_2] |_{X(\Omega,t_1)\in[x_1,x_1+dx_1]}\right] \end{aligned} \tag{14.13}$$

for $t_2 > t_1 + \Delta t > t_o + \Delta t$ and $\Delta t$ sufficiently small. The first passage time probability density function to the level $l$ for such a random process satisfies the following integral equation:

$$f_{X(t)/(x_o,t_o)}(x) = \int_{t_o}^{t} f_{L(l)/(x_o,t_o)}(\lambda) f_{X(t)/(l,\lambda)}(x)d\lambda \tag{14.14}$$

**Proof**
The proof is detailed in Appendix 14.A.

### 14.3.5 Rice Series Approach

A fundamental result dating back to Rice (1945) (see also Verechtchaguina et al., 2006) for the first passage time probability density function is stated in the following theorem.

**Theorem 14.4 Rice Series for First Passage Time**
The probability density function of the first passage time to the level $l$ for a random process $X$, comprising of differentiable signals, is given by

$$\begin{aligned} f_{L(l)}(t) = {} & f_1(t) - \int_0^t f_2(t_1,t)dt_1 + \frac{1}{2!}\int_0^t \left[\int_0^t f_3(t_1,t_2,t)dt_2\right] dt_1 \\ & - \frac{1}{3!}\int_0^t \left[\int_0^t \left[\int_0^t f_4(t_1,t_2,t_3,t)dt_3\right] dt_2\right] dt_1 + \cdots \end{aligned} \tag{14.15}$$

where

$$\begin{aligned} & f_1(t)dt = P[\text{level crossing in } (t,t+dt)] \\ & f_2(t_1,t)dt_1 dt = P[\text{LC in } (t_1,t_1+dt_1), \text{LC in } (t,t+dt)] \\ & f_3(t_1,t_2,t)dt_1 dt_2 dt = \\ & \qquad P[\text{LC in } (t_1,t_1+dt_1), \text{LC in } (t_2,t_2+dt_2), \text{LC in } (t,t+dt)] \end{aligned} \tag{14.16}$$

etc. Here, LC denotes an upward *level crossing*.

**Proof**
The proof is detailed in Appendix 14.B.

## 14.4 MAXIMUM LEVEL AND THE FIRST PASSAGE TIME

The maximum level of a continuous signal is related to the first passage time of the signal according to the following:

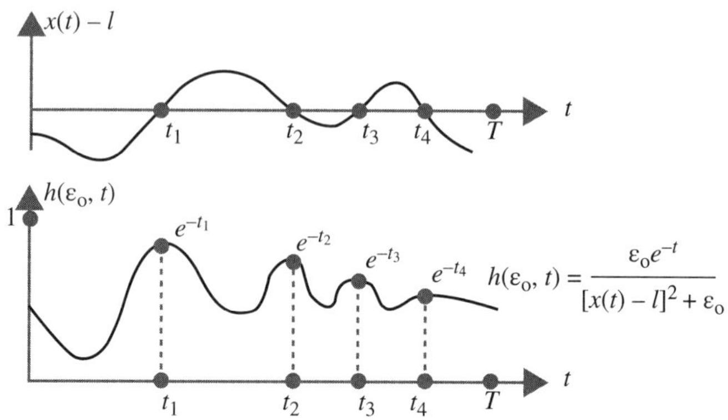


**FIGURE 14.11** Illustration of a function that transforms the first zero crossing time into the time associated with the maximum level.

If the first passage time of a signal $x:\boldsymbol{R}\to\boldsymbol{R}$ to a level $l$ occurs at the time $t_{FP}$, then the maximum level of the signal on the interval $[0, t_{FP}]$ is $l$.

or

If the unique maximum level of a signal $x:\boldsymbol{R}\to\boldsymbol{R}$ on an interval $[0, t_{FP}]$ is $l$ and this occurs at the time $t_{FP}$, then the first passage time to the level $l$ occurs at the time $t_{FP}$.

### 14.4.1 Conversion: First Passage Time to a Maximum Level Problem

Consider a bounded and continuous signal $x:\boldsymbol{R}\to\boldsymbol{R}$ on an interval commencing at $t=0$. The function

$$h(\varepsilon,t)=\frac{\varepsilon e^{-t}}{[x(t)-l]^2+\varepsilon} \tag{14.17}$$

has the form shown in Figure 14.11. The level crossing times are at the times

$$t:\quad h(\varepsilon,t)=e^{-t} \tag{14.18}$$

Such a transformation has generated a normalized signal, and the time of the first passage to the level $l$ is the time of the maximum of the signal as $\varepsilon\to 0$, that is, the transformation has converted the first passage time problem into a maximum level problem. One example of this conversion is shown in Figure 14.12.

### 14.4.2 Basic Definitions

Consider a random process $X(\Omega, t)$ where all the constituent signals $\{X(\omega,t):\quad \omega\in S\}$ are continuous. The following definitions can be made.

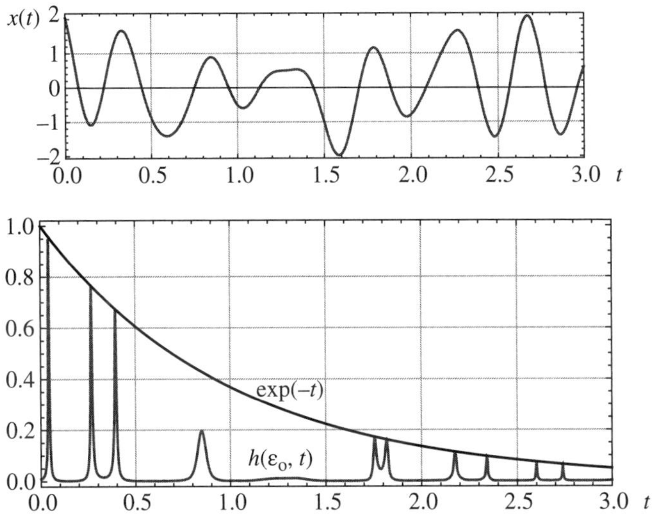


**FIGURE 14.12** Illustration of a signal and the associated function $h(\varepsilon_o, t)$ for the case of $l = 1$ and $\varepsilon_o = 0.01$.

**Definition: Maximum Random Process** The maximum random process $M(\Omega, t)$ is defined according to

$$M(\Omega,t) = \max\{X(\Omega,\lambda) : \lambda \in [0,t]\} \tag{14.19}$$

For $t$ fixed, the probability density function of $M(\Omega, t)$ is denoted $f_{M(t)}(m)$. By definition,

$$\begin{aligned} f_{M(t)}(m)dm &\approx P[M(\Omega,t) \in [m, m+dm]] \\ &= P[\max\{X(\Omega,\lambda) : \lambda \in [0,t]\} \in [m, m+dm]] \end{aligned} \tag{14.20}$$

for $dm$ sufficiently small.

***14.4.2.1 First Passage Time Random Process*** In the following, $L$ is used not $F$ for the first passage time as $F$ is used for the cumulative distribution function. $L$ denotes level in the first level crossing time, which is synonymous with the first passage time. For notational brevity, FP is used for *first passage*. Blake and Lindsey (1973, pp. 295–296) is a useful reference for first passage time theory.

**Definition: First Passage Time Random Process** Define the random process $L(\Omega, T, l)$ associated with the first passage time across the level $l$ in the interval $[0, T]$ as

$$L(\Omega,T,l) = \begin{cases} t : X(\Omega,t) = l, X(\Omega,\lambda) < l \text{ for } \lambda \in (0,t) & \text{FP of } l \text{ on } [0,T) \\ T & \text{no FP of } l \text{ on } [0,T) \end{cases} \tag{14.21}$$

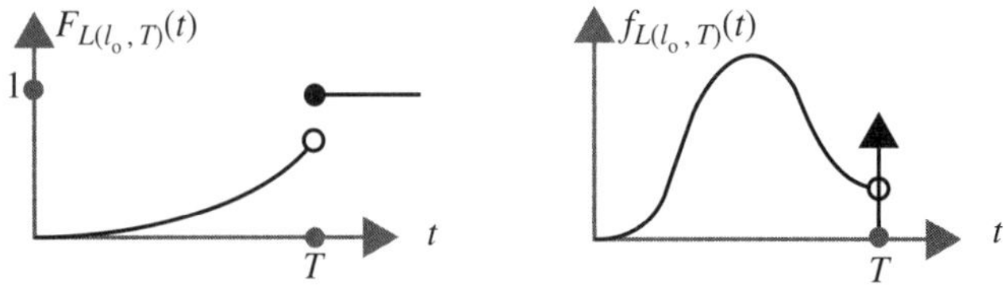


**FIGURE 14.13** Illustration of the form of the cumulative distribution function $F_{L(l_o,T)}(t)$ and probability density function $f_{L(l_o,T)}(t)$ for $T$ fixed.

Define the random process $L(\Omega, l)$ associated with the first passage time across the level $l$ in the interval $(0, \infty)$ as

$$L(\Omega,l) = \begin{cases} t: X(\Omega,t) = l, X(\Omega,\lambda) < l \text{ for } \lambda \in (0,t) & \text{first FP of } l \text{ at } t \\ \infty & \text{no FP of } l \text{ on } [0, \infty) \end{cases} \quad (14.22)$$

**Definition: Cumulative Distribution Function–Probability Density Function** For $l$ fixed with a value $l_o$, the cumulative distribution function and the probability density function of $L(\Omega, l_o)$ are:

$$\begin{aligned} F_{L(l_o)}(t) &= P[L(\Omega,l_o) \le t] = P[\text{first passage time to } l_o \text{ is prior to, or at, } t] \\ &= P[\text{one or more passage times to } l_o \text{ on } [0,t]] \end{aligned} \quad (14.23)$$

$$\begin{aligned} f_{L(l_o)}(t)dt &\approx P[L(\Omega,l_o) \in [t,t+dt] \\ &= P[\text{first passage times to } l_o \text{ occurs in } [t,t+dt]] \end{aligned} \quad (14.24)$$

for $dt$ sufficiently small. Similar definitions hold for the cumulative distribution function and the probability density function of $L(\Omega, T, l_o)$. The cumulative distribution function and probability density functions for this case are illustrated in Figure 14.13. Strictly speaking, the probability density function does not exist due to the discontinuity in the cumulative distribution function. The use of an impulse associated with the discrete outcome at $t = T$ when treated appropriately is unproblematic.

From the definition of $L(\Omega, T, l_o)$, it follows that

$$P[\text{no first passage to } l_o \text{ on } [0,T]] = 1 - F_{L(l_o,T)}(T^-) \quad (14.25)$$

Finally, the following events are equivalent assuming the random process starts below the level $l_o$:

$$\begin{aligned} &\{\text{first passage time to } l_o \text{ is prior to, or at, } t\} = \\ &\qquad \{\text{one or more passage times to } l_o \text{ on } [0,T]\} \end{aligned} \quad (14.26)$$

***14.4.2.2 Relationship between Level Crossing and Maximum Level*** Figure 14.14 illustrates the nature of the maximum of a signal on the interval $[0, t_o]$ and the

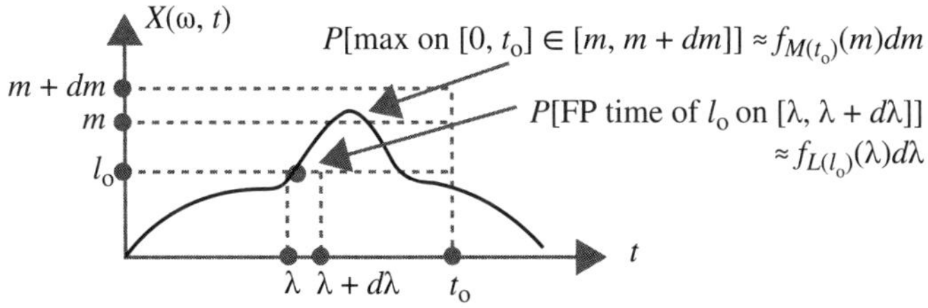


**FIGURE 14.14** Illustration of the maximum level and first passage time of one waveform $X(\omega, t)$ from the random process $X(\Omega, t)$.

first passage time to the level $l_o$. One difference is that there are, potentially, many times on $[0, t_o]$ when the signal reaches a level close to its maximum amplitude, which is assumed to be in the interval $[m, m+dm]$, but only one infinitesimally small time interval, $[\lambda, \lambda+d\lambda]$, for the first passage time to the level $l_o$.

The link between the random variables $M(\Omega, t_o)$ and $L(\Omega, l_o)$ is specified in the following theorem.

**Theorem 14.5 First Passage Time Probability–Maximum Level Probability** The maximum level and level crossing cumulative distribution functions are related according to

$$F_{L(l_o)}(t_o) = 1 - F_{M(t_o)}(l_o) \qquad F_{L(l_o)}(t_o) = \int_{l_o}^{\infty} f_{M(t_o)}(m)dm$$
$$F_{M(t_o)}(l_o) = 1 - F_{L(l_o)}(t_o) \qquad F_{M(t_o)}(l_o) = \int_{t_o}^{\infty} f_{L(l_o)}(\lambda)d\lambda \tag{14.27}$$

For the special case of a fixed interval $[0, T]$ and $t_o = T$:

$$F_{L(l_o,T)}(T^-) = 1 - F_{M(T)}(l_o)$$
$$F_{M(T)}(l_o) = 1 - F_{L(l_o,T)}(T^-) \tag{14.28}$$

The maximum level and level crossing probability density functions are related according to:

$$f_{M(t_o)}(l_o) = \frac{\partial}{\partial l_o}\int_{t_o}^{\infty} f_{L(l_o)}(\lambda)d\lambda = -\frac{\partial}{\partial l_o}\int_{-\infty}^{t_o} f_{L(l_o)}(\lambda)d\lambda \tag{14.29}$$

$$f_{L(l_o)}(t_o) = \frac{\partial}{\partial t_o}\int_{l_o}^{\infty} f_{M(t_o)}(m)dm = -\frac{\partial}{\partial t_o}\int_{-\infty}^{l_o} f_{M(t_o)}(m)dm \tag{14.30}$$

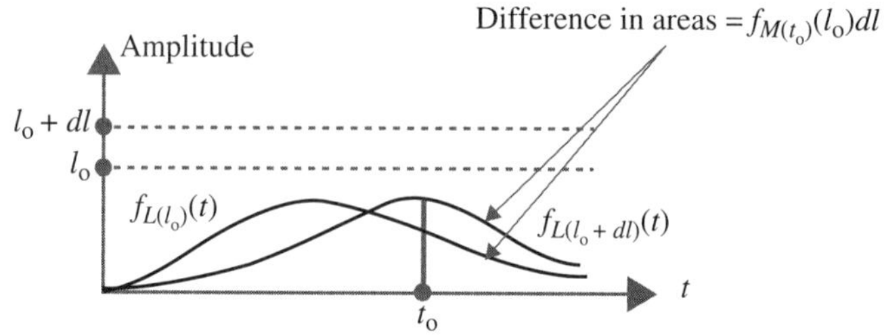


**FIGURE 14.15** Illustration of the relationship for the maximum level probability density function.

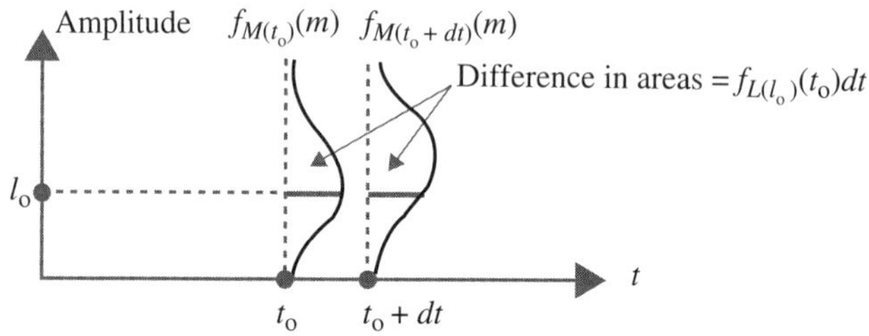


**FIGURE 14.16** Illustration of the relationship for the level crossing probability density function.

**Proof**

The proof is detailed in Appendix 14.C.

***14.4.2.3 Notes*** Equations 14.29 and 14.30 imply

$$f_{M(t_o)}(l_o)dl \approx \int_{t_o}^{\infty} f_{L(l_o+dl)}(\lambda)d\lambda - \int_{t_o}^{\infty} f_{L(l_o)}(\lambda)d\lambda \tag{14.31}$$

$$f_{L(l_o)}(t_o)dt \approx \int_{l_o}^{\infty} f_{M(t_o+dt)}(m)dm - \int_{l_o}^{\infty} f_{M(t_o)}(m)dm \tag{14.32}$$

These relationships are illustrated in Figures 14.15 and 14.16.

## 14.5 SOLUTIONS FOR THE FIRST PASSAGE TIME PDF

In general, obtaining an analytical solution for the first passage time probability density function of a random process is problematic. The following subsections detail important cases where an analytic solution to the first passage time probability density function can be obtained.

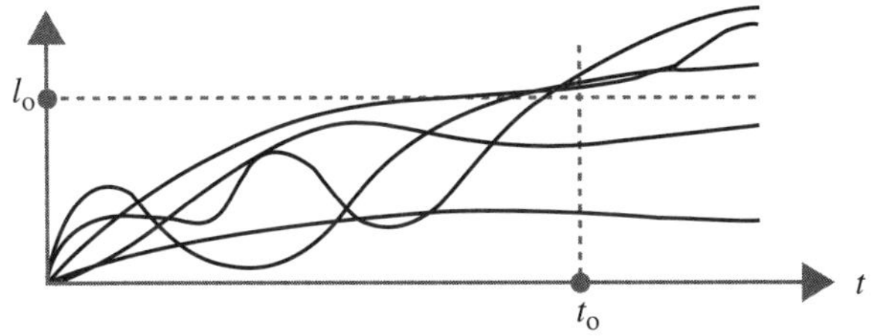


**FIGURE 14.17** Trajectories of a random process that are monotonically increasing above a level $l_o$.

### 14.5.1 First Passage Time: Monotonic Signal Case

Consider the case, as illustrated in Figure 14.17, where the signals defined by a random process are monotonically increasing, or constant, above a level $l_o$.

**Theorem 14.6 First Passage Time: Monotonic Signal Case**
If all the signals defined by a random process $X$ are monotonically increasing, or constant, above a level $l_o$, then

$$P[\text{first passage time to } l_o \text{ on } [0, t_o]] = F_{L(l_o)}(t_o) = \int_{l_o}^{\infty} f_{X(t_o)}(x)dx \tag{14.33}$$

$$f_{L(l_o)}(t_o) = \frac{\partial}{\partial t_o} \int_{l_o}^{\infty} f_{X(t_o)}(x)dx \tag{14.34}$$

Here, $f_{X(t_o)}$ is the probability density function of the random variable defined by the random process at the time $t_o$.

**Proof**
For the case of a countable number of signals, the monotonic increasing nature of the signals above thc level $l_o$ yields the result

$$P[\text{first passage time to } l_o \text{ on } [0, t_o]] = \frac{\text{number of trajectories above } l_o \text{ at time } t_o}{\text{total number of trajectories}} \tag{14.35}$$

For the uncountable case, the ratio on the right-hand side of this result is $\int_{l_o}^{\infty} f_{X(t_o)}(x)dx$.

### 14.5.2 Linear Signal plus Additive Noise

Consider the case of a random process $X$ based on a linear signal plus an additive noise signal and defined according to

$$\begin{aligned} &X: S \rightarrow S_X \qquad S=\{\omega : \omega \in \boldsymbol{R}\} \\ &X(\omega,t)=l+k(t-t_o)+N(\omega,t) \\ &S_X=\{X(\omega,t)=l+k(t-t_o)+N(\omega,t) : \omega \in S\} \end{aligned} \tag{14.36}$$

where $N$ is the noise random process. For notational convenience, the random process can be written as

$$X(\Omega,t)=l+k(t-t_o)+N(\Omega,t) \tag{14.37}$$

The first passage time of one signal from such a random process is illustrated in Figure 14.18. The jitter associated with the first passage time of an individual signal is defined according to

$$t_J=t_{FP}-t_o \tag{14.38}$$

The probability density function of the first passage time and jitter random variables is specified in the following theorem under conditions that often hold.

**Theorem 14.7 First Passage Time PDF: Linear Signal plus Additive Noise** For the case where the expected magnitude of the rate of change of the noise random process is much less than the rate of change of the signal, that is,

$$\frac{1}{k}\cdot \boldsymbol{E}\left[\left|\frac{d}{dt}N(\Omega,t_o)\right|\right] << 1 \tag{14.39}$$

the first passage time and jitter probability density functions can be approximated according to

$$f_{L(l)}(t)\approx kf_{N(t_o)}[-k(t-t_o)] \qquad f_J(\tau)\approx kf_{N(t_o)}(-k\tau) \tag{14.40}$$

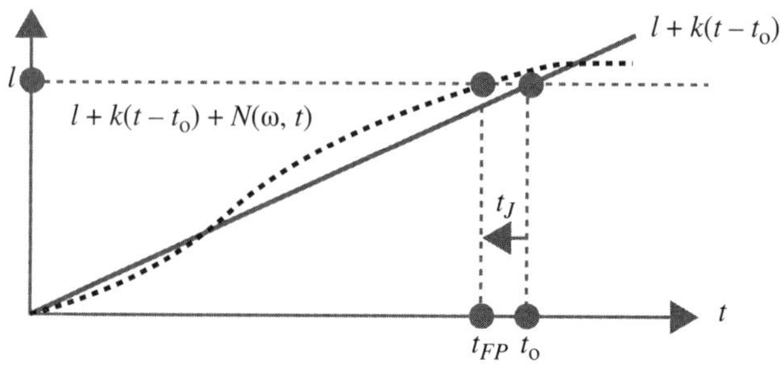


**FIGURE 14.18** First passage time of a linear signal plus an additive noise signal.

Here, $f_{N(t_o)}$ is the probability density function of the noise at the time $t_o$, and $f_J$ is the probability density function of the jitter random variable.

**Proof**
The proof is detailed in Appendix 14.D.

### 14.5.3 Random Walk

Consider an $N$-step random walk random process defined by

$$X(\Omega,t)=\sum_{k=1}^{N}A_k u(t-(k-1)) \tag{14.41}$$

where $\Omega=(A_1,\ \ldots,\ A_N)$ is a vector of identical and independent random variables with outcomes from the set $\{-1,1\}$ and with probabilities $P[A_i=1]=p$, $P[A_i=-1]=1-p$. The probability mass function for the first passage time to a level $l_o$ is specified in the following theorem.

**Theorem 14.8 First Passage Time PMF for a Random Walk**
The probability mass function for the first passage time of a random walk random process to an integer level $l_o$, for the case of $p=0.5$, is

$$p_{L(l_o)}(k)=\begin{cases}0 & k<l_o\\ \dfrac{1}{2^k}\cdot\dfrac{l_o}{k}\cdot\dfrac{k!}{\left(\dfrac{k+l_o}{2}\right)!\left(\dfrac{k-l_o}{2}\right)!} & k\geq l_o\end{cases} \tag{14.42}$$

$$k,l_o\in\{1,2,\ldots\}, k \text{ even if } l_o \text{ even, } k \text{ odd if } l_o \text{ odd}$$

**Proof**
The proof is detailed in Appendix 14.E.

*14.5.3.1 Example* The probability mass function for the first passage time to a level $l_o$ is shown in Figure 14.19.

### 14.5.4 Brownian Motion

The integral equation, as specified in Theorem 14.3, facilitates the evaluation of the first passage time probability density function of the Brownian motion.

**Theorem 14.9 First Passage Time for the Brownian Motion**
A random process $X$ that is accurately modelled as the Brownian motion, and whose probability density function evolution is given by

$$f_{X(t)/(x_o,t_o)}(x)=\frac{e^{-(x-x_o)^2/2D(t-t_o)}}{\sqrt{2\pi}\sqrt{D(t-t_o)}} \qquad t>t_o \tag{14.43}$$

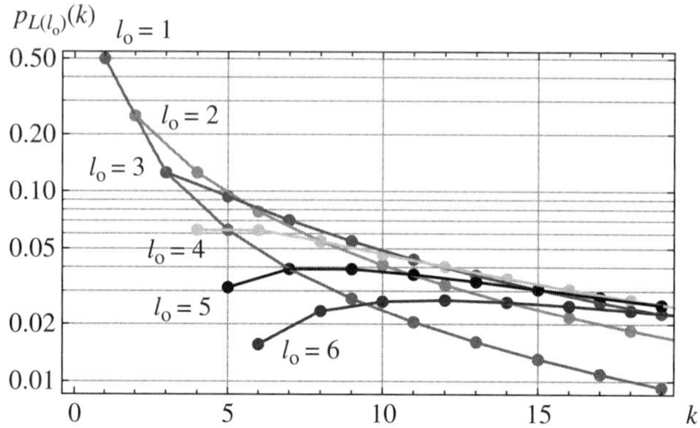


**FIGURE 14.19** Probability mass function for the first passage time of a random walk, with $p=1/2$, to a level $l_o$.

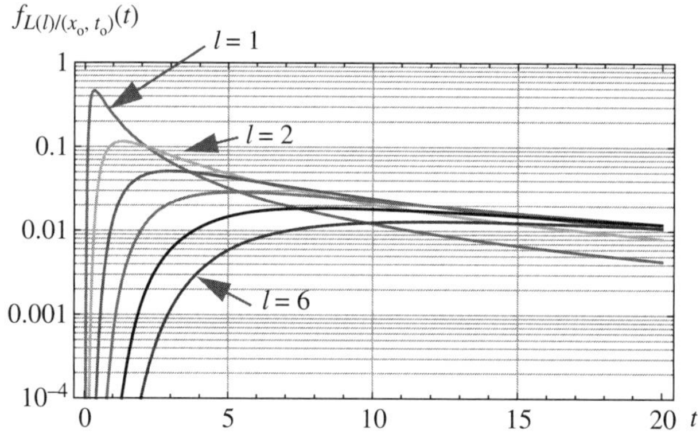


**FIGURE 14.20** Probability density function for the first passage time of the Brownian motion, with $D=1$, $x_o=0$, and $t_o=0$, to the levels $l=1$, $l=2$, $l=3$, $l=4$, $l=5$, and $l=6$.

assuming initial conditions $X(\Omega, t_o) = x_o$ and a diffusion constant $D$, has a first passage time probability density function (Blake and Lindsey 1973, p. 298) to a level $l$ of

$$f_{L(l)/(x_o,t_o)}(t) = \frac{2D(l-x_o)}{\sqrt{\pi}[2D(t-t_o)]^{3/2}} \cdot e^{-(l-x_o)^2/2D(t-t_o)} \qquad t > t_o \tag{14.44}$$

**Proof**
The proof is detailed in Appendix 14.F.

***14.5.4.1 Example*** The probability density function for the first passage time to a level $l$ of a random process that is consistent with the Brownian motion with parameters $D=1$, $x_o = t_o = 0$, is graphed in Figure 14.20.

## 14.6 PROBLEMS

**14.1** Consider the case of a random process comprising of a signal plus noise where the rate of the change of the signal ensures that all signals from the random process are monotonically increasing around a level $l_o$. The random process is defined according to

$$X(\Omega,t)=l_o+l(t-t_o)+N(\Omega,t) \tag{14.45}$$

where $N(\Omega, t)$ is the noise random process with a probability density function

$$f_{N(t)}(n)=\frac{1}{\sqrt{2\pi}\sigma_n}\cdot\exp\left[\frac{-n^2}{2\sigma_n^2}\right] \tag{14.46}$$

a. Determine the probability density function for the first passage time to the level $l_o$.

b. Graph the first time passage probability density function for the case of $k=2,\ l_o=1,\ t_o=1$, and $\sigma_n=l_o/10$.

**14.2** Consider the case of a random process comprising of a signal plus noise where the rate of the change of the signal ensures that all signals from the random process are monotonically increasing around a level $l_o$. The random process is defined according to

$$X(\Omega,t)=p(t)+N(\Omega,t) \tag{14.47}$$

where $p(t)$ is a polynomial function with $p(t_o)=l_o$ and $N(\Omega, t)$ is the noise random process with a probability density function

$$f_{N(t)}(n)=\frac{1}{\sqrt{2\pi}\sigma_n}\cdot\exp\left[\frac{-n^2}{2\sigma_n^2}\right] \tag{14.48}$$

a. Determine the probability density function for the first passage time to the level $l_o$.

b. Graph the first passage time probability density function for the case of

$$p(t)=l_o+k_1(t-t_o)+k_2(t-t_o)^2+\cdots \tag{14.49}$$

where $l_o=1$, $\sigma_n=l_o/10$, $t_o=1$, $k_1=2$, $k_2=-2$, $k_3=1$, $k_4=0$, $k_5=0.25$ and $k_6=k_7=\cdots=0$.

**14.3** Consider the case of a linear signal and additive white noise defined according to

$$X(\Omega,t)=l_o+k(t-t_o)+N(\Omega,t) \tag{14.50}$$

where $N(\Omega, t)$ is the white noise random process defined according to

$$N(\Omega,t)=\sum_{k=1}^{N}\frac{\sqrt{2}A_o}{\sqrt{N}}\sin(2\pi F_k t+\Omega_k) \tag{14.51}$$

Here, $F_1, F_2, \ldots F_N$ are a sequence of independent random variables with a uniform distribution on $[0, f_{max}]$ and $(\Omega_1, \ldots, \Omega_N)$ are independent random variables with a uniform distribution on $[0, 2\pi]$. Assume $l_o=2,\ t_o=1$ and $k=2$.

a. Specify an integral expression for the expectation of the magnitude of the rate of change of the noise random process.
b. For the case of $N=1000$, $f_{max}=1$, and $A_o=1$, generate 10,000 signals from the noise random process, and estimate the expected magnitude of the rate of change of the noise at any fixed point in time.
c. Scale the noise so the ratio $\frac{1}{k}\boldsymbol{E}\left[\left|\frac{d}{dt}N(\Omega,t_o)\right|\right]$ is 0.1. Specify an approximate expression for the first passage probability density function across the level of $l_o$ and the associated jitter probability density function. Confirm these probability density functions by results arising from simulation of 10,000 signals.

**14.4** Consider a random walk random process defined according to

$$X(\Omega,t)=\sum_{k=1}^{\infty}A_k u(t-(k-1)\Delta t) \tag{14.52}$$

where $\Omega=(A_1,\ldots,A_k,\ldots)$ is a vector of identical and independent random variables with outcomes from the set $\{-\Delta x, \Delta x\}$ and with probabilities $P[A_i=\pm\Delta x]=0.5$.

a. Specify the probability mass function for the first passage time to a level $l_o$ for $l_o\in\{\Delta x, 2\Delta x, 3\Delta x, \ldots\}$.
b. Use Stirling's approximation to the factorial to establish an approximation to the probability mass function for the case of $l_o=N\Delta x, N\in\{1,2,\ldots\}$.
c. Graph the probability mass function for the cases of $l_o=2,\ \Delta x=1,\ \Delta t=1$; $l_o=2, \Delta x=2/3,\ \Delta t=4/9$, and $l_o=2,\ \Delta x=1/2,\ \Delta t=1/4$.
d. Compare the results with that for Brownian motion with $D=1,\ x_o=0$, and $t_o=0$.

**14.5** Consider a random process that is consistent with Brownian motion with parameters $x_o=t_o=0$. Graph the probability density function for the first passage time to a level $l_o=10$ for the cases of $D\in\{1,2,3,4,5,6\}$.

**14.6** Consider a random process defined according to

$$X(\Omega,t)=\Omega t u(t) \tag{14.53}$$

where the random variable $\Omega$ has the probability density function

$$f_\Omega(\omega) = \frac{\omega}{\sigma^2} \cdot \exp\left[\frac{-\omega^2}{2\sigma^2}\right] u(\omega) \tag{14.54}$$

a. Determine the probability density function for the maximum on the interval $[0, t]$.
b. Determine the probability density function for the first passage time to the level $l_o$.
c. For this random process, confirm the relationships:

$$f_{M(t_o)}(l_o) = \frac{\partial}{\partial l_o} \int_{t_o}^{\infty} f_{L(l_o)}(\lambda) d\lambda = -\frac{\partial}{\partial l_o} \int_{-\infty}^{t_o} f_{L(l_o)}(\lambda) d\lambda \tag{14.55}$$

$$f_{L(l_o)}(t_o) = \frac{\partial}{\partial t_o} \int_{l_o}^{\infty} f_{M(t_o)}(m) dm = -\frac{\partial}{\partial t_o} \int_{-\infty}^{l_o} f_{M(t_o)}(m) dm \tag{14.56}$$

between the probability density function of the maximum and the probability density function of the first passage time.

d. Graph the probability density function for the maximum and the first passage time assuming $l_o = 2$, the interval $(0, t_o), t_o = 2$, and the case of $\sigma^2 = 1$.

## APPENDIX 14.A PROOF OF THEOREM 14.3

For $X : S \to S_X$, define the bound on the magnitude of the rate of change on the interval $[t_o, t]$ according to

$$x'_{max} = \sup\left\{\left|\frac{d}{dt_1} X(\omega, t_1)\right| : \quad t_1 \in [t_o, t],\ \omega \in S\right\} \tag{14.57}$$

It then follows, from the assumption of differentiability and a bound on the magnitude of the rate of change, that

$$|x(t_1 + \Delta t) - x(t_1)| < \Delta t x'_{max} \tag{14.58}$$

for $t_1 \in [t_o, t]$ and $\Delta t$ sufficiently small. Hence, for all $\Delta\varepsilon > 0$, there exists a value for $\Delta t$ such that

$$|x(t_1 + \Delta t) - x(t_1)| < \Delta\varepsilon \qquad t_1 \in [t_o, t] \tag{14.59}$$

Consider an upward crossing of the level $l$ by $X(\omega, t)$ on the interval $[t_o + i\Delta t, t_o + (i+1)\Delta t]$ with $X(\omega, t_o + i\Delta t) \in [l - \Delta\varepsilon, l]$. Next, consider the change

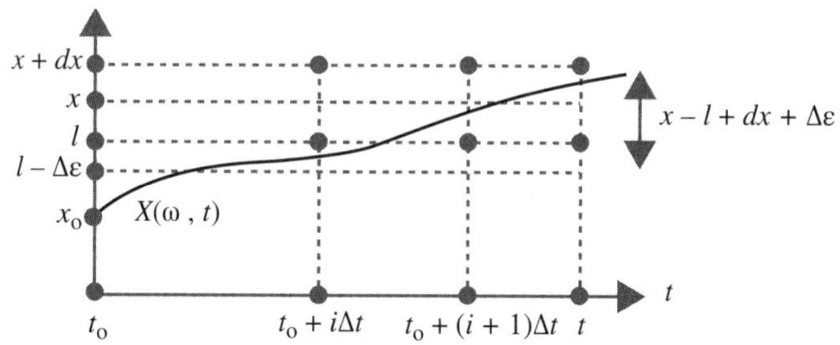


**FIGURE 14.21** Geometry for the upward transition of a differentiable signal above the level $l$.

between the signal levels at a time $t_o + i\Delta t$ and a time $t$, as illustrated in Figure 14.21 for the case of $X(\omega,t) \in [x, x+dx], t > t_o + (i+1)\Delta t$. The maximum possible change between the two levels is $x - l + dx + \Delta\varepsilon$; the minimum possible change is $x - l$.

It follows, using the notation FP for *first passage* and $I = [x-l, x-l+dx+\Delta\varepsilon]$, that

$$\begin{aligned}
&\{X(\Omega,t) \in [x, x+dx]\} = \\
&\quad \{\text{FP to } l \text{ on } [t_o, t_o+\Delta t], X(\Omega,t) - X(\Omega,t_o) \in I\} \cup \\
&\quad \{\text{FP to } l \text{ on } [t_o+\Delta t, t_o+2\Delta t], X(\Omega,t) - X(\Omega,t_o+\Delta t) \in I\} \cup \ldots \\
&\quad \{\text{FP to } l \text{ on } [t_o+i\Delta t, t_o+(i+1)\Delta t], X(\Omega,t) - X(\Omega,t_o+i\Delta t) \in I\} \cup \ldots
\end{aligned} \tag{14.60}$$

As these events are disjoint, and with $t - t_o = N\Delta t$, it follows that

$$\begin{aligned}
&f_{X(t)/(x_o,t_o)}(x)dx = \\
&\sum_{i=0}^{N-1} P[\text{FP to } l \text{ on } [t_o+i\Delta t, t_o+(i+1)\Delta t], X(\Omega,t) - X(\Omega,t_o+i\Delta t) \in I] \\
&\quad = \sum_{i=0}^{N-1} P[\text{FP to } l \text{ on } [t_o+i\Delta t, t_o+(i+1)\Delta t]]. \\
&\quad P\left[X(\Omega,t) - X(\Omega,t_o+i\Delta t) \in I|_{\text{FP to } l \text{ on } [t_o+i\Delta t, t_o+(i+1)\Delta t]}\right] \\
&\quad \approx \sum_{i=0}^{N-1} f_{L(l)/(x_o,t_o)}(t_o+i\Delta t)\Delta t. \\
&\quad P\left[X(\Omega,t) - X(\Omega,t_o+i\Delta t) \in I|_{\text{FP to } l \text{ on } [t_o+i\Delta t, t_o+(i+1)\Delta t]}\right]
\end{aligned} \tag{14.61}$$

The next step is to determine

$$P\left[X(\Omega,t) - X(\Omega,t_o+i\Delta t) \in I|_{\text{FP to } l \text{ on } [t_o+i\Delta t, t_o+(i+1)\Delta t]}\right] \tag{14.62}$$

The interval $\Delta t$ can be chosen arbitrarily small such that $\Delta\varepsilon << dx$, $X(\Omega,t) \approx l$ for $t \in [t_o + i\Delta t, t_o + (i+1)\Delta t]$ when there is an upward passage time of $l$ on $[t_o + i\Delta t, t_o + (i+1)\Delta t]$. It then follows that

$$
\begin{aligned}
& P\left[X(\Omega,t) - X(\Omega,t_o + i\Delta t) \in I|_{\text{FP to } l \text{ on } [t_o+i\Delta t, t_o+(i+1)\Delta t]}\right] \\
& \approx P\begin{bmatrix} X(\Omega,t) - X(\Omega,t_o + i\Delta t) \in [x-l, x-l+dx+\Delta\varepsilon]| \\ X(\Omega,t_o+i\Delta t) \approx l, X(\Omega,t_1) < l \text{ for } t_1 \in [t_o, t_o + i\Delta t] \end{bmatrix} \\
& \approx P\left[X(\Omega,t) \in [x, x+dx]|_{X(\Omega,t_o+i\Delta t)\approx l}\right] \approx f_{X(t)/(l,t_o+i\Delta t)}(x)dx
\end{aligned}
\tag{14.63}
$$

Here, the Markov property of the random process has been used. Thus,

$$
f_{X(t)/(x_o,t_o)}(x)dx \approx \sum_{i=1}^{N} f_{L(l)/(x_o,t_o)}(t_o + i\Delta t) f_{X(t)/(l,t_o+i\Delta t)}(x)dx\Delta t \tag{14.64}
$$

and hence, as $\Delta t \to 0$,

$$
f_{X(t)/(x_o,t_o)}(x) = \int_{t_o}^{t} f_{L(l)/(x_o,t_o)}(\lambda) f_{X(t)/(l,\lambda)}(x)d\lambda \tag{14.65}
$$

## APPENDIX 14.B PROOF OF THEOREM 14.4

Define

$$
f_1(t)dt = P[\text{level crossing in } (t, t+dt)] \tag{14.66}
$$

Using LC as shorthand notation for an upward *level crossing*, this probability can be written as

$$
\begin{aligned}
f_1(t)dt = {} & P[\text{no LC in } (0,t), \text{LC in } (t,t+dt)] \\
& + P[1 \text{ LC in } (0,t), \text{LC in } (t,t+dt)] \\
& + P[2 \text{ LC in } (0,t), \text{LC in } (t,t+dt)] + \cdots
\end{aligned}
\tag{14.67}
$$

Define

$$
f_2(t_1,t)dt_1dt = P[\text{LC in } (t_1, t_1+dt_1), \text{LC in } (t,t+dt)] \tag{14.68}
$$

where $0 < t_1 < t - dt_1$. Signals that are consistent with $f_2(t_1, t)dt_1dt$ being nonzero are illustrated in Figure 14.22.

Consider the integral $dt\int_0^t f_2(t_1,t)dt_1$ that is approximated by the sum

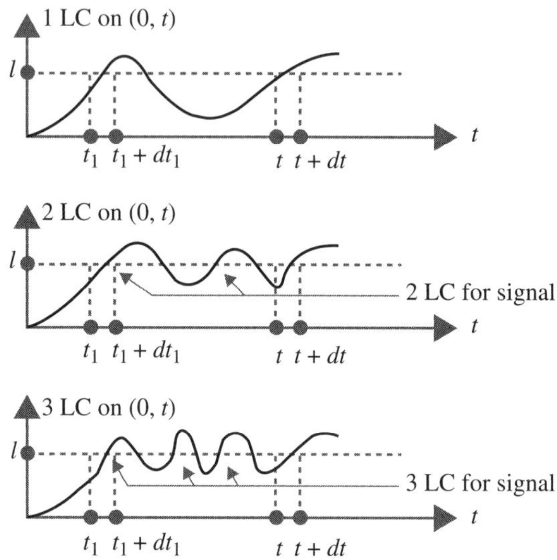


**FIGURE 14.22** Illustration of signals that contribute to $f_2(t_1, t)dt_1 dt$ being nonzero.

$$\begin{aligned} dt\int_0^t f_2(t_1,t)dt_1 \approx\, & P[\text{LCin}\,(0,dt_1),\text{LCin}\,(t,t+dt)] \\ & + P[\text{LCin}\,(dt_1,2dt_1),\text{LCin}\,(t,t+dt)] + \cdots \\ & + P[\text{LCin}\,(idt_1,(i+1)dt_1),\text{LCin}\,(t,t+dt)] + \cdots \\ & + P[\text{LCin}\,(t-dt_1,\,t),\text{LCin}\,(t,t+dt)] \end{aligned} \tag{14.69}$$

for $dt_1$ sufficiently small, $t = Ndt_1$, and $i \in \{0,\ \ldots,\ N-1\}$. The individual probabilities specified in this equation are consistent, respectively, with the events

$$\begin{aligned} & \{\text{LCin}\,(0,dt_1),\text{LCin}\,(t,t+dt)\} \\ & \{\text{LCin}\,(dt_1,2dt_1),\text{LCin}\,(t,t+dt)\}\ldots \\ & \{\text{LCin}\,(idt_1,(i+1)dt_1),\text{LCin}\,(t,t+dt)\}\ldots \\ & \{\text{LCin}\,(t-dt_1,t),\text{LCin}\,(t,t+dt)\} \end{aligned} \tag{14.70}$$

A signal with a single level crossing on $(0, t)$, and with a level crossing on $(t, t+dt)$, is captured once by these events. A signal with two level crossings on $(0, t)$, and with a level crossing on $(t, t+dt)$, is captured twice by these events. Consider level crossings, one in the interval $(idt_1,(i+1)dt_1)$ and the other in the interval $(jdt_1,(j+1)dt_1)$. Similarly, signals with three level crossings on $(0, t)$, and with a level crossing on $(t, t+dt)$, are captured three times by the events specified in Equation 14.70. Consider level crossings, one in the interval $(idt_1,(i+1)dt_1)$, one in the interval $(jdt_1,(j+1)dt_1)$, and one in the interval $(kdt_1,(k+1)dt_1)$. It then follows that

$$dt\int_0^t f_2(t_1,t)dt_1 = P[1\,\mathrm{LC\,on}\,(0,t),\mathrm{LC\,on}\,(t,t+dt)] + 2P[2\,\mathrm{LC\,on}\,(0,t),\mathrm{LC\,on}\,(t,t+dt)] + 3P[3\,\mathrm{LC\,on}\,(0,t),\mathrm{LC\,on}\,(t,t+dt)] + \cdots \tag{14.71}$$

From Equation 14.67 and this equation, it then follows that

$$f_1(t)dt - dt\int_0^t f_2(t_1,t)dt_1 = P[\mathrm{no\,LC\,in}\,(0,t),\mathrm{LC\,in}\,(t,t+dt)] - P[2\,\mathrm{LC\,on}\,(0,t),\mathrm{LC\,on}\,(t,t+dt)] - 2P[3\,\mathrm{LC\,on}\,(0,t),\mathrm{LC\,on}\,(t,t+dt)] - 3P[4\,\mathrm{LC\,on}\,(0,t),\mathrm{LC\,on}\,(t,t+dt)] - \cdots \tag{14.72}$$

To eliminate the second term in this expression, consider

$$f_3(t_1,t_2,t)dt_1dt_2dt = P[\mathrm{LC\,in}\,(t_1,t_1+dt_1),\mathrm{LC\,in}\,(t_2,t_2+dt_2),\mathrm{LC\,in}\,(t,t+dt)] \tag{14.73}$$

Signals consistent with $f_3(t_1, t_2, t)dt_1dt_2dt$ being nonzero are illustrated in Figure 14.23. The double integral of $f_3(t_1, t_2, t)dt_1dt_2dt$ can be approximated according to

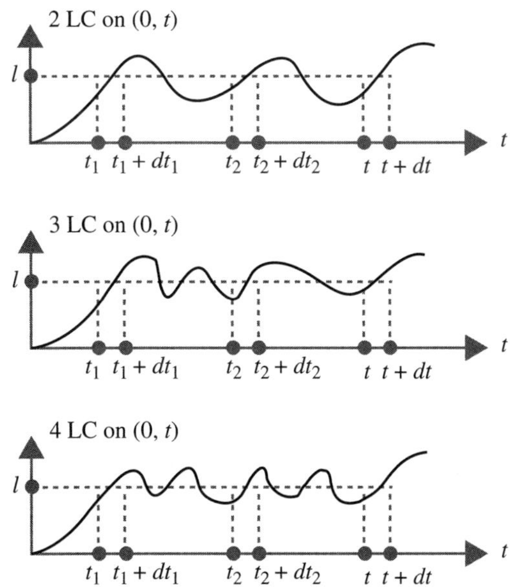


**FIGURE 14.23** Illustration of signals that contribute to $f_3(t_1, t_2, t)dt_1dt_2dt$ being nonzero.

$$dt\int_0^t\left[\int_0^t f_3(t_1,t_2,t)dt_2\right]dt_1 \approx \\ \sum_{i=0}^{N-1}\sum_{j=0,j\neq i}^{N-1} P[\text{LCin}(idt_1,(i+1)dt_1),\text{LCin}(jdt_1,(j+1)dt_1),\text{LCin}(t,t+dt)] \tag{14.74}$$

assuming the probability of two level crossings in an interval with measure $dt_2 = dt_1$ is negligible. The individual probabilities specified in this equation are consistent, respectively, with the events

$$\begin{aligned}&\{\text{LCin}(0,dt_1),\text{LCin}(dt_1,2dt_1),\text{LCin}(t,t+dt)\}\\&\{\text{LCin}(0,dt_1),\text{LCin}(2dt_1,3dt_1),\text{LCin}(t,t+dt)\}\ldots\\&\{\text{LCin}(idt_1,(i+1)dt_1),\text{LCin}(jdt_1,(j+1)dt_1),\text{LCin}(t,t+dt)\}\ldots\\&\{\text{LCin}(t-2dt_1,t-dt_1),\text{LCin}(t-dt_1,t),\text{LCin}(t,t+dt)\}\end{aligned} \tag{14.75}$$

A signal with two level crossings on $(0, t)$, and with a level crossing on $(t,t+dt)$, is captured twice by these events. Consider a level crossing in $(i_o dt_1,(i_o+1)dt_1)$ and another in $(j_o dt_1,(j_o+1)dt_1)$. These level crossings are captured by the events

$$\begin{aligned}&\{\text{LCin}(i_o dt_1,(i_o+1)dt_1),\text{LCin}(j_o dt_1,(j_o+1)dt_1),\text{LCin}(t,t+dt)\}\\&\{\text{LCin}(j_o dt_1,(j_o+1)dt_1),\text{LCin}(i_o dt_1,(i_o+1)dt_1),\text{LCin}(t,t+dt)\}\end{aligned} \tag{14.76}$$

A signal with three level crossings on $(0, t)$, and with a level crossing on $(t,t+dt)$, is captured six times by the events specified in Equation 14.75 (zero crossing pairs $(1,2)$, $(1,3)$, $(2,3)$; the double summation also captures these according to $(2,1)$, $(3,1)$, $(3,2)$). Similarly a signal with a four level crossings on $(0, t)$, and with a level crossing on $(t,t+dt)$, is captured 12 times by the events specified in Equation 14.75 (zero crossing pairs $(1,2)$, $(1,3)$, $(1,4)$, $(2,3)$, $(2,4)$, $(3,4)$ and their reverse order). It then follows that

$$\begin{aligned}\frac{dt}{2}\int_0^t\left[\int_{t_1}^t f_3(t_1,t_2,t)dt_2\right]dt_1 &= P[2\text{LCon}(0,t),\text{LCon}(t,t+dt)]\\&\quad+3P[3\text{LCon}(0,t),\text{LCon}(t,t+dt)]\\&\quad+6P[4\text{LCon}(0,t),\text{LCon}(t,t+dt)]+\cdots\end{aligned} \tag{14.77}$$

Hence,

$$\begin{aligned}&f_1(t)dt - dt\int_0^t f_2(t_1,t)dt_1 + \frac{dt}{2!}\int_0^t\left[\int_0^t f_3(t_1,t_2,t)dt_2\right]dt_1 = \\&\quad P[\text{no LC in }(0,t),\ \text{LC in }(t,t+dt)]\\&\quad+P[3\ \text{LC on }(0.t),\ \text{LC on }(t,t+dt)]\\&\quad+3P[4\ \text{LC on }(0,t),\ \text{LC on }(t,t+dt)]+\cdots\end{aligned} \tag{14.78}$$

The required result follows from continued iteration.

## APPENDIX 14.C PROOF OF THEOREM 14.5

First, from physical, or geometrical, considerations, the following events are equivalent:

$$\{\text{FP time to } l_o \text{ occurs on } [0,t_o]\} = \{\text{maximum level on } [0,t_o] \ge l_o\} \tag{14.79}$$

$$\{\text{maximum level on } [0,t_o] < l_o\} = \{\text{no first passage time to } l_o \text{ on } [0,t_o]\} \tag{14.80}$$

These two results link the random variable $L(\Omega, l_o)$ with the random variable $M(\Omega, t_o)$ according to

$$\begin{aligned} F_{L(l_o)}(t_o) &= 1 - F_{M(t_o)}(l_o) \\ F_{M(t_o)}(l_o) &= 1 - F_{L(l_o)}(t_o) \end{aligned} \tag{14.81}$$

which is the first required result. The assumptions here, consistent with continuous random variables, are

$$\begin{aligned} P[\text{maximum level on } [0,t_o] \ge l_o] &= P[\text{maximum level on } [0,t_o] > l_o] \\ P[\text{maximum level on } [0,t_o] < l_o] &= P[\text{maximum level on } [0,t_o] \le l_o] \end{aligned} \tag{14.82}$$

Second, consider the following partition of events defining the set $S_{FP}$:

$$\begin{aligned} S_{FP} = &\{\text{first passage time to } l_o \text{ on } [t_o, t_o + dt]\} \cup \\ &\{\text{first passage time to } l_o \text{ on } (0, t_o)\} \cup \\ &\{\text{no first passage time to } l_o \text{ on } (0, t_o + dt)\} \end{aligned} \tag{14.83}$$

As the events are disjoint, it follows that

$$\begin{aligned} 1 = &f_{L(l_o)}(t_o)dt + P[\text{maximum level on } (0,t_o) \ge l_o] \\ &+ P[\text{maximum level on } (0, t_o + dt) < l_o] \\ \Rightarrow \quad &1 = f_{L(l_o)}(t_o)dt + \left[1 - F_{M(t_o)}(l_o)\right] + F_{M(t_o+dt)}(l_o) \\ \Rightarrow \quad &f_{L(l_o)}(t_o)dt = -\left[F_{M(t_o+dt)}(l_o) - F_{M(t_o)}(l_o)\right] \end{aligned} \tag{14.84}$$

Hence,

$$f_{L(l_o)}(t) = -\frac{\partial}{\partial t_o} F_{M(t_o)}(l_o) = -\frac{\partial}{\partial t_o}\int_{-\infty}^{l_o} f_{M(t_o)}(m)dm = \frac{\partial}{\partial t_o}\int_{l_o}^{\infty} f_{M(t_o)}(m)dm \tag{14.85}$$

The last result follows as $\int_{-\infty}^{l_o} f_{M(t_o)}(m)dm = 1 - \int_{l_o}^{\infty} f_{M(t_o)}(m)dm.$

Third, consider

$$F_{M(t_o)}(l_o) = 1 - F_{L(l_o)}(t_o) \tag{14.86}$$

Differentiation yields

$$f_{M(t_o)}(l_o) = \frac{\partial}{\partial l_o} F_{L(l_o)}(t_o) = -\frac{\partial}{\partial l_o} \int_{-\infty}^{t_o} f_{L(l_o)}(\lambda) d\lambda = \frac{\partial}{\partial l_o} \int_{t_o}^{\infty} f_{L(l_o)}(\lambda) d\lambda \tag{14.87}$$

The last result follows as $\int_{-\infty}^{t_o} f_{L(l_o)}(\lambda) d\lambda = 1 - \int_{t_o}^{\infty} f_{L(l_o)}(\lambda) d\lambda$.

## APPENDIX 14.D PROOF OF THEOREM 14.7

The first passage time for a specific signal from the random process is defined according to

$$\begin{aligned} t_{FP} &= t: \ l + k(t - t_o) + N(\omega, t) = l \\ \Rightarrow \quad t_{FP} &= t : t = t_o - \frac{N(\omega, t)}{k} \end{aligned} \tag{14.88}$$

For the case where a first-order Taylor series expansion

$$N(\omega, t) \approx N(\omega, t_o) + (t - t_o)\frac{d}{dt}N(\omega, t_o) \qquad t \in [t_o - \Delta, t_o + \Delta] \tag{14.89}$$

is valid around the zero noise level crossing time of $t_o$, it follows that

$$\begin{aligned} t_{FP} &\approx t_o - \frac{1}{k} \cdot \left[ N(\omega, t_o) + (t_{\mathrm{FP}} - t_o)\frac{d}{dt}N(\omega, t_o) \right] \\ \Rightarrow \qquad t_{FP} &\approx t_o - \frac{N(\omega, t_o)/k}{1 + \frac{1}{k} \cdot \frac{d}{dt}N(\omega, t_o)} \end{aligned} \tag{14.90}$$

For the case of

$$\frac{1}{k} \cdot \left| \frac{d}{dt}N(\omega, t_o) \right| << 1 \tag{14.91}$$

which is consistent with the rate of change of the noise signal being much less than the rate of change of the signal, it follows that

$$t_{\mathrm{FP}} \approx t_o - \frac{N(\omega, t_o)}{k} \qquad t_{\mathrm{J}} = \frac{-N(\omega, t_o)}{k} \tag{14.92}$$

Using the result, if $Z = \alpha + \beta X$, then $f_Z(z) = \frac{1}{|\beta|} f_X\left[\frac{z-\alpha}{\beta}\right]$, it follows that

$$f_{L(l)}(t) \approx k f_{N(t_o)}(-k(t - t_o)) \qquad f_J(\tau) \approx k f_{N(t_o)}(-k\tau) \tag{14.93}$$

## APPENDIX 14.E PROOF OF THEOREM 14.8

In $k$ steps of a random walk, there are $2^k$ distinct paths. Consider the paths, around the level $l_o$, illustrated in Figure 14.24. The probability of the event of a first passage time to a level $l_o$ in $k$ steps, assuming $k \geq l_o$, then is

$$\begin{aligned} p_{L(l_o)}(k) = \frac{1}{2^k}[&\#\text{ paths that have level } l_o \text{ after } k \text{ steps} \\ &- \#\text{ paths that have previously reached the level } l_o \text{ in } k-2 \text{ steps}] \end{aligned} \tag{14.94}$$

where

$$\begin{aligned} &\#\text{ paths that have previously reached the level } l_o \text{ in } k-2 \text{ steps} = \\ &\quad \#\text{ paths that have the level } l_o + 1 \text{ in } k-1 \text{ steps} + \\ &\quad \#\text{ paths that have previously reached the level } l_o \text{ in } k-2 \text{ steps and} \\ &\qquad \text{have the level } l_o - 1 \text{ after } k-1 \text{ steps} \end{aligned} \tag{14.95}$$

The second reflection principle (e.g., Karr, 1993, p. 10) for random walks with $p = 1/2$ states that the number of paths that pass through the level $l_o$ and are at the level $l_o - 1$ after $k$ steps is equal to the number of paths that are at the level $l_o + 1$ after $k$ steps. The paths shown in Figure 14.24 illustrate this result. Thus,

$$\begin{aligned} p_{L(l_o)}(k) = \frac{1}{2^k}[&\#\text{ paths that have level } l_o \text{ after } k \text{ steps} \\ &-2(\#\text{ paths that have level } l_o + 1 \text{ after } k-1 \text{ steps})] \end{aligned} \tag{14.96}$$

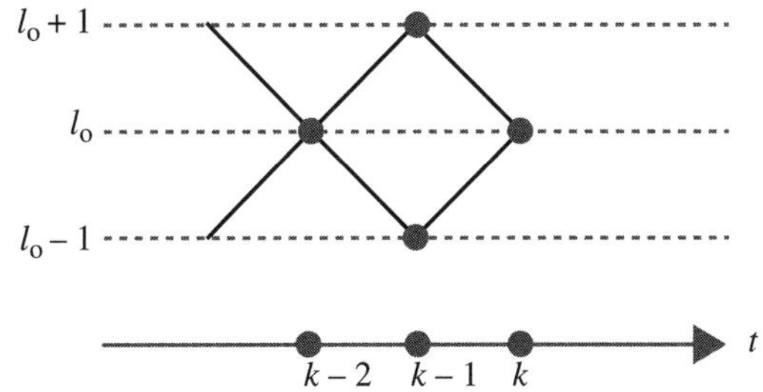


**FIGURE 14.24** Paths of a random walk around the level $l_o$.

To determine the number of paths to a set level $b$ in $n$ steps, note that such an outcome is consistent with $b_U$ up steps and $n-b_U$ down steps where $b=b_U-(n-b_U)$. This implies $b=2b_U-n$ and $b_U=(b+n)/2$. The outcome of $b_U$ up steps in $n$ steps is consistent with $b_U=(b+n)/2$ successes in $n$ trials where a success is associated with an up step. The number of such outcomes is given by

$$\frac{n!}{\left(n-\left(\frac{b+n}{2}\right)\right)!\left(\frac{b+n}{2}\right)!}=\frac{n!}{\left(\frac{n-b}{2}\right)!\left(\frac{n+b}{2}\right)!} \tag{14.97}$$

It then follows that

$$\begin{aligned} p_{L(l_o)}(k) &= \frac{1}{2^k}\left[\frac{k!}{\left(\frac{k-l_o}{2}\right)!\left(\frac{k+l_o}{2}\right)!}-\frac{2(k-1)!}{\left(\frac{k-1-(l_o+1)}{2}\right)!\left(\frac{k-1+(l_o+1)}{2}\right)!}\right] \\ &= \frac{1}{2^k}\left[\frac{k!}{\left(\frac{k-l_o}{2}\right)!\left(\frac{k+l_o}{2}\right)!}-\frac{2(k-1)!}{\left(\frac{k-l_o}{2}-1\right)!\left(\frac{k+l_o}{2}\right)!}\right] \end{aligned} \tag{14.98}$$

Simplifying yields the required result:

$$\begin{aligned} p_{L(l_o)}(k) &= \frac{1}{2^k}\cdot\frac{(k-1)!}{\left(\frac{k+l_o}{2}\right)!\left(\frac{k-l_o}{2}\right)!}\cdot\left[k-2\left(\frac{k-l_o}{2}\right)\right] \\ &= \frac{1}{2^k}\cdot\frac{l_o}{k}\cdot\frac{k!}{\left(\frac{k+l_o}{2}\right)!\left(\frac{k-l_o}{2}\right)!} \end{aligned} \tag{14.99}$$

## APPENDIX 14.F PROOF OF THEOREM 14.9

With the probability density function specified by Equations 14.43 and 14.14 (Theorem 14.3) becomes

$$\frac{e^{-(x-x_o)^2/2D(t-t_o)}}{\sqrt{2\pi}\sqrt{D(t-t_o)}}=\int_{t_o}^{t} f_{L(l)/(x_o,t_o)}(\lambda)\frac{e^{-(x-l)^2/2D(t-\lambda)}}{\sqrt{2\pi}\sqrt{D(t-\lambda)}}d\lambda \tag{14.100}$$

A standard way of solving this equation for the first passage time is to integrate both sides with respect to $x$ (Blake and Lindsey, 1973, p. 298), leading to

$$\int_{l}^{\infty} \frac{e^{-(x-x_o)^2/2D(t-t_o)}}{\sqrt{2\pi}\sqrt{D(t-t_o)}} dx = \int_{l}^{\infty} \left[ \int_{t_o}^{t} f_{L(l)/(x_o,t_o)}(\lambda) \frac{e^{-(x-l)^2/2D(t-\lambda)}}{\sqrt{2\pi}\sqrt{D(t-\lambda)}} d\lambda \right] dx \tag{14.101}$$

Interchanging the order of integration on the right-hand side yields

$$\begin{aligned} \int_{l}^{\infty} \frac{e^{-(x-x_o)^2/2D(t-t_o)}}{\sqrt{2\pi}\sqrt{D(t-t_o)}} dx &= \int_{t_o}^{t} f_{L(l)/(x_o,t_o)}(\lambda) \left[ \int_{l}^{\infty} \left[ \frac{e^{-(x-l)^2/2D(t-\lambda)}}{\sqrt{2\pi}\sqrt{D(t-\lambda)}} \right] dx \right] d\lambda \\ &= \frac{1}{2} \int_{t_o}^{t} f_{L(l)/(x_o,t_o)}(\lambda) d\lambda \end{aligned} \tag{14.102}$$

as $\frac{e^{-(x-l)^2/2D(t-\lambda)}}{\sqrt{2\pi}\sqrt{D(t-\lambda)}}$ is a Gaussian probability density function that is symmetrical around $x = l$. Differentiation with respect to $t$ yields

$$f_{L(l)/(x_o,t_o)}(t) = 2\frac{d}{dt} \int_{l}^{\infty} \frac{e^{-(x-x_o)^2/2D(t-t_o)}}{\sqrt{2\pi}\sqrt{D(t-t_o)}} dx \tag{14.103}$$

A change of variable $y = (x-x_o)/\sqrt{2D(t-t_o)}$ yields

$$\begin{aligned} &f_{L(l)/(x_o,t_o)}(t) = \\ &2\frac{d}{dt}\left[ \int_{(l-x_o)/\sqrt{2D(t-t_o)}}^{\infty} \frac{e^{-y^2}}{\sqrt{\pi}} dy \right] = -2\frac{d}{dt}\left[ \frac{(l-x_o)}{\sqrt{2D(t-t_o)}} \right] \cdot \frac{e^{-(l-x_o)^2/2D(t-t_o)}}{\sqrt{\pi}} \\ &= 2\left[ \frac{-2D(l-x_o)}{2[2D(t-t_o)]^{3/2}} \right] \cdot \frac{e^{-(l-x_o)^2/2D(t-t_o)}}{\sqrt{\pi}} \\ &= \frac{2D(l-x_o)}{\sqrt{\pi}[2D(t-t_o)]^{3/2}} \cdot e^{-(l-x_o)^2/2D(t-t_o)} \end{aligned} \tag{14.104}$$

which is the required result.

# REFERENCE MATERIAL

## 1. FOURIER TRANSFORM RELATIONSHIPS

| $x(t) = \int_{-\infty}^{\infty} X(f) e^{j2\pi ft} df \quad X \in \boldsymbol{L}$ | $X(f) = \int_{-\infty}^{\infty} x(t) e^{-j2\pi ft} dt \quad x \in L$ |
|---|---|
| $\begin{cases} 1 & 0 \le t < T \\ 0 & \text{elsewhere} \end{cases}$ | $T\text{sinc}(fT)e^{-j\pi fT}$ |
| $\text{sinc}(t/\Delta t)$ | $\begin{cases} \Delta t & -1/2\Delta t \le f < 1/2\Delta t \\ 0 & \text{elsewhere} \end{cases}$ |
| $u[t-\alpha] - u[t-\beta] \quad \beta > \alpha$ | $(\beta - \alpha)\text{sinc}[f(\beta - \alpha)]e^{-j\pi f(\beta + \alpha)}$ |
| $\begin{cases} c_k e^{j2\pi k f_i t} & 0 \le t < T \\ 0 & \text{elsewhere} \end{cases}$ | $c_k T e^{-j\pi(f - kf_i)T} \text{sinc}((f - kf_i)T)$ |
| $\begin{cases} \sin(2\pi k f_o t) & 0 \le t < T \\ 0 & \text{elsewhere} \end{cases}$ | $\frac{jTe^{-j\pi fT}}{2} \begin{bmatrix} -e^{j\pi k f_o T}\text{sinc}((f - kf_o)T) + \\ e^{-j\pi k f_o T}\text{sinc}((f + kf_o)T) \end{bmatrix}$ |
| $\begin{cases} \sin(2\pi f_C t) & 0 \le t < T \\ 0 & \text{elsewhere} \end{cases}$ | $\frac{jTe^{-j\pi fT}}{2} \begin{bmatrix} -e^{j\pi f_C T}\text{sinc}((f - f_C)T) + \\ e^{-j\pi k f_C T}\text{sinc}((f + f_C)T) \end{bmatrix}$ |

*(continued)*

**TABLE** *(continued)*

| | |
|---|---|
| $\begin{cases} \cos(2\pi f_C t) & 0 \le t < T \\ 0 & \text{elsewhere} \end{cases}$ | $\frac{Te^{-j\pi fT}}{2}\begin{bmatrix} e^{j\pi f_C T}\text{sinc}((f-f_C)T)+ \\ e^{-j\pi f_C T}\text{sinc}((f+f_C)T) \end{bmatrix}$ |
| $\frac{t^{i-1}}{\tau^i(i-1)!}\cdot e^{-t/\tau}u(t) \quad i \in \mathbf{Z}^+$ | $\frac{1}{(1+j2\pi f\tau)^i}$ |
| $\frac{e^{-(t-\mu)^2/2\sigma^2}}{\sqrt{2\pi}\sigma}$ | $e^{-j2\pi f\mu - 2\pi^2 f^2\sigma^2}$ |

## 2. COMMON RANDOM VARIABLE

| Random variable $X$ | Notation | Sample space $S_X$ | PMF/PDF | Mean | Variance |
|---|---|---|---|---|---|
| Bernoulli | | $\{0, 1\}$ | $p_X(0)=(1-p), p_X(1)=p$ | $p$ | $p-p^2$ |
| Binomial | $b(k, N, p)$ | $\{0, 1, 2, \ldots, N\}$ | $\frac{N!}{(N-k)!k!}\cdot p^k(1-p)^{N-k}$ | $Np$ | $Np(1-p)$ |
| Poisson | $\text{Poi}(\lambda)$ | $\{0, 1, 2, \ldots\}$ | $p_X(k)=\frac{\lambda^k e^{-\lambda}}{k!}$ | $\lambda$ | $\lambda$ |
| Uniform | | $[\alpha, \beta]$ | $f_X(x)=\begin{cases} \frac{1}{\beta-\alpha} & x \in [\alpha,\beta] \\ 0 & x \notin [\alpha,\beta] \end{cases}$ | $\frac{\beta+\alpha}{2}$ | $\frac{(\beta-\alpha)^2}{12}$ |
| Exponential | $\text{Exp}(\lambda)$ | $\boldsymbol{R}^+$ | $f_X(x)=\begin{cases} \lambda e^{-\lambda x} & x>0 \\ 0 & x<0 \end{cases}$ | $\frac{1}{\lambda}$ | $\frac{1}{\lambda^2}$ |
| Normal/ Gaussian | $N[\mu, \sigma]$ | $\boldsymbol{R}$ | $f_X(x)=\frac{1}{\sqrt{2\pi}\sigma}\cdot\exp\left[\frac{-(x-\mu)^2}{2\sigma^2}\right]$ | $\mu$ | $\sigma^2$ |
| Rayleigh | | $\boldsymbol{R}^+$ | $f_X(x)=\frac{x}{\sigma^2}\cdot\exp\left[\frac{-x^2}{2\sigma^2}\right]$ | $\frac{\sqrt{\pi}\sigma}{\sqrt{2}}$ | $\frac{(4-\pi)\sigma^2}{2}$ |
| Gamma | $\Gamma(\alpha, \beta)$ | $\boldsymbol{R}^+$ | $f_{\Gamma(\alpha,\beta)}(x)=\frac{\alpha(\alpha x)^{\beta-1}e^{-\alpha x}}{\Gamma(\beta)}$ | $\frac{\beta}{\alpha}$ | $\frac{\beta}{\alpha^2}$ |

## 3. NOTATION

| Symbol | Meaning |
|---|---|
| $\lfloor\ \rfloor$ | Greatest integer function |
| $*$ | Conjugation operator, outer measure operator |
| $\forall$ | For all |
| $\exists$ | There exists |
| $\emptyset$ | The empty set |
| $\beta$ | Roll-off factor in a raised cosine pulse |
| $\Gamma$ | Gamma function |
| $\Gamma_i$ | Random variable defined by *i*th ordered outcome of a Poisson point process |
| $\Delta X_{(i)}$ | *i*th ordered difference random variable |
| $\Delta X_{(j)-(i)}$ | *j*th to *i*th ordered difference random variable |
| $\delta$ | Dirac delta |
| $\delta_K$ | Kronecker delta function |
| $\rho_{XY}$ | Correlation coefficient between the random variables $X$ and $Y$ |
| $\rho_{XX}$ | Correlation coefficient of a random process $X$ |
| $\Sigma$ | Correlation matrix |
| $\tau$ | Time constant for exponential function |
| $\Phi_X$ | Characteristic function of a random variable $X$ |
| $\sigma_X^2, \sigma_X^2(t)$ | Variance of a random variable/random process $X$ |
| $\Omega$ | Sigma algebra, set of experimental outcomes |
| $\Omega_B$ | Boral Sigma algebra defined by open sets of $\boldsymbol{R}$ |
| $\lambda$ | Point rate for a Poisson point random process |
| $\lambda_k$ | Birth rate for the *k*th state of a birth-death random process |
| $\mu$ | Measure of a set |
| $\mu*$ | Outer measure of a set |
| $\mu_k$ | Death rate for the *k*th state of a birth-death random process |
| $\mu_{t_o}$ | Mean of a random variable defined by a random process at $t_o$ |
| $\mu_X, \mu_X(t)$ | Mean of a random variable/random process $X$ |
| $A \subseteq B$ | $A$ is a subset of $B$ |
| $A^C$ | Complement of the set $A$ |
| $A_i$ | *i*th amplitude random variable |
| $b(k, N, p)$ | Binomial probability mass function: Probability of $k$ successes in $N$ independent trials where the probability of success is $p$ |
| $b_i$ | *i*th basis function in a basis set |
| $c_i$ | *i*th coefficient in exponential Fourier series decomposition |
| corr, cov | Correlation function, covariance function |
| $\boldsymbol{C}$ | The set of complex numbers |
| $\mathrm{Ci}(x)$ | Cosine integral: $-\gamma - \ln(x) + \int_0^x \frac{1-\cos(\lambda)}{\lambda} d\lambda, \gamma$ – Euler function |
| $C_i$ | Random variable defined by *i*th coefficient of the Fourier series decomposition of signals of a random process |
| $C_t$ | Effective correlation time function |
| $C_{XX}$ | Covariance function of a random process $X$ |

*(continued)*

**TABLE** *(continued)*

| Symbol | Meaning |
|---|---|
| $dg$, $dg(x)$ | Integrator function in a Lebesgue–Stieltjes integral |
| $dX_C, \lvert dX_C \rvert^2$ | Cramer spectrum, Cramer power spectrum |
| $D$ | Pulse duration, diffusion constant |
| $\mathrm{erf}_1$, $\mathrm{erf}_2$ | Error function |
| $\boldsymbol{E}$ | Expectation operator |
| $E_k$ | $k$th state of a birth–death random process |
| $f^+, f^-$ | Positive and negative components of a function $f$ |
| $f_{L(l)}$ | PDF for the first passage time to the level $l$ |
| $f_{(i)}, f_{\Delta(i)}$ | PDF of $i$th ordered random variable, PDF of $i$th ordered difference random variable |
| $f_{\Delta_{(j)-(i)}}$ | PDF of $j$th to $i$th ordered difference random variable |
| $f_{(i)(j)}$ | Joint PDF of $i$th and $j$th ordered random variable |
| $f_{\Delta_{(i)}\Delta_{(j)}}$ | Joint PDF of $i$th and $j$th ordered difference random variable |
| $f_{(i)\Delta_{(j)-(i)}}$ | Joint PDF of $i$th ordered random variable and $j$th to $i$th ordered difference random variable |
| $f_{M(t)}$ | PDF for the maximum level on the interval $[0, t]$ |
| $f_X$ | Probability density function of a random variable $X$ |
| $\hat{f}_X$ | Estimator for the probability density function $f_X$ |
| $f_{XY}$ | Joint PDF of the random variables $X$ and $Y$ |
| $f_{Y/X=x}$ | Conditional PDF of the random variable $Y$ given the random variable $X$ takes on the value $x$ |
| $f_\Omega$ | Probability density function for experimental outcomes |
| $F$ | Sigma algebra of events, spectral distribution function |
| $F_i$ | $i$th ordered random variable defining the frequency in a white noise random process |
| $F_{(i)}$ | CDF of $i$th ordered random variable |
| $F_{L(l)}$ | CDF for the first passage time to the level $l$ |
| $F_{M(t)}$ | CDF for the maximum level on the interval $[0, t]$ |
| $F_X$ | Cumulative distribution function of a random variable $X$ |
| $\hat{F}_X$ | Estimator for the cumulative distribution function $F_X$ |
| $F_{XY}$ | Joint CDF of the random variables $X$ and $Y$ |
| $g_I$, $g_D$ | Increasing and decreasing components of a function $g$ |
| $G(T, f), G_\infty(f)$ | PSD of a signal, or a random process, over the intervals $[0, T]$ and $(-\infty, \infty)$ |
| $G(\omega, T, f)$ | PSD of $\omega$th signal of a random process over the interval $[0, T]$ |
| $G(\Omega, T, f)$ | Random process defined by power spectral densities, over the interval $[0, T]$, of individual signals of a random process |
| $G_I$ | Integrated power spectrum associated with the Cramer transform |
| $G_X(T, f), G_{X_\infty}(f)$ | PSD of a random process $X$ over the intervals $[0, T], (-\infty, \infty)$ |
| $G_{XY}(T, f)$ | Cross PSD of the random processes $X$ and $Y$ over $[0, T]$ |
| $h$ | Impulse response of a linear filter |
| $H_i$ | Hermite polynomial of order $i$ |
| $I$ | The interval $I$ |

**TABLE** *(continued)*

| Symbol | Meaning |
|---|---|
| $I_A$ | Indicator function for the set $A : I_A(x) = \begin{cases} 1 & x \in A \\ 0 & x \notin A \end{cases}$ |
| Im | Imaginary part of |
| inf | Infimum: greatest lower bound |
| $j$ | Imaginary unit number: $j^2 = -1$ |
| $J_i$ | $i$th jitter random variable in a jittered random process |
| $K_i$ | $i$th random variable defined by the population level of a birth-death random process after the $i$th transition time |
| $\boldsymbol{L}, \boldsymbol{L}^2$ | Set of Lebesgue integrable functions on $(-\infty, \infty)$, set of squared integrable functions on $(-\infty, \infty)$ |
| $\boldsymbol{L}[\alpha, \beta]$ | Set of Lebesgue integrable functions on $[\alpha, \beta]$ |
| $m, m_g$ | Lebesgue measure operator, Lebesgue–Stieltjes measure operator based on a function $g$ |
| min, max | Minimum, maximum operator |
| $N$ | Number of events, variables, pulses, etc. in a sample space, set, summation, etc |
| $\boldsymbol{N}$ | The set of positive integers $\{1, 2, \dots\}$ |
| $N[\mu, \sigma]$ | A Gaussian random variable with a mean $\mu$ and a variance $\sigma^2$ |
| $N_A$ | Random variable defined by the number of points in a set $A$ |
| $\binom{N}{k}$ | Combination formula: $\binom{N}{k} = \frac{N!}{(N-k)!k!}$ |
| $p$ | Probability of outcome, pulse function |
| $p_{ij}$ | Transition probability from state $i$ to state $j$ |
| $p_R, P_R$ | Rectangular pulse function, Fourier transform of rectangular pulse function |
| $p_{RC}, P_{RC}$ | Raised cosine pulse function, Fourier transform of raised cosine pulse |
| $p_X$ | Probability mass function of a random variable $X$ |
| $p_\Omega$ | Probability mass function for experimental outcomes |
| $p_{XY}$ | Joint PMF of the random variables $X$ and $Y$ |
| $p_{Y/X=x}$ | Conditional probability mass function for $Y$ given $X = x$ |
| $P$ | Probability operator, transition probability matrix |
| $\bar{P}, \bar{P}(T), \bar{P}_\infty$ | Average power of a signal, or random process, on the finite interval, the interval $[0, T]$ and the infinite interval |
| $P(t)$ | Instantaneous power |
| $\boldsymbol{Q}$ | The set of rational numbers |
| $r$ | Pulse rate, waveform rate, signalling rate |
| re, r. e. | Relative error |
| $\boldsymbol{R}, \boldsymbol{R}^+$ | The set of real numbers, the set of real numbers greater than zero |
| $\bar{R}(T, \tau), R_\infty(\tau)$ | Time averaged autocorrelation function of a signal, or a random process, evaluated on $[0, T]$ and on $[0, \infty]$ |
| $R(T, t_1, t_2), R(T, t, \tau)$ | Autocorrelation function of a signal, or a random process, evaluated on $[0, T]$ |
| $\bar{R}(\omega, T, \tau)$ | Time averaged autocorrelation function of $\omega$th signal of a random process evaluated on $[0, T]$ |

*(continued)*

**TABLE** *(continued)*

| Symbol | Meaning |
|---|---|
| $R(\omega, T, t_1, t_2)$, $R(\omega, T, t, \tau)$ | Autocorrelation function of $\omega$th signal of a random process evaluated on $[0, T]$ |
| $\bar{R}(\Omega, T, \tau)$ | Random process defined by the time averaged autocorrelation function, evaluated on $[0, T]$, of individual signals of a random process |
| $R(\Omega, T, t_1, t_2)$, $R(\Omega, T, t, \tau)$ | Random process defined by the autocorrelation function, evaluated on $[0, T]$, of individual signals of a random process |
| $R_{XX}(t_1, t_2), R_{XX}(t, \tau)$ | Autocorrelation function of the random process $X$ evaluated on $[0, T]$ |
| $R_{XY}(t_1, t_2)$ | Cross-correlation function of the random processes $X$ and $Y$ evaluated on $[0, T]$ |
| $\bar{R}_{XX}(T, \tau)$ | Time-averaged autocorrelation function of the random process $X$ evaluated on $[0, T]$ |
| Re | Real part |
| $s$ | Simple function |
| $S$ | Set of experimental outcomes |
| $S_X$ | Sample space of a random variable $X$, sample space of functions defined by a random process $X$ |
| $S_\Phi$ | Signal set |
| $\mathrm{Si}(x)$ | Sine integral: $\int_0^x \frac{\sin(\lambda)}{\lambda} d\lambda$ |
| $\mathrm{sgn}(x)$ | Signum or sign function: $\begin{cases} -1 & x<0 \\ 0 & x=0 \\ 1 & x>0 \end{cases}$ |
| $\mathrm{sinc}(x)$ | Sinc function: $\sin(\pi x)/\pi x$ |
| sup | Supremum: least upper bound |
| $t_{FP}$ | First passage time |
| $T$ | Length of interval $[0, T]$ |
| $u$ | Unit step function |
| var | Variance operator |
| $x \in A$ | $x$ is an element of $A$ |
| $x_F$ | Fourier series of a function $x$ |
| $X_{(i)}$ | $i$th ordered random variable |
| $X(f), X(T, f)$ | Fourier transform of a function $x$ evaluated on the intervals $(-\infty, \infty)$ and $[0, T]$ |
| $X, X(\Omega, t)$ | Random variable, random process |
| $X(\omega, t)$ | $\omega$th signal from the random process $X$ |
| $X(\Omega, T, f)$ | Random process comprising of Fourier transform of signals of $X(\Omega, t)$ evaluated over the interval $[0, T]$. |
| $X_C$ | Cramer transform of a signal or a random process |
| $X_{t_o}$ | Random variable defined by the random process $X$ at $t_o$ |
| $\mathbf{Z}, \mathbf{Z}^+$ | The set of integers, the set of positive integers $\{1, 2, \ldots\}$ |

## 4. ACRONYMS

| Acronym | Definition |
|---|---|
| a.e. | Almost everywhere |
| c.c | Countable case |
| iff | If and only if |
| rms | Root mean square |
| r.e. | Relative error |
| u.c. | Uncountable case |
| BJT | Bipolar junction transistor |
| CDF | Cumulative distribution function |
| CMI | Code mark inversion |
| DAC | Digital-to-analogue converter |
| LC | Level crossing |
| NRZ | Nonreturn to zero |
| PDF | Probability density function |
| PMF | Probability mass function |
| PSD | Power spectral density |
| RP | Random process |
| RV | Random variable |
| SNR | Signal-to-noise ratio |

# REFERENCES


Allen AO. *Probability, Statistics and Queueing Theory with Computer Science Applications*. New York: Academic Press; 1978.

Arnold BC, Balakrishnan N, Nagarja HN. *A First Course in Order Statistics*. Philadelphia: SIAM; 2008.

Blake IF, Lindsey WC. Level-crossing problems for random processes. *IEEE Trans Inf Theory* 1973;**19**:295–315.

Brillinger DR. *Time Series: Data Analysis and Theory*. Philadelphia: SIAM; 2001.

Burk F. *Lebesgue Measure and Integration: An Introduction*. New York: John Wiley & Sons; 1998.

Carleson L. On convergence and growth of partial sums of Fourier series. *Acta Math* 1966;**116**:135–157.

Carlson AB. *Communication Systems*. New York: McGraw Hill; 1986.

Carter M, van Brunt B. *The Lebesgue-Stieltjes Integral: A Practical Introduction*. New York: Springer; 2000.

Champeney DC. *A Handbook of Fourier Theorems*. New York: Cambridge University Press; 1987.

Cohen L. The history of noise. *IEEE Signal Process Mag* 2005; November:20–45.

Cohn DL. *Measure Theory*. Boston: Birkhäuser; 1980.

David HA, Nagaraja HN. *Order Statistics*. Hoboken (NJ): John Wiley & Sons, Inc.; 2003.

Dayley DJ, Vere-Jones D. *Theory of Point Processes: Volume I: Elementary Theory and Methods*. New York: Springer; 2003.

Debnath L, Mikusinski P. *Introduction to Hilbert Spaces with Applications*. Boston: Academic Press; 1999.

Feferman S. *The Number Systems: Foundations of Algebra and Analysis*. Reading (MA): Addison Wesley; 1964.

Feller W. On the normal approximation to the binomial distribution. *Ann Math Stat* 1945;**16**:319–329.

Feller W. *An Introduction to Probability Theory and its Applications*. New York: John Wiley & Sons; 1957.

Gardner CW. *Handbook of Stochastic Methods for Physics, Chemistry and the Natural Sciences*. Berlin: Springer; 1985.

Gradsteyn IS, Ryzhik IM. *Table of Integrals, Series and Products*. New York: Academic Press; 1980.

Grigolini P, Aquino G, Bologna M, Lukovic M, West BJ. A theory of 1/*f* noise in human cognition. *Physica A* 2009;**388**:4192–4204.

Grimmett GR, Stirzaker DR. *Probability and Random Processes*. New York: Oxford University Press; 1992.

Gross D, Harris CM. *Fundamentals of Queueing Theory*. New York: John Wiley & Sons; 1998.

Haaser NB, Sullivan JA. *Real Analysis*. New York: Dover; 1991. Chapter 10.

Hirschman II. *Infinite Series*. New York: Holt, Rinehart & Winston; 1962.

Howard RM. *Principles of Random Signal Analysis and Low Noise Design: The Power Spectral Density and its Applications*. New York: John Wiley & Sons; 2002.

Howard RM. Dirac delta and singular distributions: the general non-good function case. Proceedings of the International Conference on Engineering and Applied Science; July 24–27, 2012; Beijing, China. 2012a. p 294–312.

Howard RM. The power spectral density of a correlated and jittered pulse train. *Proceedings of the 3rd International Conference on Mathematical Models for Engineering Science*; December 2–4; Paris. WSEAS Press; 2012b. p 190–196. Available at http://www.worldses.org/online/2012.htm. Accessed December 23, 2014.

Howard RM. Modelling the PDF evolution in a CMOS inverter during switching. *22nd International Conference on Noise and Fluctuations*; June 24–28, 2013; Montpellier, France. IEEE; 2013, 4 pages. Available at http://ieeexplore.ieee.org/xpl/mostRecentIssue.jsp?punumber=6573520. Accessed December 23, 2014.

Howard RM, Raffel LA. General models for 1/*f* noise. In: Jamal Deen M, Celik-Butler Z, Levinshtein ME, editors. *Noise in Devices and Circuits*. Volume **5113**, Bellingham (WA): SPIE; 2003. Proceedings of SPIE; p 282–293.

Jain PK, Gupta VP. *Lebesgue Measure and Integration*. New York: Wiley Eastern Limited; 1986.

Karr AF. *Probability*. New York: Springer Verlag; 1993.

Kawata T. On the Fourier series of a stationary process II. *Z Wahrscheinlichkeitstheorie* 1969;**13**:25–38.

Kay SM. *Fundamentals of Statistical Signal Processing: Estimation Theory*. Englewood Cliffs (NJ): Prentice Hall; 1993.

Kleinrock L. *Queueing Systems Volume 1: Theory*. New York: John Wiley & Sons; 1975.

Koopmans LH. *The Spectral Analysis of Time Series*. New York/San Diego (CA): Academic Press; 1974/1995.

Kranz SG. *Real Analysis and Foundations*. Boca Raton (FL): Chapman Hall; 2005.

Kreyszig E. *Introductory Functional Analysis with Applications*. New York: John Wiley & Sons; 1978.

Li J, Chen JB. The probability density evolution method for dynamic response analysis of non-linear stochastic structures. *Int J Numer Methods Eng* 2006;**65**:882–903.

Li J, Chen JB. The principle of preservation of probability and the generalized density evolution equation. *Struct Saf* 2008;**30**:65–77.

Li J, Chen J, Sun W, Peng Y. Advances of the probability density evolution method for nonlinear stochastic systems. *Probab Eng Mech* 2012;**28**:132–142.

Liebovitch LS. The fractal random telegraph signal: signal analysis and applications. *Ann Biomed Eng* 1988;**16**:483–494.

Loeve M. *Probability Theory I*. New York: Springer Verlag; 1977.

Marsden JE, Hoffman MJ. *Elementary Classical Analysis*. San Francisco: W.H. Freeman and Co.; 1993.

Meyer PA. Stochastic processes from 1950 to the present. *Electron J Hist Probab Stat* 2009;**5**:1–42.

Milotti E. A pedagogical review of 1/*f* noise. Arxiv preprint physics/0204033 2002. Available at http://arxiv.org. Accessed December 8, 2014.

Montroll EW, Shlesinger MF. On 1/*f* noise and other distributions with long tails. *Proc Natl Acad Sci* 1982;**79**:3380–3383.

Neyman J, Scott EL. Statistical approach to problems of cosmology. *J R Stat Soc Ser B* 1958;**20**:1–43.

Neyman J, Scott EL. Processes of clustering and applications. In: Lewis PAW, editor. *Stochastic Point Processes: Statistical Analysis, Theory and Applications*. New York: Wiley-Interscience; 1972. p 646–681.

Papoulis A. *Probability, Random Variables, and Stochastic Processes*. New York: McGraw Hill; 1965.

Papoulis A, Pillai SU. *Probability, Random Variables, and Stochastic Processes*. New York: McGraw Hill; 2002.

Parzen E. *Modern Probability Theory and Its Applications*. New York: John Wiley & Sons; 1960.

Parzen, E., 1962a, *Stochastic Processes*, San Francisco: Holden-Day Inc.

Parzen, E., 1962b, On estimation of a probability density function and mode, *Ann Math Stat*, vol **33**, pp. 1065–1076.

Peebles PZ. *Probability, Random Variables, and Random Signal Principles*. New York: McGraw-Hill; 2001.

Priestley MB. *Spectral Analysis and Time Series*. London/New York: Academic Press; 1981.

Proakis JG, Salehi M. *Digital Communications*. New York: McGraw Hill; 2008.

Ricciardi LM. Diffusion Process and Related Topics in Biology. *Volume 14, Lecture Notes in Biomathematics*. Berlin/New York: Springer Verlag; 1977.

Ricciardi LM, Sacerdote L, Sato S. Diffusion approximation and first passage time problem for a model neuron. II Outline of a computational model. *Math Biosci* 1983;**64**:29–44.

Rice SO. Mathematical analysis of random noise. *Bell Syst Tech J* 1944;**23**:282–332.

Rice SO. Mathematical analysis of random noise. *Bell Syst Tech J* 1945;**24**:46–156.

Risken H. *The Fokker-Planck Equation*. Berlin/New York: Springer Verlag; 1989.

Ross SM. *Introduction to Probability Models*. Burlington (MA): Harcourt/Academic Press; 2000.

Saleh BA, Teich MA. Statistical properties of a nonstationary Neyman-Scott cluster process. *IEEE Trans Inf Theory* 1983;**29**:939–941.

Schwartz SC. Estimation of probability density by an orthogonal series. *Ann Math Stat* 1967;**38**:1261–1265.

Shimokawa T, Pakdaman K, Takahata T, Tanabe S, Sato S. A first-passage-time analysis of the periodically forced noisy leaky integrate-and-fire model. *Biol Cybern* 2000;**83**:327–340.

Silverman BW. *Density Estimation for Statistics and Data Analysis*. London: Taylor & Francis; 1986.

Smith PL. Stochastic dynamic models of response time and accuracy: a foundational primer. *J Math Psychol* 2000;**44**:408–463.

Spiegel MR. *Mathematical Handbook of Formulas and Tables*. New York: McGraw Hill; 1968.

Sprecher DA. *Elements of Real Analysis*. New York: Dover; 1970.

Stirling J. *Methodus Differentialis*. Londini: Typis Gul. Bowyer. Impensis G. Strahan; 1730.

Thompson JR, Tapia JR. *Nonparametric Function Estimation, Modelling, and Simulation*. Philadelphia: SIAM; 1990.

Titchmarsh EC. *The Theory of Functions*. London: Oxford University Press; 1939.

Titchmarsh EC. *Introduction to the Theory of Fourier Integrals*. London: Oxford University Press; 1948.

Tolstov GP. *Fourier Series*. New York: Dover; 1962.

van den Elzen HC. Calculating power spectral densities for data signals. *Proc IEEE* 1970;**58**:942–943.

Verechtchaguina T, Sokolov IM, Schimansky-Geier L. First passage time densities in resonate-and-fire models. *Phys Rev E* 2006;**73**:031108–1–031108–11.

Wiener N. Generalized harmonic analysis. *Acta Math* 1930;**55**:117–258.

Yaglom AM. *An Introduction to the Theory of Stationary Random Functions*. New York: Dover; 1962.

# INDEX



*A Signal Theoretic Introduction to Random Processes*, First Edition. Roy M. Howard.